Advances in
Cognitive Ergonomics

Advances in Human Factors and Ergonomics Series

Series Editors

Gavriel Salvendy
Professor Emeritus
Purdue University
West Lafayette, Indiana

Chair Professor & Head
Tsinghua University
Beijing, People's Republic of China

Waldemar Karwowski
Professor & Chair
University of Central Florida
Orlando, Florida, U.S.A.

Advances in
Cognitive Ergonomics

Edited by
David B. Kaber
Guy Boy

CRC Press
Taylor & Francis Group
Boca Raton London New York

CRC Press is an imprint of the
Taylor & Francis Group, an **informa** business

CRC Press
Taylor & Francis Group
6000 Broken Sound Parkway NW, Suite 300
Boca Raton, FL 33487-2742

First issued in paperback 2017

ISBN-13: 978-1-4398-3491-6 (hbk)
ISBN-13: 978-1-138-11654-2 (pbk)

Visit the Taylor & Francis Web site at
http://www.taylorandfrancis.com

and the CRC Press Web site at
http://www.crcpress.com

Table of Contents

Section V: Ergonomics in Product Design

Section VI: Human Factors in Aviation Systems

Section VII: Human Factors in Driving

Section X: Information Visualization for Situation Awareness

Section XI: Mental Models

Section XII: Perceptuo-Motor Skills and Psychophysical Assessment

Preface

For many researchers and practitioners working in the field of human factors and ergonomics, the *Applied Human Factors & Ergonomics (AHFE) Conference* represents a culmination of the efforts of Waldemar Karwowski and Gavriel Salvendy (along with other researchers) in organizing various topical research meetings over the years. These have included the *International Conference on Human Aspects of Advanced Manufacturing* (HAAMAHA), the *Human-Computer Interaction International* (HCII) Conference, and the *Computer-Aided Ergonomics and Safety Conference* (CAES). The meetings have provided forums for exposure of research focusing on macro- and micro-ergonomic issues in manufacturing and other production and service environments, HCI applications, and safety-critical domains. Important contributions to ergonomics science have been made through these meetings and related books and journals, such as the journal of *Human Factors & Ergonomics in Manufacturing and Service Industries*. The *AHFE Conference* now provides an even broader "stage" for macro- and micro-ergonomics research by integrating new topical meetings, including the first conference on *Cross-Cultural Decision Making*, the first conference on *Human Factors & Ergonomics in Healthcare*, the first conference on *Neuroergonomics*, and the first conference on *Applied Digital Human Modeling*.

The present book presents those chapter submissions to the "Cognitive Ergonomics" Editorial Board for the AHFE Conference; however, the contributions reflect the interests of many of the joint meetings, including ergonomics in healthcare task design, the role of cross-cultural differences in computing systems design, neuroergonomic models for describing perception, decision making and motor task performance, as well as development of models of human factors as a basis for product design. Each of the chapters appearing in this book were either reviewed by the members of Editorial Board or reflect their own research contributions. For these efforts, we extend our sincere thanks and appreciation to the Board members listed below:

G. Bedny, USA
O. Burov, Ukraine
N. Cooke, USA
X. Fang, USA
T. Gao, PR China
D. Gopher, Israel
K. Itoh, Japan
D. Kaber, USA
J. Layer, USA
J. Lewis, USA
H. Liao, USA

C. Ling, USA
Y. Liu, USA
T. Marek, Poland
A. Ozok, USA
O. Parlangeli, Italy
R. Proctor, USA
G. Song, USA
N. Stanton, UK
K. Vu, USA
T. Waldmann, Ireland

The inclusion and organization of the chapters in the book is the result of an *organic* process. Authors identified a "target" board for submission of their work. Two to three reviewers evaluated each extended abstract. In the event that critical comments were generated through the review process, these were conveyed to the authors. Final chapter submissions (including revisions) were organized according to the editorial boards for the meeting. Subsequently, the book editors reviewed the content of each chapter. For this particular volume, sets of keywords (other than those provided by the authors) were generated by the editors from the review process as a basis for coherently organizing the chapters. This led to identification of subject matter areas that represent contemporary cognitive ergonomics topics. Some subject areas are the result of an integration of several subareas in order to ensure a "critical mass" of at least three chapters per topic. The specific subject areas covered in the book include:

I. Cultural Differences in Computing Systems Design
II. Decision Making and Decision Support
III. Desktop/Mobile Interface Design
IV. Ergonomics in Design
V. Ergonomics in Product Design
VI. Human Factors in Aviation Systems
VII. Human Factors in Driving
VIII. Human Factors in Manufacturing
IX. Human Factors in NextGen Operations
X. Information Visualization for Situation Awareness
XI. Mental Models
XII. Perceptuo-Motor Skills & Psychophysical Assessment
XIII. Task Analysis
XIV. Training Technology
XV. Virtual Reality for Behavior Assessment
XVI. Virtual Reality for Psychomotor Training

The chapters in the book come from an international group of authors with diverse backgrounds in ergonomics, psychology, architecture, computer science, engineering, sociology, etc. The specific chapter topics in the various areas range from biometric systems development, military command and control, cellular phone interface design, methodologies for workplace design, medical device design, cockpit display and decision tool design for pilots, driver visual and cognitive processes, and performance of inspection tasks in manufacturing operations and extend to human-automation integration in future aviation systems, novel 3-D display technologies for enhancing information analysis, training methods for mental models, approaches to activity analysis, new research-oriented frameworks and paradigms in training, and the use of virtual reality for skill development and assessment. The implications of all this work include design recommendations for complex systems and commercial products, new procedures for operator training and self-regulation as well as methods for accessibility to systems, and specification of ergonomic interventions at the user. It is expected that this book would be of special value to practitioners involved in design process development, design and prototyping of systems, products and services, as well as training process design for a broad range of applications and markets in various countries.

April 2010

David B. Kaber
Raleigh, North Carolina
USA

Guy A. Boy
Melbourne, Florida
USA

Editors

Biometric Symbol Design for the Public - Case Studies in the United States and Four Asian Countries

Yee-Yin Choong, Brian Stanton, Mary Theofanos

National Institute of Standards and Technology
100 Bureau Drive
Gaithersburg, MD 20899, USA

ABSTRACT

The use of biometric systems has been expanded beyond traditional law enforcement applications to other areas such as identity management, access control, e-commerce, and even healthcare. With the deployment of biometric systems on the rise, the user bases are also expanding from targeted users such as police to general computer users. This phenomenon challenges biometric researchers and developers to design systems with good usability. This paper evaluated a set of symbols intended for use in biometric systems to help users better understand biometric operations. Six studies with a total of 186 participants were conducted in the United States and in four Asian countries to investigate the cultural effects on people's perception and understanding of the symbols. Some symbols show culture-free results, while some have mixed results. The cross-cultural implications of the case studies are discussed and future research is recommended.

Keywords: Biometrics, Symbols, Usability, Cross-Cultural

INTRODUCTION

The application of biometric systems will become ubiquitous in the near future. Government and industry have a common challenge in today's global society to provide more robust tools and governance principles on how to deploy these tools intelligently to meet national and international needs (National Science and Technology Council Subcommittee (NSTC) on Biometrics, 2006). The NSTC specifically challenged technology developers and researchers to develop biometrics-enabled systems that are intuitive and usable to operators and end users.

Graphic design is a form of visual communication that uses visual elements such as image, color, form, shape, and typography as a unique type of language- visual language. The use of icons and symbols to facilitate communication has been pervasive in areas such as computing technology, mobile devices, appliances, etc. Visual representations can be used as interaction widgets on a user interface, status indicators, warning signals, or to provide graphical instructions. Similar application of icons and symbols will be implemented for biometric systems.

While Horton (1994) stated that one benefit of using icons is to reduce the needs of text translation when marketing products globally, other researchers noted that visual language can be problematic when it's used to communicate across cultures. Icons can be effective in one culture but offensive in another culture (Shirk and Smith, 1994). Graphical representations have been mainly designed and tested in the West, but often times they are targeted for international use. Plocher, Garg, and Wang (1999) reported findings on Chinese users' difficulties in recognizing application tool bar icons for process control workstations designed in the United States (U.S.) and in Australia. Choong and Salvendy (1998) examined the performance differences between American and Chinese users in recognizing icons presented in different modes. A combined presentation mode was found as the best choice since the performance with a combined mode was at least as good as or better than the performance with either alphanumeric or pictorial modes. Kurniawan, Goonetilleke, and Shih (2001) reported similar findings that bimodal (text and picture) icons provide the most appropriate and best meanings for Chinese.

With the rapid proliferation of biometric systems for use by the general public in the near future, users may be unfamiliar with particular implementations and they may not understand the local language in which instructions for use are described. It is important that the symbols used have consistent significance globally and do not cause offense. As many public biometric systems are used by foreign nationals, a consistent international standard set of symbols will reduce the difficulty that the wider community experiences in finding and using biometric systems.

The international biometrics community has recognized the need and importance of establishing standards to ensure a high priority and comprehensive approach to biometrics worldwide. A new subcommittee (SC37) of ISO JTC 1 was established in 2002 to accomplish this goal. In 2008, SC37 issued a call for contributions of pictograms, icons and symbols for use within biometric systems. The National Institute of Standards and Technology (NIST) responded to the call by proposing a set of symbols for biometric systems that aims at enhancing the users' performance with these systems as well as the users' understanding of the use and goals of the systems. The objective was to help the general public understand the concepts and procedures for using electronic systems that collect and/or evaluate biometrics.

PROPOSED SYMBOLS AND PICTOGRAMS FOR BIOMETRIC USE

The proposed set of symbols are to be used to identify the biometric modality, provide instructions related to the scanning activities, display dynamic information related to the scanning process, and indicate the status of the biometric sensor.

Individual Symbols

The proposed symbols (Fig. 1) include concepts for directions and concepts for sensor activity or feedback. Some concepts have multiple variants that were evaluated to identify the best symbol for the corresponding concept.

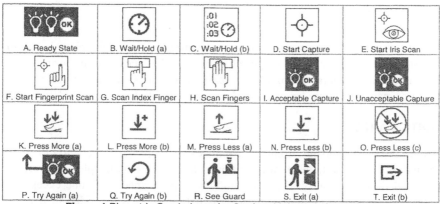

Figure 1 Biometric Symbols under Study, with intended meanings

Procedural Symbols

Although each individual symbol was designed for a concept, it is intended that the symbols be combined to fully illustrate the biometric scanning processes. For example in a customs or immigration environment, procedures constructed from the individual symbols can be presented as a series of posters while passengers are in the queue, or a series of transitional frames in a biometric booth. An example of this type of composite symbols was constructed and evaluated (Fig. 2).

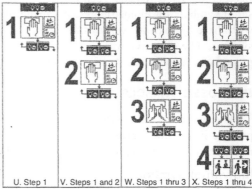

U. Step 1 | V. Steps 1 and 2 | W. Steps 1 thru 3 | X. Steps 1 thru 4

Figure 2 Composite Symbols Representing Steps in a Fingerprinting Procedure

METHOD

A study was designed to evaluate users' interpretation and comprehension against the intended meanings of the proposed symbols. The study was performed in the U.S. and, with collaboration from five research teams in Asia, was replicated in four countries, namely, China, Japan, South Korea (A), South Korea (B), and Taiwan.

PARTICIPANTS

A total of 186 participants from five countries participated in the study (Table 1). The recruitment of participants was cautioned to include only people who were born and resided in the country where the study was performed, with one exception in Korea (B) where one participant was born in Indonesia, had resided in Korea for years and was fluent in Korean. A majority of the participants had engineering or science backgrounds as the studies were conducted on campus where the research teams reside, except for the case of Korea (B) where two rounds of the study were conducted, one on campus and the other in the neighborhood where participants with various backgrounds (e.g. manager, housewife, or teacher) participated.

Table 1 Participant Demographics

Country	Number of Participants	Female (%)	Male (%)	Age Range	Age Average
China	15	53.3%	46.7%	21-29	23.6
Japan	12	0%	100%	20-23	21.2
South Korea (A)	14	50%	50%	21-36	27.5
South Korea (B)	100	49%	51%	19-58	27.8
Taiwan	30	50%	50%	20-36	25.7
U.S.	15	53.3%	46.7%	19-65	29.5

The genders of the participants were pretty balanced, except for the Japanese. The average age of participants from each team was comparable across all cultures.

APPARATUS

As stated earlier, ensuring the usability of biometric systems should be a high priority in product development as challenged by NSTC. Usability is defined as "The extent to which a product can be used by specified users to achieve specified goals with effectiveness, efficiency, and satisfaction in a specified context of use." (ISO, 1999). To develop a usable system, the context in which the biometric system will be used should be considered from the early stages of the design lifecycle.

During our pilot runs before the study, we noticed that people would interpret the symbols more closely to the intended meanings if they had a visual reminder of the context. Some of the more abstract symbols were interpreted with meanings unrelated to biometrics when the context was forgotten. In order to assist our study participants in evaluating the symbols with reasonable background information, it was important to emphasize the intended operational context of the symbols. A mockup of a fingerprint sensor was developed (Fig. 3) as an example of the type of devices that would use the symbols. Each research team in Asia also received from NIST a fingerprint sensor mockup for use in their study.

Figure 3 Mockup of a Fingerprint Sensor

All test materials were developed in English by NIST in the U.S.. Before conducting the studies in Asia, all five Asian teams translated the test materials into their local languages. Peer reviews were performed to ensure the translation was valid and comparable to the original English version.

PROCEDURE

A detailed, step-by-step, test procedure and protocol was developed by NIST and disseminated to all research teams. This ensured that all studies were carried out in a similar fashion. The studies were conducted in two parts.

Part 1 Symbol Interpretation

Part 1 was performed as a one-on-one interview. The interviewer provided each participant with the background of the study, explaining that the symbols were designed to describe biometric processing such as fingerprinting and iris scanning in a multi-lingual environment. The interviewer also described the context of use of the symbols with the fingerprint sensor mockup as a reminder. The symbols (without the intended meanings) were printed on separate sheets of paper and presented in the same order for all participants. The interviewer asked participants to look at the symbols one at a time and provide their interpretation for each symbol. The interviewer just took notes and did not provide feedback on participants' interpretations.

Once the participant had viewed and interpreted each individual symbol (Fig. 1), then the composite symbols were presented in incremental steps with one step added to the presentation each time (Fig. 2) [1]. Again each participant was asked to interpret the composite symbols intended to represent steps in a process.

Part 2 Meaning Matching

Each participant was presented with two pages of test materials. Each page contained a column of possible meanings (in Table 2) and a column of the proposed symbols (only the individual symbols in Fig. 1). Since all participants from part 1 also participated in part 2, the symbols in part 2 were randomized so that the presentation order was different from part 1 to eliminate any sequence effect. Based on what we learned from the pilot runs, words that can be possible interpretations, but not matching the intended meanings of some symbols were inserted intentionally to further investigate participants' perceptions of those symbols.

The interviewer asked each participant to examine the meaning choices and the

[1] The team from Taiwan did not perform the study on the composite symbols due to time and resource constraints.

symbols, and match the best meaning, if any, with each symbol. The participants were instructed that a meaning may be used for more than one symbol, not every meaning had to be used, and only one meaning could be chosen for each symbol.

Table 2 Choices for Intended Meanings in Part 2

1.	Ready state	2.	Wait/Hold	3.	Start Capture	4.	Go in that direction
5.	Scan fingers	6.	Start biometric scan	7.	Start fingerprint scan	8.	Scan index finger
9.	Start iris scan	10.	Look here	11.	Move hand forward	12.	Press more
13.	Press less	14.	Do not press	15.	Give up	16.	Acceptable capture
17.	Unacceptable capture	18.	Try again	19.	Exit	20.	See guard
21.	Turn Around	22.	(none of the above)				

RESULTS AND DISCUSSION

A data analysis template was developed by NIST and disseminated to each Asian research team after each team finished the study in their country. All teams performed their data analyses independently. The Asian researchers also translated the data into English and sent the results to NIST for cross-cultural analysis.

PART 1 SYMBOL INTERPRETATION

A coding scheme was developed to investigate the participants' perceptions and understanding of the proposed symbols. Each interpretation was coded into one of the following categories: "Correct"- the interpretation matched the intended meaning; "Approximate"- the interpretation is related to the intended meaning or to a concept that can lead to the intended meaning, but not exact; and "Incorrect"- the interpretation is totally unrelated to the intended meaning. Due to the subjective nature of the data, only one researcher from each team coded the data collected by his/her team so that the coding would be consistent. The correct interpretation rates are shown in Table 3, and the approximate interpretation rates and incorrect interpretation rates are shown in Table 4.

Table 3 Part 1 Correct Interpretation Rates across Countries

Country	Symbols											
	A	B	C	D	E	F	G	H	I	J	K	L
China	20 00%	46 67%	46 67%	0.00%	53.33%	26 67%	53 33%	86 67%	53 33%	40 00%	66 67%	53.33%
Japan	16 67%	0 00%	0 00%	0.00%	33 33%	50 00%	0 00%	8 33%	8 33%	16 67%	8 33%	50 00%
South Korea (A)	42 86%	50 00%	64 29%	7 14%	35 71%	21 43%	92 86%	92 86%	85 71%	85 71%	57 14%	64 29%
South Korea (B)	7 00%	4 00%	7 00%	4 00%	25 00%	18 00%	13 00%	16 00%	18 00%	17 00%	21 00%	42 00%
Taiwan	16 67%	23 33%	60 00%	13.33%	36.67%	23 33%	73 33%	83 33%	80 00%	80 00%	46 67%	86 67%
U S	26 67%	40 00%	26 67%	20 00%	40 00%	53 33%	40 00%	86 67%	73 33%	86 67%	53 33%	33 33%

Country	Symbols											
	M	N	O	P	Q	R	S	T	U	V	W	X
China	13 33%	60 00%	26 67%	20 00%	20 00%	26 67%	60 00%	53 33%	46 67%	86 67%	93 33%	46 67%
Japan	0 00%	33 33%	16 67%	16 67%	25 00%	16 67%	41 67%	33 33%	8 33%	25 00%	8 33%	66 67%
South Korea (A)	0 00%	57 14%	35 71%	50 00%	50 00%	85 71%	57 14%	57 14%	57 14%	35 71%	50 00%	78 57%
South Korea (B)	0 00%	40.00%	13 00%	4 00%	9 00%	19 00%	22 00%	17 00%	12 00%	22 00%	21 00%	17 00%
Taiwan	3.33%	83.33%	36 67%	33 33%	40 00%	63 33%	60 00%	30 00%				
U S	6 67%	26.67%	26 67%	13 33%	80 00%	66 67%	66 67%	20 00%	33 33%	33 33%	33 33%	53 33%

The correct interpretation rates demonstrate great variations cross culturally. There are a few symbols (E, L, and N with green shadings) with higher than 25% correctness across all cultures. When the approximate interpretations were also considered, participants' interpretations were getting closer (above 50%) to the intended meanings for all countries on symbol E. Two symbols, D and M (with pink shadings), have interpretation rates lower than 20% across all cultures. Specifically, if approximate interpretations were counted, participants still had problems interpreting symbol M as they did not perceive the symbol with a concept that will lead to the intended meaning. Further examination of the data indicated that the majority of participants interpreted symbol M as "lift or remove your finger" (China 73%, Japan 75%, Korea (A) 100%, Korea (B) 68%, Taiwan 93%, U.S. 93%). For symbol D, some participants interpreted it as a "target" (Korea (A) 15%, Korea (B) 21%, Taiwan 17%, U.S. 33%), while 27% of Chinese participants thought it represented the "scan area" and 17% of Japanese participants thought that it directed users to "press the target".

Table 4 Part 1 Approximate Interpretation Rates across Countries

Country	Symbols											
	A	B	C	D	E	F	G	H	I	J	K	L
China	13.33%	33.33%	26.67%	80.00%	40.00%	66.67%	33.33%	6.67%	13.33%	0.00%	26.67%	33.33%
Japan	8.33%	100.00%	66.67%	0.00%	41.67%	33.33%	50.00%	33.33%	8.33%	0.00%	66.67%	8.33%
South Korea (A)	7.14%	21.43%	28.57%	57.14%	50.00%	78.57%	7.14%	7.14%	14.29%	14.29%	28.57%	14.29%
South Korea (B)	16.00%	50.00%	28.00%	23.00%	26.00%	24.00%	45.00%	33.00%	7.00%	6.00%	48.00%	19.00%
Taiwan	13.33%	53.33%	23.33%	30.00%	13.33%	46.67%	20.00%	13.33%	6.67%	10.00%	46.67%	0.00%
U.S.	20.00%	40.00%	46.67%	46.67%	13.33%	33.33%	53.33%	13.33%	20.00%	6.67%	6.67%	6.67%

Country	Symbols											
	M	N	O	P	Q	R	S	T	U	V	W	X
China	0.00%	13.33%	33.33%	26.67%	0.00%	20.00%	13.33%	6.67%	40.00%	0.00%	6.67%	26.67%
Japan	0.00%	0.00%	75.00%	0.00%	0.00%	16.67%	41.67%	0.00%	25.00%	33.33%	58.33%	16.67%
South Korea (A)	0.00%	7.14%	42.86%	35.71%	14.29%	14.29%	35.71%	7.14%	42.86%	64.29%	50.00%	21.43%
South Korea (B)	1.00%	7.00%	52.00%	12.00%	27.00%	46.00%	39.00%	29.00%	12.00%	6.00%	11.00%	10.00%
Taiwan	0.00%	3.33%	40.00%	40.00%	13.33%	16.67%	26.67%	26.67%				
U.S.	0.00%	13.33%	20.00%	26.67%	0.00%	20.00%	26.67%	26.67%	53.33%	60.00%	60.00%	6.67%

For the composite symbols (U thru X) that represent procedural concepts, although there were no consistent interpretation results, interesting cultural differences were observed on how people described their impressions of these symbols. There are fundamental differences between East Asians and Westerners in how they perceive and think about the world. Westerners reason analytically, paying attention to objects and using logical rules to understand events, whereas East Asians reason holistically, focusing on objects in their surrounding field and the relationships among them (Nisbett, 2003). The East Asian participants in the studies tended to include relationships among objects and the field on the composite symbols during interpretation; whereas the U.S. participants tended to describe objects and their states on the symbols. For example, in interpreting symbol U, some interpretations from each culture are quoted: "First step, start when the upper two lights on. Put four fingers onto the area, press 3s. If not OK, do again; If OK, next step." (China); "if two lights are on, touch it and press it. After several seconds, the light is on. If the light doesn't turn on, go inside." (Japan); "When the light of the fingerprint scanner is turned on, scan the fingerprint during 3 seconds. If the green light is turned on, it has been completed successfully. If the red light is turned on, do it again." (Korea (A)); "In the first step, the system will start if the two lights are on. If the one light is going on after you press the button for 3 seconds with the four fingers of your right hand, you can go to the next step. Otherwise, redo the first step." (Korea (B)); "Light comes on to start. Put all 4 fingers down for 3 seconds. If not ok do it again, if ok move on" (U.S.).

PROMISING SYMBOLS

From the results of Part 1 and Part 2, seven symbols show promises that don't seem to be affected by cultures. When textual cues (part 2) are provided, majority (above

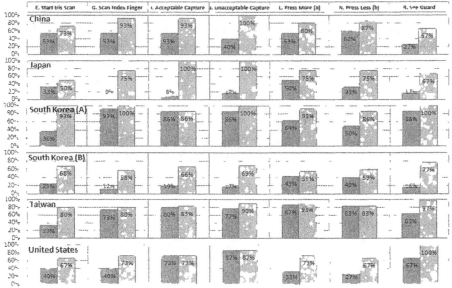

Figure 4 Promising Biometric Symbols

50%) participants were able to better recognize the symbols and their intended meanings (Fig. 4). Further improvements can be expected when the symbols are used operationally in biometric systems, e.g. as icons on a sensor.

SYMBOLS NEEDING DESIGN ALTERNATIVES

Four symbols (Fig. 5) caused confusion in participants from all cultures. With textual cues provided, results were not much better or even worse in some cases (below 50%). Design alternatives are needed for the concepts of "start scan" and "start fingerprint scan". For "Press Less," variants (a) and

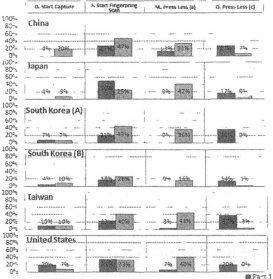

Figure 5 Symbols Needing Design Alternatives

10

(c) (symbols M and O) should be further investigated and variant (b) (symbol N) is the best choice.

SYMBOLS WITH MIXED RESULTS

The remaining symbols (Fig. 6) show mixed results among cultures. With the textual cues, only U.S. participants (73%) could match the meaning to symbol A. For symbol H, the majority of participants could match it with the intended meaning except for Korea (B) with 29% participants choosing "Start fingerprint scan" as the meaning. Majority (more than 50%) of East Asian participants reached the intended meaning for symbol K while 47% of U.S. participants chose the same answer with

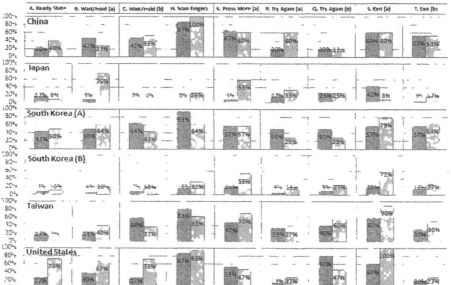

Figure 6 Symbols with Mixed Results

20% selecting "Scan fingers" as the meaning. For variants of "try again" (symbols P and Q), 60% of the Chinese found symbol P closer to the meaning, whereas 60% of the Taiwanese found symbol Q closer to the meaning, and participants from other cultures did not find either variant good for the intended meaning. For symbol S representing "Exit," the majority of participants found it plausible except for the Japanese with 75% choosing "Go in that direction" as the answer. These symbols will require further investigation to determine their feasibility of representing the intended meanings.

CONCLUSIONS

Researchers and developers of biometric technologies will soon be faced with significant challenges to provide reliable and usable systems for operators and end users who may have different cultural backgrounds. NIST has proposed a set of symbols intended to be used in biometric systems with the goals to facilitate user performance and improve usability. The symbols were evaluated in six case studies in four Asian countries and in the U.S.. Seven symbols show great promise to be culture free, four symbols did not work well, and nine symbols require further investigation to determine their utility.

While the numbers of participants of those case studies were not substantial (except for Korea (B)) for drawing statistical inferences, the results of the studies provide great insights for future research. The findings will serve well when future research is performed to investigate those symbols in operational settings for reaching the ultimate goal of an international standard set of biometric symbols.

ACKNOWLEDGEMENTS

The symbols were designed by Patrick Hofman, designpH. The authors would like to acknowledge the Asian research teams, Dr. P.L. Patrick Rau from Tsinghua University in China, Dr. Xiangshi Ren from Kochi University of Technology in Japan, Dr. Yong Gu Ji from Yonsei University in South Korea (Korea (A)), Dr. Young-Bin Kwon from Chung-Ang University in South Korea (Korea (B)), and Dr. S.F. Max Liang from National Taipei University of Technology in Taiwan, for their great collaboration and dedication. The Department of Homeland Security Science and Technology Directorate sponsored the production of this material under HSHQDC-09-X-00467 with NIST.

REFERENCES

Choong, Y.Y., & Salvendy, G. (1998). "Design of icons for use by Chinese in Mainland China." *Interacting with Computers,* 9, 417-430.

Horton, W. (1994). *The icon book: visual symbols for computer systems and documentation.* New York: John Wiley & Sons, Inc.

ISO 13401:1999. *Human-centered design process for interactive systems.*

Kurniawan, S. H., Goonetilleke, R. S., & Shih, H. M. (2001). "Involving Chinese users in analyzing the effects of languages and modalities on computer icons." In C. Stephanidis (Ed.), *Universal Access in HCI: Towards an Information Society for All* (pp. 491-495), Proceedings of the HCI International Conference August 5-10 2001, volume 3, Mahwah, New Jersey: LEA.

National Science and Technology Council Subcommittee on Biometrics. (2006), *The National Biometrics Challenge*, August, 2006, retrieved from http://www.biometrics.gov/NSTC/Publications.aspx.

Nisbett, R.E. (2003). *The geography of thought: why we think the way we do.* Free Press.

Plocher, T.A., Garg, C., and Wang, C.S. (1999). *Use of colors, symbols, and icons in Honeywell process control workstations in China.* Joint technical report prepared with Tianjin University, December 1999.

Shirk, H. N., & Smith, H. T. (1994). "Some issues influencing computer icon design." *Technical Communication*, Fourth Quarter, 680-689.

CHAPTER 2

An Empirical Study of Korean Culture Effects on the Usability of Biometric Symbols

Young-Bin Kwon[1], Yooyoung Lee[1,2], Yee-Yin Choong[2]

[1]Computer Engineering, Chung-Ang University
Seoul, Korea

[2]National Institute of Standards and Technology
100 Bureau Drive
Gaithersburg, MD 20899, USA

ABSTRACT

Biometrics is an umbrella term for methods that identify an individual based on physiological and/or behavioral characteristics such as fingerprint, face, iris, retina, vein, palm, voice, gait, signature, etc. The use of biometric systems is increasing worldwide; consequently there is a need for understanding its procedures via common biometric symbols. However, people with different backgrounds, such as native language, culture, customs, life style, education level, and religion, have various perceptions and expectations of any given symbol. We evaluated how Korean culture influences the use of biometric symbols. Our study was performed by interviewing 100 subjects residing in South Korea using 24 symbols which were developed by the National Institute of Standards and Technology (NIST) Biometrics Usability group. The results present empirical evidence of potential differences in understanding and expectations of biometric symbols due to Korean culture and user knowledge.

Keywords: Biometrics, Usability, Symbol, Cultural factors, Korean

INTRODUCTION

The collective term for methods that identify an individual based on physiological and/or behavioral characteristics (e.g. fingerprint, face, voice, etc.) is biometrics (Jain et al., 2004).

System designers and developers recognized early on that products should be user-friendly. An usability study is one way that ensures that products are both usable and intuitive, i.e. the user is able (a) to figure out what to do, and (b) to tell what is going on (Norman, 1988).

The use of biometric systems is increasing worldwide, e.g. airport, access control, etc. Consequently, there is a need for users to understand easily and efficiently the procedures of any biometric systems. The National Institute of Standards and Technology (NIST) Biometrics Usability group is devising symbols for use within biometric systems, aiming for biometric usability improvements (Mary et al., 2008).

One important aspect is that people with different backgrounds, such as native language, culture, customs, life style, education level, and religion, have varying perceptions and expectations of any given symbol. This means that a symbol might be clearly understandable for one user (or one nation), but not understandable for another (or another nation) (Nielsen, 1994).

In this paper, we present an empirical study to evaluate possible Korean cultural effects on interpreting symbols that were designed to be used in biometric systems. 100 participants residing in South Korea contributed to our evaluation via interviews and structured surveys. We illustrate our results using the usability method "As-Is-Analysis" (Mary et al., 2008) and discuss our assumptions based on the analyzed data.

This paper is divided into five sections. This section served to introduce the notion of biometrics and its usability. The second section gives an overview on our method and evaluation procedure. We present the collected data and discuss the analyzed results in the third and fourth sections. The last section contains conclusion and future works.

METHOD

PARTICIPANTS

All 100 participants were of Korean origin except for one person. We later discovered that the person was born in Indonesia, but he has been staying in Korea for some period of time. Based on the fact that this person has enough knowledge of Korean's culture we decided to include him in our experiment.

Of the 100 participants, gender was near evenly distributed with 51% of the participants being male and 49% female. The average age of the participants was 28 years old. The data was collected of people with diverse occupations such as student, associate, service worker, soldier, teacher, etc. Some of associates are related to IT field and some are not. The demographic results for age and occupation are illustrated in Figure 1.

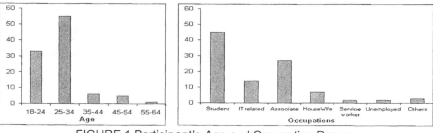

FIGURE 1 Participant's Age and Occupation Ranges

We conducted the study by interviewing two different groups. For the first group (Group-1), the data was collected from 22 participants who were either attending (or visitors) the engineering school of Chung-Ang University. The data of the second group (Group-2) was gathered by asking people on the street or by visiting commercial stores nearby Chung-Ang University. There were 78 people in Group-2. The occupations and ages of Group-2 were more diverse than those of Group-1.

APPARATUS

The symbols provided by NIST were of two types. The symbols in the first type were designed such that each symbol has only one meaning: the symbols with their intended meanings are given in Figure 2. The second type (Figure 3) was composed by combining multiple symbols to indicate sequential steps in a biometric collection process.

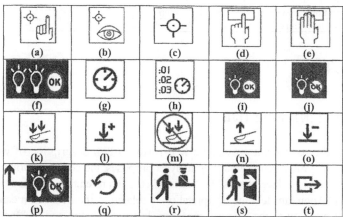

Intended Meanings: (a) Fingerprint scan, (b) Iris scan, (c) Capture, (d) Left Index finger on platen, (e) Left 4 fingers on platen, (f) Ready State, (g)-(h) Wait/hold, (i) Acceptable, (j) Unacceptable, (k)-(m) Press less, (n)-(o) Press less, (p)-(q) Try again, (r) See guard, (s)-(t) Exit

FIGURE 2 Symbol Evaluation Type 1

16

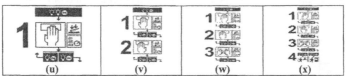

FIGURE 3 Symbol Evaluation Type 2

DATA COLLECTION PROCEDURE

As a preliminary step to the study itself, all textual materials were translated into Korean by a native speaker residing in South Korea. All participants were instructed and interviewed by a native speaker.

The study was conducted in two parts. In part one, *Interpretive*, the participants were shown 24 biometric symbols (Figure 2 and 3) one at a time and were asked to interpret them freely and to describe their impressions in words. The only instruction given to the participants was that the symbols might be used for biometric systems. In part two, *Matching*, each participant was presented the 20 biometric symbols (Figure 2) and were asked to pick, for each symbol, one answer out of the 22 choices (Table 1) which they thought matches the symbol the best. The symbol order for the two parts was different, to avoid interference between the interpretive and matching results.

Table 1 Given Meanings for Matching

Num	Meaning	Num	Meaning	Num	Meaning
1	Ready state	9	Start iris scan	17	Unacceptable capture
2	Wait/Hold	10	Look here	18	Try again
3	Start Capture	11	Move hand forward	19	Exit
4	Go in that direction	12	Press more	20	See guard
5	Scan fingers	13	Press less	21	Turn in a circle
6	Start biometric scan	14	Do not press	22	None of the above
7	Start fingerprint scan	15	Give up		
8	Scan index finger	16	Acceptable capture		

RESULTS

PART ONE: INTERPRETIVE

In this part, the participant provided a free-form written description of what he/she believed each symbol represents. For analysis purposes, the collected written responses were categorized into three classes: 1) "Correct"- the interpretation matched the intended meaning; 2) "Approximate"- the interpretation is related to the intended meaning or to a concept that can lead to the intended meaning, but not exactly; and 3) "Incorrect"- the interpretation is totally unrelated to the intended meaning. The results from this analysis are presented in Table 2. The lower-case letters in Table 2 correspond to the symbols in Figures 2 and 3. On average, across all 24 symbols, only 16% were correct, 24% were approximate, and 60% were incorrect.

Table 2 Categorized Answers (Average: correct (16.3%) approximate (24.0%) incorrect (59.6%))

Symbols	(a) fp scan	(b) iris scan	(c) capture	(d) left index	(e) left 4fp	(f) ready
Correct	18%	25%	5%	13%	16%	7%
Approximate	24%	26%	22%	45%	33%	16%
Incorrect	58%	49%	73%	42%	51%	77%

Symbols	(g) wait/hold	(h) wait/hold	(i) accep.	(j) unaccep.	(k) press ↑	(l) press ↑
Correct	4%	7%	18%	17%	21%	43%
Approximate	50%	28%	7%	6%	48%	19%
Incorrect	45%	64%	75%	77%	30%	38%

Symbols	(m) press ↓	(n) press ↓	(o) press ↓	(p) try again	(q) try again	(r) see guard
Correct	13%	0%	40%	4%	9%	19%
Approximate	53%	1%	7%	12%	27%	46%
Incorrect	34%	99%	53%	84%	64%	35%

Symbols	(s) exit	(t) exit	(u) 1 step	(v) 2 steps	(w) 3 steps	(x) 4 steps
Correct	22%	17%	12%	22%	21%	17%
Approximate	39%	29%	12%	6%	11%	10%
Incorrect	39%	54%	76%	72%	68%	73%

For symbols (a) and (b), the intended meanings are "fingerprint scan" and "iris scan". Several participants interpreted symbol (a) as "pointing at something" or "press something". Over 30% of the participants commented that symbol (b) was related to the eye rather than specifically to the iris. Of this 30%, some were categorized as correct, while others were categorized as approximate or incorrect— depending on the details of the participant's response.

Most participants struggled with symbol (c)-only 5% got this symbol correct and 73% got it incorrect; the answers ranged from "sun" to "aiming/shooting" instead of the intended meaning "capture". The symbols (f), (i), and (j) with their intended meanings of "ready state", "acceptable", and "unacceptable" were also not well understood. Most comments involved "light", e.g. "turning on/off the light."

Symbols (g) and (h) both stand for "wait/hold, both of these symbols scored poorly (4% and 7% correct, respectively). More than 30% people interpreted (g) as "clock" which we classified as "approximate".

Symbols (k) and (l) with the intended meaning of "press more", the answers for (k) were mostly related to "press", while for (l) the answers were either very accurate or very disconnected from the intended meaning (e.g. "dropping" or "hospital").

Symbols (m), (n), and (o) stand for "press less". More than 40% of the participants interpreted (m) as "do not press" which was classified as "approximate". Over 60% thought of symbol (n) as "lift your finger". Our analysis showed that symbol (o) was much clearer (40% correct answers) to participants than the symbols (m) and (n).

Symbols (p) and (q) share the same meaning: "try again", but symbol (q) turned out to be easier to understand for the participants. For (s) and (t), many participants were struggling between "exit" and "enter" and (s) turned out to be easier to grasp (+15%) for the participants.

For symbols (u)-(x) in Figure 3, if participants interpreted the symbol (u) correctly, then most of the time the answers for (v)-(x) were correct too. We noticed that some participants thought of these symbols as "too complicated" or "do not answer" because of the composite design.

PART TWO: MATCHING

For each symbol, percentages of meaning options (Table 1) selected by participants were calculated and ranked. Table 3 shows rank 1 (most common answer) to rank 5. The answers that matched the intended meanings (IM) are bold with gray background. For example, the first row in Table 3 is read as: Symbol (a) with an intended meaning of 7, "fingerprint scan", had most (26%) of the participants correctly matching it to 7, a lesser number (19%) matching symbol (a) to 8 "scan index finger", 15% of the participants identifying it as 5 "scan fingers", and so forth.

Table 3 Ranked Matching Rate for Meaning to Symbol

SB	IM	1st / %		2nd / %		3rd / %		4th / %		5th / %	
(a)	7	7	26%	8	19%	5	15%	4	8%	1	5%
(b)	9	9	68%	10	12%	2	6%	6	5%	20	5%
(c)	3	10	30%	1	17%	22	11%	3	10%	9	7%
(d)	8	8	58%	13	15%	5	8%	7	6%	12	4%
(e)	5	5	32%	7	29%	12	17%	11	4%	13	4%
(f)	1	16	33%	22	26%	1	10%	3	6%	18	4%
(g)	2	22	32%	2	20%	1	10%	4	7%	11	6%
(h)	2	22	37%	2	19%	1	8%	6	7%	18	6%
(i)	16	16	66%	22	7%	1	4%	3	3%	12	3%
(j)	17	17	69%	22	12%	15	4%	1	2%	2	2%
(k)	12	12	53%	22	12%	5	7%	11	6%	13	6%
(l)	12	12	55%	22	10%	4	5%	13	5%	3	4%
(m)	13	14	76%	22	6%	15	5%	7	4%	5	2%
(n)	13	15	27%	22	18%	13	16%	14	11%	16	5%
(o)	13	13	59%	22	15%	15	5%	17	4%	12	3%
(p)	18	17	31%	22	25%	18	14%	15	6%	19	5%
(q)	18	21	54%	18	25%	22	6%	4	4%	1	2%
(r)	20	20	77%	22	7%	11	5%	17	3%	19	3%
(s)	19	19	72%	22	7%	4	4%	9	4%	17	3%
(t)	19	22	29%	19	27%	4	13%	21	12%	15	6%

SB (Symbol)-The relevant symbol is illustrated in Figure 2; IM (Intended Meaning)-Number in Table 1

Based on Table 3, participants were still confused with some symbols. The "confusion criterion" had two components: 1) when the first choice was not the correct choice: symbols (c), (f), (g), (h), (m), (n), (p), (q), and (t); or 2) when the first choice was the correct choice, but it's less than 50%--symbols (a) and (e). For symbols (b), (d), (i), (j), (k), (r), and (s), the results are much better for this matching case than for the prior interpretive case. For example, symbol (b) had a first choice correct rate of 68% for matching, while only a 25% correct rate for the interpretive case (Table 2).

On the other hand, Table 3 also shows that the top five choices for symbol (m), "press less", were all incorrect; while for the interpretive case (Table 2) symbol (m) had a 13% correct rate. In the matching case, 76% of the participants selected "do not press" as the meaning for (m).

Interestingly for (n), "press less", 27% participants picked meaning option (15), "give up". Unfortunately, the symbol (n)'s often-given answer in the interpretive case, "lift your finger" was not provided as one of the choices in Table 1 and so no direct comparison could be done.

DISCUSSION

SYMBOL USABILITY

The graph in Figure 4 illustrates the comparison result between interpretive and matching cases, sorted by level of difficulty in understanding the symbols (from hardest (n) to easiest (l)), based on the correct rates from the interpretive case. It appears that, when options were given, the matching case has better interpretations than interpretive case, except for symbol (m). Nevertheless, the participants were still struggling with the symbols (n), (g), (p), (c), (f), (h), (q), (m), (e), (t) and (a) as shown in Figure 4, with matching and interpretive rates less than 50%. Those symbols (e.g. c, f, m) will require clearer illustration so that ordinary people would understand them better.

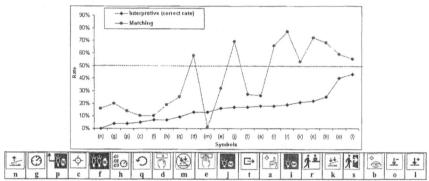

FIGURE 4 Comparisons between Interpretive and Matching cases (Sorted by Interpretive (Correct Rate) - Hardest (n) to Easiest (l))

USER KNOWLEDGE OF BIOMETRIC SYSTEMS

As stated earlier, the participants were recruited from two groups: Group-1 with 22 engineering school attendees and Group-2 with 78 street volunteers. Participants from Group-1 appeared to be more familiar with biometric systems, because the majority was students majoring in engineering where classes about biometrics are offered. Some participants in Group-2 lacked the knowledge with regards to biometrics. This section investigates the implications of users' biometric knowledge on their interpretations of the symbols. Figure 5 (a) and (b) below show comparisons of Group-1 vs. Group-2 participants for the interpretive and matching cases, respectively. Before comparing them, the symbols were sorted based on the results of Group-2 (lowest to highest) to investigate the trend patterns between the two groups.

We observed that users' comprehension of the symbols and their knowledge about biometric systems were related. In general, Group-1 participants had better understanding of the symbols than Group-2 participants did, in both the interpretive and matching cases, Figure 5(a) and (b). However, the perceived difficulties were different for Goup-1 and Group-2. For example, symbols (c) and (h) were the most difficult to comprehend for Group-2, whereas (n) and (t) were the most difficult for Group-1. When pre-defined meanings were provided, the data of the two groups follows a more similar pattern (Figure 5 (b)) than in interpretive case (Figure 5 (a)).

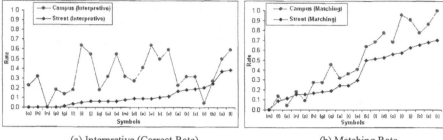

(a) Interpretive (Correct Rate)   (b) Matching Rate

FIGURE 5 Evaluation of Group-1 (campus) and Group-2 (street) –Sorted by Group-2

CULTURAL FACTORS

From the results of this empirical study, we ask the question, "Do Koreans have a difficult time to understanding a symbol because of the symbol's design or because of Korean cultural factors?"

To investigate possible answers to this question, all responses for 20 symbols (Figure 2) from the interpretive case were re-analyzed. Using our Korean background and experience to judge, we flagged responses that were not correct but had references to Korean culture. Of the 100 participants, 30 provided culturally referenced interpretations to one or more symbols while 70 participants made no cultural references in interpreting any of 20 symbols. Table 4 shows these 30 participants whose responses were with Korean cultural references for some symbols (marked in green). The 20 symbols were pre-sorted—from hardest (n) to easiest (l)—based on the overall average correct rates.

Participants made culturally misinterpretations to eight symbols with three that are worth paid attention to, namely, symbols (c), (e), and (l). Symbol (c), "capture," had the highest number of participants (20) who made culturally referenced responses. Symbol (e), "left hand on platen," had eight participants and symbol (l), "press more," had four participants who gave culturally referenced interpretations. The difficulties in understanding of the remaining 17 symbols were considered design related and culturally insignificant since those symbols had no or fewer than two misinterpretations with cultural references.

It should be noted that Korea currently has one of the world's largest troops (Anthony et al, 2006). Also, serving in the military is mandatory for male citizens of South Korea. Thus, this fact may possibly contribute to the multiple occurrences of the "aiming" and "shooting" misinterpretations for symbol (c).

Table 4 Reponses Affected by Korean Culture - Hardest (n) to Easiest (l)

ID	n	g	p	c	f	h	q	d	m	e	j	t	a	i	r	k	s	b	o	l
G1_14																				▨
G1_21										▨										
G2_1										▨										
G2_2				▨																
G2_5										▨										
G2_7				▨									▨							
G2_8										▨										
G2_11				▨																
G2_12				▨	▨															
G2_13				▨																
G2_14				▨																
G2_17																				▨
G2_23		▨																		
G2_26				▨																
G2_33				▨																
G2_34										▨										
G2_35				▨				▨												
G2_36				▨																
G2_39										▨										
G2_42																			▨	
G2_47				▨																
G2_51																			▨	
G2_54													▨							
G2_59				▨																
G2_60										▨										
G2_64				▨																
G2_67										▨										
G2_72				▨																
G2_73				▨																
G2_78																				▨
Total	0	1	0	20	1	0	0	1	0	8	0	0	2	0	0	0	0	0	2	4

For symbol (e), "left hand on platen," there were answers such as "vending machine" or "ticketing in metro". The vending machines in Korea normally have a big rectangular button for each product, unlike the vending machines common in Western countries. Participants may view the rectangular platen shown in symbol (e) as similar to their experience with the big button on the vending machines.

Korea also holds the world's highest internet access (NationMaster, 2009) and the highest rank for computer game addiction including action/violence games. This also might contribute to misinterpretations such as "aiming/shooting" for (c), and also responses of "level up" for (l) which have their source from the environment or computer games.

There are other possible cultural sources that may contribute to the confusion. For example, several participants also interpreted some symbols as a singer's name or an animation character, both part of the Korean pop culture as well. One participant interpreted the symbol (o), "press less," as "unhappy face"- (⊥-) influenced by Korean emoticon illustrations. Emoticons are different between Korean and Western countries, e.g., Korea vs. the United States:

　　　　laughing: ^o^　　vs.　:D　　　crying: ㅠ.ㅠ　vs.　:`-(

The results show that misinterpretations of some symbols were affected by the Korean culture and participants' environment. Yet, symbols (e.g. (n)) which participants had most difficulties in understanding were not affected by cultural factors (see Table 4). In these cases, it is more likely that the symbol's design, rather the user's cultural background, was causing the confusion.

CONCLUSIONS

We presented and analyzed the biometric symbol data collected in South Korea. This empirical study represents an initial step in understanding how Korean users may interpret symbols designed for use in biometric systems.

We examined three factors that potentially affected the usability of biometric symbol: 1) the symbols themselves, 2) user knowledge of biometric systems, and 3) Korean culture. In general, some of the symbols are more difficult to interpret than others for the Korean participants in this study.

We evaluated in detail how Koreans interpreted the proposed symbols in the context of biometrics. It became apparent that most participants seemed to understand a simple symbol design better (e.g. Figure 2) compared to a composite symbol design (e.g. Figure 3). Participants were able to match symbols with intended meanings better when pre-defined choices were provided which only requires users to recognize (as in the matching case), rather than have to generate meanings (as in the interpretive case). The results show that some participants were affected by the Korean culture and environment in understanding some symbols. We observed that besides cultural factors, the symbol design influences a user's understanding the most. At last, we found that a user's knowledge and experience about a system also influence symbol comprehension.

The results will provide a valuable basis to further research on investigating cultural implications in operational biometric environment.

ACKNOWLEDGEMENT

We would like to acknowledge the contributions of Ah-Young Kim.

REFERENCES

Jain, A. K., Ross, Arun, Prabhakar, Salil (2004), "An introduction to biometric recognition", *IEEE Transactions on Circuits and Systems for Video Technology 14th (1):* 4–20,

Norman, D.A. (1988), *Design of everyday things*. New York, NY: Currency Doubleday.

Mary Theofanos, Brian Stanton, Cari A. Wolfson (2008), *Biometrics & Usability*, Gaithersburg, MD: National Institute of Standards and Technology (NIST) IR 7543

Nielsen, J. (1994), Usability Engineering, Boston, *MA: AP Professional*

Anthony H. Cordesman, Martin Kleibe (2006), "The Asian Conventional Military Balance in 2006 - South Korea's Armed Forces", *CSIS (Page 24*", Website: http://www.csis.org/media/csis/pubs/060626_asia_balance_powers.pdf.

NationMaster (2009), Internet Statistics: Broadband access by country, Website: http://www.nationmaster.com/graph/int_bro_acc_percap-internet-broadband-access-per-capita

Chapter 3

Make Icons Make Sense: Solving Symbols for Global Audiences

Patrick Hofmann

designpH
101-26 Kirketon Road
Darlinghurst NSW 2010
Australia

ABSTRACT

Why are icons so often misinterpreted? Is there something in their minimalism that makes their interpretation vary so widely? Is there something in us – our upbringing, our language, our culture, our education, our gender, our age, our religion – that makes us interpret them so differently from our fellow human?

In our efforts to symbolize actions, categories, and places for global products and interfaces, we have come closer to an answer. We have encountered new challenges in creating single icons that work for the entire planet. We have witnessed the surprising reactions of various cultures and age groups on these icons, and have learned new ideas for moving forward.

Keywords: graphics, icons, symbols, internationalization, localization, cultural interpretation, visual language, visual instruction, wordless instruction

INTRODUCTION

Imagine the image of two circles, one next to the other. What are they? Set in this square frame, it may be the face of some dice, or a cutout of a square button. Is the frame a part of the symbol or object in question, or is it merely a frame that surrounds or crops the image?

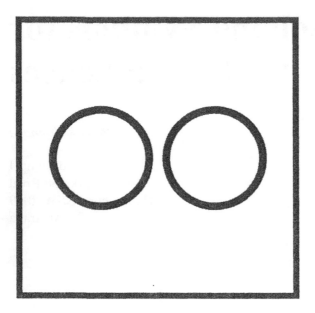

Figure 1: A symbol of simple geometric shapes is open to interpretation

So what is the answer? Irrespective of our upbringing, our language, our culture, our education, our gender, our age, our religion, the most subtle changes and additions to these two circles can dramatically change what these two circles represent.

If we place another circle or dot inside each one, they become eyes, or even breasts. If we place a big square above them, they become wheels. If we place a '1' before them, the circles become numbers.

Figure 2: Simple changes of an object's placement, position, and context have a dramatic impact on its interpretation and meaning.

Herein lies the challenge with icons: they are trimmed to such an extent as to reduce or eliminate reading, to purely provoke an instant reaction and immediate interpretation. Since icons shouldn't require 'reading', we should stop thinking of localization, internationalization, and even translation as disciplines involving the reading of words and images, but rather, as disciplines that research, design, and address the reaction and response of a user to product.

BIOMETRICS SYMBOLOGY FOR NIST

In my design of biometrics symbols for the National Institute of Standards and Technology, I learned countless lessons in icon-izing fingerprinting procedures and biometric security procedures for ports of entry.

Use symbols for simple commands and feedback, not as a detailed instruction

If symbols are meant to be instantly recognized and not read, is it possible to provide a purely visual illustration of several fingerprinting steps that are intended to be read before using the biometric device? Although it achieved some success during usability testing, users did not easily read or rely on complex multi-step visuals that involved decision-making, choosing alternatives, and troubleshooting when procedures fail.

Figure 3: An early iteration of visual instruction depicting the use of a fingerprinting device. Although some users were able to follow this detailed collection of symbols, they performed far better when integrated individually into the display of the biometric device, rather than as a separate standalone instruction.

Symbols that offer feedback can be less literal

The numerous sets of graphic image ranged from being realistic and representative, to symbolic and abstract. In Figure 3, the detailed visual procedure used more representative symbols and less abstract ones, perhaps because the entire procedure was being read in a more sequential fashion. However, as we tested the symbols as single images that dynamically appeared as feedback in the biometrics system display, we could afford to use less literal and more abstract symbols. For example, the symbol for "Press your finger down harder" changed from a very literal finger with arrows to a mere "+" sign in the display.

Figure 4: For alternative symbols expressing "Press your finger down harder", ranging from literal to abstract.

Context is everything, and then some

In creating symbols for the fingerprinting and biometric devices, we discovered that the success of a single symbol depended entirely on its context. If the 'fingerprinting' symbol was used as a wayfinding image to direct people to the biometric device, did the same symbol suffice to label the actual biometric device, and to provide feedback or instruction within the device display?

In reality, the fingerprinting symbol itself was problematic as it used the human finger, which is associated with countless gestures and representations. Does a single index finger denote pointing upward? Is it signaling importance? Is it directing movement? Is it denoting fingerprinting? To prevent confusion, arrows were used to denote direction, and stylized fingerprints were used to denote fingerprinting.

Figure 5: The raised index finger is associated with many gestures and representations. Its meaning is entirely dependent on its context and positioning.

POINTS OF INTEREST ON GOOGLE MAPS

Similar to my symbology work for NIST, my efforts to icon-ize points of interest and landmarks on Google Maps were equally as challenging. Aiming to create a single set of symbols that internationalized well for the entire planet, I encountered many scenarios where acute localization was necessary:

Hospitals

The icon to represent medicine or hospitals: a 'cross', whether it is red (as in Red Cross), or white (as in Swiss neutrality and the helpful St Bernhard), or the green of pharmacies and apothecaries in continental Europe, is not acceptable in many parts of the middle East. Likewise, the Red Crescent that was developed for many these regions is not recognized universally. In Israel, the Star of David is used to denote hospitals. In the end, tactic localization is required to prevent discomfort or offence.

Figure 6: Localized symbols to represent 'Hospital'.

Places of worship

Instead of visualizing the different icons that represent the most popular faiths on the planet, I first asked myself: is there any one icon that can represent all of them? Is there a building façade or shape, or any human shape or position that can represent all of them? As the human act of kneeling wasn't universal to all religions, I abandoned the idea of showing the human figure, and instead developed a silhouette of a place of worship, an amalgamation of a temple, mosque, synagogue, church, and chapel. Although this generated generally positive results, we discovered over time that a generic 'worship' symbol wasn't as helpful to users as the specific religious symbol, which often provided an indication of the religious makeup of a particular locale.

Figure 7: Symbols to represent 'Places of Worship'. The building at the top was an attempt to amalgamate the silhouette of a temple, mosque, synagogue, church and chapel into a single internationalized image. Eventually, we used individual symbols for better clarity.

But these two simple examples demonstrate that cultural groups can be defined by geography and religion. Are there cultural segments or groups that are defined by other criteria?

WHAT DEFINES A CULTURE?

We are continually reminded that as we design and document our products for international audiences, we must be aware of the cultural differences of our various lingual and cultural user groups. For example, if your information is destined for a handful of countries bordering the Mediterranean Sea, avoid using symbols of the hand or fingers in your documents -- like the warning "hand" or the reminder "finger", as they have very different (and very vulgar) connotations in some of those nations and cultures.

But what defines culture? Political borders, language, religion, and traditions are typically identified as factors that distinguish one culture from another. However, have we considered **age** as a distinguishing factor in our work?

CONSIDERING AGE GROUPS AS CULTURES

Due to the widespread development, accessibility, and affordability of digital consumer products – including mobile phones, music players, game systems – the spectrum of age in our audiences have expanded considerably.

Since a six-year-old child or a ninety-six-year-old adult both are users of a mobile phone, we shouldn't forget to internationalize or localize such products for multiple ages, not just languages. Using visual images and symbols as the foundation, we must recognize how different age groups interpret them.

For example, in our printed information or digital interfaces, what symbol would we use to denote a television? Maybe I'm aging myself here, but most of would think of a monitor with two channel knobs and some rabbit-ears on top. Would an 8-year old know what this is? Would a 10-year old? Even a 16-year old? If we simplified it to look like our standard TVs today, would that make it better?

According to my research, the answer is No. We have far too many appliances that look the same: microwave ovens, computer monitors, and so on. Surprisingly, to internationalize this image for most ages and countries, placing the letters TV inside this minimalistic TV set would work best, even if your native tongue's word for TV is something else.

Figure 8: Some television symbols are unrecognized by some age groups

Aside from the TV symbol, what would we use to represent the telephone? Similar to the traditional TV, the telephone has come a long way in shape, size, and complexity. Not all children, even teenagers, would immediately recognize the first two telephone symbols above. As tweens and teenagers become the next generation of big-spending consumers, we must acknowledge that their interpretation of symbols is quite different than our own. A depiction of the mobile phone may be the newest symbolic standard for "telephone" for most ages and most countries around the world, even developing nations, who have growing if not superior cellular telephone infrastructure and penetration.

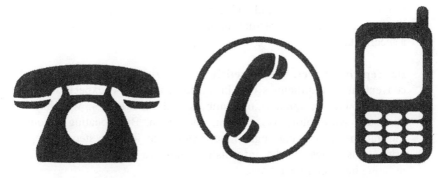

Figure 9: Some telephone symbols are unrecognized by some age groups

But for how long will these symbols last? The traditional tabletop rotary-dial telephone has been around for much of the twentieth century. Will the hand-held cell-phone symbol last for the 21st century? As our users and customers age, so too do our icons. As technology advances and consumers adopt new products, the symbolic and visual conventions that we use to instruct and inform must advance as well.

CONCLUSION

In the end, the simplicity and minimalism of icons make their interpretation vary so widely, but all the influences that we've highlighted do even more: our upbringing, our language, our culture, our education, our gender, our age, our religion, our experiences – make us interpret them so differently from our fellow human.

Yet, to aim for a correct interpretation of an icon through simplicity and minimalism, we need design them with these basic principles:

- **Reduce visual traffic, and include only what is vitally necessary.** By removing extraneous details, we help users scan and focus their attention on only the relevant areas of a visual. Highly accurate, over-detailed illustrations make the visuals realistic, but so overly detailed that they require 'reading'. Users don't like reading wordy, detailed textual instructions, so why would we want them to painstakingly read a visually wordy, detailed pictorial instruction?"

- **Illustrate the message, not the object in its anatomically perfect form.** Draftspeople and CAD artists focus on transcribing the physical object from its true form to a flat image. This means achieving realism and technical accuracy, instead of concentrating on the message of the illustration (which is to instruct the user to follow a command or complete a specific task).

- **Be repetitive; yes, be repetitive.** In writing, we repeat wording conventions and structures that bring consistency and expectation to our text. Likewise in families of symbols and illustrations, we should be consistent with color, size, weight, and so on. By repeating a common illustration and making slight modifications to it from step to step, we bring consistency to the information and we amplify the change in meaning between illustrations.

With these principles in mind, we can better achieve the simplicity and minimalism of icons: they need to eliminate reading and need to purely provoke an instant reaction and immediate interpretation. For this reason, we need to expand our thinking of internationalization and localization to accommodate how people react and respond to visual language, not just to how they read textual language.

CHAPTER 4

Usability Evaluation on Icons of Fingerprint Scanning System for Taiwanese Users

Sheau-Farn Max Liang, Po-Hsiang Hsu

Department of Industrial Engineering and Management
National Taipei University of Technology
Taipei 10608, Taiwan, ROC

ABSTRACT

A set of twenty icons for fingerprint scanning systems designed by National Institute of Standard and Technology (NIST) was evaluated through questionnaires. Results indicated that the recognition rates of approximately half of these icons were significantly lower than the standard rate of 66.7% stipulated by International Organization for Standard (ISO). Four icons with low recognition rates were selected for redesign. Three alternatives were designed for each of the four icons to access the conformability to their intended meanings against originals. Results showed that the conformabilities of three out of four original icons have been improved by replacing them with their alternative designs. The findings of this study should be able to provide relevant icon design suggestions for enhancing the usability of fingerprint scanning systems.

Keywords: Biometric System, Icon, Usability, Confusion Matrix

INTRODUCTION

Icon is one of the most important components of a Graphical User Interface (GUI) due to its aesthetic attractiveness, possibility of rapid recognition, and potential of internationalization. A well-designed icon should be visually distinctive and appropriately represents its intended meanings. That is, the discriminability and meaningfulness of icons are two important factors for icon design (Liang, 2006). An icon is usually not designed alone but together with other icons as a set. Confusion may occur when similar or common features shared among icons that represent different meanings (Wickens et al., 2004). The perceived similarity between two icons is increased by the number of common features and decreased by the number of distinctive features (Tversky, 1977).

Modified from Peirce's semiotic triangle, a dual triadic model was used to explain two components of icon interpretation: comprehension of represented meanings and familiarity of icons (Liang, 2006). For the design of computer systems, such as biometric systems, the represented meanings are usually system functions or status. They are domain-specific and task-dependent. Users can comprehend them through training or providing an appropriate context of use. On the other hand, the familiarity of icons depends not only on user's domain experience but also on his or her cultural background.

As part of the National Institute of Standard and Technology (NIST) biometric usability research, the usability of a set of twenty icons designed for fingerprint scanning systems was evaluated in Taiwan. These icons represent the functions and state status of interacting with the system, such as (1) ready state, (2) fingerprinting on the platen (scan index finger and scan fingers), (3) start capture (start capture, start fingerprint scan, and start iris scan), (4) wait/hold, (5) press more/less, (6) acceptable/unacceptable capture, (7) retry, and (8) give up/exit (see guard, give up, and exit). The discriminability and meaningfulness of these twenty icons were analyzed in this study. The twenty icons are listed in Table 1.

Table 1 Twenty Icons for Biometric Systems

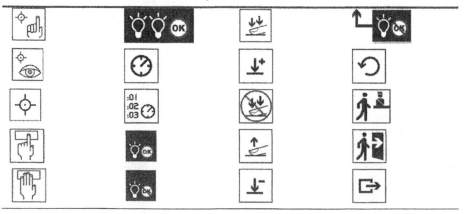

METHODS

As shown in Figure 1, in Phase 1, the data from subjects were collected through Questionnaire 1 and 2. The data collected from Questionnaire 2 were represented as Confusion Matrix for calculating correct recognition rates on these icons. A statistical test was then conducted to identify whether the rates were above or below an acceptable threshold. Results of the test were summarized. In phase 2, icons with low recognition rates were selected for redesign. Questionnaire 3 was distributed to collect data for further statistical test on examining the conformability of intended meanings between the redesigned icons and their originals. Results of the test were summarized.

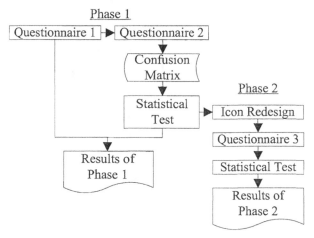

Figure 1. Flowchart of Methods

SUBJECTS

Thirty university students with various major backgrounds and without any experience of traveling abroad were recruited as subjects for answering Questionnaire 1 and 2. The subjects could represent new users of fingerprint scanning systems. More fifty-five university students with various majors were recruited as subjects for answering Questionnaire 3.

QUESTIONNAIRE 1 AND 2

Two questionnaires were designed to collect data. Questionnaire 1 was designed to list the twenty icons and ask subjects to interpret and write down the meanings of each icon. Examples of questions are presented in Figure 2 (a). Questionnaire 2 was designed to list both icons and possible meanings and ask subjects to match the best meanings with each icon. Examples of questions are presented in Figure 2 (b).

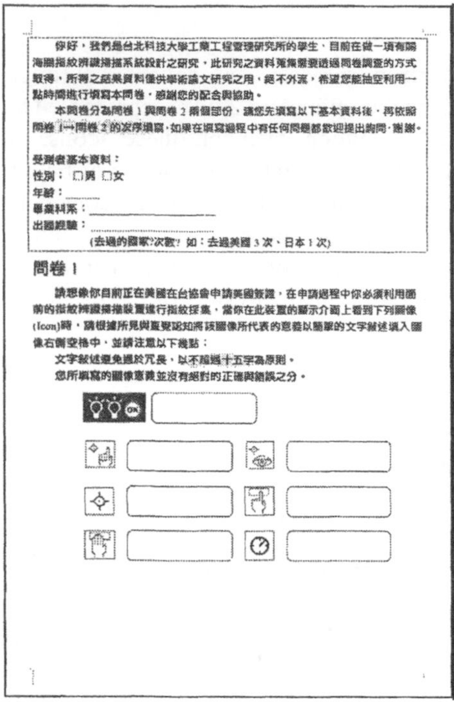

你好，我們是台北科技大學工業工程管理研究所的學生，目前在做一項有關海關指紋辨識掃描系統設計之研究，此研究之資料蒐集需要透過問卷調查的方式取得，所得之結果資料僅供學術論文研究之用，絕不外流，希望您能抽空利用一點時間進行填寫本問卷，感謝您的配合與協助。

本問卷分為問卷 1 與問卷 2 兩個部份，請您先填寫以下基本資料後，再依照問卷 1→問卷 2 的次序填寫，如果在填寫過程中有任何問題都歡迎提出詢問，謝謝。

受測者基本資料：
性別： □男 □女
年齡：_____
畢業科系：________________
出國經驗：________________
(去過的國家?次數? 如：去過美國 3 次、日本 1 次)

問卷 1

請想像你目前正在美國在台協會申請美國簽證，在申請過程中你必須利用面前的指紋辨識掃描裝置進行指紋採集，當你在此裝置的顯示介面上看到下列圖像(Icon)時，請根據所見與直覺認知將該圖像所代表的意義以簡單的文字敘述填入圖像右側空格中，並請注意以下幾點：

文字敘述避免過於冗長，以不超過十五字為原則。
您所填寫的圖像意義並沒有絕對的正確與錯誤之分。

(a)

問卷 2

請想像你目前正在美國在台協會申請美國簽證，在申請過程中你必須利用面前的指紋辨識掃描裝置進行指紋採集，當你在此裝置的顯示介面上看到下列圖像(Icon)時，請根據您所見與直覺認知將下列左側的代碼(A～U)填入到你認為意義相符合圖像的右側空格中，當你認為該圖像沒有與其相符合的意義時，請填入 X 表示以上皆非。
並請注意以下幾點：
同一個代碼可以重複填入到不同的圖像空格中。
並非每個代碼都要使用。
每個圖像空格中請勿填入超過一個代碼。

代碼 意義 圖像 圖像

A. 待命狀態
B. 等待/等候
C. 開始採集
D. 往此方向
E. 掃描手指
F. 啟動生物特徵掃描
G. 啟動指紋掃描
H. 掃描食指
I. 啟動虹膜掃描
J. 看此處
K. 將手往前移
L. 加重按壓力道

(b)

Figure 2. Examples of Questions in Questionnaire 1 (a), and in Questionnaire 2 (b)

Subjects first answered Questionnaire 1 then answered Questionnaire 2 to avoid the carryover effect. A mockup of the fingerprint scanning system shown in Figure 3 was introduced in the questionnaires and provided during answering the questionnaires to help remind the subjects of the operational context.

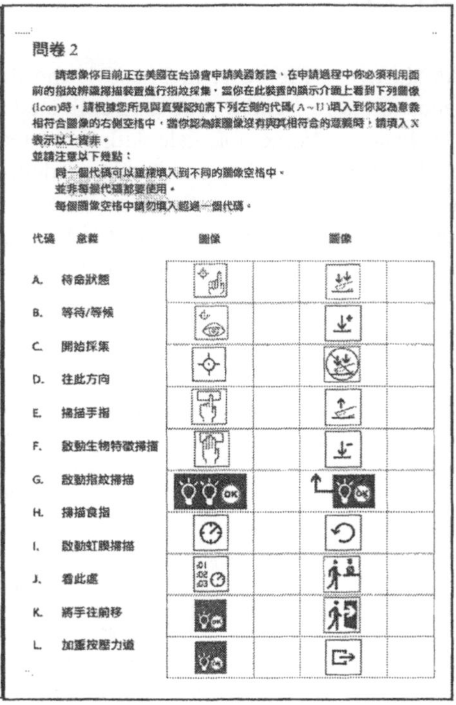

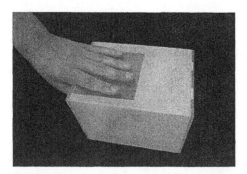

Figure 3. Mockup of the Fingerprint Scanning System

Subjects' demographic data, such as gender, age, and university majors, were collected at the beginning of Questionnaire 1. To make their answers concise, subjects were asked to write down the meanings of each icon in the fill box left to the icon within fifteen Chinese characters. Subjects were also told that there was no correct or wrong answer. In Questionnaire 2, twenty-one possible meanings with codes from A to U were listed, and the twenty icons were also listed with a fill box left to each icon (See Figure 2 (b)). For each icon, subjects were asked to write down a code associated with the meanings that they thought most matched the meanings of the icon. Subjects were asked to only write down a code for each icon, but could apply the same code to different icons. Subjects were also told that it was not necessary to apply all the codes to the icons, and if they could not find any listed meanings that matched the meanings of an icon then they could apply the code of "X" to the icon to indicating "cannot find the answer from the list."

CONFUSION MATRIX AND STATISTICAL TEST

Data collected from Questionnaire 2 were analyzed by the confusion matrix to calculate the correct percentages of icon-meanings matching. A binomial test was applied to examine whether each correct percentage meets the correct percentage of 66.7% recommended by the International Organization for Standard (ISO) (Foster, 1990).

REDESIGN

Three alternatives were designed for each of the icons with low recognition rate in the binomial test. The redesigns were based on general design principles, practical examples, and the feedback from the subjects.

QUESTIONNAIRE 3 AND STATISTICAL TEST

For each redesigned icon, the original and its alternatives were listed in Questionnaire 3. Subjects were asked to rate the original and its alternatives the degrees of conformability to their intended meanings with a 5-point Likert scale. The data from Questionnaire 3 were analyzed through an ANOVA test or similar non-parametric test.

RESULTS AND DISCUSSIONS

Results of Phase 1 (icon interpretation and icon-meanings matching) and results of phase 2 (icons conformability to intended meanings) are presented in this section with discussions.

DEMOGRAPHIC DATA

Among thirty subjects in Phase 1, half (fifteen) was female and half was male. The mean value of their age was 25.7 with the standard deviation of 2.4. Among fifty-five subjects in Phase 2, twenty were female and thirty-five were male. The mean value of their age was 24.3 with the standard deviation of 1.8.

RESULTS OF PHASE 1

A binomial test on the data collected from Questionnaire 2 showed that the icon-meanings matching rates of eleven out of the twenty icons were not significantly lower than 66.7%. These icons, their intended meanings, and matching rates are listed in Table 2.

Table 2 Icons with Acceptable Matching Rates

Icon	Intended Meanings	Matching Rate	Icon	Intended Meanings	Matching Rate
	Start iris scan	80.0%		Press more	93.3%
	Scan Index finger	80.0%		Press less	83.3%
	Scan fingers	63.3%		Retry	60.0%
	Acceptable	83.3%		See guard	96.7%
	Unacceptable	90.0%		Exit	90.0%
	Press more	70.0%			

The binominal test also revealed that the icon-meanings matching rates of the other nine of the twenty icons were significantly lower than 66.7%. These icons, their intended meanings, and matching rates are listed in Table 3.

Table 3 Icons with Low Matching Rates

Icon	Intended Meanings	Matching Rate	Icon	Intended Meanings	Matching Rate
(icon)	Start fingerprint scan	40.0%	(icon)	Press less	3.3%
(icon)	Start capture	10.0%	(icon)	Press less	43.3%
(icon)	Ready state	3.3%	(icon)	Retry	26.7%
(icon)	Wait/hold	40.0%	(icon)	Give up	0.0%
(icon)	Wait/hold	36.7%			

DISCUSSION OF PHASE 1

Results of Phase 1 showed that subjects were not familiar with the abstract symbol of "⟡" found in three icons. For example, about 73% of the subjects could not interpret the icon representing "Start capture" in Questionnaire 1. Though the percentage decreased to 20% for answering Questionnaire 2, the responses of icon-meanings match were diverse to seven different answers. This indicated the confusion of the icon-meanings match. A redesign for better meaningfulness seems necessary.

Subjects also confused with icons sharing common features, such as the icon representing "ready state" could not be distinguished from the icon representing "acceptable" since both had the same blue background color, white light bulb symbol, and green circled OK sign. Another pair of icons with the same problem was the icon representing "retry" (see Table 3) and the icon representing "unacceptable." Both had the same blue background color, white light bulb symbol, and red prohibitive OK sign. Another more abstract icon representing "retry" did not suffered from this problem and met the acceptable matching rate (see Table 2). A redesign for better distinguishability seems necessary.

Both icons with a clock symbol only and with a clock symbol with seconds to represent "wait/hold" (see Table 3) failed to meet the acceptable match rate. The data collected from Questionnaire 1 and 2 showed that it was difficult for subjects to interpret these two icons, and some of the subjects associated them with the wrong meanings of "ready state." A redesign for better meaningfulness seems necessary.

Subject also strongly associated two icons representing "press less" (see Table 3) with the wrong meanings of "stop press." However, a more abstract icon representing "press less" (see Table 2) met the acceptable rate. Results indicated that more abstract icon designs representing "press more/less" were more effectively recognized.

Finally, subject strongly associated the icon representing "give up" (see Table 3) with the wrong meanings of "exit." This might be due to the share with common feature (arrow sign) with the icon represent "exit" (see Table 2). A redesign for better distinguishability seems necessary.

RESULTS OF PHASE 2

As indicated in previous section, four icons representing "start capture", "ready state", "wait/hold", and "give up" were selected for redesign. A Friedman test was applied due to a lack of normality on the data collected from Questionnaire 3. A Wilcoxon test was applied for further post-hoc pair-comparisons. Results were showed in Table 4.

Table 4 Comparisons between Alternative Designs and Originals (Conformabilities of icons were not significant different with the same underlines at $p < 0.05$ level)

Intended Meanings	Conformability ← Higher Lower →			
Start capture				
Ready state				
Wait/hold				
Give up				

DISCUSSION

For conforming to the meanings of "start capture," three alternative designs were significantly better than the original. The icon with a blue round shape and a white triangle inside (see Table 4) had the highest conformability to the intended meanings, followed by the icon with a blue round shape and a white tick sign inside, and followed by the icon with a blue circle and a blue vertical line. Results showed that subjects more accepted the icons with familiar symbols or features found in other existed electronic appliances and devices or computer systems.

For conforming to the meanings of "ready state," three alternative designs were not significantly better than the original. The problem of the original design was the association with the wrong meanings and the confusion with the similar icon representing "acceptable" (see Table 2), so the redesign could not solve the design problem. Further research is necessary.

For conforming to the meanings of "wait/hold," two alternative designs were significantly better than the original. They were the icon with an hourglass symbol (see Table 4) and the icon with a circle composed of twelve center-emitted lines. Both icons can be found in computer systems. Again, subjects more accepted the icons with familiar symbols or features.

For conforming to the meanings of "give up," only one alternative design was significantly better than the original. That is, the icon with a blue round shape and a white cross sign inside (see Table 4) had the highest conformability to the intended meanings. The problem of the original design was the association with the wrong meanings and the confusion with the icon representing "exit" (see Table 2). New design removed the common feature (arrow sign) that originally shared with the confused icon and was more acceptable to the subjects.

CONCLUSION

Design issues regarding the discriminability and meaningfulness of the twenty icons used in biometric systems have been identified. Possible better alternative designs have been explored. To enhance the meaningfulness of icons, the icons representing "start capture" and "wait/hold" should be redesigned with symbols or features that users are more familiar with, or a training or instruction should be provided for users to comprehend the associations between the icons and their intended meanings. To enhance the discriminability of icons, the icons representing "ready state" and "give up" should be redesigned with symbols or features that can be distinguishable from other icons in the same set. Eventually, the common and different features applied for the design of icons as a set should be reviewed with care.

The findings of this study should be able to provide relevant design suggestions for the design of public icons or symbols. Further research on this issue may apply a more systematic design approach, such as the axiomatic design method, for enhancing the overall usability of the icons.

REFERENCES

Foster, J. J., 1990, Standardizing public information symbols: Proposals for a simpler procedure. Information Design Journal, 6/2, 161–168.

Liang, S. -F. M. 2006, Analyzing icon design with axiomatic method: A case study of alarm icons in process control displays, Asian Journal of Ergonomics, 7, 1-2, 11-28.

Tversky, A., 1977. Features of similarity. Psychological Review, 84: 327-352.

Wickens, C. D., Lee, J. D., Liu, Y. and Gordon Becker, S. E., 2004. An Introduction of Human Factors Engineering (2nd. ed.). Pearson Prentice Hall, Upper Saddle River, NJ.

Chapter 5

Design and Evaluate Biometric Device Symbols for Chinese

Pei-Luen Patrick Rau, Jun Liu

Institute of Human Factors & Ergonomics,
Department of Industrial Engineering
Tsinghua University
Beijing, 100084, China

ABSTRACT

The study evaluated the usability and understandability of a set of biometric symbols with 15 Chinese users through interview and symbol-meaning matching tasks. Three characteristics were found in Chinese users' comprehension of the symbols. First, Chinese users understood concrete symbols easier than abstract symbols. Second, symbols were more understandable for Chinese users when the context is provided. Third, action symbols were more easily understood than feedback symbols. The findings supported existing theories in cross-cultural studies and provided meaningful suggestions for biometric symbol design.

Keywords: Chinese user, symbol comprehension, symbol design, biometric device, usability

INTRODUCTION

It is essential to examine carefully the impacts of cultural diversity on human-machine performance, in order to develop or design user interfaces which are effective and easy to use for international users (Choong and Salvendy, 1999). With the rapid proliferation of biometric systems used by the general public in the near future, it is important that the symbols used in those biometric systems communicate consistent meanings across different cultures and do not cause offense. The U.S. National Institute of Standards and Technology's (NIST) Biometrics Usability group has developed a set of pictograms, icons and symbols for fingerprint systems with the goals to enhance user performance with these systems as well as user's understanding of the use and objectives of the systems. Through collaborations between NIST and universities from China and South Korea, a cross-cultural usability study was designed to investigate user's comprehension and understanding of the proposed symbols and icons.

The purpose of this article is to introduce the user study in China. Previous research has documented the cognitive and cultural differences between Chinese and Western populations, including cognitive styles, cognitive (verbal and visual) abilities, digit span, color association, personality and cultural patterns (Choong and Salvendy, 1998). Consequently, Chinese users' comprehension and recognitions for icons and symbols are affected by their culture and cognitive characteristics. For example, Chinese users can understand concrete symbols easier than abstract symbols, while most Westerners have the opposite tendencies. This difference is mainly caused by their different cognitive styles: Westerners have tendency to think in analytic, abstract, imaginative and linear ways, while thinking style of Easterners is synthetic, concrete, relying on periphery and parallel (Rau et al., 2004).

Due to the uniqueness of the Chinese population, the user study in China is expected to investigate Chinese users' comprehension and interpretation of the biometric symbols and to give out suggestions to improve the design according to Chinese users' characteristics.

METHODOLOGY

In the study, a set of biometric symbols and pictograms were proposed by NIST and were evaluated through user testing. This part introduces the symbols, participants, test procedure and data analysis method.

SYMBOLS AND PICTOGRAMS

The proposed set of symbols and pictograms is to help the general public understand the concepts and procedures for using electronic systems that collect and/or evaluate fingerprints. The symbols and pictograms, including both directional symbols and action or feedback symbols, are designed to be used to:

- Provide static instructions related to a fingerprint sensor
- Display dynamic information related to the fingerprint sensor
- Indicate the status of the fingerprint sensor.

In the user study, 24 symbols and pictograms were evaluated. The set of symbols included nine categories:

- Ready State
- Fingerprinting on the Platen
- Start Capture
- Wait/Hold
- Press more/less
- Acceptable/Unacceptable Capture
- Retry
- Give up/Exit.
- Demonstration for the whole scanning process

Within each category there were several alternative symbols. Figure 1 shows one sample of the proposed symbols.

FIGURE 1 Sample of the proposed symbols.

PARTICIPANTS

Fifteen Chinese participated in the study, including eight females and seven males. All participants were students from Tsinghua University in Beijing, China. Participants were specifically recruited with differing educational backgrounds and disciplines. Participants ranged in age from 21 to 29.

PROCEDURE

The user study was divided into two parts: symbols interpretation and symbols and meanings matching.

In the first part symbols interpretation, we used interviews to collect rich, detailed information about the symbols. The objective of the interviews was to identify which of the alternative symbols were most easily understood and which were more difficult to comprehend. This allowed the usability team to identify the most promising symbols.

The participants were interviewed one by one in the same lab. The interviewer first described the background of the study and the context of use of the symbols, explaining that we were designing symbols to describe biometric processing such as fingerprinting and iris scanning in a multi-lingual environment. Then, the 24 proposed symbols were shown to the participant one by one in the same order. Each of the symbols was printed on a separate sheet of paper. The participant was asked to interpret the meaning of each symbol.

The second part was symbols and meaning matching task. Each participant was presented with two pages of test materials. Each page contains a column of possible meanings (numbered and randomized) and a column of the proposed symbols. The task was to match the best meaning, if any, with each symbol. An intended meaning can be used for more than one symbol. It should be mentioned that the symbols showing the whole scanning process were not included in part 2, cause there is no difficulty to match those symbols with their intended meanings.

All participants went through both parts. During the whole process, mockup of a fingerprint scanner was put next to the participant as a visual reminder (see Figure 2).

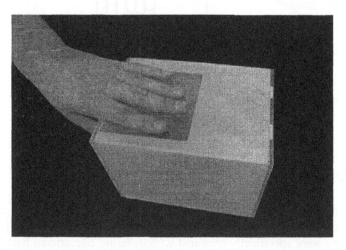

FIGURE 2 Mockup of a fingerprint scanner.

DATA ANALYSIS

The symbol interpretations collected in part 1 were compared with the original intended meaning and graded by a researcher as one of three levels: correct interpretation, approximate interpretation, and incorrect interpretation. A *correct interpretation* was when the participant's interpretation of the symbol matched the intended meaning. For example, Figure 1 shows a symbol with the intended meaning of "start fingerprint scan". A participant's response of "going to ask me to leave a fingerprint" would be marked as a *correct interpretation*. An interpretation of "touch here" would be marked as an *approximate interpretation* as the participant understood that an action involving a finger was depicted. An *incorrect interpretation* would be "go this way" as it does not involve a finger.

The matching data collected in part 2 were calculated for the correct rate of each symbol. Also the incorrect choices were analyzed together to understand users' comprehension.

RESULTS

Table 1 shows the user study results of both parts. Among the 24 symbols, the number of correct interpretations in part 1 ranges from 0 to 14. The number of correct matching in part 2 also has a large range from 1 to 15. By analyzing the quantitative data as well as the original interview records, three characteristics were found in Chinese users' comprehension of the symbols.

First, Chinese users can understand concrete symbols easier than abstract symbols. Among the 24 symbols evaluated in the study, there were 5 pairs of symbols with one concrete and one abstract having the same intended meaning. The 5 pairs were No.7 and 8, No.9 and 10, No.12 and 13, No.16 and 17, and No.19 and 20. Except No.12 and No.13, in the other four pairs, the concrete symbols all got higher number of correct interpretations than the abstract ones. While the exception of No.12 and No.13 was due to the different direction of the arrows: 12 participants got confused with the up arrow in symbol No. 12 and interpret it as "raise up the finger" rather than "press less".

Second, Chinese users can understand symbols easier when the context is provided. To demonstrate the whole scanning process, symbols No.1, 2, 3, 8, 9, 16, 18, 20 were synthesized together into symbols No.21-24. Therefore, in the synthesized symbols, each single symbol was surrounded by a context. This change resulted with easier interpretations of each single symbol by Chinese participants. When shown the symbols No.21-24, the participants often had the tendency to understand the symbols holistically with their context and their links with each other. The context and links helped their understanding a lot.

Third, Chinese users can understand action symbols easier than feedback symbols. For example, symbols No.1, 14 and 15 all feedback the status of the machine or the process. Their numbers of incorrect interpretations in the interview were respectively 14, 12, and 9, lying in a relatively higher level than other symbols. Furthermore, Chinese participants not only understood action symbols better, but also interpreted feedback symbols as indicating user actions as well. For example, symbol No.1 was misunderstood by 6 participants as "turn on the light" or "press the ok button".

Table 1: Chinese user study results

No.	Symbol	Intended meaning	Part 1: interpretation results			Part 2: matching results
			# of correct	# of approximate	# of incorrect	# of correct
Category 1: Ready state						
1.		Ready state	0	1	14	6
Category 2: Fingerprinting on the platen						
2		Scan index finger	0	10	5	14
3		Scan fingers	0	7	8	15
Category 3: Start capture						
4		Start Capture	5	5	5	3
5		Start fingerprint scan	4	11	0	7
6		Start iris scan	8	6	1	11

Table 1 (Continued): Chinese user study results

No.	Symbol	Intended meaning	Part 1: interpretation results			Part 2: matching results
			# of correct	# of approximate	# of incorrect	# of correct
Category 4: Wait/Hold						
7		Wait/hold	0	3	12	4
8		Wait/hold	3	0	12	8
Category 5: Press more/less						
9		Press more	9	3	3	9
10		Press more	6	3	6	12
11		Press less	4	5	6	1
12		Press less	2	0	13	5
13		Press less	8	3	4	13
Category 6: Acceptable/unacceptable capture						
14		Acceptable	3	0	12	14
15		Unacceptable	6	0	9	15

Table 1 (Continued): Chinese user study results

No.	Symbol	Intended meaning	Part 1: interpretation results			Part 2: matching results
			# of correct	# of approximate	# of incorrect	# of correct
Category 7: Retry						
16		Retry	1	2	12	9
17		Retry	2	0	13	2
Category 8: Give up/exit						
18		See guard	4	3	8	10
19		Exist	7	2	6	8
20		Exist	9	1	5	9
Category 9: The whole scanning process						
21		Ready, right hand,4 fingers, then press down, wait 3 seconds, if bad do again	0	6	9	-
22		1st step as before. 2nd step do as before but with left hand.	12	0	3	-

Table 1 (Continued): Chinese user study results

No.	Symbol	Intended meaning	Part 1: interpretation results			Part 2: matching results
			# of correct	# of approximate	# of incorrect	# of correct
23		1 and 2 as before. Step3 with two thumbs.	14	0	1	-
24		1, 2, 3, as before. If ok (green), then exit; if red light go see guard.	4	4	7	-

DISCUSSION AND CONCLUSION

In the usability evaluation of the biometric symbols, Chinese participants were found to understand concrete and contextually related symbols better than abstract and no context symbols. These findings supported existing results and theories in cross-cultural studies. For example, Choong and Salvendy (1998) found that Chinese users can understand concrete symbols easier than abstract symbols, while the American users had an opposite tendency. The difference can be due to their different reasoning and cognitive styles. Chiu (1972) claimed that reasoning of American children has an inferential-categorical style, whereas the reasoning of Chinese children was more likely to display a relational-contextual style. Our findings that concrete and context-related symbols are easier for Chinese users are consistent with the literatures.

Another characteristic of the Chinese user is that action symbols are more understandable than feedback symbols. In other words, Chinese users feel easier when the symbol tells them to do something than to receive something. To the best of our knowledge, this is a brand new finding in cross-cultural usability studies. More empirical studies are needed to test the finding.

All the characteristics of Chinese user discussed above allow usability and design teams to modify the proposed symbols. The findings suggest symbol designers considering three guidelines for Chinese users. First, the symbols should be concrete if possible. Second, the symbols can be shown with context. Third, feedback symbols should be paid additional attention and be as clear as possible.

In conclusion, the study evaluated the usability and understandability of a set of biometric symbols with Chinese users. The symbols that are concrete, context-related and indicating user actions are more easily understood by Chinese users than those abstract, without context, and feedback symbols. The findings supported existing theories in cross-cultural interface design and provided meaningful suggestions for biometric symbol design.

REFERENCES

Chiu, L. H. (1972). A cross-cultural comparison of cognitive styles in Chinese and American children. *International Journal of Psychology*, 7, 235-242.

Choong, Y-Y., Salvendy, G. (1998). Design of icons for use by Chinese in mainland China. *Interacting with Computers*, 9, 417-430

Choong, Y-Y., Salvendy, G. (1999). Implications for Design of Computer Interfaces for Chinese Users in Mainland China. *International Journal of Human-Computer Interaction*, 11(1), 29-46

Rau, PLP., Choong, Y-Y., Salvendy, G. (2004). A Cross Cultural Study on Knowledge Representation and Structure in Human Computer Interfaces, *International Journal of Industrial Ergonomics* 43, 117-1

International Study of NIST Pictograms, Icons and Symbols for use with Biometric Systems - The Case of South Korea

Sang Min Ko, Jong Kyu Choi, Yong Gu Ji

Department of Information and Industrial Engineering
Yonsei University
Seoul, Korea

ABSTRACT

With the importance of the protection of personal information and system security emphasized, the use of biometric system including a fingerprint scanner is being sharply extended. The biometric recognition technology can prevent the illegal use or access of ATM, Mobile Phone, Smartcard, Desktop PC, and Workstation & Computer Networks, so they are being used as an alternative to the existing forgetful PIN or password, and OTP (One Time Password). Such a biometric system is extending its scope of use to the fields requiring personal information in daily life including visa issuance and immigration. Due to the use of the biometric system on behalf of a keyboard or mouse, there comes up the necessity of making changes even to the icon and symbol that we are used to at a viewpoint of a user's interface. In other words, it is required that there should be a research work on Icon and Symbol that can prevent the violative situations in the use of biometric system by conveying a consistent meaning to the users with different cultural backgrounds. The Nation Institute of Standards and Technology (NIST) Biometrics Usability Group made plans for international cooperative work that can convey a common meaning regardless of cultural differences in the use of fingerprint scanning systems with the aim of preparing the international standard plan. This research was executed as an international cooperative work as to the results of research on icons and symbols whose usability was proved for Americans. This research conducted users' evaluation targeting Koreans in relation to the 24 symbols consisting of 8 categories in total and pictogram suggested through NIST-led research results. Selection of 14 subjects, 7males and 7 females, was proceeding with background knowledge on human engineering and different disciplines. The research was proceeded with a users' evaluation and analysis on a matching stage where subjective interpretation of individual signs & pictograms had to be matched with the individual symbols and meaning.

Keywords: Pictograms, Icons, Symbols, Biometric Symbols, Biometric Systems, Fingerprint Scanner

INTRODUCTION

Commonly, Biometric technology is defined as follow. "Biometric technologies are automated methods for recognizing individuals based on biological and behavioral characteristics." (BTAM, 2005). For the Biometric Technology, the ideal employed body part has to be Universal, retain an inherent unique characteristic for each person, have a permanent characteristic than does not alter or change, and easy collectable acquisition and quantification for the sensor. The body parts used in the Biometric Technology is fingerprints, face, iris, blood vessel patter that are physical characteristic and, signature and voice that are behavioral feature. The use of Biometric recognition technology in the modernistic form was in the 1960s, at that moment, it was used the fingerprints to set up stock trading certificate of authentication (Song, 2003). Biometric recognition technology suggest itself a technology with important increase on the usefulness and the protection and security of personal information which give as result an expansion trend on the scope of application in different areas like finance. Recently, because of the internet propagation and information technology development, different types of online service are being revitalized to utilize a safer user's recognition technology.

Biometric recognition technology can prevent the illegal use and illegal connection of ATM(Automated Teller Machines), mobile phone, smart card, desktop PC, workstation and computer network and; it is an alternative the easy to forget existing PIN (Personal Identification Number), password, OTP (One Time Password). Moreover, it is being spotlighted as an alternative for passport and driving license that has probability of robbery or lose. In the United States, the currently implemented US-VISIT (United States Visitor and Immigrant Status Indicator Technology) is the first designated overall system introduced all over the world for entry-exit and frontier administration (Wikipedia, 2009). This system gather and analyze biometric data such as fingerprint scanning and face photographing to trace and compare data base of person considered terrorists, criminals, illegal immigrants. In the early stage of implementation, it was only applied to VISA holders; however, from the 30th of September of 2004, with the Visa Waiver Program (VWP), it is equally applied to every entering person (VWP, 2009). The Republic of Korea Ministry of Justice, followed by United States and Japan, is expecting in 2010 to implement a similar management system as the US-VISIT for foreigner entry-exit named ROK-VISIT (Republic of Korea Visitor and Immigrant Status Indicator Technology) (Wikipedia, 2009). Biometric recognition technology standard is being forwarded by each country of the world standardized organization with various countries standard forms.

OVERVIEW OF NIST USER TESTING FOR SYMBOLS AND PICTOGRAMS

Technology development related to biometric recognition and followed by the accomplished standardized systematic, the application field of biometric recognition system is expected to have a rapid expansion in the near future. In this situation, to deliver consistent meaning to users with diverse cultural background, it is demanded researches of icons and symbols that would allow prevent violation situation beforehand of the direction for the use of biometric recognition system. Accordingly, NIST Biometrics Usability group has developed pictogram, icon and symbol to increase user's understanding of the objective of use and procedure of biometric recognition system utilizing fingerprints. NIST Biometrics Usability group developed symbols and icons were evaluated to users to prove how much it was able to be understood. User's evaluation was proceed with United States users as subject, and that evaluation result verified if the developed symbols were satisfactory. Next, description about the procedure and results of the content and users' evaluation of the suggested symbols and pictograph from NIST Biometrics Usability group is done.

SYMBOLS AND PICTOGRAMS

NIST Biometrics Usability group distinguished the user's fingerprint recognition system application process in 8 able to happen procedure categories from 'Ready State' to 'Give up/exit', and then constitute the 20 symbols and pictograms that can occur in each category (Table1). Through the research, it was suggested some alternative for each category and included the take effect directional situation and action that occur with the users when using the suggested symbols of the fingerprint recognition system or situation of the feedback. Moreover, by combining the 20 symbols it was developed the 'Capture Procedures' that proposed the issued overall interaction procedures visually between the user and fingerprint scanner. 'Capture Procedures' from table 1, explains the combined 11 symbols use to recognize the 10 fingerprints of the overall process.

Table 1 : The set of symbols included eight categories of symbols and capture procedure

Category	Symbols		Category	Symbols		Capture Procedures
1.Ready State			5. Press more /less		Press less (down)	
2.Fingerprint ing on the platen		Scan index finger			Press less (not prohibitive)	
		Scan fingers			Press less (abstract)	
3. Start Capture		Start capture	6. Acceptable/ Unacceptable capture		Acceptable (in instructions)	
		Start fingerprint scan			Unacceptable (in instructions)	
		Start iris scan	7. Retry		Retry (in instructions)	Full capture procedure for 10-print - dual

4. Wait / Hold		Alt 1: clock only			Retry (on device)	thumbs
		Alt 2: Clock with seconds	8.Give up /exit		See guard (after unacceptable captures)	
5. Press more / less		Press more (down)			Give up (abstract)	
		Press more (abstract)			Exit (after successful captures)	

RESULTS

A total of 13 employees of NIST have participated as subjects in the user evaluation. Each subject had different field of study and major with diverse background knowledge, and all the subjects had experience with fingerprint recognition, however, subjects associated with physical recognition area were not included. NIST researchers applied the interview methodology to acquire plentiful and more detailed information about symbols and icons. The objective of the interview is to verify that the optionally suggested symbols to the subjects are easy to understand and there is no interpretation problem.

The user evaluation was proceed in 2 steps. The first step consisted on the meaning interpretation of the combined symbols for the process with symbols and pictogram that were suggested. After the interviewer explained to the subjects about general information of use of fingerprint scanner, it was progressed with the interpretation process of the presented symbol and pictogram printed in the test material. In the second step, it was matched the previously defined meaning by the researches with each of the symbols. The same subject participated for each of the experiments; and each symbol was proceed in order.

Through the users' interview, the collected data was analyzed to then be classified in 'correct interpretation, approximate interpretation, incorrect interpretation' levels.

Correct interpretation refers that the subject interpretation matched with the researcher' intended meaning. Approximate interpretation refers to the not perfect interpretation of the meaning, however, a contextual interpretation of the symbol that includes part of the meaning intended by the researchers. Lastly, incorrect interpretation refers to the totally different interpretation of the meaning by the subject with the intended meaning of the researcher.

From the analysis of the interview, subjects having scientific background knowledge showed a better result in the identification of abstract symbols. Having this result as basis, it was identified that is important to take into consideration the background knowledge of the subjects for afterward studies. Moreover, it was verified that when visual reminder of the context for employment process of fingerprint scanner was given to the subjects for abstract symbols interpretation, the interpretation was very near to the one intended by the researchers.

PRACTICAL USE OF THE SET OF SYMBOLS

NIST Biometrics Usability group had as objective giving assistance to the user about the commonly demanded concept and procedure understanding when utilizing electronic system with the suggested symbols and pictograms from the study result that has as purpose collecting fingerprints or evaluating, Through the study, the presented symbols and pictograms can be employ with the objective of including the possible to occur directional symbols and actions or feedback symbols in the system direction of use.

- Provide static instructions related to a fingerprint sensor
- Display dynamic information related to the fingerprint sensor
- Indicate the status of the fingerprint sensor

Researches of NIST Biometrics Usability group try to constitute the different explanation for each symbols in each of the different situations that can occur when the user interacts with the fingerprint scanner. For instance, when passing through and proceeding in an entry-exit immigration office or customs office or related immigration' procedure, each symbol or pictogram can be proposed to the users in the following form.

- A series of posters while passengers are in the queue
- A series of transitional frames in a biometric booth
- An animated video or series of transitional frames while passengers are in the queue

- Instructional leaflets for passengers to read in the queue

The developed symbols, pictogram and icons by the NIST Biometrics Usability group, expect to provide to the users a high level ease of use by suggesting a consistent concept and procedure when employing biometric information system for finger prints.

USER TESTING FOR NIST SYMBOLS AND PICTOGRAMS IN SOUTH KOREA

NIST Biometrics Usability group planned an international collaborative research that would provide an international standard of the Pictogram, Icon, Symbol transmitting a common meaning about the fingerprint recognition system usage, without cultural difference. The international collaborative reseach programmed the usability evaluation process in two steps. The first one dealing with the performance of developing fingerprint recognition and related Pictogram, Icon, Symbol and the second one having as target users from different countries. In the first step, was performed by the NIST Biometrics Usability group, having as subject 13 NIST employees conducting until the stage of user evaluation and usefulness verification. The second step was programmed with subject ordinary person that researches of each country selected for the evaluation.

PREPARATION OF A TEST

The overall user evaluation performed by the NIST Biometrics Usability group was planned in two equal stages, 'Symbol Interpretation' and 'Symbols and Meanings Matching'; and it was proceed interview to gather diverse subjective data. First, the English version of the experiment content and test material was translated into Korean and   verification of the Localization operation validity was performed. Each experiment was headed with three person in total: one subject, the administrator who is responsible of the overall flow of the experiment and interview, and the note taker who is responsible of recording the comments made by the subject during the experiment. NIST presented the following conditions on the profile of the subjects participating in the user evaluation.

- It is preferred to have participants of mixed educational backgrounds.
- Participants should have limited experiences (shorter than one month) of staying in a Western country (e.g. Europe, Canada, or USA).
- Try to balance the number of male and female participants.

- There are no hard criteria on age or occupations.

There were 14 subjects participating in the user evaluation. The average age of the subjects is 27.5, and it was composed by people between twenties and thirties. Taking into consideration the proportion of male and female subjects, it was selected 7 male and 7 female subjects. The subjects had different background knowledge as Mechanical Engineering, Art History, Computer Engineering, Human Computer Interaction and more. There were not subjects with foreign resident or with long period stay outside the country, and the only subject who experienced biometric recognition technology was the 6th subject, during issuing the US-VISA.

TEST PROCEDURE AND PROTOCOL

To Follow, it is intended to describe about the overall experiment procedure and protocol. Before heading with the evaluation, Pilot testing on the previously passed through Localization process questionnaire was done so as to verify that there were no problems or mistake on the content or problem on the progress. The user evaluation about the Pictogram, Icon and Symbol presented by NIST was performed in the following 4 steps.

- Step 1. Greeting/Briefing : In the first step of the experiment, it was gathered information about the subject demography and explained the overall process and objective of the evaluation.

- Step 2. Symbols Interpretation : In this step, it was collected the opinion and the subjective interpretation of the Pictogram, Icon, and Symbol presented by NIST.

- Step 3. Symbols and Meanings Matching : In this step, it was matched the 20 symbols with the corresponding meaning previously defined by the NIST researchers.

- Step 4. Debriefing : In the last step of the user evaluation, it was identified the subjective point of view of the subject and its doubts.

ANALYSES AND RESULTS

In the Symbol Interpretation procedure, the gathered information through users' interview, was analyzed and then classified in 'Correct Interpretation, Approximate Interpretation, Incorrect Interpretation' levels. For example, in Table 3 the 7th symbol refers to 'Wait/hold'. If it was interpreted as "Wait" or "It shows the time that the finger must be position on the fingerprint scanner." with the same meaning; then, it was classified as 'Correct Interpretation'. For similar and partially correct meaning like "Do something after a set time.", it was classified as 'Approximate Interpretation"; and for meanings completely different like "6 minutes have passed.", it was classified as 'Incorrect Interpretation'. 'Correct Interpretation Rate' refers to the rate of answers by the subject about the symbols classified as 'Correct Interpretation'. The 'Interpretation Rate' is the difference between 'Correct Interpretation' and 'Approximate Interpretation'. The analysis of the Symbol Interpretation is shown in Table 2.

Table 2 Analyses Results of Symbols Interpretation

No.	1	2	3	4	5	6	7	8	9	10
Correct Interpretation	6	10	11	1	0	0	4	1	6	10
Approximate interpretation	1	3	3	13	14	1	2	3	6	0
Incorrect interpretation	7	1	0	0	0	13	8	10	2	4
CIR (%)	42.9	71.4	78.6	7.1	0.0	0.0	28.6	7.1	42.9	71.4
IR (%)	50.0	92.9	100	100	100	7.1	42.9	28.6	85.7	71.4

No.	11	12	13	14	15	16	17	18	19	20
Correct Interpretation	5	0	9	12	12	10	2	12	8	0
Approximate interpretation	0	1	0	0	0	3	5	2	5	0
Incorrect interpretation	9	13	5	2	2	1	7	0	1	14
CIR (%)	35.7	0.0	64.3	85.7	85.7	71.4	14.3	85.7	57.1	0.0

IR (%)	35.7	7.1	64.3	85.7	85.7	92.9	50.0	100	92.9	0.0

※ Note: L – Left, M – Middle, R – Right, CIR – Correct Interpretation Rate, IR – Interpretation Rate

The analyses of the Interpretation Rate demonstrate that the subject interpreted the 12 symbols (numbers 2, 3, 4, 5, 9, 10, 13, 14, 15, 16, 18, 19) similarly(more than 60%) compared to the correct meaning. On the contrary, 8 symbols (number 1, 6, 7, 8, 11, 12, 17, 20) were misinterpreted by the subjects. Only number 4 and 5 showed a high Interpretation Rate, while the rest 9 symbols were classified as 'Approximate Interpretation'. Moreover, from the symbols with low interpretation rate, number 6, 8, 12, 20 takes a rate above 70% of 'Incorrect Interpretation' while the other 5 symbols were not.

From the Symbol Interpretation analysis where Korean subjects participated in the user evaluation, it appears that they had difficulties in the interpretation of the parts Ready State, Wait/Hold, Press Less from Press More/Less Category. Moreover, 'Start capture' from Start Capture Category, 'Retry(on device)' from Retry Category, and 'Give up' from Give up/Exit Category presented difficulties for the subjects. Also, from the result analysis of Press More/Less Category, showed a rate of 42.8% (down) and 71.4% (abstract) for Press More symbols' Correct Interpretation; and a rate of 35.7% (down), 0% (not prohibitive), and 64.3% (abstract) for Press Less symbols' Correct Interpretation. Having as basis these results analysis, it is possible to demonstrate that the Korean subjects had a near interpretation of the abstract symbols. The analysis of Capture Procedure verified that the 11 symbols combined were more familiar and easy to understand its meaning for the subjects compare to when it was given separately.

In the analysis of Symbols and Meaning Matching step, by the test materials, it was calculated the numbers of correct coinciding symbols with their meaning. The 'Accuracy Rate' is the rate of correct matching between symbols and meaning that each subject had answer. The analysis of the result about Symbols and Meaning Matching is shown as follow in Table3.

Table 3 Analyses Results of Symbols and Meanings Matching

No.	1	2	3	4	5	6	7	8	9	10
AR (%)	50.0	100	64.3	42.9	92.9	7.1	64.3	42.9	57.1	92.9

No.	11	12	13	14	15	16	17	18	19	20
AR (%)	0.0	35.7	85.7	85.7	100	28.6	28.6	100	78.6	0.0

☐ Note: L – Left, M – Middle, R – Right, AR – Accuracy Rate

The analysis of the result obtained about Accuracy Rate showed that the subjects had selected more than 60% of the meaning previously defined of 10 symbols (numbers 2, 3, 5, 7, 10, 13, 14, 15, 18, 19). On the contrary, for the rest 10 symbols (numbers 1, 4, 6, 8, 9, 11, 12, 16, 17, 20), it was selected a different meaning from the previous one. The symbols with high CIR (Correct Interpretation Rate) for Symbol Interpretation had also high Accuracy Rate for Symbols and Meaning Matching, and included symbols with high IR (Interpretation Rate). However, symbol 16 that refers to Retry showed a high CIR and IR in Symbol Interpretation, the result for AR was low (28.6%) for Symbols and Meaning Matching.

CONCLUSION AND FURTHER STUDY

This research aimed to proceed an user evaluation, targeting 14 Koreans, about the 24 symbols and pictograms created by the NIST Biometrics Usability group; and then analyze the results. The analysis of the result of Symbols Interpretations demonstrated that the Korean subjects presented difficulties in the interpretation of Ready State, Wait/Hold, and Press Less from Press More/Less Category. Moreover, Symbols and Meaning Matching analysis showed that the subjects had difficulties in the interpretation of Ready State, Press Less from Press More/Less Category, and Retry. Through this analysis, it was possible to verify the interpretation of Koreans about the symbols and pictograms developed by the NIST Biometrics Usability group with the purpose of delivering a consistent meaning to users with different cultural background. This research could be utilized as basis for further studies related with symbols and pictograms development for fingerprint users to deliver a more clear understanding of the symbols and pictograms. Furthermore, it would be possible to applied when developing symbols and pictograms not only for fingerprint, but also for others biometric recognition machine.

REFERENCES

Compiled and Published by National Biometric Security Project. (2005). Biometric Technology Application Manual Volume 1 Biometrics Basics
NIST Biometrics and Usability Website. (2009). : ,
http://zing.ncsl.nist.gov/biousa

64

Wikipedia – Pictogram. (2009). :
http://en.wikipedia.org/wiki/Pictogram
Wikipedia – ROK-VISIT. (2009). :
http://ko.wikipedia.org/wiki/ROK-VISIT
Wikipedia – US-VISIT. (2009). :
http://en.wikipedia.org/wiki/US-VISIT
Wikipedia – ROK-VISIT. (2009). :
http://ko.wikipedia.org/wiki/ROK-VISIT
Song. (2003). "Application biometric authentication technology for strengthen financial security",Hyundai Information Technology.
TRAVEL.STATE.GOV. (2009). Visa Waiver Program Website :
http://travel.state.gov/visa/temp/without/without_1990.html

Chapter 7

A Cognitive and Holistic Approach to Developing Metrics for Decision Support in Command and Control

Daniel Lafond[1], François Vachon[2],
Robert Rousseau[3], Sébastien Tremblay[2]

[1]Defence R&D Canada

[2]Université Laval

[3]Neosapiens Inc.

Université Laval

ABSTRACT

We present an assessment methodology to determine to what extent command and control and crisis management systems effectively support the cognitive processes that lie at the heart of these activities. The approach is focused on cognitive functions that cover the set of processes engaged by command and control and crisis management. We put forward three cognitive functions as being key for achieving optimal cognitive support in these contexts: Situation monitoring, attentional control/management, and planning/coordination of activities. Metrics for the holistic assessment of cognitive support are derived from online measurement using well-established cognitive psychology paradigms directly related to these functions.

Keywords: Cognitive systems engineering, command and control, crisis management, decision support, metrics

INTRODUCTION

Information systems are increasingly used to support cognitive work, particularly in complex and dynamic tasks. Such tasks include Command and Control (C2) and Crisis Management (CM). C2 is defined as the exercise of authority and direction by a designated leadership over resources in the accomplishment of a mission (Keane, 2005). While C2 is predominantly associated with military operations, it also applies to many non-military activities, including CM. C2 often takes place in the context of high-reliability organizations that operate in dangerous or volatile environments under severe constraints such as high risk, time pressure, complexity, and ambiguity (e.g., the response phase in CM). The functions of C2 mainly concern planning, directing, coordinating, and controlling the employment of available resources. Executing C2 functions engages a variety of cognitive processes such as monitoring, recognition, causal learning, search, planning, judgment, and choice (Gonzalez, Vanyukov, & Martin, 2005). The ability to successfully manage cognitive resources under time constraints is key to the successful exercise of C2. Advances in information technologies provide a potential for improving quality and speed of C2 with computerized support tools known as decision support systems (DSS) or C2 systems (Alberts & Hayes, 2006). Such systems aim to augment the sensemaking, decision making and execution processes of commanders and their team. Here, we present recent progress in developing an assessment methodology to determine to what extent a DSS effectively supports the cognitive processes that lie at the heart of C2 functions.

COGNITIVE SUPPORT FUNCTIONS

A DSS is an interactive computer-based system intended to help decision-makers utilize data and models to identify problems and make decisions (Sprague, 1980). C2 systems refer more specifically to information systems designed to support C2 activities both in the military domain and in CM (e.g., Jungert & Hallberg, 2009). Research on DSS has been primarily concerned with the behavioral aspects of managerial work and has often neglected to address the cognitive aspects of decision support (Chen & Lee, 2003). Identifying means to align a DSS with the requirement of cognitive support is a formidable challenge. Some forms of cognitive support can actually hinder rather than enhance performance (e.g., Lerch & Harter, 2001). Providing cognitive support proves very challenging in dynamic situations evolving in real-time, which is typical of C2.

Recent progress in assessing cognitive support in a DSS has been achieved by using an approach called Decision Centered Testing (DCT; see Potter & Rousseau, 2010). DCT involves creating specific test conditions that exert pressure on the decision making process. Pressure points are indentified through a work domain analysis. DCT is derived from a Cognitive Systems Engineering (CSE) framework (Woods & Roth, 1988). CSE considers the human agent and the information system as a single unit: the joint cognitive system. Taking a joint cognitive system perspective implies that the purpose of a DSS is to team with the human user to achieve coordinated and effective cognitive work. DCT is based on a set of five basic requirements from CSE (see Potter, Woods, Roth, Fowlkes, & Hoffman, 2006) that are essential for a joint cognitive system to effectively conduct cognitive work in any goal-directed task like C2. These requirements are: observability (ability to form insights into a process), directability (ability to re-direct resources and activities), teamwork with agents (ability to coordinate and synchronize across agents), directed attention (ability to shift and re-orient focus) and resilience (ability to anticipate and adapt to unexpected changes or events). Although these generic requirements provide guidance for deriving cognitively relevant metrics, these support requirements are not clearly connected to underlying cognitive processes.

The basic tenet of the present research is that major progress in providing effective support to decision-makers working with a C2 system can be achieved by identifying cognitive functions that cover the set of processes engaged by C2. Metrics derived from these functions then form the basis for evaluating the effectiveness of cognitive support in a DSS. Based on an exhaustive review of the literature, previous work and Subject Matter Expert (SME) interviews, we put forward three cognitive functions as being key for achieving optimal cognitive support in C2: 1) Situation monitoring; 2) Attentional control/management; and 3) Planning/coordination of activities. *Situation monitoring* enables the development of an awareness of the situation through perception, recognition, and interpretation, and can be associated with the "observability" support function. *Attentional control/management* refers to the interplay between the voluntary deployment of attention (based on knowledge, expectations, and current goals) and reflexive attentional shifts caused by events with absolute or relative attention-attracting properties. It can be associated with the "directed attention" support function. *Planning/coordination of activities* involves dynamically engaging, interrupting, and re-engaging in multiple concurrent tasks in accordance with changing priorities. Planning/coordination can be associated with the "directability" support function. The "teamwork with agents" support function can be associated to information sharing and interpersonal coordination. Finally, the "resilience" support function is not associated to a specific cognitive function. Indeed, it is a behavioral outcome observed by directly considering performance under various conditions.

ASSESSMENT METHODOLOGY

Metrics for the assessment of cognitive support in a DSS are derived from online measurement using well-established cognitive psychology paradigms directly related to the cognitive functions previously identified as key for effective C2. Similar to the DCT approach, the present methodology consists in measuring how well the joint cognitive system accomplishes the critical cognitive functions of C2 identified by selectively exerting pressure on each of these functions. The actual tests can be performed in a laboratory setting, using microworlds or high fidelity simulations, or in the field during a scripted exercise. Provided that the necessary online measures are collected, it is even possible to apply these metrics for the evaluation of cognitive support based on data from real (non-scripted) operations. Controlled environments will lead to measurements with a greater internal validity (confounding factors are eliminated) whereas natural task settings confer a greater external validity (results may generalize better to work as it occurs in the field).

Situation Monitoring

Situation monitoring effectiveness is tested using the change blindness paradigm (e.g., Durlach, 2004). Change blindness is defined as the failure to detect changes that occur in a visual scene. In complex and dynamic environments, operators cannot remain focused on one aspect of the DSS interface, and the operators' ability to detect change when they turn their attention back to a given aspect of the DSS is crucial. Since dynamic environments typically involve numerous changes (either discreet or continuous), the key to applying this paradigm is to identify a subset of changes deemed critical for the successful accomplishment of the task. The paradigm requires a measurable response from the participant (e.g., clicking on the element that has changed) to indicate that the critical change has been detected (e.g., Smallman & St. John, 2003). The assessment of situation monitoring can be enhanced by eye-movement recordings (see Figure 1), which can serve as an on-line index of attentional allocation, or at least, of some visual processing over the display. Indeed, eye tracking allows dissociating the contribution of perceptual and attentional factors to failure in situation monitoring, for example, by contrasting undetected changes that were never fixated to changes that remained undetected despite being fixated (Vachon, Hodgetts, Vallières, & Tremblay, 2010).

Attentional Control/Management

Research on dynamic decision-making showed attention management to be a key issue for cognitive support in time-sensitive C2 tasks (Boiney, 2005; Lerch & Harter, 2001). Attentional control/management is tested using distraction as a paradigm that measures the disruptive effects of involuntary attentional shifts on task execution. Distracting stimuli can be of any modality (visual, auditory, tactile, etc.). For example, experimental findings show that the effects of auditory distraction are substantial, widespread, and pervasive (see Banbury, Macken, Tremblay, & Jones, 2001, for a review). In applied environments, extraneous speech and noise are one of the most often-mentioned sources of disturbance (Beaman, 2005). The grounds for using distraction as a paradigm to test attentional control/management lie in the property of irrelevant stimuli to attract attention away from the processing of the task (see Cowan, 1995). The effects of distraction correspond to the difference (on one or several performance indices) between conditions with and without distraction.

Planning/Coordination of Activities

Planning/coordination of activities is investigated using the interruption paradigm. This paradigm can be applied to dynamic control tasks by interrupting the on-going activity with a higher priority task in order to measure the cost of such interruptions. This situation is typical of C2 environments, where operators must manage several tasks and deal with continuously evolving situations. Task interruption is a topic of major interest in C2 research (e.g., Trafton, Altmann, & Brock, 2005). The findings that interruption is a key aspect of strategic task management in complex and dynamic environments and that it has a negative impact on goal activation and retrieval (e.g., Hodgetts & Jones, 2006) constitute key reasons for using the task interruption paradigm to put pressure on planning and coordination of activities in C2 environments. The effects of interruption correspond to the difference (in terms of performance indices) between conditions with and without interruption.

ADAPTING THE APPROACH TO THE CONTEXT

The method described herein focuses on three generic cognitive functions underlying C2 to assess the effectiveness of a DSS in terms of cognitive support. While the main aspects of this approach are intended to be applicable to all C2 tasks, it is necessary to adapt the paradigms used for testing each cognitive function to the specific work context for which a DSS is designed. Indeed, the specific changes in a situation deemed critical can only be defined by considering the specifics of the work domain. The type of distractors to use (e.g., task-irrelevant visual stimuli or auditory signals) should be of the same nature as those typically encountered in the work domain. The duration and nature of interruptions should also take into account those generally observed in natural settings.

Application to Naval Air Defense

As an example, we provide an overview of the application of this methodology to the context of naval air defense (Lafond et al., 2009). We developed a microworld called Simulated Combat Control System (S-CCS) as a testing platform. Microworlds are simulated task environments designed for the study of decision making in complex and dynamic situations and provide a good compromise between experimental control and realism (Brehmer & Dörner, 1993). The S-CCS microworld simulates the threat evaluation (i.e., threat intent and threat capability assessment) and combat power management (i.e., response planning, execution, and monitoring) processes. The scenarios developed involve multiple unknown radar contacts (called tracks) moving in the vicinity of the ship with possible missile attacks requiring retaliatory missile firing from own ship. The participant plays the role of a sensor weapon controller on a frigate. This role involves three sub-tasks: 1) determine the level of threat (i.e., hostile, uncertain, or non-hostile) of all the tracks; 2) determine the level of threat (temporal) immediacy of all tracks classified as hostile; and 3) neutralize hostile tracks. The task is performed individually on a computer, following a tutorial and practice sessions.

Three variants of a fictitious C2 system are compared (see Figure 1). DSS 0 stands for a relatively standard C2 system, simplified compared to its real-world counterparts for use in S-CCS. Its visual interface has three parts: 1) a tactical geo-spatial display screen, 2) a track parameters text list, and 3) a set of action buttons. The tactical display screen is a type of radar display representing in real time all objects moving on the screen. DSS 1 adds a temporal overview display (TOD) to the DSS-0 interface (see Rousseau, Tremblay, Lafond, Vachon, & Breton, 2007). The TOD, located to the right of the geospatial display, represents temporal information by presenting each track across a timeline. In fact, the TOD presents the same information as the geospatial display, but in a time-scaled event-based representation. This was done to provide explicit support to maintaining temporal awareness. However such a representation is more abstract in as much as it does not visually represent objects in their physical environment. DSS 2 adds a change-history table to the basic DSS 0 interface. This form of cognitive support is inspired by work from Smallman and St. John (2003) in the naval air warfare domain. The change-history DSS automatically detects important changes to the situation and logs them in a table that is linked to the geospatial display. Each time a critical change happens, the track id number and the time the change occurred (in ms) are instantly logged in a table located below the geospatial display. Information from the most recent change is always presented in the left cell and each time a new change is recorded, the information moves to the right of the screen.

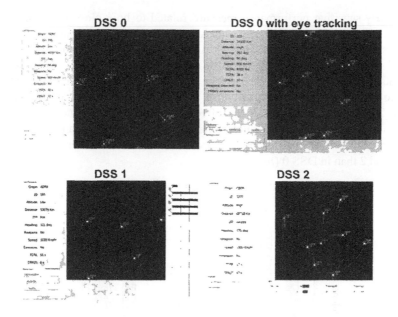

FIGURE 1. Screenshot of each DSS used in the present study.

The three paradigms proposed in the present assessment methodology were adapted to the naval C2 simulation. In the context of S-CCS, a critical change was defined as a change in track parameters that lead to a higher threat level (a non-hostile or uncertain contact becoming hostile). Participants were required to re-evaluate the threat-level of contacts following such a change. The change was considered detected if the track was selected within 15-s and then reclassified. Distraction was implemented in the simulated naval air defense task through the auditory modality as fragmented conversations that were not task-relevant. These conversations were created using a 3-D auditory recording system and played through headphones in order to increase their attention-capturing power. Interruptions consisted in the sudden occurrence of a secondary task (questions presented on-screen) related to the context of S-CCS but irrelevant in regards to the primary task. The questions were based on the QUASA technique (Quantitative Analysis of Situational Awareness) developed by McGuiness (2004) which combines both objective queries (true/false probes) and subjective self-ratings of confidence for each response. Although QUASA probes are not naturalistic interruptions, the task of a sensor weapon controller can be interrupted by similar requests from the operations room supervisor. The presence or absence of distraction and interruption was manipulated so as to yield four experimental conditions forming a 2 × 2 within-subject design (critical changes occurred in all conditions). Performance measures include threat evaluation accuracy and combat power management effectiveness (proportion of incoming missiles successfully neutralized before reaching the ship). Defensive effectiveness is a novel and particularly sensitive performance measure related to how close the ship came to being hit. This measure is defined as the sum of the time-to-ship value for all hostile contacts when either neutralized or when they hit the ship. A total of zero means that all hostile contacts that attempted to hit the own ship succeeded in doing so. The total is then divided by the number of hostile contacts in order to obtain an average time-to-ship value. Higher values indicate a greater defensive effectiveness. Figure 2 shows an overview of the effects of each DSS, of interruption and of distraction on defensive effectiveness. The effects of interruption and distraction are significant in every DSS ($ps < .05$). Critically, defensive effectiveness was lower in DSS 1 and 2 than in DSS 0 ($ps < .01$).

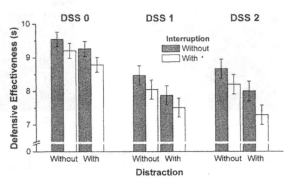

FIGURE 2. Holistic DSS comparison.

As seen from the example in Figure 2, the proposed assessment methodology can provide a holistic evaluation of various DSS. Indeed, the approach makes it possible to see the overall effects of a DSS on cognitive functioning. It can show whether a specific gain is accompanied by other gains or losses on other dimensions, allowing for a more comprehensive appreciation of the effects of a DSS. Interestingly, the example in Figure 2 shows that although DSS 1 and DSS 2 may have other benefits (e.g., in terms of change detection, threat evaluation, etc.), they actually lead to a reduced defense effectiveness compared to DSS 0.

CONCLUSION

The prevalence of surveillance and information collection technologies provides decision-makers with greater volume and complexity of information to monitor and on which to base decisions than ever before. In this ever increasing dynamic and information rich environment, the role for DSS to augment cognition is growing ever more important. However, unless a better understanding is gained of the factors that facilitate the cognitive processes underlying decision-making, the design and development of such technology may serve only to exacerbate rather than enhance the desired effect. We proposed a methodology to evaluate the efficiency of a DSS based to three generic cognitive functions that cover the set of processes engaged in C2. We described the research paradigms relevant to the assessment of each cognitive function and show how each paradigm can be applied in a specific task context by providing an example based on recent work using a simulated naval air defense task. The scenarios and conditions utilized allowed us to study cognitive functions in interaction and to make a comparative assessment of a baseline interface and two experimental DSS. This methodology provided an integrated and comprehensive assessment of cognitive support, which could not have been obtained by relying on isolated measure-specific tests.

Cognitive support may come with specific benefits and drawbacks that can be identified using different cognitive metrics and performance measures (speed vs. accuracy, number vs. types of errors, etc). For instance, some forms of support may increase the ability to detect changes but slow other aspects of task execution. In some conditions this will lead to increased performance while reducing efficiency in other conditions. Understanding these interactions will allow optimizing cognitive support for the specific requirements of the situation. Some key advantages of the present approach are that these measurements are unobtrusive and objective and can be applied online, both in controlled laboratory simulations and in the field. These metrics can provide a useful diagnostic capability for the evaluation of C2 systems and for the development of advanced adaptive decision support technologies.

The goal of this work was to provide a unique evaluation methodology to assess whether or not a DSS truly augments specific aspects of cognition. The innovation in this endeavor lies in the holistic approach adopted in deriving our metrics from concurrent rather than isolated manipulations and measurements. Future work will be directed to extending the current approach for the assessment of support to team cognition in C2 environments.

ACKNOWLEDGEMENTS

We are thankful to Julie Champagne, Sergei Smolov, Thierry Moisan, Laurence Dumont, and Benoît Vallières for assistance in programming, data collection, and data analysis. This work was supported by a R&D partnership grant from the National Sciences and Engineering Research Council of Canada with Defence R&D Canada, Neosapiens Inc. and Thales Canada, Systems Division.

REFERENCES

Alberts, D.S., & Hayes, R.E. (2006), *"Understanding Command and Control. The Future of Command and Control."* CCRP, DoD, Washington, DC.

Banbury, S., Macken, W.J., Tremblay, S., & Jones, D.M. (2001), "Auditory distraction and short-term memory: Phenomena and practical implications." *Human Factors*, 43, 12–30.

Beaman, C.P. (2005), "Auditory distraction from low-intensity noise: A review of the consequences for learning and workplace environments." *Applied Cognitive Psychology*, 19, 1041–1064.

Boiney, L. (2005), "Team decision making in time-sensitive environments."

Proceedings of the 10th international Command and Control Research and technology Symposium. MacLean, VA.

Brehmer, B., & Dörner, D. (1993), "Experiments with computer simulated microworlds: Escaping both the narrow straits of the laboratory and the deep blue sea of the field study." *Computers in Human Behavior*, 9, 171–184.

Chen, J.Q., & Lee, S.M. (2003), "An exploratory cognitive DSS for strategic decision making." *Decision Support Systems,* 36, 147–160.

Cowan, N. (1995), *"Attention and memory: An integrated framework."* Oxford: Oxford University Press.

Durlach, P.J. (2004), "Change blindness and its implications for complex monitoring and control systems design and operator training." *Human-Computer Interaction,* 19, 423–451.

Gonzalez C., Vanyukov, P., & Martin M.K. (2005), "The use of microworlds to study dynamic decision making." *Computers in Human Behavior*, 21, 273–286.

Hodgetts, H.M., & Jones, D.M. (2006), "Interruption of the Tower of London task: Support for a goal-activation approach." *Journal of Experimental Psychology: General*, 135, 103–115.

Jobidon, M.-E., Breton, R., Rousseau, R., & Tremblay, S. (2006), "Team response to workload transition: The role of team structure." *Proceedings of the 50th Meeting of the Human Factors and Ergonomics Society*. San Francisco, CA.

Jungert, E., & Hallberg, N. (2009), "Capabilities of C2 Systems for Crisis Management in Local Communities." *Proceedings of the 6th International ISCRAM Conference*. Gothenburg, Sweden.

Keane, M. (2005), "C2." *Dictionary of Modern Strategy and Tactics*. US Naval Institute Press.

Lafond, D., Champagne, J., Hervet, G., Gagnon, J.-F., Tremblay, S., & Rousseau, R. (2009), "Decision analysis using policy capturing and process tracing techniques in a simulated naval air-defense task." *Proceedings of the 53rd Meeting of the Human Factors and Ergonomics Society* (pp. 1220–1224). San Antonio, TX.

Lerch, F.J., & Harter, D.E. (2001), "Cognitive support for real-time dynamic decision making." *Information Systems Research,* 12, 1, 63–82.

McGuinness, B. (2004), "Quantitative Analysis of Situational Awareness (QUASA): Applying Signal Detection Theory to True/False Probes and Self-Ratings." Bristol, UK.

Potter, S., & Rousseau, R. (2010), "Evaluating the Resilience of a Human-Computer Decision-Making Team: A Methodology for Decision-Centered Testing." In E.S. Patterson and J. Miller (Eds.), *Macrocognition Metrics and Scenarios: Design and Evaluation for Real-World Teams*. Aldershot: Ashgate.

Potter, S.S., Woods, D.D., Roth, E.M., Fowlkes, J., & Hoffman, R.R. (2006), "Evaluating the effectiveness of a joint cognitive system: metrics, techniques, and frameworks." *Proceedings of the 50th Annual Meeting of the Human Factors and Ergonomics Society* (pp. 314–318). Santa Monica, CA.

Rousseau, R., Tremblay, S., Lafond, D., Vachon, F., & Breton, R. (2007), "Assessing temporal support for dynamic decision making in C2." *Proceedings of the 51st Annual Meeting of the Human Factors and*

Ergonomics Society (pp. 1259–1262). Baltimore, MD.

Smallman, H.S., & St. John, M. (2003), "CHEX (Change History EXplicit): New HCI concepts for change awareness." *Proceedings of the Human Factors and Ergonomics Society 46th Annual Meeting* (pp. 528–532). Santa Monica, CA.

Sprague, R.H. (1980), "A framework for the development of decision support systems." *Management Information Sciences Quarterly*, 4, 1–26.

Trafton, J.G., Altmann, E.M., & Brock, D.P. (2005), "Huh, what was I doing? How people use environmental cues after an interruption." *Proceedings of the Human Factors and Ergonomics Society 49th Annual Meeting*. Orlando, FL.

Vachon, F., Hodgetts, M.M., Vallières, B.R., & Tremblay, S. (2010), "Detecting Transient Changes in Dynamic Displays: When Sound Makes You "More" Blind." *Poster presented at 22nd Annual Convention of the Association for Psychological Science*, Boston, MA.

Woods, D.D., & Roth, E.M. (1988), "Cognitive Systems Engineering." In M. Helander (Ed.), *Handbook of Human-Computer Interaction* (pp. 3–43). New York: North-Holland.

Chapter 8

A Cognitive Systems Engineering Approach to Support Municipal Police Cognition

Edward J. Glantz, Michael D. McNeese

The Pennsylvania State University

ABSTRACT

Municipal police officers perform cognitive activities, such as decision making, judgment and problem solving, while facing real world constraints that include large, socially distributed problem spaces with potential for high risk outcomes. The goal of this paper is to improve cognitive effectiveness in the municipal policing domain by successively applying cognitive systems engineering methodologies to 1) understand this police domain and related cognitive constraints, 2) analyze sources of complexity and needs in the cognitive activities, and 3) based on these findings develop technology interventions to reduce officer task complexity within domain constraints. Findings suggest it is possible to improve the use of technology to enhance cognitive effectiveness, and that there are strengths in combining methodologies, such as concept mapping and the critical decision method, when identifying interventions.

Keywords: cognitive systems engineering, municipal police, cognition

INTRODUCTION

In their performance of duties, police officers need to make effective decisions, show good judgment, remember details and react quickly to evolving situations. This is not always accomplished, however, as illustrated in these scenarios:

- As it gets close to bar closing time, Officer Stevens finds herself behind a suspicious car. As they slow to a traffic light, she decides to reach over and enter the car's license into her mobile data terminal. Although her attention was only diverted for a moment, she still ran into the car she was looking up.
- Officer Corl is responding to a silent bank alarm. Unfortunately, there are so many new banks in that part of town that he went to the wrong one. The second bank was not right either. Frustrated and delayed, he finds the bank on the third try.
- Officer Walker approaches a house to serve a warrant. He wishes he had more information from the caseworker. As the door opens, his fears are confirmed – it is now a barricaded gunman situation.

The officers in these and other scenarios were observed and interviewed as they worked in an effort to understand and improve the cognitive aspects of their work. Police work involves uncertainty, evolving scenarios, collaboration with other officers and agencies, and information derived or mediated by information systems. This combination of complex work, officers and system dependence often results in suboptimal results. In several situations, the information systems that were intended to support officer cognitive activities, such as judgment and decision-making, actually had an opposite effect. For example, poorly designed interfaces, partial system access or limited system availability resulted in cognitive breakdowns that could distract officers in the performance of their duties. In other situations, support tools that could have helped were noticeably missing, also creating an unnecessary cognitive burden on officers.

An understanding of this problem is important to domain practitioners, cognitive engineers and researchers. Effective cognition and collaboration is important to police who are faced with increasingly high stakes and risks that result from the outcomes of their decisions.

The goal of this paper is to improve cognitive effectiveness in the policing domain through three objectives: (a) develop an understanding of the police domain, and the effect domain constraints place on the cognitive activities of officers; (b) analyze the cognitive tasks undertaken by police officers to identify sources of complexity and specific officer needs; and (c) explore technology interventions to reduce task complexity and accommodate domain constraints. The first and second objectives reflect analytical efforts, while the third objective extends this analysis into design intervention.

Contributions from this paper include development of the MAKADAM methodology that facilitates identification of cognitive support interventions by integrating ethnography, knowledge elicitation and active participatory design, design of a prototype that supports the cognitive activities of police officers during critical incidents, collection of policing data over a two-year period, and analysis of policing data and eight analytical findings.

LITERATURE REVIEW

The first analytical objective is to develop an understanding of cognitive constraints within the police domain, and the second is to develop insight into the complexity of cognitive police tasks. These understandings can then be used to explore a technology intervention objective to reduce task complexity while accommodating domain constraints.

There are several methodological and theoretical foundations available to study the design of digital artifacts used in work, including cognitive systems engineering (CSE), human computer interaction (HCI), and computer-supported cooperative work (CSCW), for example. CSE is a multidisciplinary field with scientific conceptual foundations originating in the 1980's (Hollnagel & Woods, 1983; Norman, 1986; Rasmussen, Pejtersen, & Goodstein, 1994; Woods & Roth, 1998). Since then it has evolved from a scientific endeavor to also include tools and methods to inform engineering design (Dowell & Long, 1998; Eggleston, 2002). CSE now represents both a theoretical science and practical engineering within the context of system design for complex systems. Furthermore, CSE "toolkits" or frameworks have been developed for the analysis of work domains (McNeese, Zaff, Citera, Brown, & Whitaker, 1995; Potter, Roth, Woods, & Elm, 2000; Vicente, 1999). Each framework represents multiple techniques, methods and approaches for knowledge elicitation, knowledge capture and design support.

CSE is useful to study work environments that find "users in an information-rich world with little time to make sense out of events surrounding them, make decisions, or perform timely activities" (McNeese, 2002, p. 80). CSE takes a user-centered and problem-based focus to designs that support users in complex settings. This means creating designs that are not just usable, but also useful in supporting decisions and other cognitive activities within time constraints at varying levels of information richness.

The relevance of information technology, and the information it processes, is highly dependent on the user's ability to use and understand the technology (McNeese,

2002; Woods, 1998). Cognitive systems engineering expands the level of analysis beyond the user or the computer-user dyad. The unit of analysis in CSE is a cognitive system (Roth, Patterson, & Mumaw, 2002; Woods & Roth, 1998). This is a triad consisting of the user, the system and the work context. As such, CSE is a method suitable for analysis of collaborative activities and represents a significant strength relative to traditional human-computer research which tends to focus on either the user (e.g. solitary cognition, problem-solving, sensemaking) or the user with the user's machine (e.g. human-computer interface issues and key stroke models. A CSE approach emphasizes environment and context and allows for a distributed unit of analysis that measures the complex interdependencies between the user and the user's artifact, and in socially distributed work, the other users and their artifacts. These studies are useful in answering questions such as: (a) who needs to communicate with whom, when and how (workspace design); (b) what information needs to be shared (artifact features); and (c) who has what information and when (system design)?

Current work practices are dependent on current artifacts and represent actions used by workers to complete tasks with the tools at their disposal. Unexplored possibilities represent another form of inefficiency resulting from intrinsic work constraints that are not currently part of the system design. Although they may represent very productive ways to complete a job, they are not exploited. The identification of these functional actions, workaround activities and currently unexplored possibilities represents an important reason to conduct field studies of work under naturalistic conditions (Hutchins, 1995; Vicente, 1999, p. 96; Xiao & Milgram, 2003). This is important since new designs should not necessarily support work in its currently articulated form, but rather should attempt to overcome current system inefficiencies and exploit new opportunities.

This paper uses the MAKADAM framework, which is a modified version of the McNeese AKADAM (Advanced Knowledge and Design Acquisition Methodology) framework (McNeese et al., 1995; Zaff, McNeese, & Snyder, 1993). AKADAM and MAKADAM explore the ability of a user-centric aid to support the cognitive needs of police officers in dynamic environments. This begins with an artifact-independent analysis of the cognitive work domain resulting in an abstraction hierarchy diagram, followed by a task analysis resulting in scenarios and concept maps. The substitution of functional decomposition methods should not change the outcome, although domain experts may find it easier to work with the abstraction hierarchy. This provides an opportunity to explore decomposition techniques within the participatory MAKADAM framework.

APPROACH AND METHODS

Insight into design of cognitive systems can be obtained by studying municipal police officers and the environment within which they work. Similar to emergency medical and fire responders, municipal police officers exemplify experts facing dynamic work environments. Situations include large problem spaces, ill-defined problems and high risk outcomes. Police information systems should support making sense of these situations thereby improving decision-making, judgment, problem-solving and perception.

Three Pennsylvania municipal police departments are included in this paper. The headquarters of the three police departments are located within 10 miles of each other. Although managed by different municipalities, the police departments in this paper participate in a county-based law enforcement consortium. They work closely together, are dispatched by a common 911 center and share radio frequencies and computing technology. Municipal police department A (MPD-A) includes a borough with retail and student housing, as well as two residential townships. MPD-B and MPD-C provide police support for neighboring townships that include residences, student housing, rural farmland and developed commercial sectors. Approximately 130 square miles are covered by the three police departments, with a total population served of 81,183.

Pennsylvania police statistics track crime based on seriousness of offense, which can be used to compare these departments. More serious Part I offenses include murder, rape, aggravated assault and arson. Part II offenses include lesser assaults, forgery, drug abuse, drunkenness, and disorderly conduct. Table 1 reveals that the three departments are consistent with state averages for the more serious Part I Incidents per officer. An active student population in the area covered contributes to an above average rate of Part II Incidents per officer, especially in the central and southern borough area of MPD-A.

Table 1 Crime Statistics for the Three Included Municipal Police Departments ("Pennsylvania State Police Uniform Crime Report," 2004)

	Officers	Part I incidents	Part I incidents per officer	Part II incidents	Part II incidents per officer
MPD – A	63	1,008	16	5,680	90
MPD – B	18	186	10	733	41
MPD - C	15	230	15	528	35
Pennsylvania	24,646	326,961	13	625,008	25

The first analytical stage uses ethnographic techniques to reveal the backdrop or landscape within which officers perform their cognitive activities. Both semi-structured and unstructured discussions occurred between the researcher and the officers during these ridealongs. To better understand the cognitive domain, the probes focused on the use and manipulation of information by officers to make decisions and solve problems.

Outcomes from the cognitive domain analysis in the ethnography stage include evaluation of intrinsic work constraints and current work practices, with an emphasis on currently unexplored possibilities and workaround activities. This includes identification of cognitive design leverage points, or generalized areas of cognitive police work not effectively supported by current systems. Design leverage points from the ethnography stage focus the identification of cognitive design seeds during knowledge elicitation in the second analytical research stage. Design seeds represent specific cognitive tasks not effectively supported by current artifacts. Design seeds then form the basis for prototype development during the intervention part of this paper. Other researchers conducting cognitive task analysis have used design seeds as well. Patterson, Woods, Tinapple and Roth (2001) triggered development of modular design concepts to assist intelligence analysts suffering from data overload. Hoffman, Coffey, Carnot and Novak (2002) generated 35 leverage points in a study of weather forecasters. The efforts of this paper add to that body of literature.

Knowledge elicitation in the second analytical stage uses the critical decision method (CDM) (Klein, Calderwood, & MacGregor, 1989) as a cognitive task analysis technique to elicit cues, knowledge requirements and decision points used in solving real world problems by police experts. The first four CDM questions ask the respondent to describe a critical incident, its timeline, the decision-making heuristics, and a post hoc analysis of decision-making effectiveness. A fifth question was added to the traditional critical decision method asking the officer to consider system support and possible enhancements that might have improved the decision-making effectiveness. This fifth question helped to focus development of an intervention in the third research stage.

The CDM is adapted to take advantage of the knowledge elicitation and representation benefits from concept mapping as well as to facilitate expert exploration of leverage points identified in the first analytical stage to identify design seeds. For example, concept mapping by officers eliminates the need to transcribe interview notes allowing the researcher and expert to simultaneously share in the representation of knowledge with immediate benefits of interpretation and discussion. Using this technique, the police expert represents his or her perspective directly without interpretive bias by the researcher.

The third and final research stage in this paper applies knowledge gained from the analytical stages by using participatory design methods to develop a prototype capable of enhancing cognitive police work. Using the MAKADAM methodology which links cognitive analysis to cognitive design, interventions based on design seeds improve cognitive police work by both complying with domain constraints and reducing task complexity. In the third research stage, one of the identified design seeds is developed into a prototype using participatory design.

The design research stage applies cognitive domain and task knowledge in a scaled world to explore a design intervention capable of improving cognitive work in dynamic police environments. This scaled world includes a command center scenario derived from police experiences with a real world barricaded gunman. The researcher and domain experts work together using participatory design methods to develop a prototype. The prototype is based on a design seed identified in the knowledge elicitation stage and is developed in an iterative bootstrapping process that includes mapping information and decision needs with visual concepts. Bootstrapping is the use of multiple, converging techniques where each step provides knowledge and insight used to guide the next step (Potter et al., 2000). Scenarios are then created to supplement the use of the prototype as a police training tool in a scaled world setting. It should be noted that the participatory design methods used in this paper are similar to those associated with Scandinavian participatory design (Greenbaum & Kyng, 1991; Kyng, 1994).

RESULTS

Analytical results from ethnographic methods, including domain training, secondary data analysis and observation of officers working in the field, provide insight into the domain and its constraints for cognitive work. Results from the knowledge elicitation methods based on the critical decision method provide insight into the complexity of cognitive tasks. Intervention results begin with the selection of a design seed from knowledge elicitation for prototype development. This is followed by domain and task modeling that includes representations of the work domain structure, work activity processes, information flows and decision-ladders used in decision-making. In addition to guiding prototype development, these models are used to create scaled world training scenarios. The scenarios can be used to train officers in real world decision-making.

An outcome of this paper is an evaluation of the effectiveness of this prototype for use in the mobile command center shown in real world settings. This is now possible since officers in the participatory design session implemented the prototype (e.g. a portable laptop, projector, magnetic icons and a magnetic whiteboard) in a mobile command center. Plans have been made to train officers from six county-based municipal police departments in the use of the prototype during field exercises.

DISCUSSION AND CONCLUSIONS

Recent events, including the 2001 World Trade Center attacks and the 2005 Hurricane Katrina, have made clear the need to support the cognitive activities of emergency planners and responders. Police officers working in dynamic environments make decisions that involve large problem spaces, time constraints, uncertainty, ill-defined situations, conflicting goals and socially distributed responses. The identification of challenges to enhance the cognitive effectiveness of police officers began during the literature review. The findings continued to evolve during the subsequent analytical and intervention research stages that include ethnography, knowledge elicitation and participatory design. Eight findings were identified and substantiated during this process.

First, cognitive demands in scope of policing are increasing. Legal, social, and technological change has contributed to an increase in the cognitive requirements associated with policing. Police are expected to show good judgment and quick response under challenging circumstances. In addition to mastery of a diverse set of skills, police are also expected to be active users of information systems both to record and support their activities.

Second, increasing police autonomy increases cognitive challenges, including challenges for information sharing, coordination and monitoring. The autonomy stems from a combination of policing reforms and the introduction of mobile technologies. Reforms that require a more visible community police presence support administrative decisions to distribute officers in the field for patrol and report writing. Cell phones, wireless networks and mobile computing help keep these officers in the field while they perform office-type duties that otherwise would require a return to the station. However, even though police work autonomously, they still need to monitor each other and be ready to provide support when needed. This creates new system design challenges for information sharing and situation monitoring.

Third, current system solutions encourage workarounds by officers. Gaps in the support of current work practices are an outcome from present police information system planning, use, maintenance and training. Since so many technologies are becoming available for police use, the importance of workarounds cannot be underestimated. From a design perspective, the workarounds that result from system inefficiency not only consume resources, but also may impact the quality of the data needed for cognitive actions.

Fourth, system interoperability gaps in the sharing of information between agencies reduce decision-making quality. These gaps are visible during daily police operations, but have also become quite visible during recent national disasters. Potential interoperability exists between first responders, dispatch, and other groups including health services. Since a shared commitment from several organizations may be required, design opportunities to close this gap may prove more difficult than others.

Fifth, police officers consciously weigh benefits and costs of use before adopting technology solutions. Officers will choose the technology that best accomplishes the task while consuming the least amount of cognitive resources. From a system design perspective, this suggests that an intervention with lesser capabilities may be preferred in different contexts due to the user's perceived efficiency of performance. There are benefits, then, for maintaining the multiple communication opportunities currently available to police (e.g. chat, radio, cell, email, and face-to-face), as long as each has a demonstrated effectiveness during certain tasks.

Sixth, achieving the benefits of user-centered interventions requires design insight as the tool is both conceived and used. There are benefits of actively involving the users in the design conception, as opposed to just surveys, for example. Active participation can be facilitated by combining users and artifact components in a scaled world setting.

Seventh, not all solutions require a high tech approach. Designers must balance the high cognitive loads associated with situations and tools against anticipated information provision and needs. This suggests that the best design may not be the most technical. In some contexts, targeted, simple appliance-like solutions have great potential.

Eighth, historical roots of policing that emphasizes cases, learning through storytelling and experiential development of decision-making capabilities cannot be fully replaced by integrated, automated solutions. To this end, cognitive designers may use process modeling to highlight non-obvious information exchanges and value.

REFERENCES

Dowell, J., & Long, J. (1998). Target paper: Conception of the cognitive engineering design problem. *Ergonomics*, 41(2), 126-129.

Eggleston, R. G. (2002). Cognitive systems engineering at 20-something: Where do we stand? In M. D. McNeese & M. A. Vidulich (Eds.), *Cognitive systems engineering in military aviation environments: Avoiding cogminutia fragmentosa* (pp. 15-78). Wright-Patterson Air Force Base, OH: Human Systems Information Center.

Greenbaum, J. M., & Kyng, M. (1991). *Design at work: Cooperative design of computer systems*. Mahwah, NJ: Lawrence Erlbaum Associates, Inc.

Hoffman, R. R., Coffey, J. W., Carnot, M. J., & Novak, J. D. (2002). An empirical comparison of methods for eliciting and modeling expert knowledge. In *Proceedings of the 46th Meeting of the Human Factors and Ergonomics Society*, Santa Monica, CA: HFES

Hollnagel, E., & Woods, D. D. (1983). Cognitive systems engineering: New wine in new bottles. *International Journal of Man-Machine Studies*, 18(6), 583-600.

Hutchins, E. (1995). *Cognition in the wild*: MIT Press.

Klein, G. A., Calderwood, R., & MacGregor, D. (1989). Critical decision method for eliciting knowledge. *IEEE Transaction on System, Man, and Cybernetics*, 19(3).

Kyng, M. (1994). Scandinavian design: Users in product development. Paper presented at the *Proceedings of the SIGCHI conference on Human factors in computing systems: celebrating interdependence,* Boston, Massachusetts

McNeese, M. D. (2002). Discovering how cognitive systems should be engineered for aviation domains: A developmental look at work, research, and practice. In M. D. McNeese & M. Vidulich (Eds.), *Cognitive systems engineering in military aviation environments: Avoiding cogminutia fragmentosa* (pp. 77-116). Wright-Patterson Air Force Base, OH: HSIAC Press.

McNeese, M. D., Zaff, B. S., Citera, M., Brown, C. E., & Whitaker, R. (1995). Akadam: Eliciting user knowledge to support participatory ergonomics. *The International Journal of Industrial Ergonomics*, 15(5), 345-363.

Norman, D. A. (1986). Cognitive engineering. In D. A. Norman & S. W. Draper (Eds.), *User centered design: New perspectives on human-computer interaction* (pp. 31-61). Hillsdale, NJ: Erlbaum.

Patterson, E., S., Woods, D. D., Tinapple, D., and Roth, E. M. (2001). Using cognitive task analysis (CTA) to seed design concepts for intelligence analysts under data overload. In *Proceedings of the Human Factors and Ergonomics Society 45th Annual Meeting*. Santa Monica, CA: HFES

Potter, S. S., Roth, E. M., Woods, D. D., & Elm, W. C. (2000). Bootstrapping multiple converging cognitive task analysis techniques for system design. In J. M. Schraagen, S. F. Chipman & V. L. Shalin (Eds.), *Cognitive task analysis* (pp. 317-340). Mahwah, NJ: Lawrence Erlbaum.

Rasmussen, J., Pejtersen, A. M., & Goodstein, L. P. (1994). *Cognitive systems engineering*. New York: John Wiley & Sons, Inc.

Roth, E. M., Patterson, E. S., & Mumaw, R. J. (2002). Cognitive engineering: Issues in user-centered system design. In J. J. Marciniak (Ed.), *Encyclopedia of software engineering* (2nd ed., pp. 163 - 179). New York: Wiley-Interscience: John Wiley & Sons.

Vicente, K. J. (1999). *Cognitive work analysis: Towards safe, productive & healthy computer-based work*. Mahwah, NJ: Erlbaum.

Woods, D. D. (1998). Designs are hypotheses about how artifacts shape cognition and collaboration. *Ergonomics*, 41, 168-173.

Woods, D. D., & Roth, E. M. (1998). Cognitive systems engineering. In M. Helander (Ed.), *Handbook of human-computer interaction*. Amsterdam: Elsevier.

Xiao, Y., & Milgram, P. (2003). Dare I embark on a field study? Toward an understanding of field studies. *The Qualitative Report* 8(2), 306-313.

Zaff, B. S., McNeese, M. D., & Snyder, D. E. (1993). Capturing multiple perspectives: A user-centered approach to knowledge acquisition. *Knowledge Acquisition*, 5(1), 79-116.

Chapter 9

Decision Support for Option Awareness in Complex Emergency Scenarios

Mark S. Pfaff[1], Jill L. Drury[2],
Gary L. Klein[2], Loretta D. More[3]

[1]Indiana University – Indianapolis

[2]The MITRE Corporation

[3]The Pennsylvania State University

ABSTRACT

To enhance the decision support of emergency responders, we are examining the ability of decision space visualization tools to enhance option awareness and support more robust decision making. The current study extends prior work to detail the impact of the decision space information provided to users, relating the correctness of decisions to the levels of complexity represented in the events and the affordances for understanding alternative actions. This research provides further insight into the value of decision space information and option awareness for users working in complex environments.

Keywords: Decision making; emergency management; situation awareness

INTRODUCTION

Emergency response is an especially challenging domain for decision support, as life-or-death decisions must be made very quickly, often with high levels of uncertainty in the information available about the situation. While model-based decision support tools have been used extensively in emergency planning, their high sensitivity to input conditions has limited their utility during ongoing emergency events.

Because of the sensitivity of critical decisions to conditions outside of the decision maker's control, a course of action (COA) that appears to be optimal at one moment may indeed lead to failure due to one sensitive piece of information being inaccurately reported. In response to this, Lempert et al. (2003) and Chandrasekaran (2007) describe a robust decision making (RDM) methodology for identifying COAs that are less susceptible to these discrepancies in the situation space information.

This RDM approach is well-suited to dynamic work environments in which the information available to describe a given situation unavoidably falls short, to some degree, of the precise information required to determine an optimal solution to the problem. This gap is referred to by Hall et al. (2007) as the Situation Space-Decision Space gap. The "situation space" consists of whatever descriptive information is available from sources such as reports, surveillance, or sensors. This information, when well-presented, helps develop *situation awareness* (Endsley. 1988) among system users. Decision makers with that situation space information must then map the information into a "decision space" of potential options and possible consequences. The outcome of this decision space is *option awareness*.

The overall goal of this research program is to enhance the support of emergency responders using model-based decision aids grounded on these principles of robust decision making. In the current study, the goal was to examine the ability of decision space visualization tools to enhance option awareness and user confidence in scenarios of varying levels of situational complexity.

BACKGROUND

Preceding studies (Drury et al., 2009a; Drury et al., 2009b; Klein et al., 2009) have supported the core principles of this research. We developed a decision-space visualization aid that displayed box plots representing the distribution of costs (in damage to property, loss of life, projected future events, and other factors) over thousands of simulated possible futures, each representing outcomes related to tiny variations in these uncertain pieces of situation space information. Users with this tool displayed increased correctness in selection of COAs as well as higher confidence in their decisions in a series of emergency management events. Users with the decision-space visualization also rated the system significantly more supportive for decision making than those with only situation space information. However, this tool was entirely automated, with no actual user input. In Drury et al. (2009c) some users entered their own estimates of values used in the forecast model, yet all users saw the same recommendations from the decision aid. Consistent with Williamson and Shneiderman (1992), those who believed they were controlling the model had higher confidence in their decisions and more highly rated the degree of decision support that was provided to them.

To explore the impact of complexity on decision making using the aids, we developed four specific types of trade-off conflicts between cost components; each showed significantly different impacts on decision-making speed. For the current study, situation complexity was increased by introducing scenarios including temporal patterns of activity and geospatial conflicts. This increased complexity was expected to reduce intuitive situation awareness, which consequently would reduce their confidence in an unaided choice of options. This reduction in situation awareness may also impact reliance upon the decision-space visualizations, changing their effect on option choice and confidence.

This study further extends our prior work by giving users actual control over model underlying the decision-space visualization aid. The recommendations made by the decision aid and the metrics of successful resolution of events now directly reflect the user's assessment of the emergency event's characteristics.

VISUALIZING COSTS

To enable decision makers to compare the relative desirability of alternative emergency COAs, a cost metric is needed that takes into account both immediate and future costs. Therefore, the cost metric needs to be a multi-attributed utility (MAU) function (Chatfield et al., 1978; Keeney and Raiffa, 1993). In our cost metric, immediate costs consist of the cost to send resources, property damage, and a dollar value assigned to both the injuries and loss of life (known as casualties). Future costs are calculated based on maintaining sufficient resources to handle emergencies in the near future. Only the costs of the *extra* damage that might occur in the future due to insufficiently reserved resources are charged against the current emergency.

We based our visualization on Tukey's (1977) box plots to provide a simple affordance for comparing the cost-distributions of the COAs because they are a common visualization of distributions that typical research subjects can be readily trained to read. We further simplified the box plot visualization by excluding outlier data points (see Figure 0.1).

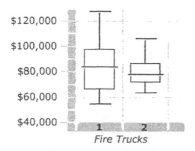

Figure 0.1 Box plot visualization showing the distribution of costs of sending one fire truck versus two, assuming a range of possible futures

Though all of the COA permutations for an event were calculated, only the box plots of the six most algorithmically-determined "robust" COAs were displayed. To determine the most robust courses of action, we used a "best three out of five parameters" algorithm based on the five distribution parameters of the box plots: the five parameters being the median, the 25th and 75th percentiles (the bounds of the box), and the best and worst outcomes (the upper and lower whiskers). Although such filtering may result in a theoretical reduction in option awareness (Drury et al., 2009c), additional comprehension practically would be minimal at best from visualizing additional less-robust options, at the cost of information overload and distraction.

METHODOLOGY

The experiment employed a mixed design.

Between subjects: One group of participants (the SS group) received only situation space information that consisted of a textual description of an emergency event and a map showing location(s) of interest. The other group (the DS group) received the situation space and a box plot visualization of the decision space.

Within subjects: Each participant, regardless of group, was asked to make decisions in ten complex scenarios and in ten simple scenarios. The order of presentation was random.

Complex scenarios included a log of two or three previous events that had occurred earlier in the day. Together with the current event, the incidents formed a pattern that had one or more spatio-temporal characteristics, such as incidents that were spread along a route in a particular direction. Furthermore, each of the spatio-temporal patterns could be subject to different types of conflict. For example, when there are resources located at two different stations, there can be a conflict between sending resources from the closest station and reserving resources by sending them from the farther station that has many more resources available.

Simple cases did not include a log (participants could assume that the current event was the first one of the day), and resources were available from only one station. No patterns or conflicts were present in these scenarios.

PROCEDURES

All participants were asked to read a paper copy of a one-page introduction to the experiment, which included Institutional Review Board (IRB) information. Next they received a copy of a training manual and a list of Frequently Asked Questions (FAQ). Next, they were given ten training scenarios in the computerized test bed so that participants could become familiar with its interface and the types of decisions they were being asked to make. Participants received feedback on the parameter estimates during the training phase. Training gave participants a chance to practice using the interface and think about the types of questions being asked of them, and achieve a consistent calibration to the scenario elements. After the training, they completed 20 scenarios on the computer test bed during which participants were asked to play the roles of police or fire/rescue commanders.

Each scenario contained a short textual situation-space description of the emergency that included information that suggested the likely cost of the incident (e.g., a fire at a jewelry shop) and the likelihood of other incidents occurring in the near future. It also contained a map that showed the location of the current event, any previous events if noted in the log, and the locations of the two fire/rescue stations or police stations. Each scenario was completely independent of the others.

After reading the textual description and viewing the map, each participant was asked to estimate three input parameters: the current magnitude of the emergency incident (from a low value of 1 to a high value of 5), the likely impact of the event in terms of property damage and injuries/loss of life that could result (low, medium low, medium, medium high, or high), and the potential for future events occurring (less than usual, same as usual, or more than usual). Participants were then asked to rate their confidence after each of these estimations using a scale from a low of 0 to a high of 7.

The input parameters entered by participants dynamically fed the event forecast model. Because there could be varying amounts of combinations of emergency response vehicles (police squad cars, ambulances, or fire trucks) available from two stations, the six most robust options were displayed in order by location and quantity, rather than rank, to avoid leading DS participants straight to the normatively correct answer.

After entering the three input parameters, the DS group was shown the six box plots of the decision space, but the SS group was not. Participants were then asked to select from all the possible permutations how many resources to send. Resource allocation was made from pull-down menus of resources available at each station. Immediately after choosing a resource allocation, participants were asked to rate their confidence in that decision.

After completing the scenarios, participants answered survey questions, including questions probing their subjective assessment of the decision support provided to them. As part of the post-test survey, we assessed three different information processing traits that might interact with using the decision aid: where participants fell on the spectra of risk taking versus risk aversion (Blaise and Weber, 2006), visual versus verbal information processing (Childers et al., 1985), and vivid versus non-vivid imaging (Sheehan, 1967). We also surveyed participants' prior experience with emergency response and with box plots.

HYPOTHESES

We defined five a priori hypotheses.

H1: The DS group will make decisions that will result in more positive outcomes than the SS group. *Rationale:* Since the box plots take into account a number of factors that affect decision outcome, to the degree that they make the outcome quality clear, participants in the DS group will be able to choose the normatively correct COA more often.

H2: The DS group will be more confident in their decisions than the SS group.

Rationale: The box plots will make it clearer which COAs will lead to more positive outcomes, so to the extent that this clarity is communicated to participants, they will be more confident in their decisions.

H3: Simple decisions will be made faster than complex decisions by participants in both groups. *Rationale:* Simple decisions should be easier to make than complex decisions, and therefore all participants should make them faster.

H4: The DS group will give higher scores for the quality of the information provided about events than those receiving only the situation space information *Rationale:* The box plots should help in framing and interpreting the situation space information, helping the DS group to engage in more effective sense-making than the SS group, which only receives situation information.

H5: Participants in both groups who believe there is a large chance of future events occurring will under-allocate resources to the current event. *Rationale:* Some of the scenarios include hints that another emergency is highly likely to happen in the near future. If participants use those hints to develop a belief that another emergency event is likely to happen soon, we hypothesize that they will want to conserve resources to address that possible future event.

RESULTS

We recruited a total of 24 participants (15 male and 9 female, ranging from 18 to 65 years old) from two locations of a not-for-profit corporation and a university.

Preliminary stepwise regression determined which confounding variables, if any, were acting as covariates in the following analyses. Decision times were log-transformed for a normal distribution (back-transformed means of decision times appear in parentheses).

H1 (the DS group will make decisions that will result in more positive outcomes) was supported. The optimality of the selected COA was measured according to the ranked list of options produced by the model based on the participant's parameter estimates (see Table 0.1).

Table 0.1 Percentage of selected COAs by rank

Condition	Rank of Selected COA						
	1st	2nd	3rd	4th	5th	6th	Other
Situation Space	30.56%	15.00%	12.78%	12.78%	11.11%	6.11%	11.67%
Decision Space	42.67%	19.00%	11.67%	8.67%	8.33%	4.00%	5.67%
$X^2 = 14.42, p < .05$							

Participants in the DS condition selected the first-ranked COA 42.67% of the time, as compared to 30.56% of the time for the SS group. Participants without the decision space information selected options not among the top six 11.67% of the time, as compared to 5.67% of the time for the DS group.

Wilcoxon rank sum tests of the selected COA rank showed higher ranked choices in the DS group ($M = 2.55$, $SE = 0.11$) than the SS group ($M = 3.24$, $SE = 0.14$), $z = 3.58$, $p < .001$, $r = .16$, and a significant main effect and interaction for gender, $z = 4.68$, $p < .001$, $r = .19$, (see Figure 0.2).

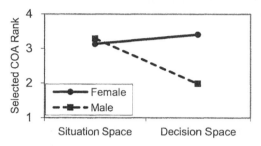

Condition	Female	Male
Situation Space	3.13	3.29
Decision Space	3.41	1.99

Figure 0.2 Interaction between condition and gender for selected COA rank

This result was explored with additional rank sum tests to identify whether event complexity was underlying this difference between genders in the DS condition. The effect of gender, though significant for both levels of complexity, showed a significantly greater difference for the complex events ($M_{female} - M_{male} = 1.88$) than simple events ($M_{female} - M_{male} = 0.96$), $t(148) = 2.36$, $p < 05$.

H2 (the DS group will be more confident in their decisions) was not supported. There was no difference in confidence levels between the two conditions.

H3 (simple decisions will be made faster than complex decisions by all participants) was supported. A within-subjects factorial ANOVA of decision time by event type, controlling for familiarity with box plots, confirmed that decision times for simple events were much faster ($M = 9.42$ (12.45 seconds), $SE = 0.07$) than those for complex events ($M = 9.63$ (15.16 seconds), $SE = 0.07$), $F(1,22) = 14.75$, $p < .001$. Those familiar with box plots also showed faster decision times ($M = 9.35$ (11.55 seconds), $SE = 0.09$) than those who were not ($M = 9.70$ (16.35 seconds), $SE = 0.10$), $F(1,22) = 6.52$, $p < .05$. There was no interaction between the terms, nor was there any main effect for condition.

H4 (the DS group will give higher scores for the quality of the event information than those only receiving situation space information) was supported. The assessment of event information quality was aggregated from four semantic-differential survey questions about the quality of the information describing each event, reported on a scale of 1 to 7. A mixed ANCOVA of information quality by condition, controlling for risk aversion and self-reported usefulness of the map, showed higher ratings of event information in the DS condition ($M = 5.46$, $SE = 0.22$) than the SS condition ($M = 4.37$, $SE = 0.29$), $F(1,20) = 8.10$, $p < .05$. Event information quality was strongly correlated with the rating of map usefulness ($\beta = .70$, $t(20) = 4.10$, $p < .001$) while risk aversion was inversely correlated with event information quality, $\beta = -.36$, $t(20) = 2.05$, $p < .10$.

H5 (participants will under-allocate resources in anticipation of future events) was supported. Resource accuracy measured over- or under-allocation by subtracting the total resources allocated in the first-ranked COA from the total resources allocated by the participant. A within-subjects ANOVA of resource accuracy by the level of future likelihood of events showed that for events rated as having "less than usual" likelihood of future events were only slightly under-allocated ($M = -0.09$, $SE = 0.17$), while under-allocation was much more pronounced for those rated "same as usual" ($M = -0.50$, $SE = 0.16$) and "more than usual" ($M = -0.55$, $SE = 0.15$), $F(2,476) = 3.50$, $p < .05$.

DISCUSSION

The primary focus of this study was to extend prior results regarding decision space visualization by examining the effects of providing users with direct control over the visualization. Consistent with previous studies (Drury et al., 2009a; Drury et al., 2009b), the results above demonstrate that a visualization that helps identify the most robust options has the ability to guide decision makers more often toward options with more positive outcomes, while improving the perceived quality of the event information, and without introducing any additional decision time.

The most surprising result was the effect for gender on the impact of the decision-space visualization, which had not been seen in prior studies. However, many studies have shown significant gender differences in risk assessment (Gustafson, 1998) and technology usage (Taylor, 2004), both of which are in play simultaneously here. The difference appears to be accentuated by the complexity of events, which also is a new factor introduced in this study. Unfortunately, the present experiment does not provide enough resolution to further unpack this result, but possible explanations to be tested further include whether additional training on the tool could balance out the difference, or whether this difference is only apparent among naïve users such as the ones used in this study and would not be found among experienced emergency managers.

Prior studies showed an increase in confidence for participants with the decision support, which did not occur here. In the previous study (Drury et al., 2009b), the user inputs were not actually controlling the decision space visualization. All participants saw the same recommendations regardless of their input, perhaps making the tool seem stable and robust, providing sensible recommendations for all of the events. In the present study, with the tool now very sensitive to the participants' skill in estimating the event characteristics, the consistency of the recommendations may vary significantly from event to event, counteracting any boost in confidence from having the visualization. This effect may diminish with additional training.

Another significant difference is the more positive evaluation of the event information by DS participants, when previously there had been no significant difference between conditions. Given that the DS group did not show significantly greater confidence in their choices, this higher positive evaluation is a curious result. It may be that they appreciated having additional information, their training was sufficient for them to more often select the right option, but not sufficient for them to comprehend implications of the visualizations well enough to affect their confidence in those options.

Lastly, the tendency for users to under-allocate resources in anticipation of future events had not been supported previously, but we believed this to be related to constraints in how such conflicts had been written into scenario events (Drury et al., 2009b). Scenarios in this study were more varied and benefitted from extensive pre-testing to verify their perceived characteristics.

The results of this study revealed compelling trends describing the impact of the decision space information provided to users, relating the correctness of decisions and the user's perceptions of the process to the complexity of events and a variety of individual differences not observed in prior studies. This research provides additional insight into the value of decision space information and option awareness for users working in complex, emerging, and uncertain task environments.

FUTURE WORK

In the current experiment we demonstrated the effectiveness of decision spaces in aiding decision making in scenarios of increased complexity; where a single decision maker in a single department must coordinate resources across multiple sub-units. The reciprocal interdependence among the stations introduced the complexity of synergistic joint action. However, because actions occur within a single department the outcomes were scored from a single perspective.

Our future work will be to increase both scenario and organizational complexity by developing decision spaces to support collaborative decision making across agencies. Building upon the current results, we will introduce collaborative decision spaces based upon a super-ordinate scoring model to support inter-departmental decision making between two decision makers. Scenarios for decision making can be designed so that in some the most robust option for joint action will conflict with the most robust options in individual decision spaces (as is often the case in the real world). In other scenarios, the collaborative and individual decision spaces will agree. In this way, we can assess the impacts of providing collaborative decision space, individual decision spaces, and no decision space on correct decisions and confidence. In addition, we will assess decision space impact on collaboration, both on the amount and on the type of information exchanged between the participants in joint actions.

REFERENCES

Blais, A., & Weber, E. (2006). A domain-specific risk-taking (DOSPERT) scale for adult populations. *Judgment and Decision Making, 1*(1), 33-47.

Chandrasekaran, B. (2005). From optimal to robust COAs: Challenges in providing integrated decision support for simulation-based COA planning, *Laboratory for AI Research, The Ohio State University*.

Chatfield, D. C., Klein, G. L., Copeland, M. G., Gidcumb, C. F., & Schafer, C. (1978). *Multiattribute utility theory and conjoint measurement techniques for air systems evaluations*. Technical Report, Pacific Missile Test Center, N6126-78-1998.

Childers, T. L., Houston, M. J., & Heckler, S. E. (1985). Measurement of individual differences in visual versus verbal information processing. *The Journal of Consumer Research, 12*(2), 125-134.

Drury, J. L., Pfaff, M., More, L., & Klein, G. L. (2009a). A principled method of scenario design for testing emergency response decision-making. *Proceedings of the 2009 International Conference on Information Systems for Crisis Response and Management (ISCRAM 2009)*, Goteborg, Sweden.

Drury, J. L., Klein, G. L, Pfaff, M., & More, L. (2009b). Dynamic decision support for emergency responders. *Proceedings of the 2009 IEEE Technologies for Homeland Security Conference*, Waltham, MA.

Endsley, M. R. (1988). Design and evaluation for situation awareness enhancement. *Proceedings of the Human Factors Society 32nd Annual Meeting*, 97-101.

Gustafson, P. E. (1998). Gender differences in risk perception: Theoretical and methodological perspectives. *Risk Analysis, 18*(6), 805-811.

Keeney, R. L. and Raiffa, H. (1993). *Decisions with multiple objectives: Preferences and value tradeoffs*. New York, NY: Cambridge Univ. Press.

Klein, G. L., Pfaff, M., & Drury, J. L (2009). Supporting A Robust Decision Space. *Proceedings of the 2009 AAAI Spring Symposium on Technosocial Predictive Analytics*, Stanford, CA.

Hall, D. L., Hellar, B. & McNeese, M. (2007). Rethinking the data overload problem: Closing the gap between situation assessment and decision making. *Proceedings of the 2007 National Symposium on Sensor and Data Fusion (NSSDF) Military Sensing Symposia (MSS)*, McLean, VA.

Lempert, R. J., Popper S. W., & Bankes, S. C. (2003). *Shaping the next one hundred years: New methods for quantitative, long-term policy analysis*. Santa Monica, Calif., RAND MR-1626.

Sheehan, P. (1967). A shortened form of Betts' questionnaire upon mental imagery. *Journal of Clinical Psychology, 23*(3), 386-389

Taylor, W. A. (2004). Computer-mediated knowledge sharing and individual user differences: An exploratory study. *European Journal of Information Systems, 13*, 52-64.

Tukey, J. W. (1977) *Exploratory data analysis*. Reading, Mass: Addison-Wesley.

Willamson, C. & Shneiderman, B. (1992). The dynamic HomeFinder: Evaluating dynamic queries in a real-estate information exploration system. *Proceedings of the 15th Annual International ACM SIGIR Conference on R&D in Information Retrieval*, Copenhagen.

Chapter 10

Cognitive Task Analysis for Maritime Collisions

Angela Li Sin Tan[1], Martin Helander[1], Kenny Choo[2], Soh Boon Kee[2]

[1]Center for Human Factors & Ergonomics
Nanyang Technological University
50 Nanyang Avenue N3.1-B1a-02 Singapore 639798

[2]Cognition & Human Factors Research Lab
DSO National Laboratories
27 Medical Drive #11-17 Singapore 117510

ABSTRACT

This study used Cognitive Task Analysis (Crandall, Klein, & Hoffman, 2006) to elicit expertise in collision avoidance at sea. The knowledge elicited was used to develop a decision-support tool. Watch officers are constantly fed with vast amounts of information and at the same time, they are required to draw conclusions about the time-critical situations at sea. The mental workload experienced can be reduced with a decision-support tool, which factors in relevant data, assesses the situation and computes the best course of action. Five experienced naval officers were interviewed and their expertise was elicited using the Critical Decision Method (Klein, Calderwood, & MacGregor, 1989). This was combined with information from a training facility, as well as information from (1) literature on collision avoidance and (2) manuals for navigation at sea. The information was organised and programmed into the iGEN Cognitive Agent, a decision-support architecture developed by CHI Systems (Zachary, Ryder, & Hicinbothom, 2000). The decision-support tool was designed to read key information from sensing devices, primarily the radar, about the position, speed and course of surrounding vessels. It then interpreted the situation. The recommendations from iGEN were then validated (these results are reported separately in another paper).

Keywords: Cognitive Task Analysis, Cognitive Systems, Maritime Collision Avoidance,

INTRODUCTION

A recent study of navigation accidents identified several situational factors related to human, task, system, and environment. It was found that task requirements related to anticipation, perception, criticality, and diagnosis were common factors (Gould, Roed, Koefoed, Bridger, & Moen, 2006). Chauvin, Clostermann, and Hoc (2008) also found that 55% of the young watch officers who participated in a study of situation awareness at sea performed a maneuver that was against regulations, and 34% performed an unsafe manoeuvre.

A large amount of information is constantly fed to the watch officers in a naval bridge team, and they will need to make critical decisions while navigating. To do so, they need to filter information and draw conclusions about time-critical situations as they assess the risk of collision with surrounding vessels, and determine the best course of action. Their mental workload can be reduced by using a decision-support tool, which factors in the relevant data, assesses the situation and computes the best course of action.

The objective of this study was to elicit the expertise in collision avoidance at sea, and use this knowledge to develop a decision-support tool for aiding decision-making. The elicited material was organized into an information flow and an advisory system was programmed using the iGEN software developed by CHI Systems (Zachary, et al., 2000). The advisory system was designed to read key information from sensing devices, primarily the radar, about the position, speed and course of the surrounding vessels. It then interprets the situation and suggests an action using the information flow reported in this study.

iGEN

The iGEN software is an integrated development environment that facilitates development, testing and deployment of cognitive agents (Zachary et. al., 2000). Cognitive agents, such as iGEN, are machines that mimic human cognition. CHI Systems, through their understanding of human expertise and experience in developing advisory systems, designed iGEN to support quick prototyping and development of advisory systems. Expert cognitive systems, which were developed using iGen, have been applied as training agents (Ryder, Santarelli, Scolaro, & Zachary, 2000; Zachary & Ryder, 1997), decision aids (Zachary, et al., 2000), and as a human substitute in hazardous environments (Zachary, et al., 2005).

The iGEN Cognitive Agent is modeled using COGNET, a human cognition model that integrates models from cognitive science, decision science and psychology. The COGNET model mirror concepts in the Human Information Processing model (Wickens & Hollands, 2000). The software engine, Blackboard Architecture for Task-Oriented Networking (BATON), interprets cues and places them in its memory. The information is then processed by various sets of cognitive rules before an action is decided and executed. The information processing mechanism can broadly be classified as declarative knowledge and attention management. Both are known to be exhibited by experts.

The iGEN software is intended to be used by a team consisting of a cognitive analyst and a software engineer. The software engineer is responsible for building the shell that translates information between the virtual world and the physical world. The parameters to be communicated are specified by the cognitive analyst. In addition to these parameters, the cognitive analyst builds the declarative knowledge and attention management in the BATON. The subsequent sections focus on how the cognitive analyst collects the information and designs the advisory system.

KNOWLEDGE ELICITATION

In this study, the requirements were elicited using Cognitive Task Analysis (CTA). The objective was to determine the tasks that the agent would support for the collision avoidance task. Specifically, the analyst identified declarative knowledge that was used in the situation; perceptual cues for sensing the external environment; thought processes expressed as GOMS-like compiled goal hierarchies; and actions exhibited by the Subject Matter Expert (SME). The knowledge was analysed and the challenges in building the advisory system were identified.

Five experienced naval officers (two active and three retired) were interviewed and their expertise elicited using the Critical Decision Method (Klein, et al., 1989). All participants had more than 10 years of experience in managing naval bridges. This was coupled with document reviews of (1) existing literature for the development of collision avoidance and (2) manuals for navigation at sea.

DOCUMENT REVIEW

Two themes were explored in the document review: collision avoidance systems and collision avoidance computation. The first theme investigated whether there were existing collision avoidance systems to avoid the reinvention of the wheel. It was found that there was no system that kept track of both higher and lower levels of cognitive activity. Grabowski and Wallace (1993) built an advisory system that supported the lower level cognitive skills (e.g. manoeuvring), but not the track-keeping of the vessel. Yang, Shi, Liu, and Hu (2007) developed a multi-agent-based decision-making system for solving the problem of collision avoidance. Each vessel was an independent agent with a set of operating rules that can negotiate the collision situation. They were working on a risk-evaluation model, which had little in common with this study where the advisory system would be tested empirically.

The second theme investigated techniques for computing collision avoidance. The Admiralty Manual of Navigation: BR 45 Volume 1 Chapter 17 (1987) offers some guidance on computation of relative velocity and collision avoidance. The chapter presents the use of "relative track" to compute the closest point of approach. If the closest point of approach is zero, there will be a collision. It also discusses the presentation and limitations of the radar of judging collision.

The International Regulations for Preventing Collisions at Sea, also known as the Rules of the Road, documents useful principles for collision avoidance during navigation, which can be expressed into collision avoidance computation

(*Admiralty Manual of Navigation: BR 45*, 1987, Volume 4 Chapter 9; Allen, 2004). This contains 38 collision avoidance rules, which are adopted by 97% of the world's shipping tonnage (Allen, 2004). It was first written during the Industrial Revolution in the 19th Century, when collision and near-collision rates were increasing.

INTERVIEWS

A semi-structured interview technique, Critical Decision Method, was used to interview the experts (Klein, et al., 1989). They were asked to recall critical incidents of collision avoidance that they had experienced. They then broke the task into three to six steps and briefly discussed each sub-task. The interviewer identified the sub-tasks that imposed high cognitive load and probed deeper for (1) goals and shifts in goals; (2) perceptual cues needed in making the decisions; (3) difficult mental tasks in the situation; and (4) actions taken to overcome the challenges.

Seven stories that capture the essence of the expertise in avoiding collision at sea were collected. In all stories, the situation and the action taken by the SME were first described. The goals, cues and mental tasks in handling the situation were elaborated. Table 1 presents one of the stories where a naval ship was investigating a trawler, and the trawler unexpectedly charged at the naval ship. Based on the content analysis of the stories, we then derived the knowledge and decision rules that were used by experts in collision avoidance, which were later programmed using iGEN.

ANALYSIS

The knowledge elicitation process identified five main challenges in designing an advisory system for the naval bridge. They are: (1) knowledge management, (2) shifts in priorities, (3) situation interpretation, (4) self-awareness, and (5) level of automation. The iGEN cognitive agent was assessed for its ability to address these issues, and it was concluded that the first four challenges can be addressed using the iGEN cognitive agent.

We noted that members in the bridge team were manually processing data. However, in many cases the procedures can easily be automated. As an example, the radar automatically detects contact within a specified range, but it does not compute the risk of collision, unless the radar specialist physically acquires the contact by clicking on it. Although the radar can automatically acquire targets, this is not useful for piloting in Singapore waters, as the large number of vessels would quickly overload the system without helping the watch officer. Hence, an automated cues management system would be useful to help the watch officer focus on the information that has priority.

KNOWLEDGE MANAGEMENT

The blackboard capability in the iGEN cognitive agent allows the designer to specify the relationship between objectives. There are two main relationships:

hierarchy and links. Hierarchies mimic human information processing, while the links store associations that human creates between objects. The hierarchy supports quick programming. During programming, the designer can select from a pre-defined drop-down list (generated from the blackboard). The links allow fast information retrieval when running the advisory system. When the advisory system receives a new cue, there is no need to run through all objects to find an item. The program starts searching for items that has a relationship with the cue.

Table 1. A story on a near-collision situation

> **Challenging situation and action(s) taken** – During a patrol mission the SME was instructed to investigate a vessel of interest (VOI), a suspicious trawler which was ahead of the SME's ship. The bridge team was tasked to get closer to the VOI, so that the Commanding Officer could question their intent by shouting across. As his team approached the VOI at about 12 knots, the VOI turned and headed towards his vessel. The SME noticed that the VOI might come into a light collision with his vessel and ordered his bridge team to reverse at 10 knots to avoid the collision.
>
> **Goals & sub-goals shift in goals** - The initial goal was to ensure the territorial integrity of Singapore by preventing illegal entry into Singapore's waters. This was to be achieved by investigating any potential threats. When the VOI charged at the naval ship unexpectedly, the SME was concerned with the safety of his vessel. At this stage, safety became the active primary goal. Safety was previously a dormant goal, but was activated when the vessel and its occupants' safety was compromised.
>
> **Cues** – Three cues helped the SME to identify the danger and make his decision. First, the movement of the VOI was unusual. Typically, vessels will cooperate and slow down when stopped by the naval authority. Second, the distance to the VOI as visually assessed by the SME to be relatively near. This meant that the SME would need to act quickly. Third, the SME judged that the relative speed of the VOI heading towards his ship was about 15 knots. At this instance, the naval ship had slowed down to 6 knots. The combined speed was judged to be too fast for a starboard turn.
>
> **Strategies and mental tasks** - The SME made use of his knowledge of his vessel. He was aware of that the vessel was capable of reverse propulsion and he applied this knowledge appropriately. A novice watch officer may not be as competent and choose to turn port or starboard. This decision would have resulted in a collision because his vessel had slowed down to 6 knots.
>
> Also, a novice watch officer may not know that in situations pertaining to safety to the ship, he can override an order, if he sees that the ship is in danger. The last order given by the Commanding Officer was to close up to investigate. A novice watch officer may seek the Commanding Officer's approval to mitigate the collision situation instead.

The iGEN cognitive agent can also support simple computations. Simple computations can be performed by the iGEN cognitive agent, while difficult mathematics can be passed on to the external environment for computation. The program can also retrieve information from databases. Every vessel has a record of its own capability such as maximum speed and performance at different speed. Such information can be programmed to be used by BATON for making decisions. Vessels that might pose dangers can be computed by iGEN based on the information derived from the ship's sensors and other databases. Such vessels can then be highlighted to help watch officer in processing the most important data.

SHIFTS IN PRIORITIES

The inherent goal for collision avoidance is safety. When safety is not compromised, bridge teams set their attention to navigation and missions. For navigation, there is a planned route and the bridge team operates the vessel from point to point based on the planned route. But should safety concerns arise, e.g. an incoming vessel on collision course, the bridge team must alter the route, and safety becomes a top priority.

In addition to achieving safety, naval vessels execute missions. Three types of naval missions were reported in the interviews: convoy operations, investigation, and live firing exercises. The navigation in each of these missions is different. In one story, the expert led the last vessel in a convoy and experienced a near collision with a vessel that was laying cables. Following a convoy may still be considered as point-to-point navigation but there are additional requirements imposed on the bridge team. Conflicting goals might arise and usually, safety will take priority over other goals.

The iGEN cognitive agent possesses good modules for supporting goals. Goals are triggered when a combination of conditions are met. There is a "trigger condition" module that facilitates the input of these conditions. The iGEN software also supports the shifts of goals using a "priority" feature. Each goal has a "priority", which is a number. The various goals will all be "calling for attention". The goal with the highest priority gets the attention. This is in analogy with Selfridge's Pandemonium model (Selfridge & Neisser, 1995).

SITUATION INTERPRETATION

In the story presented in Table 1, the expert thought that the sudden turn of a vessel was irregular and could be a hostile act. A sudden turn can be detected by analyzing differential velocities from radar plots. Challenging incidents that experts face can be analysed, interpreted and programmed into the advisory system. In addition to the incidents, the characteristics of the vessels can also be taken into account.

Situation interpretation refers to processes which deal with tacit information. These are the processes which we (operators) are well familiar with, but do not necessarily think about, because we have become so used to them. One good example is tying your tie or shoe laces, or similar. We know how to do it, but since the procedure has become automated in our motor skills, it is difficult to describe in words.

The iGEN cognitive agent can mimic our cognitive abilities if it is fed the perceptions one receives from the environment. Features of challenging situations can be programmed into the system. The difficulty is to know whether the information from the sensors is good enough, complete, and timely so that the implications are clear with sufficient time to make decisions. The advisory system is good only if the inputs are complete, accurate, and timely. If the sensing technology produces better results than the human, the advisory system will be able to perform better than the human. The contrary is also true.

SELF-AWARENESS

It was noted that the quality of the decisions made by the SMEs hinge on their knowledge of the capabilities and limitations of the vessel they operate. The SME, who knew that his vessel is capable of doing reverse propulsion, was able to quickly move away from the hostile vessel, see story in Table 1...

In designing the advisory system, one requirement would be self-awareness. The advisory system should be aware of its maximum speed, advance and transfer at different speeds, and stopping distance. Such information can be stored in the iGEN cognitive agent and used for computation of "what-if" scenarios.

The database can also be used for diagnosing problems in the vessel. For example, one SME spotted a rotational radar drift. The interpretation would be that the vessel was spinning on the spot. But it was not! Something was wrong with the radar. Self-diagnostic features can be added by checking the data against other sources of data and computing if these data, as a whole, conform to natural effects and/or those in normal operations.

LEVEL OF AUTOMATION

The most difficult issue in this study was related to the level of automation. Sheridan (2002) discussed varying degrees of automation ranging from no technology aid to an aid that performs the task automatically and informs the human only when asked. The advisory system can possibly have an auto-pilot mode, where the commander is informed only about the decisions made by the advisory system. On the other extreme, the advisory system could be used to only provide the expert with information; and all decisions are then made by the human commander.

Currently, the watch officer has to manually take charge of situations, control the bridge team and decide the own ship's actions at sea. The navigation task is at the lowest degree of automation. In the story in Table 1, the watch officer did not seek the Commanding Officer's approval before executing his decision. Could we likewise allow the automation to take control and steer the ship away from danger?

This is one issue which iGEN was unable to address. iGEN can be used only to a specified level of automation. There are no features or modules in the iGEN software that allow the operator to select the level of automation. It is also difficult to design for the advisory system to take over when the bridge team is incapacitated.

When iGEN was conceived, the focus was mainly on expert thinking. Most expertise literature addressed the individual. In addition, to facilitate operation at various levels of automation is typically difficult to achieve. The current iGEN requires one to determine the level of command and autonomy given to the advisory system at the early design stage. While it may be possible to include the selection of different level of automation, this requires more complex programming.

To conclude the evaluation, we would like to highlight that the advisory system is only as good as the rules that are programmed into the system. Despite having identified the potential of using iGEN for building our advisory system, it is important to note that advisory system works based on the procedural rules identified. While these rules work most of the time, it is still susceptible to most failure like all other automation. In one of the stories, the SME decided on an action that did not comply with the navigation traffic rules, but was a safer decision. The question then arises if there are categories of scenarios which are better not to be controlled by automation. Rigorous validation and testing should be conducted.

DESIGN OF THE ADVISORY SYSTEM

The Cognitive Task Analysis led to the development of an information flow map, which was subsequently programmed into the iGEN cognitive agent. The advisory system was designed to read the measurements from other navigation tools such as the radar, sonar, AIS (Automatic Identification System), and ECDIS (Electronic Chart Display and Information System). It interprets the data (position, speed course, class, etc) for nearby vessels and assesses if there will be a risk of collision. The conclusion of the assessment is communicated to the watch officer in the form of alerts and suggestions.

The Advisory System will interpret the information and make sense of the situation. The interpretation is based on the findings derived from the Cognitive Task Analysis. Figure 1 shows the information flow for assessing the collision situation. When a vessel of interest (VOI) is picked up by the radar, the tool determines if the VOI is within a tracked zone. The concept of having a tracked zone is to sort out VOIs that will not be in the path of the own vessel in the near future. This reduces the information processing load. The remaining VOIs, which are within the tracked zone, are evaluated for their risk of collision. This is done by examining historical data of each vessel, their closest point of approach (CPA), and time to CPA.

If a VOI is judged to have a potential risk of collision, the system assessed if there was a prior assessment of the vessel. If the VOI has been evaluated, and a corrective action has been taken, the CPA should be increasing. This is performed to ensure the VOI is conforming to the expectations.

If the VOI has not been evaluated, the system alerts the watch officer of the vessel. It then computes if the watch officer has the right of way. The watch officer does not have the right of way when (1) the VOI is a big vessel with low maneuverability; (2) the VOI is not a power-driven vessel; (3) the VOI is on the starboard/right; and (4) the VOI is ahead and both vessels have to take action to mitigate the situation. If the watch officer has the right of way, the system would check if the CPA is increasing and advise if the VOI is giving way. Otherwise, the system would inform the watch officer that he needs to prepare to give way. The system would then evaluate the course of action and advise the watch officer to take a safer route for avoiding the collision. The corrective action will be determined by analyzing the position of the VOI (ahead, behind, left, or right), the obstacles in the path, and the time to make the maneuvers.

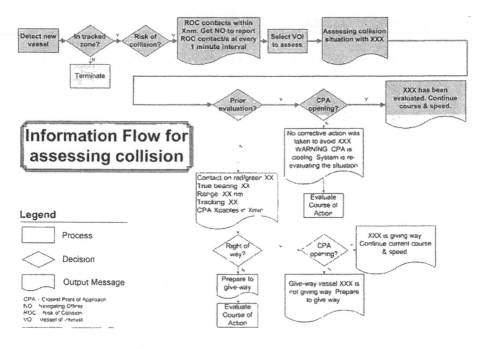

Figure 1. Information flow for assessing the risk of collision.

CONCLUSION

This paper demonstrates the application of Cognitive Task Analysis for the design of a first-of-its-kind system for avoiding collision at sea. Five naval officers were interviewed using the Critical Decision Method and the goals, cues, mental strategies, and actions considered in the stories were analyzed. The findings were used to develop an advisory system using the iGEN software.

Five main challenges for the construction of the decision-support tool were identified, and iGEN was assessed for its ability to address these issues. The five challenges are: (1) level of automation; (2) knowledge management; (3) shifts in priorities; (4) situation interpretation; and (5) self-awareness. Only the level of automation could not be mitigated using the iGEN cognitive agent.

The findings from the Cognitive Task Analysis were also translated into an information flow map, which was subsequently programmed into an advisory system. The advisory system has been validated and will be reported in a separate paper. In the validation, the decisions generated by the tool were first compared with the decisions made by officers in a training simulation and later critique by a panel of experts. It was found that the tool was able to identify imminent dangers and the situations that the officers should attend to (83% similar identifications). The decisions made were rated to be better than the decisions made by the trainees. However, further studies need to be conducted to improve the resolution of the decisions and the scope of decision scenarios.

ACKNOWLEDGEMENTS

We sincerely acknowledge the help of MAJ Adeline Heng, Republic of Singapore Navy.

REFERENCES

Admiralty Manual of Navigation: BR 45 (1987). (2nd ed. Vol. 1). Great Britain: Ministry of Defence, Directorate of Naval Warfare.

Allen, C. H. (2004). *Farwell's rules of the nautical road* (8th ed.): US Naval Institute Press.

Chauvin, C., Clostermann, J. P., & Hoc, J.-M. (2008). Situation awareness and the decision-making process in a dynamic situation: Avoiding collisions at sea. *Journal of Cognitive Engineering and Decision Making, 2*, 1-23.

Crandall, B., Klein, G. A., & Hoffman, R. H. (2006). *Working minds: A practitioner's guide to cognitive task analysis*. Cambridge, Mass: The MIT Press.

Gould, K. S., Roed, B. K., Koefoed, V. F., Bridger, R. S., & Moen, B. E. (2006). Performance-shaping factors associated with navigation accidents in the Royal Norwegian Navy. *Military Psychology, 18*(3), 111-129.

Grabowski, M., & Wallace, W. A. (1993). An expert system for maritime pilots: Its design and assessment using gaming. *Management Science, 39*(12), 1506-1520.

Klein, G. A., Calderwood, R., & MacGregor, D. (1989). Critical decision method for eliciting knowledge. *IEEE Transactions on Systems, Man, & Cybernetics, 19*(3), 462-472.

Ryder, J., Santarelli, T., Scolaro, J., & Zachary, W. (2000). *Comparison of cognitive model uses in intelligent training systems*. Paper presented at the *Human Factor and Ergonomics Society 44th Annual Meeting* Santa Monica, CA.

Selfridge, O. G., & Neisser, U. (1995). Pattern recognition by machine *Computers and thought* (pp. 237-250): MIT Press.

Wickens, C. D., & Hollands, J. G. (2000). *Engineering psychology and human performance*. Upper Saddle River, NJ: Prentice Hall.

Yang, S., Shi, C., Liu, Y., & Hu, Q. (2007). *Application of multi-agent technology in decision-making system for vessel automatic anti-collision*. Paper presented at the *2007 IEEE International Conference on Automation and Logistics*.

Zachary, W., & Ryder, J. (1997). Decision support systems: Integrating decision aiding and decision training. In M. Helander, T. K. Landauer & P. Prabhu (Eds.), *Handbook of Human-Computer Interaction* (2nd ed., pp. 1235-1258). Amsterdam: Elsevier.

Zachary, W., Ryder, J., & Hicinbothom, J. (2000). Building cognitive task analyses and models of a decision-making team in a complex real-time environment. In J. M. Schraagen, S. F. Chipman & V. L. Shalin (Eds.), *Cognitive Task Analysis*. Mahwah, NJ: Lawrence Erlbaum Associates.

Zachary, W., Ryder, J., Strokes, J., Glenn, F., Le Mentec, J. C., & Santarelli, T. (2005). A COGNET/iGEN cognitive model that mimics human performance and learning in a simulated work environment. In K. A. Gluck & P. R. W (Eds.), *Modeling human behavior with integrative cognitive architectures: Comparison, evaluation, and validation* (pp. 113-175): Routledge.

Chapter 11

Modelling Decision Making in the Armed Forces

Laura A. Rafferty[1], Neville A. Stanton[1], Guy H. Walker[2]

[1]Transportation Research Group
School of Civil Engineering and the Environment
University of Southampton
Southampton, UK

[2]School of the Built Environment
Heriot- Watt University
Edinburgh, UK

ABSTRACT

This research explores the way in which military teams make decisions in naturalistic environments. An initial review of the literature surrounding both teamwork and decision making allowed for the development of a model of team decision making; the F3 model. Through the application of the EAST methodology to case studies of decision making the processes involved in team decision making have been explored. This investigation allowed for the F3 models factors to be measured, with quantitative figures provided for a number of the factors as well as qualitatively exploring the way in which the factors interact with one another.

Keywords: Teamwork, EAST, Naturalistic Decision Making, Military

NATURALISTIC DECISION MAKING IN TEAMS

Naturalistic Decision Making (NDM) is the study of how people make decisions in real world settings (Klein, 2008). The discipline explores the strategies people use to make difficult decisions under poor conditions such as uncertainty, time pressure, risk, multiple decision makers, high stakes, vague goals and unstable conditions (Schraagen, Militello, Ormerod and Lipshitz, 2008). Klein states that people in these situations are not generating and comparing different courses of action, as believed by traditional decision making research. Rather, these people are using their past experience and schemata (mental model) to rapidly categorise situations and make judgements (Klein, 2008).

Naturalistic decision making within teams is an important concept and recently interest in the topic has increased. A thorough review of the literature revealed some interesting findings. A frequency plot of the literature was carried out in order to indentify key factors. Factors affecting teamwork and decision making were drawn from over 80 pieces of literature spanning a date range of over 30 years. Each factor affecting teamwork was given a quantitative value based upon the number of times it occurred within the literature. The factors were then organised with respect to their contribution to total variance, in descending order. The frequency plot identified five factors that accounted for most of the discussion, the five most prominent factors affecting teamwork and decision making, as prescribed by the literature. These factors were communication; cooperation; coordination; situation awareness and schemata. Links between these factors were also drawn out of the literature with quantitative values being derived for the strength of the links. The thickness of the lines represents the number of papers discussing relationships between the concepts. From this a model of team decision making was developed, shown below in figure 1.

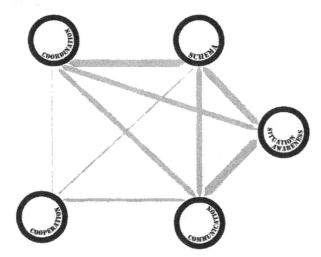

Figure 1. Famous Five of Fratricide (F3) model (Rafferty et al., 2009)

In line with the NDM paradigm this is a formative, descriptive model exploring the way in which people actually make decisions, rather than a normative, prescriptive model which states how people should make decisions. Situation awareness and schemata are key aspects of individual decision making. Stanton et al (2006) define situation awareness as 'activated knowledge for a specific task, at a specific time within a system.' Situation awareness is an awareness of the external environment. Schema can be defined as small scale models of reality that the mind constructs and uses to anticipate events (Craik, 1943). Schemata represent an internal set of 'theories of the world' built up through past experience which set expectations and direct searches for information, whereas situation awareness represents the processing of information from the external environment. Situation awareness is seen as a systemic property (phenotype), it is the product of, rather than the sum of, each individuals schema based 'theory of the world' (genotype) (Stanton et al., 2009). The importance of schemata is apparent in the study of expert decision makers who often exhibit intuitive competence. They are able to quickly identify critical cues and diagnose a situation almost instantaneously based upon their schemata built up from past experiences (Mosier, 2008).

Once we begin to move away from individual decision making and toward team decision making communication becomes an important concept. Communication up dates schemata, which subsequently affect situation awareness. Team decision making is then affected by two further concepts – coordination and cooperation, as levels of coordination and cooperation affect the level and quality of communications that flow throughout a team. Coordination and cooperation can also directly impact schemata, for example adequate coordination ensures that everyone holds the right expectations of where everyone else is, others roles and so on.

In order to explore team decision making within the military, as well as to validate and build upon the F3 model, two case studies of team decision making were undertaken. Research in NDM is generally based around case studies in order to ensure that the decision maker is immersed in their real life context, that the decision making processes they use are those used on a day to day basis. NDM research tries to identify differences in abilities and subtle skills so that they can be understood and supported in training and decision support systems as well as informing the development of recommendations for effective decision making.

METHOD

Two case studies are discussed in this paper, the Black Hawk case study and the tank crew case study. The first case study was undertaken on the accidental shoot down of two U.S. Black Hawk helicopters by two U.S. F15 fighter jets in northern Iraq during the Persian Gulf War in 1994 (USAF, 1994), (see Rafferty et al., 2009 for a full account). The second case study involved the observation of a combined battle group undergoing pre deployment training in a training facility of vehicle specific simulators. The specific mission analysed was a quick attack in which one Challenger II tank crew accidentally engaged a friendly Recce vehicle (see Rafferty et al., 2009 for an in depth account).

The case studies were analysed using the EAST methodology. The Event Analysis for Systemic Teamwork (EAST) (Stanton, Baber and Harris, 2008) methodology provides a systems level methodology that explores emergent properties arising from the interactions within complex, multi agent systems. The EAST methodology consists of the integration of six individual Ergonomics methods. These methods are: Hierarchical Task Analysis (HTA: Annett, 2005), Coordination Demand Analysis (CDA: Burke, 2005), Communications Usage Diagram (CUD: Watts & Monk, 2000), Social Network Analysis (SNA: Driskall & Mullen 2005), Information networks (IN: e.g. Ogden, 1987) and an enhanced form of Operation Sequence Diagram (OSD: Kirwan & Ainsworth, 1992). Through the use of multiple methods EAST is able to explore the: who, when, where, what and how of a scenario, as illustrated below in **Error! Reference source not found.**.

Table 1 Data analysis methods within EAST method

	HTA	CDA	CUD	SNA	OSD	IN
Who	░		░		░	
When	░		░		░	
Where	░		░			
What	░	░			░	
How	░					
Why	░					░

The use of numerous methods validates the results and allows for multiple perspectives within the analysis. Such multiple perspectives are needed in order to explore the emergent properties that arise in complex socio technical systems. Previous applications of EAST (Walker et al., 2006; Salmon et al., 2008) have allowed for a clear illustration of such interplay, depicting the information space in which decisions and interactions between decision makers took place. The six EAST methods allow for the exploration of each of the F3 models five factors, as shown below in table 2.

Table 2 EAST Method and F3 Model Factors

Famous Five Factor	EAST method
Communication	Social Network Analysis
	Communication Usage Diagram
	Operation Sequence Diagram
Coordination	Coordination Demands Analysis
Cooperation	Social Network Analysis
Schema	Information Networks (Genotype)
Situation Awareness	Information Networks (Phenotype)

The results of the EAST analysis of these two case studies are discussed in the following section.

RESULTS

FIVE FACTORS

The HTA analysis identified differences between the effective and less effective teams in both case studies. Within the Black Hawk case study the HTA identified that the effective system completed their tasks as a networked system undertaking tasks collaboratively, whereas the less effective system undertook their tasks as a linear chain of command with tasks remaining very separate. In the tank crew case study the HTA revealed a greater level of preparation within the effective team when compared to the less effective team. In addition to this there was also a greater level of non mission related tasks in the less effective team compared to the effective team. All of these points highlight a higher level of coordination within the effective teams of both case studies.

The SNA results highlight a higher level of sociometric status in the effective decision making scenarios of both case studies when compared to the less effective decision making scenarios. The greater the value of sociometric status an agent has the greater the contribution an agent makes to the communication flow of the network. From the analysis it is also proposed that the two effective decision making scenario systems had a greater number of agents making a greater contribution to the communication flow when compared to the two less effective decision making scenario systems. This would mean that information could, and did, flow with much greater ease. The SNA also revealed differences in the communication structures. Within the Black Hawk case study the SNA highlighted the importance of three key players in the effective system – the three high level organisations. In the less effective system the communication structure was much more convoluted, without the hierarchical structure of the effective scenario. Looking at the tank crew case study, again, the SNA highlighted the importance of a hierarchical communication structure present in the effective scenario but not in the less effective scenario. In addition to this the SNA allows a value to be derived for cooperation. Within the tank study there was no difference in the cooperation value between the effective and less effective tank crews. Within the Black Hawk case study there was an increased level of cooperation in the less effective crew (0.1) when compared to the effective crew (0.007).

The information networks highlight the higher number of incorrect schemata within the less effective scenarios of both case studies. This data allows us to identify a higher level of correct schemata and a lower level of incorrect schemata for the effective teams in both case studies. The data also reveals a higher level of information was present in the effective teams in both case studies.

The results of the CDA analysis for both the Black Hawk case study and the tank crew case study reveal a higher level of coordination in the two effective scenarios than the two less effective scenarios. Within the tank crew case study the overall coordination value was higher in the effective team (3) compared to the less effective team (1.9), similarly in the BH case study the effective team value was higher (2.7) when compared to the less effective team (1.9). These average coordination figures are based on a scale of 1 – 3, with 3 representing a high level of coordination.

The results of the CUD analysis reveal that for both case studies there were lower levels of communication in the less effective scenario compared to the effective scenario. In addition to this in the Tank crew study it was found that there were higher levels of non mission related communication acts in the less effective scenario (31% of all communication acts) than in the effective scenario (14% of all communication acts).

LINKS

The links for the initial model, the model derived from the literature, were the result of a frequency plot as discussed earlier. Each link has a quantitative value representing the frequency with which the link was mentioned in the literature. For the Black Hawk and tank crew case studies the links were defined through analysis of the transcripts surrounding the incidents. The F3 model was populated by stepping through the mission performance and coding for the links in the model as well as identifying breakdowns in these links. Two examples of how these links were identified in the tank crew case study can be seen below:

"Guys there's friendly dismounted troops where we are going now so just take it easy"

The first extract is taken from the effective tank crews' transcript and has been labelled as representing the link *communication – situation awareness*. The Commanders effective communication of an update he had received to the tank crew enabled the tank crew to have a correct situation awareness of their surroundings.

"Um 12, is that 02..... and um 0 I don't know who that one is over there"

The second extract is taken from the less effective tank crews' transcript and has been labelled as representing a breakdown in the link *coordination – situation awareness*. Due to a lack of coordination between the friendly call signs involved in the mission the less effective tank crew have a poor situation awareness regarding the identity of the tanks around them.

In this way the mission transcripts enabled a representation of the effective links between factors within the model, as well as identifying breakdowns in the links between the factors in the model.

If we compare the final model of the less effective tank crews breakdowns to the original model of team decision making breakdowns derived from the literature we can see a number of interesting differences in the models

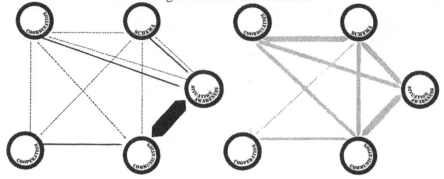

Figure 2. Tank crew case study model   Figure 3. Model derived from the literature

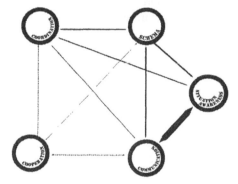

Figure 4. Model derived from the BH case study

Figure 3 above represents the initial F3 model derived from the literature. The link between communication and situation awareness has the highest frequency of breakdowns in both models, however this rate is far higher in the model derived from the Tank study, figure 2.

We can also see that a number of links – the links in grey dashed lines – are not broken in the Tank study model. This is as expected; we did not anticipate that every possible breakdown would occur in every act of less effective decision making. It is worth noting, however, that those breakdowns not represented in the tank crew study were the weakest links in the model derived from the literature.

Figure 4 represents the model derived from the Black Hawk case study (see Rafferty et al. under review). If we compare the model from the Black Hawk case study to the model derived from the tank crew study we can see a number of similarities. The largest link in both models is that between communication and situation awareness, although this is double the size in the tank crew study example.

The link between schema and situation awareness is also very similar in both models, as too is the link between coordination and situation awareness. From this we can suggest that situation awareness is the most important factor in decision making within teams, followed by communication.

The differences between the models also provide some interesting findings. In the tank study a new link has been formed between cooperation and situation awareness – this is a link previously unidentified in the literature we have reviewed and requires further analysis. To summarise the comparison of the three populated F3 models, the most frequently broken relationships in all three models is the link between communication and situation awareness. Cooperation appears to be an important factor in the tank crew model as well as in the model derived from the literature. Cooperation did not appear to play a role in the Black Hawk case study. It is put forward that cooperation may not have appeared to be a factor in the Black Hawk case study due to the nature of the data. The data used in the analysis was taken from the Aircraft Accident Investigation report rather than being based on data gathered by the analyst. It is posited that the accident investigators who collected the initial data may not have been aware of the importance of cooperation and may have missed its impact on the scenario. The importance of cooperation was further emphasised in the tank crew study with the identification of a new link breakdown between cooperation and situation awareness.

SUMMARY

Results from the EAST analysis identified that levels of coordination and communication were highest in the effective teams of both case studies and lowest in the less effective teams. This provides strong evidence for the F3 model and provides evidence for the supposition that low levels of coordination and communication may act as causal factors for less effective decision making.

The results also show that cooperation within the less effective tank crew was at a low level, again providing support for the F3 model. The results also highlighted the importance of the correlation of incorrect schemata with decision making. The research has shown that the level of incorrect schemata is highest within the less effective teams of both case studies and lowest in the effective team providing support for the importance of schemata in decision making. The same was found with respect to situation awareness.

Table 3 Levels of F3 Factors for less effective and defective decision making

F3 factors	Less effective		Effective	
	Tank study	BH study	Tank study	BH study
Communication	low levels of communication acts	low levels of communication acts	higher levels of communication acts	higher levels of communication acts
Communication	lower sociometric status	lower sociometric status	higher levels of sociometric status	higher levels of sociometric status
Communication	non hierarchical	convoluted structure	hierarchical	3 key agents dominate
Coordination	lower levels of coordination	lower levels of coordination	higher levels of coordination	higher levels of coordination
Cooperation	no difference	higher levels of cooperation	no difference	lower levels of cooperation
Schema	lower levels of correct schema	lower levels of correct schema	higher levels of correct schema	higher levels of correct schema
Situation awareness	lower levels of information	lower levels of information	higher levels of information	higher levels of information

The application of EAST to two scenarios allowed for data to be inputted into each section of the F3 model. This allowed for the comparison of levels of the F3 models factors across different scenarios. The results revealed that the link between communication and situation awareness is consistently the most frequently broken link in less effective team decision making. A new link breakdown between cooperation and situation awareness was identified during the tank crew, this along with the identification of breakdowns in other cooperation links provided support for the importance of cooperation breakdowns.

CONCLUSIONS

The results of this comparison indicate that the five factors of the F3 model are indeed prominent in team decision making, and that there are differences in the levels of these factors with respect to teams that made effective decisions and teams that made less effective decisions. In addition to this the F3 model has enabled examination of the relationships between these factors and breakdowns that can occur. Again key differences in the links and broken links between factors are found between effective and less effective decision making.

The results of this research have provided a model and diagnostic metrics exploring naturalistic decision making performance within teams, in order to enhance understanding of how teams make decisions along with the differences between effective and less effective decision making. The research has also led to the identification of key areas for effective decision making thus informing recommendations, training procedures and interventions to foster more effective decision making by experts in complex situations.

REFERENCES

Annett, J. (2005). "Hierarchical Task Analysis". In N. A. Stanton et al. (Eds), *Handbook of Human Factors and Ergonomics Methods*. London: CRC.

Burke, S. C. (2005). "Team Task Analysis". In N. A. Stanton et al. (Eds.), *Handbook of Human Factors and Ergonomics Methods*. London: CRC.

Craik, K. (1943). *The Nature of Explanation*. Cambridge University Press: Cambridge.

Driskell, J. E., & Mullen, B. (2005). "Social Network Analysis". In N. A. Stanton et al. (Eds.), *Handbook of Human Factors and Ergonomics Methods* (pp. 58.1-58.6). London:CRC.

Klein, G. (2004). *The Power of Intuition*. Bantam Dell Pub. Group.

Klein, G. (2008). "Naturalistic Decision Making." *Human Factors, 50(3) 456-460.*

Klein, G. & Armstrong, A. A. (2005). "Critical Decision Method." In N. A. Stanton et al. (Eds.), *Handbook of Human Factors and Ergonomics Methods*. London: CRC.

Klein, G. and Hoffman, R. (2008), "Macrocognition, Mental Models and Cognitive Task Analysis Methodology." In Eds. Schraagen, J.M., Militello, L.G., Ormerod, T. and Lipshitz, R. *Naturalistic Decision Making and Macrocognition*. Ashgate:

Mosier, K.L. (2008). "Technology and Naturalistic Decision Making: Myths and realities." In Eds. Schraagen, Militello, Ormerod and Lipshitz. *Naturalistic Decision Making and Macrocognition*. Ashgate: Aldershot.

Rafferty, L. A., Stanton, N. A. and Walker, G. H. (2009). "FEAST – Fratricide Event Analysis of Systemic Teamwork". Under review. *Theoretical Issues in Ergonomics Science.*

Salmon, P.M., Stanton, N.A., Walker, G.H. and Jenkins, D.P. (2009). *Distributed Situation Awareness*. Ashgate: Aldershot.

Schraagen, J.M., Militello, L.G., Ormerod, T. and Lipshitz, R. (2008). *Naturalistic Decision Making and Macrocognition*. Ashgate: Aldershot.

Stanton, N.A., Baber, C. and Harris, D. (2008). *Modelling Command and Control: Event Analysis of Systemic Teamwork*. Ashgate: Aldershot.

Stanton, N.A., Salmon, P.M., Walker, G.H. and Jenkins, D. (2009). "Genotype and phenotype schemata and their role in distributed situation awareness in collaborative systems." *Theoretical Issues in Ergonomics Science,* 10 (1,) 43-68.

USAF Aircraft Accident Investigation Board. U.S. Army Black Hawk Helicopters 87-26000 and 88-26060: Volume 1, Executive Summary: UH-60 Black Hawk Helicopter Accident, 14 April 1994. Available from www.schwabhall.com/opc report.htm.

Walker, G.H., Gibson, H., Stanton, N.A., Baber, C., Salmon, P. and Green, D.

(2006). "Event analysis of systemic teamwork (EAST): a novel integration of ergonomics methods to analyse C4i activity." *Ergonomics,* 49(12), 1345-1369.

Chapter 12

Is Analytical Thinking Useful to Technology Teaching?

Raymy K. O' Flynn, Thomas Waldmann

Faculty of Science and Engineering
Manufacturing and Operations Department
University of Limerick
Limerick, Ireland

ABSTRACT

The Iowa Gambling Task is a famous and frequently-used neuropsychological task that is designed to simulate real-world decision-making. In a study conducted by Evans, Kemish and Turnbull (2004) results reported a significant difference in the performance of educated and less well educated participants in the IGT. This study looks at the effects of a student's course choice and the effect it has on their everyday decision making. 128 students took part in a computerised version of Bechara, Damasio, Anderson, and Damasio (1994) and Maia and McClelland's Iowa gambling Task (2004).

Keywords: Iowa Gambling Task, Decision- making, Somatic Marker, Education.

INTRODUCTION

In life people make many important decisions regarding their career, money etc. These decisions are often of great importance and do not allow for bad judgment calls. It is therefore in our interest to develop good decision making skills as early as possible. The current study was undertaken to investigate if the type of education received has an effect on decision making. A computerised version of The Iowa Gambling Task (Bechara, Damasio, Anderson, Damasio 1994) and Maia and McCelland's (2004) somatic marker questionnaire was then used to determine the decision making skills and knowledge of participants in different college courses.

Decisions vary in the degree to which they rely on intuitive and analytical processes (Hammond, Hamm, Grassia, and Pearson 1987). The IGT looks explicitly at how real world decisions are achieved, and Demareea, Burns, and DeDonnoa (2009) found that IQ was a significant predictor of IGT performance suggesting that the IGT is a cognitive task, as has also been suggested by Maia and McCelland (2004). Hammond, Hamm, Grassia and Pearson (1987) pointed out that recognitional decisions are more common in everyday decision making than an analytical style of decision making. Recognitional decision making was found to occur more frequently when the decision maker was more experienced, environmental conditions were less stable and pressure or time constraints were greater. With colleges offering work experience, co-operative work placements and internships, this would suggest the more interactive courses that allow for these placements and encourage projects facilitating decision making and use of intuition may benefit students' decision making in the work place. Advantageous decision making is necessary in order to perform under time uncertainty and pressure constraints. Deciding advantageously on the IGT may suggest that participants select options that lead to a positive outcome.

Evans, Bowman, and Turnbull (2005) demonstrated that education has a paradoxical effect on the Iowa Gambling task. Less-well-educated people adopt a more advantageous strategy in the task than their university educated counterparts. Mulderig (1995) suggested that tertiary level establishments encourage students to rely solely on documented evidence when forming arguments. The idea that people's level of education can have an effect on decision making requires further investigation. Hooper, Luciana, Conklin, and Yarger (2004) found that skills aiding learning to make advantageous decisions occurs from adolescence to adulthood "14- 17 year olds in this study generate a pattern of performance that is different from healthy adults, suggesting that even older adolescents do not have the same capabilities as adults to make decisions that will be advantageous in the long run". As most attend second level schooling from the age of 10-19 and tertiary level education from as young as 17 years old during these formative years it is important to explore decision making skills and exercise them when necessary. Evans et al. (2005) found that performing the IGT did in fact meet many of the criteria for intuitive operations. Brand, Heinze, Labudda and Markowitsch (2008) pointed out that a strategic approach to decision making in the IGT may be impossible within the first few trials as participants must explore and figure out contingencies. When these have been discovered cognitive strategies or a more analytical approach may be favoured.

The IGT is constructed in such a way that participants can freely pick from four decks of cards, during the 100 trials participants will experience wins and losses in all of the decks. Two decks are considered to be good, and two considered to be bad. The "bad decks" have high gains and high losses associated with them resulting in an over all net loss when choosing only from these decks, while the good decks have smaller rewards participants will ultimately make a net gain on these decks. The IGT is thought to simulate real world decision making with participants undergoing periods of reward, punishment and uncertainty. (Bechara, Damasio et al. 2000) Thus a successful completion of the IGT can said to be a predictor of advantageous decision making.

METHOD

Participants

128 participants took part in a computerized version of the Iowa Gambling Task (IGT). Male and female participants ranged from 17- 42 years of age and were students at the University of Limerick. They did not receive any monetary reward for taking part. Participants were asked to provide personal information such as their age, course taken at college, level of education attained so far, and relevant information pertaining to work experience, or extra education availed of by the participants during their education.

Iowa Gambling Task

The study used a computerised version of the Iowa Gambling Task (Bechara et al., 1994) and a computerised version of Maia and McCellands IGT accompanying questionnaire. Facsimile rather than real money was used. The nature of reward on the task (i.e facsimile versus real reinforce) has been shown not to affect performance on the IGT (Bowman and Turnbull 2003). Bechara (2000) also found that IGT results were unaffected by the use of manual versus computerised testing systems. Participants were allowed to select cards, in any order, from any of four decks (A, B, C, and D) as in Bechara et al. (1994). Decks A and B were disadvantageous decks and decks C and D were advantageous. As per instructions in Bechara et al. (1994) Losses were more frequent on selection of decks A and B, and the task was terminated after 100 selections. A direct replication of Maia and McCelland's (2004) questionnaire was presented on computer. Participants' knowledge on the task was assessed as outlined by Maia and McCelland's (2004).

RESULTS

Tests were conducted using a computerized version of Bechara et al. (1994) IGT and a computerised version of Maia and McCelland's (2004) somatic marker questionnaire. Based on Maia and McCelland's (2004) definition participants were tested on their behaviour (selecting from more advantageous decks C and D on average during a trial), Level 0, Level 1, Level 2 and Level 2 calculated knowledge. As in Bechara et al. (1994) the 100 card selections were sub-divided into five blocks and each block was calculated by subtracting the number of good from bad card selections ((C + D) – (A + B)). A net score of above zero implied that the participants were selecting cards advantageously, and if participants scored a net of below zero it was considered disadvantageous.

Round	Block
1	1
2+3	2
4+5	3
6+7	4
8+9	5

Table 1- Round Numbers and corresponding Blocks

No significant difference between groups was found for the time at which Level 1 stage of knowledge was reported. See Figure 1 for a stem and leaf plot of level 1 knowledge and course studied.

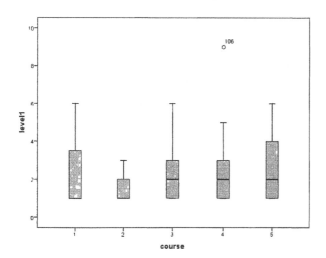

Figure 1- Level 1 knowledge by Course: (1= Digital Media and Design, 2= Product Design and Technology, 3= Woodwork, 4= Metalwork, 5= Engineering)

Digital Media Design Students were significantly slower than Product Design and Technology Students in reporting Level 2 knowledge. (Mann-Whitney U=34.5, p=0.008) and Woodwork student teachers (Mann- Whitney U= 100.5, p= 0.037) See Figure 2 for stem and leaf plot of level 2knowledge by course.

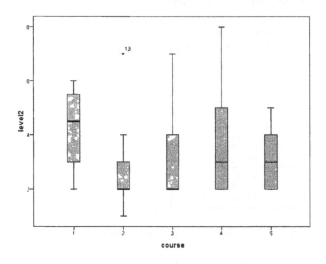

Figure 2- Level 2 knowledge by Course: (1= Digital Media and Design, 2= Product Design and Technology, 3= Woodwork, 4= Metalwork, 5= Engineering)

At Level 2 calculated knowledge Product Design and Technology students performed significantly better than Digital Media Design students (Mann-Whitney U =29, p= 0.028). Metalwork students also performed better than Digital Media and Design students (Mann- Whitney U=29, p=0.036) See Figure 3 for stem and leaf plot of level 2 calculated knowledge by course.

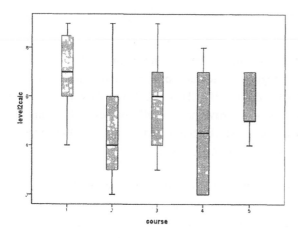

Figure 3- Level 2 Calculated knowledge versus course: (1= Digital Media and Design, 2= Product Design and Technology, 3= Woodwork, 4= Metalwork, 5= Engineering)

In blocks 3, 4 and 5 Product Design and Technology students made significantly fewer advantageous decisions than Woodwork students. Block 3 N=45 p=0.015. Block 4 N=46 p=0.049. Block 5 N=50 p=0.060 and Metalwork student teachers in Block 3. Block 3 N= 37 p=0.071.

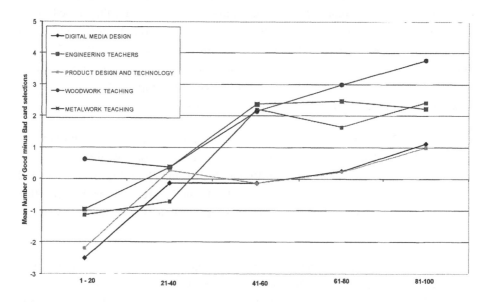

Figure 4- Mean number of good minus bad card selections per block for participants. (Digital Media and Design, Product Design and Technology, Woodwork 1st yr, Woodwork 4th yr, Metalwork 1st yr, Metalwork 4th yr, Engineering)

CONCLUSIONS

This investigation replicates many findings of the existing literature (Bechara et al. 1994; Bechara et al. 2000) with regular systematic progress and improvement in performance across the five blocks of the IGT. This effect was shown in the five university courses tested. The rate of improvement in all courses fell within the range of performance previously reported (Bechara and Damasio 2002) Steady learning was seen across the blocks in all courses. See figure 4.

Product Design and Technology students demonstrated a high Level 2 understanding and Level 2 Calculated understanding with most students reaching both levels ahead of other courses, Product designers on average reporting both understandings between round 2 and 4 respectively (see Table 1 for related Blocks). However they did not show a high level of advantageous behaviour across Blocks 3, 4 and 5 where Level 2 and Level 2 calculated knowledge were previously reported. Product Designers reached only a mean score of 1 in their net advantageous choices by Block 5, although positive net scores reflect advantageous performance. However, Product Design and Technology students' mean score was below that of other students. This suggests that even though Product Design and Technology students could report the advantageous strategy early they failed to act on such knowledge. Digital Media and Design students reported Level 2 and Level 2 calculated knowledge later than students from other courses, reaching Level 2 knowledge at round 5 (Block 3) and reported Level 2 calculated knowledge at round 7 (Block 4). Their net advantageous choices only improved towards a mean score of 1 similar to Product Design students' scores.

Woodwork, Engineering and Metalwork students all reported Level 2 and Level 2 calculated knowledge of the advantageous strategy at roughly the same time. Reporting level 2 knowledge between Round 2 -3 (Block 2) and Level 2 calculated knowledge early in Rounds 5-6 (Block 3-4). From Figure 4 it can be seen that reporting on such knowledge and acting on it occurred simultaneously for students from all three courses.

In the IGT average success (advantageous decisions) in Blocks 3-5 was lower than reported in previous studies. This may be due to the analytical nature of cognition taught in engineering courses. A study using the Cognitive Styles Index (CSI) (Allinson and Hayes 1996) rreported that in the case of engineering courses the average student's CSI score was 44 (scores above 38 are considered "analytical learners") (Cosgrave 2004). Engineers tested therefore had a tendency towards analytical learning and thinking allowing more logical, compliant thought and more structured systematic approaches to problems and decisions. This type of cognitive style may reduce IGT net score as participants from engineering backgrounds decision make with a more analytical approach rather than the explorative style necessary at the start of the task. Further investigation into the effects of cognitive styles on IGT performance will be undertaken in future studies. The process of deciding advantageously is not only logical but emotional as well. (Bechara and Damasio 2004)

REFERENCES

Allinson, C. W. and J. Hayes (1996). "The Cognitive Style Index: A Measure of Intuition-Analysis For Organizational Research." Journal of Management Studies 33(1): 119-135.

Bechara, A. and A. R. Damasio (2004). "The Somatic Marker Hypothesis: A Neural Theory of Economic Decision." Elsevier: Games and Economic Behaviour 52: 336-372.

Bechara, A., A. R. Damasio, et al. (1994). "Insensitivity to future consequences following damage to prefrontal cortex." Cognition 50: 7-15.

Bechara, A. and H. Damasio (2002). "Decision-making and addiction (part I): impaired activation of somatic states in substance dependent individuals when pondering decisions with negative future consequences." Neuropsychologia 40(10): 1675-1689.

Bechara, A., H. Damasio, et al. (2000). "Emotion, Decision Making and the Orbitofrontal Cortex." Cereb. Cortex 10(3): 295-307.

Matthias Brand, M.,Katharina Heinz, K., Labudda,K. and Markowitsch, H.J. (2008). "The role of strategies in deciding advantageously in ambiguous and risky situations." Cognitive Processing 9(3): 159-173.

Cosgrave, E. (2004). MSc Thesis. Manufacturing and Operations Engineering. Limerick, University of Limerick.

Demareea, H. A., Burns,K.J and DeDonno, M.A. (2009). "Intelligence, but not emotional intelligence, predicts Iowa Gambling Task performance." Elsevier Intelligence 38(2): 249-254.

Evans, C. E. Y., K. Kemish, et al. (2004). "Paradoxical effects of education on the Iowa Gambling Task." Brain and Cognition 54(3): 240-244.

Hammond, K. R., Hamm,R.M., Grassia,F. and Pearson,G. (1987). "Direct comparison of the relative efficiency on intuitive and analytical cognition." IEEE Trans. Syst. Man Cybern. 17(5): 753-770.

126

Hooper,C.J Luciana, Conklin, and Yarger. (2004). "Adolescents' Performance on the Iowa Gambling Task: Implications for the Development of Decision Making and Ventromedial Prefrontal Cortex." Developmental Psychology 40(6): 1148-1158.

Maia, T. V. and McClelland J. L. (2004). "A Reexamination of the Evidence for the Somatic Marker Hypothesis: What Participants Really Know in the Iowa Gambling Task." Proceedings of the National Academy of Sciences of the United States of America 101(45): 16075-16080.

Mulderig, G. P. (1995). The Health Handbook. Lexington, MA, CENGAGE Learning.

Chapter 13

Ergonomic and Control Rooms: A Case Study of a Hydroelectric Enterprise

Christianne S. Falcão e Vasconcelos, Marcelo Marcio Soares

Departamento de Design
Universidade Federal de Pernambuco
Recife, PE, BRAZIL

ABSTRACT

This paper presents the ergonomic participation in a project to modernize the control room in a hydroelectric company from Brazil. In this design, the ergonomics interferes from the viability studies, in the early stages, until the workplace definition, furniture setting, computerized equipment development, setting screens, the staff recruitment and the training and organization of the work. The control room under study has been undergoing several steps of modernization owing to the shifts of the information media,when the synoptic panels/screens where changed. The limits of the ergonomic proposal, its contribution to the analysis of the work situation, the use of an ergonomic methodology as well as the results will be discussed.

Keywords: Control room, Ergonomic Evaluation, video wall.

INTRODUCTION

Ergonomics can be defined as the application of the knowledge of human characteristics to the systems project. Such systems are inserted within a context that may affect its performance. Following Karwowski (2006), the aim of ergonomics is to contribute to the project and to evaluate tasks, works, products, environments and systems making them compatible with the human needs, habilities and limitations.

From the application of an ergonomic methodology, a series of elements that modulate work activities in an determined environment are identified. Thus, the present article intends to present and discuss the analysis of the modernization of the physical environment of a control room of the hydroelectric sector, based on the Ergonomizing Intervention (Moraes & Mont'Alvão, 2003). In this article, we identify conflicts between operators, work environment and equipment resulting from lacking or inadequate elements through opinions and suggestions of the users involved acquired through Ergonomics research tools.

Due to the constant changes in the selling and distribution of electric energy, that generates a fairly competitive market (Nobrega & Filho, 2002), the company under study has been investing in the modernization of control rooms, whose function is to control the transmission of electric energy generated by the company. The control room analyzed herein is located in the city of Recife, Brazil, and contains complex systems monitored by operators through a computerized system, SAGE, composed of hardware and software, that allows command activities and the execution of maneuvers in the electric substations.

Operator activities present a high degree of complexity requiring from them safe and optimized system piloting. In this framework, system operator capacitance is of extreme importance. A simple error can cause losses since the company is penalized by contingency occurrence.

In the conception of the project of the control room, ergonomics interferes from viability studies, in the initial phases, until the implementation of the designed situations, passing by the definition of work environments, the furniture, the computerized equipment, the configuration of screen, the recruitment of personnel and by the training and work organization. The earlier ergonomics is introduced the better will the adaptation of the future project be.

METHODS

As presented before, the company analyzed is within a new phase characterized by changes in the sector. For this reason, it has been undergoing various steps of reforms to adapt the new information system adopted, when synoptic panels where totally replaced. Besides, and certainly for this same reason, the management showed interest in an ergonomic study of work conditions.

The analysis was based in the methodology Ergonomizing Intervention ("Intervenção Ergonomizadora" in portuguese) proposed by Moraes & Mont'Alvão (2003). It encompasses all the steps, since the definition of the problems through the final detailing and optimization of the solution adopted. The method uses a systemic approach, through which are characterized the dysfunction that cause constraints to man, negatively affecting within System Human Task Machine.

In the first research step, data collection was done through observations and interviews with operators and management. The second began from the acquisition by the company of a video wall panel to substitute the preexisting plasma one. During all the steps of the research, there was a personal monitoring of all the process, especially during the main activity periods: peak hours, shift change and training.

ANALYSIS OF THE USE OF THE COMPUTERIZED SYSTEM

In all developed or developing countries, activities of a control room influence directly our daily lives. With economic development, new technologies arise, causing a constant transformation in the means of organization of work and requiring a better preparation of operators. In the automation era, characterized by Wickens et al. (1998), where the machine assumes a task which in some cases is performed by a human operator, the relationships between the users and the production means are substantially altered. Modern communication systems and powerful computers have allowed the search for costs cuts in large scale, centralizing various activities within an unique control center.

The control room under study benefits from this technological evolution, where the transmition is achieved at distance, by grouping the majority of commands and measures in a single place. To warrant a normal functioning of the system, punctual regulation must be executed automatically or by human operators.

Some years ago, the synoptic panel was used which, with the advent of microchips, was replaced by the software system SAGE – Open System of Energy Management. Thus, communication means, information and control of the operators were modified causing a significant change in the organization of the work of these rooms.

Since the shift from analogical to digital systems, designers of the automated systems subdivided the information through monitors in the consoles of the control rooms. Therefore, what was possible with a direct conduction and an easy access to indicators and commands of the synoptic panels, which allowed a general view of the system, is presently done by successive screen calls, offering a fragmented visualization of the process.

WORK STATIONS OF THE OPERATORS

With the introduction of the computerized system, the first stage of research, the control room had two work stations of the operators (Operators 1 and 2) and one of the supervisor. The workbench of the operators was composed of a "U" shaped modular system. Over it, we found the following elements: two 21" monitors, one to operate the SAGE system and another for internal communication and reports, a telephone central, a printer, a mouse and keyboard set for each system, a 15" monitor and CPU for the smart system, a software used to decode system alerts when contingencies occur (Figure 1).

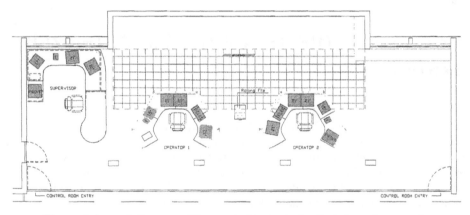

Figure 1. Layout diagram of the control room in the first step of research.

With only 57cm of depth, the workbench was insufficient for the quantity of existing equipments which hampered the filling of documents, the handling of folders with procedures and other needs. To view the monitors, the operator rotated on his body and moved with his chair constantly, especially when they where answering a telephone call and needed to query data or to operate the system (Figure 2).

Figure 2. Work Station of the operator and plasma panel with 52"

The space under the desk was adequately dimensioned, but complaints of the operators were registered related to the quantity of free cables and apparent electric sockets causing incidents like the brownout of equipment with feet or even accidents with electric shocks.

Between the workstations of operators two rolling file chariots are placed with suspended folders. The first file has norms and instructions of the control activity and the other one the report with the procedures and maneuvers to be realized during the shift.

In the second step of the research, the furniture of operators 1 and 2 was not modified and another workstation was created, of Operator 3, which occupied the former desk of the supervisor. In this way, the modules of the furniture were divided and one part was adapted to be used by the supervisor at the opposite side of the room (Figure 3).

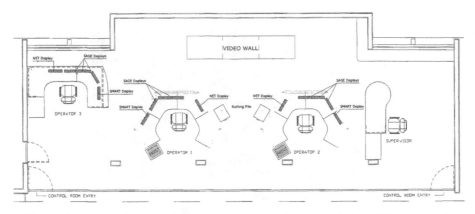

Figure 3. Layout diagram of the control room in the second step of the research

DISPLAYS AND PANELS

In control centers, much of the activities carried out by the operators involve the use of video displays. In some centers other types of displays may be available, analogical indicators, alert lights, digital displays, plasma screens, video walls, etc.

In the control room under study, besides CRT displays (cathode-ray tubes), mentioned in the previous item, there is also a 52" plasma display located in front of the two workstations of the operators, as shown in Figure 2. Because of its small size for visualization of the data of the SAGE system, it was almost not used by the operators, being one of the most criticized points. Evidence of this fact allowed a discussion for the installation of a video wall (Figure 4).

Figure 4. Work Station of the operator and video wall

Despite insufficient height between the floor and the roof for a video wall, its installation allowed the visualization of the whole network of substations served by the company, giving the proper conditions for the operator to adjust the configuration of screens to their needs.

The CRT displays (Cathode-ray Tubes, until then the most used) were replaced by LCD (Liquid Crystal Display) because of their flat screen, smaller space occupancy, lighter weight, lower energy consumption, smaller heat emission, higher visual comfort for prolonged use, 40% superior lifespan and larger effective area compared to a CRT (ISO 9241-6, 1999).

Another factor appointed by the operators was the insufficient number of displays in the work station for the visualization of the system. This was solved by installing an additional. The work station of operators now harbors the following equipments: three 21" displays for the SAGE system, one 21" display for networked communication operations and report filling, one 15" display for the SMART system.

The wallpapers of the screens of the system shows color contrasts and light intensity differences. Such factors, following Grandjean & Kroemer (2005), can offer daze risks by the light sources, damaging the vision, as well as to make difficult the distinction of details in the darker zones.

RESULTS AND DISCUSSION

Based on the data presented, we established that some modifications were considered satisfactory. However, we verified that the furniture and equipment are not fully adequate to the activities developed and that new interventions are necessary.

In the next section, the main interferences identified and suggested improvements are presented.

- **Displays** - One factor pointed out as satisfactory by the operators was the introduction of an additional monitor on the workstation for the operation of the system. However, the screen background deserves special attention because of the presence of color contrast and brightness differences. Such factor, as well as the use of inadequate colors for data presentation can be corrected by modifications of the SAGE system.

- **Workstation Layout** - As observed by the operators, the separate placement of the workstations makes the communication between them difficult, especially with Operator 3, whose position does not allow the visualization of the panel.

We advise the need to group the three operators in an unique cell. The workbench was criticized for the lack of space to house the equipment and by its "U" shaped configuration, forcing the operator to take inadequate postures.

Thus, as seen in the norm ISO 11064-3 (2002), the new placement of the workstations must be thought from the location of the video wall panel and the work surface dimensioned as a function of the quantity of equipments it will harbor. The distribution of monitors in the workbench must be made in such a way that these are placed facing the operators, preventing constant movements of the later between the screens. The mouse and keyboard must also remain in the same position, constantly adapting to the different forms of outreach.

A space for the passage of the cables of the equipments must be planned. The rolling file chariots must be adapted to the new furniture and located at the center of the cell, next to the Operators 2 and 3, such as to ease the access to folders.

CONCLUSIONS

The adaptation in control rooms of the computerized system, replacing the synoptic panels, demands an integration between the equipment and the furniture, to ease its operation with safety, reducing the workload not only physical but also psychological and cognitive.

The replacement of the plasma panel by a video wall, despite the physical dimensions of the room not allowing a proper installation, provided a significant improvement of the visualization problem of the SAGE system. Such solution is reinforced by the introduction of an additional monitor on the workstation.

As presented, much of the activities realized by the operators, involve the use of monitors of the workstation and the video wall panel, despite the persistence of the use of paper when the former query norms/procedures and fill reports. Such activities are influenced by the lack of conformity of elements that make up the room, such as the furniture and the equipments of the computerized system.

As a result, the systemic research approach allowed to verify that the modernization processes in the control room call forth the need of adaptation of the work to the new situation and that for a correct planning of the equipment and furniture the detailed knowledge of the process and especially of the user is necessary.

REFERENCES

Grandjean, E. Kroemer, K. (2005). Manual de Ergonomia: adaptando o trabalho ao homem. Trad. Lia Buarque M. Guimarães. Porto Alegre: Bookman.

International Organization for Standartization. (1999) ISO 9241-6 – Ergonomic requirements for office work with visual display terminals (VDTs). Gevève.

International Organization for Standartization. (2002) ISO 11064-3 – Ergonomic design of control centres – Part 3: Control room layout. Gevève.

Karwowski, Waldemar. (2006). International Encyclopedia of Ergonomics and Human Factors – CD Room. Florida: Taylor & Francis.

Moraes, Anamaria. Mont'Alvão, Claudia. (2003).Ergonomia: Conceito e Aplicações. Rio de Janeiro: 2AB Editora.

Nobrega, A. Filho, L. (2002) "A comercialização da energia elétrica das estatais federais por meio de Leilões Públicos", proceedings of 3° Simpósio de Especialistas em Operação de Centrais Hidrelétricas – SEPOCH. Foz do Iguaçu, PR.

Wickens, C. et al. (1998). An Introduction to Human Factors Engineering. United States: Longman.

Wood, John. (2001) "Control Room design", In: People in Control: Human factors in control room design. Noyes, J. Bransby, M. (Ed.). United Kingdon: IEE.

Chapter 14

Relation Between Kawaii Feeling and Biological Signals

Michiko Ohkura, Sayaka Goto, Asami Higo, Tetsuro Aoto

Shibaura Institute of Technology
Tokyo, 135-8548
Japan

ABSTRACT

In the advanced information society of the 21st century with its communication infrastructure of computers and networks, it is crucial to enhance software that utilizes these technologies, that is, digital content. Among such content, various Japanese kawaii characters such as Hello Kitty and Pokemon have become popular all over the world, suggesting that the allure of Japanese cuteness can be attained worldwide. However, since few studies have focused on the kawaii attributes of the interfaces of interactive systems or other artificial products, we focus on a systematic analysis of kawaii interfaces themselves: kawaii caused by the attributes of such interfaces as shapes, colors, textures, and materials. Our aim is to clarify a method for constructing a kawaii interface from the research results. Kawaii is one of the important kansei values for future interactive systems and industrial products.

We previously performed experiments and obtained valuable knowledge on such kawaii attributes as kawaii shapes and kawaii colors. However, only questionnaire was employed for the evaluation of kawaii feeling in those experiments. Although questionnaires are the most common form of kansei evaluation, they suffer from some demerits such as linguistic ambiguity, possibility of mixing intension of experimenters and/or participants to the results, and interruption of system's stream of information input/output.

Thus, to compensate for these questionnaire demerits, we examined the possibility of using biological signals. This paper describes a trial to detect comfortable feeling and/or excitement caused by seeing kawaii objects by changes in biological signals.

Keywords: Kawaii, kansei, feeling, interface, biological signal

INTRODUCTION

Recently, the kansei value has become crucial in manufacturing in Japan. The Japanese Ministry of Economy, Trade and Industry (METI) determined it to the fourth highest characteristic of industrial products after function, reliability, and cost. According to METI, it is important not only to offer new functions and competitive prices but also to create a new value to strengthen Japan's industrial competitiveness. Focusing on kansei as a new value axis, METI launched the "Kansei Value Creation Initiative" in 2007 (METI, 2008 and Araki, 2007) and held a kansei value creation fair called the "Kansei-Japan Design Exhibition" at Les Arts Decoratifs (Museum of Decorative Arts) at the Palais du Louvre, Paris in December 2008. Launched as an event of the "Kansei Value Creation Years," the exhibition had more than 10,000 visitors during its ten-day run and was received favorably (METI, 2009).

From several years ago, we focused on the kawaii attributes of industrial products, because we considered kawaii one of the most important kansei values. Various Japanese kawaii characters such as Hello Kitty and Pokemon have become popular all over the world. However, since few studies have focused on kawaii attributes, we systematically analyze the kawaii interfaces themselves: kawaii feelings caused by such attributes as shapes, colors, and materials. Our aim is to clarify a method for constructing a kawaii interface from the research results. Kawaii might become one important kansei value for future interactive systems and industrial products of Asian industries.

We previously performed experiments and obtained valuable knowledge on kawaii attributes (Murai et al., 2008, Ohkura and Aoto, 2007, Ohkura and Aoto, 2008, Ohkura et al., 2008, and Ohkura et al., 2009). For example, curved shapes such as a torus and a sphere are generally evaluated as more kawaii than straight-lined shapes. However, a discrepancy exists in kawaii colors among the experimental results (Murai et al., 2008). Warmer colors tended to be chosen as the most kawaii in one experiment, while in another experiment most male participants chose blue objects as the most kawaii. Thus, to solve this discrepancy, we performed a new experiment that employed the Muncell Color System, where a color has three elements: hue, saturations, and brightness (Ohkura et al., 2009). The following are the obtained results:

- Brightness is effective for kawaii colors.
- Saturation is also effective for kawaii colors.
- Although all hues can be chosen as the most kawaii color, purple and yellow tend to be chosen most often.

- The comparison results between pure color and the most bright and saturated candidates for each hue in the experiment depend on hues.

These results suggest that the discrepancy between the results of our previous experiments may be caused by the condition that only red, blue, and green were candidates of hues in some experiments.

Meanwhile, only questionnaires were employed to evaluate kawaii colors in all the above experiments. Although questionnaires are the most common form of kansei evaluation, they suffer from the following demerits:

- Linguistic ambiguity
- Possibility of mixing the intensions of experimenters and/or participants intention with the results
- Interruption of system's stream of information input/output

Thus, to compensate for these questionnaire demerits, we examined the possibility of using biological signals that offer the following merits and can supplement the above demerits:

- Can be measured by physical quantities.
- Difficult to be controlled by intensions of experimenter and participants.
- Can be measured continuously without interruption.

This paper describes our trials to clarify the relation between kawaii colors and biological signals and the relation between kawaii sizes and biological signals.

KAWAII COLORS AND BIOLOGICAL SIGNALS

EXPERIMENTAL METHOD

The first experiment addressed kawaii colors. We performed a preliminary experiment to select color candidates from 381 colors in the color table (Azo, 2010) employing four male and two female students in their twenties as participants. As the experimental results, pink and colors basically identical to pink were selected as kawaii colors, while dark brown and dark green were selected as non-kawaii colors. Thus, we selected pink (5R 7/10 from the Munsell Color System (Emori et al., 1979)), blue (10B 7/6), brown (5R 4/6), and green (between 2G 4/4 and 3G 4/4) for the following evaluation experiment, where green represented a middle color between kawaii and non-kawaii colors.

FIGURE 1 shows the experimental setup. Participants watched a large, 100-inch screen whose surface was covered by one of the four above colors and evaluated its kawaii degree on a 7-scale evaluation.

The experimental procedures were as follows:

1. Participants sat on chairs.
2. The experimenter explained the experiment.
3. Participants wore electrodes.
4. Participants remained quiet for 30 seconds.
5. Participants watched the screen displayed by a color for 30 seconds.
6. Participants answered the questionnaire.

Steps 5 and 6 were repeated four times for the four colors, which were displayed in random order. Such biological signals as heart rate, Garvanic Skin Reflex (GSR), breathing rate, and Electroencephalogram (EEG) were measured both before and while watching. The biological signals were measured by BIOPAC Student Lab (BIOPAC Systems, Inc.), except for EEG which was measured by the Brain Builder Unit (Brain Function Research Center).

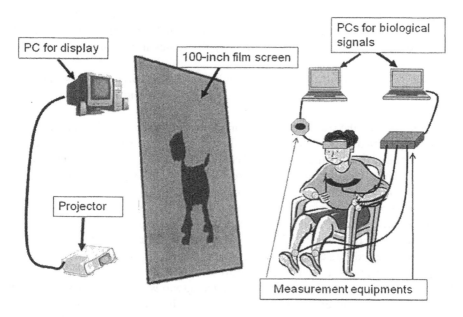

FIGURE 1 Experimental setup

EXPERIMENTAL RESULTS

Experiments were performed with eight female and eight male student volunteers in their twenties. Based on our previous experiments (Aoki et al., 2006, Ohkura et al. 2, 2009, Aoto and Ohkura, 2007, and Aoto and Ohkura 2, 2007), we selected the following physiological indexes for analysis:

- Average heart rate

- Variance of heart rate
- Average R-R interval
- Variance of R-R interval
- Average GSR
- Variance of GSR
- Number of breaths
- Variance of number of breaths
- Average breath magnitude
- Variance of breath magnitude
- Ratio of power spectrum of Theta, Slow alpha, Mid alpha, Fast alpha, and Beta waves
- Ratio of dominant duration of Theta, Slow alpha, Mid alpha, Fast alpha, and Beta waves

The R-R interval, defined as the time interval between the two R waves of the ECG, is the inverse of the heart rate, the number of heart beats per minute. All indexes described above were normalized by the values in the quiet state for each participant.

FIGURE 2 shows the questionnaire results, where the horizontal axis shows the participants (male: a- h, female: i-p) and the vertical axis shows the kawaii degree of each color.

Table 1 shows the results of a 2-factor analysis of variance. The main effect of color is significant at the 1% level, and the main effect of gender is significant at the 5% level, and a significant cross effect exists between color and gender at the 5% level. Thus, we successfully analyzed the biological signals by dividing the participants into two groups by kawaii scores.

The data of the biological indexes were divided into the following two groups: the kawaii group where scores were above 0, and the non-kawaii group where the kawaii scores were below 0. The data with 0 score were omitted from analysis. From the unpaired t-test results of the difference of the mean value of the two groups, the heart rates, the numbers of heart beats per minute, showed a significant difference (Table 2). The heart rate of the kawaii group was significantly faster than that of the non-kawaii group. This result suggests that watching a kawaii color is more exciting than watching a non-kawaii color, because a decrease of the heart rate is considered an index of relaxation.

Moreover, from the unpaired t-test results of the difference of the mean value of the two groups by gender, the ratio of the dominant duration of the mid alpha wave showed a significant difference. The ratio of the dominant duration of the mid alpha wave of the kawaii group was significantly larger than that of the non-kawaii group. This result suggests that watching a kawaii color is more exciting than watching a non-kawaii color, because a decrease of the ratio of the dominant duration of the mid alpha wave is also considered as a index of relaxation.

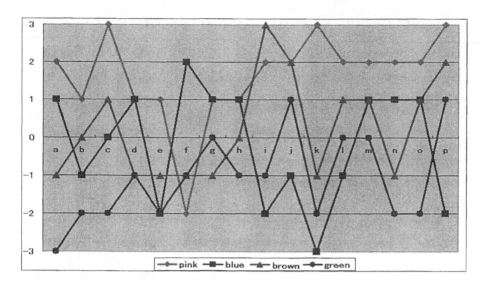

FIGURE 2 Questionnaire results of first experiment

Table 1 Analysis of variance

FACTOR	SUM OF SQUARED DEVIATION	DOF	MEAN SQUARE	F-VALUE	P-VALUE
COLOR	60.31	3	20.10	13.94	0.000**
GENDER	6.25	1	6.25	4.33	0.042*
C x G	17.13	3	5.71	3.96	0.013*
ERROR	80.75	56	1.44		
TOTAL	164.44	63			

Table 2 Unpaired t-test of difference of mean value of heart rate

FACTOR	KAWAII	NON-KAWAII	DIIFFERENCE	EQUALITY OF VARIANCE	T-TEST
NUMBER	32	26		F:1.74	T:1.75
AVERAGE	3.14	0.88	2.26	DOF1:25	DOF:56
UNBIASED VARIANCE	17.92	31.11		DOF2:31	0.08
SD	4.23	5.58	1.34	P:0.15	0.04*

KAWAII SIZES AND BIOLOGICAL SIGNALS

The second experiment addressed the kawaii sizes of objects. The experimental setup resembled that shown in FIGURE 1, except the display used two projectors with polarized filters and the participants wore polarized glasses to watch the objects on the screen stereoscopically. Participants watched an object on the large screen and evaluated its kawaii degree. The shape and the color of each object were set as torus and yellow, 5Y8/14 in MCS, as shown in FIGURE 3, based on the results of our previous experiment described above. The size, which means visual angle, of each object was set as one of the four sizes shown in Table 3 based on our preliminary experiment.

Experiments were conducted with twelve female and twelve male student volunteers in their twenties. FIGURE 4 shows the questionnaire results, where the horizontal axis shows the participants (male: m1-m12, female: f1-f12) and the vertical axis shows the kawaii degree for each object size. The results of a 2-factor analysis of variance show that the main effect of size is significant at the 1% level and the main effect of gender is not significant (Table 4). The biological signal data were divided into two groups similar to the first experiment. From the unpaired t-test results of the difference of the mean value of the two groups, the heart rates showed a significant difference (Table 5). Since a higher heart rate shows the unrelaxed state, the mental state when feeling kawaii is considered more exciting than not feeling kawaii. In addition, the results of the similar difference test of the heart rate for object C showed a significant difference at the 5% level, and the results of the similar difference test of the heart rate for object D showed a significant difference at the 1% level. These results show that the mental state when feeling kawaii is probably more exciting than not feeling kawaii even if the size of the object being watched is the same.

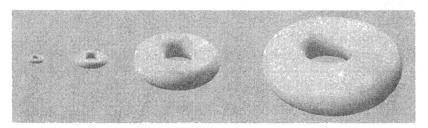

FIGURE 3 Objects employed for second experiment

Table 3 Analysis of variance

OBJECT	A	B	C	D
SIZE	1	2	6	10
VERTICAL DEGREE	3.7	7.5	22.0	35.4
HORIZONTAL DEGREE	5.0	10.2	29.5	47.3

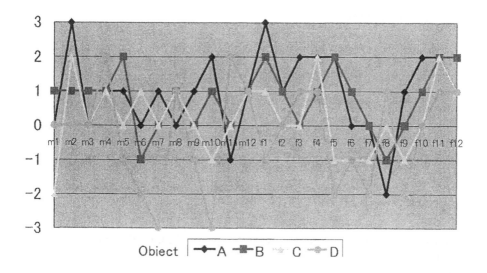

FIGURE 4 Questionnaire results of second experiment

Table 4 Analysis of variance

FACTOR	SUM OF SQUARED DEVIATION	DOF	MEAN SQUARE	F-VALUE	P-VALUE
SIZE	23.5	3	7.83	5.46	0.002**
GENDER	0.67	1	0.67	0.46	0.497
S x G	0.83	3	0.28	0.19	0.901
ERROR	126.3	88	1.44		
TOTAL	151.3	95			

Table 5 Unpaired t-test of difference of mean value of heart rate

FACTOR	KAWAII	NON-KAWAII	DIIFFERENCE	EQUALITY OF VARIANCE	T-TEST
NUMBER	50	20		F:2.99	T:2.555
AVERAGE	3.45	0.60	2.85	DOF1:49	DOF:68
UNBIASED VARIANCE	18.98	14.61		DOF2:19	0.013
SD	4.36	3.82	0.534	P:0.542	0.006**

CONCLUSION

In the 21st century, the kansei values of industrial products will probably be considered crucial. In this study, we focused on kawaii as one important kansei values for future industrial products.

In this article, we focused on the relation between kawaii attributes and biological signals. We performed two experiments to clarify the relation between kawaii colors and biological signals and the relation between kawaii sizes and biological signals.

ACKNOWLEDGMENTS

This research was partly supported by the Grant-in-Aid for Scientific Research (C) (No. 09017360), Japan Society for the Promotion of Science and by the SIT Research Promotion Funds. We thank to the students of Shibaura Institute of Technology who contributed to this research and served as volunteers.

REFERENCES

METI (2008), "Kansei Value Creation Initiative"
   http://www.meti.go.jp/english/information/downloadfiles/PressRelease/080620KANSEI.pdf
Araki, J. (2007), "Kansei and Value Creation Initiative," *Journal of Japan Society of Kansei Engineering,* Vol. 7(3), pp.417-419. (in Japanese)
METI (2009), Announcement of the "Kansei Value Creation Museum"
   http://www.meti.go.jp/english/press/data/20090119_02.html
Murai, S. et al. (2008), "Systematic study for "kawaii" products (The third report) - Comparison of "kawaii" between 2D and 3D-," *Proceedings of VRSJ2008.* (In Japanese)
Ohkura, M., and Aoto, T. (2007), "Systematic Study for "Kawaii" Products,"

Proceedings of the 1ˢᵗ International Conference on Kansei Engineering and Emotion Research (KEER2007), Sapporo, Japan.

Ohkura, M., and Aoto, T. (2008), "Systematic Study for "Kawaii" Products," *Proceedings of Workshop of DIS2008*, Cape Town, South Africa.

Ohkura, M. et al. (2008), "Systematic Study for "Kawaii" Products (The second report) -Comparison of "kawaii" colors and shapes-," *Proceedings of SICE2008*, Chofu, Japan.

Ohkura, M. et al. (2009), "Systematic Study for "Kawaii" Products –Study on Kawaii Colors using Virtual Objects-," *Proceedings of HCI International 2009*, San Diego, California.

Azo (2010), Color guide http://www.color-guide.com/e_index.shtml.

Emori, Y. et al. (1979), "Colors – Their Science and Culture –," Asakura publishing Co. Ltd., Tokyo. (in Japanese)

Aoki, Y. et al. (2006), "Assessment of relief level of living space using immersive space simulator(Report 4) –Examination of physiological indices to detect "Uneasiness"," *Proceedings of VRSJ the 10th Annual Conference,* Tokyo, Japan, (CD-ROM), Vol.11, 2B1-4. (in Japanese)

Ohkura, M. et al. 2 (2009), "Evaluation of Comfortable Spaces for Women using Virtual Environment -Objective Evaluation by Biological Signals-," *Kansei Engineering International*, Vol.8, No.1, pp.67-72.

Aoto, T., and Ohkura, M. (2007), "Study on Usage of Biological Signal to Evaluate Kansei of a System," *Proceedings of the 9ᵗʰ Annual Conference of JSKE 2007,* Tokyo, Japan, (CD-ROM), H-27. (in Japanese)

Aoto, T., and Ohkura, M. 2 (2007), "Study on Usage of Biological Signal to Evaluate Kansei of a System", *Proceedings of the 1ˢᵗ International Conference on Kansei Engineering and Emotion Research (KEER2007)*, Sapporo, Japan, (CD-ROM), L-9.

CHAPTER 15

Impact Factors of Three-Dimensional Interface

Yu-Chun Huang, Chun-Chin Su, Fong-Gong Wu, Chien-Hsu Chen

Institute of Industrial Design
National Cheng Kung University
Tainan 70101, Taiwan (R.O.C)

ABSTRACT

In this study, we discussed the present researches about the interaction of the three-dimensional (3D) interface. However, we found that participants presented weaker performances associated with the depth direction (z-axis). Three possible factors could be summarized to the display mode in 3D presentation, the degree of freedom in input devices and the compatibility between the display mode and the input devices. Therefore, we tried to design an augment reality (AR) 3D interface base on those possible impact factors and to test whether the participants' performance could be improved by this design. However, the result shows indeed there were significant differences between the three moving directions. But, there were also found the interference effect induce by the 2D cues. Those interferences may be the key solutions to development the 3D user interface.

Keywords: 3D User Interface, Augment reality, Human-Computer Interface (HCI)

INTRODUCTION

Human always desire to reach the reality, and it seems without the end. Now a day as the evolution of display technology progressed, we experience the colorful, exquisite, and even stereo vivid images or animations in our lives. Especially the 3D display is the most interesting structure. Although most interactions with computers today are in a two-dimensional environment, three-dimensional technologies have lots of potential applications in many different fields, such as medication, aviation, and information visualization (e.g., Ellis & McGreevy, 1983; Gaunt & Gaunt, 1978; Pepper, Smith, & Cole, 1981). Moreover, it is expected that the two-dimensional interface will be taken place by the three-dimensional interface in some areas in the near future. Therefore, it is an important task to understand the quality of interaction and make the new interactive 3D applications possible.

THE FINDINGS OF WEAKNESS PERGFORMANCE IN THE DEPTH DIRECTION (Z-AXIS)

In the previous studies, researchers had used different ways of 3D presentation to execute visuo-motor task and to see users' performances with those interfaces. The 3D presentations can be easily divided into 3D perspective display with monocular depth cues and binocular disparity features. What concerned about the most is that the interactions with computer are in 2D display; however, the 3D graphic computer has been opening a virtual 3D environment into a 2D display system. Also, many researches began to use this technology to create a 3D perspective view as their experimental conditions.

Ellis and McGreenvy (1987) had selected a 3D perspective format to compare with the original plan-view (the vertical information was presented by text display) to see the pilots' decision time in traffic avoidance. The 3D perspective view was based on the visual cue of liner perspective that Ellis believed that its dimensionality matches with the pilots' visualizations of their situations in three-dimensional space, and this condition really improved the timing of their decision making.

Massimino et al. (1989) also used the liner perspective cue as the 3D perspective background. In order to determine how movements in each of the six degrees of freedom (DOF) devices affected operator's tracking performance. They found that tracking motion got much higher error rate in the z-direction than in the x and y directions. Therefore, they also recommended that a compensatory display maybe displayed an error for the z-direction might be helpful.

In order to enhance the depth effect, Liao and Johnson (2004) used 3D perspective view to test users' ability and strategy in target acquisition. They used the droplines both on the target and the cursor to enhance the relative position which can resolve the height/distance ambiguity. Their results also found both the mean horizontal (x-axis) and the mean vertical (y-axis) movement times were significantly lower than the mean depth (z-axis) movement time. Hence, the presence of droplines led to poorer performance in the depth domain, and this finding was the opposite of the dropline effect in the other domains.

As propagating the faster computers and displays, it comes with the possibility of real-time interaction within a reasonably sophisticated 3D visual environment. Those offspring researches had develop in many areas, such as the stereopsis perception, illusion, attention, virtual reality (VR) and virtual environments (VEs). For those researches, scientist also found out some asymmetries of the weakness performance in the depth space.

Ware and Balakrishnan (1994) had asked participants to perform a volume location task (1cm cube) with targets to the left or behind the starting location of the cursor using a head-tracked field sequential stereo display and found that the performance in z-direction was to be 10% slower than the performance in the x-direction. Those participants not only performed more slowly in the depth-direction but also get higher error rate along this direction in the tracking task (Zhai et al., 1997). They tried to decompose the tracking performance into six components (three in translation and three in rotation). Tests revealed that participants' tracking errors in the depth dimension were about 45% (with no practice) to 35% (with practice) which was larger than those in the horizontal and vertical dimensions. They also found out participants had initially larger tracking errors along the vertical axis than the horizontal axis, and it was likely due to attention allocation strategy.

Unlike most experiments involving 3D tasks, Grossman et al. (2004) performed on a true 3D volumetric display which provided with reliable perception of all three spatial dimensions. They observed that moving forwards and backwards in depth is slower than moving left and right in selecting targets; the width was more critical than the height and depth of that target. Thus, the effect of the width and depth were dependant on the moving angle, while the effect of the height was constant regardless of the moving angle.

IS 3D USER INTERFACE BETTER?

In the previous 3D perspective or 3D display conditions, the experimental evidences seemed directed that it took more times and error rates to move in depth direction. Nevertheless, there were many researches to compare user's performances in different 3D display conditions, and their result showed that 3D interface can ease the differences from the other two dimensions, even better. But, what exactly the difference of these two kinds of interfaces, that would be the crucial factors to improve in 3D interface design. Therefore, we try to follow the previous studies to clarify the differences.

Takemura et al. (1988) had recruited participants to perform in the volume location task (1cm and 2cm cubes) using a field sequential stereo display and a large number of target locations (x, y, z). They found no significant difference between target positions in different dimensions. Boritz (1999) thought that it may have a reduced accuracy requirement of the volume location task and result in a reduced sensitivity to position difference.

The evidence also demonstrated that using 3D view can improve the performance in the pointing task (Sollenberger & Milgram, 1993). They studied the abilities of participants to identify the visual network to which specified point belonged in monoscopic and stereoscopic modes with and without rotation of the network. They found that rotation and stereoscopic viewing improved performance both individually and as a combination.

Arthur et al. (1993) studied a similar task using monoscopic and stereoscopic displays with and without head-coupling. They found that the head-coupled stereoscopic mode resulted in the shortest response time and lowest error rate.

Zhai studied a serial studies to evaluate how the different display modes and input control devices affect different motion tasks (Zhai, 1995; Zhai & Milgram,

1997). They have set several studies in 3D docking, tracking and target acquisition tasks using a variety of controllers and control types in conjunction with monoscopic and field sequential stereoscopic displays. Consequently, they found that the stereoscopic display was superior to that of the monoscopic display, and asymmetries in the amount of error existed across the x, y and z axes.

Boritz (1997) also found that the stereoscopic performance was superior to the monoscopic one. He investigated 3D point location using a six degree of freedom input device. Then, test by using four different visual feedback modes: monoscopic fixed viewpoint, stereoscopic fixed viewpoint, monoscopic head-tracked perspective, and stereoscopic head-tracked perspective. The results indicated that stereoscopic performance was superior to that of the monoscope, and that asymmetries existed both in across and within axes.

THE POSSIBLE IMPACT FACTORS

The evidences showed that in the 3D display environment, users can eliminate the weakness of their performances in depth direction. Furthermore, in some comparison studies, stereoscope seemed to provide user more effective cues to accomplish those task quickly and precisely than the monoscope condition; however, what exactly the differences between this two display mode are still unknown. We analyze the differences from the studies to find out possible factors. Then, we could improve the 3D user interface base on users' needs.

First, in the 3D perspective display the size and distance could be ambiguous. The accuracy of depth perception was limited by the pictorial cues. Thus, the visual angle of the visual perspective was an important factor.

The second is focusing on the compatibility between the display modes and motor control devices (DOF). As Norman's (2002) control-display compatibility principle, the designed relationship between controls and displays should be mapped naturally. In the previous 3D perspective studies, although the depth can be perceived by the pictorial cues, the motion direction was limited along the oblique direction. Hence, the motor control was also constraint by the lower degree of freedom input device.

Furthermore, the competing compatibility and familiarity of the motion directions in 3D space were also important factors. For instance, when the computer mouse moved from the left to the right, the cursor would follow the same route. When we forwarded the scroll and tried to move from the nearest site to the farthest site, the cursor would move the near to the far; However, this kind of motion could also be the up and zoom functions in the 2D interface, which were frequently used. This kind of misleading mapping familiarity could be the effect of motor control concept in 3D environment. Therefore, based on this reason, we can predict that in the 3D space, user performs better in horizontal than in the vertical, and the depth.

EXPERIMENT: TARGET ACQUISITION TASK

Base on the present works, an intuitive and high performance 3D user interface should contain a stereo visual output and a higher degree of freedom input. Therefore, we build up an augment reality of three-dimensional environment that could interact with the 3D objects by hand and physical marker which was identified as virtual reality coordinates. And, in this experiment we would like to see whether the new 3D interface of target acquisition task will be the results of the previous studies.

METHOD

Apparatus

In order to present a vivid and stereo environment. We use the augment reality environment to display the virtual workspace. The experiment setup, we use a static webcam (as the participants' visual point) to detect the goal and initial target (physical marker). The initial target which is held by the participant's right index finger, the goal target is the place that participants have to hit by their finger. All the conditions and experimental procedure setup were drawn and recorded by the Flash action script 3.0.

In these virtual workspace (size was 22 cm) it contains five positions on each axis (total visual angle were 30.75°) that center appears as target box (size of 2 and 4 cm; visual angle of each box were 1.43°, 2.86°). The participants have to move from the initial box (size as target box) to the location of the target by moving his/her index finger. In these positions, three kinds of moving axes will appear.

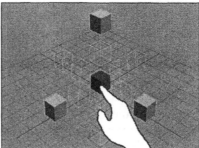

FIGURE 1. The experimental setup and illustration of the virtual workspaces

Participants

There were 5 male and 5 female unpaid volunteers, age ranging from 23 to 30, served as participants in this experiment. We measure the participants' hand length in which they could reach to the farthest location in the virtual location. And, all participants were right handed and used their right hand to control as an input device.

Procedure

A 3D static target acquisition task was used for the experiment. The scene was consisted of an initial target and a goal target. The initial target was at the middle point of a single axis moving route, while the goal target was rendered as a yellow cube and presented at the adjacent four positions. In order to complete the task, participants have to move from the initial target to the goal target whether the different size, position or direction of the goal target. The position of the initial target and the goal target were positioned in a random location, but the distance between the initial and goal targets was manipulated in the range of 5 or 10 cm in 3D space.

The experiment was divided into six blocks, each block presented one kinds of target size and one kinds of moving direction. The sequences of those blocks would be manipulated in the random order. And, each of the block contains 40 trials (total trials would be 240) limited in 5 seconds, the inter-stimulus interval (ISI) was fixed as 2 seconds. Between each of the block there was 2 minutes break.

Experimental Design

A three factors repeated measures of within-participant design was applied as experimental design. Three independent variables were: two levels of goal target size (2, 4 cm); distance between initial and goal targets (4, 8 cm); the direction along moving axis (x, y, z axis). The dependent variable was recorded the completion time that the time between participants to move their finger from the middle point to the goal target.

RESULT

Our interface has three major factors to observe the performance of this virtual workspace. The participants' performances were recorded as the trial complete time. The trial completion time was counted when participants moving their finger from the middle position to the target box. However, if the participants could not complete the motion in time, that were count as error. And, three factors repeat measure analysis of variance (ANOVA) was used to analysis in trial completion time.

The results of the ANOVA are presented in Table 1 and Figure 1. The three main factors, both directions and distance show significant effect, and there were no significant effect of moving distance. But, it contains interaction effect between the direction and distance. It presented the reverse performance of moving along with the z axis to x and y axes, and the further target seems have better depth guideline to the participants.

Table 1 Statistic results of repeat measure ANOVA

Variables	SS	df	MS	F	p
Direction	2.58	2	1.30	5.82	.014
Size	.29	1	.29	5.83	.046
Distance	.19	1	.19	2.47	.160
Direction*Size	.52	2	.26	1.86	.192
Direction*Distance	1.82	2	.91	10.06	.002
Size*Distance	.06	1	.06	.79	.403
Direction*Size* Distance	.02	2	.01	.16	.851

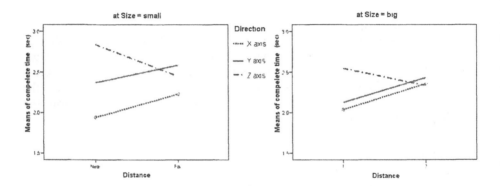

FIGURE 2. Interaction effect between direction and distance

DISCUSSION

As the result shown, we found the different moving direction presented significant effect. Associated to the present works, movement the in z direction actually had poor performance. And, according to the Fitts' law (Fitts, 1954) prediction that lager distance or the smaller size of target width would take more time to execute the task. In this study, we also found the same tendency in the x and direction. But, the movement along z axis got an opposite result interference by the different distance.

We observed and conclude the possible factors could be summarized as the 2D occlusion effect and level of depth perception. The 2D occlusion effect was come from the special character of augment reality that stacking up virtual images on the real environment. It presented the virtual images always on the top of the participants' finger. However, when the near target appears, the occlusion effect led the interference to the participants. They seem need more times to aware the relative depth position. On the contrary, when participants perceive further targets they got more depth cues (linear perspective, size consistency) to help them judge to depth position at the star. Hence, we could see the performance of far distance condition in those three axes as figure 2 shown. Participants seem to perform an equal effect on the further movement.

CONCLUSIONS

In this paper we had presented a basic finding of the weak performance in depth direction. Base on the comparison of different tasks and 3D presentation modes, we develop an augment reality 3D interface to see whether the 3D interface would induce user's depth perception and how they move to the target position.

Major findings were the different moving direction and target size presented significant effect, and the interaction effect of direction and distance. It means, in this 3D interface the depth perception that induce by real environment could improve the performance in virtual interactions. And, the depth cues induce by 2D pictorial cues were also important to participants. Moreover, according to the Fitts' law prediction the lager distance and the smaller size of target width would take more time to execute the task. In this study, we also found the same performance in the depth direction.

In future development, we may try to test the actual 3D pointing movements under the manipulation of target size, the distance to the target and the direction to the target. And, we can modify this interface to the Fitts' law for 3D interface. Murata and Iwase (2001) had attempted to extend Fitts' law to an actual three-dimensional movement (pointing) task to enhance its predictive performance in this domain.

Furthermore, in this study participants were only been asked to move the cursor in a single direction. It would be worthy to observe the moving path when the cursor and target are in multi-direction. Therefore, moving across different dimensional spaces should be discussed as well.

Last, multi-modal of feedback was also an important factor to enhance human performances with virtual interface. Therefore, comparison of visual, auditory, haptic feedback would be another issue on the 3D interaction interface.

REFERENCES

Arthur K. W., Booth K. S. & Ware C. (1993). Evaluating 3D Task Performance for Fishtank Virtual Worlds, ACM Transactions on Information Systems, 11(3): 239-265.

Boritz J. & Booth K. S. (1997). A study of interactive 3D point location in a computer simulated virtual environment. In ACM Symposium on Virtual Reality Software and Technology (VRST), 181–187.

Boritz J. & Booth K. S. (1998). A study of interactive 6DOF docking in a computerized virtual environment. In Proceedings of IEEE Virtual Reality Annual Internationl Symposium (VRAIS), 139–146.

Ellis S. R, McGreevy M. W. & Hitchcock R. J. (1987). Perspective traffic display format and airline pilot traffic avoidance. Human Factors, 29, 371-382.

Fitts, P. M. (1954). The information capacity of the human motor system in controlling the amplitude of movement. Journal of Experimental Psychology, 47, 381-391.

Fiorentino M., Monno G., Renzulli P. A. & Uva A. E., (2003) "3D Pointing in Virtual Reality: Experimental Study", XIII ADM - XV INGEGRAF International Conference on Tools And Methods Evolution In Engineering Design, Napoli, June 3th and June 6th.

Groen J. & Werkhoven P.J. (1998). Visuomotor adaptation to virtual hand position in interactive virtual environments. Presence: Teleoperators and Virtual Environments, 7, 429-446.

Grossman T. & Balakrishnan R., (2004) "Pointing at Trivariate Targets in 3D Environments", Proceedings of the 2004 Conference on Human Factors in Computing Systems, p.447- 454, April 24-29, Vienna, Austria.

Johnson T. E. (1963). Sketchpad III, a computer program for drawing in three dimensions. In Proceedings of the Spring Joint Computer Conference. Reprinted in Tutorial and Selected Readings in Interactive Computer Graphics, Herbert Freeman ed., IEEE Computer Society, Silver Spring MD 1984, 20-26.

Takemura H., Tomono A. & Kobayashi Y. (1988). An evaluation of 3D object pointing using a field sequential stereoscopic display. In Proceedings of Graphics Interface, 112-118.

Kim, W. S., Ellis, S. R, Tyler, M. E., Hannaford, B., & Stark, L. W. (1987). Quantitative evaluation of perspective and stereoscopic displays in three-axis manual tracking tasks. IEEE Transactions on Systems, Alan, and Cybemetics, SMC-17, 61-71.

Liao, M. & Johnson, W. W. (2004). Characterizing the effects of droplines on target acquisition performance on a 3D perspective display. Human Factors, 46(3), 476-496.

MacKenzie J. S. (1992). Fitts' law as a research and design tool in human-computer interaction. Human-Computer Interaction, 7, 91-1 39.

Massimino, M. I., Sheridan, T. B., & Roseborough, 1. B. (1989). One handed tracking in six degrees of freedom. In Proceedings of the IEEE International Conference on Systems, Alan, and Cybemetics, 498-503, New York: IEEE.

Murata A. & Iwase H. (2001). Extending Fitts' Law to a three-dimensional pointing task. Human Movement Science, 20, 791–805.

Norman D. (2002). The design of everyday things. Published: Perseus Books Group. U.S.A.

Sollenberger R. L. and Milgram P. (1993). Effects of stereoscopic and rotational displays in a three-dimensional path-tracing task. Human Factors, 35(3):483-499.

Sutherland I. E. (1968). Head mounted three dimensional display. In Proceedings of the Fall Joint Computer Conference, 757–764.

Van Erp, J. B. F. & Oving A. B. (2002). Control performance with three translational degrees of freedom. Human Factors, 44, 144-155.

Ware C. & Balakrishnan R. (1994). Target acquisition in Fish Tank VR: The effects of lag and frame rate. In Proceedings of Graphics Interface.

Ware C. & Osborne S. (1990). Exploration and virtual camera control in virtual three dimensional environments. In 1990 Symposium on Interactive 3D Graphics, 175–183.

Zhai S. (1995). Human Performance in Six Degree Of Freedom Input Control, Doctoral Dissertation, Department of Industrial Engineering, University of Toronto.

Zhai S., Milgram P. & Rastogi A. (1997). Anisotropic human performance in six degree-of-freedom tracking: An evaluation of three-dimensional display and control interfaces. IEEE Transactions on Systems, Alan, and Cybernetics - Part A: Systems and Humans, 27, 518-528.

Zimmerman T. G., Lanier J., Blanchard C., Bryson S. & Harvill Y. (1987). A hand gesture interface device. In J.M. Carroll and P.P. Tanner, editors, Proceedings of Human Factors in Computing Systems and Graphics Interface, 189–192.

Chapter 16

Helping the User Fail: Ergonomic Modalities and Distraction

M. Hancock[1], K. Wood[2], M. Rowthorn[2], B. Badger[2],
A. Abouzahr[2], A. Woods[2], C-A. Sturt[2], K. Koudijs[2], M. Kinkey[2],
R. Greveson[2], R. Myers[2], S. Wheeler[2], W. Brasher[2]

[1]Celestech, Inc.
4505 E. Chandler Blvd., Suite 155
Phoenix, AZ 85048, USA

[2]Project Sirius
406 Dartmouth Ave. West
Melbourne, FL 32901, USA

ABSTRACT

Performance on cognitive tasks is negatively correlated with subject stress levels. Instrumenting real-time performance feedback (e.g. elapsed time, current "accuracy") into the user- interface (UI) of a computing system affects stress levels, cognitive load, and thereby performance.

The purpose of this study is to assess the relative effects of the modality of instrumentation (e.g. graphical, textual, combinations) on performance in the presence/absence of distracters. In this study, distracters are textual messages providing information unrelated to task performance. It is hypothesized that some feedback modalities are more greatly affected by the presence of distracters than others.

A trial constituted a computer program presenting two visual patterns to the subject on a computer display. The first one (the labile pattern), can be modified by the subject through the manipulation of simple keyboard controls. The other, the target pattern, is fixed. A static sequence of parameter settings is used, so that each subject encounters identical tasks. Each subject sees exactly the same labile-target pattern pairs. The subject uses the controls to "morph" the labile pattern onto the target pattern. Subject performance is a numerically calculated accuracy envelope of the morph, which measures the subjects' accuracy as a function of time-on-task. The computer program collects data as the trials are performed.

The display supports a variety of performance user feedback modalities (graphical and textual). The specific goal of the experiment (not known to the subjects) is an assessment of the relationship between performance changes and the mix of feedback modes when distracters are added to the instrumentation. It is hypothesized that the presence of distracters will affect user performance due to increased cognitive loading. It is also hypothesized that individual differences in performance will correlate with gross demographic factors (e.g. gender, age, handedness, cultural background, computer literacy, etc.). Subjects are current residents of four western countries on three continents.

Characterization of these effects provides insight into how control interface designers can design performance metrics to "best condition" user performance.

Keywords: User interface, complexity, cognitive overload, cognitive ergonomics, interface design, web application.

Acknowledgement: This research was funded in part by Celestech, Inc., Phoenix, AZ, http://www.celestech.com

PROBLEM STATEMENT

Cognitive ergonomics includes considerations of the mental and perceptual load placed on a user by feedback instrumentation. The relationship between aspects of the user demographic and cognitive load were investigated by taking participants through a series of on-screen tasks. Each user was given a randomly selected suite of gauges. Some of which provided useful feedback on the task and some did not. The user was not given any documentation describing the instrumentation. Participants were required to work out how to use the gauges provided in their own time. Users had no idea whether there was a time limit since there were no instructions. This enabled users to determine their own limitations and pressures. Human performance in advanced, complex systems can be expected to differ. It is possible to gather, quantify, and assess these differences.

We can quantify whether too much information is bad information. The goal is to determine the relationships between demographic features such as nationality, age, gender, computer literacy, and performance. In particular, the following questions are addressed: Does performance vary consistently with UI complexity for both genders? Does being married or single affect performance?

It was hypothesized that the number and utility of the gauges would affect specific performance variables (speed and accuracy) by inducing conditions that could result in varying levels of stress (e.g. performance anxiety and increased cognitive load.) In particular, it was hypothesized that more obtrusive instrumentation would be associated with reduced performance for most demographic strata.

It was hypothesized that the effect of obtrusiveness would combine with attributes of the subject creating differential results among groups as well as individuals within groups.

BACKGROUND

Human Factors and Ergonomics (HFE) seek to improve the efficiency of interactions between the user and an interface. The applications of HFE have included Integrated Hazard Displays in aircraft (Alexander & Wickens, 2006), the integration of protective systems in automobiles to prevent crashes (May, Baldwin, & Paraguayan; Inagaki, Itoh, and Nagai, 2006), and the affect of team collaboration on tasks (Tollner et al., 2006). HFE has led to the advent of Decision Support Systems, which increase the operator's ability to make quicker and/or better quality decisions contingent on the operator's skill level (Hayes & Anderson). Research in the field of HFE has also been applied to limiting the negative ramifications of increased technological assistance to an operator to increase the safety and efficiency of the operator. An example of this is lazy driving developing as a result of Advanced Driver Assistance Systems along with night vision (Torbjörn, Kovordáni, and Ohlsson, 2006), and observing human complacency relative to automated systems. The military has taken advantage of HFE research, with its applications in such fields as the management of Unmanned Aerial Vehicles (UAVS) (Levinthal & Wickens, 2006; Wilson, Russell and Davis, 2006).

METHOD

A computer program was written that presents two visual patterns to the subject on a computer display. The display has a graphical interface. The display also had set of gauges with erroneous information to confuse the user. A website was created for the distribution of the program. The data was collected for analyst purposes.

The Project Sirius team was made up of a group of people whom had never met and most live on different continents around the globe. A team project website and emails are the center focus for communication and collaboration.

SAMPLING METHODOLOGY

Each Team Member was giving 500 or more login keys to distribute to each volunteer test participant they found. Login codes for the experiment application were distributed mostly by personal e-mail. E-mail attachment was also the primary method of distributing the experiment application itself.

A login key allows a participant to access and execute the software program for the experiment. This program performs the following functions:

1.) thanks the participant
2.) gives brief instructions on using the software
3.) presents a demographic survey

4.) takes the participant through a trial run

5.) takes the participant through four collection runs while collecting metric data

6.) presents a feedback survey

7.) uploads data to the project server

PROJECT WEBSITE

In order to acquire participants for the study, a website was created (http://bit.ly/cerL3k) where possible participants can download the JAVA program and a JAVA update if needed to run the program , request a login code via email (sirius.project@yahoo.com), which every member of Project Sirius has access to, to begin the trials and to ultimately send their data to the server, and view the instructions from installing the program to uninstalling it in nine descriptive steps and pointers all on the same front page.

The website was as helpful as the emails sent to the possible participants. Almost the first thing you see is the link to download the JAVA program used for collecting data from the participants as well as the JAVA sun update side-by-side to work the program if needed. There was also an easily recognizable link to contact the team for requesting a login code. Furthermore, due to its simple and easy to follow layout, you can view the instructions on the front page from installing the program to uninstalling it in nine descriptive steps and pointers.

APPARATUS

A static sequence of parameter settings is used so that each subject encounters identical tasks (that is, exactly the same labile-target pattern pairs.) Both patterns are closed parametric curves whose shape and location is determined by the values of numeric parameters that modulate its x and y terms using constant, linear, quadratic, and higher order monomials.

The subject uses the controls to "morph" the labile pattern onto the target pattern. Changing the control settings immediately redraws the labile curve using the modified parameters. There is no time limit; the software terminates the run when the "error of fit" between the two curves falls below a threshold value.

No documentation of the gauges or description of them other than the on-screen annotations was provided. This was done to avoid suggesting any particular strategy for their use. The user was also provided several different amounts of gauges on each task.

DATA COLLECTION

Subject performance is measured by collecting time to complete a sequence of four on-screen pattern-matching tasks, after conducting a trial run for training. This translates into a numerically calculated accuracy envelope of the morph, which measures the subjects' accuracy as a function of time-on-task; data are collected by the computer program as the trials are performed.

Data from each test participant's performance was collected by the experiment application as the subject performed the experimental trials. At the end of the program, the collected data was written to a file, and that file was sent via FTP to a data collection server.

The test data was originally collected in an HTML-like form with tags signifying the meaning of different data points. Afterwards, the data files were parsed into a CSV format for ease of use with analyzing programs.

TEST VARIABLES

The test variables are the presence or absence of various categories of visual instrumentation providing on-screen feedback to the user during performance of the task.

Each user saw a specific variety of visual gauges with various levels and types of feedback that were relevant to or irrelevant to task performance. The goal of the experiment (unknown to the subjects) was an assessment of the relationship between the number and utility of the gauges and subject performance.

GAUGES

A variety of UI forms and gauges were used:

Values: This is a display of the numerical and bar-graph representations of the two parameters the test subject can manipulate, along with the buttons for changing them.

Target: This is a display of the numerical and bar-graph representations of the parameter values which would cause the trial task to succeed, shown only in a Control test scenario.

Diff: This is a display of the numerical and bar-graph representations of the difference between the two parameters which the test subject can manipulate.

Mag: This is a display of the numerical and bar-graph representations of the magnitude of the subject-manipulated shape, measured as the length plus the width.

Area: This is a display of the numerical and bar-graph representations of the area of the subject-manipulated shape, measured as the length times the width.

Prop: This is a display of the numerical and bar-graph representations of the proportion of the subject-manipulated shape's area to the target shape's area.

Time: This is a display of the numerical and bar-graph representations of the time the test subject has taken on the current trial.

Error: This is a display of the numerical and bar-graph representations of the error margin between the subject-manipulated shape and the target shape, measured as the average of the absolute difference between all the points in the subject-manipulated shape and the target shape.

The following figure shows the relative visual complexity of the UI, for each of the two extremes, of our four test modes.

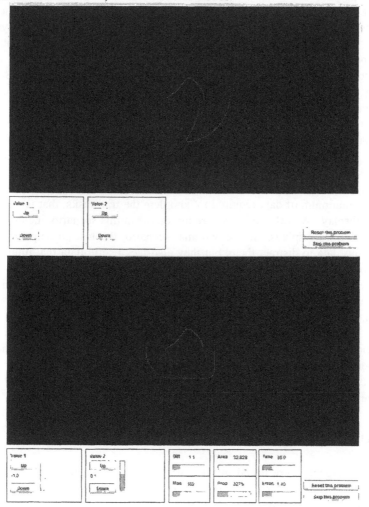

FIGURE 1 Minimal and Maximal

TEST SCENARIOS

The experiment application has four basic test scenarios. Each time the application is run, a test scenario is selected randomly, with some scenarios being more likely than others. All of the trial tasks are then run under the conditions of this same test scenario. The four basic test scenarios and their likelihoods are:

- (10%) Control
- (25%) Maximal
- (25%) Minimal
- (40%) Random

The purpose of the **Control** scenario is to provide a "base case" in which the test participant is given exactly the information needed to complete the trial tasks in the most efficient way possible. Both number and bar displays are shown for the Current Values and Target Values. This allows the test participant to directly compare the Current Values to what they should be in order to complete the trial task. The only other display shown is the Shapes display, which is shown in all test scenarios. The Control scenario is the only test scenario in which the Target Value displays are shown.

The purpose of the **Maximal** scenario is to overwhelm the test participant with data, much of which is useless. All displays are shown except for the Target Values display. This creates a situation in which the test participant must search for relevant data amongst the displays.

The purpose of the **Minimal** scenario is to give the test participant the absolute minimum of data required to complete the trial tasks, that is, only the Shapes display. All other displays are turned off in this scenario, including the Current Values displays. In the Minimal scenario, the test participant has no extraneous data to distract them, but they also lack useful data that might help them to complete the trial tasks more efficiently.

The purpose of the **Random** scenario is to create a variety of data display combination test cases. Each individual display has a 50% chance of being shown. The number and bar displays for each value may be shown or not shown independently. In addition to providing varied test cases, the Random scenario can showcase the effect of an disorganized, senseless, user interface on user performance.

Each subject performed four task runs with their particular gauge suite; the software automatically collected and uploaded the performance metrics.

SUBJECT DEMOGRAPHICS

Forty participants were tested in total, with an optional demographic survey, measuring Gender, Marital Status, Number of Children, Age, Handedness, Computer Literacy, Education Level, Job, Work Experience, Spare Time Activities, and Handicaps.

ANALYSIS

The principal statistic used for analysis of the data is the Pearson Correlation Coefficient. The correlation coefficient measures the strength of any linear relationship between two columns of data. It is denoted by the Greek letter "rho" (Coolican, 1990).

Rho is always between -1 and +1, inclusive. The larger the absolute value of rho, the stronger the correlation. A positive correlation means that the two columns tend to move in the same direction, that is, when one increase, the other does; and when one decreases, the other does. A negative correlation means the data tend to move in opposite directions, that is, when one goes up the other goes down.

Just because two columns have little linear correlation does not mean that they are not "related" in some manner; it just means that the relationships that exist are non-linear.

Correlation does not mean "causality". Two columns can be correlated because they arise from a common cause, or other reasons.

Before mathematical analysis, the first column (filename), and last column (participant comments) were deleted. The results of five participants were removed because they did not provide demographic data.

There were several instances for which it appears that people logged in multiple times, but filled out the demographic survey only once. Data from their survey was propagated into the gaps based upon their login ID.

The collected data were numerically coded to facilitate mathematical analysis. The coding is described in the following table:

The following table shows the impact of the presence or absence of each gauge on the performance. A correlation between the presence and absences of the gauge was measured against trial times. Negative correlation values mean that when the gauge was present, participants took longer to complete the trial; Positive correlation values mean that when the gauge was absent, the participants took longer to complete the trial.

CONCLUSIONS

SOME SIMPLE OBSERVATIONS

Five of the Forty subjects were left handed, and four out of five of these left handed subjects rated their computer experience as Extremely Experienced, one was moderately experienced. Age did not appear to be a factor in computer experience, with many of the older participants (those above the age of forty), rating themselves as very experienced or extremely experienced.

There is a high positive correlation (0.85) between participants' ages and the years of work experience they have, and that subjects deeming themselves to be computer-literate participants tend to have fewer children than those deeming themselves to be less computer literate.

Very low rho values suggest that "gender" is not strongly related to the amount of time required for a participant to complete a trial.

Differences in rho values suggest that married people do significantly worse on the training trial, but by trial four, the performance gap between married and unmarried participants has disappeared.

Differences in rho values suggest that single people are generally able to complete the trials more quickly than married people. This might be related to the fact that married participants tend to be a little older than unmarried participants.

Data shows that the performance of participants improved as they proceeded throughout the trials (One through Four). This chart shows the percentage performance improvement from trials One to Four.

THE IMPACT OF INSTRUMENTATION COMPLEXITY ON PARTICIPANT PERFORMANCE

Practice Improved Performance (Improved = Shorter Trial Time)

The performance of most people improved with experience (Trial zero to Trial four). Only seven percent of the participants had a final trial time that was longer than their initial trial time. The two participants with decreased performance were both young United States working males. There was a correlation between Age, Job, and performance level in these cases. For most participants, the improvement was over sixty percent.

The Mean Time of Trial Times is affected by the number of gauges on the User Interface

When the number of gauges is zero, users must rely entirely on geometric intuition. For a small number of gauges (four or five) average trial times are smallest (best performance). When the number of gauges is six, average trial times is greatest (poorest performance). When the number of gauges is seven through to nine, performance is better but not as good as with four or five gauges. When the number of gauges is ten or more, trial times are large (poor performance).

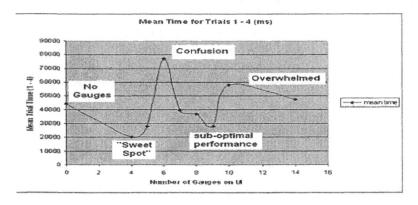

FIGURE 2 A Process Hypothesis

The tripartite nature of the time vs. gauge count graph suggests a process:
1. When there are too few gauges, users must rely on their intuition
2. With a small number of gauges, users successfully incorporate gauge use into their strategy.
3. At some point, the number of gauges becomes daunting, but not overwhelming. Users attempt to use them, but cannot do so

effectively; the end result is degraded performance.

4. When the number of gauges becomes overwhelming, the user ignores them, and the level of performance returns to virtually the same level it had when no gauges.

BIBLIOGRAPHY AND REFERENCES

Alexander, A.L. and Wickens, C.D. (2006). "Integrated Hazard Displays: Individual Differences in Visual Scanning and Pilot Performance." *Proc. HFES, 50th Annual Meeting 2006.*

Coolican, H. (1990). *Research Methods and Statistics in Psychology.* London: Hodder and Stoughton.

Dietrich, M., Bahner, J.E., and Hueper, A.D. (2006). "Misuse of Automated Aids in Process Control: Complacency, Automation Bias, and Possible Training Interventions." *Proc. HFES, 50th Annual Meeting 2006.*

Funke, G.J., Dukes, A.W., Galster, S.M., and Nelson, W.T. (2006). "Evaluation of Interface Types in an Adversarial Team Based Environment." *Proc. HFES, 50th Annual Meeting 2006.*

Hayes, C.C. and Anderson, R.A. (2006). "Benefits of Decision Support Tools for Users With Differing Levels of Domain Expertise." *Proc. HFES, 50th Annual Meeting 2006.*

Inagaki, T., Itoh, M., and Nagai, Y. (2006). "Efficacy and Acceptance of Driver Support Under Possible Mismatches Between Driver's Intent and Traffic Conditions." *Proc. HFES, 50th Annual Meeting 2006.*

Levinthal, B.R. and Wickens, C.D. (2006). "Management of Multiple UAVS With Imperfect Automation." *Proc. HFES, 50th Annual Meeting 2006.*

May, J.F., Baldwin, C.L. and Parasuraman R. (2006). "Prevention of Rear-End Crashes in Drivers with Task-Induced Fatigue Through the Use of Auditory Collision Avoidance Warnings" *Proc. HFES, 50th Annual Meeting 2006.*

Tollner, A.M. et al. (2006). "Change Blindness In Teams: are Three Pairs of Eyes Better Than One?" *Proc. HFES, 50th Annual Meeting 2006.*

Torbjörn, A., Kovordányi, R. and Ohlsson, K. (2006). "Continuous Versus Situation Dependent Night Vision Presentation in Automotive Applications" *Proc. HFES, 50th Annual Meeting 2006.*

Wilson, G.F., Davis, I., and Russel, C.A. (2006). "The Importance of Determining Individual Operator Capabilities When Applying Adaptive Aiding." *Proc. HFES, 50th Annual Meeting 2006.*

Chapter 17

Workload-Based Evaluation of Supervisory Control Interfaces for Life Science Automation

Sang-Hwan Kim[1], Rebecca S. Green[2], David B. Kaber[3], Matthias Weippert[4], Dagmar Arndt[4], Regina Stoll[4] & Prithima Mosaly[3]

[1]Department of Industrial and Manufacturing Systems Engineering
University of Michigan-Dearborn
Dearborn, MI 48128-1491

[2]SA Technologies, Inc.
Marietta, GA 30066

[3]Edward P. Fitts Department of Industrial & Systems Engineering
North Carolina State University
Raleigh, 27695-7906

[4]Institute for Preventive Medicine
University of Rostock
Rostock, Germany 18055

ABSTRACT

High demands for new drug development and advances in robotic technologies have led to automation of lab-based compound screening processes. This advancement has also led to a change in the human operator role from direct process controller to monitor and process supervisor. The objective of this research was to compare two supervisory control interfaces prototyped based on different conceptual design approaches, including an ecological interface design (EID) and conventional engineering interface design (ENG), for supporting operator decision making on screening process errors and managing mental workload. Based on examination of an existing supervisory control system, two virtual supervisory control interfaces (VSCIs) were developed to present the EID and ENG design approaches. Results indicated that the EID approach lowered workload, as measured using heart rate (a decrease in beats per minute), compared with the ENG design. The EID approach also increased operator SA, particularly in terms of perception of elements in the task. The results of the study provide a basis for future interface development in the life science domain and underscore the need for interface design guidelines to improve human performance and increase productivity.

Keywords: Life science automation (LSA), supervisory control interface, and ecological interface design.

INTRODUCTION

The life sciences domain presents many interesting opportunities in terms of human-automation interaction research. In particular, compound screening processes, identifying biological and/or chemical agents for drug-derivative development, have become increasingly automated since the early 1990s (Thurow and Stoll, 2001). Operators that once worked with individual automated tools are now involved in programming, monitoring and managing compound screening process as well as data analysis (Thurow, Göde, Dingerdissen and Stoll, 2004). In such processes, lab operators typically interact with multiple robotic systems for materials handling and liquid transfer, as well as automated analytical measurement systems. Operator task requirements are twofold: (1) plan and program screening experiments based on the design of previous manual assays; and (2) monitor automated systems for errors, such as robot positioning and material handling errors (Kaber et al., 2009). The latter task often requires human supervisors to make decisions to correct abnormal automation processing or to determine the acceptability of results of a completed process. This occurs through operator rule- and knowledge-based behaviors. Error identification in such systems is critical because system failures are very costly for drug discovery labs in terms of time and resources.

Ecological interface design (EID) is a theoretical framework for the design of interfaces for complex human-machine systems (Vicente & Rasmussen, 1992). It originated with the work of Rasmussen and Vicente (1989) and Vicente and Rasmussen (1992) who sought to create an interface design methodology that would support skilled users in coping with unanticipated events. The framework focuses on three general design principles (Vicente, 2002; Vicente & Rasmussen, 1992) to support:

- skill-based behavior (automated behavior) - users should be able to directly manipulate the interface;

- rule-based behavior (cue-action associations not involving cognitive processing) - the interface should provide a one-to-one mapping between work domain constraints and perceptual information. Object displays that integrate several directly measurable variables into a single, more meaningful variable are an example of the application of this principle; and
- knowledge-based behavior (analytical problem-solving) - the work domain should be represented in the form of an abstraction hierarchy (AH) that would serve as an external mental model

In general, an EID and specific interface features should promote the substitution of knowledge-based behaviors with use of lower levels of cognitive control (skill- and rule-based behavior) that involve fast, effortless processing and fewer errors. An EID should also support knowledge-based behavior in the case of less experienced users and for experts in managing unexpected problems (Vicente and Rasmussen, 1992). The EID approach has been used successfully in diverse application domains including process control, aviation, software engineering, and military command and control (Vicente, 2002). In most domains, EID has been found to reveal information requirements that were not captured by the existing system interfaces. When empirical evaluations have been conducted, ecological interfaces have been shown to improve user performance over existing system interfaces (Burns et al., 2008).

With this in mind, we expected that the EID approach might be beneficial for supervisory control (SC) interface design for automated drug discovery processes in order to facilitate effective problem detection and decision-making. We also expected the EID approach, presenting a clear process model for operators, would yield lower demands in cognitive workload and greater operator process situation awareness (Endsley, 1995; Kaber and Endsley, 1998). The existing SC interfaces for such processes are text-based and pose high visual loads. We refer to these as conventional "engineering (ENG) interfaces." The focus of this study was on comparing two SC interfaces, prototyped based on different conceptual design approaches, including EID vs. ENG, but with a similar goal of supporting operator decision making on process errors in terms of mental workload and SA responses.

METHOD

PROTOTYPE INTERFACES

Prototypes of the SC interfaces for the automated compound screening process were developed using a systematic procedure. Initially, information about the actual process was recorded and used to develop a running discrete event simulation application. The process under study consisted of four major automated workstations, including a microculture plate incubator, a barcode printer and reader, a robot for pipetting liquids into plates, and a microplate reaction (fluorescence) reader. Microplates were run through 15 steps in this process. The system manipulated one or two plates at the same time and each plate required different processing times at each station. An error log recorded on actual system performance showed that most errors occurred in the pipetting (liquid transfer) and microplate reading steps. When a pipetting error was observed from the SC interface, an operator had to decide whether the error could be corrected and the system could continue by manual replacement of a robot liquid dispensing fixture or if a plate should be scraped because of limited remaining processing time. Similarly, for microplate reading errors, operators were required to make a decision between manual repositioning of the plate at the reader or discarding the plate. When an operator makes a decision on a plate error, they need to project the total elapsed processing time of the plate, including the error correction time, and compare it with an acceptable processing time (or compound death time). In the process under study, it was possible for biological test materials to expire before process completion, particularly, if errors occurred.

Based on this process information and simulation, two Java-based prototypes of virtual supervisory control interfaces (VSCIs) were developed and used in experimental test trials. One was developed based on the EID concept. Work domain constraints were mapped through AH models. These models were used as a basis for the EID, which presented the process and automated systems according to their physical form along with the specific domain constrains to support operator diagnosis of system states. Figure 1 shows a screenshot of the EID. The interface presents the physical location of stations and current position and status of plates during the process. Information is also provided on current station times, mean processing times, and elapsed time for each plate.

Another prototype was developed based on an existing, standard engineered interface design (ENG) for this type of system. The interface presented text-based information to operators on a screen with no physical process model or specific operating constraints (Figure 2).

In the simulation of both interfaces, the plate processing time at each step was randomly generated following a normal distribution with the mean time and variance based on actual process data. The two types of error occurrences (microplate pipetting and misreading) were, however, manipulated in the simulation based on a predetermined schedule.

PARTICIPANTS AND TASK

Five expert biochemists at the University of Rostock (Germany), Center for Life Science Automation, who were highly familiar with automated compound screening systems, participated in this study. Access to experts in this field is extremely limited because of current demands on labs. Although the study sample size was relatively small, the background of the subjects was expected to promote the generalizability of results versus using a novice population. The experimental task required participants to monitor 11 microplates running through the simulated screening process. The microplates were processed in three epochs or groups (plates 1-4, 5-8, and 9-11). In each test trial, two types of errors were scheduled to occur in plate processing and the chemists were required to decide whether the simulated plates should be discarded or the errors corrected as quickly and as accurately as possible.

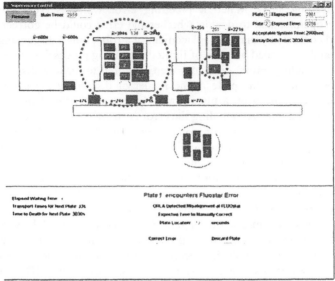

FIGURE 1. Ecological Interface Design (EID).

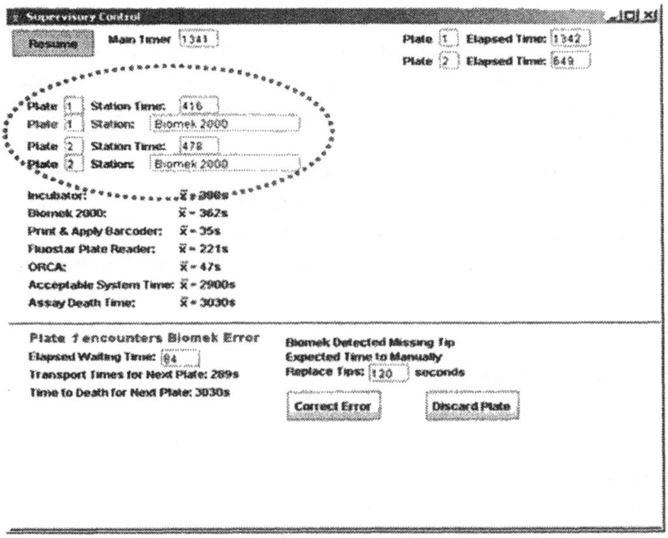

FIGURE 2. Conventional Engineering Interface Design (ENG).

EXPERIMENT DESIGN AND DEPENDENT MEASURES

During the course of the experiment, each participant completed two trials with each interface type (EID and ENG) under two different levels of task complexity for a total of four test trials. Participant exposure to the conditions was randomized. A high complexity trial involved two epochs in which plate errors occurred (Epoch 1 and 3) and for which total plate processing time was mentally challenging to project. A low complexity trial included only one epoch (Epoch 2) in which errors occurred and operators could easily project processing time. Two types of workload measurement were conducted, including subjective workload ratings using the NASA-TLX and physiological workload on the basis of heart rate. The workload measures were expected to be significantly lower for the EID interface. Operator situation awareness (SA) was also measured using the situation awareness global assessment technique (SAGAT) (Endsley, 1995). Simulation freezes were conducted in between processing of groups of micro-plates and chemists were quizzed on the current state of the process simulation without having access to the interface. Chemist responses to SA questions were graded in terms of the ground-truth of the simulation. The SA scores were expected to be higher for the EID interface than the ENG interface.

ANOVAs were used to determine the significance of the interface design approach, level of task complexity, and occurrence of errors on the physiological (heart rate) and subjective measures of workload (NASA TLX) as well as SAGAT scores. The significance criterion for the study was $\alpha = 0.05$. Further, correlation analyses were also conducted on the response measures to identify any significant linear relations.

PROCEDURE AND SETUP

The study was divided into three parts: (1) a 10-minute setup and familiarization of participants with the equipment and material; (2) two 15-minute training sessions; and (3) four trials lasting approximately 30-minutes each. Participants were trained on scenarios consisting of eight microplates proceeding through the assay. A standard PC computer with a monitor and mouse was used to compare the two virtual interface designs in test trials. A Polar watch and chest strap were mounted on the participants to measure heart rate (HR) before and during their exposure to the experimental trials. The heart rate measure was collected during each epoch in every trial and averaged at the end. The SAGAT questions were also posed at the end of each epoch during the simulation freeze. NASA-TLX measures were taken at the end of each trial. The total experiment duration for each participant was approximately 3.5 hours.

RESULTS

Since there were individual differences in heart rate (HR) and perceived mental workload (measured using the NASA-TLX), both measures were normalized to provide a consistent mechanism for interpreting results. The data were converted to standardized (z) scores for each participant across the trials. Graphical analysis and diagnostic tests on residuals for models revealed no ceiling or floor effects and the standardized data conformed with a normal distribution.

ANOVA results indicated that there was a significant effect of epoch (F(11,59)= 29.5, p<0.0001) and level of task complexity (F(1,59)=6.48, p=0.014) on HR. The interface type also approached significance (F(1,59)=3.79, p=0.058). Post-hoc analysis using Duncan's test revealed that regardless of the sequence of presentation of the type of interface and levels of complexity, participant HR was higher at the start of the experiment and gradually stabilized over time (increase in epoch number). ANOVA results also revealed that the EID (mean standardized HR=72.06 bpm) reduced HR compared to the ENG (mean standardized HR=74.25bpm) interface design (p=0.058) (see Figure 3). Surprisingly, HR was significantly higher in the low-level complexity condition (p=0.014). It is possible that subjects experienced some anxiety in anticipation of additional errors in the simulation. The absence of errors was also associated with lower HR, but this effect was only marginally significant (F(1,59)=3.06, p=0.087).

ANOVA results on the normalized NASA-TLX scores revealed the interface design to be significant (F(1,59)=6.79, p=0.013). The epoch and level of task complexity were not significant. Surprisingly, the EID (mean scores=62.37) was associated with higher subjective workload than the ENG interface (mean scores=57.10) (see Figure 4). It is possible that subjects were more accustomed to the ENG interface based on their prior experience and felt less comfortable (or frustrated) in using the new EID. However, correlation analysis using Pearson product-moments revealed no significant association between the objective and subjective measures of workload (p=0.43).

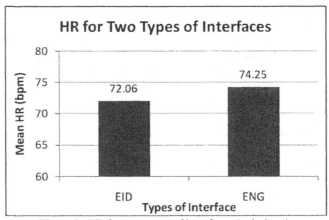

Figure 3. HR for two types of interfaces (unit: bpm).

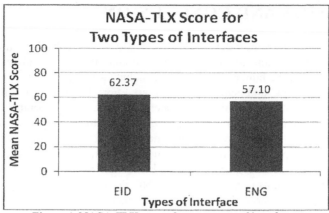

Figure 4. NASA-TLX score for two types of interfaces.

ANOVA results on overall SA score revealed only the effect interface design to be marginally significant (F(1,35)=2.88, p=0.0997). Figure 5 shows that the EID resulted in a higher mean SA score (*M*=44%) than the ENG (*M*=36%) interface. Among the three levels of SA defined by Endsley (1995), including perception (level 1), comprehension (level 2) and projection (level 3), the interface design also had a marginally significant affect on level 1 SA score (F(1,11)=4.87, p=0.063). The EID was associated with higher level 1 SA (*M*=57%) than the ENG (*M*=38%) interface. The other two levels of SA (comprehension and projection) were not affected by the interface design or other experimental manipulations (epoch, complexity). The fact that the results on SA were marginally significant is likely due to the small sample size for the experiment and lower sensitivity for rejecting the null hypothesis of no difference among the interface conditions.

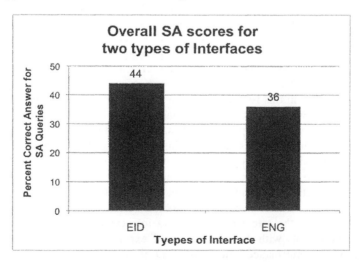

Figure 5. Overall SA scores for two types of interfaces.

DISCUSSION AND CONCLUSION

Due to demands for new drug development and advances in robotic technology, lab-based biological screening processes have been automated. This advancement has created a change in the human operator role including removal from direct process control to monitoring and supervision. In an effort to minimize the cognitive workload on operators, complex interfaces for supervisory control have been developed for compound screening processes. In this study, we tested two prototypes including an EID and conventional engineering interface design (ENG). In line with our expectations, the EID reduced user mental workload, based on a physiological measure (HR), and increased situation awareness in terms of user perception of features, as compared to the ENG design. These findings support the contention of prior research that interfaces designed following the EID framework aim to reduce user mental workload (Vincente, 1999) and to enhance situation awareness (Nielson et al., 2007) when dealing with unfamiliar and unanticipated events, which are attributed to increased psychological pressure. However, we suspect that the small sample of experts used in the study led to the marginal statistical differences among the two interface designs. That said, the expertise of the operators in the study substantially contributes to the validity of any findings.

Contrary to our expectations, HR was significantly higher under the low task complexity condition. It may be possible that the errors in the simulation in the first epoch (period 1) led participants to be more attentive to the display. This eventually made it easier for subjects to make decisions on errors in the third epoch under the high complexity condition. On the other hand, errors introduced in the second epoch for the low complexity condition may have disturbed a smooth monitoring process and led to an increase in HR. Contrary to our expectations, subjective workload (NASA-TLX) was significantly higher while using the EID compared to the ENG interface design. Since the participants were expert biochemists and had used an ENG interface for an actual screening process, introduction of the new EID may have led to a subjective bias in the ratings. Although training was provided before the experiment on both interface designs, the participants' prior experience in using an ENG design may have masked some of the critical improvements/advantages in using the EID prototype.

The main caveat in this research was the study sample size. Expert biochemists with experience in working with ENG interface designs in screening processes were used. This is a very small population, as is, with limited time available to spend outside normal work activities. Beyond this, the study did not investigate other performance, decision accuracy, or reaction time measures that might provide further insights into the interface design impact on process supervisory control.

With these limitations in mind, two directions of future investigation in the life science automation domain need to be pursued by researchers. First, it would be interesting to assess the effect of a new EID for screening process using a large sample of participants with different levels of expertise and work characteristics. Secondly, other performance effects of EID in screening process needs to be investigated to better understand the impact of interface design principles. Ultimately, it can be expected that a series of studies would lead to the development of interface design guidelines for improving human performance and increasing productivity in the domain of life science automation.

ACKNOWLEDGEMENTS

This research was completed while the first and second authors, Sang-Hwan Kim and Rebecca Green, worked as research assistants at North Carolina State University. This research was supported by a National Science Foundation (NSF) Information Technology Research Grant (no. 046852). Ephraim Glinert was the technical monitor for the NSF. The views and opinions expressed in this paper are those of the authors and do not necessarily reflect the views of the NSF. We thank Dr. Kerstin Thurow of the University of Rostock and CELISCA for providing us with access to high-throughput biological screening systems and supporting the research through allocation of biochemist and process engineer time to the effort.

REFERENCES

Burns, C. M, Skraaning, G., Jamieson, G. A., Lau, N., Kwok, J., Welch, R., and Andresen, G. (2008), "Evaluation of Ecological Interface Design for Nuclear Process Control: Situation Awareness Effects." *Human Factors*, 50(4), 663-679.

Endsley, M. R. (1995), "Measurement of situation awareness in dynamic systems." *Human Factors*, 37, 65-84.

Kaber, D. B. and Endsley, M. R. (1998), "Team situation awareness for process control safety and performance." *Process Safety Progress*, 17(1), 43-48.

Kaber, D. B., Stoll, N., Thurow, K., Green, R. S., Kim, S-H and Mosaly, P. (2009). "Human-automation interaction strategies and models for life science applications." *Human Factors & Ergonomics in Manufacturing*, 19, 601-621.

Nielson C. W., Goodrich, M. A. and Ricks, R. W. (2007), "Ecological interface for improving mobile robot teleoperation," *IEEE Trans. on Robotics*, 23, 927-941.

Rasmussen, J. and Vicente, K.J. (1989), "Coping with human error through system design: Implications for ecological interface design." *International Journal of Man-Machine Studies*, 31, 517-534.

Thurow, K. and Stoll, N. (2001). "Aktuelle Tendenzen in der Laborautomation." In *Praxishandbuch Laborleiter* (Chap. 11.3, pp. 1-22).

Thurow, K., Göde, B., Dingerdissen, U. and Stoll, N. (2004), "Laboratory information management systems for life science applications." *Organic Process Research & Development*, 12.8, A-M).

Vicente, K. J. (1999), "Ecological Interface Design: Supporting operator adaptation, continuous learning, distributed, collaborative work." *Proceedings of the Human Centered Processes Conference*, 93-97.

Vicente, K. J. and Rasmussen, J. (1992), "Ecological interface design: Theoretical foundations." *IEEE Trans. on Systems, Man, and Cybernetics*, 22(4), 589-606.

Vicente, K. J. (2002), "Ecological interface design: Progress and challenges." *Human Factors*, 44(1), 62-78.

CHAPTER 18

The Elder's Discrimination of Icons with Color Discrimination on Cell Phone

Jo-Yu Kuo, Fong-Gong Wu

Department of Industrial Design
National Cheng Kung University
No.1 University Rd., Tainan City 701, Taiwan (R.O.C)

ABSTRACT

The elder's discrimination of icons with color discrimination will influence their quality of future life. Observing from the current market, we found that cell phone is the most popular and essential technology products. Owing one can represents the user's characteristics has already become a fashion symbol. But the elder cannot enjoy the mobility and convenience brought by cell phone, since the changes of physiological function, the designer enlarge the size of button and screen resulting a stupid and ugly features of cell phone. Moreover, human received 85% of environment stimuli by visual ability, which indicated it will affect the most part of their daily life. With increased recognition of the degree of cell phone color icons, helping the elder to enhance the reliability for technology product, purchase intention and product competitiveness. With a rule for cell phone interface color icons, we can find it will be more suitable and close to the human needs, especially for a super-ageing society. Therefore, people do not be forced to change their familiar e-life due to physical degradation.

Experiments were aimed at investigating the discrimination of color icons of cell phone. A total of 16 healthy participants (9 male, 7 female, averages age 62) were tested on the experiment device which is a 3.2" screen touch phone with 640 x 480 pixels resolution and the brightness of about 40 ~ 50cd. In this research, the selected function of main menu icons include: messages, contacts, call records, tools, camera, game, Internet, multimedia folder and application. We use the confusion matrix to analyze the top three recognized icons for each function, then re-design three set of figures (set A, set B, set C). Set A represents the simple and 2D icons, set C represent the complex and 3D icons. According to the literature, a good icon should design with clear figure-ground perception, so all experimental icons displayed in a square background.

There are three different experiments: discrimination of icon size, discrimination of different color combination and subjective preference. First, we investigated the discrimination of color and no-saturation icons for the elders. Icons began to show in 40 x 40 pixels with different matrix numbers from 1x1 to 3x4. This pre-experiment is assigned for supporting the next task about appropriated icon size.

The second part of the experiment focused on the color combination between icons and interface background. The icons were composed of three different set of icons and 19 different background colors. Thus, for reducing the participant's eyestrain and increase test efficiency, the icon cataloged into 4 colors: red, blue, yellow, green. The color of background took 30 for the pitch for color hue (H). Thus, Lightness (S) and saturation (B) is divided into two kinds of 50,100. The participants were asked to search the screen and click the same icon offered at random. The other experiment of this session is a select-multi-icon task. Both of those were measured the reaction time, correct rate and analyze the data by Duncan's test. Finally, the participants were asked to rank their preference after each task in the test.

The result showed the first experiment that the size of color icons should be designed to at least 80 pixels, and recognizing the color icons is easier for elders then no-saturation icons. However, the elders are lack of recognized association for simple figures (e.g. SetA), but after learning, the neat lines of icons are helpful to recall their memory and perform better on reaction time. Although the elder need a high brightness of background color, when it's brighter than icons and too dazzling, it will affect the correct rate significantly. Green and blue are vague color for the elder, the interface designer should aware of those combinations. There are three recommend designed-icons for all color combinations (background /icon color): blue/yellow, white/green and gray/red.

Keywords elder, icon, color discrimination, cell phone

INTRODUCTION

GRAPHIC INTERFACE DESIGN

Research have showed that no matter recall memory (Shepard, 1967),free memory(Paivio and Csapo,1973), or semantic memory work (Potter and Faulconer,1975), picture superiority effect. Since the advance in LCD technology, there's more and more research on the small-screen interface for portable electric product. Swierenga (1990) found that there's no specific different for searching section election of main menu between 12 lines and 24 lines screen on handy electric products.

Weiss (2002) pointed out that the number of menu election shouldn't not over 9 and the size of screen display at least 4x12 character which is between 5x7 and 20x20 pixels. Buyukk(2001) also think the participant's performance of menu searching task for handy electric products are influenced on small-screen interface.

SCREEN AND COLORS

Tufte(1992) showed gray as a multi-color be used on unobvious screen, it will emphasize the other 2-3 kinds of color effects. Designer should avoid strong brightness and contrast colors for background, but increase the contrast to gray background for selection of high brightness and contrast color as window boarder, and Tufte(1992) said yellow is the only appropriated color. Durrett and Trezona(1992) showed red and white windows boarder or white text on black background have same effect. Overall, while increasing the color contrast to increase readability, changing in brightness must be more cautious than the changing in hue.

Pastoor (1990) found that in the two kinds of same luminance contrast, the participants prefer to cool (blue and green blue) as the background color for the subjective measurement about 800 colors combination. Saturation is the most important factor affecting rates, non-saturated color combination has a better rating.

EXPERIMENT DESIGN

EXPERIMENT ENVIRONMENT

16 participants (9 female and 7 male, mean 62 years old)were recruited for this research. In addition to have presbyopia due to age, growth characteristics, all of them have no color blindness, cataracts, glaucoma, macular degeneration and other eye diseases, and corrected visual acuity of 0.8 or above. Before the measurements, participants must no staying up late , no excessive use of eyes, usually have the habit of using mobile phones, and had used two or more functional mobile phones.

Indoor experiment environment with illumination 300~500lux, temperature 26 ~ 30 degrees, the noise is less than 45dB. Users was a natural sitting position, adjust the cell phone away from the eyes fixed on the distance of about 30cm, cell phone screens are measured perpendicular to the line of sight by placing on desktop. Experimental equipment is the Nokia 5800 touch screen mobile phone, 3.2 "screen, resolution 640 x 480 pixels, and the brightness of about 40 ~ 50cd.

First, collect the interface icons in past five years, and take the results from "The Main Menu Icon Design on Color Display of Cell Phone"(Yao Meiling, 2004) into consideration which indicated the following common functions in main menu through the method of focus group and questionnaire survey: Contact, log, settings, messages, profile, camera, apps, game and internet. Then compare with the icon name and exact function, finally named menu included: messages, contact, log, tools, camera, game, internet, multi-media, and application.

Confusion matrix used to select the top three good degree of icon pattern recognition, and then change the combination of color between icons and background. Considering the number of trials is so large that the elderly would feel tired easily, re-deign the exacting icons to three sets of icons with four kinds of colors(Red, blue, green and yellow). The background colors were selected by every 30 Hue, 50 or 100 for saturation and brightness. At last, select the experiment color by eliminating the colors which are hue-close, low subject preference with poor discrimination in previous literature and taking into consideration of HSB color resolution on cell phone.

The pattern design for set A, set B and set C are from simple to complicated, and set A is 2D icon, set B is 2D/3D icon and set C is 3D icon.

Previous studies showed that the specific figure-ground effect is required for a good graphic interface design, so icons are all surrounded by square boarder for induce the interference of pattern details.

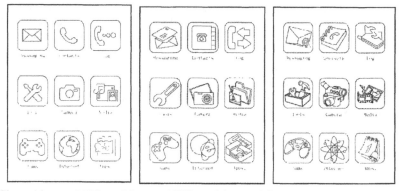

Figure1.icon set AFigure2.icon set BFigure3.icon set C

THE DISCRIMINATION OF ICON SIZE

In order to understand the recognizable size of the existing icons, two group of non-saturate icons and color icons are set to explore the smallest identifiable size in different numbers of icons. To avoid the individual oral semantic difference, the participants were asked to point out the icons displayed on screen on a comparison form with large pixel. The smallest icon is 40 pixel x 40 pixels, 5 pixels as a range for length and width, increasing the size until the participants recognize it. This trial cost 10 minutes.

Independent variables:the number of icons, saturation (black and white / color)
Dependent variables:size of icons。
Control variables: square, same hue, brightness and saturation.

THE VISIBILITY OF CELL PHONE ICONS

This experiment investigated that whether the elders' reaction time decreased and enhance the effeteness or not under different color combination for icons and background. Each participants all take set A~C with four colors combined with different background colors.

a. search time:

Before staring the test, there's a attention alarm and then a non-saturation target icon last 2 seconds, then a 3x3 set icons randomly. The participants asked to click the icon. A trial is finished when participants test three target icon for Each color of background and icon. The participants completed this section experiment with 76 trials. Take a minute break every 10 minutes, and not over 1 hour a day.

Independent variables: Color of icons, color of background
Dependent variables: Searching time
Control variables: size of icons, numbers of icons. (Side length 80 pixels square with 3x3 matrix in the center of screen.)

b. Task-analysis:

The participants were asked to finish a simple task in this section. Make sure the participants understand the function of all icons and considerate the elders' experienceof usage. The task is click"log"icon for checking missed call, click "message" to send and click"alarm"to set reminder.

Independent variables: Colors of icons
Dependent variables: reaction time
Control variables: size of icons, numbers of icons. (Side length 80 pixels square with 3x3 matrix in the center of screen.)

CONCLUSION

The rate of discrimination for icon pattern higher than 80% (setA: 7, setB: 6, setC: 6). since the complicated icons (setC) are all perspective type and make the participants think too much leas the result of incereaing error and reaction time. There's no specific different between the average rate of discrimination, (setA: 81.25%, setB: 81.25%, setC 80.56%). After re-design the easy-confused icons then do the recognizable experiment.

THE SIZE OF RECOGNIZABLE ICONS

The results show that the average size of discrimination of the color icons (83.96pixel)is smaller than non-saturation(94.65pixel), when the number of icons on screen increased, the smallest size trend to decreased. Presumably due to some reasons that the participants want to see the details of the expectations for single icon. Cautiously, take 80 pixels as the standard of icon side length instead to 50 pixels existed.

The significant value of non-saturation is 0.022(<0.05), and the color icons is 0.121(>0.05). It represents there is no significant difference, that is, the number of color icons on cell phone screen is not the factor of icon size discrimination.

Table 1. Duncan test for the numbers and size of B/W icons

number of icons	B/W icon size (pixel)	Grouping
3x4	78.75	
3x3		
3x1		
2x4		
3x2		
2x2		
2x3		
2x1	100.625	
1x1	119.0625	

The results showed that the participants are more cautious for the first targrt icon, so there's only grouping for single icons.And the colors do help the participants since average searing time of colors icon(6.09 s) is fast than non-saturation icons(12.68 s).

THE DISCRIMINATION OF COLOR ICONS

Table 2 show the Duncan test for A~C set icons on different colors. There's no grouping and specific different between three set with green color which is middle-color. In addition, the participant had better performance on set A with red and yellow colors.

Table 3 show the Duncan test for A~C set icons with 4 colors. The participants had better performance(less searching time) on set A icons and set B icons with green colors, but not for set C icons

Table 2. Duncan test for A~C set icons on different colors

Red			Blue		
Icon set	time (sec.)	Grouping	Icon set	time (sec.)	grouping
A	1.5069		B	1.4961	
C	1.6996		C	1.7426	
B	1.8065		A	1.9084	
Green			**Yellow**		
Icon set	time (sec.)	grouping	Icon set	time (sec.)	grouping
A	1.414		A	1.4254	
B	1.512		B	1.7312	
C	1.6009		C	1.7707	

Table 3. Duncan test for A~C set icons with 4 colors

Set A icons			Set B icons		
icon colors	time(sec.)	grouping	icon colors	time(sec.)	grouping
GREEN	1.4140		GREEN	1.5384	
Yellow			BLUE	1.6519	
RED	1.5069		Yellow	1.7312	
BLUE	1.6994		RED	1.8065	
Set C icons					
icon colors	time(sec.)	grouping			
GREEN	1.6009				
RED					
Yellow					
BLUE	1.7957				

The obvious grouping happen on background color which is (255,255,255)、

(0,127,255)、 (0,0,127)、 (63,0,127)、 (127,0,63)、 (127,127,127) and show on Table 4. Three of those have close-hue which is not suggested by previous studies, it's possible that the participant can't distinguish it with background, and due to the re-design icon for this experiment is center-white, the menu icon become no boarder for them. Therefore, enhance the contrast accidently and speed the searching time. In addition, gray is middle-color, it's predicated the most obvious grouping.

Table4. Duncan test for the icon set and background color

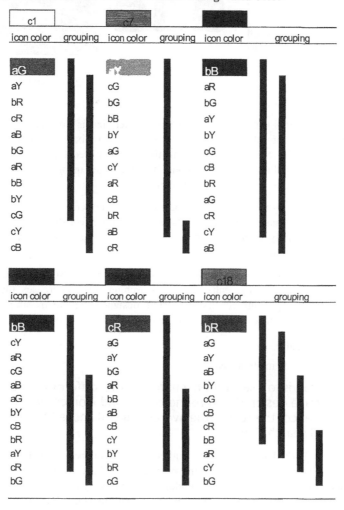

DISCUSSION AND SUGGESTION

This research discussed the pattern and colors combination of icons on cell phone for the elderly, and the results help designers get a more objective basis with visual constraints not the personal preferences. The result suggested three sets (Figure 4~6) and divided into patterns, colors in two parts.

Figure4.icon design1   Figure5.icon design2   Figure6.icon design3

- Icon pattern:

(1) Existing mobile phone interface icons pixel size is not suitable for the elderly. It should be increased to at least 80 pixels. In the smallest recognizable icon size experiment, the area required for color icons are smaller than the black- white icons. it is suggested using color scheme In the small-screen icon interface design.

(2) There's no significant different between the numbers of color icons and smallest recognizable size. but the numbers do effect the searching time for the elderly which should be taken into consideration while design small-screen interface.

(3) The elderly had bad association with simple icons at first , but it helps the participants memory, enhances the searching time, decreases the error after learning. This results suggested designer the clean, simple icon which is benefit for interface performance.

- Icon color:

(1) This research investigated

two kinds of saturation changes. the different polarity of interface design (background colors and icon color saturation difference) the result, the average age of those little effect on visual search time.

(2) Although the background with higher brightness is better for the elderly, it doesn't work while the dazzling brightness is higher than icon and effect the error of discrimination. In addition, the elderly have fewer error rate when the brightness of icon higher than interface background. The halftone such as blue and green were influenced by background color to searching time obviously, so the designer should pay more attention when choosing those colors.

REFERENCES

Buyukkokten, O., Garcia-Molina, H., and Paepcke, A., 2001, Accordion summarization for end-game browsing on PDAs and cellular phones, Special Interest Group on

Durrett J. and Trezona J., 1982, How to use color displays effectively: The elements of color vision and their implications foe programmers, Pipeline,7(2), 13-16

Paivio A and Csapo K. Picture superiority in free recall: Imagery or dual coding? Cognitive Psychology. 1973; 5; pp176-206.

Pastoor, S., "Legibility and subjective preference for color combinations in text," Human Factors, pp. 157-171, 1990.

Potter, M. C. and Faulconer, B. Time to understand pictures and words. Nature; 1975; 253; pp437-438.

Shepard, R. N.. Recognition memory for words, sentences, and pictures. Journal of Verbal Learning & Verbal Behavior. 1967; 6; pp156-163.

Weiss, S., 2002, Handheld Usability, John Wiley & Sons, New York, USA.

Yao Meiling, 2004, color-screen mobile phone main menu image symbol design research, National Chiao Tung University Institute of Applied Arts, Master's thesis

Chapter 19

Utilizing the Keyboard to Issue Commands: Do Self-Efficacy, Goal Orientation, and Computer Anxiety Impact Usage of Efficient Techniques with Software?

S. Camille Peres, Jo Rain Jardina, Courtney Titus

University of Houston-Clear Lake

ABSTRACT

Even though using the Keyboard to Issue Commands (KIC) is a more efficient software utilization technique than other techniques (e.g., using the mouse to click on the icons or menus) even experienced users still do not make the switch to this more efficient technique. A study found that if a participant observed a peer using efficient techniques, they were more likely to increase their use of those efficient techniques (KIC), but the observation of a peer using KIC only accounted for a small percent of the variability. To further investigate why people adopt KIC, the current study looks to see if goal orientation, self-efficacy with computers, and computer anxiety is related to why some people adopt KIC after the observation of this technique. To investigate this, we manipulated whether participants saw someone using KIC with Microsoft Word® and then had them complete several questionnaires about their goal orientation, self-efficacy with computers, and computer anxiety. The results indicate that the change in self-efficacy relates to the observation of KIC, and goal orientation relates to KIC usage.

Keywords: efficiency, software, goal orientation, computer anxiety, and self-efficacy

INTRODUCTION

Currently, there are two primary ways in which users typically interact with software, using the mouse or keyboard. Research has found that using the Keyboard to Issue Commands (KIC) is the more efficient method of these two (Lane, Napier, Peres, & Sándor, 2005). However, both novice and experienced users often fail to make the switch to this efficient technique (Lane et. al, 2005). Our goals with the current study is to investigate what variables may be involved in users' decision to adopt (or not) KIC.

Previous research conducted on this topic found that certain user demographics (e.g., experience with the software, age, training methods) were unrelated to the adoption of KIC (Lane et. al, 2005). However, research conducted by Peres, Tamborello, Fleetwood, Chung, and Paige-Smith (2004) revealed that users were more likely to adopt KIC if they worked with someone who used the technique. Similar findings were provided by a study conducted by Peres, Tamborello Fleetwood, and Nguyen (in review) in which the observation of a peer using the efficient technique increased the likelihood of that participant's usage of KIC. However, the observation of a peer using efficient strategies only accounted for 9% of the variability associated with participants' change in efficient method utilization (Peres et al., in review).

In addition to experience with computers and peer observation, a user's weighting of costs and benefits of using KIC have also been investigated in relation to KIC adoption. Peres, Fleetwood, Yang, Tamborello, and Paige-Smith's (2005) study found that participants' weighting of costs and benefits of using KICs was related to KIC usage. A later study conducted by Powell, Peres, Nguyen, Bruton, and Muse (2009) yielded similar results. However, studies have also found that costs and benefits were not related to participants' change in their KIC usage (Powell et al, 2009; Jardina, Peres, & Titus, in review).

The current study further investigates the adoption of KIC by examining its relationship between goal orientation, self-efficacy with computer use, and computer anxiety. For instance, it is possible that users with a performance goal orientation, marked by a need for external indicators of success, are less motivated to adopt efficient techniques. Users with a mastery orientation, on the other hand, may be more motivated because of their desire to become proficient in a topic. Further, many studies have found a relationship between an individual's performance and self-efficacy (Smith, 2001). It is reasonable to think that those users with a low self-efficacy with computer use may be less confident with adopting alternative techniques for issuing commands. Similarly, people's anxiety levels with computers may also affect their motivation to use efficient techniques.

METHODS

PARTICIPANTS

Twenty-two University of Houston-Clear Lake students participated in the experiment (3 males) (Age, $M = 31.3$, $SD = 8.31$). All participants received course credits for their participation.

SETUP

The experiment was carried out on a PC with Windows XP, a 3GHz Pentium 4 processor, 2.0 GB RAM, and 17" monitor. Observations were recorded with Noldus Observer XT 7.0 and uLog software. Two similar computer workstations were set up in adjoining rooms, one for the participant in an observation room and one for the confederate in a control room.

DESIGN AND MEASURES

There were five constructs examined in this study: KIC usage, self-efficacy with computers, anxiety with computers, goal orientation, and efficiency observed. Efficiency observed was a between subject variable and had four levels: Inefficient, Intermediate, Efficient, and Control. Participants were randomly assigned to one of these four conditions.

Participants' KIC usage was measured by calculating the proportion of times they used KIC for each command (i.e., cut, copy, paste, bold, italicize, underline, and find) and then averaging these proportions. This was done for the Microsoft Word® tasks done both before and after the intervention phase (phase when efficiency observed was manipulated—see description below).

Participants' anxiety with computers was measured using the Computer Anxiety Trait Scale (CATS: Gaudron & Vignoli, 2002), which consisted of 15 items related to the level of anxiety experienced with computer usage. Participants rated the items on a 5-point Likert scale with 5 being "very much" and 1 being "not at all." An example item from the survey is "When you are in situations where you use or you are about to use a computer you feel relaxed." A higher score on the CATS indicated a higher computer anxiety, while a lower score indicated a lower computer anxiety.

Self-efficacy (SE) was measured using a survey consisting of 30 items related to the level of confidence participants had with their ability to perform various computer tasks. Participants rated the extent to which they agreed or disagreed with each item on a 5-point Likert scale with 5 being "strongly disagree" and 1 being "strongly agree." An example item is "I am confident I can apply font formats (e.g., size, style, bold, italic, underline)." A higher mean indicated a higher self-efficacy with computer tasks, while a lower mean indicated the opposite.

The three measures just described (KIC, SE, and CATS) were completed two different times during the study—once before the intervention and again after the intervention (see description below). Participants' *change* in these variables was calculated by subtracting the "before intervention" phase from the "after intervention" phase. Thus there were three different scores for KIC, SE, and CATS—a "pre" score, a "post" score, and a difference (or change) score.

To measure participants' goal orientation (GO), participants completed a survey which consisted of 15 items related to performance orientation and 15 items related to mastery orientation. All 30 of the goal orientation items came from a survey used by Button, Mathieu, & Zajac's (1996). Participants rated the extent to which they agreed or disagreed with each item on a 5-point Likert scale with 5 being "strongly disagree" and 1 being "strongly agree." An example of a performance orientation item is "When working with a team, I like to be the one who performs the best." An example of a mastery orientation item is "The opportunity to learn new things is important to me." A mean rating was calculated for both performance and mastery orientation giving a separate score for each dimension. A higher score indicated a higher orientation for that dimension and vice versa for a lower score. The participants' score on the performance scale is an indicator of how motivated they are by performance measures like their ability to demonstrate competence to others or wanting to feel superiority towards others. Their score on the mastery scale indicates how much they are motivated by learning, competence, and goals they make for themselves. Because goal orientation does not vary remarkably from situation to situation, the participants completed this measure only once—after the last phase of the study.

Participants also completed a 15-item demographics survey after the last phase of the study. The survey included items related to the occupation and age as well as their level of computer usage (e.g. "How many years have you used a personal computer?").

PROCEDURES

With the exception of the surveys and the addition of the intermediate group, the procedures for this experiment were identical to the Powell et al (2009) and the Peres et al. (in review) studies. In order to participate in the study, all participants had to complete a qualifying survey in the university's psychology participant pool, prior to arrival to the study. The items in the CATS and the computer self-efficacy survey were just a small portion of the surveys included in the qualifying survey, the remainder of the surveys belonged to other psychology experiments.

After arrival to the study, the participants read and signed the consent form and filled out the demographics survey. The experiment lasted approximately an hour and a half and took place in three phases.

There were three phases to the study and during the first phase the participant completed a set of editing tasks in Microsoft Word®. All sets of editing tasks consisted of 15 tasks that would ask the participant to *find* certain phrases and *underline* that phrase or *find* a word and *italicize* that word. The second or intervention phase involved the participant reading instructions out loud and observing the confederate complete similar editing tasks (or navigating a website a for the control condition). All participants were randomly assigned to one of the four intervention conditions. For the final phase of the experiment, participants completed a set of editing tasks, similar to the ones completed in the first and second phases, as the confederate read the instruction out loud and observed the participant.

There were four conditions utilized in this study, three experimental and one control. The confederate issued the commands for the tasks by issuing efficient or inefficient techniques, depending on which of the three experimental conditions the participant was randomly assigned to. The confederate issued all commands, like *cut, copy*, and *paste*, in the efficient condition with KIC, all commands in the inefficient condition with icons, and all commands in the intermediate condition with a mixture of KIC and icons. The control condition consisted of the confederate carrying out tasks on a website as the participant read the instructions out loud to the confederate.

After the participant completed the final phase, the participant was given the CATS, computer self-efficacy survey, and the goal orientation survey. When the participant completed the surveys, he or she was debriefed and dismissed

RESULTS

RELATIONSHIP BETWEEN INITIAL KIC USAGE AND INDIVIDUAL DIFFERENCES

Pearson correlations were calculated to identify if there was a relationship between the participants' initial KIC usage and their responses on the GO measures and their scores for the "pre" CATS and SE (Note: Participants' initial KIC scores were considered an approximation of their KIC usage before the intervention.). There was no relationship found between participants' initial KIC usage and their pre SE, pre CATS, and performance GO. There was, however, a significant negative correlation, $r(19) = -0.487$, $p = 0.025$, between initial KIC usage and mastery GO. Specifically, the higher someone's mastery goal orientation, the lower their initial KIC proportion. See Figure 1.1.

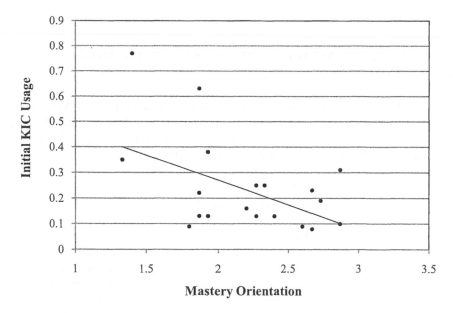

FIGURE 1.1 The figure shows the relationship between initial KIC usage and a participant's mastery goal orientation.

RELATIONSHIP BETWEEN CHANGES IN KIC USAGE AND INDIVIDUAL DIFFERENCES

To examine if participants' changes in KIC usage could be predicted based on individual difference measures in this study, we calculated a step-wise multiple linear regression. The change in KIC usage was the dependent variable and their scores on the GO scales (mastery and performance) and the "pre" CATS and SE scores were used as the predictors. None of these variables loaded into the model and the p values and Beta weights are listed in Table 1.

Table 1.1 Beta weights, t, and p values for multiple regression predicting the change in KIC.

Variables	Beta	t	p
Pre Anxiety with computers	0.073	0.169	0.869
Pre Self-Efficacy with Computers	0.324	0.805	0.438
Mastery Goal Orientation	0.426	1.192	0.258
Performance Goal Orientation	-0.141	-0.373	0.716

EFFECTS OF EFFICIENCY OBSERVED ON CHANGES IN INDIVIDUAL DIFFERENCES

In order to determine if there was a change in participants' ratings of CATS and SE, and if so, whether this change differed by the amount of efficiency they observed, two repeated measures ANOVAs were calculated—one for CATS and the other for SE. For the CATS ratings, there was no change in their scores, $F(1, 12) = 0.012$, $p = 0.913$, there was not a main effect of group, $F(1, 12) = 0.582$, $p = 0.638$, nor was there an interaction by group $F(3, 12) = 1.968$, $p = 0.173$.

For the SE ratings, there was no significant difference between the before and after ratings, $F(1, 17) = 3.697$, $p = 0.071$, nor was there a main effect of group, $F(1, 17) = 1.559$, $p = 0.216$. However, there was an interaction between group and change in SE, $F(3, 17) = 3.575$, $p = 0.036$. As seen in Figure 1.2, participants in the efficient and intermediate group decreased their SE and those in the inefficient and control group either had no change or increased in their SE respectively. A Tukey's Post Hoc analysis was conducted to examine this interaction further. The control group was significantly different from the intermediate group ($p = 0.04$) and approached a significant difference from the efficient group ($p = 0.07$). All other comparisons were not significant ($p > 0.33$).

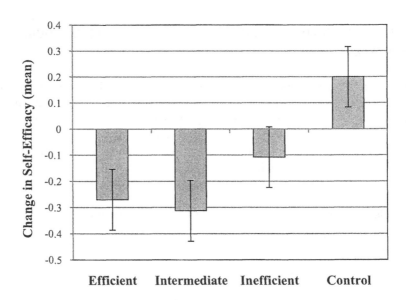

FIGURE 1.2 The mean self-efficacy change for each group. Error bars represent the standard error of the mean.

EFFECTS OF EFFICIENCY OBSERVED ON CHANGES IN KIC USAGE

To explore the effects of participants' observed efficiency on their change in KIC usage a one-way ANOVA was conducted. There was no significant difference found in their change of KIC usage, $F(1, 18) = 2.354$, $p = 0.106$, but the results of the ANOVA are in the direction we expected (See Figure 1.3). Given the small sample size used in this study for this preliminary data, and a previous study's effect size ($r^2 = 0.09$) (Peres et al, in review), this study did not have sufficient enough power to find a significant effect.

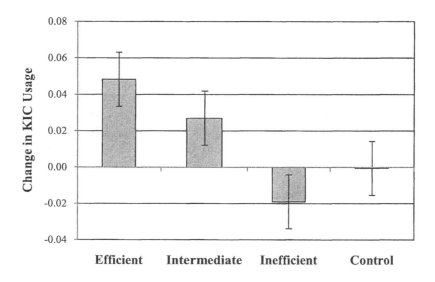

FIGURE 1.3 The change in the mean proportion of KIC usage per group. Error bars represent the standard error of the mean.

DISCUSSION

The overall goal of this research is to see what additional variables contribute to the adoption of efficient techniques. The preliminary data show that self-efficacy with computers, computer anxiety, and performance goal orientation do not predict whether a participant will use KIC in the initial task. The preliminary data did find a negative relationship between initial KIC usage and mastery goal orientation. This is fascinating because it seems counterintuitive that mastery goal orientation would be negatively related to KIC usage. It seemed likely that mastery goal orientation would be positively related to KIC usage because it was thought that people with a higher mastery goal orientation would be more concerned with learning the software and therefore be more likely to progress to efficient techniques. Perhaps, people with a mastery goal orientation are focused on learning all the functions of a software, such as Microsoft Word®, and are less concerned with efficient techniques to carry out these functions. Additionally, self-efficacy with computers, computer anxiety, and performance and mastery goal orientation were not predictive of whether someone would change their KIC usage.

Another intriguing result was the lack of significance in the change of computer anxiety within subjects and an absence of change in computer anxiety between the four groups. The authors thought computer anxiety would increase for participants in the efficient group because seeing a new and more efficient technique would make participants feel uncomfortable or inadequate. During the debriefing, most participants reported witnessing others utilize KIC in their daily lives, so maybe participants were familiar enough with KIC to not feel anxious after observing someone use an efficient technique. Furthermore, maybe a new technique is not enough to change a person's computer anxiety, but the utilization of a new function, which the participant is unfamiliar with, would be enough to change their computer anxiety. Equally, maybe a person's computer anxiety does not change after the observation a new technique because a person's computer anxiety remains stable over time or after the accumulation a certain amount of experience with computers.

Even though group did not have an effect on computer anxiety, group did have an effect on people's change in their self-efficacy. Specifically, those participants in the efficient group reported a lower self-efficacy with computers after the intervention phase. Perhaps, after seeing someone use an efficient technique a participant felt less self-efficacy with computers because they were unsure about how to issue this new technique and therefore started to question themselves on their own knowledge with the software.

The results support previous findings on the impact of watching peers utilize efficient techniques and the subsequent adoption of these efficient techniques (Peres et al, in review). The results, when considered in conjunction with results presented in previous studies (Peres et al, in review), show that the individuals differences associated with goal orientation, self-efficacy with computers, and computer anxiety, are not related to KIC usage or change in KIC usage. It is, however, likely that people's adoption of efficient techniques relates more to their environment and training than any specific demographic characteristic or personality trait.

REFERENCES

Button, S. B., Mathieu, J. E., & Zajac, D. M., (1996). Goal orientation in organizational research: A conceptual and empirical foundation. Organizational Behavior and Human Decision Processes, 67, 26-48.

Gaudron, J., Vignoli, E. (2002). Assessing computer anxiety with the interaction model of anxiety: Development and validation of the computer anxiety trait subscale. *Computers in Human Behavior*, 18, 315-325.

Jardina, J.R., Peres, S.C. & Titus, C. (In Review). Keyboard shortcut usage: Benefits, costs, and the impact of observing others use efficient techniques.

Lane, D. M., Napier, H. A., Peres, S. C., & Sandor, A. (2005). Hidden costs of graphical user interfaces: failure to make the transition from menus and icon toolbars to keyboard shortcuts. *International Journal of Human-Computer Interaction*, 18(2), 133-144.

Peres, S. C., Tamborello, F. P., Fleetwood, M.D., Chung, P., & Paige-Smith, D. L. (2004). Keyboard shortcut usage: The roles of social factors and computer experience. *Proceedings of Human Factors and Ergonomics Society 48th Annual Meeting*, New Orleans, LA, pp. 803-807, Human Factors and Ergonomics Society.

Peres, S. C., Fleetwood, M. D., Yang, M., Tamborello, F. P., & Paige-Smith, D. (2005). Pros, cons, and changing behavior: An application in the use of the keyboard to issue commands. *Proceedings of Human Factors and Ergonomics Society 49th Annual Meeting*, Orlando, FL, pp. 637-641, Human Factors and Ergonomics Society.

Peres, S. C., Tamborello, F. P., Fleetwood, M. D., & Nguyen, V. D. (In Review). Observing efficient behavior leads to the adoption of efficient behavior: An investigation of the influence of peer observation on the adoption of efficient software interaction techniques.

Powel, C., Peres, S. C., Nguyen, V., Burton, K. E., Muse, L. (2009) Using the keyboard to issue commands: The relation of observing others using efficient techniques on the weightings of costs and benefits. *Proceedings of Human Factors and Ergonomics Society 53rd Annual Meeting*, San Antonio, TX, pp. 637-641, Human Factors and Ergonomics Society.

Chapter 20

A Study of How Users Customize Their Mobile Phones

Wei-Jen Chen, Li-Chieh Chen

Interaction Design Laboratory
Tatung University
Taipei 104, Taiwan

ABSTRACT

Today's mobile communication technology is growing more rapidly than ever before. In order to be competitive, mobile phone makers incorporate as many functions into their mobile phones as possible. However, with the growing multitude of functions, its user interface design, integration, and maturity all will affect users' expectation and perceived complexity. To address this, many mobile phones allow for "User Customization" as a means to let users freely configure interface or functions, which leads the user's experience more logical, "friendly" and personalized per the individual user. Our research objective seeks to analyze user behavior patterns in customizing mobile phone, such as layout logic, menu hierarchy, and function priority. In the "Work" mode survey, we found out that both genders set "Contact" as their first customization choice and "Calendar" the second. In the "Leisure" mode, "Contact" was still the most popular choice and "Multimedia", "Game" that followed. So, as we can see, despite the various functions or programs provided, the initial function "The ability to connection people" still remains significant.

Keywords: Mobile customization, User interface personalization, Customizable user interface, Self-As-Entertainment

INTRODUCTION

With the rapid growing technology in telecom industry, the service providers and mobile phone makers provide many add-on values or services for users. Recently, mobile phones and relevant devices have various potentialities in multimedia (Marco et al., 2009). Mobile devices have great functionalities not only in mobility but also can process more digital data than ever before (Anderson and Blackwood, 2004). These digital contents such as video, photos…etc are now all can be customized as user's personal identity in mobile phone. Therefore, with mobile service and digital contents available, most users to some degree customize their mobile phone for personal identity and/or convenience. Children and young adults think mobile phones not only for communication, but for chat and gossip (Davie et al., 2004). Also, mobile devices were wildly used as business equipments in recent years (Attewell, 2006).

In this study, we developed a survey to examine existing user customization services and gender differences regarding if achieve users' expectations and the types of customization users covet. The purpose of this study would like to provide a better understanding of whether the services deliver on users' needs and customizable functions that suit. The next section describes the methodology used in this study. Afterwards, the results are analyzed. Finally, some findings and conclusions are drawn.

METHODOLOGY

We conducted the questionnaire using three formats: scale, multiple-choice and open descriptive. In order to specifically investigate customizable user interface, we divided the questionnaire into two sections which are: 1. Customizable User Interface, 2. Work/Leisure.

In the first section of Customizable User Interface, two categories are included which are "General Profile" and "Personalization". The first categories: General Profile, respondents are asked to choose (one) their "Most Frequently Customized Item" from some options, e.g. "Main Menu", "Template Style", "Wallpaper", "Contacts", "Text Message", and "others". Based on the respondent's choice of "Most Frequently Customized Item", the respondent is then asked to complete the questionnaire which uses a Likert scale and open descriptive. Under the "Visual Presentation", attributes includes "Layout Logic", "Item Focus Emphasis", "Visual Consistency", "Visual Legibility", "Icon Representation", "Information Density" and "Visual Satisfaction". Under "Ease of Operation", attributes includes

"Customizable Items", "Flow Logic", "Ease of Use", "Operational Complexity" and "Operational Satisfaction". Under "Overall User Experience" asks respondents to describe the pros and cons of "Most Frequently Customized Item" and how it may be improved.

In the second categories: Personalization, there were the following five multiple-choice questions:

1. What mobile phone accessories do you use to add personal identity?
2. What type of content(s) do you transfer from computer to mobile phone?
3. Which item(s) do you customize in your mobile phone?
4. When do you usually customize your mobile phone?
5. What is your purpose of customizing mobile phone?

In the section of Work/Leisure, the following three multiple-choice questions are included:

1. What function(s) do you customize while working?
2. Which program(s) do you setup as shortcut while working?
3. Which program(s) do you setup as shortcut while leisure?

In order to investigate if any gender difference existence, the questionnaires are classified separately into males' and females' answers. Pursuant to the findings of this research, we created separate tables with percentages based on above-mentioned multiple-choice questions to identify the differences or similarities between two groups' choices. Also, the independent-samples t-test was used to test statistically if any significant difference between genders and preferences.

RESULTS

GENERAL PROFILE

The results of the initial questionnaire utilize Likert scale, we used mean value to compare with each attribute and the values are then rounded to the first decimal point.

A 31.8% majority of respondents chose "Main Menu" as most frequently customized item. These users wanted "Main Menu" interfaces to be easy to use and instinctive. Respondents desired the ability to add/update applications, widgets and new functionalities (not just visuals like wallpapers and icons) via an online store or service. In order to fully demonstrate personal style, respondents wished for updates from the Internet (their websites/blogs, widgets, personalized "home" spaces) to reflect on "Main Menu" directly.

Of respondents that chose "Contacts", these users indicated instead of using default cataloging, they preferred customizable multiple hierarchy management of their contacts. This group of users wanted customizable search capabilities to simplify the search process and save time.

Of respondents that chose "Template Style", they wanted greater flexibility in their ability to manipulate template layout elements and they wanted manipulations be intuitive. Additionally, some users want music or video integrated and become customizable options.

Of respondents that chose "Wallpaper", respondents indicated that this was common practice from their experience with personal computers; therefore they expect the experience to be similar on their mobile phones. In order to have better visual legibility, respondents wanted the ability to scale, crop, and adjust the transparency of the wallpapers.

Only 9.1% of respondents chose "Message" as the most frequently customized item. They indicated that, regardless of phone model, the message interface of mobile devices they've used all look similar. To improve messaging efficiency, users wanted message's interface able to add personal presets or customized shortcut.

PERSONALIZATION

PHONE ACCESSORIES

Both genders would somehow add accessories on their mobile phones. More females (50%) than males (27.2%) would use "Phone Strap" to add personal identity (Fig. 1). In general, females tend to be more willing to add accessories on their phones than males do.

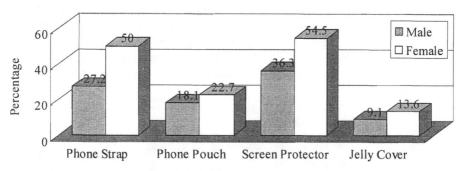

Figure 1. Add accessories on mobile phone as personal identity

TRANSFER COTNENT FROM COMPUTER TO MOBILE PHONE

As it is obvious, "Music" and "Photo/Video" are the main contents that both genders would transfer to their mobile phones (Fig. 2). More females (72.7%) than males (45.4%) would transfer "Music" from computer to mobile phones. However, about 81% of males more than 54% of females would transfer "Photo/Video". As for the "Software", there is a difference between genders, males (27.2%) vs. females (4.5%). Also, there are more males (45.4%) would transfer "Game" from computer to mobile phones than females (13.6%) do.

Regarding "The ease of use to transfer content from computer to mobile phone", the independent-samples t-test indicated that there was no significant difference between genders (t=1.875, df=31, p=0.070). In the "Purpose of connecting mobile to computer", there were more males (54.5%) than females (31.8%) chose "Download to Mobile". In contrast, there were more females (50.0%) than males (36.3%) chose "Upload to Computer". Of the response of how often do they connect mobile to the computer, about 45.4% males and 36.3% females answered that they would do it once every six months.

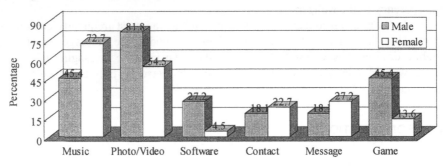

Figure 2. Transfer content from computer to mobile phone.

ITEM THAT CUSTOMIZE IN MOBILE PHONE

General, females seems more interested in customization them males do (Fig. 3). The majority of males (90.9%) and females (95.4%) change their phone's "Ringtones". Also, the majority of males (81.8%) and females (95.4%) customize their "Main Menu". Both genders, about 75%, would change "Theme". More females (68.1%) than males (45.4%) would customize their mobile phones' "Contact" and there are much more females (68.1%) customize "Message" than males (18.1%) do.

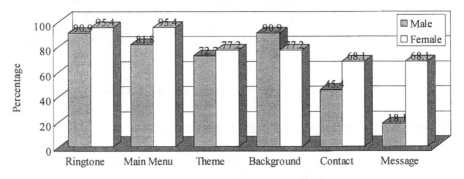

Figure 3. Item that customize in mobile phone.

WHEN TO CUSTOMIZE THE MOBILE PHONE

Comparing with "Sitting in the Traffic" (males 27.2%, females 40.9%) and "Before Sleep" (males 36.3%, females 22.7%), there were noticed differences that almost both genders customize their phones when they were "Bored" (males 90.9%, females 81.8%). Regarding "The satisfaction when customizing their mobile phones", the independent-samples t-test indicated that there was no significant difference between genders (t=1.057, df=31, p=0.299).

THE PURPOSE OF CUSTOMIZING THE MOBILE PHONE

The majority of both males (72.7%) and females (59.1%) customize their phones for "Convenient" (Fig. 4). Also, the majority of both males (63.6%) and females (68.1%) are for "Change Mood". More females (27.2%) than males (9.1%) would like to "Share (show off)" their mobile phone after customization. Of the response of "How often do they customize the mobile phone", about 45.4% males and 40.9% females answered that they would do it once per month.

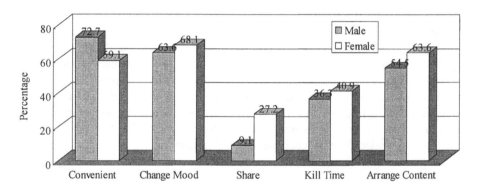

Figure 4. The purpose of customizing the mobile phone

WORK / LEISURE MODE

SET SHORTCUT WHILE WORKING

In "Work" mode, about 100% majority of males chose "Contact" as must have shortcut item and so did females (90.9%) do (Fig. 5). These users wanted "Contacts" interfaces to be easy to customize and instinctive, due to the rapid pace of working environment. Male respondents (72.7%) also desired the direct access to the "Calendar", and the ability to synchronize content via own computer or online service. Nevertheless, only males (18.1%) and females (4.5%) would set "Note" as shortcut due to the limited screen size, inefficient input device or some other reasons.

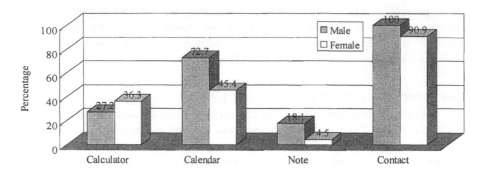

Figure 5. Program that setup as shortcut while working

SET SHORTCUT WHILE LEISURE

As same as "Work" mode, "Contact" is considered a must have shortcut in "Leisure" mode. Males (63.6%) and females (54.5%) chose "Contact" as the important shortcut in "Leisure" mode (Fig. 6). Respondents indicated that, regardless of any circumstance, the ability to access "Contact" immediately is their initial demand. Besides "Contact", "Multimedia", "GPS" and "Game" are the following three programs that respondents would like to set as shortcut while "Leisure". Nevertheless, it seems beyond ordinary expectation that more females (40.1%) than males (36.3%) would set "Game" as shortcut.

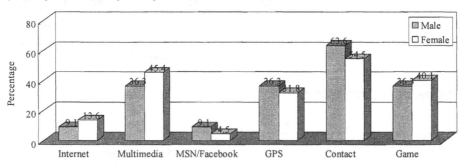

Figure 6. Program that setup as shortcut while leisure

CONCLUSIONS

Although mobile phone's menu hierarchy system has been great improved, there are still a lot obstacles for developers to overcome in user interface (UI) aspect (Young, 2006). Therefore, based on this research, in the stage of developing UI architecture for mobile phones, the UI designers can refer to which functions are users frequently used and provide appropriate flexibilities for users in customizing these functions. Moreover, based on the findings of this research, service providers can provide specific functions to targeted groups and facilitate creating customization-friendly interfaces and make customization become delighted for most users.

With in depth research, we noticed that in customization, there are two different mindsets of users who are so called technophobe and techno-geek. Our research will assist UI designers have clear understandings of what these users needs and design customizable interfaces that suit. The research revealed that approximately 85% of both genders customized their phones when they were "Bored". Also, both genders that transfer from computer to mobile are mainly multimedia contents: Music and Photo/Video. This result can be explained when people customize their mobile phones not merely change the settings but treat customization self-as-entertainment. Hence, continuous research is needed to study users' psychological demand such as: hedonic motivation, perceived playfulness…etc. We are yet to determine if it is a psychologically activity for user to customize the mobile phone for entertainment.

In the "Work" mode survey, we found out that both genders set "Contact" as their first customization choice and "Calendar" the second. Interestingly, in the "Leisure" mode, "Contact" was still the most popular choice and "Multimedia", "Game" that followed. So, as we can see, despite the various functions or programs provided, the initial function "The ability to connection people" still remains significant. Besides "Contact", "Multimedia" and "Game" play important roles among many users. This result mirrored mobile phone user's self-as-entertainment psychologically activity.

REFERENCES

Anderson, P., Blackwood, A., (2004). Mobile and PDA technologies and their future use in education. JISC Technology and Standards Watch 04–03 (November).

Attewell, J. (2006). Mobile technologies and learning: a technology update and m-learning project summary, Learning and Skills Development Agency.

Davie, R., Panting, C., Charlton, T., (2004). Mobile phone ownership and usage among pre-adolescents. Telematics and Informatics 21, 359–373.

Marco, Sa., Luis, C., Luis, D., Tiago R., (2009). Supporting the design of mobile interactive artifacts. Advances in Engineering Software, Volume 40, Issue 12, 1279-1286.

Young S., Sang H., Tonya S., Maury N., Kei., (2006). Systematic evaluation methodology for cell phone user interfaces. Interacting with Computers, Volume 18, Issue 2, 304-325.

CHAPTER 21

Ergonomics for Elderly People: A Brazilian Case Study in Mass Public Transport

Jairo José Drummond Câmara, Lívia Galvão Fiuza

Center for Research and Development in Design and Ergonomics (CPqD)
Universidade do Estado de Minas Gerais / Escola de Design
Belo Horizonte, MG, Brazil

ABSTRACT

The technological development of our society has allowed major changes, as well as the break of new and varied barriers. One of it is the advance in life expectancy. Never in our history has been recorded such high rates of life expectancy. This advance brings new perspectives to be taken within the society.

One of the most important factors for the development of any large urban center is a good public transport system. Brazil, despite showing good levels of development in the past 15 years, still has major shortcomings in this sector. And as regards the elderly, the mass public transport is poor and there is not, in most cases, a system's suitability for that specific audience, who suffers various physical constraints due to this inadequacy.

According to studies released by IBGE (Brazilian Institute of Geography and Statistics), the elderly population in Brazil tends to grow steadily, exceeding 30 million people over the next 20 years. As a consequence, studies' focusing on this segment of the population is necessary. Yet, the lack of data directed to that specific audience is still noticeable.

In this scenario, it's of great importance to improve the ergonomic studies in order to minimize the exclusion of this significant segment of population in society, minimizing their dependence on other individuals.

The fact of elderly people do not be considered in most product projects, seriously compromises the quality of their lives. The obstacles imposed by the products that surround them, impede or compromise the performance of everyday simple tasks. Unsuitable products can discourage the elderly to perform a certain task just as cause accidents by improper handling.

To get a complete analysis, which was able to meet the needs of the specific public, it was necessary to make analyses that fitted the entire frame to be studied. According to HENDRICK (2006) this study is called macroergonomics, which is the third generation of ergonomic studies, which seeks recognition of the organizational context and can interfere with the activity of work, in a systematic way.

This article has as main objective to present a case study of ergonomics facing the elderly, more specifically in the public transport system. This study was made through a detailed literature review on the theme.

Key words: macroergonomics, elderly people, public mass transport system.

INTRODUCTION

According to the World Health Organization (apud Magalhães, 1987), people are considered "seniors" when they reach the age of 65 years old in developed countries and, in underdeveloped or countries in development, when they reach the age of 60.

Sociologically, people transit in three stages: the first one is the training time, where the individual acquires skills to produce in the society. The second one is the production time, when the individual is at the peak of its potential, living in the modern society on its most reinforcing moment. The third one is the time of non production, where there is a decrease of their physical strength and activities that are a result of this strength. In the third phase, the situation also begins to revert in the biological, social and psychological level, as their physical body goes into a declination and they do not have available anymore, the means they used to have (physical strength, speed of reasoning, etc.). Socially, their professional space is extinguished and in most cases, their social death is pronounced with the advent of retirement. Their social and family roles start to disappear as children grow up and leave home and their role as father (or mother) ceases.

The Brazilian population of elderly is increasing its longevity and, therefore, the elderly society is looking for better living conditions, in a way to increase their presence and activities in urban centers. This fact leads to the discussion and the need for new public policies, broader, with specific laws that provide protection throughout the life course, where the elderly have access to all services and public places without restrictions or barriers that might difficult the movement independently.

The Accessibility, defined by the Brazilian Association of Technical Standards (ABNT) as: "Possibility and range condition for use, with security and independence, of buildings, spaces, urban furniture and equipment" (NBR 9050, 1994, p. 2), is of great importance for the elderly in view of their new active participation in society. Therefore, accessibility should already be considered at the beginning of the project design to be implemented, respecting and complying with the directive of the laws and standards, and therewith, be able to develop and resolve all interfaces of living spaces with users, respecting the socio-cultural differences of the actors involved, and thus avoiding the numerous design errors caused by the construction of buildings and urban infrastructure that have become major architectural barriers, restraining, in an arbitrary manner, the access of elderly and disabled people of all kinds.

ERGONOMICS AND ELDERLY PEOPLE

In 1949, the Ergonomics science arises, defined by Alain Wisner (1991) "as a set of scientific knowledge relating to mankind and necessary for the design of tools, machines and devices that can be used with maximum comfort, safety and effectiveness", searching to rescue the meaning of the Greek word ERGON *(work)*.

Thereby, Ergonomics aims to adapt the work to mankind, giving back its dignity. It has a globalizing approach and, of course, turns back to the problem of individuals who are withdrawn from the labor market by the mandatory retirement by age, when they still had great potential to share with society.

Statistics show a profile of the population where, increasingly, the portion corresponding to the elderly will be greater and more meaningful. If in one hand they are the main consumers of the future; the pure and simple removal of the same group of productive people should also be questioned.

The tasks management is affected by aging, influencing the performance of tasks. The increasing difficulty in managing multiple tasks simultaneously also increases with age. But experience plays a fundamental role in the management of simultaneous tasks.

Ergonomics, according to Bustamante (1996), worries about the adequacy of the work to the person, considering the anthropometric and physiological diversity of the worker, just as the variability of the efficiency over time. It also considers the need for a suitable place where to work. It is the elderly in their relationship with work.

According to these sciences, the formation of wrinkles and baldness are not substantive changes in the senile process. More important are the changes that cause the decrease in the intensity and rapidity of the reactions of the human body to external stimuli, the reduction in the speed of separation of tissues after injury, the greater possibility of nutritional deficiency and the very gradual involution of the functional capacity of the whole body. The elderly, due to react less violently to infections and other injuries, presents more discreet symptoms and signs of diseases. Often, diseases are already in an advanced stage in their bodies when they request for medical care, often struggling to overcome them. Feeding little or too much, suffering from dehydration or excessive moisturizing, tends to have far more serious consequences of climate changes than the young person. Many factors promote a deep insecurity in seniors, needing an adjustment to their new image.

THE MASS PUBLIC TRANSPORT IN BRAZIL

We live in a moment where there is a crisis of mobility with increased congestion, air pollution, the deaths in the traffic and the time spent for travel, in cars or buses. The road infrastructure is a determinant factor of physical and territorial planning, and with the pressure on the tremendous growth in the fleet of cars, most part of the public investment is allocated to this mode of private transport. It's understood that this is the result of an exclusive mobility model focused on the car, which has received over the past year more than R\$ 12 billion of Reais in tax incentives from the federal government for the production and sale of cars, benefiting manufacturers, concessionaires and vehicle owners.

With so many public resources for car expenses only this year, it would be possible to significantly improve the public transport system, which today does not receive the attention it should of the government as an essential service, as required by the Federal Constitution: the investment is inadequate and the users pay the full costs of operation of the transportation system, including gratuities and discounts (students, seniors and persons with disabilities, for example) and all local and federal taxes totaling more than 40% of the rates.

With the increasing competition in the market for the urban transport, characteristic of Brazilian cities from the 90's, the performance of services is becoming increasingly important.

The inefficiency, in a generally plan, and the poor quality of services in a more specific focus may charge a high price to operators and government agencies linked to public transportation, dramatically reducing their participation in meeting urban mobility. And that is a high cost to society.

The urban transport system must be oriented to greater rationality and efficiency. It doesn't fit, in its planning process, a straightforward view of the market. Not talking about a competition for passenger but for a service cheaply, safely and comfortably. Each technology has its role and importance in this structure (...). (Source: Estado de Minas Newspaper, Luis Francisco Tomazzi Prosdocimi, 26/09/2008).

METHODOLOGY

This study presented more analytical than practical characteristics. So, the methodology was structured in order to allow the team to identify the products of relevance to the study. This work has allowed a deeper understanding of the relationship between the mass public transportation and the elderly.

In general, the project was organized in two main phases. The first was about the raising of the main features and the flaws in the Brazilian mass public transportation system, especially regards the elderly. The second phase concerned about the analysis and discussion of the data collected. This system required a detailed survey of information from official web pages, technical files, libraries and queries to database information.

CONCLUSIONS

Through the analysis of the collected information and data, it was concluded:

Considering the work in its broadest form, like any activity performed by a person or machine, comes the conclusion: To perform any activities, it's necessary that the built environment offers certain facilities to enable the implementation of these activities, in other words, equipment and access elements should be designed with proper sizing and positioning according to the needs of all users. In these elements are included: Stairs, ramps, lifts, corridors, ceramic and metal sanitary, cabinets, handrails, guardrails, and commands to drive the devices. For people who can stand, and have medium height, 1.60 m to 1.80 m high, it can be applied the standards already established by the use for residential projects. Now, for those who can not stand, it is necessary to locate the devices that are manually operated, within manual reach of a seated person.

The promotion of accessibility is a structuring element and guarantees of non-exclusion. Simultaneously, a higher quality of the transport system (in terms of inclusive design of vehicles and infrastructure, but also of driver training and information available), translates into a more balanced system, contributing to the accessibility as a key guarantee of social sustainability of the transportation sector.

A friendly environment results in the provision of physical and psychosocial of compensatory nature to promote physical health, functionality and the psychological well-being of older people.

The analysis and design of systems for universal access has been over the last decade the subject of several national and international initiatives, which have occurred more or less autonomously. The universal design seems to be the most suitable design to be considered for the improvement of the quality of mass public transport system regarding elderly people.

The Universal Design means to design products and environments to be usable by all people within the limits of the possible, without the need for adaptation or specialized design. (Wright, 2001:55). According to the Center for Universal Design at the University of the State of North Carolina there are seven principles of universal design:

• **Drawing equitable** - can be used by people with diverse skills, avoid segregating or stigmatizing any users and has an attractive design for all.

• **Flexibility of use** - accommodating a wide range of individual preferences and skills, allows the left-and right-handed use, facilitates the accuracy and precision of User: also adapts to the rhythm of any person.

• **Simple and intuitive use** - easy to understand, regardless of the user experience or her knowledge, language proficiency, or current concentration level.

• **Perceptible Information** - effectively communicate the necessary information to the user, regardless of environmental conditions or sensory abilities of the same.

• **Tolerance to errors** - contains elements that reduce the danger of mistakes. Requires little physical exertion - can be used efficiently and comfortably with the minimum expenditure of energy.

• **Size and space suitable for approaching, reaching, manipulation and use** --

are guaranteed, regardless of the size of the user, posture (sitting or standing) or their mobility.

THE AUTHORS WOULD LIKE TO THANK TO CNPQ (NATIONAL COUNCIL FOR SCIENTIFIC AND TECHNOLOGICAL DEVELOPMENT) AND FAPEMIG FOR THEIR SUPPORT.

REFERENCES

HENDRICK, H. & KLEINER, B. (2006) *Macroergonomia- Uma Introdução aos projetos de Sistema de Trabalho.* Rio de Janeiro: Editora Virtual Científica.

Associação Brasileira de Normas Técnicas. *Acessibilidade de pessoas portadoras de deficiências a edificações, espaço, mobiliário e equipamentos urbanos - NBR9050/94.* Rio de Janeiro ABNT/ Fundo Social de Solidariedade do Estado de São Paulo,1994. 59p.

MAGALHÃES, Dirceu Nogueira. *Invenção da velhice.* Rio de Janeiro: Editora do autor, 1987.

BUSTAMANTE, Antônio; MENÉNDEZ, Concha. *Una ergonomia en evolución.* Apostila. Barcelona, 1996.

IIDA, Itiro. *Ergonomia: Projeto e Produção.* – 2ª. edição rev. e ampl. São Paulo: Edgar Blücher, 2005. 609p.

CAMARA, J. J. D. - *Levantamento Ergonômico do Sistema do Metrô de Superfície de Belo Horizonte - O Design das Instalações Físicas e do Material Rodante.* In: 49 REUNIAO ANUAL DA SBPC, 1997. CD Rom com os anais da 49/50 Reuniões da SBPC. BELO HORIZONTE: SBPC. v. Único. p. 0-0.

WRIGHT, Charles. *Facilitando o transporte para todos.* Washington, D.C.: Banco Interamericano de Desenvolvimento, 1ª ed., 2001.

LARICA, Neville Jordan. **Design de Transportes: Arte em Função da Mobilidade**. Rio de Janeiro: 2AB / PUC-RIO, 2003. 216p.

Chapter 22

Ergodesign: A Systemic and Systematic Approach: How the Task Activities Determine the Product Conception and Configuration

Anamaria de Moraes

PUC-Rio Pontifical Catholic Universty of Rio de Janeiro - Brasil
LEUI – Laboratory of Ergonomics and Usability of Interfaces
moraergo@puc-rio.br

ABSTRACT

Ergodesign in project development allow a design process with ergonomics since the beginning till the evaluation. Products – physical equipments and cognitive systems better fit the operators/users/consumers. Schematic model shows the appliance of ergodesign, and finally the design process including the project activities.

Keywords: Ergodesign, ergonomics, product development

INTRODUCTION

This paper deals with the concept of Ergodesign, considering an ergonomic approach of the product design process. The stages and phases to be accomplished are also described in detail. The ergodesign observe systemic and systematic development beginning with the problem delimitation, the ergonomic diagnosis and design, ergonomic tests and evaluations and ergonomic validation. The analysis of the task activities provide the identification of all possibilities of operation and use and, consequently, provide fitted design.

THE PROBLEM

Designers - architects, product designers, engineers, screen designers - do not agree about the participation of the ergonomists during the development of products, since definition of the problem, the generation of alternatives till the selection of configurations and layouts. They say that ergonomics must be applied at the end of the process and do not consider ergonomic evaluation and validation.

ERGODESIGN

Accorder Cushman & Rosenberg (1991) product design is the process of creating newest and better products for people to use. The ergonomists is responsible for the product usability focusing in the comfort, learning, efficiency and safety.

The design of interactive products point out the necessity of the compatibility with physical, psychic and cognitive characteristics of the operators and users. Further attention shall be given to the ergonomic features of the product. Ergonomics brings to the design process a systematic approach to the analysis, configuration, and evaluation of usability requirements. The ergonomists know methods, techniques and references that increase the designer ability to develop interfaces.

After Chapanis (1995), Ergonomics or human factors is a body of knowledge about human abilities, human limitations and other human characteristics that are relevant to design. Ergonomic design or human factors engineering is the application of ergonomics information to the design of tools, machines, systems, tasks, jobs and environments for safe, comfortable and effective human use. "The significant word in those definitions is *design* because it separates Ergonomics from such purely academic disciplines as anthropology, physiology and psychology. We study people but we study them not because we merely want to add to our store of basic information. We study people in special circumstances because our aim is to apply what we know or what we find out to design of practical things - of things that we have to do or want to use because of our inclinations."

METHODS AND TECHNIQUES

The ergodesign observes the following systematic development: the problem delimitation and the system comprehension (appreciation; similar of Francophile demand analyses); the ergonomic diagnosis (task analysis, activities behavioral analyses, macroergonomics analysis); the ergonomic design; ergonomic tests and evaluations. The task exigencies and constraints determine the system conception, product configuration in terms of functions to be performed by the man or by the machine.

Meister (1989) says that when performed correctly, system design is akin to system engineering problem solving. "It begins with determination of the goals for design, specification of functions to be performed by the system, conceptualization of alternative ways of performing these functions (alternative problem solutions), analysis and comparison of these alternatives, and selection of the most effective alternative. In human factors, the equivalent of system analysis consists of the mission, function and task analysis with which the system development begins. (...) System thinking considers the function, purpose, and goals of the system and how the goals of the system can be reconciled with those of the "suprasystem" of which it is a part and with the subsystems that form part of it. The approach emphasizes input-output features and a purpose orientation, aspects that are particularly appropriate for sciences with a high degree of application." (Meister, 1989)

According to Stammers & Shephard (1990) whilst a simple definition of a task may be possible for everyday use, a context of ergonomics a more complex view must be taken. In particular, three interacting components of tasks must be considered:

Task requirements: Refer to the objectives or goals for the task performer. For example, a word processor user, having completed a document, may be required to save the document on to a permanent storage device.

Task environment: Refers to factors in the work situation that limit and determine how an individual can perform. This may be through euther restricting the types of action that can be taken and their sequencing, or by providing aids or assistance that channel user/operator actions in a particular way.

Task behaviour: Refers to the actual actions that are carried out by the user within the constraintas of the task environment in order to fulfill the task requirements. Behaviour of the user may be limited by inherent psychological or physiological factors, or a lack of appropriate knowledge os skill. The actions employed may also have been developed through experience to optimize efficiency and to minimize effort.

The first two aspects are determined by the system context; this incorporates the organizational context, the operating requirements and the limitations of the technology involved, the prescriptive elements of training, the structure of the interface, the operating procedures, the environmental conditions and the influence of other externally connected events.

Almost all these elements can be observed, recorded or predicted accurately. Thus the basic framework of task activity, in so far as it is determined by its context, can be accurately documented. Task behaviour, on the other hand, can vary greatly between individuals and with experience. A consensus may have to be reached then on what optimal performance will involve. This is particularly the case when a task is largely cognitive in nature.

RESULTS

A flow diagram of activities practiced for more than twenty years in product development in design under graduation and pos- graduation courses is presented and explained. All techniques performed in each phase are explicated.

The ergonomic intervention can be divided in the following major stages (Figure 1):

- Ergonomic appreciation;
- Ergonomic diagnosis;
- Ergonomic design;
- Evaluation, validation and ergonomic tests;
- Ergonomic refine and optimization.

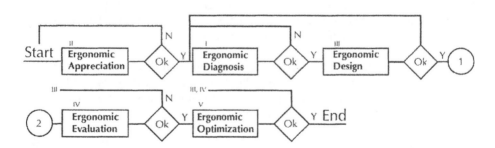

Figure 1 Flowchart of Ergodesign

The ergonomic appreciation is an exploratory phase that includes mapping the ergonomic problems. Consists in the systematization of human-machine-task-system and in the delimitation of ergonomics problems – postures, information, actions, controls, cognitive, communications, interface human-screens, displacements, manual material handling, environment, and physical variables (illumination, noise, vibration, and temperature), work team and design, management. This is the moment of field observations and non-structured interviews with workers, supervisors and managers.

Photo, tape and video and draw registers are made. This phase ends with the ergonomic report: illustrated (photos and sketches), modeling and human-machine-task-system dysfunctions. The report also presents: the hierarchical presentation of the problems, considering human work constraints and heath costs. Gravity, urgency and tendency determine the priority of work stations and situations to intervention and modification. Prediction with cause-consequence problems guides future hypotheses to be treated during the diagnosis. Some preliminary suggestions of ameliorations are welcome.

The ergonomic diagnosis permits go into details of problems that took priority and test predictions or experiment hypotheses. According the emphasis of the research and assuming the manager demands, it is time of macroergonomics analyses or task behavioral activities analyses. Measurements of physical variables, environmental factors, organizational aspects and work design are made. Methods and techniques: Systematic observations; behavioral register in the field:; video records; structured interviews; questionnaires; rating scales; verbalizations; think aloud; heuristics; participatory evaluation, usability tests; and computer programs.

We implement reports of frequency, sequence, duration of posture assumptions, uptake of information, controls' activations, communications, displacements, manual material handling. The levels of amplitude and detailing of the analysis depend of the scope determined in the contract as well as the time and budget available by the manager. This phase ends with the ergonomic diagnosis-ratification or refutation of predictions and hypothesis. An overview of what literature points out and the research conclusions establish the ergonomic recommendations to system and subsystems changes, considering physical, cognitive and satisfaction contents.

The ergonomic design adapts the existing situation to the whole of recommendations. The team generates concepts, alternatives, configurations, dimensions, details for each subsystem and component: panels, screens, furniture, layout, environment, activities, instructions, communications by software, hardware or humanware The objectives are emotional satisfaction, comfort, safety and health of operators, consumers, and users, and maintenance group, supervisors and managers. Reports, schemas, graphics, draws present the results and conclusions they are part of ergonomic project.

The ergonomic evaluation or validation give back to operators, consumers, and users, and maintenance group, supervisors and managers the obtained results They were the actors of the research, so they are the best subjects to evaluate the new propositions. At this moment with the real operators/ consumers/ users we evaluate prototypes, simulators and models. Consists in simulations, evaluations using test techniques simulations and models. Satisfaction questionnaires, rating scales, focus group are applied. The objective is the participation of the actors in all decision of ergodesign team.

The <u>ergonomic refine and optimization</u> implies the project revision, after its evaluation. What is available in the market and the demand of the contractor considering technological changes, price and time for implementation are discussed. Meetings with the workers participation are recommended. After all this consideration it is time to ergonomic specifications for system, subsystems and components mentioned above. All the phases are presented in Figure 2.

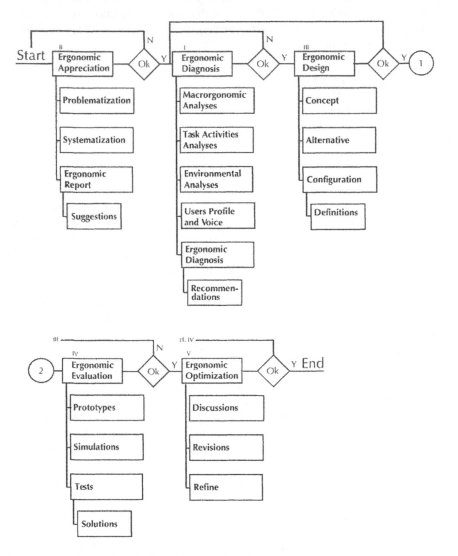

Figure 2 Flowchart of Ergodesign Going into Details

According Yap et al. (1997), the ergodesign approach is creative macroergonomics approach that has as objective the human and system attributes simultaneously with the design conceptualization and development. As a technology the ergodesign has a design orientation and in this way an important tool in the efficiency and purpose of ergonomics implementation in design and products development, equipments and systems.

McClelland (1990) affirms that influence individually the attitudes and experiences of designers is necessary but not sufficient to assure that ergonomics recommendation receive the desirable attention during the project activity. We may emphasize the understanding and contribution to design management, be conscious that ergonomics is part of project development. To allow this objective the ergonomist must knows about design and a change of perspective – forget the traditional role of correction, evaluation an criticism and have a way of think more prospective and process oriented. Operationally the ergonomist must be less an applied scientist and more a designer. So must see the role of the ergonomist as a partner in the design process not as outsider. This implies accept the division of the responsibilities for the project decisions.

"Many ergonomists seem to think that their task is ended when they have provided general guidelines and turned them over to engineers and designers to interpret and implement. I disagree. After all, we, not engineers and designers, are or should be experts on systems interfaces on reach limits, on control-display movement relationships, on subjective responses to colors, in cognitive aspects, and acceptable time lags. When we are engaged in product or information development or are members of a design and development team is our responsibility to translate general guidelines into specific design recommendations to the project. That means using some special techniques and methods and some conventional techniques and methods in special ways." (Chapanis, 1995)

After Porter (1999) we shall mention that "It is essential that ergonomics input to a product takes place throughout the design process but nowhere is it more important than at the concept and early development stages of design. Basic ergonomic criteria such us the adoption of comfortable and effective postures need to be satisfied very early on. If these criteria are not thoroughly assessed then there is usually only very limited scope for modifications later on as all the other design team members will have progressed too far to make major changes without considerable financial and time penalties.

A sequence flow of step by step to be followed during the iterative product development considers the exchanges between ergodesign and product developer. (Figure 3).

218

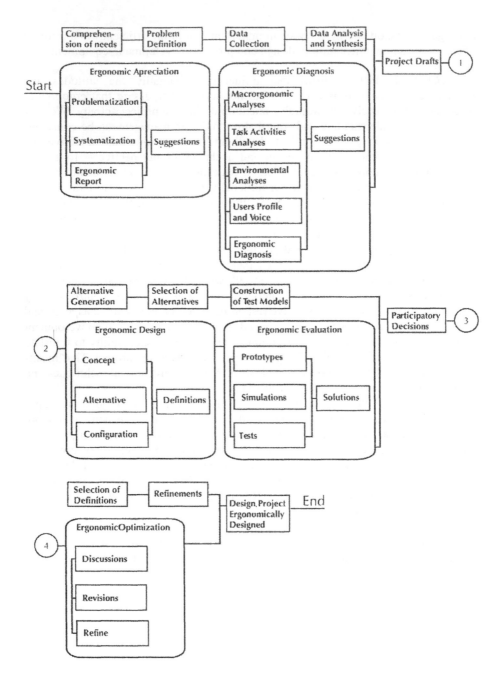

Figure 3 Flowchart of Product Development

CONCLUSIONS

To well march in all the steps of the model designers and ergonomists must change ideas and work together. The product developers need to know ergonomics their methods and its applications. It is important to recognize that about user centered design this is the principal task of ergonomists over the past fifty years. By the other side the ergonomists may be conscious their work is not finished when they propose some general recommendations. The ergonomist is the one that better understand how to adapt the system, subsystems and components to the physical, psychic and cognitive limits and potentialities of operators /users/consumers. It is his obligation participates of all discussions since the definition of the problem till the solutions' evaluation. He can not be afraid of the responsibilities related to the project development. He is part of project team never mind if someone says the product will be a success or a fiasco. He is an ergodesigner.

REFERENCES

Chapanis, A. (1995), "Ergonomic in product development a personal view." *Ergonomics*, 38 (8), 1264-1638.

Cushman, W.H., Rosenberg, D. (1991). *Human factors in product design*. Elsevier, Amsterdam.

McClelland, I. (1990). "Marketing ergonomics to industrial designers." *Ergonomics*, 33 (4), 391-398.

Meister, D. (1989). Conceptual aspects of human factors. Johns Hopkins, Baltimore..

Porter, S., Porter, J. M. (1999). "Designing for usability; input of ergonomics information at an appropriate point, and appropriate form, in the design process." *Human factors in product design: current practice and future trends*. Taylor & Francis, London. 15-25.

Stammers, R.B., Shephard, A. (1995). "Task analyses." *Evaluation of human work*, Taylor & Francis, London. 144-168.

Yap, L., Vitalis, T., Legg, S. (1997) "Ergodesign: from description to transformation." *Proceedings of the 13th triennial congress of the international ergonomic association*. Finnish Institute of Occupational Health, Helsinki, 2, 320-322.

Chapter 23

Light, Colour and Surface as Ergonomics Aspects in Space Recognition and Urban Orientation: *Azulejo's* (Glazed Tiles) as Paradigm

Carla Lobo

ESAD Escola Superior de Artes e Design das Caldas da Rainha, Portugal
CIAUD | Faculdade de Arquitectura da Universidade Técnica de Lisboa

ABSTRACT

The aim of this study is to assess glazed tile's cladding (*azulejos*) qualities in urban spaces, to identify its perceptive potential as an ergonomic aspect in urban environment, allowing a more intentional and consistent use of the *azulejo*.
Sight sense in human interpretation of physical reality is crucial. In the mental process of spatial organization, differences in the visual perceptive field have an important role. Perception variation, synchronic or diachronic, is crucial in this interpretation process. A psychometric measurement of glazed tile's cladding is being conducted in order to produce conclusions that improve the product design (of glazed tiles), and environmental design process.

Keywords: Human factors, perception, ergonomics, *azulejo*, light, colour, surface, spatial orientation, design process

CONTEXT

A vital urban environment is one in which the visual elements - light, colour, and architectonic form - signify and express civic functions. (Swirnoff 2000, p.IX)

The sense of spatial appropriation by the citizen largely depends on its legibility, symbolic content, safety and wellfare, which are linked with the spaces contributing to an harmonious relation between them and the external world (Lynch 2002; Pallasma 2005).

Public space should be designed as an organizing structure of the territory, an area of continuity and differentiation, an axis of urban environment.

Creating guidance systems based on structures that promote safety, and comfort, will contribute to a positive environment. In urban areas survival has become not only a physical need, but also a psychological and emotional condition. These areas provide "information overdoses", diminishing our capability of visually isolate elements. In order to organize our visual spatial memory we need to focus on visual references that allow similarities and differences recognitions, in a permanent comparative process as we move in urban spaces.

USER | SPACE RELATIONS

Framing and identifying our environment is a vital activity for all movable animals, as Lynch said. Elaborating a wayfinding system through identity structures in order to facilitate life and in urban areas survival, has been a constant issue on living beings history.

While walking through the city, we go through personal mind maps created on basis of meaningful perceptual/spatial/emotional/socio-cultural features. They succeed one another in our journey, creating rhythms, defining time perceptions. According to Lynch we build our image through a collage of successive images, rarely seeing the city as a whole (2002).

In urban areas it is difficult to isolate elements, dissociate them from visual noise. What catch our attention are the similarities and the differences, the rhymes and rhythms, as stated by Humphrey, "likeness tempered with difference"(1980). This comparative process enables us to find common principles on diversity, recognizing structural principles of knowledge, making possible for us to walk trough a mutant reality.

COLOUR, LIGHT AND TEXTURE IN THE URBAN ENVIRONMENT

Colour in the urban environment is an expression of collective identity built by the will of an evolving society, constantly changing, becoming a cultural reference, an important factor for the humanization of public spaces.

The loss of building surface plasticity, motivated by modernist principles, has contributed to vision supremacy on spatial perception. Without tactile elements, which claim physical proximity and interaction with the user /viewer, the architectural structure becomes flat, inhospitable, unfriendly.

People appreciate colour and texture variations in their environments (Mahnke 1996; Swirnoff 2000, 2003). They consider them pleasant for the eyes and touch, a sign of abundance, a return to Nature, which somehow softens the subject/object relationship: texture appeals to the touch, to an interaction between built and user, not only as functional feature, but also as tactile pleasure, linking us to built environment.

Diversity in the urban landscape is considered as a vital quality (Humphrey 1980; Lancaster 1996; Lynch 2002). Colour, and chromatic diversity, enhanced by light variations, are positive elements in our space image composition, not only for their emotional and psychological value, but also by its sensory capabilities that can alter our spatial perception. Highlighting details, breaking the monotony by introducing rhythm and proportion, increasing spatial readability by differentiating volumes, establishing figure/ground clarification, hierarchizing spaces (Porter 1982; Merwein, Rodeck and Mahnke 2007), improving wayfinding and wayshowing tasks (Mollerup 2005) are processes to achieve this diversity.

AZULEJO'S AS PARADIGM

Size, shape, colour and location are part of the mental process of objects and space organization (Swirnoff 2003). These are fundamental factors in user's efficiency, as they promote environment recognition and categorization (Friedman & Thompson 1976).

Considering these variables, glazed tiles are important elements of space orientation in urban landscape, with great perceptive prominence due to its light reflection capacities and perceptive changes (Lobo & Pernão 2009).

Because of its intrinsic qualities (clay's plasticity, glaze's characteristics), extrinsic qualities (colour, gloss, texture), and emotional characteristics (easily recognized, familiarity), *azulejo's* cladding can provide aesthetic pleasure, human comfort and reflecting as well ergonomic concerns. They create surfaces in permanent mutation allowing different forms of visual communication in the object/environment/observer system (Lobo & Pernão 2009).

AZULEJOS AS POTENTIAL EMOTIONAL LANDMARKS

Since Ancient Times, the use of ceramic in claddings and structural elements has integrated architecture in a comprehensive way. Glazed or natural its presence is atavistic, with a socio-cultural significance, expressing the action of the human hand, embodying customization as opposition to the cold impersonal and flawless materials of mass production.

Azulejo's claddings because of their inherent characteristics, resulting from the nature of the material and its production process, make the perceptual experience a prominent event through the significant variations that occurs with different light conditions, viewing proximity or distance, or angle of vision.

Sensory-perceptual features make *azulejo* eligible not only as functional protection material, but also as differentiation element, with a social function: guiding element and landmark. Its surface qualities – colour, texture and gloss - consequence of the material nature; the variety, stability and longevity of glaze's colour, even in high saturated hues compared to other materials; the shifting brightness due to the reflectiveness of glossy glazes; the possibility to create random colour patterns or graphic designs on the surface; the chance of texturing the surface, give glazed tiles is distinctiveness.

FIGURE 1 Azulejo's social-cultural significance | light reflection and perceptive changes.

AZULEJOS COLOUR AND TEXTURE AS DETACHMENT FACTORS

The *azulejo* laying process results in an irregular surface where incident light bounces and scatters, creating multiple colour and light reflection perceptions. *Azulejo's* colour variations in terms of chroma, are consequence of glaze's thickness and evenness: more thickness results in an higher saturation. These colour variations are further strengthened by the effect of emboss motifs on the glaze: relief "opens" the glaze, which becomes less thick on prominent lines, losing colour saturation; on the contrary, on "lower" lines of the surface, due to glaze's accumulation, the colour saturation increases.

When light interacts with glazed tiles cladding, it scatters in different directions. Depending on the viewing angle one can "see" a mirrored surface, a specular reflection or a sum of coloured squares of the same hue, but with different value and chroma.

Figure 2 Glazed colour variations due to the surface emboss and glazed characteristics.

The reflectivity of a glossy glaze, affects the colour of the surrounding buildings through the reflection of light rays, and brightness reflection. This apparent colour modification of the tiles as a consequence of the reflection of surrounding colours, and the quality of the light reflected by *azulejo's* surfaces, create an emotional atmosphere (Zumthor 2006), which stand out from its environment. For this reason an *azulejo's* surface detach from other adjacent areas, painted or plastered. Glossy surface colour reflection is much stronger than that of a matte surface (Lancaster 1996), and colour saturation perception changes, as specular reflection depends the observer situation.

These shifts in specular reflection, associated to brightness variations, can also help to identify the relative position of the viewer, and building, in regard to the sun.

Colour and texture can merge with distance, but the reflectiveness of a glossy glaze will contribute to the detachment of that surface, over the background of the other buildings, increasing spatial legibility.

FIGURE 3 *Azulejo's* colour and reflectiveness as an emotional feature, a detachment factor, and an integration aspect.

AZULEJO'S CLADDINGS AS SPATIAL REFERENCE

Wayfinding is a structured process in which we try to establish relations among multiple stimuli. This process of assimilation is a continuous succession of classifications, through which we mentally "ordered" rhythms and rhymes, setting visual and emotional harmonies in space (synchronic), and time (diachrony) (Humphrey 1980), developing an internal representation of perceived environmental features (Golledge 1999).

Glazed ceramic cladding provides visual and tactile richness, allowing diversity in the perception of the object at different distances and viewing angles, making them easy to recognize, and to remember.

The relation between observer and object determines *azulejo's* claddings perception: if we consider the observer in motion, the variation of the distance will result in different visual stimuli that attract our attention, from the overview vision to detail proximity, triggering new visual sensory-perceptual experiences. Due to the optical mixture phenomenon, at an urban scale we are able to see a uniform colour, as we move closer this impression is converted into brightness and colours; at an even closer distance an array of colours and textures is revealed, with tactile qualities and patterns.

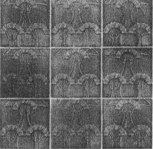

FIGURE 4 Distance and viewing angle provide perceptual diversity, from an uniform colour veil to an array of colours and textures.

The way we perceive *azulejo's* cladding patterns, their texture or the grid created by the tile laying, is fundamental for referencing our position in relation to that surface. If they are clearly perceived, the distance is proximal and the viewing angle is close to 90°. If texture, pattern and grid merge into homogeneous colour, then the viewer is at a greater distance, or the viewing angle is smaller, or both.

Proximity or distance leads to different visual patterns. This enables us to feel the pulse of facade rhythms, leading the eye and body to discoveries, in this place of experiences: the public space.

Unity and complexity conveyed by ceramic glazed tiles provide balance to the environment. Diversity tends to arouse visual perception (Foster 1976), which can lead to a clearer understanding of urban space (Friedman & Thompson 1976).

226

FIGURE 5 The pattern merging into colour, increases depth perception.

AZULEJOS AS AN ERGONOMIC FACTOR

In an urban environment, our attention needs to be stimulated; colour can act as an information finding system (Mollerup 2005). People with normal vision are more receptive to hue contrasts; visually impaired persons are more sensitive to differences in value than in hue. Although the brightness is generally a negative factor in terms of visual communication for people with reduced vision, in the case of *azulejo* it might work as visual a reference.

The reflections caused by the incidence of light on its surface, which vary with the angle of vision and with the inclination of the sun, will provide specific differences in brightness, allowing movement references for those with reduced vision, without becoming a visual uneasiness. Even if you have similar hues in contiguous façades, *azulejo's* mutability is recognized as a reference, helping us when moving.

FIGURE 6 Value perception variations in glazed tile's surfaces, due to viewing position, and light change.

Given the plasticity of the ceramic body it is also possible to add textures and reliefs, mechanical or manual, to the surface of the tiles, giving them an haptic value, which enables its use as guidance systems, or even as signage for the visually impaired.

Human reliability on visual, tactile and proprioceptive information (Allen 1999), turns glazed tiles into a reference material in the environmental design process. Its structural, colour and brightness longevity allows an environmentally responsible use, without losing visual or tactile qualities during its lifetime.

Figure 7 Colour schemes, as well as tactile features can contribute to spatial readability, and as wayfinding

METHODOLOGY

In order to assess the relevance of *azulejo's* claddings in urban environments a cross methodology was designed, making possible to interconnect a theoretical approach with data brought from the author professional and academic experience and knowledge in glazed tiles production and use, and in colour studies on urban rehabilitation, with perceptual evaluations on the effect of the use of this material on urban spaces.

This methodology is based on a multidisciplinary bibliographic review in the areas of colour, light, perception, architecture and design, expert's interviews, and on a psychometric measurement of glazed tile's cladding ergonomic potentials.

Psychometric measurement proceedings:

1. Case study selection: selection based on different cladding examples and specific characteristics, both intrinsic and extrinsic.

2. Definition of perceptual variation parameters to be measured: distance, light conditions, viewing angles, surrounding colours.

3. Measurement: as colour perception depends on several variables, by measuring it we can assess the importance of several perceptual features.

3.1. Inherent colour directly from surface

3.2. Apparent colour measured with device

3.3. Photographic contextualization of the study stating the circumstances of each observation: pictures taken from the predefined viewing points, at the same moment of observation

3.4. Data registration and analysis: identification form developed for this research

4. Surveys

4.1.The Human Factor: evaluation of perceptual variations due to circumstances of observation in context (synchronic) compared with differences, in distance, light condition, viewing angles, and time (diachronic), and ergonomics abilities: balance, order, orientation, perception levels, comfort levels.

4.2. Photographic contextualization of the study stating the circumstances of each observation: pictures taken from the viewing points at the same moment of observation

4.3. Data registration and analysis

5. Results analysis

CONCLUSIONS

Colour, texture, size, and quality surface are fundamental to the understanding of spatial context. Ceramic claddings, as visual and physical aspects of architecture, interact with the environment, changing with light, inviting to the Human contact, actively participating in the construction of a vital and stimulating urban space.

In this study we demonstrated and underlined the important role of *azulejos* in the emotional and functional quality of urban spaces.

Azulejos visual and haptic qualities - color, texture and gloss – can be a reference in ergonomic features, functioning as an anchor point, supporting our visual memory construction, contributing to wayfinding and urban identity.

The sensory-perceptual and spatial features of this material, qualify it as an improving element in spatial legibility and readability, allowing us to characterize, organize, and hierarchize objects and spaces.

The adoption of a cross study methodology enabled a transversal approach to the subject, including different areas of knowledge, enriching the scope of the study.

The findings of this study aim to be a contribution to a sustainable and meaningful (re)innovation of glazed tile's claddings, in his design process – from thinking to production and application, as well as a opportunity to establish perception features as an important element in colour planning methodologies.

REFERENCES

Allen, G. (1999). Cognitive Maps, and Wayfinding: bases for individual differences in spatial cognition and behavior. In Golledge, R. (ed.). *Wayfinding behavior: cognitive mapping and other spatial processes*. The Johns Hopkins University Press, Baltimore.

Foster, N. (1976). On the Use of Colour in Buildings . In Porter, T. & Mikellides, B. (ed.). *Colour for Architecture*. Studio Vista, London.

Friedman, S., Thompson, S. (1976), Colour, Competence, and Cognition: Notes towards a Psychology of environmental Colour, In Porter, T. & Mikellides, B. (ed.) (1976), *Colour for Architecture*. Studio Vista, London.

Golledge, R. (ed.). (1999). *Wayfinding behavior: cognitive mapping and other spatial processes.* The John Hopkins University Press, Baltimore.

Humphrey, N. (1980). Natural Aesthetics. In Mikellides, Byron (ed.). *Architecture for People.* Studio Vista, London.

Lancaster, M. (1996). *Colourscape.* Academy Editions, London.

Lobo, C., Pernão, J. (2009), Glazed Tiles as an Improving Element for the Environmental Quality in Urban Landscape. *11th Congress of The International Colour Association - AIC Sidney2009 Proceedings.*

Lynch, K. (2002). *A Imagem da Cidade.* Edições 70, Lisboa.

Merwein, G., Rodeck, B. and Mahnke, F. (2007), *Color – Communication in Architectural Space.* Birkhauser, Basel.

Mollerup, P. (2005). *Wayshowing - A Guide to Environmental Signage. Principles & Practices.* Lars Muller Publishers, Baden.

Pallasmaa, J. (2005). *The eyes of the Skin – Architecture and the Senses.* John Willey & Sons Ltd.,Chichester, UK.

Porter, T. (1982), Architectural color, Whitney Library of Design, New York.

Swirnoff, L. (2000). *The Color of the Cities – An International Perspective.* McGraw Hill,New York.

Swirnoff, L. (2003). *Dimensional Color – Second Edition.* W. W. Norton & Company, New York.

Zumthor, P. (2006). *Atmosferas,* 1st Edition, Editorial Gustavo Gili, SL, Barcelona.

Chapter 24

Accessibility and Inclusion: The Case Study of the Faculty of Architecture of Lisbon

Fernando Moreira Da Silva

CIAUD – Research Centre in Architecture
Urban Planning and Design
TU Lisbon
Lisbon, 1349-055, PORTUGAL

ABSTRACT

In 2006, three faculties of the Technical University of Lisbon started a joint research project, trying to identify the most frequent tasks performed in learning activities at university level, school services and campus mobility; the groups of people with disabilities presenting limitations to perform those activities; a set of priorities and tools, guidelines and checklists contributing to solve the identified problems.

This project intended to obtain an evaluation and a methodology to promote qualitative change to those terms on the grounds of an analysis of the efficiency of current practices and the experimentation with alternative solutions for physical environment design and learning tools. The design of such a methodology at its dissemination among schools and universities would be a major contribution to reverse current situations, making university accessibility, and in consequence the attendance, more universal and inclusive. The aim was to reach out to minority groups with specific needs, improving their citizenship fulfilment.

This paper presents the Case Study of the Faculty of Architecture of the Technical University of Lisbon, one of the schools that were chosen for the project, focusing the accessibility for wheel chair users and people with impaired vision, enhancing the methodology that was used and the achieved results.

Keywords: Accessibility, Ergonomics, Inclusive Design, Disability

INTRODUCTION

During the last decade, concerning people with disabilities, accessibility was considered one of the most relevant aspects to achieve social participation and a better quality of life. Accessibility plays also an important role in inclusive schools and work integration. The accessibility achievement is related with services and urban space and with more and better functionality security and comfort.

Limitations faced by People with disabilities performing activities related with the use of school equipment, instruments, materials and limitations to move around usually leads to restrictions to their social participation and to their inclusion in school activities (Covington et al., 1997; Clarkson et al., 2003).

The study of social participation restriction imposed by social organization, beliefs and values and other environmental factors like interaction with services terminals and equipment has become progressively more important putting the focus on the external factors and how they contribute to produce barriers instead of facilitators to people with disabilities social participation taking into account their activity limitations. As environment factors have an important role on the performance level of those people, to study possible solutions to creating the least restrictive environment as possible is then a priority (Goldsmith, 1998; 2001; 2005).

In spite of having subscribed documents like the World Program of Action Concerning Disabled People (ONU 1981) and the Standard Rules on Equalization of Opportunities for Disabled People (ONU 1996), where is possible to find clear objectives to be achieved in the near future concerning accessibility, Portugal has experienced difficulties to implement them (Preiser et al., 2001).

Those difficulties are expressed in the proposal for the National Plan to Promote Accessibility. With the diagnosis made in those documents it is possible to observe that Portugal doesn't have enough data to characterize the majority of the situations presented.

PROJECT PRESENTATION

Currently society social participation is the main focus of studies on people with disabilities. Their abilities, performance levels, activity limitations and social participation restrictions are the most relevant aspects to be studied. Activity limitations may be defined as the gap between a person performance level and task requirements. The least restrictive environment possible is in general the most accessible one and has an important role on the performance activity level of those persons. The needs of an individual or of a group can be characterized by their ability to do, to decide and to adapt themselves to a certain task taking into account its accessibility (Imrie, 2001; Lidwell et al., 2003; Lehmann, 2004).

Accessibility studies need to consider the definition of user groups and their demographic estimates. Those definitions are usually profoundly influenced by impairment and functional abilities definitions, task requirements considered, and social expectations on the quality of services (Barker et al., 1995; Holmes-Siedle, 2003).

The first main problem is the identification of groups and their definition in terms of functional abilities and activities limitations. The application to the Portuguese population of the proposed new classification is still under study and as not yet been applied to learning, communication and movement activities, because evaluation methods and criteria remain to be defined.

On the other hand, new products are introduced into the market everyday. Such products promote new ways to fulfil a task without taken into account users with disabilities thus creating new accessibility problems causing social exclusion of those persons. This project tries to find solutions to fulfil the gap between mainstream and assistive technology and activities limitations imposed by what is generally considered the "normal way to perform a task" (Kohn, 1997). So, it was very important the identification of the more frequent tasks performed in learning activities at university level, school services and campus mobility at the Technical University of Lisbon; the definition of groups of people with disabilities presenting limitations to perform those activities; and the establishment of a set of priorities and tools, guidelines and checklists contributing to solve the problems identified.

During the project the group of researchers tried to develop a set of methodologies and tools based on modern education/(re)habilitation theories, information technology and inclusive design contributing to support the bio psychosocial disability model and to objective description of human functioning in activities related with moving around, communication and learning, taking into account the built environment and the use of information technology.

This project started in October 2005 having as partners three faculties of the Technical University of Lisbon, Portugal, and it was funded by FCT – The Foundation for The Science and Technology, of Portugal (POCI/AUR/61223/2004). Performed as external consultants, Professor Marcus Ormerod from the University of Salford, and Professor Klaus Miesenberger from the University of Linz.

The team of research was composed by Fernando Moreira Da Silva (Faculty of Architecture), Maria Leonor Moniz Pereira and Carla Espadinha (Faculty of Human Kinetics), and João Brisson Lopes (Technical Superior Institute). Several master students from the same identified schools took part in the research team.

This paper presents the Case Study of the Faculty of Architecture (FA).

OBJECTIVES

To collect standards, legislation and existing studies on accessibility in university campus; to identify the Faculty of Architecture as one of the places to develop the study at the university campus; to identify students with special needs depending on the existence of accessibility conditions to participate in academic life without restrictions; to identify the most frequent services and equipments used in academic and social tasks at FA; to define the network of routes more frequently used by students at FA; to define scenarios and evaluation methodologies related with those routes and equipments interaction regarding its accessibility and usability; to develop a software to help data compilation and treatment; to create a set of guidelines and to propose future work.

PROJECT DESCRIPTION

Between 2005 and 2008, the project tried to enlarge and consolidate a network with national and international partnerships, which had in common the subject of Inclusive Design, Accessibility and Social Inclusion in Higher Education.

Having this objective in mind, the project, through the Faculty of Human Kinetics, started to integrate a group of 29 entities which formed an Erasmus Mundus program (Action 4: Enhancing Attractiveness). This network was established during a former project HEAG (Higher Education Accessibility Guide) whose result may be consulted in http://www.european-agency.org/heag/. These candidates belong to 28 countries, being 21 Universities, 8 governmental supporting services for people with special needs and 1 enterprise of Internet services.

This was fundamental to develop the Case Study of the Faculty of Architecture.

For a better understanding of the project, we have divided the project's description by the different tasks performed.

TASKS

This project was composed by several tasks:
Literature Review based on Standards, Legal Documents and Existing Studies
Target Spaces and Group Identification:
Building and Validation of Instruments to Carry out the Studies:
Realisation of Field Studies:
Development of the informatics' tool: *Access Tek*

During the first task a data collection and literature review was implemented, focusing in standards, legal documents and published studies on the accessibility offered to disabled people to university physical spaces (buildings and other facilities) in view of disability type and disability degree (Ostroff, 1990; United

States A.T.B.C.B., 1991; Terry, 1997; Martins, 1999; Nicolle, & Abascal, 2001; Adler, 2002; Harber, et al., 2004).

The review of published studies was focused on obtaining the "state-of-the-art" of the university physical spaces that are used by disabled people and identifying the types of spaces which cause barriers. The review also assessed the methodologies used in such studies and drew conclusions on existing accessibility and/or desirable practices, examples of good practice and negative effects (e.g., course rejection on the grounds of no accessibility). The existence and efficiency of technical help were reviewed together with the impact of such helps in providing better accessibility (Mace, 1991; Tilley, 2005).

The second task main objective was the identification of the physical spaces (faculty of architecture buildings, accessibilities and attached facilities) that were to be the object of the studies that were carried out by the following tasks, including the identification of characteristics of the spaces, type and frequency of use and architectural barriers to their use by disabled people. The Faculty of Architecture was selected for the study, and this selection was founded in the diversity and type of structures, period of construction and complexity of problems; identification of the disabled population based on the disability type, degree and characteristics, including identification of the population that is able to use the existing spaces and the ones who are barred to use them; identification of the types, specifics and efficiency of help which are available in these physical spaces; selection and further characterization of the physical spaces that will be the subject of the following studies and of the population groups whose interaction with the selected spaces will be studied, including the definition of representative groups and sample populations.

This task began by the surveying of the physical spaces of the Faculty of Architecture (buildings and related facilities). The spaces characteristics were classified according to parameters such as use, its frequency and the existence or absence of architectural barriers or the existence of accessibility promoting facilities. The aim was to select the buildings and exterior spaces to be studied by this project. At the same time, the population of disabled people was identified according to characteristics such as type and degree of impairment, need, ease or impossibility to use some spaces, with the aim of identifying representative groups of the impaired population whose interaction with the selected spaces were studied. Such groups were set according to specific accessibility requirements.

The task also addressed the existence, type, frequency and efficiency of use of any existing technical help in the buildings and exterior spaces (including accessibility to FA). Then there was a sample selection of the physical spaces that were the subject of the present study. This way, the target spaces and user groups were selected.

The next step was the "Building and validation of instruments to carry out the studies". This task main objective was to select the methodologies that were applied to carry out the proposed studies by building and testing procedural guidelines, protocols and all other instruments needed to correctly carry out the above studies.

The first result of this task took the form of proposals for protocols, experimental instruments and guidelines for the fieldwork developed in the next task.

The project developed two methods of analysis (one of check-list type), obtaining data concerning the physical and communicational accessibility to a certain place, according to two different adopted criteria: the most recent Portuguese legislation about inclusive design and disabled people (163/2006) and the international references (UFAS and ADA) (Ostroff, 1990; United States A.T.B.C.B., 1991; Terry, 1997; Harber, et al., 2004). The second method is the SAP – Synthesis of Analyses of Paths (figure 1).

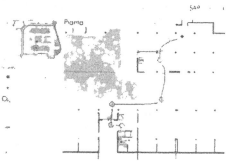

Figure 1. Example of SAP formulary (Synthesis of Analysis of Paths)

These elements were then validated and corrections were introduced. For this, a pre-testing took place. 16 students with disabilities took part in the project as important users of the spaces (fig 2).

Figure 2. Student in an electrical wheelchair, taking part in the project's research.

Pre-testing data results helped the evaluation of the experimental hypothesis, instruments and protocols and their correctness, mainly what concerns the physical barriers and the cognitive and communicational aspects. Corrections and amendments to the instruments and protocols were introduced as shown by evaluation and the new versions were then tested again. In this phase there was an independent review performed by the consultant Professor Marcus Ormerod. At the end of this task, the final version to implement the experiments, protocols and instruments were validated and available for the realisation of field studies.

The large-scale experiments according to the methodologies selected in the task "Building and validation of instruments to carry out the studies" took place and involved all representative user groups that were previously identified and were carried out in the selected physical spaces used by such groups of users or the spaces they are unable to use.

The layout of this task depended on the results of the previous one. Therefore, the realization of, e.g., surveys, interviews and field observations, their sequence and combination depended on the above results. A questionnaire was produced to answer the main questions of the research project, which are: to know and understand the main access and integration difficulties at the faculty of architecture by users with special needs, which can be of physical, communicational or social-economic order, among other. After each experiment, both users and team of research were debriefed.

The aim for this task was to collect the largest possible data on physical space usage as well as the quality and efficiency of any technical help that are available. It was expected to identify opportunities for accessibility and technical help improvement.

The raw data collected was subjected to preliminary processing, including statistical processing, to guarantee the significance and validity of the data itself.

There was a definition of scenarios and evaluation methodologies about accessibility of the paths and interaction with the services and equipment. Then it was very important the definition of the more used services and the creation of specific paths for usage.

After analyzing all the data collected, and in function of the priorities, the research group decided as a methodology to establish a circular preferential path for each building at the faculty of architecture (exterior and interior), allowing the derivation in small secondary paths, obtaining, this way, better results and better use of the spaces. To validate the option and the project's impact, a group of disabled people in wheelchairs were included in the study, which helped to widen the scope and to improve the accessibility methods and results that were taken, and to clarify how and why these were adopted in this access audit.

All the physical and communicational barriers were also recorded and Braille information was also produced.

At this moment the research group experienced a real need to create a new tool for the treatment of the enormous and complex data, collected and analyzed during the project working. Based on the analyses of the methods, drawings and questionnaires, the project developed a software program.

The informatics' tool (*Access Tek*) was developed by Gonçalo Semedo at Superior Technical Institute, as a Master Course project supervised by professor João Brisson Lopes, in a way to facilitate the manipulation of all the research data, which is normally in a large quantity and handled in paper; and to allow a more correct and global analysis and representation of the physical paths used by the students, teachers and staff in general in the faculty's buildings, the surrounding environment and, by example, the bus stops.

The software that was created, using the KML language, defined and used by Google Earth product, has a multimedia character because the information is mainly visual, (image and video), but also supporting schemes and sketches and sound (noisy environments, by example, able to promote disorientation).

This informatics' tool presents a very simple and intuitive way of working, functioning in a way similar to a geo-referencing system, supporting paths inside the buildings, based on technical drawings, providing information about passages, corridors and doors (fig 3).

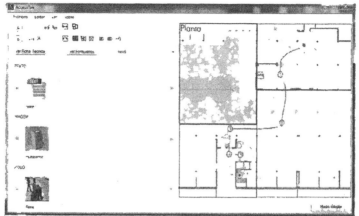

Figure 3. *Access Tek* tool representing the building's plan, sub-areas and related multimedia information.

The software also integrates an existent language for the association of the multimedia data with the different locals in the paths. This tool allows the easy addition, modification and removal of multimedia elements with the places in the paths, being a guarantee of the internal coherency of the data.

It is a descriptor of the spaces/paths/accessibility through a map, which tries to solve all main problems of the analyses put by the research group, being present the large and some times inappropriate inquiries. The idea of this software is after a map being able to have defined areas more restricted, about which there is more information about all the possible types (images, text, film and sound). It's a technical program which allows the creation/edition/removing of inquiries, maps, descriptions. It possesses classification of the locals and paths, identifying founded difficulties by different levels of solution (fig 4-5).

238

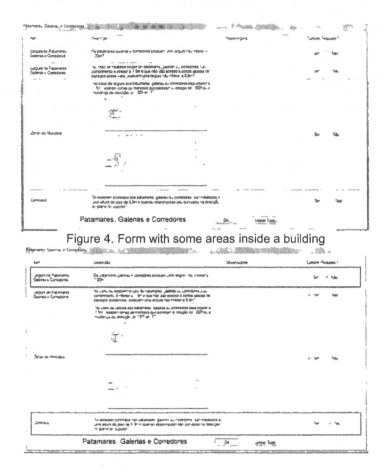

Figure 4. Form with some areas inside a building

Figure 5. The same form presented in fig 4, after being completed,
showing in red the not existent prerequisites.

CONCLUSIONS

The terms for attending Higher education, as for accessing information and knowledge, have until now been determined by models which are exclusively designed to satisfy the needs of the supposedly "normal" or "ideal" user, making it difficult and even excluding growing sectors of the population with university potential from university attendance.

This project, as all, intended to obtain evaluation and a specific methodology to promote qualitative change to those terms on the ground of an analysis of the efficiency of current practices and the experimentation with alternative solutions for physical environment design and learning tools, which we think we have achieved.

The case study of the Faculty of Architecture allowed experiencing a new approach, a new methodology, to study the accessibility and inclusion of disabled people which we believe may contribute to the reflection and discussion of this issue, and the implementation of similar studies in this and other research areas.

The group of researchers feels that the design of such a methodology and its dissemination among schools and universities will be a major contribution to reverse current situations, making university attendance more universal. The aim is to reach out to minority groups with specific needs, improving their citizenship fulfilment. This is important success factor for universities since the weight of such groups in university attendance is constantly increasing.

Although this project appears to be confined to the space of the Technical University of Lisbon, and in this paper specifically to the Faculty of Architecture as a Case Study, there is regional significance. The effects of the dissemination of this project and its results should in fact promote new ways to discuss and analyse the theme of this project and, particularly, to provide the tools that will allow the implementation of processes for countrywide inclusion: inclusive design and social inclusion.

REFERENCES

Adler, D. (2002). *Metric Handbook – Planning and Design Data*. London: Architectural Press, Second Edition.

Barker, P. et al. (1995). *Building Sight – A handbook of building and interior design solutions to include the needs of visually impaired people*. London: Royal National Institute for the Blinds, United Kingdom.

Clarkson, J. et al. (2003). *Inclusive Design: Design for the Whole Population*. London: Springer-Verlag.

Covington, G. et al. (1997). *Access by Design*. New York: Van Nostrand Reinhold.

Goldsmith, S. (1998). *Designing for the Disable – The new paradigm*. Butterworth-Heinemann.

Goldsmith, S. (2001). *Manual of Practical Guidance for Architects*. Butterworth-Heinemann.

Goldsmith, S. (2005). *Universal Design*. Oxford: Architectural Press

Harber, L. et al. (2004). *UFAS Retrofit Guide*. North Carolina: Wiley

Holmes-Siedle, James (2003). *Barrier-free Design*. Oxford: Architectural Press

Imrie, R. & Hall, P. (2001). *Inclusive Design: Designing and Developing Accessible Environments*. London: Spon Press

Kohn, J. (1997). *The Ergonomic Casebook*. US: Lewis

Lehmann, K. (2004). *Surviving Inclusion*. US: Scarecrow Education.

Lidwell, W. et al. (2003). *Universal Principles of Design*. Massachusetts: Rockport Publishers

Mace, R. (1991). *The Accessible Housing: Design File*. US: Wiley

Martins, J. V. (1999). *Normas Técnicas de Construção e Acessibilidade*. Lisbon: Edição de Autor.

Nicolle, C. & Abascal, J. (ed) (2001). *Inclusive Design Guidelines for HC*. London: Taylor & Francis

Ostroff, Elaine (1990). *UFAS Accessibility Checklist*, supervised by Ron Mace e Lucy Harber. USA: produced by Barrier Free Environments and Adaptive Environments Center for U.S. Architectural & Transportation Barriers Compliance Board.

Preiser, Wolfgang F.E. & Elaine Ostrof (2001). *Universal Design Handbook*. USA: McGraw-Hill nProfessional.

Terry, E. (ed) (1997). *Pocket Guide to the ADA*. Birmingham: Wiley

Tilley, A. (2005). *As medidas do Homem e da Mulher: fatores humanos em design*. São Paulo: Bookman

United States A.T.B.C.B. (1991). *ADA Accessibility Guidelines – Checklist for Buildings and Facilities*. USA: BNI.

Chapter 25

Sign Systems for Touristic Information

João Neves, Fernando Moreira da Silva

CIAUD – Research Centre in Architecture
Urban Planning and Design
TU Lisbon
Lisbon, 1349-055, PORTUGAL

ABSTRACT

The highest affluence of people to certain places such as airports, commercial areas, events, public services, tourism installations, etc., led to the need to guide these persons in unknown places and to communicate basic messages with a language understandable by everyone. This mobility brought road development associated with a growing flux of people that have to move from one side to another for different reasons. That movement has developed the need to learn new rules, which ones will be formalized through signs that make the access or the circulation to certain places easier.

This project addresses Touristic Information and aims to give a contribution to a more legible, understandable, inclusive and ergonomic approach in the research area, through the validation of the existing signage, mainly at Iberian Peninsula level, the design of new methodologies applied to the conception of symbols and to conceive a valuable tool for the conception of specific sign systems for tourism information.

Keywords: Sign Systems, Pictograms, Ergonomics, Touristic information, Wayfinding

INTRODUCTION

Signage is constituted by a multiplicity of signs that require a deep and systematic study in which, because of the quantity and diversity of symbols, their characteristics aren't always understood, sometimes causing disrespect and transference for signage. Vertical signage is constituted by several signs, classified in different categories according to their characteristics, meaning that they are constituted by signs or panels of signage transmitting a visual message, thanks to their localization, their form, colour and type, and also through symbols and alphanumeric characters (Diário da República Nr. 227 - October 1st. 1998).

Regardless of the spoken or written language, the need for a universal visual communication language is underlined as fundamental. Symbols' usage to represent touristic attractions is a decisive step in that direction.

The level of demand in information transmission through pictograms compels to conceive concise signs, simple, rapidly understandable; to achieve this we have to search for elementary graphic structures, to be easily perceived. Generally, conceptual models (having in mind the conception of pictograms) must present information in a simple and clear way, limiting possible ambiguities.

It happens that signage systems developed for information (in the tourism area or in traffic), are often empirically conceived and, most of the times, disrespecting the rules already contemplated by law, originating multiple, incoherent and illegible systems, thus making difficult their decoding.

For the present research project a concrete problem was identified: the sign systems for tourism information in Europe, and in the entire world, are totally different from one another, having no graphic relationship, being incoherent from the graphic point of view: there isn't a normalized system.

Strategically, there is a concrete answer in the communication design, to obtain research methodologies applied to the conception of symbols for tourism information. So, design harnesses innovation and differentiation, creating concrete answers to an identified problem. Design projects different objects or means of communication for human use, being a discipline or activity that is intimately related with the conception, planning and production of signage equipment.

The main aim of this research is to contribute for the reflection and knowledge in the signage design area (as a discipline of the communication design), trying to conceive a valuable tool for the conception of specific sign systems for tourism information, designing uniform sign systems which may help to evolve not only designers but multidisciplinary teams, communicating clear and unequivocal messages for the user.

USER, MOBILITY AND TERRITORY

The development of the railway networks, the advent of the automobile industry and the growth of the aerial fluxes, allied to a growing world scaled globalization, brought a greater individual social mobility coming from different regions and continents. Commerce, industry, leisure and other activities caused the abolition of borders, whether they are physical, linguistic or even cultural, in order to make the circulation of people and goods easier.

"Social mobility assumes the fluxes of individual groups, from different geographical proveniences and different socio-cultural characteristics, moving from one point to another based on very distinguished reasons. This social dynamic implicates the circumstantial basics, which means that the passage through determined places is sporadic as a result of a naturally itinerant activity. Therefore, it generates new situations, morphological and organizational unawareness of these places, and consequentially, it presumes a high level of intelligibility or indetermination, which raises dilemmas in the individuals' actuation necessities and even risks" (Costa, 1989).

Social mobility supposes a displacement from a place to other places in a certain territory. If accessibility is considered the access conditions destined to mobilely handicapped and special educational needed people, on the other side, accessibility is understood as the easiness in accessing or displacing between two points. Although both notions of accessibility are related, especially in the way reduced mobility or special educational needs are a conditioning aspect in accessing or displacing in a pre-determined space, it is considered in the present study that accessibility in the access or displacement on a territory may or may not be known or pre-determined.

The greater affluence of people to such places as airports, commercial areas, events, public services, etc., has defused the necessity of giving those people an orientation in an unknown place and to communicate basic messages through an understandable language. On another hand, that mobility brought along traffic developments associated to a growing flux of individuals that displace themselves from a point to another. That displacement often performed in unknown spaces, defused the need to learn new rules, which become normalized through signs that facilitate the access/circulation to/in determined places.

Signage and signalectics are constituted by multiple signs which require a profound and systematic study of a code in which, by the quantity not always are their characteristics apprehended, sometimes causing disrespect and alienation regarding the delivered message. Signage and signalectics are constituted by multiple levels, categorized in different categories regarding their characteristics, constituted by signs or panels which transmit a visual message, thanks to their location, shape, color, type and even through symbols and characters.

SIGN SYSTEMS

A system can be seen as a set of interrelated, interacting or independent elements forming, or considered to form, a collective entity (Heskett, 2005). The purpose of a system is to provide clear information on the consequences of choosing a route or a particular direction, but letting users decide exactly where to go.

In design areas, the collective quality manifests itself in several ways. Different elements can combine in functionally as in transport systems. A system requires principles, rules and procedures to ensure a smooth and orderly interaction in the interrelationship of ideas with shapes. This means having systematic thought qualities which infer in the methodical, logical and determined procedures.

Each signal offers very specific information, coded in a way that may be simultaneously linked with all the others. The growing importance of systems design, in contrast with the design focused on shapes, can be attributed to a globalization that affects the collective activity. Being the signal a physical object, with a self-image and to which was assigned a meaning, then we are before a sign.

Each one of us, in various situations, have encountered difficulties in accessing certain physical space, either by the ineffective signalectics, or by its improper use or even due to the illegibility of the graphism. Signalectics contribute effectively in the orientation of people and goods in a given territory. It is a discipline of the visual communication science who studies the functional relationships between signs of orientation in space and behavior of individuals (Costa, 1989). At the same time, it is the technique that organizes and regulates these relationships.

Each signal contributes to form the system, which means, signals have characteristics that differentiate them, forming the whole system which, nevertheless, requires the seizure of its own signification rules. The signal is an artifact with different meanings and unique features which makes it, different from the others and at the same time, related to the system.

Design as a project activity that involves creativity, proposes the adaptation of means to ends. Design projects create various media for human use, therefore being a subject or activity that is closely related to the conception, planning and production of equipment such as traffic signs. The graphism of signs, as the preferred means of information transmission needs other disciplines to contribute to the attainment of its objectives: to convey clear and unequivocal messages to the user, contributing to the improvement of accessibilities.

Sign systems for tourist information have developed slowly, looking to solve specific problems in each moment and relying mostly on international agreements and protocols. Despite the impossibility to adopt a universal unified sign system, there were moments where the ratified protocols were without a doubt an important impetus for example in the standardization of traffic signing.

The advent of the car became a lever that triggered the evolution of sign systems, creating greater social mobility and generating the apprehension of new rules through signs of orientation in space, which communicate and transmit information constituting a sign system - signalectics. The increased traffic movement has brought along the problem of the international regulation of signals, which began to be examined at European level since 1908. The global standardization of traffic signs was attempted in a United Nations conference in 1968, achieving only a partial match of the European and North American systems.

Traffic signaling systems (which include several times signs for tourist information) are not uniformed worldwide, existing different systems of signs which, by their distribution, lead us to consider now two fundamental systems with different shapes, colors and graphics. One of them is the European system, based on pictograms and ratified by several countries through the 1949 'Geneva Convention', implemented in most European countries, much of Africa (according to colonizing countries) and almost the whole of Asia. The other is the American system, based primarily on the use of spelling applied to squares or rectangles and based on the "Manual on Uniform Traffic Control Devices for Streets and Highways" of the United States of America, published on 1948. This system is currently used in Anglo-Saxon countries (territories of the Commonwealth) in the American continent, Australia and other countries of Oceania, fundamentally.

Signage consists of several signs, classified into different categories according to their characteristics, which means that, it is constituted by signals or panels that convey a visual message, thanks to its location, its shape, its color and its type and still through symbols. Signals are composed by several elements that contribute to the final appearance of the artifact.

Signage can be defined as a system composed of independent elements (which transmit certain information or an obligation to act) that are interrelated with the function of communicating messages. Therefore, signage is a system composed by interrelated elements (signs), simultaneously independent (by their classification) forming a collective entity - a sign system. Each artifact unit (sign) contributes to form a whole (the system), which means, the signals (objects built by man) are not individually designed, but having in mind the collective entity that unites them. The sign (the unit that belongs to a whole) is therefore a physical object with different meanings and unique characteristics which makes it different from the others, and at the same time related to the system. Being the signal a physical object, with a self-image and to which was assigned a meaning, then we are before a sign.

Iconicity includes several degrees of analogy and fidelity to the model, varying from hyperrealism to schematics or extreme abstraction. For signage as for signalectics, the maximum iconicity corresponds to pictograms (representing objects and people), and the minimum iconicity to what it is called "ideograms or non-figurative symbols" (Costa, 1989).

To convey messages, sign systems use pictograms, which are not more than simplified figurative signs that represent forms and objects in the environment. Pictographic system is a term introduced in this paper to define elements of an inter-related system, making use of figurative signs which represent things and objects of the environment (pictograms). Pictographic system is then a set of descriptive signalectic elements that interrelate to form a whole, involving the use of pictograms.

SIGNS

The sign is composed by its physical form and a mental concept associated with it, and this concept is, actually, an apprehension of external reality. The sign only relates to reality through concepts and people who use it.

In relation to the study of signs and taking into account the two main streams ('semiology' starred by the linguist Ferdinand de Saussure and the 'semiotics' of the philosopher CS Peirce), to this study the development evidences the three major areas of study covered in semiotics:

1. The sign itself. It is the study of different varieties of signs, the different ways in which these convey meaning, and the ways they relate to the people who use them.

2. Codes or systems in which signs are organized in. This study covers shapes developed by a variety of codes to meet the needs of a society or culture, or to explore the communication channels available for their transmission.

3. The culture within which these codes and signs are organized and, in turn, depends on the use of these codes and signs in relation to their existence and shape.

It appears crucial for this research, which aims to obtain a detailed study of sign systems for tourist information, a careful analysis of all aspects related with the three areas mentioned above: Sign and its meaning; The system or how the signs are organized; Culture or users to whom the signs are made for, and at this level the ergonomics is very important, in terms of achieving a total social inclusion.

SIMPLIFIED FIGURATIVE SIGNS: PICTOGRAMS

Pictograms are requested to transmit critical information to large numbers of people from a different language, having in common social and cultural traits, and to who are not supplied any teachings to decode these messages. This type of images (pictograms) are a good support in the orientation in public or private spaces and services.

Although pictograms appear to be absolutely self-explanatory and universal, in fact, they possess cultural limitations. Joan Costa (1998) defines pictogram as a figurative simplified sign representing things and objects in the environment. The term pictogram absorbs other variants of the iconic sign: ideogram and badge, despite their fundamental differences, because if the pictogram is an analogical image, the ideogram is an outline of an idea, a concept or a non-visual phenomenon and the emblem a highly institutionalized conventional figure. Pictogram was the generalized noun to refer to all of them.

A pictogram represents an object in a simplified way, which may be more or less iconic (more or less the same as the real model), but what matters most of all is that it is visible to the largest possible number of users. It also required a comprehensive understanding of the system to develop, and then conceive pictograms individually, consistently and contributing to the overall uniformity.

Any image that contributes to form a pictogram tends to take on the characteristics and convey the sense of the total category of objects belonging to the object in question (Massironi, 1983). This means that an image to be represented by a pictogram tends to regulate the design of other pictograms that are contained in the same category. The requirements for an information transmission through pictograms undertakes to create concise, simple, quick to understand signs; to achieve this, one has to look for elementary graphic structures, to do justice to a certain type of perception (Aicher, 1995). In general, the conceptual model (taking into account the design of pictograms) should present information in a more simple, clear and unambiguous as possible (Mijksenaar, 2001).

CONTRIBUTION AND PRELIMINARY RESEARCH CONCLUSIONS

The current research aims to develop the contribution to knowledge in the area of design, specifically in signalectics (as a discipline of communication design), trying to be a valuable tool for designing specific sign systems for information. The proposal and aim for this research project is to design uniformed sign systems that involve not only the designer but also multi-disciplinary teams that communicate clear and unambiguous message to the user.

This study contributes to a broader understanding of sign systems and the interrelationship of its components. The research will also demonstrate the importance of other disciplines and studies for the development of signalectics as a system of information transmission and pictographics as a sign system that conveys messages, among which ergonomics and inclusive design.

Design, as an eminently creative activity, requires the design of artifacts that respond to expressed needs in a projectual logic with the purpose of an industrial production, taking into account social, cultural, economic and environmental aspects. In this sense, design enhances innovation and differentiation, creating added value. Design works on several levels, such as industrial design, equipment design, graphic design, among others, contributing decisively to the improvement of the signage system, looking for new solutions in terms of safety, shape, graphism and materials used.

From the deep study of several sign systems and their constituents, the conclusion is that much of the legislation or manuals that regulate the production of signs is sufficient for its proper performance, noting however that such rules and standards are not always respected by the entities in the process, from conception to placement in space.

Signalectic systems developed for information (whether in tourism or traffic) are often designed empirically, disregarding the majority of the rules contemplated by law, resulting in multiple inconsistent, illegible systems, making it difficult to decode.

Even though there are definite rules for sorting and organization of systems, ignoring the hierarchies and imposed rules is abundant, causing a detachment from the imposed rule, embarrassing users and sometimes causing danger. Also, it was verified that signs' dimensions is not always respected, even if significant improvements are experienced.

Regarding the use of colour, the legislation is largely followed and well applied. However, that legislation is yet to be standardized, because in the classification of signs and taking into account the same class of signs, different colours are applied, which implies a memory and understanding effort by the user which could be reduced.

The evolution of printing in signs denotes a quantum leap in terms of readability, although it is notoriously difficult to follow the legislation by those who design signs, not respecting sizes, the font in use, spacing, etc. The arrows used on signage and plaques are very different, which could and should be standardized in order to simplify their shape and improve their usefulness, verifying that in determined signage, reading and comprehension is deficient.

The applied pictograms lack of uniform criteria, both in terms of shape and colour, because not all pictograms are easily perceptible. The iconic level used in the design of pictograms is very uneven, with some being extremely simplified and easily perceptible and other requiring a higher level of decoding by the user, caused by the complexity of the sign.

In order to improve sign systems for public or tourist information, it would be important that all the involved agents understand the problem that the system is directly related to the security of society, which is why there should be more governmental action with all entities, regarding levels of awareness, training and even supervision of the same, in order to further the continuous improvement of sign systems in place and those which will be developed from now on.

REFERENCES

AIGA, American Institute of Graphic Arts (May 8, 2008) Symbol Signs 1979. AIGA Website: http://www.aiga.com

AICHER, Otl e Krampen – Sistemas de signos en la comunicación visual. 4.ª ed. México: Gustavo Gili, 1995.

BESSA, José Pedro Barbosa de – Representações do masculino e do feminino na sinalética. Aveiro: Departamento de Comunicação e Artes da Universidade de Aveiro, 2005. (356 p.). Tese de doutoramento.

Capítulo I - Artigo 2.º do Regulamento de Sinalização do Trânsito aprovado pelo Artigo 1.º do Decreto Regulamentar n.º 22-A/98 de 1 de Outubro, com as alterações impostas pelo Decreto Regulamentar n.º 41/2002 de 20 de Agosto.

Capítulo I, Artigo 2.º do Regulamento de Sinalização do Trânsito aprovado pelo Artigo 1.º do Decreto Regulamentar n.º 22-A/98 de 1 de Outubro, com as alterações impostas pelo Decreto Regulamentar n.º 41/2002 de 20 de Agosto.

COSTA, Joan – La esquemática: Visualizar la información. 1.ª ed. Barcelona: Paidós, 1998.

COSTA, Joan – Señalética. 2.ª ed. Barcelona: Ediciones CEAC, 1989.

Diário da República - I Scrics-B Nr. 227 - October 1st. (1998), Chapter I - Article 2. Regulamento de Sinalização do Trânsito.

MASSIRONI, Manfredo – Ver pelo desenho: aspectos técnicos, cognitivos, comunicativos. 1.ª ed. Lisboa: Edições 70, 1983.

MIJKSENAAR, Paul – Diseño de la información. 1.ª ed. Mexico: Gustavo Gili, 2001. ISBN 968-887-389-6

Chapter 26

The Importance of Sustainability in the Communication Design Practice

Maria Cadarso, Fernando Moreira da Silva

CIAUD – Research Center in Architecture, Urban Planning and Design
Faculty of Architecture, TU Lisbon

ABSTRACT

Until recently, "eco" and "sustainable," have been subjects mainly for the industrial designer to research. However, gradually Graphic and Communication Designers, especially since the "First Things First Manifesto," have become increasing aware of their need to raise a conscience in fundamental issues like ethics, environmental and social responsibility. Paper is a sensitive issue nowadays, not so much because of the number of trees that are cut down (they are a renewable source), but more to do with the transformation process of wood into paper. For Communication Designers paper is one of the main supports, but they use many other that also need to be researched. However this environmental concerns should not be seen just an altruistic duty from designers, they much rather are a recent but increasing demand from companies. To give an example "Sustainable Reports" have increased 100% in one year alone, comparing 2002 with 2003. For designers the knowledge about sustainability can be an empowering situation. AIGA points out the need for designers to have a proactive attitude, showing to clients that they are informed, and able to advise then. What gives designer more responsibility, but also new market opportunities, and more control over is work.

Keywords: communication design; cognitive ergonomics; design principles; sustainability

INTRODUCTION

This paper is part of an ongoing PhD research, which is expected to bring relevant contribution in defining Sustainable Communication Design principles and code of practice. It starts with a general overview about what is Sustainable Development, and where the concept began. Next, we acknowledge the most relevant contributions from Industrial Designers to "eco" and "sustainable design". And finally we analyse the most significant aspects from Communication Design evolution. Equally important, are the efforts from some institutions, designers, and writers, among others, already engaged in working towards sustainability. However, the PhD research is wider and could not be completed without analysing the markets and the interaction from fields such as cognitive ergonomics as a way to uncover more respectful forms of connecting with the consumer / individual, to whom Communication Designers mostly work for, but it will not be the focus of this paper.

We live in a world that faces unprecedented social and environmental challenges. The differences between north and south countries are still far from being overcome. Even in industrialized countries social differences are a growing problem. This is a problem that concerns us all, and individuals, as citizens and as professionals. How Communication Designers can contribute, to proactively work towards a common solution, is what this PhD research is set to uncover.

In 1983, under the chairmanship of Mrs Brundtland, The World Commission on Environmental and Development (WCED) was created. This commission task was about "identifying and promoting the cause of sustainable development" (O'Riordan 2000) and the first definition of sustainable development was: "For development to be sustainable, it must take account of social and ecological factors, as well as economic ones; of the living and non-living resource base; and of the long-term as well as the short term advantages and disadvantages of alternative action." (WCED 1983 cited in O'Riordan 2000)

Many important conferences and documents came later on. Nowadays, it is estimated that there are 64 sustainable development definitions and many interpretations of the term "sustainability", and the number will continue to grow as the global debate on the topic widens. For some, it means maintaining the status quo. For others it is equated with notions of responsibility, conservation and stewardship. However for a growing number of people, sustainable development is a "triple bottom line" activity, based in economic, social and environmental impacts. (AIGA – American Institute of Graphic Arts 2003). For the PhD research, and therefore for this paper, we are going to use the concept of "Sustainable Development" as the triple bottom down balance between economical, social and environmental issues

STEPS TOWARDS A DESIGN FOR SUSTAINABILITY

Already in 1971 Papanek, in his book "Design for the real world", challenged designers to act upon a social responsibility. He wrote that designers could propose from simple solutions, to products, or services to be used by the community and the society (Papanek 1985). In 1971 the world faced the first energetic crisis, and in 1974 the petroleum barrel was costing more then ever before. Since then, there has been a rising environmental awareness. It was facing the need to produce eco-solution that eco design first emerged. In the early 90s TuDelf University and Philips designers' created a method for Life-Cycle Assessment (LCA), which is the evaluation of the environmental impact cost product: during production, life and disposal, in what concerns resources, and energy (Faud-luke 2002).

Taking into account that we need to consume fewer resources the model "Factor 4" was first proposed in 1995 by Weizsacker, Lovins, H, Lovins, A (cited in Wuppertal Institute 2009), as a model that would hold the key to sustainability. However "Factor 4", was soon proved not to be sufficient and the Factor 10 was recommended, meaning we need to reduce by 90% the use of natural resources, in a global scale by 2050 (Schmidt-Bleek 2000 cited in Factor 10 Institute 2009).

Facing the need to reduce the resources consumption, analysing product's cycle and environmental impact just was not enough. Han Brezet (1997) proposed a four-step model, that represented a paradigm shifting that initiated the transition from "eco" to "sustainable" design. The four steps are: product improvement; product redesign; function innovation; and system innovation. Also Ursula Tischner (2005) provided a great contribution by asking designers to consider social and economic issues, while seeking a greater balance between countries, people, and wealth.

Ezio Manzini and Carlo Vezzoli (2002) took Design for Sustainability a step further. They proposed an interconnected system of services instead of physical objects for a sustainable quotidian. Also, Ezio Manzini with François Jégou (2003), have been researching in sustainable lifestyle options. The concept is based on communities that engage themselves in finding sustainable solutions for their daily problems, like taking care of children, older people, cooking for the community, or lift sharing. By proposing this, Ezio Manzini, wishes to empower people to find the necessary solutions by using the available resources at hand (Manzini, Collina, Evans 2004). When Manzini is proposing shared products or services as an alternative, he is obviously looking for less production and less consumption of resources, but he also brings design to a new levels: service, social, community design, just a few to mention. And those are a very interesting new area for designers to research and work.

COMMUNICATION & GRAPHIC DESIGNERS TOWARDS SUSTAINABILITY

Parallel actions were emerging among graphic and communication designers. Already in 1964, twenty-two Graphic and Communication designers signed the first version on the "First Things First Manifesto". The second, and latest version led by Max Bruinsma, in 2000, essentially claimed that Graphic Designers should be able to work independently and regardless the marketing, and the advertising, to pursuit more valuable causes.

Although the "First Things First Manifesto" was a call for action, involvement and engagement to graphic and communication designers, it had less impact then the desired one. Apart from some individual designers and initiatives with a more acute perception of the state of the world, it is fairly reasonable to say that most communication designers (including graphic and visual) feel that they work for the client, under their directives and briefing, as a response of the market.

In order to establish a common ground, for the PhD research and therefore in the paper, we use the term "Communication Designer" as the discipline that communicates visually but also in more abstract ways. That is capable of sending the required message, in an explicit or implicit form (for example, an event, an experience, a perceived feeling). And when we refer to "Communication designers", in the context of this research, it includes "graphic designers", and "visual designers" as well. As well, "Sustainable Communication Design", is used as the discipline that communicates in visual but also in more abstract ways. That is capable of sending the required message, in an explicit or implicit form; that consciously, and within its possibilities, reduces adverse environmental impact and takes in consideration the social aspects.

In matters of Sustainable Graphic Design, one of the most relevant contributions is being given by American Institute of Graphic Arts (AIGA 2003) with their Sustainable Centre, where designers can find information on resources, technical advise and information on materials, such as paper. Using data from the United States, we can understand the relevance of Communication Designer working towards Sustainability: "Americans receive over 65 billion pieces of unsolicited mail each year, equal to 230 appeals, catalogues and advertisements for every person in the country. According to the not-for-profit organization Environmental Defence, 17 billion catalogues were produced in 2001 using mostly 100 percent virgin fibre paper. That is 64 catalogues for every person in America" (AIGA 2003).

Paper is a sensitive issue nowadays, not so much because of the number of trees that are cut down (they are a renewable source), but more to do with the transformation process of wood into paper (AIGA 2003). For Communication Designers paper is one support, but they use many other that also need to be researched.

However this environmental concern should not be seen just an altruistic duty from designers, they much rather are a new but increasing demand from companies.

To give an example "Sustainable Reports" have increased 100% in one year alone, comparing 2002 with 2003 (AIGA 2003). For designers the knowledge about sustainability can be an empowering situation: "This increased attention to environmental responsibility can be an opportunity for designers to be seen as critical advisors to corporations on how to reduce their negative impacts without compromising the imperative for product differentiation and promotion through design and printing" (AIGA 2003).

What AIGA is pointing out, is the need for designers to have a proactive attitude, showing to clients that they are informed, and able to advise then. What gives designers, for one hand more responsibility and more control over his work, and on the other side there are new market opportunities.

The Designer Accord, represent an important step. For designers to adopt the accord they must undertake some steps, which are quite open and easy to follow, so they are intended to be a kind of motivation for beginners. On the other hand, if these guidelines were too restrictive they would be felt as discouraging. It is also reasonable to expect that most of the 170 000 members are probably just using the accord as a way to promote their image, or as a light commitment.

Specifically committed to Sustainable Communication Design, is the Society of Graphic Designer of Canada (GDR 2009), that in April 2009, during the annual general meeting, proposed the first definition for Sustainable Communication Design: "Sustainable communication design is the application of sustainability principles to communication design practice. Practitioners consider the full life cycle of products and services, and commit to strategies, processes and materials that value environmental, cultural, social and economic responsibility". This definition came with a statement of values and principles to guide the GDC's members during their design practice (GDR 2009). The statement has three parts; the first one is assuming responsibility in this interconnect world. The second is about the in-house changes that can be done. And the third part is a set of guidelines for the design practice and client advising.

Brian Dougherty is the book "Green Graphic Design" author, and also an experienced Communication Designer, partner in is studio "Celery Design Collaborative". Based on his eleven years of experience, he says, "the message designers make, the brand we built, and the causes we promote can have impacts far beyond the paper we print on", and adds "In addition to seeking our better material and manufacturing techniques, designers can craft and deliver messages that have a positive impact on the world" (Dougherty 2008).

Although the presented cases are relevant they demonstrate that the paradigm is still the same as conventional communication design and advertising. They think how to use eco supports, or less pollution printing, or going carbon free, however there is much more to be done.

THE RESEARCH LAYOUT

Research objective and reasons

The PhD research aims to define Sustainable Communication Design as a way to add value to Communication through the regular practice of Communication Design, and using the triple bottom down (economy, social and environmental) principles of sustainability.

Reason why this research is being undertaken: first, when we study to the Designer Accord that has been adopted by 170 000 designer (or agencies) it is clear that there is a growing interest among practitioners. Secondly, as it was presented earlier in this paper, more then one institution has provided guidelines, but none are consensual or even connected. Third, the information available not always is accurate or reliable. Fourth, there is a growing interest in the consumer and market for sustainable options. Finally, it can be an opportunity to empower designers, to adopt other practices, and to explain then to clients.

Research questions

To guide through the hypothesis validation, four research questions should be bring relevant contributions:
- How can a code of practice with ethic and values, be define?
- How can layers of "sustainability" be inserted during project methodology?
- How can Communication Designer help consumers overcoming sustainable illiteracy?
- How to add value to communication through sustainability?

Research strategy

With the research process evolution other research methods may be needed, but for the moment, these are the ones being taken in consideration:
- Data collecting and literature review;
- Survey by interview with a panel of experts

Research techniques

Data collection
Apart from the information collecting mentioned before, other forms of survey or data collecting are not been foreseen, for now. If later, it proves to be necessary, the research will adopt the necessary techniques.

Delphi methodology

This research aims for a consensus in defining Sustainable Communication Design, as a concept and as a way to improve the design practice. For this reason a focus group of experts agreeing in same sort of definition, or guidelines, or procedures would be most relevant for Sustainable Communication Design to reach full acceptance and maturity. Within this perspective, and as it is foreseen now, from all the research techniques, Delphi Methodology seems to be the more adequate, in this case.

In their book "The Delphi Method: techniques and applications" Linstone and Turrof (2002) state that "Delphi may be characterized as a method for structuring a group communication process, so that the process is effective in allowing a group of individual, as a whole, to deal with a complex problem".

The way Delphi works is: a monitor (or a team) carefully selects a group of people and provides then with an initial questionnaire, that should completed alone, without the others participants, neither knowledge or influence. When all the participants have replied, the monitor (or the team) looks for a consensus in their replies. In the end of this phase the participants are allowed to review their replays. Completed the first round, they go for the second one, and follow the same procedure. The rounds are repeated as many as necessary to reach a general consensus (Linstone and Turrof 2002).

Delphi methodology has two basic models; one is the "conventional" Delphi, where the questionnaires, the replays and the consensus, are done by written hand; the other one is "Real-Time Delphi" that explores the computer technology, as a way to facilitate the process and make it quicker. (Linstone and Turrof 2002).

For now we are considering Real-Time Delphi, the possibilities provided by the computer, can allow, speed, real time discussion between persons from different countries; but also structure and systematization when collecting information, and while looking for consensus. For this research a site is being prepared www.sustainablecommunicationdesign.net and that will host a platform with restrict access, specially design to work with the Delphi Method (see ahead in tools).

Approach analysis

The data analysis will be prominently qualitative, because as it can be read in the objectives, this research deals more with concepts, definitions and guidelines. In what concerns, materials, techniques and tools, the goal is not to experiment, but to take advantage from the existing data.

Research tools

Site | www.sustainablecommunicationdesign.net (under construction)

For this research, the investigator is expected to create a website to share the contents and evolution of the research with the community in general (academia, other designers, and society in general. The site will be specially designed as an investigation platform to host the Delphi Method.

Blog | http://sustainablecommunicationdesign.blogspot.com/
The existing blog is soon expected to be transfer into the site, to improve discussion between the researcher and the community in general (academia, other designers, and society in general).

CONCLUSION

Looking in perspective the growing need to have a sustainable pattern is evident. Associations such as AIGA, Design Accord, or Society of Graphic Designer of Canada, just to mentions the most relevant ones, demonstrate the relevance to research in this area. However, the lack of interconnected information between these associations, or the inability to speak as one, brings even more pertinence to the proposed PhD research. As the research, first outcome, it is evident the need to collect the emerging data related to Sustainable Communication Design, in order to validate it, and to be used as common ground data for all designers. Through the methodology here presented, we expect to be able to understand how sustainability can be applied to de Communication Design practice, and how it can be used to add value to the end result. This means breaking down the intervenient parts, the stakeholders, but above looking for a new paradigm shift in how Communication can be defined and applied. That also means to find a balance between social, nature and economic aspects, and how it reflects in the designer's studio, project, the client and the final consumer. Also, in looking for a more respectful ways for communication, not just environmental or socially speaking, but also in terms of content, we try to establish a different relation with the final consumer or end user. For that, we expect to interact with fields, such as ergonomic cognitive, as a way to relate in a more adequate manner to the individual. In conclusion the positive aspects are, the awareness, the rising interest, and definitely the first attempts to define Sustainable Communication Design, because they set ground for this research to evolve in a structured way, and hopefully be able to contribute to a better design practice.

REFERENCES

AIGA Centre for Sustainable Design (2008), *Our Mission*, accessed in 30 August 2009, <http://sustainability.aiga.org/sus_mission

AIGA (2003), Print Design and Environmental Responsibility, *The Woodside Institute for Sustainable Communication*, New York, accessed in 12 June 2006,<http://sustainability.aiga.org/sus_paper?searchtext=print%20design%20

258

and%20environmental%20resp&curl=1>

Brezet, H. (1997), Dynamics in ecodesign practice'. *UNEP Industry and Environment*, 20(1-2)

Bruinsma, M. (2000), *First Things First 2000 a design manifesto*, accessed in 30 August 2008, <http://www.xs4all.nl/~maxb/ftf2000.htm>

Carter, N. (2007), *The politics of the Environment*, Cambridge University Press, Cambridge

Creswell, J. (2003), *Research Design: qualitative, quantitative and mixed methods approaches*, 2nd ed., Sage Publications, California

Designer Accord (2009), *Mission*, accessed in 30 August 2009, <http://www.designersaccord.org/mission/>

Design for Sustainability (1997), *Part I What and Why D4S – Introduction*, accessed in 30 August 2009, <http://www.d4s-de.org/>

Dougherty, B. (2008), *Green Graphic Design*, Allworth Press, New York

Factor 10 Institute (2009), *The factor 10 Institute*, accessed in 12 December 2008, <http://www.factor10-institute.org/about.html#factor10_institute>

Fuad-luke, A. (2002), *The eco-design handbook*, Thames & Hudson, London

Hickman, L. (2008), *A Good Life: the guide to ethical living*, Transworld Publishers, London

Jackson, T. (2005), Live Better by Consuming Less? Is There a "Double Dividend" in Sustainable Consumption?, *Journal of Industrial Ecology*, 9(1-2), **pp.**19-36.

Linstone, H. and Turoff, M. (2002), *The Delphi Method Techniques and Applications*, accessed via the authors' site in 28 August 2009 <http://is.njit.edu/pubs/delphibook/>

Manzini, E. (2006), *Design, ethics and sustainability: guidelines for a transition phase*, Politecnico di Milano, Milano

Manzini, E. and Vezzoli C. (2002), *Product-Service Systems and Sustainability: opportunities for Sustainable Solutions*, United Nations Environmental Programme, Milano

Manzini, E. and Jégou, F. (2003), *Quotidiano Sostenibile - Scenari di Vita Urbana*, Edizione Ambiente, Milan

Manzini, E., Collina and Evans, S. (2004), *Solution Oriented Partnership - How to design industrialised sustainable solutions*, Cranfield University, Bedfordshire

Mcdonough, W. and Braungart, M. (2002), *Cradle to cradle*, North Point Press, New York

O'Riordan, T. (2000), *Environmental Science for Environmental Management*, Pearson Education Limited, Essex

O'Rourke, D. (2005), Market Movements: nongovernmental organization strategies to Influence global production and consumption. *Journal of Industrial Ecology*, 9(1-2), pp. 115-128

Ottman, J. (1999), *Green Marketing: Opportunity for Innovation*, 2nd Edition, Greenleaf Publishing Limiteted, Sheffield

Papanek, V. (1985), *Design for the real world*, Thames & Hudson Ltd, London

Robson, C. (2002), *Real World Research*, 2nd ed., Blackwell Publishing, Oxford

Schelling, T. C. (2006), *Micromotives and macrobehavior*, W. W. Norton &

Company, Inc, New York

Sherin, A. (2008) *SustainAble a handbook of material and applications for graphic designers and their clients*, Rockport Publishers, Inc, Massachusetts

Steffen, A, ed. (2008), *World changing. A user's guide for 21st century*, Abrams, New York

Society For Environmental Graphic Design (2009), *About us*, accessed in 30 August 2009, <http://www.segd.org/#/about-us/index.html>

SustainCommWorld (2009), *About us*, accessed in 30 August 2009, http://www.sustaincommworld.com/general/about_us.asp>

US Environmental Protection Agency (2008), *Life-Cycle Assessment (LCA)*, accessed 15 August 2009, <http://www.epa.gov/nrmrl/lcaccess/>

Tishener, U (2005), *Eco-Design – Methods and Tolls*, Econcept, accessed in 06 February 2005<www.econcept.org/data/tools.pdf>

The Global Development Research Centre (2009), *Sustainability Concepts: Factor 10*, accessed in 6 September 2009,< http://www.gdrc.org/sustdev/concepts/11-fl0.html>

The Institute for Sustainable Communication (2009), *Our Mission, Vision and Values*, <http://www.sustainablecommunication.org/home/mission-and-values>

The Society of Graphic Designers of Canada (2009), *World Graphic Design Day 2009: GDC Adopters Sustainability Principles*, accessed in 30 August 2009, <http://www.gdc.net/designers/index/articles659.php>

Worldwatch Institute (2008), *State of the World 2008: innovations for a Sustainable Economy*, Worldwatch Institute, Washington

Wuppertal Institute (2009), *Factor Four*, accessed in 12 August 2009 <http://www.wupperinst.org/FactorFour/index.html>

Chapter 27

Developing a Sustainable System for Attending the Basic Needs of the Brazilian Population at the Bottom of the Pyramid (BOP)

Lia Buarque de Macedo Guimarães

Graduate Program in Industrial Engineering
Federal University of Rio Grande do Sul.
Porto Alegre, RS 90.035-190, Brazil

ABSTRACT

This paper presents a project based on the Sociotechnical Design method (Guimarães, 2010) for product and process design, which follows the Cradle to Cradle (C2C) approach to system design (McDonough and Braungart, 2002) the ZERI (Zero Emissions Research & Initiatives) methodology for sustainable chain development (Pauli, 1998) and is in accordance to the sustainability concept of the Triple Bottom Line or TBL (the interaction among people, planet and profit) proposed by Elkington (1999). Sociotechnical Design considers the basic needs of the population at the bottom of the pyramid (BOP) and the residual materials available in the regions in order to transform them into high value products that attend the identified basic needs and was used for transforming a poor Brazilian town into a sustainable one. The project is being prototyped in the town of Palmares do Sul, RS, to make possible the social, environmental and economic analysis for further improvements.

Keywords: Basic Needs, BOP Population, Product and Process Development, Sustainability

INTRODUCTION

Brazil is an outstanding country: the richer country in the world for its biodiversity and one of the poorest in terms of social disparity. 29% of the Brazilian population, according to the latest census of 2000 (IBGE, 2002), earns R$ 80 a month or U$ 44, noting that at that time, a minimum wage was R$ 151 or U$ 84 (Fundação Getúlio Vargas, 2001) in jobs requiring a very low level of school education. The same census showed that 63.2% of the Brazilian population had less than eight years of schooling (an incomplete first grade level of education); only 16.3% have second grade, and 6.4% a university level.

According to Barros et al. (2007) based on the Gini coefficient, only 10 countries in the world (seven in Africa and three in South America) are worse than Brazil in terms of income concentration. When we add to this a high level of unemployment (10% on average), deficiencies in social protection and labor laws, we get an economic and social situation that facilitates the adoption of deficient product and process development along with deficient jobs and explains the overall low quality of life of most Brazilians today.

This underdeveloped condition is only explained by the political retarding of the country, that does not show much improvement since Brazil was discovered in 1500: the country was built to deliver natural resources for the colonizers and industrialization was never an issue for the country's development because this would have prepared the country to compete with the more advanced ones. This situation did not change even after independence in 1822. Furtado (1959) pointed out that the many limitations imposed on the installation of a strong industrial system in Brazil had a deep negative influence on the establishment of its economic structure, defining an indirect development dependent on the industrialized countries. As stated by Jaguaribe (2006), even nowadays the Brazilian dependence on external capital and technology is so intense as to leave no room for native technology to develop. Because the majority of the population has low level of education, the situation does not change.

In this context, looking for increasing the opportunity of new jobs and a better quality of life mainly for the BOP Brazilian population, a project, financially supported by the National Research Board (CNPq), is under development, based on the Cradle to Cradle (C2C) approach for system design, the ZERI (Zero Emissions Research & Initiatives) methodology proposed by Pauli (1998) for sustainable chain development, under the major concept of the Triple Bottom Line (based on the interaction among people, planet and profit) proposed by Elkington (1999) what should be the essence of sustainability.

THE IDEA BEHIND THE PROJECT WITH ITS SIX PRODUCTIVE CHAINS

The idea behind the project, which is being prototyped in the town of Palmares do Sul –RS, is the use of raw materials which are abundant in the surrounding regions, and discarded (either for their low commercial value or because ways of reusing such materials were not sought). They will be processed in low/medium technology factories, which will use recycled oil and rice shell/casing as fuel, with low environmental impact. The processes were designed for reducing, reusing and recycling materials and energy, what is inherent to the C2C policy (McDonough and Braungart, 2002). C2C takes into account the process of production as a whole, from the selection and extraction of components, to the processing, production, utilization and use. At the final stage of their life cycle, the products will be renovated either by re-processing in the technological or biological metabolism.

For the project to reach its end, six productive chains were planned for the usage of residual raw material from the surrounding regions, according to the Zeri methodology for chain development which main goal is: no liquid, no gaseous and no solid residuals; all of the inputs are utilized during the production; in case of residual waste, it is used by other industries. The criteria of the selection in the Zeri methodology (Pauli, 1998) are: assess the potential to the aggregated value; establish the necessity for energy; determine the investments of capital; revise the necessities of physical space; calculate the opportunities for the creation of new jobs.

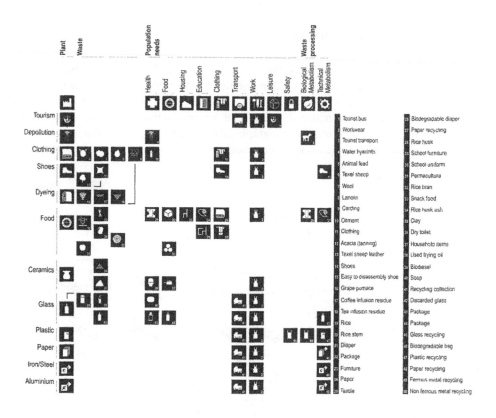

Figure 1. First step of the Sociotechnical Design method: crossing the basic needs and available materials in the matrix result in the product and process opportunities that appear in the cells

The general method for product and process development used in the project was the Sociotechnical Design (Guimarães, 2010) which develops in three steps: 1) the identification of demands and residuals by the use of a matrix (as in Figure 1) that crosses the basic needs of the population (health, food, housing, education, clothing, transport, work, leisure, safety) with the residue materials available in the region. Solution opportunities (i.e. the fulfillment of the matrix cells) like foodstuffs, clothes, shoes and health products were obtained mainly by brainstorming; 2) identification of the qualities of the products and processes according to the potential users, using a participative design tool: the Macroergonomics Design proposed by Fogliatto and Guimarães (1999). The final users who will get prime focus in this specific project will be the students from the Brazilian public schools (about 48 million people) and daycares; 3) evaluation of the cost and benefits of the solutions, by using a matrix that considers the sustainability, quality and cost of each proposal, what is consistent with the Triple Bottom Line (Elkington, 1999).

THE PRODUCTIVE CHAINS

There are six productive chains: the rice (food) chain, the rice case/celulose/silicon chain, the ceramics chain, the clothing chain, the footwear chain, products from the selective collect chain: glass, plastic, oil, mate and coffee leftover, paper, iron, aluminum and organic material waste. The six productive chains are structured as shown in Figure 2 and may be geographically defined, regarding the origins of the raw material and the regions with the possibility for processing.

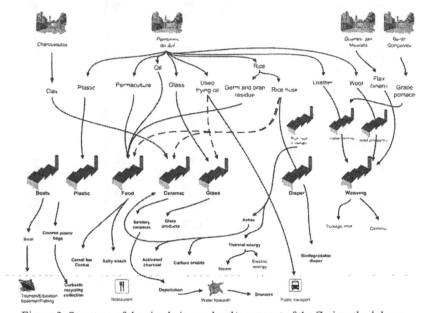

Figure 2. Structure of the six chains under the concept of the Zeri methodology

The objective of the rice chain is to formulate food products, which are nutritionally balanced and sensorial adequate, using the bran of rice, a sub product of the rice producing industry, and other sub products, The products (such as cakes and cereal bars made with rice and vegetables from permaculture, molasses and "dulce de leche") have already been prototyped and analyzed for its physical-chemical, nutritional and sensorial composition in 2008, by the Institute for Science and Technology of the Federal University of Rio Grande do Sul. Fruits and vegetables to be added in the food produced in the factory will come from permaculture, which is being developed in Palmares do Sul to generate more jobs, although it is also possible to establish a partnership with small producers for the collection of fruits and vegetables and also molasses and "dulce de leche", abundant in the region, they cannot sell.

The idea behind the rice casing chain this is the eco-innovation in the obtainment of cellulose, silicon and coal by using rice casings. The obtainment of either cellulose or silicon/coal will depend on the supplier for the chain, for there are two possibilities for the processing of rice in the region: plain or parbolized. The parbolized rice producers burn the casings to obtain energy during the process itself since this type of rice is heated to almost 60°C. resulting in the residue of ashes. The silicon obtained from the ashes is in a process of characterization but the asses themselves are suitable for ceramics. The plain rice producers do not burn the casing during the process because the rice is merely peeled and cleaned and, as such, the residue is the casing, which contains sufficient cellulose to form a cellulose pulp at a large scale. There is a special group studying the possibility of transforming this pulp into biodegradable diapers.

The differential of the ceramics chain is that the energetic input of the stoves the of the factory will be the residue casings from the plain rice producers. Through the well executed burning (with no environmental impact) this biomass generates an ash that will be combined with the residual clay moved in the coal mines of the Copelmi Company from the region of Charqueadas, The most important product of this factory is the ecological toilets to be distributed for the BOP population who have no sanitary system at home, According to the 2000 IBGE Census (2010), 52% of the Brazilian population has no sanitary treatment. Specifically in the south region, only 38,9% of the towns have some sort of treatment: 61,1% do not collect the residue, 17,2% only collect it and 21,7% collect and treat. The lack of sanitary infra-structure is, without any doubt, one of the major responsible for diseases and low quality of life of the population, rising the costs of the Brazilian Health System. Besides the eco-toilets, other products like utilitarian, Lego type bricks, tiles etc. are also being considered in the project.

The objective of the clothing chain is the eco-innovation in the production of basic clothing (pants, shirts and sweathers) through the use of fibers from:
1) Rice straw which will be supplied by the rice producers of Palmares do Sul and the surrounding towns already partners to the research;
2) flax/linen fiber, residue from the linen which is planted in the region of Guarani das Missões, only for the extraction of oil from the seeds: fibers are thrown away;
3) Wool from the sheep Texel from Palmares do Sul but also from other surrounding regions (Caçapava, Viamão, Barra do Ribeiro). The wool will be cleanned in a system that reuses the water, and the residual lanolin will be separated for further use in pharmacos (i.e. ointment).

The chain for clothing is to be structured taking into consideration the fact that the fibers will be rowed in Sapucaia do Sul who already has the adequate technology. The threads will be sent to the factory in Palmares do Sul for either weaving or knitting. Clothing will be dyed with natural ink (mainly with coffee and mate from the selective collect system).

The objective of the footwear chain is the eco-innovation in shoe production using Texel sheep leather (which skins are actually discarded) treated in a new biologically process of tanning with low environmental impact (using tannin from acacia tree or residual grapes available in the vineyard of the south of Brazil). As to get a clear picture, the region of wine production makes 104.251.438 (kg) of residue (full of tannin) which is completely thrown away. The shoe is being designed for material re-use: following the design for environment (DfE) and the cradle-to-cradle criteria, the shoes will be distributed in the schools and will get back to the factory after six months of use by the children, in order to be disassembled: the sole will be reused as material in the technological system (to produce more soles) while the leather will be transformed in the biological system.

Brazil has tradition in shoe manufacturing: it is the third largest producer of footwear in the world market after China and India, with 8.4 thousand industries, which produce approximately 725 million pairs per year (of which 189 million, are for export) with its productive sector installed in different regions. The State of Rio Grande do Sul is the main producer at the national level, followed by the State of São Paulo. The destined markets of the Brazilian production of traditional shoes are the United States, which absorbs 50% of the volume exported. This industry is the prime generator of employment in the country. In 2007, there were 312 thousand 579 people working directly in the shoe industry (Abicalçados, 2008).

The sociotechnical footwear factory being designed for Palmares do Sul is the one that will employ more people: 400 people working in two shifts in order to keep the plant area small. This same 400 people will work in the clothing factory, since leather, flax/linen and wool are seasonable.

A garbage selective system is being developed to collect paper, iron, aluminum, plastic, glass, oil, mate and coffee, and the citizens will be paid for selecting at home and delivering these residues. All the collected waste but glass, plastic, oil, coffee and mate, will be prepared in a garbage center. Glass and plastic (but PET) will be re-manufactured (in the technological metabolism) and the other materials (PET, paper, iron, aluminum) will be sell. The selective collecting will also focus on the retrieving of the used oil for the energetic input for the glass factory and the public bus. Brazilians have the habit of consuming large amount of coffee. In special, gauchos (Brazilians from Rio Grande Sul and also Argentineans and Uruguayans) drink mate all the time. Both beverages are made from plants and their leftovers can be used for tanning the fibers of the clothing factory. The selective collection will generate around 50 jobs.

The same center will be responsible for the organic waste also collected selectively: Organic residue from the kitchen will be processed in the biodegradable brown bags to avoid manual contact (what is a significant gain for the health and safety of the workers) and treated to generate organic material for permaculture or plantation of trees of the streets and parks of the town. Bathroom waste, like diapers, absorbents and used paper (in white bags) will be sent to the municipal sanitary garbage plant near the town. The center will be also responsible for collecting the waste from the eco-toilets in order to help cleaning the river .

The plastic factory will re-manufacture the different types of plastic (but PET) recovered from the selective garbage collect to produce the brown and white biodegradable bags for the organic waste. The idea is to sell the bags to the stores and supermarkets, so the population will not need to by the bags.

The glass factory will be the first to be implemented because is the easiest and cheapest and has the potential to attract tourism. Glass is easily re-manufactured, abundant, has no commercial value therefore is not much of an interest to collectors The cost of the ovens are very low (U$100), the energy is practically free (the population will provide the used oil that will be the oven fuel).

Another important point which adds to the glass factory is the heat it will provide (around 300°C) which will be used as energy for the food factory. The products from the glass factory will basically be utilitarian, and for decoration. The same glass could be used in the verification of the ceramic produced from the rice casing ash. This factory will work more as a touristic attraction, since employs only a few people (around 10).

THE MAJOR PRODUCT: THE SUSTAINABLE TOWN

Fishing from the river around the town is an import source of income to the citizens but it is polluted due to the lack of sanitary system. Therefore, a cleaning system based on the eco-toilets and aquatic plants (Eichhornia crassipes) will be installed by the river, and its functioning will be in charge of the fishermen who will get extra money from this new activity. The recovery of the polluted Palmares river and Lagoa dos Patos (a beautiful lake with ecological appeal) can be a source of interest for ecological tourism. Actually, in order to generate more jobs, tourism is an important part of the project and will be based on;
1) holding adequate environmental and social opportunities offered by region's innate (such as the Lagoa dos Patos lagoon, fishing);
2) the creation of new areas of interest using the region's natural resources (such as sustainable fishing);
3) the creation of new areas of interest using the artificial resources of the region (including the food, clothing and shoes factories to be installed, but mainly the glass one), the creation of parks and restoration of historical heritage (monuments and buildings);
4) the increase of quality of life (by the installation of an ecological sanitary system in places with no assistance) and the generation of jobs; and
5) the symbolic value of the project by showing what can be done with what is considered junk. Thus, the plan is to improve the infrastructure of hotels and restaurants in the town that will receive the tourists and therefore, generate more jobs and revenue. A bus system, also fuelled by used oil will be installed to transport tourists and all citizens anytime and for free. The residue of this oil will be remanufactured into soap. The PET plastic, paper, iron and aluminum that will be sold will generate part of the money to pay for the town maintenance (i.e., the bus driver and general maintenance of parks). This project aims to demonstrate the technical feasibility and economic cultural development of towns, from the implementation of socio-cultural development based on tourism. The recovery of historical and local culture (boats fabrication, for example) is under review.

RELEVANCE OF THE SUSTAINABLE TOWN AND THE PRODUCTIVE CHAINS: IMPACTS ON THE SOCIETY

The relevance of the project is the structuring of a sustainable town, which depends on the project of the arrangement of the six chains, can be summarized as follows:

1) by the generation of high socially aggregated value: the fabrication of clothing, shoes and foodstuffs as to meet the needs of schools means raising the quality of life of at least 48 million and 700 thousand people who are registered in the Brazilian public learning system;

2) by the generation of jobs and financial resources for poor regions: the town of Palmares do Sul that was considered to pioneer the Project is closed to Porto Alegre (distant 80 km), capital of Rio Grande do Sul, and has a population of less than 10,000 inhabitants. According to the IBGE Demographic Census of 2000 (IBGE, 2010) at least 60% of these inhabitants do not have good quality of life because of a lack of employment. Many of them were expelled from the field according to the mechanization of farming and rice and have no training in other areas. The town offers few opportunities because, besides a few services (bank, pharmacy etc) there are only rice processing plants that employ little hand labor (on average 6 people / plant). Despite the fact that job generation "upsizing", is mentioned in the Zeri methodology, Pauli (1998) stresses the creation of jobs but do not talk about the quality of the jobs. This project amplifies the idea of upsizing as includes the quality of work life, besides being inclusive since is being designed (Bittencourt, 2008; Brod Jr, 2010) to absorb people with disabilities. It is an important goal to enhance the human development index by promoting jobs (in the factories, in the logistics, in the tourism and in the de-pollution system handling and treatment) and a sanitary system for the town. With the implementation of this project, the expectation is human, economical and environmental development with both monetary and social gains along with the protection of the environment (the purpose of the TBL).

3) by the reduction of the environmental impact of products. In the case of the rice casing, when it is not burned (and generally they are wrongly burned, generating high levels of $CO2$), the casing is deposited anywhere, representing a serious environmental problem. Also, it is as if money had been thrown away, for the raw material can be useful as silicon, coal and cellulose. In the case of the shoe production chain, a vegetal tanning of low cost may serve as an example of a positive environmental impact for the Brazilian shoe producing sector, which is decidedly pollutant.

4) by the increasing of exports of new products: the shoe chain, due to the ecological appeal of a cleaner have the potential to transform the leatherier-footwear sector and increase the exportation. China is the greatest competitor in regards to Brazil in the area of footwear, and Brazil is one of the few able to make a front to the low costs achieved by China, exporting products with environmental appeal. Also, the clothing chain has the possibility to export socially and ecologically strong products to Europe as well as to South America;

5) by exporting of technology: Brazil is, in the majority of areas, an imitator of products and a buyer of technologies developed in the first world, at a high cost for the entire country. The innovations in alternative products and technologies are an escape route from the technological dependence ever since the discovering.

CONCLUSIONS

The project model is now being tested in Palmares do Sul, a small town in Rio Grande do Sul, closed to the capital (Porto Alegre) which deals with tons of residuals from the agro industry. The products are being designed mainly to fulfill the needs of the students enrolled in the Brazilian public schools and daycares: food, clothing and shoes; ceramic eco-toilets for the population with no sanitary system. Residues from the collect system such as glass for utilitarian products to attract touristim, and plastic for producing degradable bags to be used in the collect system.

An educational course was developed to prepare the citizens for the change. They will be taught about the importance of depollution, better sanitary habits and their participation in the improvement of the town by selecting the garbage, reducing and re-using resources and the gains they will get (like being paid for garbage selection, getting free transportation, more jobs and quality of life)

The implementation of this project in the first town would be an example that innovative processes and products are possible and that ideas should be endogenous, not imported with no questioning. Brazil is big, poor, with lots of social and economic problems but at the same time rich artistically (for its music, dance and arts) and environmentally (by its biodiversity). So far, Brazil has being performing the role of feeding the developed world with raw materials. This innovative Project might be a door for changing this situation, as the country can start exporting sustainable products with high aggregated value, demonstrating that it is possible to enhance the environment and the society with economic advantages.

It is also important to note the importance of this Project in relation to the interaction among the academy, the enterprises and the society. Two banks are interested in financing the project: one federal bank is interested in the factories and a statal bank is interested in the selective collect system This interaction will favor the exchange of information, and in the near future, more towns, in all parts of Brazil, using the same model, may develop their own factories with their native materials and in accordance with their needs and regional culture.

REFERENCES

Abicalçados (2008) Resenha Estatística anual 2007. http://www.abicalcados.com.br.
Barros, R.P., Foguel, M.N. and Ulyssea, G. (2007) Desigualdade de renda no Brasil: uma análise da queda recente. Rio de Janeiro: IPEA.
Bittencourt, R.S. (2008) Proposta de um modelo conceitual para o planejamento de instalações industriais livre de barreiras. PHD Thesis, Escola de Engenharia, Universidade Federal do Rio Grande do Sul, Porto Alegre.
Brod Jr, M. (2010) Linguagem de produção inclusiva: a linguagem gráfico-verbal,

gráfico-visual e gesto-visual para atividades de produção. PHD Thesis, Escola de Engenharia, Universidade Federal do Rio Grande do Sul, Porto Alegre.

Elkington, J. (1999) Cannibals with forks: the triple bottom line of the 21st century business. Oxford: Capstone.

Fogliatto, F. S. and Guimarães, L. B. de M. (1999) Design macroergonômico: uma proposta metodológica para o projeto de produto, Produto e Produção, Volume 3 No. 3 pp. 1-15.

Furtado, C. (1959) Formação econômica do Brasil. Rio de Janeiro: Fundo de Cultura.

Guimarães, L. B. de M. (2010) "Design sociotecnico", in: Design Sociotecnico, Guimarães, L. B. de M. (coord.) pp. 4.1-4.83.

Instituto Brasileiro de Geografia e Estatística (IBGE) (2000) Censo demográfico 2000. Características da população e dos domicílios: resultados do universo. http://www.ibge.gov.br.

Instituto Brasileiro de Geografia e Estatística (IBGE) (2008) Pesquisa nacional de amostra de domicílios (PNAD) 2006. http://www.ibge.gov.br.

Jaguaribe, H. (2006) Atual problema do desenvolvimento brasileiro. Revista Brasileira de Ciências Sociais, Volume 21 No. 60.

McDonough, W. and Braungart, M. (2002) Cradle to cradle: remaking the way we make things. New York: North Point Press.

Pauli, G. (1998) Upsizing: como gerar mais renda, criar mais postos de trabalho e eliminar a poluição. Porto Alegre: Fundação Zeri Brasil / L&PM.

Chapter 28

Ergonomic Pattern Mapping: A New Method for Participatory Design Processes in the Workplace

Marcello Silva e Santos, Mario Cesar Rodriguez Vidal,
Paulo Vitor Carvalho

GENTE Laboratory –
Federal University of Rio de Janeiro (COPPE/UFRJ)

ABSTRACT

The practice of ergonomics lacks instruments to either map or regulate cognitive, organizational and environmental related impacts. When ergonomic transformations through design – such as a new workstation or facility design – is at play, ergonomists tend to yield terrain to design professionals and engineers. Empirical evidence shows that participation of ergonomists as project facilitators, may contribute to an effective professional synergy amongst the stakeholders in a multidisciplinary venue. When that happens, eventual conflicts are dissipated in exchange for more convergent design alternatives. In addition, ergonomics professionals are supposed to be neutral participants, even though they tend to assure proper work conditions are guaranteed. This paper presents a participatory design method, in which users are encouraged to participate in the design process by sharing their work activities with the design team. The results inferred from the ergonomic action and translated into a new design, are then compiled into a "Ergonomic Pattern Manual".

Keywords: Ergonomic Design, Cognitive Maps, Project Management

INTRODUCTION

Human Factors professionals have long missed effective instruments to either map or regulate comprehensively and effectively, cognitive and organizational related impacts, especially the environmental ones. As a consequence they have been put aside when it comes to facilities design or even more specific workstation design in organizations. As a result, chances are lost for HFE professionals to employ their knowledge in order to improve overall worker performance. of an office layout or an entire new. This happens either for their lack of specific knowledge or because they do not know the appropriate tools to serve both architectural, engineering and ergonomical needs of the built work environment undergoing such transformation processes – or new ones that will be set up.

According to PIKAAR & DE LOOZE (2008) it may not be sufficiently clear for the general public the notion of what human factors & ergonomics is, or what functions an HFE professional really performs. Once human factors & ergonomics outreach is unveiled in the course of a given project management, surprise reactions are generally a positive collateral effect of that realization. One of the reasons for that outcome is the fact that applied HFE works, through participatory actions, tends to empower work system's users to develop their own project solutions. In addition, those professionals are always impartial participants, not taking any particular side, even though they tend to make sure proper work conditions are guaranteed.

After years of experience in applied ergonomics projects and programs in a multitude of multinational organizations, we have found a staggering amount of evidence showing that most "ergonomic problems" people and organizations face in their operations are due to design flaws (lack of physical space, environmental comfort, poor accessibility and ambience, etc.). It seems clear that several of those could have been prevented, if better planning and project management had taken place. In fact, over one third of all work environment's design transformation processes (remodeling, renovations, layout reorganizations) happen in the first year of operation. The following graphs (Figure 1) illustrate the importance of both careful planning of work facilities and the close relationship between quality in the production of the Built Work Environment (BWE) and Ergonomics.

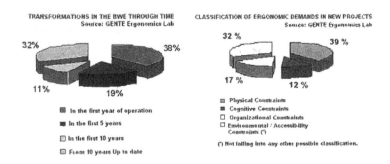

FIGURE 1. Graphs of Timeline of Changes and Nature of Ergonomic Demands

Although not explicit, the results to which the figures refer to can unveil other classes of problems. The more important ones are related to the mishandling of data uncovered in the course of ergonomic actions and its subsequent financial losses. The mishandling of data is usually due to the classification objective itself – as to differentiate aspects and impacts of ergonomic task analysis, which made us create a separate category for environmental constraints as can be seen on the graph to the right. In terms of the direct economical impact, the cost of ergonomic actions may raise several times from the earliest design stage to the date when the facility becomes fully operational. In practical terms, human factors and ergonomics actions are economically more effective when employed in the earlier stages of a given work system's design. (Hendrick, 2002).

BACKGROUND

In spite of comprising what must be the most widely explored of any group of ergonomic variables (Wilson, & Corlett, 2005), the work environment is not considered a separate set of human factors & ergonomics domain. When one asks in which category the work environment should fall in regard of its appropriateness to each one of the usual and accepted ergonomic domains – physical, organizational or cognitive, as its is currently classified by IEA (International Ergonomics Association) – it would be probably more prudent to simply answer: none. After all, the environment, whether built or not, cannot be a part of a domain for the sole reason that all those different domains actually take place *within* boundaries or environments. It is not merely a linguistic or semantic matter as one may assume, but rather a epistemological one, if we prefer to reassure terminology precision. Every single ergonomic related analysis, actions, processes or projects are utterly attached to a given scenario, which alters those contexts as well as it is altered by it.

In other words, we can infer that, environmental inadequacies – rather than design and project flaws – must be both studied as a separate dimension, with its particularities and peculiarities and, concurrently, as part of each dimension depending what variables are involved. For example, a poorly ventilated room needs to be analyzed according to its compliance to environmental comfort levels (physical domain). However, the root cause for the inadequacy must be addressed broadly, within the entire *spectrum* of ergonomics and *also* in the context of the planning, design and construction of that work facility that is under scrutiny. In order to better characterize and classify all possible variables, we suggest the addition of a new ergonomic "dimension", a study locus capable of adequately encompass technical and conceptual limits and contents. Not only it would enable a better understanding and treatment of those questions, but also become a tool of sorts for preventing the imbalance caused by all different variables or factors that may impact the desirable appropriateness of the work environment (Figure 2).

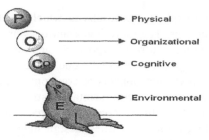

FIGURE 1. The PhOCoE Model for Workspace Design and Treatment (Santos, 2008)

When ergonomics fails, organizations do loose, no doubt about it. However, when that happens, workers also suffer, become sick, work badly or simply miss work. The "PHOCOE" Model (Foca or *SEAL*, in Portuguese, what explains the sea mammal "supporting" the model), derived from that idea, but it shifts the focus, because of its design intentions, as it underlines the need for considering the environment as an independent – as well as interdependent – domain for the study and practice of ergonomics, both in terms of processes and (design) projects. We propose the work environment being considered as both a separate entity for analysis and, at concurrently and isolatedly, a "shield" covering every dimension of a work system. Regardless how effective is the organization in providing good ergonomics for its workers, everything falls if they are forced to work in a inadequate workspace. In addition, there is strong evidence to point out to a corresponding cause-effect relationship between work environment efficacy and overall quality at work life (QWL) standards.

In sum, design actions like planning a new layout, redesigning routines or evaluating a new facility, for example, cannot be performed dissociated from a "macro-organizational" stand point. Instead, it must be dealt with as a multi-faceted demarche, in which we have multiple issues permeating through various possible domains – ergonomically or beyond. It must be carried out as a "whole event" , even though it is strongly recommended situating the question (s) as much within a context as possible.

In the real world of corporate action development, organizations and professionals are formally hired to conduct a transformation project, thus becoming subject to contractual deeds in which their competencies can be accordingly evaluated and measured individually. In other words, if we have 10 different project "entities", we will have 10 different identities, standings, behaviors and moreover inclinations towards the global project. If no integrating mechanism for those different processes and viewpoints is set in place, there will be a potential risk for this project to become more like a patchwork quilt rather than an combined, consolidated result.

DISCUSSIONS FOR A METHODOLOGICAL FRAMEWORK

The use of applied ergonomics approach to project management is a plausible alternative to mediate what some call "project synapses" (Santos, 2009), the various turning points and key moments that occur when a project is carried out. The main reason for that is because ergonomics has clearly a facilitation nature: it is a process of managing interfaces, both either through a project or a process. A project is also carried out by some kind of due diligence action, hence, there is always some kind of process to be managed. When we further investigate in the course of an ergonomic action, the demands that are usually called "ergonomic problems" are, in reality, engineering problems, design problems and so forth. Similarly, we find that a lot of those problems happen because ergonomic parameters were not given similar priority as metric ones, like how much equipment one can get into an allotted space (Kohn, 1999). Those problems tend to lead to deeply embedded ergonomic inadequacies if not properly dealt with. Empirical evidence has shown that participation of ergonomists as project facilitators, may contribute to an effective professional synergy amongst the various stakeholders in a multidisciplinary venue. Naturally, limits and outreach depend on the establishment an ergonomic maturity scale within organizations (Santos & Vidal, 2009).

HF & Ergonomics is also known as the science of managing interfaces from the individual level all the way up to global influence, as it is observed by recent links to the concept of sustainable development (Zink et. al, 2008). Because those professionals are not "specialists" in any particular process or participatory action project, it becomes usually easier for them to mediate eventual controversies arising in the midst of a management process or project flow of events. On the other hand, multidisciplinary actions and participatory projects presents a real challenge in terms of accommodating questions such as yielding budget and timeline together. We present below a list of the main problems that were found in a selection of major projects we were involved in:

- Autonomy and Power of Authority – it is crucial to identify the false prevalence caused by "expertise dominance". The designer often thinks his solutions were accepted because were good, while in truth they became good just because were accepted.
- Project to Product Transformation – Differently from a chair or a desk, in which you can follow specs "mirroring" the project into the final product, in building construction it is impossible to replicate all the details.
- Unavoidability and Variability – Being able to control the variability – not the unexpected events – is the key for a successful project.

The first step of a well suited solution for all those issues is to try using tools and methods that are cross-culturally interchangeable. In other words, to use techniques that have been successfully tested before, but not necessarily for that particular objective (successfully plan, design and build the work environment). In order to situate our quest for a suitable method, we particularly looked into participatory approaches for dealing with transformation projects. One such method, Concurrent Engineering, appeared in the 1970's and received innumerous worldwide contributions for at least 2 decades, until it became almost an standardized project management procedure, even though it never really became a solid overspread project methodology (or even an overall adopted terminology).

Other project management models, techniques and methodologies encouraged some paradigmatic changes. It all reflected the new times that were characterized by gradual loosening of corporative authoritarian rules and a move towards more participatory actions within organizations (Pasmore, 1994). It is important to emphasize that only recently, project management processes began to be professionally handled outside the boundaries of corporation's functional structure (Alsene, 1999). That means added opportunity for ergonomics professionals to act as project mediators, interpreting major trends, conversational actions and directions, managing eventual conflicts and so forth. However, in order for that to happen, it is essential that those professionals acquire not only the ability towards cross-communication, but the agility to understand an respond to multidisciplinary interactions and eventual conflicts that might arise from them. After all, it should be more important how effectively a project is carried out – its result – than the project methodology itself – its "recipe".

PROPOSING A METHOD FOR DEVELOPING AND SUSTAINING ERGONOMIC PROJECTS– THE ERGONOMIC PATTERN MAPPING TECHNIQUE

Every single work environment – understood in here as a man-built environment designed for labor activities – should be studied as a *Built* Work Environment. If this is a true hypothesis, then it makes no sense to evaluate any given work facility only as a design accomplishment, an object to be used as an equipment or a tool. The BWE is in reality a complex scenario in which work actions take place because there are actors to play specific parts. Therefore, if the scenario is not well set, so will not be the "acts" at play. This analogy serves the purpose of emphasizing the need for organizations to plan ahead their new work facilities and offices, as well as their remodeling or renovations of any kind. In addition, in order for this planning to work, it is necessary for designers, managers and other stakeholders to analyze the work as it is – or will be – performed by the users of those BWE's. In Ergonomics, we use the term "comprehension", stressing the need of understanding the meaning of *real* work, not the *prescribed* work as it is relayed to designers and builders.

Christopher Alexander, an architect and mathematician , developed a design theory based on what he calls *"Pattern Language"*, an attempt of establishing a design method for laymen to exercise their creative power. It all originated because of Alexander's realization that traditional design has not been capable to produce adequate houses, parks and cities for men to live in, use and enjoy. Traditional design, according to him, is characterized by a distant relationship between designers and users. Because of that, the customer of any given design product stand as a rigid interface between the creator (the designer) and the creature (the houses, buildings, offices, etc.). One reason for that may be that – more specifically in the case of architecture – in modern societies architects do not have much of a true connection with their clients, like they did in the past. As a result, their ideas and project decisions are often misinterpreted – it goes both ways – and since usually there is no evaluation in the aftermath, bad ideas are some times perpetuated. Alexander called those ideas "anti-patterns" because they became standards for repetition instead of appropriateness.

As a pattern is intended to describe a solution, it may also be used to replicate them by analogy, even though a pattern may not be exactly repeated. Every conception or creation act is governed by some kind of *pattern language* – in a sense that they constitute a blend of intuitive codes and implicit perceptual-cognitive elements – conversely, they share patterns with worldly artifacts, derived themselves from universal patterns and languages different people use. In addition, not only the form, shape or style of our buildings follow similar cultural insights, but also their quality or lack thereof. Once those cultural codes are no longer shared, subjacent processes are terminated and, consequently, we loose the ability of seeing the essence and feeling our lives in our "new temples" (Alexander, 1977). In short, using "a" pattern language is more than a form of participatory design, rather is an empowerment technique, enabling users to formulate their "codes" in accordance to their "core" beliefs.

With all that in mind we envisioned a possibility of taking advantage of this participatory and empowering approach for the ergonomic design of the BWE. Naturally, it does not mean that ergonomists will take architect's or designer's jobs. As "managers of interfaces", ergonomics professionals are well fit to act as "drivers" in this intricate relationship between users and providers of the work environment, mediating talks and balancing expectations and needs, better yet without having to take sides. Once the product that originated from a pattern – or a collection thereof – is tested and approved by users, it can be used recurrently, consolidating the a *true* pattern . Since ergonomics works by extracting information from a context to establish its standards, we see the use of a pattern language as a coherent and natural way for ergonomics to become involved into the BWE design and project.

Our experience to date using ergonomic pattern mapping – the adopted terminology for this method – have shown preliminary results that are extremely satisfactory. In sum, we achieve true operating models to be used for improving work conditions, as well as QWL indicators that can be used for registering, modeling, disseminating and consolidating good practices and ergonomic solutions. In the following figure (Fig. 3) we present an example of one of those patterns, developed within the scope of an participatory ergonomics program in a major oil company. The text is partially scrambled for contractual security reasons. A design pattern does not have a rigid method of construction. However, we sense our choice of using cognitive maps – for opening up broad understanding within the work group – is quite compatible with the original idea set by Alexander and his followers. In terms of the actual pattern configuration, it starts with a number and

278

title above the picture or figure that states the situation. Then a synthetic situation description leading to the pattern is outlined. After that, some hypothesis may be described or a list of justifications for a pre-diagnostics is set to address a solution. The solution is then summarized and illustrated with a scheme, picture or any type of diagram that will express a general notion for the concept, making it also easier in terms of its broadcasting. In the final part, normative standards and links to other interconnected patterns are established, so that a sequential design flow may come along.

CONCEITO 8 – A Distribuição Cartesiana e Desorganizadora

FIGURE 3. Ergonomic Pattern – Layout Distribution (GENTE, 2007)

In essence, patterns are concepts that convey a generalization of essential ideas (*what* is it) either conventionally written in a knowledge database – as in books, documents, portal and so forth – or through primary representations. The prepositions, emerged from the core concept, are variables that determine the form (*how* is it) from which primary and secondary concepts are connected. In a sense, a pattern is a cognitive roadmap, graphically inducing a collective creative action to take place. As a result, a pattern assumes the position of a common representation of a specific concept within a context and possibly beyond that. It is recommended for the ergonomics team and the workgroup, in charge of creating patterns, to have someone familiarized with project management as well as design actions.

Whenever we are establishing ergonomic patterns, it is also important to emphasize one fundamental characteristics of this idea. A pattern will always serve as a guideline, a pathway towards an objective not the objective itself. That is what HFE professionals try to do: rather than showing *our* solution we let individuals and organizations decide on *their* ideal solution by presenting a set of guidelines and examples.

Social inclusion of workers within organizations, initially timid in nature, gradually gained momentum until it evolved to the sociothechnical model (Trist, 1981). By understanding organizations as live organisms with people in its core, the sociotechnical model, along with the widespread use of ergonomic principles in the workplace, established parameters for studying and evaluating those complex systems as a whole. In other words, it means that everything is influenced by the environment, which in turn influences the entire organization. In this context, organizations need an adequate design, in order to become flexible enough to withstand the various constraints they need to overcome in order to survive (Hartley, 1998).

CONCLUSION

Organizations, their managers and people in general are becoming more and more convinced that management models that encourage participation and empower the various stakeholders in the course of projects – regardless of types – are far more successful to deliver their goals than traditional ones with all their regulatory protocols and excessive compartmentalized functional structure. One of the main reasons may be that methodological tools designed with participatory characteristics in mind tend to convey relative stability to "unstable scenarios", such as those in multidisciplinary environments. In the case of design projects and other facility transformation processes, they also stimulate creativity and deliver a more sustainable built work environment for their direct and indirect users.

Sustainable work environments depend not only on how they are built – what technique, materials they use – but also on the way they are planned and designed – what "philosophy" was embedded into it. After all, what is the use of all those certified green buildings, that conserve energy and save money and natural resources, if they still present problems in their operations and, moreover, if people having to work inside them suffer?

The use of ergonomic pattern mapping – derived from the PhOCOE project management model – offers a real opportunity, as initial empirical evidence has demonstrated its validity in terms of "amalgamating" multi-professional theories and practices, ideas and concepts that are most likely to be employed separately, limited by individual expertises. Numerous future applications can be foreseen, but its use in interdisciplinary actions, under conflict situations depending on efficient decision-making strategies, is probably the best possible scenario to practice it. We should add that a built environment may not be necessarily inadequate in its essence. But if it is unfit or inadequate for the work activities their users perform, it does become unfit or inadequate as a Work Environment *for those people.* Consequently, compliance enforcement is also enhanced by the guidance of ergonomic patterns. Therefore, with enough planning and the correct design methodology, we might put as much effort to deliver a good BWE as we do for producing an unfit one that will not serve its purpose effectively. As a bonus, when we work towards an ideal situation, we actually reduce overall efforts and, consequently, engineering, production and management costs involved in the planning, designing and production of those BWE`s.

REFERENCES

Alexander, C. et.al. (1977), *A Pattern Language: Towns, Buildings, Construction.* New York: Oxford University Press.

Alexander, C. (1979), *The Timeless Way of Building*, New York: Oxford University Press.

Alsene, E., (1999), *Internal changes and project management structures within enterprises*, Project Management, Vol 17, Issue 6, December, Pages 367-376.

GENTE/COPPE Laboratory (2008), PETROBRAS E&P SERV Project Report., Rio de Janeiro, GENTE Ergonomics Lab.

Hartley, J.R. (1998), Concurrent Engineering - Shortening Lead Times, Raising Quality, and Lowering Costs, Portland: Productivity Press.

Hendrick, H, (2004), Handbook of Human Factors and Ergonomics Methods, London: Taylor & Francis.

Hendrick, H. ; Kleiner, B. M. (2002), Macroergonomics – Theory, Methods and Applications, Boca Raton, CRC Press.

Kohn, P.J. (ed.), (1999), Ergonomic Process Management – A Blueprint for Quality and Compliance, New York: Lewis Publishers.

Novak, J. (1990), Concept mapping: A useful tool for science education, Res.Sci. Teaching, 27, (10), p. 937–949.

Pasmore, W. (1994), Creating Strategic Change: Designing the Flexible, High-Performing Organization , New York: Wiley, John & Sons.

Pikaar R.N., Koningsveld, E.A.P., Settels, P.J.M. (2008), Meeting Diversity in Ergonomics, Amsterdam: Elsevier.

Santos, M.(et.al) , (2009), Can we really opt in terms of ergonomic methodologies and/or approaches? Annals of XVII IEA Congress, Beijing.

Santos, M. (2008), Conception Ergonomics for Prevention of Environmental Inadequacies, Ergonomic Action Journal.(in portuguese), Rio de Janeiro: EVC, Vol. 2, No 6, p. 44-57.

Trist, E., (1981), The Evolution of Socio-Technical Systems – A Conceptual Framework and an Action Research Program, Paper # 2, June, Toronto: QWL Centre

Vidal, M. & Bonfatti, R. (2003), Conversational Action: an Ergonomic Approach to Interaction in: Grant P., Rethinking communicative interaction, Amsterdam, JB, p.108-120.

Vidal, M., Santos, M., (2009), The Ergonomic Maturity of a Company Enhancing the Effectiveness of Ergonomics Processes, Annals of XVII IEA Congress, Beijing.

Wilson, J & Corlett, N. (eds.), (2005), Evaluation of Human Work, 3rd. Edition, Boca Raton, CRC Press.

Zink, K.(Ed), (2008), Corporate Sustainability as a Challenge for Comprehensive Management, Kaiserlautern, Physica-Verlag.

Principle Driven Design: Ergonomic Requirements of Multimedia as the Main Driver in the Development of a Tool for Prevention in Occupational Health and Safety

Denis A. Coelho

Human Technology Group
Dept. of Electromechanical Engineering
University of Beira Interior
6201-001 Covilhã, Portugal

ABSTRACT

A software application was conceived aiming at two goals: contributing to raise awareness about the importance of improving health and safety at work in a preventive approach, and supporting the conduction of a preliminary diagnosis of working conditions in Small and Medium Enterprises, once the user has gained knowledge. The whole development was triggered directly by these two goals, broken down into actions and objectives seen from the users perspective and seeking compliance with ergonomic recommendations and guidelines for multimedia. The software makes use of a set of innovative interface tools, both in what concerns user tools for orientation, access and assimilation of the knowledge base and in terms of the structure of the software contents as well as its graphic design. A requirement analysis was carried out, that led to the design of the conceptual structure of the software application, alongside user interaction sketches, and the identification of illustration and animation spots. The concept of user level was instrumental to focus user motivation for learning and progression in the application's knowledge base, and also led to the creation of quizzes within each content category. Development started with literature review and acquaintance with the applicable legislation and regulations, aiming at meeting the needs of Small and Medium Enterprises (SMEs) with an industrial character, in terms of continual improvement of their practices in the realm of Health and Safety at Work (HSW). Systematization of hazard and risk categories was pursued. The interface is structured in nodes of information, interconnected, and endowed with graphical elements that have interactive features. The structure of the software supports a hypertext organization of the information, with a view to enable the user to design her, or his, own path on the way to attain a deepened and enlarged understanding of concepts and greater conscience leading to increased knowledge about the subject matter. The user is given several settings options, enabling the program to assume a configuration that is intended as more suitable to the user. This process also entails a quiz that is the basis for attributing a user level. The possibility of bypassing this quiz is also foreseen, with immediate transfer to what is considered the second part of the software interface. This area, can be explored starting from the main console, by selecting one of the four hazard categories. A detailed depiction is found in each of them, illustrated and with games, meant to test knowledge acquired, in a playful way. Once each of the categories is explored, users gain access to the third phase of the program, which is evident from the change in the central console. Here a structured evaluation of the company is carried out by the user. The program then offers recommendations, warnings and preventive measures that aim at mitigating the problems detected in the guided diagnosis and promote the improvement of HSW. The software makes use of a set of user interface tools, including a dynamic map and the possibility of user annotations appended to the contents as well as drop-down menu based navigation, derived from ergonomic guidelines for multimedia. Both a search tool and a searchable glossary are provided.

Keywords: ergonomic guidelines, ergonomics of multimedia, health and safety software

INTRODUCTION

As computing technology penetrates our workplaces, homes, schools, community organizations, automobiles, and aircraft, literally every element of our daily lives, having technology that is useful and usable is of paramount importance (Olson and Olson, 2003). Donald Norman proposed, as early as in 1993, a checklist of the characteristics of the environment leading to an optimal experience, which assisted in guiding the development reported in this paper, and consists of the following principles:

a - providing a high intensity of interaction and feedback;

b - having specific goals and established procedures;

c - motivating: providing a continual feeling of challenge, one that is neither so difficult as to create a sense of hopelessness and frustration, nor so easy as to produce boredom;

d - providing a sense of direct engagement, producing the feeling of directly experiencing the environment, directly working on the task;

e - providing appropriate tools that fit the user and task so well that they aid and do not distract;

f - avoiding distractions and disruptions that intervene and destroy the subjective experience.

The "Preventor Industrial SMEs" edutainment software application under development is approaching conclusion. It has been conceived aiming at two goals: contributing to raise awareness about the importance of improving health and safety at work in a preventive approach, and supporting the conduction of a preliminary diagnosis of working conditions in Small and Medium Enterprises, once the user has proven that she or he has gained knowledge. The whole development was triggered directly by these two goals, in accordance with principle (b), broken down into actions and objectives seen from the users perspective and seeking compliance with ergonomic recommendations and guidelines for multimedia. The software application makes use of a set of innovative interface tools, both in what concerns user tools for orientation, access and assimilation of the knowledge base and in terms of the structure of the software contents as well as its graphic design. A requirement analysis was carried out, that led to the design of the conceptual structure of the software application, alongside user interaction sketches, and the identification of illustration and animation spots. The concept of user level was instrumental as means of focusing user motivation for learning and progression in the application's knowledge base, and also led to the creation of quizzes within each content category. User levels and quizzes were the main means sought to implement principle (c) afore-mentioned. The software was developed with the support of two multimedia designers and a software engineer, working closely with an ergonomics expert and with subject matter experts.

APPROACH PURSUED IN THE DEVELOPMENT PROCESS

Development started with literature review and acquaintance with the applicable legislation and regulations, aiming at meeting the needs of Portuguese Small and Medium Enterprises (SMEs) with an industrial character, in terms of continual improvement of their practices in the realm of Health and Safety at Work (HSW).

From this point, the systematization of hazard and risk categories was pursued based on Coelho and Matias (2006). The summary of the outcome of this process is shown in Fig. 1, which also enables appreciation of the graphical design of the software.

The underlying structure of the software was designed to provide support for the information the software is meant to convey. The interface is structured in windows, or nodes of information, interconnected, and endowed with graphical elements that have interactive features. The structure of the software supports a hypertext organization of the information, with a view to enable the user to design her, or his, own path on the way to attain a deepened and enlarged understanding of concepts and greater conscience leading to increased knowledge about the subject matter of the software.

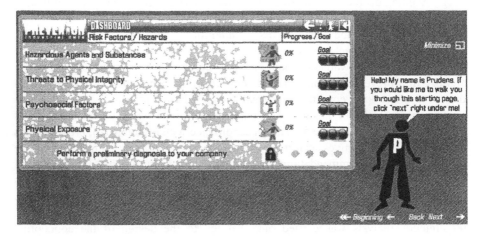

FIGURE 1. Hazard categories considered and graphical design of the Preventor industrial SMEs computer software.

OPERATION OF THE SOFTWARE AND ORGANIZATION OF INFORMATION

In first contacting with the software, the user is shown several options concerning the user type (see Fig. 2). These options enable the program to assume a configuration that is intended as more suitable to the user, according to the user choices. This process also entails a quiz that is the basis for attributing a user level. The possibility of bypassing this quiz is also foreseen, with immediate transfer to what is considered the second part of the software interface. This area, can be explored starting from the dashboard (as shown in Fig. 1), by selecting one of the four categories shown. A detailed depiction of the information is found in each of them, illustrated (examples in Fig. 3) and with games, meant to test knowledge acquired, in a playful way. The knowledge test is depicted in Fig. 4.

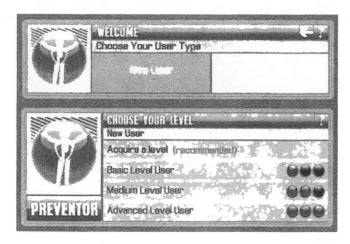

FIGURE 2. Welcome panel and selection panel for new users.

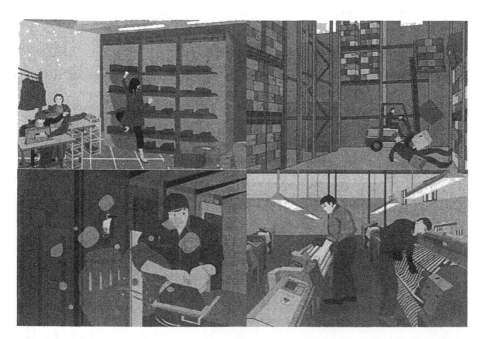

FIGURE 3. Example of illustrations from the Preventor Industrial SMEs software.

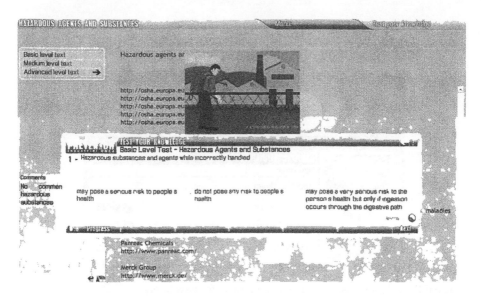

FIGURE 4. Advanced level questions about physical exposure in the Preventor Industrial SMEs software.

Once each of the categories is explored, the user gains access to the third phase of the program, which is evident from the change in the dashboard (Fig. 5). Here a structured evaluation of the company is carried out by the user (Fig. 6). The program then offers recommendations, warnings and preventive measures that aim at mitigating the problems detected in the guided diagnosis and promote the improvement of HSW (Fig. 7).

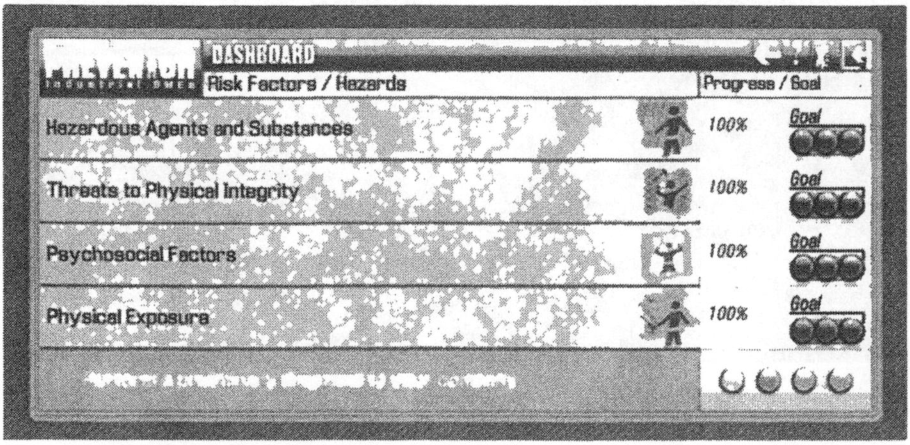

FIGURE 5. Dashboard with the diagnosis functionality enabled.

The purpose of providing information on user progression and recommendations as well as warnings and preventive measures is seeking compliance with principle (a), afore-mentioned, which is also attained directly through the high level of interaction and abundant feedback provided.

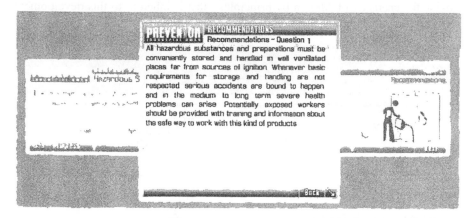

FIGURE 6. Diagnosis questions with recommendations given for each question.

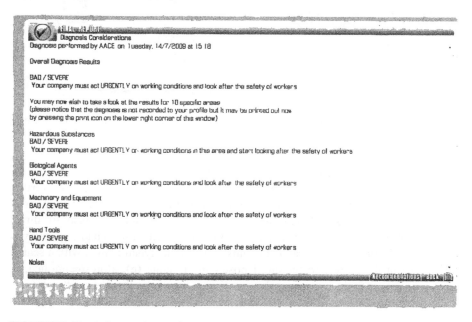

FIGURE 7. Final diagnosis results.

NAVIGATION TOOLS PROVIDED

The software that was created makes use of a set of user interface tools, including a dynamic map (Fig. 8) and the possibility of user annotations appended to the contents as well as drop-down menu based navigation (Fig. 9), derived from ergonomic guidelines for multimedia (Chignell and Waterworth, 1997). These guidelines essentially complement the principles presented earlier, and combine to provide the character of "driven by principles" that is inherent to this development leading to the software presented herein.

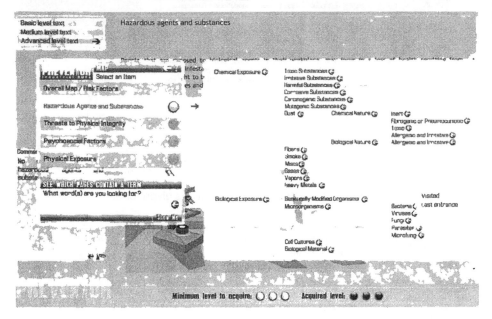

FIGURE 8. Dynamic annotated map with user specific history in the knowledge base.

The user level (basic, medium, advanced) is used as a means of motivating for learning and progression in the knowledge database, with category tests as a further means towards this goal. Animations are dispersed throughout the user experience within the program. Animations are a means of providing a sense of direct engagement, in the lines of principle (d), afore-mentioned. A searchable glossary is also provided (Fig. 10). The searchable glossary and the dynamic map with a search tool represent the answer to the quest posed in principle (e), afore-mentioned, to provide tools that fit the user and task so well, that they aid rather than distract.

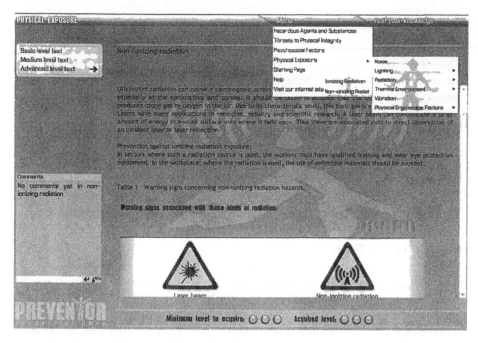

FIGURE 9. Drop down menu based navigation and user insertion of comments.

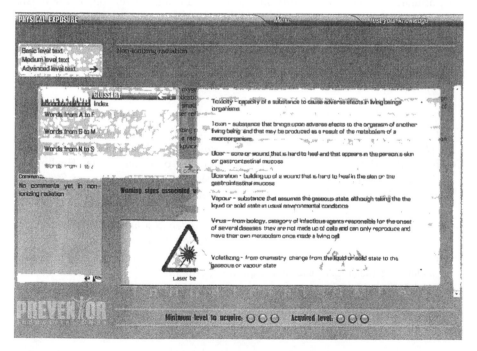

FIGURE 10. Preventor industrial SMEs glossary.

CONCLUSION

While there are a number of approaches to creation of tools to support cognitive activities, principle driven design is deemed applicable to applications where learning is paramount. This notwithstanding, user centered design should always be pursued, as is exemplified in this case. This paper showed how a set of principles was implemented in the design of a software application. All the afore-mentioned principles were directly implemented, except for principle (f) which is a negative principle, since it postulates avoiding distractions and disruptions. The software was designed in order to support a pleasant subjective experience, and hence, no disruptive or distractive functions or modes were purposefully built in the application.

ACKNOWLEDGEMENT

The software reported in this paper was developed with the support of the Portuguese Autoridade para as Condições de Trabalho (ACT) (Authority for Work Conditions) and is the result of a collaborative effort from Dr. Isilda G. Barata, André B. Silva, M.Sc., Andreia S. Campos, M.Sc., Dinis O. Nunes, M.Sc., João C. O. Matias, Ph.D. and Isabel L. Nunes, Ph.D., led by Denis A. Coelho, Ph.D.. There is a Portuguese version of this software, as well as an English version. The software is available on-line at http://preventor.ubi.pt

REFERENCES

Chignell, M., Waterworth, J. (1997) "Multimedia"; in G. Salvendy (editor) *Handbook of Human Factors and Ergonomics*, New York: Wiley.

Coelho, D.A., Matias, J.C.O. (2006) "The Benefits of Occupational Health and Safety Standards"; in Waldemar Karwowski (editor) *Handbook of Standards and Guidelines in Ergonomics and Human Factors*, New Jersey: Laurence Erlbaum Associates.

Norman, D. (1993) *Things that make us smart – defending human attributes in the age of the machine*, New York: Addison-Wesley Publishing Company.

Olson, G. M.; Olson, J. S. (2003) "Human-computer interaction: Psychological aspects of the human use of computing", *Annual Review of Psychology*, 54: 491-516.

Chapter 30

Usability Evaluation of Different Door Handles

Luis Carlos Paschoarelli[1], Raquel Santos[2]

[1]Department of Design – Faculty of Architecture, Arts and Communication
Univ. Estadual Paulista,
Bauru, SP 17033-360, BRAZIL

[2]Ergonomics Department – Human Kinetics Faculty
Technical University of Lisbon,
Cruz Quebrada, 1495-688, PORTUGAL

ABSTRACT

Ergonomic design of manual instruments aims the product usability. Frequently in different projects the variables analyzed involve force and gender influence. This study intended to evaluate maximum hand torque force and the associated effort perception of a simulated manual activity, namely, the simulation of a door opening with different kinds of door handles. 180 subjects have participated in the experiment (90 males and 90 females). Torque forces were collected with a force gauge and a static torque transducer. Data analysis was based in descriptive statistics and mean comparison analysis. The results indicated that male subjects performed mean torque forces significantly higher ($P \leq 0.05$) that female subjects. At effort perception level the results showed that on two types of door handles males presented a higher value and that in the other three types of door handles women presented a higher value of mean effort necessary to produce the maximum force torque. These results must be considered in the door handle design since, besides aesthetics, the shape must also privilege usability for all users facilitating in this way its daily tasks.

Keywords: Ergonomic Design, Usability, Door Handles,

INTRODUCTION

The objective of ergonomic design is to optimize the human/system interaction namely through usability evaluation methodologies applied in product design phase. One kind of interfaces between humans and its environment are manual devices. Its lack of usability can result in biomechanical problems, discomfort and dissatisfaction for the user. Knowing that the designer must accommodate user capacities and limitations, and that these can vary considerably among people, one of the main constraints that must be considered to a good suitability between products and humans is the necessary force for product operation.

The study of handle capacities has begun with grasping analysis. According to Napier (1956) grasping can be divided in two types: force grasping (palmar grasping) and precision grasping (digital grasping). Beyond grasping forces (that measure the hand force applied to handle an object), there's a demand to new approaches, trying to reproduce in laboratory some interfaces found in occupational tasks or daily activities like torque or pronosupination rotational movements performed by upper limbs.

Within this subject some studies have analyzed different kinds of interfaces, including tools (Mital, 1986; Mital and Channaveeraiah, 1988), cylinders (Pheasant and O'Neill, 1975; Imrhan and Jenkins, 1999), lids (Berns, 1981; Rohles et al., 1983; Nagashima and Konz, 1986; Imrhan and Loo, 1989) and knobs (Deinavayagam and Weaver, 1988; Adams and Peterson, 1988).

These approaches contribute to ergonomic design of manual devices through manual forces patterns generation and other references that contribute to a better knowledge of human biomechanics in this specific area.

The great complexity of this kind of studies results from the several variables involve as the individual aspects like, for example, the genre influence in force application.

The differences between male and female, mainly in force, are largely recognized. Among individual characteristics that can influence force, genre registers the larger difference of mean values – females produce forces of two thirds (67%) regarding male force, and it can vary from 35% to 89% depending of the muscular group evaluated (Sanders and McCormick, 1993). Besides this fact there's no consensus about how stronger male are. To palmar grasping there's evidence that females produce forces between 50% and 60% of males (Edgren et al., 2004; Imrhan, 2003; Crosby et al., 1994; Härkönen et al., 1993), whereas other authors found greater values that vary between 71% and 74% (Caporrino et al., 1998; Fransson and Winkel, 1991). To manual torque force there are studies indicating that females produce between 49.1% and 51.5% of male force (Imrhan and Jenkins, 1999; Kim and Kim, 2000), whereas other authors present values between 62% and 66% (Shih and Wang, 1996, 1997; Mital, 1986; Mital and Sanghavi, 1986). This difference it's not constant and can vary between individuals (there are many females stronger than males) and according age. For example, Kong and Lowe (2005) haven't found significant differences between genres in their sample. Imrhan and Loo (1989), Ager et al. (1984) and Fullwood (1986) have demonstrated that in childhood the difference between genders is reduced and that it augments with age, but even in lower age males where significantly stronger that females. Force differences between genders increase in adult age and decrease in old age indicating that force loss is more intense for male (Imrhan and Loo, 1989).

The present study intended to evaluate the maximum hand torque force and its associated effort perception in a simulated manual activity, namely, the simulation of a door opening with different kinds of door handles by Portuguese male and female subjects. We intended to gather data for design since, until now, this kind of information regarding Portuguese population is very scarce or even non-existent.

MATERIALS AND METHODS

All methodological procedures of this research were realized in a laboratory using human subjects. Therefore all subjects signed a permission to accomplish the required tasks for the research.

180 Portuguese adults have participated in the experiment, being 90 males (mean age=43.54 years, standard deviation (sd)=5.84; mean stature=1.71m, sd=5.44; mean weight=77.33kg, sd=9.43) and 90 females (mean age=43.34 years, ds=5.36; mean stature=1.58m, sd=5.85; mean weight=62.72kg, sd=11.91).

In this study the used equipments were according the recommendations of Caldwell et al. (1974), Chaffin and Anderson (1990) and Mital and Kumar (1998), assuring a greater credibility in results. Thus to measure force it was used the Advanced Force

Gauge AFG 500 (Mecmesin Ltd., UK) to register torque forces by means of a serial connection with Static Torque Transducer – STT, Mod. ST100-872-003 (Mecmesin Lda., UK). The torque force was measured with the subject standing in a static position with exception of upper limb movement (prono-supination of hand regarding the static interface). It was also used a digital scale (Korona, Mod. NS 5818008) to collect subjects' weight and an anthropometer (Mod. GPM) to collect anthropometric measures.

To task simulation it was used a fixed support to place the Static Torque Transducer and door handles at 950 mm high, where door handles are generally placed. This support was fixed in a metallic base of 700 x 700 mm, where the subject stands still to perform the simulated task. This metallic base had two superficial marks; one transversal at 150 mm from the door handle projection and one longitudinal along the axe projection of the torque transducer. Five door handles were used with different shapes (see figure 1).

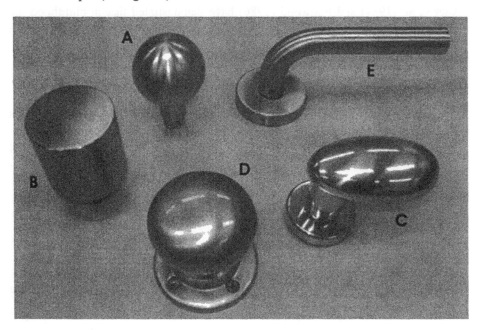

Figure 1. Different door handles (A, B, C, D and E) used to torque force evaluation

All procedures were accomplished in the Ergonomics Laboratory of the Human Kinetics Faculty of the Technical University of Lisbon. Some previous tests were realized to evaluate equipments stability, procedures standardization, verbal instructions and the time needed to the experiment.

Individually, it was told to each subject the study goals, the main procedures to perform and then the permission was signed. In sequence the subjects weight and stature were collected. Then the subject was standing in front of the support where the door handles were placed, with the front side of their shoes on the transversal line limit and with a leg aperture to proportionate a similar foot distance between the longitudinal line also marked in the support base (see figure 2).

After, standardized instructions were provided to subjects, namely the need of keeping trunk straight and static during simulation (avoiding trunk movement and body weight influence on performed force). For each door handle the subject should realize its maximum hand torque force, clockwise, relaxing afterwards during 60 seconds. The five different door handles were placed randomly on the static torque transducer to each subject simulation.

Figure 2. Position of subject in task activity simulation.

After the torque evaluation of all door handles and with the subject in a free posture the associated effort perception was collected. Thus, a standardized verbal instruction was given to solicit the door handles hierarchization in an increasing order regarding the effort needed to realize the task. To aid the subject response to this solicitation it was also presented a paper base to place the door handles.

The statistical analysis was done using SPSS Statistica 17.0 software. The *T-Student* Test to independent samples was applied to verify the presence of significant differences ($P \leq 0.05$) in torque force comparison of both genders, considering normality and variance homogeneity. To non-parametric data, or effort perception values, it was applied the *Mann-Whitney* non-parametric test, also to verify the existence of significant differences between both genders.

RESULTS

The torque force results for all door handles presented in figure 3 shows that male subjects performed significant higher torque forces ($P \leq 0.05$) than female subjects.

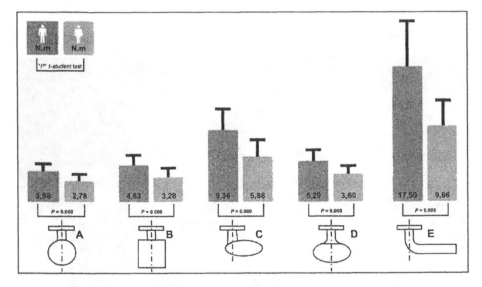

Figure 3. Torque mean and standard deviation values for the five door handles divided by gender and "*P*" value to *t-student* test.

The effort perception results by gender demonstrated that in door handles "A" and "B" males have experienced a greater mean effort to perform torque force despite *Mann-Whitney* test has showed that only door handle "A" had a significant difference ($P \leq 0.05$). For other door handles ("C", "D" and "E") females have experienced a greater mean perception effort to perform maximum torque forces, despite *Mann-Whitney* test has showed that only door handle "E" had a significant difference ($P \leq 0.05$) (see figure 4).

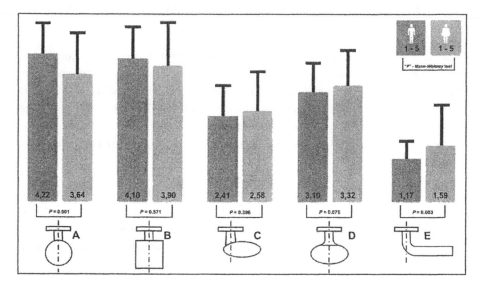

Figure 4. Associated perception effort mean and standard deviation values for the five door handles divided by gender and "P" value to *Mann-Whitney* test.

DISCUSSION AND CONCLUSIONS

The difficulties faced by subjects using manual instruments produce constraints and discomfort even in a simple torque movement to open a door. Some individuals like, for example, women are susceptible groups to those difficulties, because products are usually designed for adult males.

In this study we've tried to understand two groups of independent variables: torque forces and associate effort perception to perform a simulated activity and understand how these variables behave according both genders.

The methodology has respected ethical aspects of scientific research with humans. 180 adult subjects of both genders performed maximum torque in a simulated door open task. The torque force results showed that male subjects performed a significant greater force ($P \leq 0.05$) than females. In door handles whose handle was near rotation axe, "A", "B" and "D", this index has vary between 43% and 47%, and in door handles whose handle was far from rotation axe, "C" and "E", this index has vary between 59% and 77% respectively. Sanders and McCormick (1993) indicate that male exert more muscular force than female and that can vary between 35% and 89%. Particularly, regarding torque force, studies from Imrhan and

Jenkins (1999), Kim and Kim (2000), Shih and Wang (1997), Mital (1986) and Mital and Sanghavi (1986) have also indicated that male perform greater forces than females, which confirms the results found in our study.

Regarding effort perception levels, the data obtained must be considered exploratory and no other similar data are known to enable discussion. The comparison between male and female data showed that male had a greater effort perception to perform the activity in smaller door handles. This tendency was inverted proportionally with the door handle shape change, and it was in door handle "E" that men had a smaller effort perception ($P \leq 0.05$) if compared with women.

In general, the torque and effort perception data analysis indicate an inverse relation between exerted force and perception effort. In this sense, it can be concluded that door handle design influences directly these two variables. This means that in future designs of manual interfaces, these results must be taken in account. Besides aesthetical arguments, the facility of performing forces must be considered in order to facilitate daily activities.

REFERENCES

Adams, S. K. and Peterson, P. J. (1988) Maximum voluntary hand grip torque for circular electrical connectors. Human Factors 30 (6): 733-745.

Ager, C. L. Olivett, B. L. and Johnson, C. L. (1984) Grasp and pinch strength in children 5 to 12 years old. The American Journal of Occupational Therapy 38 (2): 107-113.

Berns, T. (1981) The handling of consumer packaging. Applied Ergonomics 12(03): 153-161.

Caldwell, L. S. Chaffin, D. B. Dukes-Bobos, F. N. Kroemer, K. H. E. Laubach, L. L. Snook, S. H. and Wasserman, D. E. (1974) A proposed standard procedure for static muscle strength testing. American Industrial Hygiene Association Journal 35: 201-206.

Caporrino, F. A. Faloppa, F. Santos, J. B. G. Réssio, C. Soares, F. H. C. Nakachima, L. R. and Segre, N. G. (1998) Population study of strength grip with Jamar dynamometer. Orthopedics Brazilian Journal 33 (2): 150-154. (in Portuguese)

Chaffin, D. B. and Anderson, G. B. J. (1990) Occupational Biomechanics. 2nd Ed. New York: John Wiley & Sons.

Crosby, C. A. Wehbé, M. A. and Mawr, B. (1994) Hand strength: normative values. The Journal of Hand Surgery 19A (4): 665-670.

Deinavayagam, S. and Weaver, T. (1988) Effects of handle length and bolt orientation on torque strength applied during simulated maintenance tasks. In: Aghazadeh, F. (Ed.) Trends in Ergonomics / Human Factors. Amsterdam: Elsevier, pp. 827-833.

Edgren, C. S. Radwin, R. G. and Irwin, C. B. (2004) Grip force vectors for varying handle diameters and hand sizes. Human Factors 46 (2): 244-251.

Fransson, C. and Winkel, J. (1991) Hand Strength: the influence of grip span and grip type. Ergonomics 34 (7): 881-892.

Fullwood, D. (1986) Australian norms for hand and finger strength of boys and girls aged 5 – 11 years. Australian Occupational Therapy Journal 33 (1): 26-36.

Härkönen R. Piirtomaa, M. and Alaranta, H. (1993) Grip strength and hand position of the dynamometer in 204 Finnish adults. Journal of Hand Surgery (British and European volume) 18B (1): 129-132.

Imrhan, S. N. (2003) Two-handed static grip strengths in males: the influence of grip width. International Journal of Industrial Ergonomics 31: 303-311.

Imrhan, S. N. and Jenkins, G. D. (1999) Flexion-extension hand torque strengths: applications in maintenance tasks. International Journal of Industrial Ergonomics, 23: 359-371.

Imrhan, S. N. and Loo, C. H. (1989) Trends in finger pinch strength in children, adults, and the elderly. Human Factors 31 (6): 689-701.

Kim, C. and Kim, T. (2000) Maximum torque exertion capabilities of Korean at varying body postures with common hand tools. In: International Ergonomics Association, 14., 2000a, San Diego. Proceedings of the International Ergonomics Association. San Diego: IEA.

Kong, Y. K. and Lowe, B. D. (2005) Evaluation of handle diameters and orientations in a maximum torque task. International Journal of Industrial Ergonomics 35: 1073-1084.

Mital, A. (1986) Effect of body posture and common hand tools on peak torque exertion capabilities. Applied Ergonomics 17 (2): 87-96.

Mital, A. and Channaveeraiah, C. (1988) Peak volitional torques for wrenches and screw drivers. International Journal of Industrial Ergonômics 3: 41-46.

Mital, A. and Kumar, S. (1998) Human muscle strength definitions, measurement, and usage: Part I – Guidelines for the practitioner. International Journal of Industrial Ergonomics 22: 101-121.

Mital, A. and Sanghavi, N. (1986) Comparison of maximum volitional torque exertion capabilities of males and females using common hand tools. Human Factors 28 (3): 283-294.

Nagashima, K. and Konz, S. (1986) Jar lids: effect of diameter, gripping material and knurling. In: Proccedings of the Human Factors Society. 30th Annual Meeting, pp. 672-674.

Napier, J. R. (1956) The prehensile movements of the human hand. The Journal of Bone and Joint Surgery 38B (4): 902-913.

Pheasant, S. and O'Neill, D. (1975) Performance in griping and turning: a study in hand/handle effectiveness. Applied Ergonomics 6 (4): 205-208.

Rohles, F.H. Moldrup, K.L. and Laviana, J.E. (1983) Opening jars: an anthropometric study of wrist-twisting strenght of children and the elderly [Report No. 83-03]. Kansas: Kansas State University.

Sanders, M. S. and McCormick, E. J. (1993) Human factors in engineering and design. New York: McGraw-Hill.

Shih, Y. C.; and Wang, M. J. J. (1996) Hand/tool interface effects on human toques capacity. International Journal of Industrial Ergonomics 18: 205-213.

Shih, Y. C.; and Wang, M. J. J. (1997) Evaluating the effects of interface factors n the torques exertion capabilities of operating handwheels. Applied Ergonomics 28 (5): 375-382.

Chapter 31

The Uses and Manufacture of Wheelchairs – An Emotional Approach

Paulo Costa, Fernando Moreira Da Silva

Polytechnic Institute of Guarda, Portugal
CIAUD – Research Centre in Architecture
Urban Planning and Design
TU Lisbon
Lisbon, 1349-055, PORTUGAL

ABSTRACT

The wheelchair is a key element in one's integration. Without it, it would not be possible to reduce the differences and allow a larger autonomy facing the environment, in the various activities of the daily life, in one's home or out of it. Professional, leisure or sporting tasks would be jeopardized if the wheelchair didn't exist. Besides the fact that the wheelchair socially represents a symbol of an explicit deficient condition, it establishes, exactly because of that, a "bridge" of communication between the person and the environment. With two main slopes, one referring to the user and the other referring to his environment, it is, in its origin, a laboratory of inquiry where the new technologies adapted to the biomechanical demands of dominant gesture allow a better integration of the disabled person that uses the wheelchair.

The purpose of this work is the study of the "relation" between the wheelchair and its user, in the most various forms, to have the possibility to identify the main points of a wheelchair project. There are inquiries that permit to identify some principles on the manufacturing of a wheelchair that would guide us to important improvements in the concept itself. The introduction of new materials and new ways of approaching the concept of project, the production and selling of wheelchairs should bring some advantages to a big number of people that are completely dependent of its use. New design, lighter materials and an assembly of colours in the several components, associated to a large range of changeable pieces would allow its user to build such an individual wheelchair as determined by his state of mind. The "destruction" of the standard wheelchair or of its social symbol through a strong individual characterization of the wheelchair by its user, when choosing its components, could be implemented in many ways, not neglecting the internet, whose importance is growing in our lives.

The access to programmes online that would allow us to choose the several modules of the chair, according to the preferences of the user and his abilities (functional, cardiovascular conditioning, cognitive ability, vision, perception and motivation), beyond a set of others that would involve the environment conditions, the predicted durability of the chair among other factors. All the documentation of these factors would serve as a reference to the construction, by the market, of "new" wheelchairs.

The result will be a chair that fits the needs of the majority of users, allowing a huge range of combinations in the choice of its components through the internet or through a store to buy those components. It would be important a normalization as a principle that would permit the several manufacturers to work under the same concept, giving the final consumer the benefit of a wider choice of materials on a cheaper price.

Keywords: Ergonomics, Emotional, Customization, Inclusive Design

INTRODUCTION

Since the antiquity that the wheelchairs are represented and constructed for the use of people with motor inabilities. Its evolution has been very slow, with the introduction of new materials in the last few decades but keeping almost unchanged its design. There are ceramic painted from the IV century AC where we can observe Hefesto (Hephaistos), the Greek God of the Metallurgy, the holy blacksmith, seated in a wheelchair pushed by white swans. In accordance with Greek Mythology, his father Zeus launched him of Olimpus Mountain, because he presented defects in the legs and was unformed. He is portrayed in his wheelchair during parties and social meetings. In this good example of social inclusion the wheelchair is seen as a more-value mechanics to help the movement, of pleasure and comfort.

In 1933 Herbert Heverest asked Harry Jennings to build a lighter wheelchair with a structure in a way to be closed. More recent chairs can be seen in any site of specialized manufacturer, with only some alterations in relation to this model of 1933 remaining equal in its essence. If we compare the evolution of the babies strollers during the last 10 years with the evolution of the wheelchairs is not possible to establish a chronological evolutive parallelism, although technically they are similar products. So, we can ask why we have not invested on the wheelchairs design, or in the same way, by comparison, the strollers have evolved so much.

The approach between people and object of use in part lays on its aesthetic component, but this one doesn't have a known mathematical formula of success. All of us "feel" what we see in a different way, as a result of a life experience and formed personality. The emotion is part of our life in everything we touch, helping us when opting what hardly can be rationally justified. When we buy a baby stroller this will have a functional but also aesthetic and emotional component. In the market there are baby strollers four times more expensive than an economic wheelchair and, although the period of use is completely different, the companies continue to produce and to sell luxury baby strollers. The emotional/aesthetic association of the set baby and stroller is so attractive that people spend a lot of money to be able to have the best.

This paper presents a research project focusing in wheelchairs, in a more ergonomic point of view, including, among other variables, the emotional design.

PROJECT PRESENTATION

In the case of a wheelchair it isn't an object to be used for little time, being in most of the cases for the rest of the user's life. It isn't an object of transition, it is an imposition to our body, to our physics involving and to whom surround us, the society. Wouldn't it be logical to invest much more money in its purchase in comparison with a baby stroller? Of course it could be, in the case the state of the art of the wheelchair being so advanced aesthetic/functionally as it is the baby stroller. Although in the market there are many types of chairs, still isn't implemented in a practical and normal way the notion of the customization. To offer the public a modular chair with different colours, materials and textures it isn't yet generalized among brands. It is rather difficult to acquire a personalized wheelchair without being through a special order, being therefore inaccessible to most of the people for being extremely expensive. Besides the price factor, the fact of being able to customize a wheelchair provides the user a chance of being able to feel the chair as an extension of his own body.

This possibility is of an enormous importance because by itself and in many cases the wheelchair is reason for social exclusion. The fact of not feeling well with our own body, or one its extensions; it isn't the best principle to face society. All the wheelchairs that actually exist, in a way or another, are technically well prepared to perform the function of the user's transport, but the question is if they are the best solution for the user's adaptation and integration in society. It must be not only something necessary and indispensable to the displacement but also pleasant to the sight and adapted to the user and never more an exclusion element.

The selection of a wheelchair passes, many times, by a complex process for the common user. There are performance variables to qualify a wheelchair, being resistance to the rolling, control and manoeuvrability, the easiness of placement and transport, and the security, the most common. When one of these criteria is maximized many times one can verify a loss of performance by another criterion (Thacker, 1994). In general we can say that a good performance of a wheelchair can be seen by the way it reduces the constraints of its user and by the adequacy to the different day to day situations (Rodrigues, 1996).

In the resistance to rolling one of the criteria that more money involves in its implementation will be the little weight of the wheelchair. However, duplicating the weight of the wheelchair and maintaining the weight of the user, the increase of the resistance to the rolling does not exceed 8% (Rodrigues and Silva, 1998). Other factors of production and sizing of the structure, such as the weight distribution between the traction wheels and the casters or the unalignment of the wheels, in spite of being the ones that more contribute for the resistance to the rolling reduction, they are not factors that will increase by themselves the price of the wheelchair, more connected with the correct structural dimension. It matters, therefore, to choose a chair which might give confidence to the user, what in most of the cases is impossible to quantify.

The control and the manoeuvrability increase when the weight of the wheelchair/user sets mainly on the traction wheels, becoming however the wheelchair more unstable (Trudel, 1995; Bruno, 1998). So, we can affirm the control and the manoeuvrability is inversely proportional to stability. This is another selection criterion which is not quantifiable because it depends on the user's physical form and on his easiness to surpass obstacles with the wheelchair. Larger agility and physical ability will allow using a more unstable chair with the same security, usufructing of a bigger associated manoeuvrability.

The easiness of placement and transport is nowadays an important factor for who needs mobility and uses the own car. In this point, the weight of the chair is a basic factor, as well as the dimension of the wheelchair when closed.

The security is one of the most important factors when choosing a wheelchair (Bertocci, 1997). Factors that characterize it are the static resistance to the impact, the inflammability of the construction materials and the effectiveness of the brakes.

The components selection for the wheelchair will influence directly its performance (Thacker, 1994; Silva and Silva, 1996). We can consider three main groups: the structure, the wheels and the casters. If we consider the structure, we essentially have chairs of vertical or horizontal latch and rigid structure. As rule we can affirm that the articulated wheelchair is heavier (Rodrigues, 1996), being also less rigid, what will increase the necessary effort for its propulsion. The used materials as the graphite in epoxy resin or other composite materials are nowadays widely used because of its lightness and resistance, being however not cheap.

This paper addresses the manual wheelchairs, prescribed for clinical situations with injuries of neurological level until C7. The regulations ANSI/RESNA or ISO can help when choosing a wheelchair. The commercial catalogues supply several measures of its products, from the width of the seat, depth, angle of lean, among others. This information is essential for correct choice of the product under the ergonomics point of view. The correct use of the wheelchair depends on the observation of these factors when buying the wheelchair. Many are adjustable with defined maximum and minimum measures what makes possible an evolution of the user with the chair throughout the years, without being necessary to invest again in the purchase of another one. Some measures to have in account are however the recess of the seat between the thighs and the lateral part of the chair, the inclination of the back/seat and of the seat in itself, the height of the supports for the arms and for the feet, the angle between the arm and forearm when the hand grasps the highest part of the propeller rings, the height of the back of the seat and the distance between the front edge of the seat and the posterior part of the knees. There are so many variables that it becomes necessary to know our own ergonomic measures to choose correctly.

Sometimes one centimetre more or less may make the difference for a correct use (Britell, 1990). There are several pathologies associated to the use of the wheelchair whose effects can be "amplified", if there wasn't an adequate choice of the wheelchair (Rodgers, 1994; Taylor, 1995), for example in the shoulders the conflict and muscle-tendons injury frames. Other pathologies are possible to be verified in the elbows, fists and hands, canal neuropathies, tendon injuries and the development of pressure ulcers.

Summarizing, we have some selection criteria which pass for technical questions, ergonomics and of performance widely described and studied in literature.

Other important factors to consider are the customization character of the wheelchair, its aesthetic or its emotional design. Ergonomic issues as psycho-sociology, psychophysiology, anthropometry and biomechanics, among others, must be taken in consideration when one develops a project for a wheelchair. All these issues must be addressed in a sustainable and integrated way, with the only intention of being able to provide one better quality of life to whom makes use of these physical supports, being by the intervention in the own mechanical systems, being by a different look to the person who uses a wheelchair. Under the mainly commercial point of view it also the companies interest to invest in all these research areas in a equitable way. As Thomas J Watson would say: "*Good Design is Good Business*". It is necessary not to look only to the mechanical or anthropometric interventions but also to think about the emotional issues, in a way to not stopping the progress. Also according to Thomas J. Watson, "*Whenever an individual or business decides that success has been attained, progress stops*".

Developed by students of the course of Equipment Design at the Polytechnic Institute of Guarda, Portugal, and guided by the researcher in the scope of a project course, we are able to present one of many possible solutions for a new image of the wheelchair. Studied and ergonomically adapted to each user, the prototype will be able to contain customizable solutions, with the introduction of new materials, colours and textures where the emotional component will be able to play a new role because of the different aesthetic of the wheelchair.

Figure 1. Prototype of the developed manual wheelchair

The adaptation of several devices to the chair, wheels of fast removal, debatable seat, rigid structure for a bigger stability and rigidity among others factors, allow a new project solution.

CONCLUSIONS

The possibilities in the modulation of a wheelchair are immense and we believe, considering the example of what happened with other products as the baby strollers, it is also possible to imprint a new dynamic to the wheelchairs' industry, for the wellbeing of all who cannot live without them. We have to think about them as an extension of the body, and being thus, psychologically, the emotional component will have a major importance in its choice and use. The wheelchair must be transformed in an element of integration and social inclusion and not of segregation.
In the next phase of the project we intended to achieve the production of the chair itself and to essay it in terms of ergonomic optimization. With this research project we aim to be able to give a contribution to knowledge through an innovative point of view of the wheelchair, more adapted to now a day's user, customizable, cheaper and inclusive.

REFERENCES

ANSI/RESNA (1998) - Wheelchairs Standards, Vol1 and 2. Resna Press
Bertocci, G., Karg, P., Hobson, D. (1997) - "*Wheeled Mobility Device Database for Transportation Safety Research and Standards*", Assistive Technology, RESNA Press

Bruno, C., Hoffman, A.H., (1998) - *"Modeling the Dynamic Stability of an Occupied Wheelchair"*, Rehabilitation Engineering and Assistive Technology Society of North America - RESNA - 98 Minneapolis

Buckley, S.M., Bhambhani, Y.N., Madill, H.M. (1995) - *"The effects of Rear Wheel Camber on Physiological and Perceptual Responses During Simulated Wheelchair Exercise"*, Rehabilitation Engineering and Assistive Technology Society of North America - RESNA - 95, Canada

Davis, J., Growney, E., Johnson, M., Iuliano, B., An K-N (1998) - *"Three-Dimensional Kinematics of the Shoulder Complex During Wheelchair Propulsion: a Technical Report"*. J. Rehab Res Dev

Kottke, F., Lehmann, J. (1990) - Handbook of Physical Medicine and Rehabilitation, 4th ed. Philadelphia. W.B. Saunders

Rodgers, M., Cayle, W. (1994) - Biomechanics of Wheelchair Propulsion During Fatigue. Arch Phys Med Rehab

Rodrigues, P.E.L.B. (1996) - *"Cadeira de Rodas Manual e o seu Desenvolvimento"*, I Jornadas Tecnológicas de Apoio e Reabilitação Soluções e Serviços, IV Feira Internacional de Ajudas Técnicas e Novas Tecnologias para Pessoas com Deficiências, Porto

Rodrigues, P.E.L.B., Silva, A.F. (1996) - *"Relationship Between Frame Deformation and Misalignment for Manual Wheelchairs"*. The Canadian Seating and Mobility Conference, Toronto

Rodrigues, P., Silva, A. (1998) - *"Cadeira de Rodas Manual"*, INR

Silva, A.F., Silva, P.C. (1996) - *"Cadeiras de Rodas Ultra-Leves para Deficientes Motores - Da Ideia ao Protótipo"*, I.P.B.

Taylor, D:, Williams, T. (1995) - *"Sports Injuries in Athletes with Disabilities: Wheelchair Racing"*. Paraplegia

Thacker, J.G., Sprigel, S.H., Morris, B.O. (1994) - *"Understanding the Technology When Selecting Wheelchairs"*, Resna Press

Trudel, G., Kirby, R.L., Bell, A.C., (1995) - *"Experimental Location of the Axis of Rotation for Rear Stability of Wheelchairs with and Without Camber"*, Rehabilitation Engineering and Assistive Technology Society of North America - Resna - 95 Vancouver, Canada

Chapter 32

Ergonomic Design of Diagnostic Ultrasound Transducers: Analysis, Redesign/Modeling, Simulation/Assessment and Product

Luis C. Paschoarelli[1], Helenice G.Coury[2], Ana B.Oliveira[2], José C.P.Silva[1]

[1]Department of Design – Faculty of Architecture, Arts and Communication
Univ. Estadual Paulista, Bauru, SP 17033-360, BRAZIL

[2]Department of Physical Therapy, Federal University of Sao Carlos
São Carlos, SP 13565-905, BRAZIL

ABSTRACT

The development of new technology has resulted in more precise medical ultrasound diagnostics; however, the intensive use of poorly designed equipment has been closely linked to musculoskeletal problems among physicians working with ultrasound. This study presents procedures for the analysis, redesign, modeling and simulation/assessment of diagnostic ultrasound transducers. The assessment was performed in four phases: 1 – evaluation during simulated activity and analysis of movement (electrogoniometry) with two commercial transducers; 2 – "redesign of the product", in which the aspects of "concept product", "detailed product" and "modeling of the mock-ups" were considered; 3 – evaluation during simulated activity with one mock-up and with two commercially available transducers. 4 – project revision and the development new mock-ups and a subjective assessment (perception scale) of all tested equipment. The project revision resulted in two new mock-ups that required smaller range of motion ($P \leq 0.05$), better load distribution and achieved better acceptability. Systematic methodology of ergonomic design can be used to develop safer and more comfortable products.

Keywords: Ergonomic Design, Usability, Redesign, Modeling, Simulation.

INTRODUCTION

Product design includes methods that vary according to technological evolution and human necessity. In recent decades, ergonomic aspects of product design have gained precedence, maximizing safety, functionality and usability. However, many products, including medical and hospital equipment, are still produced without such considerations. According to Akita (1991), although this type of equipment is characterized by high technology, the complexity of its handling and use results in low usability. One example is that the use of ultrasound examination equipment is associated with musculoskeletal problems in physicians (Wihlidal and Kumar, 1997). Their occupational activities involve static postures of the trunk and one upper limb, while the other upper limb grasps and performs repetitive movements with a transducer. Studies associating the activities of physicians who work with ultrasound and the development of work-related musculoskeletal disorders (WRMD's) have been carried out in the USA (Vanderpool et al. 1993 and Necas 1996), Canada (Russo et al., 2002), Australia (Gregory, 1998), Israel (Schoenfeld et al., 1999), Italy (Magnavita et al., 1999), the United Kingdom (Ransom, 2002) and Brazil (Barbosa and Gil Coury, 2003; Paschoarelli et al. 2008). Some of these studies propose redesigning the work environment and/or the equipment to minimize the associated problems.

Due to the association of WRMD's and the biomechanical effort exerted by the upper limbs of physicians working with ultrasound, wrist posture should be the main variable considered in the redesign of transducers. Paschoarelli and Gil Coury (2002) and Paschoarelli et al. (2008) found great amplitude of movement (27.7°) during simulated breast ultrasound examinations, and Barbosa and Gil Coury (2003) reported that such tasks are characterized by awkward postures and repetition of movement, which can be a risk for the musculoskeletal system.

Another important biomechanical aspect to be considered in the ergonomic design of ultrasound transducers is pressure imposed on the soft tissues surrounding the phalanges during occupational activities. Tichauer and Gage (1977) stress the importance of analyzing the pressure at frequent points of hand-instrument interface because arteries, veins and nerves may be compressed, resulting in inflammation, calluses, and other types of damage to the hand.

According to Hall (1997), the design of handheld instruments exerts great influence on the pressure points of the hands, and Putz-Anderson (1988) states that excessive or continuous pressure on the palm and/or phalanges indicates that a handheld instrument is badly designed. Muralidhar et al. (1999) report that the distribution of force applied by the hand is not uniform in any manual activity. According to Tichauer and Gage (1977), the satisfactory design and evaluation of handheld instruments depends on an understanding of the gripping method and the intensity of pressure at those points. To measure pressure on the tissues of the hand, force sensing resistors (FSR's) are widely used, as acknowledged in Fellows and Freivalds (1991), Radwin and Oh (1992) Bishu et al. (1993), and Hall (1997).

The ergonomic design of handheld instruments may also be enhanced by subjective feedback from usability tests (Motamedzade et al., 2007) or evaluations of other redesigned instruments (Li, 2002). Borg (1998) has suggested the use of perception criteria for the ergonomic assessment of products. Such evaluation can be carried out using visual scales, which are considered to be the most sensitive type (Collins et al., 1997). Criteria such as the perception of positive attributes (such as acceptability) or of negative attributes (such as discomfort) can be included, preferably in unipolar scales (which are more precise).

Considering that the occupational activities of physicians working with ultrasound involve ergonomic risks, and that the ultrasound transducer contributes to muscle overload of upper limb extremities, ergonomic design procedures were used to redesign the ultrasound transducer in order to evaluate the methodological phases of analysis, redesign, simulation and product.

METHODOLOGY AND RESULTS

The redesign of the ultrasound transducer according to ergonomics principles was executed in four phases: 1 – analysis of simulated activity using two commercial transducers; 2 – identification of problems and redesign of the product; 3 – evaluation of the new product; 4 –project revision and development of proposals for the final design. Specialized equipment was used and specific procedures were developed for each phase of this study, which was carried out under controlled conditions. The participation of volunteers during usability evaluations was approved by the Ethics in Research Committee (Protocol 03/2003).

PHASE 01 - ANALYSIS OF COMMERCIAL TRANSDUCERS

In this first phase, the activities of a physician working with ultrasound were simulated in a laboratory in order to analyze two different commercial transducers. Eighteen subjects participated (9 men and 9 women), with mean age of 23.72 (± = 3.01 years). For the usability tests, a specially developed activity simulator (see Figure 1 - left) was used. According to Gawron et al. (1996), a simulator is capable of demonstrating interface problems. A Gaumard Scientific mannequin (Model S230.4) was used for the simulation of clinical breast examinations, facilitating the development of a contact pattern for the transducers and eliminating ethical questions. A biaxial electrogoniometer (Bimetrics, XM65) and a torsiometer (Biometrics, Z110) were used for the kinesiological evaluation. The simulation began by familiarizing subjects with the prescribed task (ultrasound breast examination) (Figure 1 - right). The sequence of transducers used in the assessment was randomized. Detailed results of the movement evaluation were reported by Paschoarelli and Gil Coury (2002).

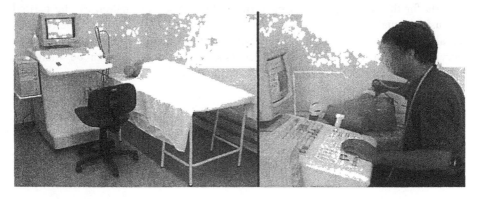

Figure 1. Ultrasound activity simulator (right). Simulation of the activity (left).

PHASE 02 - REDESIGN OF THE PRODUCT

Based on the ergonomic problems identified in previous studies and on the simulated analysis with commercial transducers described above, a project method was used corresponding to that of Norris and Wilson (1997). In this manner, the usability requirements for the new design were defined (minimizing extreme wrist postures, defining an appropriate size for grasping, avoiding sharp edges, and maximizing balance and grip adherence) based on the recommendations of Konz (1979), Meagher (1987), Putz-Anderson (1988), Mital and Kilbom (1992), McCormick and Sanders (1992); Lewis and Narayan (1993), Hedge (1998) and

Cacha (1999). Subsequently, the "product concept" stage started with a Brainwriting session, because according to Vangundy (1999) this can result in more ideas than classic brainstorming. After this, "product detailing" began, using an adaptation of the "combination of elements" technique described by Kaminski (2000) and producing satisfactory results (see Figure 2 – A). During this step, anthropometric modifications as well as digital product modeling were completed (see Figure 2 – B) in order to verify the functional systems. Finally the physical modeling took place, using rubber molds (see Figure 2 – C) and polyurethane foam for the mock-ups (see Figure 2 – D), because according to Säde et al. (1998), this is an efficient means for detecting usability problems.

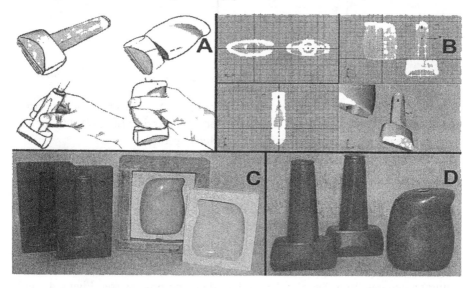

Figure 2. Results of the product concept/detailing (A). Digital modeling for functional analysis (B). Physical modeling using rubber molds (C). Results of the modeling in polyurethane foam: Mock-up 1(D).

PHASE 03 - ANALYSIS OF MOCK-UPS AND COMMERCIAL TRANSDUCERS

In this phase, the same simulation procedures described in Phase 1 were repeated in order to analyze and compare the ergonomic conditions of commercial transducers and Mock-up 1(see Figure 3 - left).

Ten subjects (5 men and 5 women) with mean age of 22.25 (± 1.49 years) participated in this stage. As well as the equipment described in Phase 1, force sensing resistors (FSR) (Interlink Eletronics – Mod. #402) were used on the palmar face of the thumb, the index, the middle and the ring fingers (see Figure 3 – center). Detailed results of the wrist movement evaluation are reported in Paschoarelli et al. (2008). Data collected from the FSR (see Figure 3 – right) indicated that Mock-up 1 presented a better and more homogenized load distribution, because superior loads were registered on the distal phalanges of the middle and ring fingers, whose participation in this type of palmar grip is normally small.

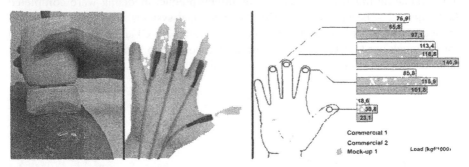

Figure 3. Simulated use of Mock-up 1 (left). Arrangement of FSR on the palmar face of the thumb, index, middle and ring fingers (center). Load results (in kgf/1000) for each sensor, for each one of the transducers and for Mock-up 1.

PHASE 04 - PROJECT REVIEW, DEVELOPMENT AND EVALUATION

After establishing that Mock-up 1 had attained a better load distribution, the product review phase began with the development of two new mock-ups (2 and 3) using procedures from Phase 2: the product detailing process (see Figure 4 - left), digital modeling (see Figure 4 - center) and physical modeling (see Figure 4 - right).

A new evaluation process was started for Mock-ups 2 and 3 in which 10 subjects (5 men and 5 women) with mean age of 24.00 (± 2.05 years) took part. The procedures for evaluating kinesiological aspects and palmar pressure distribution were the same described in Phase 3, as well as the statistical analysis of the results. The test subjects' perceptions of the transducer models were also analyzed, considering a negative attribute (discomfort), as well as a positive attribute (acceptability). After a session of simulated ultrasound task performance similar to that of previous phases, perceptual evaluation forms were filled in. In order to describe their perceptions, the subjects drew a vertical line on graphic scales, which represented the levels of discomfort and acceptability that they sensed. The Wilcoxon test ($P \leq 0.05$) was used to detect differences among the values obtained for each of the transducers.

Ten experienced physicians who work with ultrasound, having mean age of 41.9 (±
8.1 years) and mean ultrasound experience of 15.3 (± 6.2 years), participated
exclusively in the evaluation of perceived discomfort and acceptability. These
physicians used the transducers in a simulated examination and classified their
acceptability and order of preference. This group also presented suggestions for
additional improvements to the transducer.

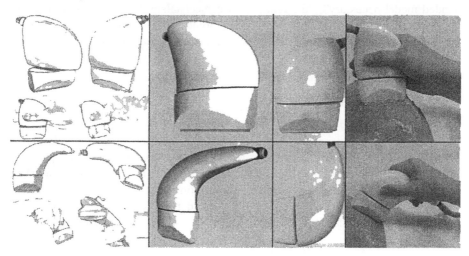

Figure 4. Development of Mock-up 2 (top) and Mock-up 3 (bottom). Conceptualization
(left), digital modeling and modeling in polyurethane foam (center), simulated use
(right).

Kinesiological evaluation with electrogoniometry demonstrated that Mock-up 2 and
Mock-up 03 facilitated better posture patterns for the upper limb than the
commercially available transducers (detailed results of this evaluation were
reported by Paschoarelli et al., 2008). As for the FSR results, Mock-up 02
presented a higher load on the index finger and smaller loads on the middle and ring
fingers due to its extended handle, which causes both a better load distribution in
the palm and a reduction to the load on the phalanges of the middle and ring
fingers. Mock-up 03, however, produced higher loads on the other fingers due to
its grip format, which accentuated the phalanges region. Although numerically
higher, the loads verified for Mock-ups 02 and 03 confirm that there was a better
load distribution in the new tranducer designs.

The results of the perceptual evaluation revealed that acceptability was higher for
Mock-ups 2 and 3, decreasing progressively for Mock-up 1, Commercial 1 and
Commercial 2. The mean discomfort levels followed suit; that is, they were higher
for Commercial 2 and Commercial 1 and decreased progressively for Mock-up 1,
Mock-up 3 and Mock-up 2. The Wilcoxon test results demonstrated statistically

significant differences ($P \leq 0.05$) among the transducers, except Mock-ups 2 and 3.

The physicians' evaluation of the transducers resulted in the following preference order: Mock-up 3, Mock-up 2 and Commercial 1. There was consensus among all the physicians that mock-ups 2 and 3 provided a better and more adequate grip. The physicians made the following observations and suggestions: "the position of the handle (parallel to the sensor) is a positive aspect"; "Mock-up 3 (in particular) facilitates movement during examinations, but its right-handed design makes a left-handed model necessary"; "the rotating base mechanism is controversial, and can generate more functional problems than solutions". Other suggestions for additional improvements included increasing the grip area and adding grooves to better accommodate the fingers.

CONCLUSIONS

This design project demonstrated that the systematic application of methodological procedures for evaluation and analysis during the development phase of a product, in this case an ultrasound transducer, provided alternatives that diminished biomechanical loads and increased the perception of positive attributes. These methodological procedures exemplify a safe and reliable evaluation method for ergonomic design based on statistical analysis, resulting in safer and more comfortable products.

This study was partially funded by FAPESP (Proc. 99/12147-7), and was awarded the Werner von Siemens Prize for Technological Innovation – Brazil/2008 (2nd Place – Health Modality).

REFERENCES

Akita, M. (1991) Design and Ergonomics. Ergonomics 34 (06), 815-824.

Barbosa, L.H., Gil Coury, H.J.C. (2003) Analysis of wrist movements in the activity of ultrasonographers. Brazilian Journal of Physical Therapy. 7 (2), 179–185.

Bishu, R.R., Wang, W. and Chin, A. (1993) Force distribution at the container hand/handle interface using force-sensing resitors. International Journal of Industrial Ergonomics. 11(03): 225-231.

Borg, G. (1998) Borg's Perceived Exertion and Pain Scales. Human Kinetics, Champaign, 101pp.

Cacha, C.A. (1999) Ergonomics and safety in hand tool design. New York: Lewis Publishers.

Collins, S. L.; Moore, R. A. and McQuay, H. J. (1997) The visual analogue pain intensity scale: what is moderate pain in millimetres? Pain. 72 (01-02): 95-97.

Fellows, G. L. and Freivalds, A. (1991) Ergonomics evaluation of a foam rubber grip for tool handles. Applied Ergonomics. 22 (04): 225-230.

Gawron, V. J., Dennison, T. W. and Biferno, M. A. Mock-ups, physical and eletronic human models, and simulations. In: O'Brien, T. G. and Charlton, S. G. Handbook human factors testing and evaluation (1996). New Jersey: Lawrence Erlbaum Associates Publishers, pp. 43-80.

Gregory, V. (1998) Musculoskeletal injuries: na occupational health and safety issue in sonography. Sound Effects [Educational Supplement].

Hall, C. (1997) External pressure at the hand during object handling and work with tools. International Journal of Industrial Ergonomics. 20(03): 191-206.

Hedge, A. (1998). Design of hand-operated devices. In: Stanton, N. (Ed.), Human Factors in Consumer Products. Taylor & Francis, London, pp. 203–222.

Kaminski, P.C. (2000). Developing products with planning, creativity and quality. Rio de Janeiro: Livros Técnicos e Científicos. (*in Portuguese*)

Konz, S. (1979) Work Design: industrial ergonomics. New York: John Wiley & Sons.

Lewis, W.G. and Narayan, C.V. (1993) Design and sizing of ergonomic handles for hand tools. Applied Ergonomics. 24 (05): 351-356.

Li, K.W. (2002) Ergonomic design and evaluation of wire-tying hand tools. International Journal of Industrial Ergonomics. 30, 149–161.

Magnavita, N., Bevilacqua, L., Mirk, P., Fileni, A., Castellino, N. (1999) Work-related musculoskeletal complaints in sonologists. Journal of Occupational and Environmental Medicine. 41 (11), 981–988.

McCormick, E.J. and Sanders, M.S. (1992) Human Factors in Engineering and Design. New York: MacGraw Hill.

Meagher, S. W. (1987) Tool design for prevention of hand and wrist injuries. Journal of Hand Surgery. 12A (05): 855-857.

Mital, A. and Kilbom, A. (1992) Design, selection and use of hand tools to alleviate trauma of the upper extremities: Part I – guidelines for the practitioner. International Journal of Industrial Ergonomics. 10 (1-2): 1-5.

Motamedzade, M., Choobinehb, A., Mououdic, MA., Arghamid, S. (2007). Ergonomic design of carpet weaving hand tools. International Journal of Industrial Ergonomics. 37, 581–587.

Muralidhar, A., Bishu, R. R., Hallbeck, M.S. (1999) The development and evaluation of an ergonomic glove. Applied Ergonomics 30 (06): 555-563.

Necas, M. (1996) Musculoskeletal symptomatology and repetitive strain injuries in diagnostic medical sonographers: a pilot study in Washington and Oregon. Journal of Diagnostic Medical Sonography.. 12 (6), 266–273.

Norris, B., Wilson, J.R. (1997) Designing Safety Into Products—Making Ergonomics Evaluation a Part of the Design Process. Institute for Occupational Ergonomics/University of Nottingham, Nottingham, 30pp.

Paschoarelli, L.C., Gil Coury, H.J.C. (2002) Preliminary evaluation of the movements of wrist movements in the activities simulated of ultrasound examination breast. In: Proccedings ABERGO – VI Latin American Congress of Ergonomics. Recife: ABERGO. (*in Portuguese*)

Paschoarelli, L.C., Oliveira, A.B. Gil Coury, H.J.C. (2008) Assessment of the ergonomic design of diagnostic ultrasound transducers through wrist movements and subjective evaluation. International Journal of Industrial Ergonomics 38 (00): 999-1006.

Putz-Anderson, V. (1988) Cumulative trauma disorsders: a manual for musculoskeletal diseases of the upper limbs. London: Taylor & Francis.

Radwin, R. and Oh, S. (1992) External finger forces in submaximal five-finger static pinch prehension. Ergonomics. 35 (03): 275-288.

Ransom, E. (2002) The Causes of Musculoskeletal Injury Amongst Sonographers in the UK. Society of Radiographers, London, 31pp.

Russo, A., Murphy, C., Lessoway, V., Berkwitz, J. (2002) The prevalence of musculoskeletal symptoms among British Columbia sonographers. Applied Ergonomics. 33 (5), 385–393.

Säde, S., Nieminen, M., Riihiaho, S., 1998. Testing usability with 3D paper prototypes—case Halton system. Applied Ergonomics. 29 (2), 67–73.

Schoenfeld, A., Goverman, J., Weiss, D.M., Meizner, I. (1999) Transducer user syndrome: an occupational hazard of the ultrasonographer. European Journal of Ultrasound. 10 (1), 41–45.

Tichauer, E.R.; Gage, H. (1977) Ergonomic principles basic to hand tool design. American Industrial Hygiene Association Journal. 38 (11): 622-634.

Vanderpool, H.E., Friis, E.A., Smith, B.S., Harms, K.L. (1993) Prevalence of carpal tunnel syndrome and other work-related musculoskeletal problems in cardiac sonographers. Journal of Occupational Medicine. 35 (6), 604–610.

VanGundy, A. B. and Naiman, L. (2007) Orchestrating Collaboration at Work: Using Music, Improv, Storytelling, and Other Arts to Improve Teamwork. BookSurge, 278pp.

Wihlidal, L.M., Kumar, S. (1997) An injury profile of practicing diagnostic medical sonographers in Alberta. International Journal of Industrial Ergonomics 19 (3), 205–216.

Chapter 33

Usability: Timeline and Emotions

Alexandre Barros Neves, Maria Lúcia L. R.Okimoto

Universidade Federal do Paraná
Curitiba, Paraná, Brazil

ABSTRACT

The objective of this study is to identify the changes in emotional relationship in different stages of the product interaction. The premises are based on the grounds of Woolley (2003), which links the different types of pleasure elicited by products at each stage in time, promoting different interaction. To evaluate these differences, subjects were selected according to the desired user profile. This group is undergoing evaluations in four different moments of contact with the new washing machine. The results are displayed by a vector representation timescale containing the different stages of interaction and the corresponding emotions outsourced.

Keywords: Design and Emotion, Usability, Semantic Differential

INTRODUCTION

Domenico de Masi (2000), in his studies points to the upward trend in contemporary objects that have the ability to interact with the User in the emotional sphere. The author's argument is centered on the study of the most important values of modern societies: globalization, subjectivity, femininity, aesthetics and emotion.

Damázio and Meyer (2008) also confirm the perception that people demands have advanced from the physical field to the emotional field.

Researches conducted in this area are trying to evaluate and measure the emotions that objects are able to evoke in individuals. The evaluation methods proposed in the thematic design emotion initially sought the tacit knowledge of the interactions, through the formulation of semantic representative with the Semantic Differential presented by (Osgood, Suci and Tannenbaun, 1957). Some studies using photographs of products as a stimulus for the evaluation (DESMET, 2003, Silva et al. 2008; MCDONAGH et al. 2002; JORDAN, 2002; MONDRAGON et al. 2005). Several authors such as Norman, Hekkert, Nagamashi, Helander, etc. contributed to leverage the research in the subject of Design and Emotion, and inspire us in the search of some deeper questions about the emotions caused by products regarding temporal aspects, considering the dynamic changes of emotions in the time line.

We aim to present a proposal to these issues that concerns us, such as: How to maintain the positive emotions elicited by the products over time?

What type of interaction can occur between User and product during the use of the product? Therefore, we sought to investigate aspects of how emotions change in time between individual and object. The objective of this proposed study is to identify the changes in emotional relationship applied in different stages of interaction with a household appliance.

The premises of this study are based on the grounds of Woolley (2003), which links the different types of pleasure elicited by the product at each stage in time, promoting different types of interaction. The author notes that even for products with greater durability, they are easily replaced and discarded when they do not provide pleasure during use any more.

Woolley (2003) also emphasizes that the pleasure of using the products is related to time, and summarizes this relationship through a cycle of pleasure and satisfaction over time, which occurs at five levels: In the pre-order when there is an anticipation pleasure to see or have information about the product and the consumer comes to believe that the acquisition would extend the pleasure; Post-purchase product, Novelty period is the period of excitement after buying the consumer product, operates by first time and discovers its operation. In the period of use, during which there is a reduction of pleasure because of the dominance of the operation of the product. In this critical phase can occur with the purchased product which can occur at the transition from pleasure to dissatisfaction stage, where the pleasure to use turns into dissatisfaction, by factors such as damage, wear, poor performance, opportunities for purchasing new product, or even boredom on the long-familiar product; Phase high satisfaction, a rare situation in which the product achieves a high level of satisfaction for a long time, and passes to be considered a "product for life", as objects of heritage and antiques. At this point, the object transcends its practical function.

METHODS

This study is being developed along with a manufacturer of household appliances. The proposal is to evaluate a washing machine for the middle class population. Users receive the products in their homes to carry out the proposed tests and these are the provision of these for their own use. To perform the evaluation experimental subjects were selected according to the desired User Profile, product focus. This group is undergoing evaluations in four different moments of contact with the new washing machine. These moments are: a) emotional image of the product category, without any visual stimulus, b) presentation of a visual stimulus of a new product, c) first physical contact with the product and d) after a couple of months of frequent use product. A total of 10 individuals are evaluating the products, in order stay slightly above the recommendation of Nielsen(1993), for the optimization of number of evaluators and the amount of usability problems in a given system. The experiment is being carried out with individuals that took part on the field test of a new model of washing machine. The field test is a normal procedure used by the manufacturer with the purpose of evaluating pre-production products and identifying possible problems before they go in production. Using the field test as a source for this kind of experiment has a number of advantages: as the test is done with a new product and this assures that none of the users has had any previous experience at any level with the product; the users are screened in advance by the manufacturer, assuring they match the target profile; the steps of the experiment can be better controlled as the manufacturer controls the delivery of the product to the homes of the users; the products are tested in the real context of use.

The following variables were controlled in the development of this study: the analysis of the various steps were performed with the same individuals; tests were made with a new product in order to eliminate any possibility of previous user's experiences with that product, and people were previously selected base on specific parameters (socio-economic status, age, number of family members, etc..). The target audiences for the study are women aged between 25 and 55 years, economic class B1 and B2, as the criterion Brazil Economic Classification, version 2008.

INSTRUMENTS FOR DATA COLLECTION

We used the semantic differential to assess the responses to emotional stimuli generated by the product. It was necessary for data collection to develop a questionnaire with a scale of semantic differential (Osgood, Suci & Tannenbaum, 1957), with the intent to measure the significance of emotion in every step of User Interaction with the washing machine.

Phase 1. Construction of the list of adjectives

The first phase was devoted to gather the adjectives to compose the semantic differential scales. Adjectives must meet the essential presuppositions of design, integrating concepts of functionality, usability and pleasure. This phase was composed of several steps: discrimination, selection, filtering and consolidation of adjectives. This process is represented graphically in the flowchart in Figure 1.

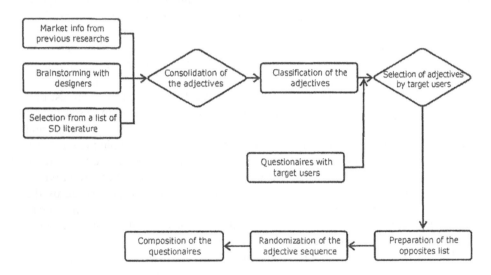

Figure 1.Flowchart for the development of semantic differential adjectives.

At the beginning of this study, we sought to collect as many concepts for the product and similar adjectives as possible. We attempted to exhaustively list a great number of concepts in order to expand the possibilities of characterization of a washing machine. These adjectives have been refined in subsequent phases, selecting the most representative for the target audience of the product in question. We used three sources for the preparation of adjectives: Source 1-Electrolux, 2009; Source 2 - Brainstorming with experts and Source 3 - PEREIRA, 1986

The first source of adjectives was extracted from market surveys of washing machines run by Electrolux. These surveys were conducted between the years 2007 and 2009 in São Paulo city, and to target consumers in various socioeconomic classes, according to the type of washing machine. We obtained from this study a list of 37 adjectives from the consumers' views on the concept of an ideal washing machine.

The second source for the list of adjectives was a divergent thinking session, using the technique of brainstorming. Took part in this exercise, eight designers with experience in home appliances, with different specialties: product, interaction, graphics, and finishes. The designers were instructed to list positive and negative adjectives relevant to the characterization of a washing machine. In the first stage of the process 233 adjectives were obtained. After a filtration of the adjectives to eliminate repeated, inappropriate or out of context terms, 120 adjectives remained.

The third source used for the selection of adjectives was Pereira which presents a study of 200 adjectives most "significant" held in Rio de Janeiro in the 1980s (Pereira, 1986). According to the author, these adjectives were selected using the Shannon index H which is a measure of the degree of importance of information, through an evaluation of both the frequency and diversity of use of adjectives (...) "(Pereira, 1986). Six designers directly involved with the design process of washing machines analyzed these adjectives. The designers were instructed to mark the adjectives, which they considered relevant in a washing machine. A total of 96 adjectives were selected in this third step, thus creating the third list of adjectives

Phase 2 - Consolidation of Adjectives

To obtain a single list of adjectives, we compiled the data of the three lists of adjectives. Repeated adjectives were identified, leaving only a single list of distinct adjectives. The adjectives were grouped by meanings, similarity, and proximity.
Thus, for each cluster, an adjective was taken as representative of the group. The representative adjectives were used to compose a new list of adjectives. Most adjectives mentioned by individuals were considered by experts as positive adjectives. We then decided to convert all adjectives in their positive form, and list their opposites only when the final phase of development of the semantic differential scales.

The consolidation exercise of adjectives was a whole day workshop conducted by a group of five people, three women and two men, all were graduates. Throughout the process, there were extensive discussions, with solutions being found through consensus among all participants, as Figure 2.

Figure 2. The consolidation exercise of adjectives

As a result of this step, 52 adjectives were obtained. Next, a questionnaire was administered to a group of 10 women consumers. The task of consumers was to list the characteristics of an ideal washing machine. At this stage, we had 6 new relevant adjectives. These new adjectives were added to the list. thus amounting to a total of 58 adjectives.

For the following phase, participants were individually asked to quantify on a scale of 1 to 5, the degree of importance of each of 58 adjectives. The goal of this step was to select the 10 most relevant adjectives for each of the following dimensions: functionality, usability, and pleasure. Table 2 shows the scores of the adjectives selected. After this phase, the opposites of the 30 selected adjectives were listed, to create the semantic differential scales. This exercise was undertaken with the aid of a dictionary of synonyms and antonyms, and has also used the negative adjectives mentioned in the early stages.

Table 1. Selection of 10 adjectives with the highest score

| | ADJECTIVES | | |
	+ POSITIVE	**- NEGATIVE**	**Average**
Functionality	**durable**	fleeting	4,9
	resistant	fragile	4,9
	technological	rudimentary	4,0
	efficient	inefficient	4,9
	economic	wasteful	4,8
	ecologic	anti-ecological	4,9
	automatic	manual	4,5
	silent	noisy	5,0
	strong	weak	4,9
	multifunctional	basic	4,3
Usability	**simple**	complicated	4,0
	easy to use	difficult to use	4,9
	controllable	uncontrollable	4,3
	safe	unsafe	4,5

	easy to clean	difficult to clean	4,6
	practical	laborious	4,9
	organized	disorganized	4,1
	ergonomic	anti-ergonomic	4,5
	smart	stupid	3,8
	comfortable	uncomfortable	4,0
Pleasure	meticulous	sloppy	4,6
	beautiful	ugly	3,6
	modern	old fashioned	4,0
	pleasant	unpleasant	3,5
	amazing	commonplace	3,7
	honest	dishonest	3,7
	stimulant	boring	3,6
	innovative	traditional	4,4
	clean	dirty	4,8
	Dynamic	Still	4,1

Application Questionnaire

Were structured four different sequences of adjectives, in order to avoid any tendency of the users in their responses, such as fatigue at the end of the fill, remembering what was answered earlier or insecurity in early returns. Each participant received in each step a different sequence of adjectives, as shows in table 3.

Table 3. Distribution of questionnaires

	Phase 1	Phase 2	Phase 3	Phase 4
Participant 1	Quest. A	Quest. B	Quest. C	Quest. D
Participant 2	Quest. B	Quest. C	Quest. D	Quest. A
Participant 3	Quest. C	Quest. D	Quest. A	Quest. B
Participant 4	Quest. D	Quest. A	Quest. B	Quest. C
Participant 5	Quest. A	Quest. B	Quest. C	Quest. D
Participant 6	Quest. B	Quest. C	Quest. D	Quest. A
Participant 7	Quest. C	Quest. D	Quest. A	Quest. B
Participant 8	Quest. D	Quest. A	Quest. B	Quest. C
Participant 9	Quest. A	Quest. B	Quest. C	Quest. D
Participant 10	Quest. B	Quest. C	Quest. D	Quest. A

When this article was written, the questionnaires were being applied to users of the field test with a particular model of washing machine, according to the phases described above. The first three phase had already been carried out. The manufacturer, according to the profile defined previously, selected participants. The questionnaires from the first phase were applied, TIME 1. The first phase was intended only to evaluate the emotional image that users have of a washing machine.

At this stage, it was not offered any visual stimulus to the participants. Participants also answered questions in order to identify the level of education and economic class.

Next, participants were instructed on how to fill the semantic differential scales. They used the semantic differential scales to associate them to "washing machine". Each participant individually completed the questionnaire

Two weeks after the first phase, started the second phase, TIME 2. The second phase took place before participants receive the washing machine. A photograph of the product was shown to participants. Participants reviewed carefully at the picture, and they were instructed to fill out the questionnaire semantic individually.

Two weeks after the second phase the respondents received the washing machine in their homes. The delivery of the washers in homes was monitored.

Participants received questionnaires corresponding to the third phase, TIME 3, before unpacking the washing machine. The subjects were instructed to complete the questionnaire as soon as they had the first physical contact with the product before using it effectively, or install it. At this stage, beyond the questionnaire with the semantic differential scales, participants were given another sheet containing a photograph of the product. They were asked to indicate which part of this photograph was the washing machine in the selection of the value assigned to each of the scales. They should note on the specific region of the image, the number of the corresponding scale. The objective of this analysis was to extract more precise information about which product feature is related to a particular concept, with the vision of the User.

After two months of intensive use of the washing machine, these users should evaluate the product again, with the same type of tool, phase 4, TIME 4.

RESULTS AND DISCUSSION

The results are displayed by a timescale containing the different stages of interaction during product use and the corresponding emotions outsourced, , as shown in the figure 2. Allowing the visualization of the emotional changes as it increases the level of interaction with the product for the User Experience.

Before applying the final filter in the adjectives for the construction of the semantic differential, it was believed that two situations could occur. The first with respect to the order of importance assigned to each dimension. The hypothesis was that there would be a degree of importance of the dimensions established by the consumer, as proposed by Jordan (Jordan, 2002). The functionality dimension would receive the highest score among the three, followed by usability and finally by Pleasure. This hypothesis is partially confirmed: the dimension of Pleasure received an average score, being 24% lower than the other two averages. This suggests a trend for a stronger appreciation for the rational and functional aspects of the product over those emotional, less tangible ones, when participants are asked about. The surprise, however, was in relation to the dimensions of functionality and usability. Both got exactly the same score, showing a great concern of the users with questions related to the efficient use of the product.

A second hypothesis was that, due to possible ambiguities in the interpretation of the less tangible pleasure related adjectives, the deviation on the scores of this dimension would be greater than on the other two. This hypothesis was confirmed. The average standard deviation of the scores given to the adjectives of size Functionality was 0.79 while Usability was 0.98 and 1.23 of Pleasure - 55% higher than the first and 25% higher than the second.

326

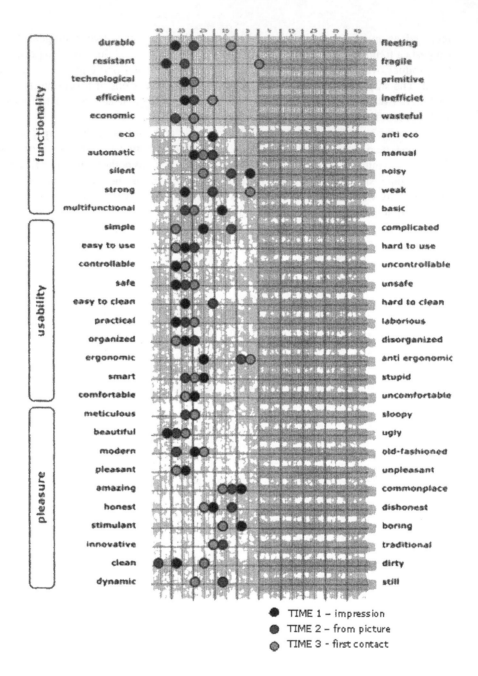

Figure 2 – Results of the experiment

In the first phase of the evaluation of the emotional image that users have of a washing machine there was a strong tendency of the respondents have mentioned earlier experiences with similar products.

Due to the introduction of new technologies in the design of this washing machine, the manufacturer has developed a product with a higher capacity for the external dimension of the product when compared to previous models.

The appearance of the compact washer reflected significantly in the responses of the questionnaires in phase 2. For some users the improvement was seen with good eyes. For other users are not very sensitive to problems of space, there was a trend of association of proportions of thin washer with a possible weakness of the product.

We hope that the user experience is reflected in the fourth phase, after 2 months of using the product. We hope that the sense of User is different, which probably will be reflected in the assessment that will make the product.

The result of the experiment shows the general trend of first impression in the following phases. Some more significant results are related to the concepts of ergonomics, strength and resistant. These concepts in the first contact were below initial expectations of users.

REFERENCES

Damazio, V. M. ; Lima, J. ; Meyer, G. C. . . In: Claudia Mont'Alvão; Vera Damazio. (Org.). Design, ergonomia e emoção 1a ed. Rio de Janeiro: Maud X, 2008, p1.

De Masi, D. O Ócio Criativo. Rio de Janeiro: Sextante, 2000.

Desmet, P.. A multilayered model of product emotions. The Design Journal, p.1-13, 2003.

Desmet, P.M.A., & Hekkert, P. (2002). The basis of product emotions. In: W. Green and P. Jordan (Eds.), Pleasure with Products, beyond usability (60-68). London: Taylor & Francis.

Jordan, P.W. Designing pleasure products. London: Taylor and Francis, 2002.

Mcdnagh et al. 2002; MCDONAGH, D. et al. Design and emotion. London: Taylor and Francis, 2004.

Mondragon, S.; Company, P.; Vergara, M.. Semantic Diferential applied to the evaluation of machine tool design. International Journal of Industrial Ergonomics, v.35, n.11, p.1021-1029, nov. 2005.

Nielsen, J. Usability engineering. Boston: Academic Press, 1993.

Osgood, C.E.; Suci, C.J. Tannenbaum , P.H. The measurement of meaning. Urbana: University of Illinois Press, 1957.

Silva, T. B. P. et al. Teste de personalidade dos produtos. In: Congresso Brasileiro de Pesquisa e Desenvolviento em Design, 8, 2008. Anais. São Paulo: AEND/Brasil, 2008. p. 3038-3043.

Woolley, M. Designing Pleasurable Products And Interfaces .Proceedings of the 2003 International Conference on Designing pleasurable products and interfaces. Pittsburgh, PA, USA, 2003.

Chapter 34

Usability of the Industrial Firing: An Anthropometric Analysis of the Traditional and the Smart System

Alexana Vilar Soares Calado, Marcelo Márcio Soares

Center of Art and Communication – Department of Design
Federal University of Pernambuco
Recife - Brazil

ABSTRACT

This article refers to the study of equipment used for cooking in commercial kitchens in restaurants which use the system of traditional cooking and kitchen by smart cooking system. Investigates the issues concerning sizing equipment studied, which constitute an important point to be considered, since the existing anthropometry variation among employees in kitchens researched was regarded as one of the main factors of discomfort inherent equipment used in cooking, particularly as regards stove industry. For this reason the survey presents a study of relations anthropometry griddle industrial Combined Oven.

Keywords: Ergonomics, usability, cooking industrial.

INTRODUCTION

After the conclusion of ergonomic studies conducted in commercial kitchens located in the city of Recife, the equipment used in these environments needed adjustments to meet ergonomically to its users. The industry of cooking was chosen to be investigated, since the equipment used for the achievement of the task (stove industrial) presented according to the research said, with a series of difficulties of use given by users. The interest on the part of the researcher came up in conducting a comparative study between: (I) the performance of the new cooking equipment used in kitchens more modern (combined oven) which operate the system with Cook and Chill, the so-called intelligent kitchens; (ii) and the industrial stove, assessing whether these new equipment would have solved the usability problems observed in the traditional system and which would be the aggregate contributions handy to these new products.

This study considered as industrial kitchen of traditional firing the cuisine that uses primarily the stove industrial as equipment to perform the task of cooking food and industrial kitchen cooking smart kitchen that uses primarily the furnace combined for carrying out the same task. One aspect of great importance to the functioning optimized a kitchen is related to the scale, management and security of industrial kitchen equipment, which interfere directly in the usability of these products (JORDAN, 1998; CUSHMAN & Rosenberg, 1991). Such equipment, typically have fixed dimensions, which were determined from pre-established standards, this is the reason that hardly ever is suited to anthropometric variation that exists among users who work inside a kitchen (PANERO & ZELNIK, 2002) and for the reasons already described is causing a number of difficulties in handling, bringing about occupational diseases.

For looking into anthropometric products searched, it was registered graphically and examined the incompatibilities between the cooking equipment and industrial users of extreme dimensions (percentile 2.5% and 97.5%). From this analysis was possible to propose recommendations dimensional obeying requirements of task activities and physiological constraints. This assessment was used the technique of applying two-dimensional anthropometrics Dummies. These dummies representing men and women of percentiles extremes – maximum and minimum – are project tools that function as if they were fixtures used to define the relationship dimensional man x machine, analyzing existing situations, setting proposals, evaluating compromise solutions or by taking the necessary revisions. Grandjean (1998) recommends for the sizing of benches for work aimed at a better production performance, adapt the arm abduction entry for angles from 8 to 23 degrees. The author also suggests a great reaching area for horizontal surfaces for forearm around the elbows with arms touching the surface between 35 and 45 degrees. Maximum reaching area of arm around his shoulders describing an arc of 55 to 65 cm. for scaling height benches for tasks performed, the surface of the bench must be set between 5 to 10 cm below the elbow height. This way the tasks which are accomplished in higher frequency shall be held within an area of great reaching. A rotation of 20 degrees of the head was taken in this job, around its axis (angle considered within the of the head), around its axis (angle considered within the limits of comfort).

We point out that this research is part of a wider search usability which includes other phases of study such as: ergonomic assessment, problems with of SHTM, analysis of failure modes and effects, which are not objects of this article.

STUDY FIELD

SELECTION VARIABLES ANTHROPOMETRY

According to Panero & Zelnik (2002), the data anthropometry of most importance for the sizing of spaces and equipment for kitchen are: Stature; Height of the eyes; elbow Height; Height sitting erect; height of eyes sitting; free space for the thighs; buttocks-knee Length; Front vertical reach; seizure; maximum body Depth; Width body Maxima. It was noted that by its own positioning adopted to accomplish the task (stood up) these variables do not fit the jobs examined, because the measures selected by Panero & Zelnik (op cit) consisted of the seating position (Calado, Barros and Almeida 2005). Due to this observation, the authors got into the following adaptation of the original table kitchen for industrial kitchens along the firing task, Stature – AP 04; height of eyes on foot – AP 05; shoulder Height on foot – AP11; elbow Height – AP 16; length of upper – C 11 (including operated front and vertical); maximum width of the thorax – LP 09; maximum width of pelvis – LP 18; maximum depth protusion abdominis – PP 15; maximum depth protusion pelvic – PP16; depth of buttocks to the knee in patella) sitting – PS 10 variable 10 would be parsed as a parameter to the depth required to allow the employee bend in the safe space. This is justified, due to the scale of industrial equipment, and the fact that officials had constantly bend moving.

CRITICISMS AND RECOMMENDATIONS ZONE INTERFACE

industrial stove

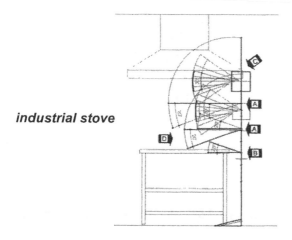

Figure 01 - intersection view of woman 2,5% and man 97,5%

A. Vision sight out of reaching area for a shorter and taller user.

B. The height of the cooker requires elbow elevation above the position of comfort for the shortest user

C. Height of stove hood makes difficult the maintaining of proper posture, requiring the tallest user a postural kyphosis, and also forces the user to work with the legs bent during cooking activity.

D. Work surface at the lowest level of the elbow height for the tallest user. Action area of two users out of the area of comfort

Vertical Combined Oven

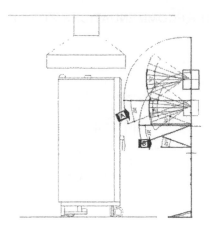

Figure 02 - intersection view of woman 2,5% and man 97,5% - Vertical Combined Oven

A. Vision sight out of reaching area for the shortest user. Action area out of the area of comfort.

B. The height of shelves (Mobile) oven structure requires elevation of shoulders and upper limbs, as well as bending vertebral column, out of the area of comfort for the shortest user.

C. The height of the oven shelves requires bending of the vertebral column out of the position of comfort for the tallest user.

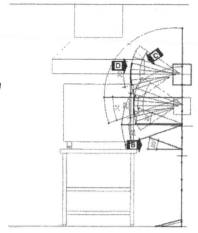

Bench combined Oven

Figure 03 - Intersection side view of woman 2,5% and man 97,5% - Bench Combined Oven

A. Vision sight out of reaching area for the shortest user.

B. Action Area out of the area of comfort for the shortest user.

C. The height of the oven shelves requires elevation of shoulders and upper limbs out of the reaching area for the shortest user.

D. Vision sight out of reaching area for the tallest user.

E. Height of stove hood makes difficult the maintaining proper posture, requiring the tallest user a postural kyphosis.

CONCLUSION

From the use of dummies anthropometry two-dimensional (shorter woman and tallest man) and consideration of biomechanical angles of comfort, interfaces parameters were defined for the rendering of extreme users. We note that the Visual requirements and requirements visibility, as well as the action requirements and biomechanical handling requirements of the arms and legs, could only be defined from the analysis of the task. To obtain the ideal situation, by a dimensional point of view, it is recommended the use of adjustments to the following elements:

Industrial stove

The sizing up of the work face of Industrial stove

- Adjust the height of the industrial stove
  Justification: allow the shorter and taller user to use it in a comfort way.

 The sizing up of the height of the support shelf for the pans

- Repositioning of cookware
Justification: avoid the constant inadequate postures of the users.

Sizing up of stove hood

- Adjust the height of stove hood
  Justification: prevent tallest users have to perform task of cooking adopting a postural kyphosis and keeping their knees bent.

Combined Oven

Sizing up of mobile structure

- Adjust the maximum and minimum height of the structure's mobile combined oven.
  Justification: prevent tallest and shortest users have to perform the task of filling the shelves of mobile structure, adopting inadequate postures.

Bench combined Oven

• In this case the height adjustment can be done on the bench that supports the oven, because the use to the user 2.5% percent requires the same have to elevate shoulders and upper limbs beyond the area of comfort. The analysis of the intersection of Visual areas and actions, obtained with the aid of dummies anthropometry two-dimensional highlights the costs arising from the use of a furniture or equipment badly positioned. The inadequacy of the height of the area, particularly as regards stove industrial demands extreme actions, discovered users from checking the angle of comfort to the coasts, highlighted the necessity of adopting a stance that inadequate a with the trunk tilted with damage to lumbar and column. Note that this same attitude had already been registered in this search in a problematic of human Task-machine system.

REFERENCES

CALADO, Alexana V. S., BARROS, Helda e ALMEIDA, Maria de Fátima X. M. (2005). Ergonomic Analysis of three restaurant kitchens; a comparative analysis. Monograph. Recife.
CUSHMAN, W. H & ROSENBERG, D. J. (1991) *Human factors in product design*. Elsevier, Amsterdam.
DUL, J. WEERDMEESTER B. (1995) *Ergonomics pratice*. Edgar Blücher,
GUIMARÃES, L.B.M. (2004) *Ergonomics Product*. Vol 1, 5a. ed. FEEng,
JORDAN, P.W. (1998) *An Introdution to Usability*.Taylor &Francis.
MORAES, Anamaria de. (1983) Application Anthropometry data: scaling of man-machine interface – Federal University of Rio de Janeiro. COPPE / UFRJ, Product Engineering.
MORAES, Anamaria de. MONT'ALVÃO, Claudia. (2003) Ergonomics: aplications and Concepts. iUsEr. Rio de Janeiro.
PANERO, J. ZELNIK, M. (2002) Human Sizing for Interior Spaces. GG. Barcelona.
SOARES, Marcelo Márcio. (2000) Ergonomics, reliability and safety of the product: in search of the total quality of the product. *ABERGO/RJ. Ergonomics Association of Rio de Janeiro*.
SOARES, M. M. & BUCICH, C.C. (2000) Product safety: reducing accidents through design. Studies in design..

Chapter 35

The Effects of Explicit Alternative Generation Techniques in Consumer Product Design

Fabio Campos, Marcelo Soares, Walter Correia,
Remo Ferreira, Eliana Melo, Arlindo Correia

UFPE – IFET-PE – Capes – CNPq

ABSTRACT

Every product design implies in a step of alternatives generation where creativity is employed. Creativity is one of the more active skills of the human beings; by it is possible to solve ergonomic problems and develop innovative ideas. The techniques able for exercising the right hemisphere of the brain, responsible for symbolic thinking and creativity, can be used in an "ad hoc" or "explicit" way. This paper presents evidence about the efficiency of their "explicit" use when compared to the "ad hoc" one, and also about which conditions can impact and be impacted in this use.

Keywords: Product design, alternative generation techniques, design methods, creativity

INTRODUCTION

Consumer Product Design, as any kind of product design, involves at least 5 steps: data gathering, alternative generation-selection, implementation-prototyping and alternative evaluation. These steps are neither necessarily done in sequence nor used only once along the design process (Bürdek, 2006).

According to Cushman and Rosenberg, the ergonomists should be involved in activities which pervade all these 5 steps such as allocation of functions and tasks, task analysis, identification of user interface requirements, development of technical specifications, preliminary and detailed design, and limited user testing with muck-ups (Cushman and Rosenberg, 1991). By other perspective, the act of solving ergonomic questions about the design of a consumer product can, itself, makes use of alternative generation techniques.

All the diverse design branches (graphic design, product design, service design, etc.) necessarily make use of these steps, but do this in a continuum ranging from all steps made intuitively (in an "ad hoc" way) (Melo et al, 2006) to all steps made explicitly ("explicitly" in the sense of being done by means of a formal, clear, technique, and not by non-structured ways). For example, a designer may choose not to gather additional information about a particular product to be designed, if he works all the time with this kind of product and feels that he already has enough data.

Good designs are often associated with innovative or creative solutions (Baxter, 2000), and the "insights" for theses creative solutions happen during the step of alternative generation, the main focus of this paper.

Although the alternative generation techniques are around for about 100 years and one intuitively knows that the use of some technique tends to be able to produce better results (Alves, Campos and Neves, 2007), they are not widely used by designers.

This paper presents evidence that the explicit use of these techniques can produce better alternatives for consumer product designs than an "ad hoc" alternative generation step, and explore some conditions where these better results occur.

Two experiments were done to try to elucidate these questions. In the first one the alternative generation with and without the use of explicit techniques are compared. In the second experiment, the results of the application of the explicit techniques by veteran designers are tested. The conclusions of both experiments generate evidence in favor of the use of explicit techniques to increase the quantity of useful alternatives generated and quality of consumer products designs.

EXPERIMENT I – WITH X WITHOUT TECHNIQUES

The aim of this experiment was to investigate if it is possible to note some advantage with the use of explicit techniques for alternatives generation versus

338

alternatives generated without any structured technique.

Sixty people from five diverse educational backgrounds (Graphic Design, Environmental Management, Civil Engineering, Radiology and Tourism) were chosen at random and divided in a control group (which used no technique) and a "technique group" (using the explicit technique) from each background, therefore resulting in 10 groups (a control group and a technique group from each background) each one with 6 people (Bomfim, 1995).

None of these 10 groups communicated with the others during the experiment.

The same briefing was delivered to each group, but to the "technique groups" were taught previously the "Brainwriting" technique (Warfield, 1994). All groups had the same time to develop their alternatives, 30 min (Bomfim, 1995).

To evaluate the generated alternatives a specialist in the field of the briefing (game design) was hired, and the "De Bono Scale" employed (De Bono, 1997).

The "De Bono Scale" categorizes the alternatives in several groups, which are in diminishing order of importance or usefulness: "Idea directly usable" (IDU), "Good idea but not to us" (GBN), "Good Idea but not now" (GNN), "Should be improved" (SBI), "Strong but not usable" (SBU), "Interesting but not usable" (INU), "Little value" (LV), and "Unusable" (U).

The alternatives were shuffled before the evaluation thus the specialist did not know from which group were the alternatives.

The results of each group and the final result of all groups summed up were plotted in bar-graphs (figures 1 to 6).

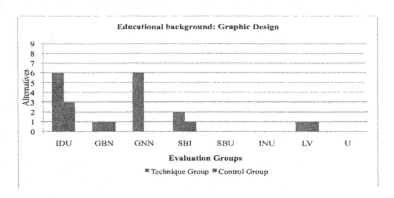

FIGURE 1 Results from the "Graphic Designers" group. (Font: authors)

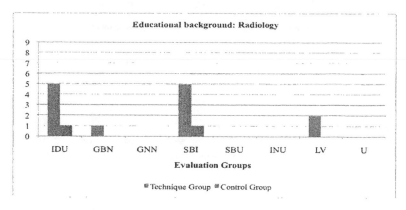

FIGURE 2 Results from the "Radiology" group. (Font: authors)

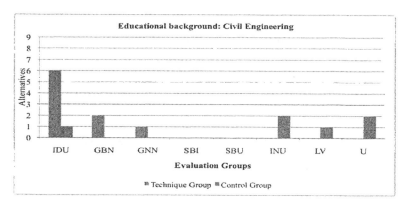

FIGURE 3 Results from the "Civil Engineering" group. (Font: authors)

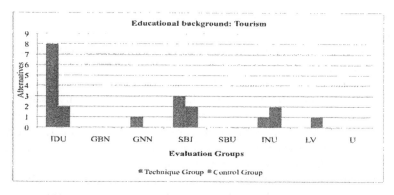

FIGURE 4 Results from the "Tourism" group. (Font: authors)

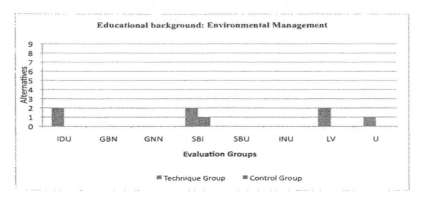

FIGURE 5 Results from the "Environmental Management" group. (Font: authors)

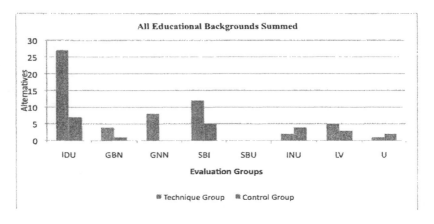

FIGURE 6 Results from all groups summed up. (Font: authors)

Even without further statistical elaboration, the graphs make clear the effects of the explicit techniques use.

Some conclusions can be drawn from these figures:

- In a real world design process, the set of "Ideas directly usable", IDUs, would be the most valuable one, regarding the conception of an actual product or the solution of a real-world problem. In Figure 1, Results from the "Graphic Designers" group, there is the double of IDUs coming from the team that used the explicit techniques.
- Comparing the results from the graph at Figure 1, with the results from the others ones, it is relevant to observe that it was, among the "educational backgrounds", the one able to generate more IDUs by the control group, evidencing a possible effect of the educational background in the "spontaneous" creativity.
- All the graphs (figures 1 to 6) shows that the use of explicit techniques

result in more IDUs than without. In the less favorable case for their use, the Graphic Designers group, Figure 1, was generated the double of IDUs; and the total results from all groups evaluated, Figure 6, shows more than 4 times IDUs using the techniques.

EXPERIMENT II – VETERAN DESIGNERS

In Experiment I people were chosen at random regarding their professional experience. In this experiment the "Graphic Design group" (vide Figure 1) exhibited the better result in the quantity of IDUs generated, thus Experiment II was devised to further explore this fact.

The main aim of Experiment II was to evolve the investigation to test the effect of the techniques with professional, veteran, designers. The idea now was to evaluate not the quantity of IDUs anymore, but the quality of consumer product concepts generated by these veteran designers.

Four groups of 6 veteran designers with relevant professional experience were chosen based in a previous analysis of their portfolio. A scale of creative product evaluation, CPSS (Creative Product Semantic Scale) (O'Quin and Besemer, 2006), was used along with an evaluation method, CPAM (Creative Product Analysis Matrix) (Besemer and Treffinger, 1981), were used by a panel of 3 specialists to evaluate the concepts. From this evaluation model we plotted the results of the opinions regarding how much the concepts of the products rated as being: surprising, original, logical, useful, worthy, and comprehensive.

Three briefings for diverse products were delivered to each group. One of the groups, the "Control Group", generated the product concepts to these 3 briefings in an "ad hoc" way. The other 3 "Technique Groups" worked upon these briefings using three different techniques: brainstorming, brainwriting and provocation (Bomfim, 1995). A combination was made where each of the 3 "technique groups" worked upon a briefing with a different technique to try to equalize the specific differences among the teams. All these concepts of consumer products were shuffled and then evaluated by the panel of specialists, without any communication among them.

The results of these evaluations were plotted and their bar-graphs shown in figures 7 to 10.

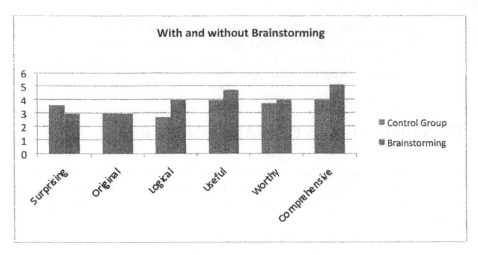

FIGURE 7 Veteran Designers with and without using the Brainstorming Technique. (Font: authors)

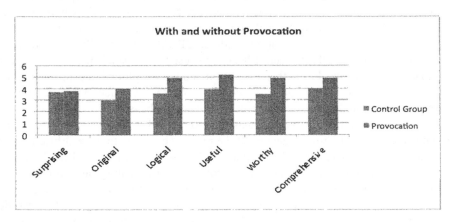

FIGURE 8 Veteran Designers with and without using the Provocation Technique. (Font: authors)

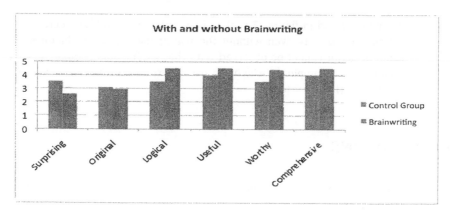

FIGURE 9 Veteran Designers with and without using the Brainwriting Technique. (Font: authors)

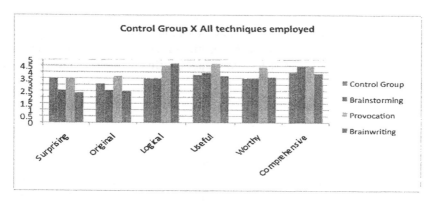

FIGURE 10 Veteran Designers – comparison among all techniques and the control group. (Font: authors)

Analyzing these graphs and taking in account the "Experiment I" it is possible to drawn some additional conclusions:

- It is possible to detect better qualities in the consumer products concepts generated with the use of explicit techniques.
- The use of the explicit techniques enhance both the quantity and quality of the alternatives generated and also the qualities of the products concepts.
- Even veteran designers can have benefits in the qualities of their designs using the explicit techniques.
- Some techniques, like "provocation" in Experiment II, Figure 8, can exhibit consistent better results than others even with different briefings and teams. The experiments done were not enough to identify which conditions interfere or cause this behavior.
- Some qualities of the consumer product concepts generated can be more improved by the use of the explicit techniques than others, for example, as

can be observed in figures 7 to 9, it is possible to get a high degree of the "surprising" quality even without the use of the explicit techniques. The same behavior cannot be observed with the "useful" and "comprehensive" qualities, for example, in which the use of explicit techniques excel, in all techniques (figures 7 to 9).

CONCLUSIONS

The two experiments provided evidence in favor of the use of explicit techniques to generate consumer product concepts.

Both the number of useful alternatives of design and the qualities of the design concepts are enhanced with the explicit techniques use. Even veteran designers can create better designs concepts using them.

Although the experiments produced strong evidence in favor of the use of explicit techniques, they also elicited some questions to be investigated like which conditions make a technique have better performance over others, and which conditions, besides the educational background, interfere with the creative potential.

REFERENCES

Alves, H. A., Campos, F., Neves, A. (2007), *"Aplicação da técnica criativa brainstorming clássico na geração de alternativas na criação de games"*. Anais do VI Simpósio Brasileiro de Jogos para Computador e Entretenimento Digital, São Leopoldo.

Baxter, M. (2000), *"Projeto de produto"* Makron Books, São Paulo.

Besemer, S.P., Treffinger, D.J. (1981), "Analysis of creative products: review and synthesis." Journal of Creative Behavior, 15, 158–78.

Bomfim, G. A. (1995), *"Metodologia para desenvolvimento de projetos"* Universidade Federal da Paraíba, João Pessoa, Paraíba.

Bürdek, B. E. (2006), *"História teoria e prática do design de produtos"* Blücher, São Paulo.

Cushman, W.H. and Rosenberg, D.J. (1991), *"Human factors in producut design"*. New York, Elsevier.

De Bono, E. (1997), *"Criatividade levada a sério: como gerar idéias produtivas através do pensamento lateral"* Pioneira, São Paulo.

Melo, E., Neves, A., Campos, F., Pimentel, H., Moura, E. (2006), *"Matriz e rede de restrição: uma proposta inicial de aplicação na fase de concepção de jogos eletrônicos"*. proceedings of the V Brazilian Symposium on Computer Games and Digital Entertainment, Recife.

O'Quin, K., Besemer, S. P. (2006), "*Using the creative product semantic scale as a metric for results-oriented business*" Creativity and Innovation Management, Vol. 15, No. 1, pp. 34-44.

Warfield, J.N. (1994), "*A science of generic design: managing complexity through systems design*". Iowa State University, Iowa.

Chapter 36

The Safety of Consumer Products Under the Perspective of Usability: Assessment of Case Studies Involving Accidents with Users in Recife / Brazil

Walter F. M. Correia, Marcelo M. Soares, Marina de Lima N. Barros, Fábio Campos, Viviane M. Gallindo

Design Department
Pernambuco Federal University - UFPE
Recife-PE, Brazil

ABSTRACT

Overall the consumer products of today have reached a high degree of complexity that often the users are not accustomed or habituated to using at a first time. This seems true when we realize that the instruction manuals often do not provide a readable and easy to understand, sometimes with a technical language, sometimes with a very extensive content and lowercase letters. It is interesting to say that although the high degree of sophistication and technology stemming with the years has maintained a strong pull from the standpoint of strategic advertising and marketing, this may produce serious frustration to users, causing problems either physical and/or cognitive. Thus, this paper presents a confluence of data pertinent to a field study conducted in Recife-PE, Brazil on consumer products, aligned to the influence of usability on the use of these products. The study runs through the stages commonly used in product development and how the testing directed to the area of usability can make a product have a higher acceptability by the general public, making it safer to use. These are shown through illustrations and testimonials, case studies of accidents involving consumer products that could have been avoided if in the project some aspects of usability have been aggregated. At the end of the article it's possible find the main findings of the study within the binomial: security and usability.

Keywords: Usability Tests, Design, Consumer Products

INTRODUCTION

According Chiozzotti (1995), the case study is a comprehensive characterization to describe a variety of surveys that collect and record data in a particular case or several cases of individuals to organize an orderly and critical report of an experience, or evaluate it analytically, aiming to make decisions about it or propose a transformation. Thus, for the submission of this article, was performed a case study with a sample taken from the two accidents questionnaires submitted by Correia (2001) in a universe of eight accidents.

The selection of the sample who composed the accidents studied was made randomly from contact with some respondents who had registered their e-mail in the questionnaires. The objectives of this case study were: a) detailed analysis of each accident, b) to analyze the activity through a flowchart of task c) perform a careful analysis of the causes of accidents and d) assess the level of usability of products. The products that were part of the study and their accidents were (i) a common drill, leading to a cut and bruise, and (ii) a hot shower, causing a shock. The case study involved the users of accident victims.

They were asked to: a) respond to a previous interview about how the accident happened, b) make a recorded simulation of such accidents; c) answer questions in the SUS - System Usability Scale. The latter aimed to evaluate the degree of usability of each product mentioned, according to Stanton and Young (1999).

CASE STUDY PROCEDURE

During the interview stage with each of the users, the victim of some kind of incident, followed the roadmap proposed by Weegles (1996), and thus took into account several aspects of the accident such as how this happened, the User's opinion about it, what is the most relevant for the product and the place where the incident happened, where the conditions in which the environment was at the time of the event, and what was the condition of psycho physiological User upon the occurrence of accident.

The accidents simulations were made from the reports sent by users, since it is aimed to reconstruct the situation for future analysis. They were asked to describe the accident, repeatedly, to obtain a recovery as close as possible to the fact. The considerations were based on (i) the "User's voice," which were recorded their descriptions and opinions about the accident, and (ii) the author's own insight to replenish the same. The evaluation of each of the SUS questionnaire was made subjectively, it depends on the User's own opinion, and its quantification and calculations were made based on parameters provided by the SUS.

ACCIDENT ANALYSIS

As mentioned previously, accidents collected from questionnaires were recorded and analyzed in detail in accordance with what was reported by users. What is presented in this session comes down to presentation of two of the accidents, an analysis of actions taken by users and, finally, its Fault Tree.

DRILLING ACCIDENT

The accident occurred when the User tried to make a hole in the wall for placement of a bush (Figure 1). The photos, shown in Figures 2 and 3, are simulating the use of the drill at the time of the accident. The drill is the same with which the User had his accident.

Figures 1, 2 e 3 – Images of the drill and its use for accident 'simulation

The related accident happened due to an excessive resistance from the wall, and then the User decided to tilt the drill during operation. At this point, the drill broke and hit the face of the User. There was a bruise and a small cut. Part of the drill remained attached to drill.

For a better understanding of the drilling steps activity, was developed a flowchart of the task, as can be seen in figure 4.

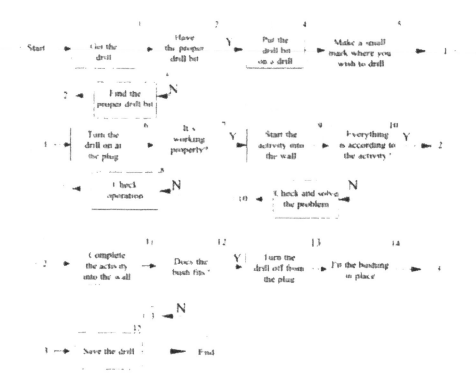

Figure 4 – Flowchart of the tasks performed by a drill

The actions that caused the accident were described as follows by User:

- Choice of equipment;
- Search the drill bit ideal for the bush in question;
- Using the reference manual for the drill bit;
- Start the activity using both hands;
- Slightly inclination to try to break the barrier found (strongest part of the wall);
- The drill latch and part of the drill bit goes against the face of the User;
- Cut and bruised in the face of the User.

The manual recommended that the User should always use the drill in a perpendicular position to the action used, in this case is the wall, which should be 90 ° with respect to it. The angle adopted by the User may have resulted in the breaking of the drill bit. In the Fault Tree on the drill (Figure 5) These considerations can be checked detailed in accordance with the User-run.

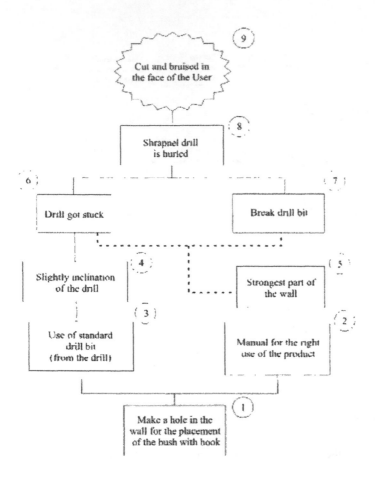

Figure 5 - Fault Tree for the accident with the drill

If the User had some more experience with the activity performed (which he affirmed that he didn't), he would know that a wall requires a stronger type of drill bit that, when it is thicker (using an bush to a thicker wall below) the drill bit should be stronger (other material).

ACCIDENT WITH AN ELECTRIC SHOWER

The shower and the simulation use from the accident analysis are presented in Figures 6 to 8, and the shower shown in those pictures is the same that has caused the accident. And at the time of the interview the user had not fixed it. In describing the accident, the user said the regulating thermostat knob got stucked and she tried to release him, holding the shower with one hand and forcing it, harder, with the other. Right at this moment the user had a slightly electrical discharge of low intensity, causing only fear, without further damage.

Figures 6, 7 and 8 - Electric shower and the accident simulation from its use

Even after the occurred, the User claims to still be using the same electric shower, but without heating the water, no longer using the thermostat knob. Following what the user said a flowchart was made including all the steps that she took using the electric shower. Those steps are described at the figure 9 as follow:

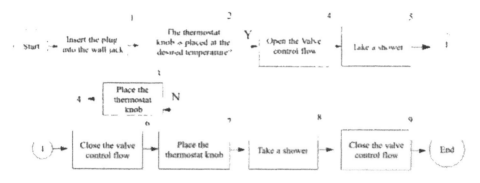

Figure 9 – Flowchart for the use of electric shower

This flowchart was developed during the moment of the interview and helped the user to remember details of the accident. The steps explained by the user are briefly described as following:

- Insert the plug into the wall jack;
- Tried to place the thermostat knob to the "summer";
- knob hard to be placed;
- Using both hands to obtain further support for reach the knob;
- The knob is activated abruptly;
- The user takes an electric shock.

The user said she tried to find the instructions to solve the problem with the knob. As she didn't find, she decided to try to solve the problem by herself. This follow action is shown in the image below at the Fault Tree (Figure 10).

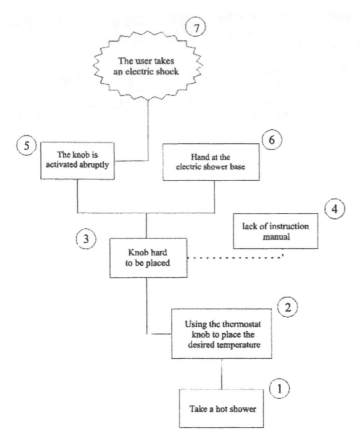

Figure 10 - Fault Tree for the incident with the electric shower

The failure of the knob was caused by problems in the product due to prolonged use of it. The key temperature was possibly damaged after having been swapped position several times, due to the preference of individual users (residents in the house). Although this fact must be taken into account, such activity should be regarded as normal and the thermostat knob's material should have been thought to work consistent with that function. However, the shock protection on the basis of the product should have been developed and placed on the product, considering the risks that the User is tended to. The main objective of this step was analyze the causes of accidents. It is observed that, for each of the accidents was found that a portion of the blame, either through carelessness or neglect, belonged to the User.

However, these faults are to be related since all the accidents could have been avoided if the products had been designed with an emphasis on the caution and safety items, moreover had been paid particular attention to the prevention of misuse of them. It can also be seen that the products analyzed did not show characteristics of a "friendly" product. This fact must consider (i) physical composition (ii) components, (iii) manual of instruction and (iv) warnings.

QUESTIONNAIRE WITH THE USERS: APPLICATION OF SUS

SUS shows to the respondents some few questions to be answered by marking on a satisfaction scale according to the level of agreement or disagreement for each question. After answered and encoded the answers are calculated, through a coefficient given by the SUS, the degree of usability of a given product.

IMPLEMENTATION PROCEDURES OF SUS

According to the recommendations made by Stanton and Young (1999) were followed all the procedures for the implementation of SUS, ranging from the explanation on that it was the questionnaire to their weighting and analysis. Through this procedure, it was possible to obtain acceptable results for the study with users. The calculations are presented in the following tables (tables 1 and 2) were made based on questionnaires answered by the SUS users followed the order presented by the authors. For a proper understanding of users' responses, it is the "0" (zero) as the lowest score for the level of usability and "100" a higher score.

SUS ANALYSIS FOR DRILL

For the first case, the drill, the calculations are presented as follows in table 1:

Table 1 - Calculation for the level of usability of the drill

Score of odd number of items = Rating of - 1	Score of even-numbered items = Rating of - 5
Item 1 $2 - 1 = 1$	Item 2 $5 - 2 = 3$
Item 3 $4 - 1 = 3$	Item 4 $5 - 3 = 2$
Item 5 $2 - 1 = 1$	Item 6 $5 - 2 = 3$
Item 7 $5 - 1 = 4$	Item 8 $5 - 5 = 0$
Item 9 $3 - 1 = 2$	Item 10 $5 - 4 = 1$
Sum of odd numbers (ON)	$= 11$
Sum of even numbers (EN)	$= 9$
Total sum of the items = NP + EN	$= 20$
Total score of usability for SUS Total of items x 2.5	$= 20$ x 2,5 $= 50$

In the case of the drill, it became a level of usability in the 50 value. Considering the scale from 0 to 100 showed, note that a level of 50 for usability in consumer products can now be considered below the moderate.

SUS ANALYSIS FOR ELECTRIC SHOWER

At the beginning the User of the electric shower, demonstrated that had a hard time by answering the questionnaire, but after a while and having the doubts clarified, we obtained the results shown in Table 2.

Table 2 - Calculation for the level of usability of electric shower.

Score of odd number of items = Rating of - 1	Score of even-numbered items = Rating of - 5
Item 1 $5 - 1 = 4$ Item 3 $5 - 1 = 4$ Item 5 $1 - 1 = 0$ Item 7 $4 - 1 = 3$ Item 9 $4 - 1 = 3$	Item 2 $5 - 5 = 0$ Item 4 $5 - 2 = 3$ Item 6 $5 - 5 = 0$ Item 8 $5 - 5 = 0$ Item 10 $5 - 2 = 3$
Sum of odd numbers (ON)	= 14
Sum of even numbers (EN)	= 6
Total sum of the items = NP + EN	= 20
Total score of usability for SUS **Total of items x 2.5**	= 20 x 2,5 = 50

It is evident that the electric shower in this study, presents a level of usability of 50. What, similarly to the previous item, should be considered low to moderate. In general can be seen that the products are presented in a not acceptable scale of usability. This scale should be considered as very low, since the product must meet and satisfy their users (that means 100 points). SUS has through the points where users complained more, try to develop solutions that seek to address gaps in certain product area. As an example, if a User strongly disagree that a product or system is easy to learn to use, you should think about how a more friendly interface can be applied to such product or system, as quoted by Stanton and Young (1999) and Stanton and Barber (2002).

CONCLUSIONS AND LESSONS LEARNED

It may be noted that that despite the Users' reluctance to remake the incident by claiming difficulty of remembering what happened, or saying that they no longer had the products, the results were satisfactory. The simulations took place at the injured Users' home, bringing him closer to the real in relation to what had happened. This was asked directly to each user that had any kind of accident. The methodology of Weegles (1996) was followed step by step, for the most part, for a correct analysis and evaluation of the study. Through the construction of the Fault Tree some lacks were observed, from both the product and the User. It is also clear that, despite the attempt at reconstruction (simulation) of the accident, all were aware that some detail or details might have been forgotten. SUS used in the interviews was the most important to certify that the products analysed in this essay have relatively low level of usability, though one can contest the fact that all the opinions given in here came from the use of products that ended up in an accident.

The authors of the SUS does not define the contents as 40, 50 or 60 are low, but show that a scale from 0 to 100, those two would be the extreme negative and positive end respectively, which can ensure that levels below 70 to 80, attest, in general, products with low usability, or they could be better studied in their conceptions. It can be affirmed that the accidents analyzed occurred through the fault of the product, and when the lack of the User, attests to the insufficient attention given to some norms of security and usability. According to Jordan (1998) and Vendrame (2000), is a mistake of designers and drafters develop a product for themselves thinking that users will act and think like them. Designing for children, the elderly, people with some type of disability or limitation, male or female, tall or short, requires a different degree insight in each case, and this is where is the difference between a good product and a bad product.

It must be clear that not only the aesthetic factor, or meet some of its functions would be sufficient, it is necessary to perform it well and properly, have a formal design and aesthetics in line with the proposal that is intended mainly to the public and to be achieved . The above cases show only a part of a random sample taken from a Brazilian capital. Many cases were found in only a small portion that is not representative. Estimate that the numbers representing meaningful and are as high as those that were found. According to the testimony of one of those interviewed: "We have at home products and appliances that are like white weapons, or rather invisible, that can hurt us at any time."

REFERENCES

Abbott, H. & Tyler, M. Safer by Design. A guide to the management and law of designing for product safety. England, Gower, 1997.

Chiozzotti, A. Pesquisa em ciências humanas e sociais. 2 ed. São Paulo, Cortez, 1995.

Correia, W. F. M. Segurança do Produto: Uma Investigação na Usabilidade de Produtos de Consumo. Dissertação de Mestrado, PPGEP / UFPE, 2002.

Jordan, P.W. An Introduction to Usability. London, Taylor & Francis, 1998.

Stanton, N A. & Barber, C. Error by design: methods for predicting device usability. Design Studies. Vol. 23, N° 4. London, Elsevier Science Ltd, July. p. 363 -384, 2002.

Stanton, N A. & YOUNG, M. S. A Guide to Methodology in Ergonomics. Designing for human use. London, Taylor & Francis, 1999.

Vendrame, A. C. Acidentes Domésticos - manual de prevenção. São Paulo, LTr Editora. p. 9-13, 2000.

Weegles, M. F. Accidents Involving Consumer Products. Doctorate's Thesys. University of Delft, 1996.

CHAPTER 37

Through the Looking Glass: Reflecting Upon the Acquisition of Expertise

Richard C. McIlroy[1], Neville A. Stanton[1], Bob Remington[2]

[1]Transportation Research Group
School of Civil Engineering and the Environment
University of Southampton
Southampton, SO17 1BJ
UK

[2]School of Psychology
University of Southampton
Southampton, SO17 1BJ
UK

ABSTRACT

The purpose of this study was to investigate cognitive skill acquisition on an individual basis in the context of learning to use a complex computer based application. Novice participants were trained in the application and subsequently engaged with the tool across three more sessions. During all test sessions participants verbalized their thoughts. Objective measures of time to complete each task, amount of mouse travel and number of mouse clicks indicated all participants acquired a level of skill across sessions. Not only did the verbal data suggest individuals' cognitive processes changed as a function of skill acquisition, but that these within-subjects changes mirror some of the between-subjects expert/novice distinctions described in the literature.

Keywords: Skill acquisition, expertise, verbal protocol analysis.

INTRODUCTION

COMMUNICATIONS PLANNING IN THE MILITARY

The ability to pilot an aircraft is a complex skill that is only acquired after an extensive training program, and indeed not all who apply themselves are ultimately successful. This is particularly true for military pilots, as these individuals have the added complexity of system dynamism; where commercial pilots are supplied with set routes, altitudes and communications requirements that are unlikely to change en-route, military pilots must consider these variables separately for each mission, be it training or operations, whilst remaining cognizant of the possibility of unforeseen circumstances arising. It is for this reason that military pilots must consider not only the planned route, but a number of planned diversions also.

The sustained high level of cognitive demand placed on military pilots arises from the need to remain aware of time keeping, hazard perception and navigation through three dimensional space whilst retaining the ability to communicate effectively with a variety of different organisations, services, and individuals, both on the ground and in the air. These communications requirements can be broadly divided into three major sub-categories; the need to alert ground authorities to the presence of aircraft passing through or travelling in close proximity to any controlled airspace; the need for effective and secure voice communications with other aircraft concurrently airborne; and the need to transfer data effectively and securely between concurrently airborne aircraft.

The Mission Planning System (MPS) software tool was developed in order to reduce the pilot's workload when in flight (Jenkins *et al.* 2008) and allows for pre-flight planning of the majority of these communications requirements. It is primarily used by military helicopter pilots and allows for the pre-programming of a number of radio frequencies relating to different organisations and individuals on the ground and in the air that pilots are able to access within the aircraft without the need for reference documents or the manual tuning of their radios.

Although the MPS software tool has proved effective in reducing the in-air workload of the military pilot (Jenkins *et al.* 2008), the tool itself is complex. The process of pre-flight communications planning is conceptually quite straightforward, however it is the requirement of the individual to learn the basic programming skills necessary for the effective and error free insertion of data into the relevant fields that is cognitively demanding. The complex cognitive skills required to operate the application must necessarily be acquired; it is this skill acquisition that is the focus of this paper.

The acquisition of cognitive skill can be broadly conceptualized as the process by which an individual learns to problem solve through the use of rational and critical thinking strategies (VanLehn, 1996) and is argued to consist of three primary phases; cognitive, associative, and autonomous (Fitts, 1964). Although this three-stage separation was originally put forward to explain the acquisition of manual skill (Fitts, 1964), the description is equally useful for describing the acquisition of cognitive skill (VanLehn, 1996). First, the *cognitive stage* describes the initial encoding of available information, often through verbal mediation. The learner begins to form a loose approximation of the target skill. In the *associative stage* the learner repeatedly attempts to perform the skill through the application of knowledge acquired in the previous phase. This is done in order to detect and eliminate errors. The final stage, the *autonomous stage*, is characterised by the increased speed and efficiency with which the skill is performed. It is in this final stage that the learner approaches expert performance; processing of information and integration of knowledge becomes fast, effortless, autonomous and unconscious (Logan, 1988).

At the heart of the three stage skill acquisition process lies the concept of knowledge conversion. It has been argued that as an individual progresses through the stages of skill acquisition domain knowledge is converted from a declarative form ('knowing that') (Anderson, 1987; Ryle, 1949) to a procedural ('knowing how') form (Anderson, 1987; Ryle, 1949). Declarative knowledge refers to individual pieces of information that are stored separately, for example theories, facts, and events. It is the time-consuming and effortful way in which knowledge is integrated that shapes the slower more error prone nature of novice behaviour (Anderson, 1993). In contrast, procedural information represents knowledge of how an individual does things, for example cognitive skills and strategies. Individual pieces of information have already been amalgamated to form coherent concepts, thus online information integration is not required to drive problem solving. Procedurally guided information retrieval is argued to be unavailable to inspection by reflection due to its fast, automatic nature (Pirolli and Recker, 1994).

Typically, novices rely on declarative knowledge to guide behaviour in a controlled manner, whereas experts can apply procedural knowledge relatively automatically. Differences between individuals at the two ends of the expertise spectrum have been studied in relation to many complex problem solving tasks, including chess (e.g., Chase and Simon, 1973), managerial decision-making (Isenberg, 1986), and programming (e.g., Pirolli and Recker, 1994). Although there is a wealth of literature on the distinctions between novices and experts (of which only a tiny proportion has been cited here) there is a distinct lack of research investigating individual change. Although it is indeed interesting and important to build our understanding of individuals' differing levels of expertise, there is also a need to investigate the origin of these differences and how it is that individuals progress through the early stages of skill acquisition. Furthermore, there is little research on how novices familiarize themselves with a new cognitively challenging application that requires training to use effectively. The current study was designed to address this issue in the context of learning to use MPS.

Across four sessions, one expert and eight novice participants engaged with the MPS computer application, preparing communications for a different hypothetical route during each session. Novice participants received one training session, followed by three test sessions. The expert, included to control for task differences, completed the only first and third test session. During all test sessions, Ericsson and Simon's (1993) 'think aloud' technique was used to capture the thought processes of all participants. To measure performance standards, the time to complete each task, number of errors made, mouse travel, and number of mouse clicks were also recorded. It was hypothesised that the acquisition of skill and improvements in problem solving ability would be reflected in error, time, and mouse data. The verbal data were analysed in an exploratory fashion to see what changes individuals experienced between their first and third attempt at communications planning.

METHOD

PARTICIPANTS, APPARATUS, AND SETTING

Novice participants were 8 post-graduate students studying at the University of Southampton. Their ages ranged from 22 to 28 (mean = 24.4, standard deviation = 2.1) and of the eight, four were male, four were female. The expert was a 40 year-old male native English speaker recruited through his participation in an ongoing research project at the University of Southampton. All were computer literate. The expert participant had 21 years of flying and mission planning experience and 10 years of MPS experience, whilst none of the novices had ever had any prior experience with MPS, and none had any experience of airspace regulations or communication procedures.

All sessions were held in the same room where a Fujitsu Siemens© E Series Lifebook laptop was situated at a desk. The laptop supported the Aerosystems International© Mission Planning System BL7 Build 0058 software. Participants were recorded using a Sony® Handicam HDR-SR12 digital video recorder capable of recording video and audio whilst their actions (mouse travel and number of clicks) on the laptop were recorded by the Jan Lellman/Fridgesoft© OdoPlus Version 1.6 Build 281 mouse tracking software. Participants were supplied with a number of different information sheets relating to each hypothetical mission, along with a number of Royal Air Force flight publications detailing airspace communications information. Participants were also supplied with a pen and paper.

PROCEDURE

Novice participants received a two hour one-to-one training session on the basic need for communications and the process by which planning takes place on MPS. They were subsequently subjected to three sessions in which they planned for a different hypothetical route each time. The sessions each lasted between 1 and 1.5 hours and were separated by one week. These three sessions will be herein referred to as the test sessions.

During the test sessions the participants were asked to work alone on planning the communications for each route. To ensure progression through each plan the participants were instructed that they may ask questions should they not know how to continue. Participants were asked to 'think aloud' constantly for the duration of each session. They were advised not to censor their thoughts, to be unconcerned with grammar, and to verbalize their thoughts no matter how trivial they believed them to be. Along with recording participants' verbalizations, mouse travel, number of clicks, and the length of time to complete each plan was recorded. The expert participant was recruited to control for task differences and was treated to the same conditions as the novices, with the exception that he only completed the first and third test sessions.

DATA REDUCTION AND STATISTICAL METHODS

Completed plans from test sessions one and three were inspected with the aid of the expert participant and errors and omissions were recorded. Participants' utterances from the first and third test sessions were transcribed verbatim and segmented based upon identifiable single units of speech. The segmentation was based upon content, reference and assertion. This was informed by studying the recordings for intonation and studying the videos to see where utterances were related to significant actions.

The number of words spoken by the participant and the experimenter in each session, and the number of experimenter inputs (a passage of speech from the experimenter preceded and followed by a question or statement from the participant) were recorded along with the mouse travel, mouse clicks, completion-time, and error data. The Wilcoxon Signed Ranks Test was then applied to this data to assess differences between sessions. Verbal data were analysed in a descriptive fashion only.

RESULTS

QUANTITATIVE DATA

All except one of the novice participants took significantly longer to complete the plan in the first session than in the third ($z=-2.20$, $p<.05$). In addition, all except one of the novice participants recorded significantly less mouse travel in the third session than in the third ($z=2.24$, $p<.05$). Mouse click data was mixed; although on average the number of clicks dropped from session one to session three, the difference was not significant ($z=-.34$, $p=.735$).

Data from the expert participant revealed some differences between the tasks in sessions one and three, in that he recorded higher values for mouse travel, mouse clicks and completion-time in the third session. Given his expert status no further skill acquisition should have taken place between sessions, hence any differences in results are likely to indicate task differences more so than differences in skill. Therefore a ratio of data from session one to session three was calculated for each measure. Expected session three values for novices were calculated from this ratio. These values would be expected if there was no improvement across sessions.

Upon comparison of the expected results with the observed results three significant differences were observed; time taken to complete the task ($z=-2.52$, $p<.05$), amount of mouse travel ($z=-2.52$, $p=<.05$), and the number of mouse clicks ($z=-2.52$, $p<.05$). In each case, the observed results were significantly less than the results expected had the novices shown no improvement.

Error data was mixed across participants, with no significant difference between the number of errors made in the first and third test sessions. Novices uttered significantly more words per minute in the first session than in the third ($z=-1.96$. $p<=.05$). Novices also required significantly fewer experimenter inputs in the third session compared to the first ($z=-2.52$, $p<.05$). The expert participant uttered a comparable number of words per minute in the first (82.86) and third (86.44) test sessions and required no experimenter inputs in either session.

VERBAL DATA

An initial coding scheme was developed *a priori* from the skill acquisition and expert/novice literature. This was applied to two of the protocols. The coding scheme was refined in an iterative fashion guided by the content of the protocols. The final coding scheme contained 14 codes. These are summarized in Table 1.

Table 1 Final codes with descriptions

CODE	DESCRIPTION
Generate Strategy	Statements giving overall description of what the participant intends to do, unspecific to any one action
Select Action	Statements describing what single action the subject is doing, or is about to do
Evaluate	Statements evaluating a previous action, choice, or piece of information
List information	Reading out or listing information directly, without interpretation
Select information	Statements referring to the selection of suitable or appropriate information for input
Confirm	Confirmation of previous action or choice, or confirmation of information either already present or previously entered in
Select input	Statements referring to the field of input, or the particular tab into which information must be entered.
Justify	Statements justifying or giving reason for a certain choice or action
Recognize	Statements about the recognition of information, the meaning of that information, and the resulting necessary actions or choices
Describe	Description of the route and the information displayed, or the current state of the plan
Queries	Statements reflecting self directed questions and times at which the subject is unsure of an action or choice

Software related	Statements of direct reference to the application, how the application affects performance, and how the subject interacts with the application
Task irrelevant	Full statements that are not directly related to the mission planning task or the software
Other	Incomplete statements and other unclassifiable statements

Prior to applying the coding scheme, all protocols were inspected with attention paid to the number of experimenter inputs and the total number of words spoken by the participant and the experimenter. Data from two participants were excluded from the analyses as a result of this inspection; one had a disproportionately high number of experimenter inputs while the other required more words to be spoken by the experimenter than they spoke themselves. Although guidance was given to all participants, the level of help required by these two individuals indicated a lack of task understanding. Skill acquisition could not be investigated as the experimenter had simply guided the two individuals through the task.

The remaining seven protocols (6 novices, 1 expert) were encoded, and the average relative frequency of each code's occurrence was calculated as a percentage of the overall code-count. To assess interrater reliability, Cohen's Kappa was calculated. The value of 0.63 indicated a good agreement between raters (Fleiss, 1981) and was significantly better than chance ($p<.001$).

Although the pattern of results for expert and novice participants was similar, three clear differences between the protocols were observed. First, the expert participant recorded fewer 'confirm' statements in both sessions than did the novices. Second, the change in the proportion of 'select information' statements across sessions was in the opposite direction for the expert participant compared to the novice participants. Novice participants produced more 'select information' utterances in the third session when compared to the first; this effect was in the opposite direction for the expert participant. Third was the difference in change across sessions in the proportion of 'select action' statements. On average, novices gave more of these reports in session one than in session three, whereas the expert participant gave more of these reports in session three rather than session one.

There were a number of notable differences when comparing the novices' protocols from the first and the third test sessions. The proportion of statements encoded as 'generate strategy', 'select action', 'confirm', and 'describe' dropped from session one to session three for all but two of the subjects (note that it was not the same two subjects that bucked this trend for each statement). The proportion of 'recognise' statements dropped from session one to session three in all but one of the subjects, and the proportion of 'query' statements dropped from session one to session three for all participants, with no exceptions. All participants uttered a higher proportion of statements encoded as 'list information' and 'select information' in the third session compared to the first.

DISCUSSION

The quantitative data recorded supported the hypothesis that participants would acquire some level of skill in using the MPS application. The decrease from session one to session three in the time required to perform the task and the amount of mouse travel taken suggest an improvement in efficiency and hence an increase in skill (Logan, 1988;). Although the number of errors recorded by each novice participant for each route did not significantly change from session one to session three, the expert data suggest that this result may be misleading. The expert required more time, a higher number of mouse clicks and a greater distance of mouse travel to plan for the route in session three than in session one. As this participant should have acquired no additional skill across sessions (he was already at expert status through his extensive experience with the application (Anderson, 1993)) this result was taken to indicate that the mission plan was either more complex or required more data considerations and inputs in session three than in session one. With more data to be considered and more data entry requirements necessarily comes more task steps, and as each task step carries with it a potential opportunity for error (Kirwan, 1994) it can be reasonably concluded that the task in session three gave participants more chance for error than did the task in session one. Taking this indication into account, along with the differences between the novices' observed session three results and those expected based on the expert's session on to session three ration, suggests that the novices acquired more skills specific to the mission planning process than is suggested by the error, observed mouse data, and time data alone.

Differences in novices' verbal reports from session on to session three also supported the hypothesis that participants would acquire communication planning skill across sessions. It was established that the task in session one required the participant to engage with less information than session three's task. It is unsurprising therefore that the codes 'list information' and 'select information' were higher in the final session; there was simply more information to list and select. Following on from this argument, one might expect the number of 'query' reports to increase across sessions. With more information to consider comes the possibility for a greater number of self-directed queries and questions regarding the successful insertion of that information. The decline in this code's occurrence across sessions points to an increase in task understanding or familiarity prevailing over any effects of increased task difficulty (Chit et al. 1989). Furthermore, although there was more information for the participant to 'recognise', the frequency with which participants made these reports was actually less in session three than in session one, again suggesting an effect of task familiarity or practice (Pirolli & Recker, 1994).

Some of the differences observed between the novices' protocols from the first and third test sessions mirrored some of the expert/novice differences described in the literature. Work by Chi et al. (1981), Glaser (1984), and Isenberg (1986) lead to the argument that experts are able to make inferences and draw meaning from the materials they are given, while novices rely on a more literal translation of the task or instructional materials. In the present study, the statements 'recognize' and 'describe' referred to surface features of the tasks, and these statements were more commonly produced in the first test session than in the third. This decline in the frequency with which novices produced these statements is indicative of a move from a characteristically novice way of engaging with a task to a method of processing more akin to expert behaviour (Chi *et al.,* 1981; Glaser, 1984; Isenberg, 1986).

Another notable difference between the novices' protocols collected from the first and third test sessions was in the frequency of statements encoded as 'select action'. First, it is important to bring to attention the task differences highlighted by data from the expert participant. Not only did the expert participant produce a higher proportion of 'select action' statements in test session three than in test session one, the overall picture gained from assessing the quantitative data indicated that the task set out in the final session required more actions from the participants than the session one task. Although the number of actions required by the task increased across sessions, the frequency with which novices produced 'select action' statements did not. In fact, novices produced fewer of these statements in the third session compared to the first. In conclusion, as task differences would suggest an increase across sessions in the need to 'select action' it is contended that the observed decrease displayed by the novices is an artifact of the acquisition of skill. It is quite possible that this effect arose from the early stages of the conversion of knowledge from a declarative to a more procedural nature (e.g. Anderson, 1989).

On final point to make is related to the literature surrounding the availability of knowledge for reflection by introspection (Pirolli & Recker, 1994). Anderson (1993) argues that the retrieval and integration of declarative knowledge is time consuming and effortful and as knowledge begins to be converted from declarative to procedural, retrieval and integration becomes more automatic. This is reflected in a reduction in the number of verbalisations; the less time and effort an individual spends thinking about a given task or situation, the less chance that individual has to verbalise those thoughts and the fewer thoughts they have to verbalize. In the current research, the number of words spoken by the novice participants declined across sessions, as did the number of words spoken per minute. This result, taken with the reduction in 'select action' statements, suggests that novice participants' knowledge became more procedural in nature as they became more familiar with the task and with the software (Anderson, 1993; Pirolli & Recker, 1994).

In conclusion, this study showed how cognitive processes change as a function of skill acquisition on an individual basis. Through applying verbal protocol analysis it has been demonstrated that the change in cognitive processes individuals experience as they acquire complex cognitive skills do reflect some of the distinctions made between individuals at the two ends of the expertise spectrum. Additionally, because many complex applications - especially those with user interfaces that are not intuitive - are essentially similar, applications of this type afford valid opportunities for further expertise research.

ACKNOWLEDGEMENTS

This research from the Human Factors Integration Defence Technology Centre was part-funded by the Human Sciences Domain of the UK Ministry of Defence Scientific Research Programme. Any views expressed are those of the author(s) and do not necessarily represent those of the Ministry of Defence or any other UK government department.

REFERENCES

Anderson, J.R. (1987). Methodologies for studying human knowledge. *Behavioural & Brain Science, 10,* 467–505.

Anderson, J.R. (1989). A theory of the origins of human knowledge. *Artificial Intelligence, 40,* 313-351.

Anderson, J.R. (1993). *Rules of the Mind.* Hillsdale, NJ: Erlbaum

Chase, W.G., & Simon, H.A. (1973). Perception in chess. *Cognitive Psychology, 4,* 55-81.

Chi, M.T.H., Bassok, M., Lewis, M., Reimann P., & Glaser R. (1989). Self-explanations: how students study and use examples in learning to solve problems. *Cognitive Science, 15,* 145–82.

Chi, M.T.H., Feltovich, P.J., and Glaser, R. (1981). Categorization and representation of physics problems by experts and novices, *Cognitive Science, 5,* 121-152.

Ericsson, K.A., & Simon, H.A. (1993). *Protocol analysis: Verbal reports as data.* (Revised ed.). Cambridge, MA: The MIT Press.

Fitts, P.M. (1964). Perceptual-motor skill learning. In *Categories of Human Learning,* A.W. Melton (ed.), pp. 243–85. New York: Academic.

Fleiss, J.L. (1981). *Statistical methods for rates and proportions.* New York: Wiley.

Glaser, R. (1984). Education and thinking: The role of knowledge. *American Psychologist, 39,* 93-104.

Isenberg, D.J. (1986). Thinking and managing: A verbal protocol analysis of managerial problem solving. *Academy of Management Journal, 29,* 775-788.

Jenkins, D.P, Stanton, N.A., Salmon, P.M., Walker, G.H., & Young, M.S. (2008). Using cognitive work analysis to explore activity allocation within military domains. *Ergonomics, 51,* 798-815.

Kirwan, B. (1994). *A guide to practical human reliability assessment.* Basingstoke, UK: Burgess Science Press.

Logan, G.D. (1988). Toward an instance theory of automatization. *Psychological Review, 95,* 492–527.

Pirolli, P., & Recker, M. (1994). Learning strategies and transfer in the domain of programming. *Cognition and Instruction, 12,* 235–75.

Ryle, G. (1949). *The Concept of Mind.* London, UK: Hutchinson.

VanLehn, K. (1996). Cognitive skill acquisition. *Annual Review of Psychology, 47,* 513-539.

CHAPTER 38

Cognitive Ergonomics: The Adaptation of the Brazilian Pilots to the Aircraft F-5EM

Homero Montandon

Brazilian National Civil Aviation Agency
São José dos Campos, SP, 12.246-870, Brazil
Márcio da Silveira Luz - PhD, Taubaté University, SP
Selma Leal de Oliveira Ribeiro - PhD, Estácio de Sá University, RJ

ABSTRACT

This work aimed at investigating the process of how the Brazilian pilots adapted themselves to the updated aircraft F-5EM, from the cognitive ergonomics point of view, once this airplane had adopted the *glass cockpit* concept in its cabin, among others modification suffered. The motivation factor of this research was the identification in the civil aviation that this concept of cockpit, with elevate level of automation and computer resource, had changed the pilot work nature, with a lot of benefits but bringing, also, some issues related to safety, with the pilots having difficult to deal with the new technology. However, the amount of information to be processed and the mental workload added, as the consequence of the installed system, with its new functionalities, imposed a challenge to the pilots, which had been trying to go through by a great amount of study and an intensive use of the simulator.

Key-words: Cognitive Ergonomic, Cognitive Psychology, Learning, Attention, Memory,

INTRODUCTION

The F-5E *Tiger II* is a light tactical fighter jet supersonic aircraft, manufactured by the Northrop Company (USA) and in service at the Brazilian Air Force (FAB) since 1975. This aircraft was submitted to an update process in order to modernize the avionic system and to provide a better capacity of navigation, attack and interception missions. The modernized model, designated as F-5EM, begun to operate in 2005.

Besides other changes, it was adopted a *glass-cockpit* style for the pilot station. This kind of cockpit is so called because the old and traditional analogical instruments were substituted by digital instruments, displayed in electronic screens.

It was also adopted the HOTAS concept, what means *hands on throttle and stick*, which permit the pilot fly and operate the systems without moving his hands off the controls, and the *head-up display* (HUD), with the most important information being displayed in a transparent screen, in front of the pilot, at his eyes line

What is significant in such modification into the cabin is the accretion in amount of information available to the pilot, and the number of new functionalities with the modern systems that the pilot has to deal with. A study made by Rodrigues (2006) shows that the volume of information was multiplied by 3 times, at least.

This research with the Brazilians' F-5 pilots was motivated by the fact that the civil aviation had suffered a similar evolution in the flight deck of its aircrafts since 1980's, where the old style of cabin, like that ones from the Boeing 707, were modified to those news ones like the Airbus A-320 *glass-cockpits*.

THE PROBLEM

This change brought a lot of benefits in terms of precision in navigation, situations awareness and physical workload decrease, due to the amount of the information allowed and the automated systems adopted. But this new technology had introduced others problems also, such as the tendency to become dependent of these automated resources, sometimes not understanding the system logic of operation, other times having surprise due to the aircraft unexpected behavior, as well as not being involved in the task during the flight and becoming somewhat apathetic or complacent, having as consequence some catastrophic accidents.

Such issues related to the pilots-aircraft relationship, which were affecting the safety in aviation, became the focus of attention to the aviation authorities, aircrafts manufactures, aviation psychologists and aviation safety professionals.

Such preoccupations were object of a multidisciplinary report, sponsored by the FAA, the US aviation authority, with the title of *The interface between flight crews and modern flight deck systems* - Report of the FAA Human Factors Team

(ABBOTT *et al,* 1996). Such as this one, others books and reports of the aeronautical community had claimed by knowledge related to human factors in aviation in order to better understand how the man acting in this type of environment and how to improve its performance in this new flight deck, aiming to improve safety (WIENER & NAGEL, 1988; WIENER 1989; REASON, 1990; BELLINGS, 1997; DEKKER & HOLLNAGEL, 1999; ENDSLEY & GARLAND, 2000).

In the civil aviation, the transition from one type of flight deck to a new one had occurred gradually, with the pilot's input becomes distant from the aircraft response, initially with the introduction of the automatic pilot (AP), which permits the airplanes keeps the altitude and heading automatically, until, finally, the operations of actual *Flight Management Computers* (FMC), which allows the complete flight planning and management by computer.

As the Brazilians' F-5 pilots were making a cliff transition between an old style of cockpit to the new generation and once this aircraft does not have an auto pilot and its controls, but having a sophisticated FMC, HOTAS, HUD and a modern multifunction radar, it was considered important to study their behavior in this adaptation process because it could bring some important knowledge of how the men adapt themselves and act in face of such king of technological environment.

METHODOLOGY

It was conducted a field research with the Brazilians' pilots. The research had a qualitative character, getting the data based on a structured interview and using the deductive method as analytical tool. The sample had the participation of 14 pilots in the universe of 40, corresponding to 35% of the total.

The theory which supports the analyses of the research data was the cognitive psychology, which propose itself at getting scientific knowledge about how people think, learn, judge, decide and employ their mental resource of perception, attention and memory.

RESULTS

The results demonstrated that the pilots were very well impressed in relation to the introduction of the HUD system, the quantity and quality of the available information, the precision in the navigation and armament employment, as well as to the new capacities offered by a better radar range, and the new operational capabilities.

But the focus was to know how they deal with the challenger of having adapted themselves, using their cognitive capabilities, and the first thing was to get their primary impression of the new cabin, what had got their attention in their first contact. The amount of information and the HUD utilization were the most impressive response, and the great number of new commands and the modified layout were the second ones.

After the first impression, the next question was to know what their opinion about the workload imposed by the new cabin. The mental workload was the greatest modification, when comparing with the old cabin, in their opinion, due to the amount of information and news functionalities to deal with.

Such demand for their attention made them concentrated on the systems' management, instead of external visualization, which is a big concern for a fighter pilot, as much as for operational reasons as well as for safety issues.

The absence of an automatic pilot (AP) as an auxiliary tool for basic flight, and the necessity to manually insert the data of the flight plan into the system also were reported as important factors that had increased the cabin workload. But now, the latest version of the system allows inserting the plan data by memory cartridge.

It was reported that, as the training and familiarization with the HOTAS and the new functionalities were getting progress, the workload was decreasing, which becomes around 10 hours of flight experience in the new cockpit operation. The next figure shows their opinions in percentage.

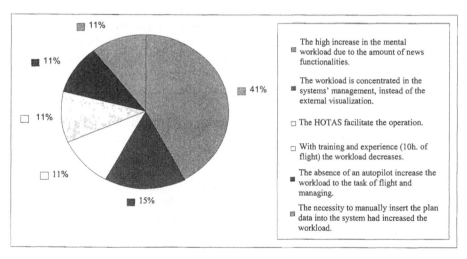

Figure 1: The workload imposed by the new cabin.

The cognitive psychology explains that such familiarization is part of a mental process that involves the memory capability, which is divided, for theoretical propose, in "long term memory" (LTM), "short term memory" (STM), and "work memory" (WM). The LTM, for the scientist opinions, has an infinitive capability of data storage or, at least, a capability great enough to be measurable.

The STM, for the other hand, has a low capacity of storage, seven plus/minus two items, according a famous article written by George Miller, in 1956. The WM is the activate part of the memory, which utilizes as much as LTM storage data as STM data to deal with a specific in charge task, depending on the specific necessity of the moment (Several authors, *apud* STERNBERG, 2000).

In the beginning of a familiarization process, when there is a few, or perhaps, none data related with the matter into the LTM, there is the necessity of a great effort of the STM, occupying almost the full WM capacity, even for very single tasks, in a sequential mental processing of information, making an use of what is called *conscious attention*. Such phenomena makes the apprentice movements or actions very slowly and, in most of the time, wrong or, at least, with a big amount of mistakes (STERNBERG, 2000).

But, as much as the experience is being accumulated, the *long term memory* begins to have more and more storage data and the memory process begins to be in parallel, with a lot of data being processed in the same time, relieving the *working memory* and the *short term memory* for other tasks. Such mental process is called *automating,* which consist in act in a desired manner at the proper time, involuntarily, self-acting, automatically (Several authors, *apud* STERNBERG, 2000). Such theoretical explanation aids to understand what the pilots reported as the increase of the mental workload at the beginning and a decrease as soon as the experience was being accumulated, around 10 hours of flight, as reported.

In respect to the difficulties found, it was reported that the demand for *time dedicated to study the system*, due to the great amount of new information to be assimilated, was the greatest obstacle to win. In 2006 the pilots missed a full simulator available to give them the possibility of training the procedures, mainly in relation to the HOTAS system.

The fact that it was possible to get the same information or to access the same operation mode by different ways was reported as another difficulty. They had trouble also to understand the *Technical Order* (TO) and assimilate the new technology due to the number of new terminology and acronyms. The location of the instruments into the cabin, in the new layout, was another point, mainly for the veteran pilots. Some pilots also mentioned the difficulty to understand the operation logic, even being surprised by the aircraft response to some specific commands. The figure below resumes their opinions.

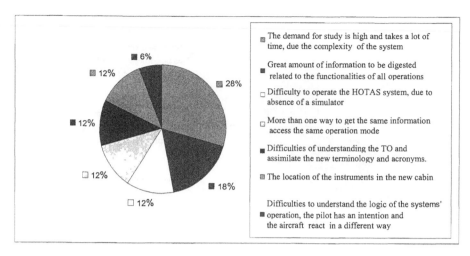

Figure 2: The pilot's difficulties to operate the system.

The Cognitive Psychology explains that these difficulties are normal and occur within the learning process, when the *conscious attention* are focused to process the new information, in a sequential way, and using significant part of the *work memory* (Several authors, *apud* STERNBERG, 2000).

Once they reported such difficulties of adaptation, it was asked to them what kind of personal strategy each one had adopted for better deal with such challenge. The answers were concentrated in three distinct aspects, complementary between each other: the first, the necessity of a lot of study and simulator training, due the great amount of information to be assimilate and the systems complexity itself, which demand a high degree of practice in order to optimize the understating of systems´ logic operation; the second, the learning process should occur in a gradually manner, step by step, divided by phases of the flight, beginning by simple tasks, such as departure and landings, up to the more advanced and complex missions, such as air combat, armament employment and interception modes radar operation; and, the last, the recommendation to utilize the screens configurations as much as possible similar to the instruments layout adopted in the old cockpit.

The figure bellow retracts their personal strategy of learning.

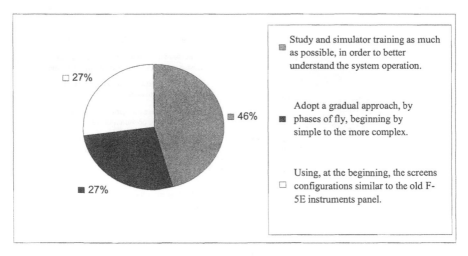

Figure 3: The personal strategy adopted to facilitate the adaptation.

Based on such strategies, the F-5EM pilots made some recommendations as *suggested training process* for the new pilots: The most emphatic opinion was the utilization of *intensive simulator training* sections, with several basic missions, aiming at the *procedures conditioning*. The conditioning, theoretically speaking, is the process to learn association, as well as is a way which each animal learn to adapt and survive in its environment (MYERS, 1999).

So, such adaptation using simulator training is considered a cheaper, safe and efficient way to realize such learning, or *conditioning*, what means, in other words, to turn the conscious mental process in *automated* process, making the pilots acts automatically, quick and safely while performing the basic normal and emergency procedures, relieving the *work memory* for operational tasks. The next stronger suggestions were the intensive systems' study, preceding the simulator utilization in order to correct understand its characteristics and operation logic, due the complexity and amount of functionalities. Other emphatic suggestion was to use the screens configurations similar to the old F-5E instruments panel.

Their suggestions has adherence and are coherent with the learning theory, which states that we learn, basically, by *association*, according to concepts that came to us since the philosophers as the ancient Greek Aristotle, passing by John Locke and David Hume. Our minds link naturally the events that occurred sequentially: we associate them, although there are others way of learning, such as by *observation* and by *imitation* (MYERS, 1999).

Such proposition aids to understand why the recommendation to utilize the screens configuration as much as possible similar to the old cockpit: it is an association technique aims at facilitate the adaptation.

Another point is the suggestion to learn gradually, step by step, from the most simple to the more complex: it find support at the theories about memory storage, which states that the key-technique to storage data into the memory is the *repetition*, so called the *practice's effects*, and that it is preferable a *distributed practice* instead of *agglomerated practices* (BAHRICK and PHELPS, 1987, *apud* STERNBERG, 2000), due to the *spacing's effect* (GLENBERG, 1977, and LEICH & OVERTON, 1987, *apud* STERNBERG, 2000), which helps the data storage, facilitating the elaborative process of mental association.

CONCLUSIONS

This research shows some important characteristic of the learning process when it was revealed the Brazilian pilots' behavior in face to the challenger to adapt at the modernized cockpit. All data got from the interviews and analyzed under the cognitive psychology corroborate its theoretical propositions about memory, attention and learning. Due to the cliff transitions between an old cockpit style to a modern one, it was natural and expected that they had found some difficulties to assimilate the new environment of work.

The tendency is, according to the theory, that, with the *practice* and the *experience*, the procedures become automatic, occurring without the use of the conscious attention, with the information being processed in parallel, in a agile and quickly way, liberating the work memory and the attention to others purposes. It is necessary to know, however, with how much of practice and experience the automatic process will occur.

The training method that had been utilized is in concordance with theoretical principles, which predict that the learning occurs in a *gradual* and *accumulative* way, being recommendable the *spaced* and not accumulated practices, and that the *repetition* is the technical-key to the learning retention in the memory, using the *practice's effect*. For that, the study and the simulator have been demonstrated as fundamental tools in this process.

So, the cognitive psychology theories and the cognitive ergonomics should be explored more deeply by aircrafts manufactures, airworthiness authorities and pilots' training managers in order to use their propositions and concepts to improve aviation safety.

REFERENCES

Abbott, K., Slotte, S., Stimson, D. K.*et al.* (1996). *The interface between flight crews and modern flight deck systems.* (Report of the FAA Human Factors Team).Washington, DC: Federal Aviation Administration.

Billings, C. E. *Aviation Automation: The Search for a Human-Centered Approach.* Mawah, NJ: Lawrence Erlbaum Associates, 1997.

Dekker, S. & Hollnagel, E. (Ed.). (1999). *Coping with Computers in the Cockpit.* Burlington, VT: Ashgate.

Endsley, M. R. & Garland, D. J. (Ed.) (2000). *Situation Awareness Analysis and Measurement.* Mahwah, NJ: Lawrence Erlbaum Associates.

Myers, D. (1999). *Introdução à Psicologia Geral.* [Introduction to General Psychology]. Rio de Janeiro: LTC SA.

Reason, J. (1990). *Human Error.* Cambridge: Ed. Cambridge University Press.

Rodrigues, F. W. (2006). *Sistemas Automáticos de Controle de Vôo em Aeronaves Modernas de Alto Desempenho.* [Flight Controls Automatic Systems at High Performance Modern Aircrafts]. (Monograph). Rio de Janeiro: Escola de Aperfeiçoamento de Oficiais da Aeronáutica.

Sternberg, R. J. (2000). *Psicologia Cognitiva.* [Cognitive Psychology]. Porto Alegre: Artmed.

Sumwalt III, R.t L. (Capt.). (2004). *Considerations about human factors in aviation safety.* In: II Aviation Human Factors International Seminar. São José dos Campos: CTA.

Wiener, E. L. & Nagel, D. C. (Ed). (1988). *Human Factors in Aviation.* Los Angeles: Ed. Academic Press.

_____. NASA Contractor Report 177528. (1989).CONTRACT NCC2-377: *Human Factors of Advanced Technology ("Glass Cockpit") Transport Aircraft.* Moffett Field, CA: Ames Research Center.

CHAPTER 39

Designing an Efficient Trajectory Clearance Information Format for Air Traffic Controllers' Displays

Miwa Hayashi[1], Richard C. Lanier[2]

[1] University Affiliated Research Center
NASA Ames Research Center
Moffett Field, CA 94035, USA

[2] Federal Aviation Administration
NASA Ames Research Center
Moffett Field, CA 94035, USA

ABSTRACT

Some of the air traffic control decision-support tools currently being developed require controllers to issue complex trajectory information as clearances to pilots. If traditional voice communication, instead of a data link, is to be used, the trajectory information must be presented to the controller in a way that facilitates accurate clearance reading. The trajectory information should also be as compact as possible so the chance of obstructing critical traffic information is minimized. The present study examined the effects of three trajectory-clearance information formats—A) most abbreviated text, B) less-abbreviated text, and C) graphical format—on controllers' clearance-reading performance. The results showed tradeoffs between clearance readability and the amount and type of displayed information. The results also indicated importance of training if more-abbreviated format is to be used.

INTRODUCTION

Recently, a number of new trajectory-based operation (TBO) tools have been proposed under the Next Generation Air Transportation System (NextGen) initiative to increase air-space capacity and reduce the environmental impacts of flights. Advanced computer technologies enable the TBO tools to calculate complex flight trajectories to achieve these goals. For such applications, a Controller-Pilot Data Link Communication system (henceforth *data link*) would be ideal for conveying the complex trajectory clearance to the pilot (FAA, 1995). However, if a TBO tool is to be deployed in the absence of a data link, the clearance would need to be issued via traditional voice communication by the human controller. Even if a data link is already available, if a transition period is expected, where there will be a mixture of data-link-equipped and -unequipped aircraft, or if the system needs to have a voice backup option, clearances will still need to be given occasionally through voice communication. If that is the case, care must be taken in designing the display of new TBO clearance information.

The controller's display for a busy sector is already very crowded. To lower the display clutter level and reduce the chances of obscuring critical traffic information, a compact format is desired for the clearance information. Yet, too much abbreviation or compression of the information could impede accurate and smooth reading of the clearance and/or increase the controller's cognitive workload.

Studies on the readability of abbreviated texts have been conducted in various domains, such as instant messaging (e.g., Kleen & Heinrichs, 2008). The current study differs from previous work by focusing on a specific air-traffic-control application, where an expert operator is required to quickly reconstruct the correct phraseology from the abbreviated information and read it accurately and smoothly under high-workload situations.

In an effort to identify efficient ways to present complex trajectory-clearance information to the controller, the present study compared three display formats: A) most abbreviated text, B) less-abbreviated text, and C) graphical format. The last format presents the trajectory clearance information directly on the map display. Some TBO tools depict the clearance route graphically on the map display. If all the other information associated with the clearance, such as speed, altitude, etc., is presented next to the graphical route, this may serve as the clearance display. Such a graphical format eliminates the need to present separate text information elsewhere, and thus helps to reduce display clutter. The added graphical information might also improve awareness of the cleared trajectory.

For the complex trajectory-clearance phraseology, the phraseology developed and used by one of NASA's proposed TBO tools—Efficient Descent Advisor (formerly, En Route Descent Advisor; EDA) (Coppenbarger, et al., 2004) was used. In this study, air-traffic controller participants were asked to read trajectory-clearance information presented in one of the three formats. No air-traffic management task was simulated. Instead, the participants were asked to perform a simple secondary task concurrently with the clearance-reading task.

METHODS

THREE CLEARANCE TYPES AND PHRASEOLOGY

To investigate the effects of the formats, the phraseologies for three EDA clearance types, denoted CT1 (descent speed), CT2 (cruise and descent speeds), and CT3 (path stretch), were used in this experiment. Examples are as follows:

- CT1 (descent speed): "American 123, EDA clearance, descend via the SHARK SIX arrival, transition at 260 knots in descent."
- CT2 (cruise and descent speeds): "United 456, EDA clearance, maintain mach .81, descend via the LUNAR FIVE arrival, transition at 280 knots in descent."
- CT3 (path stretch): "Continental 789, EDA clearance, maintain mach .77, revised routing when ready to copy," "Continental 789, at Bowie (BOW), proceed direct to the HERON 146 bearing 108 mile fix, then direct HERON," "Continental 789, descend via the SHARK SIX arrival, transition at 260 knots in descent."

Note that CT3 clearance is issued via three separate radio transmissions and is the longest and most complex clearance of the three types. The pilot's read back was not included in this study. All waypoints and Standard Terminal Arrival (STAR) names used in the experiment were fictional.

DISPLAY FORMATS

Format A is the most abbreviated text format, containing the bare minimum text information necessary to reconstruct the required phraseology. Figure 1 shows an example of a CT3 clearance in format A. (The purpose of the CLOSE button will be explained in the Tasks section.) The "C/" and "D/" indicate the cruise speed and the descent speed, respectively.

Format B, the less-abbreviated text format, presents more textual information than format A. If format A resulted in degraded clearance-reading performance compared to format B, that implies that the aggressive text abbreviation comes at a cost. Figure 2 shows the same CT3 clearance example in format B. Each line corresponds to a single radio transmission. The title bar shows "EDA CLEARANCE," allowing the participant to simply read it following the aircraft ID.

Format C is a graphical format. Figure 3 shows the same CT3 clearance example in format C. The diamond represents the aircraft, the circle represents the navigation aid, and the crosses represent the waypoints. The cleared routes are shown in cyan color. The amount of text information in format C is exactly the same as in format A. The intention was to measure the effects of the graphical information by comparing formats A and C.

FIGURE 1 Example of CT3 clearance in format A. (Window size: 7" × 4.5")

FIGURE 2 Example of CT3 clearance in format B. (Window size: 7" × 4.5")

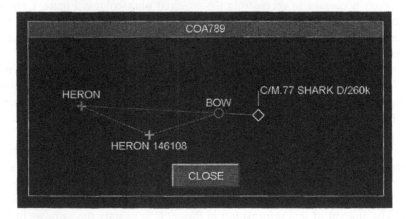

FIGURE 3 Example of CT3 clearance in format C. (Window size: 9" × 4.5")

PARTICIPANTS

Five retired and one current en-route air traffic controller participated in the study. All of the retired controllers retired within the past five years. Two were female and four were male. The ages ranged from 32 to 62 (mean: 51, std: 11), and their en-route-sector control experience ranged from five to 33 years (mean: 18, std: 11). All demonstrated sufficient visual acuity to view the displayed information used in this experiment (some used glasses).

APPARATUS

A 15-inch laptop computer was placed in front of the participant to present the clearance information. A mouse was provided on the right-hand side of the laptop for the participant's use. Another 15-inch laptop computer was placed on the left-hand side of the first laptop angled to face the participant to present the secondary-task visual stimuli. Both the clearance-information displays and the secondary-task visual stimuli were generated using ActionScript 2.0.

TASKS

The primary task was reading the trajectory clearance. Clearance information was presented on the monitor in front of the participant in one of the three formats. The participant reconstructed the proper phraseology and read it aloud. When the reading was complete, the participant clicked the CLOSE button using the mouse. The next clearance information then appeared. The process was repeated.

In addition, a secondary task was administered to assess the participant's spare attention level. The side monitor presented a yellow vertical bar whose height reduced at a constant speed. The participant was asked to press the space bar as soon as he/she noticed that the top of the bar had reached the bottom. This task mimicked monitoring for another flight to attain a certain point or altitude at a known speed. The bar size was 1.25 × 4 inches at its full height. The bar speed was randomly chosen from three values: 6.8, 8.5, and 10.3 cycles per minute. If the space bar was pressed prematurely, it was inactivated for two seconds and the vertical bar turned red. The participants were instructed that they should perform both tasks well, but if maintaining the dual tasks became difficult, the secondary task could be unattended temporarily.

PROCEDURE

Prior to the experiment date, an information packet was sent to the participants, and they were asked to read it and memorize the phraseology, the waypoint names, and the STAR names. On the experiment date, a briefing was held to review the information in the packet, and was followed by a practice session. Once the participant and the experimenter both felt comfortable, the data collection started.

The same format was used within a single trial. Each participant performed nine trials, containing three trials for each of the three formats. The orders of the formats were counterbalanced within and among the participants. Each trial consisted of four CT1, four CT2, and nine CT3 clearances (thus, 17 clearances in total), presented in a balanced order.

The data collected were the audio recording of the clearance reading and the timestamps of the participants' presses of the CLOSE button and the space bar. After completion of all the nine trials, the participants filled out a questionnaire that asked questions regarding their subjective preferences and comments.

RESULTS

SPEED OF CLEARANCE READING

The clearance reading times, measured as the time between the opening of new clearance information and the pressing of the CLOSE button, were analyzed with a four-way mixed-model analysis of variance (ANOVA). The main effects were Participant, Trial Block (the first three, second three, and the third three trials), Format (A, B, and C), and Clearance Type (CT3 and others). The Participant was treated as a random effect (Lindman, 1974), which led to more conservative statistical-inference results. The results showed statistically significant effects of the Participant ($F_{5, 810} = 157.5, p < 0.01$), Format ($F_{2, 10} = 6.5, p = 0.02$), Clearance Type ($F_{1, 5} = 409.8, p < 0.01$), Participant × Trial Block ($F_{10, 810} = 9.1, p < 0.01$), Participant × Format ($F_{10, 810} = 2.13, p = 0.02$), Participant × Clearance Type ($F_{5, 810} = 31.1, p < 0.01$), Trial Block × Clearance Type ($F_{2, 10} = 4.43, p = 0.04$), and Participant × Trial Block × Format ($F_{20, 810} = 2.21, p < 0.01$).

To visualize the Format effects, Figure 4 plots the means and standard errors of each format. Format B resulted in the fastest clearance reading, and format C resulted in the slowest. Planned comparisons showed a statistically significant difference between formats A and C ($F_{1, 10} = 4.80, p = 0.05$), but not between A and B. For the formats B and C, a more-conservative Fisher's *post hoc* comparison was applied since testing this pair was not planned originally: The results showed a statistically significant difference between these two formats ($F_{1, 10} = 13.4, p = 0.01$).

Figure 5 plots the means of clearance-reading times for each combination of the trial blocks and formats. It shows large learning effects in formats A and C, but not in format B. Later in the trials, format A resulted in even faster clearance reading than format B. However, the Trial Block × Format effect did not result in a significant difference in the above ANOVA, mainly due to the presence of the random effect. Instead, the Participant × Trial Block × Format effect was found to be statistically significant. That implies the impacts of the Trial Block × Format effects depend on the individual participant.

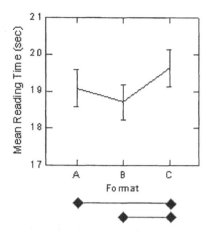

FIGURE 4 Format effects on clearance-reading time. Pairs of diamonds indicate statistically significant differences.

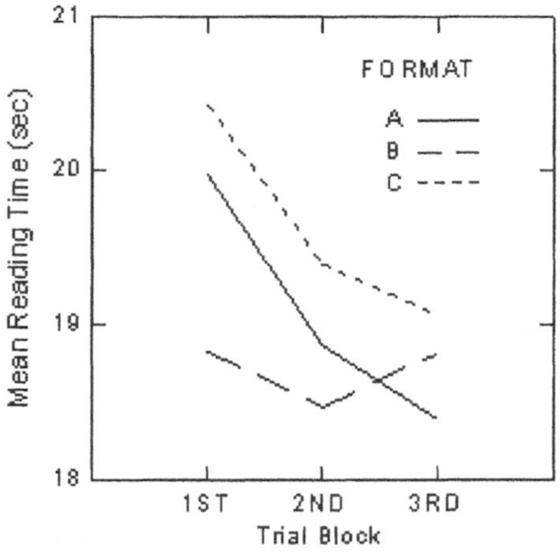

FIGURE 5 Trial Block × Format effects on clearance-reading time.

ACCURACY OF CLEARANCE READING

Errors made in clearance reading were categorized as *format neutral* or *format dependent*. The format-neutral errors were those related to the information either displayed in an identical form in all of the formats (i.e., the aircraft ID, numbers, and units) or not displayed in any format (e.g., "descent," "arrival"). The remaining errors were categorized as *format dependent*. Then, the format-dependent errors were analyzed in a three-way mixed-model ANOVA with Participant, Trial Block, and Format as the main effects. No interaction effect was included as there was only one frequency data per cell. The results showed a statistically significant Participant effect ($F_{5, 44} = 4.52$, $p < 0.01$) and a marginally significant Format effect ($F_{2, 10} = 3.55$, $p = 0.07$). A planned comparison of the Format effects revealed a marginally significant difference between formats A and B ($F_{1, 10} = 4.70$, $p = 0.06$).

Table 1 lists the frequencies of the format-dependent errors corrected and not corrected by the participants in each format. In the table, these errors were further divided into two groups: omission or wrong-word errors. A chi-square test was applied to a 2×2 table that was generated by adding the three 2×2 tables in Table 1. The test resulted in a statistically significant difference ($\chi^2_1 = 22.1$, $p < 0.01$). That means the omission errors were significantly harder to detect and correct than the wrong-word errors.

The format-neutral errors included 76 aircraft-ID errors (about 4% of all of the aircraft IDs called in this experiment; of which 35 were corrected), 41 number errors (i.e., wrong, transposed, or missing digits, excluding the aircraft-ID and the STAR numbers; 22 corrected), 33 unit errors (such as "miles" vs. "knots"; 23 corrected), and 13 noun errors (such as "descent," "arrival"; 10 corrected).

Table 1 Frequencies of format-dependent errors in clearance reading

	Format A		Format B		Format C	
	Corrected	Not Corrected	Corrected	Not Corrected	Corrected	Not Corrected
Omission	4	30	0	1	7	11
Wrong Word	22	10	8	2	29	26

RESPONSE DELAYS IN THE BAR-MONITORING TASK

The response delay in the bar-monitoring task was analyzed in the same four-way mixed-model ANOVA as in the clearance-reading task performance. However, no noteworthy effect of Format or Trial Block was found.

QUESTIONNAIRE RESULTS

Head-to-head comparison results on each pair of formats showed that the participants preferred format B the most, format A second, and format C the least. Five out of six participants remarked that they liked format B presenting each radio transmission on a separate line. However two commented that format B may be too much on the air-traffic controller's display. Three commented that they liked format A for its brevity. One participant commented that she liked format B first, but later, as she got more used to the clearances, format A became her preference. Five participants remarked that they did not think adding graphical information helped reading the clearance. One participant commented that she did not like format C because the sequence of the information was different from that in the phraseology, which forced her to look back and forth.

DISCUSSION

The results showed that controllers can read the trajectory clearance faster with the text formats (formats A and B) than with the graphical format (format C). Between the two text formats, format B (the less-abbreviated text format) generally yielded a faster reading completion time than format A (the most abbreviated text format), though the difference was not statistically significant. Format B tended to incur the fewest format-dependent errors, and therefore the least amount of time used to correct them. This explains the resulting faster reading speed using format B. The participants may also have needed to pose and recall the subsequent words least frequently with format B. Format B was also the most preferred format among the participants, though some were concerned that the large footprint of format B may be too much for the controller's display.

Large learning effects in the clearance-reading speeds were observed with formats A (the most abbreviated text format) and C (the graphical format), but not format B (the less-abbreviated text format). Note that all participants received about the same amount of training for each format before the data collection session. That means the controllers need additional training to be fluent in more-abbreviated formats, such as formats A and C. The amount of initial and recurring training required for these more-abbreviated formats may vary largely by individual controller (based on the large Participant × Trial Block × Format effect observed in the reading speed). Moreover, the format-dependent error counts suggest that omissions were particularly hard to detect and correct. Hence, the instructor may want to emphasize preventing omission of critical words.

After a sufficient amount of (on-the-trial) training with format A, the participants performed as well as, or in some cases even better than, they did with format B. Format A requires smaller footprint than format B. One way to take advantage of both formats while also reducing the need for controller training is to design the display so that the controller can select either format A or B. If the controller is relatively new to the TBO tool or has not been using the tool for a while, the complex trajectory-clearance information is shown in format B. Once he/she becomes used to reading the information, the display can be switched back to format A.

Even after some (on-the-trial) training with format C, the participants did not achieve the performance level attained with formats A and B. The graphical information of the path-stretch route offered little help in clearance reading (though it may be helpful for the purpose of route planning). Thus, display designers should not depend on graphical information of the cleared trajectory. However, the route graphics was merely added information, which the participants could have ignored if they wanted. The real disadvantage of format C appeared to be the text information being displayed out of sequence. Thus, if graphical format is to be used, the display designer needs to devise a way to present the text information in an order consistent with the phraseology.

The study found a number of format-neutral errors along the way. For instance, about 4% of all of the aircraft IDs were misspoken. (In reality, many such errors are quickly corrected by either the controller or the pilot. This still takes up precious radio time, however.) What this tells us is that no matter how much displays are improved, errors in spoken words will never completely disappear. It is still important for us to make every effort to assist the controller by designing better displays. However, the ultimate solution may be a data link.

Lastly, in this experiment, the bar-monitoring secondary task did not indicate any obvious performance interference caused by the Format or Trial Block effects. Instead, the interference appeared in the primary task performance. It seemed that all the participants managed to perform the secondary task very well. If a researcher wishes to make the secondary task a more sensitive measure of the workload level, he/she may need to raise the task intensity (e.g., faster bar speed) or select a task that requires mental resources similar to those required for the clearance-reading task, such as text search or mental arithmetic.

CONCLUSION

The study provided empirical evidence of the tradeoffs between readability and the level of abbreviation when the air traffic controllers (i.e., experts) needed to reconstruct the correct phraseology for a TBO clearance from the abbreviated information and read it aloud. The results showed that displaying the information in the less-abbreviated text format (format B) yielded the fewest clearance-reading errors and the fastest clearance-reading speed. This format was also preferred most by the participants; yet, the larger footprint required for its display may be a concern for this format. The more abbreviated and more compact text format (format A) could attain similar performance as format B (or even better performance than format B) with sufficient controller training. An idea for possible adaptive display design was discussed. This design would allow the controller to switch between formats A and B at will, to take advantage of the different levels of abbreviation without increasing the need for controller training. The graphical format (format C) caused the slowest reading-completion time as well as the greatest number of reading errors, probably due to the text information being arranged inconsistently with the clearance phraseology. Though the experiment used the phraseology for a particular TBO tool, the results obtained in this experiment are generic enough to be applied to other TBO tools' trajectory-clearance information display designs as well.

ACKNOWLEDGEMENT

The study was funded by FAA Research Solutions (AJP-64). The authors express great appreciation to all of the six controller participants for their time, effort, and patience, as well as many insightful comments. The authors are also thankful to the Flight Research Associates for their support.

REFERENCES

FAA Data Link Benefits Study Team. (1995). *User benefits of two-way data link ATC communications: aircraft delay and flight efficiency in congested en route sirspace* (Report No. DOT/FAA/CT-95/4): FAA Technical Center.

Kleen, B. A., & Heinrichs, L. (2008). "A comparison of student use and understanding of text messaging shorthand at two universities." *Issues in Information Systems*, IX(2), 412-420.

Coppenbarger, R. A., Lanier, R., Sweet, D., & Dorsky, S. (2004, Aug 16-19). "Design and development of the En Route Descent Advisor (EDA) for conflict-free arrival metering." Paper presented at the AIAA Guidance, Navigation, and Control Conference, Providence, RI.

Lindman, H. R. (1974). *Analysis of variance in complex experimental designs*. San Francisco, CA: W. H. Freeman and Company.

Chapter 40

Learning to Monitor Aircraft Datablock with Different Complexity Levels

Chen Ling, Lesheng Hua

University of Oklahoma

ABSTRACT

One of the information sources for air traffic controllers to monitor the traffic is aircraft datablocks. Such datablocks are typically composed of many fields depicting various aircraft information. The objective of this study was to understand how the objective datablock complexity and datablock exposure time affected the monitoring task learning process and performances. We trained 10 college students for six days to monitor datablocks presented briefly on a computer screen. We arbitrarily assigned a normal or abnormal status for each field. The participants' task was to indicate the status of the datablock as normal or abnormal after viewing the datablock. We varied the objective complexity by changing the number of datablock fields from 6 to 12, and the exposure time of the datablock from 3s to 0.3s. Participants' performance was measured as proportion correct of monitoring task. We found that individuals were able to learn to quickly monitor DBs, and the learning effect existed during the six days of training. The datablock with six fields led to lower level of experienced complexity, whereas the datablocks with 9 and 12 fields led to higher level of experienced complexity. This differing level of experienced complexity was demonstrated with differences in the average proportion correct in monitoring task, number of training trials needed, and the subjective mental workload. When the exposure time was longer than or equal to 1s, the monitoring task was easier to learn and perform. But when the exposure time was shorter than 1s, the task became more challenging. The capacity of learning was constrained by both the DB complexity and the exposure time. The results implicated that designing DBs for air traffic control displays should consider both the number of items in the DB and the time that users (e.g, controllers) actually spent on viewing a DB in monitoring air traffic situations.

Keywords: Training, Aircraft Datablock, Monitor, Complexity, Exposure Time

INTRODUCTION

Air traffic controllers are responsible for ensuring safe and efficient flow of air traffic. As the traffic situation evolves, information on aircraft status should be conveyed to the controllers in the most effective way through the air traffic control (ATC) displays. Datablock (DB) is the basic unit to display such information on the displays (Xing & Manning, 2005). Many types of information are relevant to the aircraft status such as flight altitude, destination, and conflict alert. However, it may not be advisable or feasible to include all relevant information in the DB. A frequently asked question by the display designers is "how many fields should be included in a DB?" More fields in a DB imply increased complexity. To answer this question, we need to understand the impact of DB complexity on the controller's task performance.

First, a good understanding of how controllers use the DB for their tasks is needed. Cognitive task analysis identified four ATC cognitive tasks: monitoring, controlling, checking, and diagnosing / problem solving. Monitoring task is one of the primary tasks for controllers for future Air traffic control system (Metzger & Parasuraman, 2001). It becomes more important when technology takes over tasks performed by human. In the monitoring task, the controllers need to monitor the aircrafts on the display and judge whether the aircrafts' status are normal. Normal status for a DB implies that all fields in the DB are in the normal status and no unusual event is present, therefore controller let the normal DB pass. On the other hand, if one or more DB fields present any unusual event, then a DB is in an abnormal status. Examples of unusual event could be that the altitude is too high or low; the speed is too fast or slow; or the separation from another nearby aircraft is lost, etc. When unusual event occurs, controller need to take some action. With the information from the monitoring task, controllers make decisions on what to do next. They could let the normal DB pass, or perform anticipated controlling actions on the abnormal DB, or check for unexpected information from the abnormal DB fields. If the situation implies a problem after checking, then further diagnosis and problem solving need to be performed.

The complexity of DB design could have an impact on the monitoring task performance. Researchers examine complexity through two lenses, the objective task complexity and experienced task complexity (Campbell, 1988). When a DB contains more fields, it presents an increased objective complexity. When controllers need to monitor DB with different levels of objective complexity, they experience different levels of task complexity. Understanding the relationship between the objective and experienced complexity help us to know what becomes too complex for human.

The impact of complexity on the task performance might also be affected by the amount of time available to perform the task. If the exposure time for controllers to view the DB on the screen is brief, then the controller's visual attention needs to be distributed among all the fields within a short period of time. Hence, the controller is under greater temporal pressure. The impact of increased complexity in the stimuli could be even more pronounced when time to view the stimuli is very brief. Studying monitoring task performance of different levels of DB complexity with different DB exposure time could facilitate deriving practical design guidelines for the ATC display.

To derive the optimal design of DBs for future ATC displays, we need to gain basic understandings on the complexity issues associated with DB. It is also helpful to understand whether controllers could learn to monitor DBs with varying levels of complexity, and what the learning process is like. Specifically, we aimed to answer the following research questions: 1) Did learning of the DB monitoring task occur? 2) How did the objective DB complexity affect the learning of monitoring task? 3) How did the exposure time affect the learning of monitoring task?

METHOD

Participants
Ten male participants participated in this experiment. They were college students on the campus of University of Oklahoma. Their ages ranged from 18 to 25. All participants had normal vision of at least 20/20 and normal color vision conditions.

Experiment Stimuli
Each participant was trained to perform the monitoring task with three types of DBs, including six-field datablock (6F-DB), nine-field datablock (9F-DB), and twelve-field datablock (12F-DB). The DBs partly mimicked a DB prototype for future en route radar display DBs with datalink. All DBs had three lines. The total numbers of fields that needed to be monitored were 6, 9, and 12 for 6F-DB, 9F-DB, and 12F-DB respectively. An example of DBs is shown in Figure 1. The meaning and location of each data fields in 12F-DB –the DB with most fields are shown in Figure 2. It needs to be noted that in this experiment, the field "flight speed" was included in the DB stimuli but its values stayed the same and were not monitored. The information in each field indicated two possible statuses: normal or abnormal (see Table 1). If all DB fields were normal, then the DB was normal. If any DB field was abnormal, then the DB was abnormal. In this study, the abnormal DB only had one DB field with abnormal status. Participants learned to monitor the DB statuses of normal or abnormal through training.

CA UAL1234 330 UP 46 H250 R292	CA UAL1234→3 330C UP290# 46 H250 R292	CA ^UAL1234→3 330C [UP]290# 46 H250 R292 K300
(a) 6F-DB	(b) 9F-DB	(c) 12F-DB

Figure 1. Example of Three Types of DB

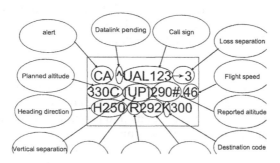

Figure 2. Meaning of the DB Fields in 12F-DB

Table 1. Description of DB Fields in Three Types of DBs

Field Name	Normal	Abnormal	6F-DB	9F-DB	12F-DB
Alert	--	CA (Conflict alert) LA (Low-altitude)	x	x	x
Datalink pending	-	^			x
Call sign	UAL123	<u>UAL123</u>	x	x	x
Loss of separation	---	→3		x	x
Planned altitude	<500 >100	>500 or <100	x	x	x
Vertical separation	----	[] (magenta)	x	x	x
Flight status	-	UP(climbing) DW (descending)	x	x	x
Reported altitude	290	290#		x	x
Heading direction	0-360	>360	x	x	x
Aircraft type	R, J	Other or no letters		x	x
Computer ID	3 digits (e.g. 292)	2 or 4 digits			x
Destination code	A-V	any of symbols @%$&			x

Note: x means that the DB type contains the corresponding DB field.

Equipment

An Optiplex GX620 Dell computer with 2GB of RAM and a Pentium D Smithfield (3.2GHz) processor was used to control the presentation of DB stimuli. The stimuli were presented on a 19-inch Dell color monitor with resolution of 1024x768. MATLAB program was used for visual stimuli presentation and data collection. The experiments were conducted in a quiet room.

Procedure

Every participant first took a visual acuity test to ensure that their vision conditions were normal. Then the participants learned the meaning of all DB fields as well as the normal and abnormal statuses for each DB field with a set of PowerPoint slides. The participants were subsequently tested with a short written quiz. They were required to achieve 95% correct in understanding the meanings and statuses of the DB fields. If they failed to achieve 95% correctness, they studied the PowerPoint slides again and took the quiz again until 95% correctness was achieved. This was to ensure that the participants learned the necessary knowledge about the task. But having the knowledge did not guarantee applying the knowledge correctly in the actual tasks. The participants were then trained to perform the computer-based monitoring tasks with multiple training sessions.

The training took place on six consecutive days for three types of DBs. 6F-DB were trained on day 1 and day 2; 9F-DB were trained on day 3 and day 4; and 12F-DB were trained on day 5 and day 6. The participants' task was to judge whether the status of the presented DB was normal or not. If the participant judged the DB to present any unusual event, they pressed the "action" button; otherwise, they pressed the "pass" button (see Figure 3). For each type of DB, each participant learned the task through four training sessions over two days. Participants spent around one hour on the two training sessions each day.

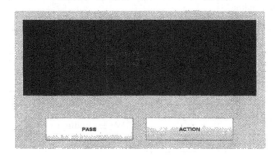

Figure 3. Screen Capture of the Monitoring Experiment

Within each session, the exposure time of DB was varied from 3s, to 2s, 1s, 0.5s, and 0.3 s consecutively. For each exposure time, the participants were provided with up to three trials to improve their performances, and each trial consisted of 12, 17, or 22 DB test cases for 6F-DB, 9F-DB, and 12F-DB respectively. The normal or abnormal statuses were randomly assigned to the DB fields. The proportion correct of DB monitoring task for each trial was calculated by dividing the number of DBs that was correctly judged of their statuses by the total number of DB test cases in the trial. This value was kept track with the Matlab program. If the participants achieved 90% correctness for the first trial, they moved on to the next session. Otherwise, they continued to learn with another trial within the same session. If the participants went through all three trials, then they had to move on to the next session no matter whether 90% correctness was achieved. At the end of each day's training, the participants filled out the NASA-TLX to measure the subjective mental workload that they experienced.

Experimental Design

The experiment was a repeated-measure design with two factors: the DB complexity level, and the exposure time. The dependent variables included the proportion correct in monitoring task, the number of training trials to learn the monitoring task, and the subjective mental workload. The hypothesis was that the more complex DB and shorter exposure time would result in lower proportion correct, more trials to train, and higher mental workload.

RESULTS

The results of the current study were arranged into three sections. First, the learning processes of the DB monitoring task were reviewed in terms of the average proportion correct achieved. The effects of DB types and exposure time on the monitoring task performance were analyzed. Second, the total numbers of training trials used were compared. Third, the subjective mental workloads experienced by the participants during the monitoring tasks were compared.

Learning Process of DB Monitoring Task

The proportion correct of monitoring task was calculated for all training sessions. Because the participants were allowed to take up to three trials in each session to learn the monitoring task, the number of trials that the participant actually took in each session could be 1, 2, or 3. The average proportion correct of trials in each session represented the level of learning for that session. The learning trends of DBs during the six day training are shown in Figure 4.

Effect of training session on monitoring task performance. ANOVA on the proportion correct showed significant differences among the four training sessions, $F(3, 27) = 11.73$, $p = 0.0001$. There was a gradual increase in average correctness over the four sessions. Post Hoc Tukey test showed that the proportion correct of the first session (M=0.92, SD=0.063) was significantly lower than those of the second (M=0.93, SD=0.066), the third (M=0.94, SD=0.053), and the fourth (M=0.95, SD=0.050) session. Such increase suggested learning effect existed for all the participants.

Effect of training session on monitoring task performance. ANOVA on the proportion correct showed significant differences among the four training sessions, $F(3, 27) = 11.73$, $p = 0.0001$. There was a gradual increase in average correctness over the four sessions. Post Hoc Tukey test showed that the proportion correct of the first session (M=0.92, SD=0.063) was significantly lower than those of the second (M=0.93, SD=0.066), the third (M=0.94, SD=0.053), and the fourth (M=0.95, SD=0.050) session. Such increase suggested learning effect existed for all participants.

Effect of DB complexity on monitoring performance. ANOVA results on the average proportion correct indicated that there was significant differences among the three types of DB, F (2, 18) =6.48, *p*=0.0076. Post Hoc Tukey test showed that the proportion correct for 6F-DB (M=0.95, SD=0.066) was higher than those of 9F-DB (M=0.93, SD=0.053), and 12F-DB (M=0.93, SD=0.057). There was no significant difference between 9F-DB and 12F-DB. This result suggested that 6F-DB had lower level of complexity, whereas 9F-DB and 12F-DB had higher level of complexity.

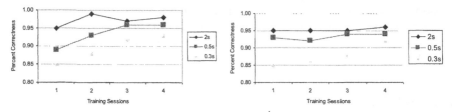

(a) Learning Trend of 6F-DB during day 1 and 2   (b) Learning Trend of 9F-DB during day 3 and 4

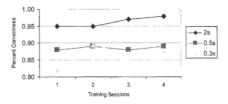

(c) Learning Trend of 12F-DB during day 5 and 6

Figure 4. Learning Trends of DBs during Six days (n=10)

Effect of Exposure Time on Monitoring Performance. The exposure time of DB varied from 3s to 2s, 1s, 0.5s, and 0.3s for each DB type. This created different levels of temporal demand and task difficulty for the participants. When the exposure time was shorter, the participants needed to make judgments faster. It was observed that as the exposure time decreased the proportion correct also decreased gradually for all three types of DBs. ANOVA results showed significant difference among the five exposure time in terms of proportion correct, F (4, 36) =43.78, *p*=0.0001). The post-hoc Tukey analysis indicated that there was no significant difference among exposure time of 1s (M=0.95, SD=0.046), 2s (M=0.96, SD=0.040), and 3s (M=0.96, SD=0.038). But the proportion correct for 0.5s (M=0.92, SD=0.055) was significantly lower than that of 1s, 2s and 3s. And proportion correct for 0.3s (M=0.88, SD=0.070) was even lower than all the other exposure time. This result suggested that when the exposure time was greater or equal to 1s, the task difficulty level was similar. But when the exposure time was less than 1s, the task difficulty level was much higher.

Total Number of Trials Took to Learn the Monitoring Task

The total numbers of trials in all four training sessions to train each type of DB were calculated and compared (see Figure 4). Because participants could take up to three trials in each session, and they went through four training sessions, the maximum number of trials for them to learn was twelve.

Effect of DB complexity on number of trials. In terms of total number of training trials, the ANOVA results showed that there was a significant difference among the number of trials for the three DB types, F (2, 18) =3.49, *p*=0.05. Post Hoc Tukey test showed that it took less trials to train 6F-DB (M=5.84, SD=1.93) than 12F-DB (M=6.48, SD=2.75) and 9F-DB (M=6.74, SD=2.50). There was no significant difference between 9F-DB and 12F-DB in number of trials.

Effect of exposure time on number of trials. There was also a significant difference in the total number of trials among the five exposure time, F (4, 36) =35.10, *p*=0.0001. Post Hoc Tukey test showed that it took the larger number of trials to train the participants to monitor DB with exposure time of 0.3s (M=8.80, SD=2.17) and with 0.5s (M=7.46, SD=2.47). It took less trials to train the participants to monitor DB with exposure time of 3s (M=5.67, SD=1.04), 2s (M=5.90, SD=1.36), 1s (M=5.93, SD=2.02) than 0.5s and 0.3s.

Subjective Mental Workload

In addition to the objective task performance, another impact of complexity was on the participant's mental workload. A common way to measure the subjective mental workload is with the NASA- Task Load Index (TLX) instrument (Hart, 1987). Mental workloads perceived by the participants were measured with NASA-TLX at the end of each day's training sessions. Mental workload ratings from the first participant were missing, leaving us with data from nine participants. The possible NASA-TLX workload rating ranged from 0 to 100. The result showed that the mental workload experienced by the participants increased as the DB complexity increased. ANOVA results on the mental workload scores revealed significant differences among the three types of DB, F (2, 16) =15.12, *p*=0.0002. Post hoc Tukey test indicated that the mental workload associated with the more complex 12F-DB (M=65.04, SD=18.72) and 9F-DB (M=59.61, SD=21.06) were higher than the less complex 6F-DB (M=47.89, SD=16.23). There was no significant difference between the 12F-DB and 9F-DB.

Additionally, the differences between the mental workload during the first and second day of training were also significant, F (2, 18) =15.93, *p*=0.004. The mental workloads experienced by the participants were generally higher on the first day (M=59.84, SD=19.50) than on the second day (M=55.19, SD=20.16). It might be because that the participants further mastered the tasks during the second day, thus they experienced less mental workload on the second day. Another interesting thing to note was the nearly identical mental workload for 12F-DB for two days of training. Such results suggested that the monitoring task for 12F-DB became so demanding that even an extra day of training did not alleviate the mental workload.

DISCUSSION

Did learning of the DB monitoring task occur?

Understanding the learning process is important for designing training programs for the controllers. In this experiment, the participants learned the DB with progressively increasing complexity. For each type of DB, they learned through four training sessions on two days. The proportion correct of monitoring tasks for each DB increased gradually across the training sessions. According to Speelman and Kirsner (2005), the reasons for such improved performance with practice were the strengthening of the knowledge structure, faster and more reliable activation of the knowledge structures, and more efficient knowledge access and information processing. The observed improvement in performance suggested that the knowledge structure related to the meaning and statuses of the DB fields were strengthened through training.

To characterize the training process, both the performance achieved and the lengths of training were important. Length of training was described by the number of trials that participants took to learn the DB monitoring task. In our analysis, two sets of objective measurements: the average proportion correct and the number of trials in the training were collected to describe the training process. Subjective measurements of the mental workload were collected as well. Interestingly, the ANOVA analysis revealed similar patterns among the measures for the DB type and exposure time duration, providing corroborating evidences for each other. The 6F-DB led to lower level of experienced complexity, whereas the 9F-DB and 12F-DB led to higher level of experienced complexity. When the exposure time was long than or equal to 1s, the monitoring task was easier to learn and to perform. But when the exposure time was shorter than 1s, the task became more challenging.

How did the complexity affect learning of the monitoring task?

The three types of DBs had different numbers of fields. Such differences in DB objective complexity affected the DB learning process. ANOVA on the monitoring performance, number of training trials needed, and the mental workload all revealed significant differences among the three DB types. In particular, the performances with 9F-DB and 12F-DB were not significantly different from each other, but both were worse than that of 6F-DB. It meant that participants spent relatively more efforts to learn and perform monitoring tasks with 9F-DB and 12F-DB than 6F-DB. The achieved level of performance was also higher for the less complex 6F-DB than the more complex 9F-DB and 12F-DB.The data indicated a breaking point in terms of experienced complexity. When the DB had 6 fields, the task complexity is lower. With DBs of 9 and 12 fields, the experienced task complexity became higher. Such results suggest designers to limit the number of DB field to around six.

It needed to be noted that the proportion correct in this study was calculated as the average value of all DB fields. Because each DB fields were encoded differently, it was easier to determine the DB status when certain fields presented an abnormal status than other s. Such differences were explored by Ling and Hua (2008) where significant differences between types of DB fields were identified.

How did the exposure time affect learning of the monitoring task?
The effect of exposure time for the DB on the learning process of monitoring task
was also studied. Again, similar patterns emerged from ANOVA analysis on both
the number of trials taken to learn the task and the average proportion correct of the
task. Significant differences existed for the five exposure time. In particular, the
exposure time of 1s, 2s, 3s were not significantly different, but they were different
from the duration of 0.5s. And exposure time of 0.3s was also significantly different
from the rest of exposure time. When the exposure time was 0.5 or 0.3s, it took
participants significantly more trials to learn the monitoring task, and the average
proportion correct was also lower. This result suggested a breaking point in terms of
task difficulty level. When the exposure time was reduced to 0.5 and 0.3s, the
monitoring tasks became much harder than to perform than the exposures time of
1s, 2s, and 3s. It suggested that in order for the controllers to be able to reliably
perform monitoring tasks, the exposure time should be at least 1s.

In conclusion, this study indicated that individuals could learn to quickly
monitor DBs. Differences in the average proportion correct in monitoring task
performance, length of training, and subjective mental workload indicates a
breaking point in experienced task complexity at DB field of six. 6F-DB led to
lower level of experienced complexity, whereas 9F-DB and 12F-DB caused higher
level of experienced complexity. When the DB contains more than six fields, the
monitoring task became more demanding for the human operator. In terms of task
difficulty level, when the exposure time was greater or equal to 1s, the monitoring
task was easier to learn and to perform. But when the exposure time was shorter
than 1s, the task became more demanding and challenging. The capacity of learning
was constrained by both the DB complexity and the exposure time. The results
implicated that designer of DBs for air traffic control displays should consider both
the number of fields in the DB and the time that users (e. g, controllers) actually
spent on viewing a DB in monitoring the air traffic situations.

ACKNOWLEDGMENT

This research was supported by Federal Aviation Administration (FAA) Civil
Aerospace Medical Institute (CAMI), Oklahoma City, with grant entitled as
"Investigating Information Complexity in Three Types of Air Traffic Control
(ATC) Displays" grant number FAA 06-G-013. The FAA grant monitor, Dr. Jing
Xing, initiated the study, designed the experiment, and contributed to data analysis.

REFERENCES

Campbell, D.J. (1988), "Task complexity: a review and analysis". *Academy of
Management Journal*, 13 (1), 40–52.

Ling, C., & Hua, L. (2008), "Monitoring task performance with datablock on air traffic control display". *Proceedings of Applied Ergonomics International 2008*, Las Vegas, NV, July 17-22.

Metzger, U., Parasuraman, R. (2001), "The role of air traffic controller in future air traffic management: an empirical study of active control versus passive monitoring". *Human Factors*, 43(4), 519-528.

Speelman, C. P., & Kirsner, K.(2005), *"Beyond the Learning Curve"'*. Oxford university press Inc., New York.

Xing, J., & Manning, C. A. (2005), "Complexity and automation displays of air traffic control: literature review and analysis". Washington, DC: Federal Aviation Administration, No: DOT/FAA/AM-05/4.

Chapter 41

The Study of Human Haptic Perception for Vehicle Haptic Seat Development

Wonsuk Chang[a], Kwangil Lee[a], Yonggu Ji[a],
Jongkweon Pyun[b], Seokhwan Kim[b]

[a]Information and Industrial Engineering
Yonsei University, Seoul, Korea

[b]Seat System Engineering Team
Hyundai-Kia Motors R&D Division, Korea

ABSTRACT

This study aims to provide practical and comprehensive guidelines for designing the haptic (or vibrotactile) interface in vehicles. Twenty participants took part in the experiments that were conducted at driving simulation environments. This study recommended the proper intensity (approximately, 26~34Hz and 2.0~3.4G), position (seat pan or back support), direction (horizontal or indirect), and inter-vibration distance (8~9cm), and showed how the characteristics of drivers, such as gender and age, had effects on setting the design variables of the haptic interface in the vehicle seat.

Keywords: Haptic seat, Vibrotactile, Guideline, Human factors

INTRODUCTION

A recent study for the improvement of the interaction between the user and the system at the Human-Computer-Interaction (HCI) is in progress. The earlier studies of the interaction method mainly depend on visual or auditory interface. However, nowadays a variety form of studies for interaction method is actively attained due to increasing quantities and the level of complexity of the information. In the vehicle field, likewise, through the development of the entertainment technique and In-vehicle Information System (IVIS) such as the navigation information, much information are delivered to the user in a complicated form and the preexisting visual and auditory information transmission method is indicated to have many problems (Tijerina, et al., 2000; Van erp & Van Veen, 2004). The perceptual overloads and its interference phenomenon within the vehicles lead to interests in a new perceptual channel called haptics, and research for applicability of haptic-based information to the interface design for vehicles along with visual and auditory-based information. The word 'haptic' originally comes from the Greek word, 'Haptesthai' that means 'touch', and is currently referred to as perception by all human tactile organs, including hand, arm, abdomen, hip, foot and other skins (Iwata, 2008). The haptic interface uses all parts of the body come under (hands, skin, body, feet, etc) for the transmission of the information. This distinctive feature could deliver much more information than the visual or auditory method and has a higher possibility to deliver information to the different parts of the body at the same time. Because of this distinctive feature of haptic interface, the importance is augmenting and the scope of the study is being various (Robles-De-La-Torre 2008). However, the haptic method used for transmitting the information by the seat lacks guidelines than the visual and auditory. Therefore, we proposes the guidelines that need to be considered when designing the haptic seat by applying the haptic interface technique for the advancement of the interaction between the user and the system and for the safety.

HAPTIC INTERFACE IN VEHICLE SEAT

There have been two streams of studies that help develop the haptic (or vibrotactile, more precisely) interface in vehicles seats. One is the group of studies that show applicability and superiority of the vibrotactile interface in vehicles seats compared to the visual and auditory interfaces in vehicles, and the other is the group of studies that give information about design elements of the tactile interface, such as types and characteristics of design variables, which can be applied to vehicles seats. In the former group of studies, Tan et al. (2003) investigated the effectiveness of using a haptic back display based on a 3-by-3 vibration tactors array for delivering attention- and direction-related information to users, and concluded that the haptic cueing system could be beneficial to users who conducted attention- and direction-related tasks with complex visual display. Lee et al. (2004) reported that the collision warning system through the visual and haptic interface outperformed through the visual and auditory interface in terms of driver's response time to collision situations and driver's attitude. The vibration was generated as the haptic cues by the actuators on the front edge and in the thigh bolsters of the seat. In addition, Riener (2008) justified the usage of haptic feedback through the vibration in the vehicle seat by conducting the experiments, which resulted in the superiority of haptic feedback compared to visual and auditory feedback in terms of reaction time to the conditions for the selected driving activities. Thus, all of the three studies show the effectiveness of using vibrotactile stimuli in vehicles seats as the haptic cues for delivering the vehicle-related information to drivers. In the latter group of studies, Morioka & Griffin (2008) reported the effect of input location (the hand, the seat without back, and the foot), frequency, and direction (fore-and-aft, lateral, and vertical) on absolute thresholds for the perception of vibration, and concluded that under the frequency of 80Hz the seat was more sensitive than the hand and the foot and more sensitive in the vertical direction than in the other directions of vibration. Although, from this study, we can get the information about the sensitivity of vibration perception from the seat without back, we still need more information, such as the perception of vibration from the seat back, the satisfaction of vibration from the seat, and information from various types of vibration, in order to design the vibrotactile interface in the vehicle seat. In the meantime, Self et al. (2008) summarized nine characteristics of the vibrotactile interface, which can be utilized as the design variables of the haptic interface. They suggested size, shape, orientation, position, moving patterns, frequency, amplitude, rhythm, and waveform, as a set of tactile characteristics, assuming vibration tactors are located on the torso of humans in the aircraft. Although they specified nine candidates of design variables for the vibrotactile interface, since all of them were originally for the military aircraft, each of the nine design variables needs to be selectively applied for the vehicle seat.

EXPERIMENT SETUP

PARTICIPANTS

The twenty participants were chosen to practice the experiment. The participants' age ranged from 22 to 58(means 37.8, standard deviation 12.3, 9 females and 11 males). Average weight was 65.4kg, and the average height was 167.3cm. They all possessed the driving licenses and their average driving experience was approximately 8 years. They had variety in jobs.

APPARATUS

For this study, we have chosen the seat that is being used at the high-classed vehicle produced at the company H. To produce the seat vibration we using the result of the pressure measurement of the seat to define the position of the seat where it shows the highest pressure distribution and attached the custom-made motor in the interior. In this study defines the motor that generates the vibration through the rotation of the eccentric weight which is attached to the general rotating motor, as an eccentric motor. Each experiment result in the standard of voltage and for the use of other eccentric motor, the characteristic of the vibration of the used eccentric motor was measured. Through the analysis of the mutual relation between the voltage(V) and frequency(Hz), we derived the equation of $y= 8.1405x - 2.4247$ (x=Voltage(V), y =Frequency(Hz)), also the relationship between voltage(V) and amplitude(G) can be represented by the equation $y= 0.1884x^2 - 0.1418x + 0.2058$ (x=Voltage(V), y =Amplitude(G)) .

STIMULUS

We considered the vibrotactile characteristics suggested by Self et al. (2008) as the candidates of design variables. As a result, four design variables, such as frequency, amplitude, position, and moving pattern, were chosen from Self et al. (2008). We additionally considered the excitatory direction of vibration as one of the design variables because the sensitivity of vibration was different according to the excitatory directions (Morioka & Griffin, 2008). The implemental conditions for each design variable can be identified through the experiments with the vehicle seat and the vibration actuator. We planned three experiments: i) the first experiment for finding proper intensity of the vibration, which is related to the design variable of frequency and amplitude; ii) the second experiment for finding minimum distance between vibrations that is distinguishable to drivers, which is related to the design variable of moving pattern, because the moving pattern is perceived when spatially distinguishable multiple vibrations are consecutively generated; and iii) the third experiment for finding proper position and direction of vibration that satisfy drivers, which is related to the design variable of position and excitatory direction.

EXPERIMENT AND RESULTS

EXPERIMENT I: FINDING PROPER INTENSITY

To present the vibration stimulus to the user, it is necessary to examine of appropriate level of the vibration intensity. Therefore, this study is to make an experiment for measuring the level of user's recognition of various intensity of the vibration stimulus.

Method

In this experiment, the vibration intensity was examined at six experimental conditions: 2 positions (seat pan, back support) × 3 directions (vertical, horizontal, indirect). Position is referred to as the place where the vibrotactile stimuli were perceived, such as seat pan and back support. Direction is referred to as the excitatory directions of vibration, such as vertical, horizontal and indirect directions. Vertical direction means that the rotated axis of the eccentric motor is perpendicular to the surface of the vehicle seat, whereas horizontal direction means that the rotated axis of the eccentric motor is parallel to the surface of the vehicle seat. Indirect direction means that the vibration generated from the eccentric motors shakes a pole in the bolster and then indirectly delivered to the drivers. Each condition presents into 11 levels of classified intensity of the vibration stimulus ranging from 2V to 7V into 0.5V. The participant responded in 'Not detected', 'Weck', 'Moderate', Strong', Too strong' in 5 points as in standard unit.

Results

First of all, as the stimulus intensity increased, it was shown to be increased in the recognized intensity in the both seat pan and back support. The intensity that is recognized 'Moderate', distributed the voltage of 3.5 to 4.5V. Also, in the case under 2.5V, since it showed the reaction of 'Not detected' up to 50%, it should restrict using the stimulus under 2.5V when designing the seat. In the comparison between the positions, in the case of the seat pan, generally, it brought a similar result even though the average of indirect direction was low. And in the case of the back support, all the intensity of the vertical direction has recognized to have strong stimulus than others. Which means, as a result, the appropriate vibration intensity for transmission of the general information is shown to be between 3.5 to 4.5V (see Figure 1).

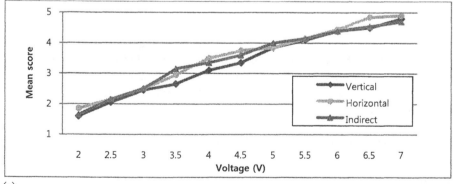

(a)

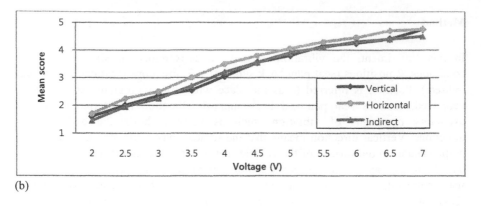

(b)

Figure 1 Mean recognized intensity for vibration intensity. (a) is result of seat pan and (b) is result of back support

In the comparison between the gender groups, at the most of intensity condition, it did not show significant difference between the male and female most of the conditions, but it has shown that females have recognized lower in the intensity of 2V to 2.5V. In the comparison between the gender groups, At the most of the intensity condition, it was shown that the 20s and 30s group (12 people) was more sensitive than the 40s and 50s (8 people).

EXPERIMENT II: FINDING MINIMUM DISTANCE

In order to design the haptic interface, the user must know how to differentiate more than two vibrations to express moving pattern. Therefore, the purpose of this experiment II is to measure the minimum distance between the stimuli by classifying the two vibrations.

Method

We investigated the minimum distance for spatially distinguishable vibrations in the seat pan and the back support. For this, the distance between the motors has changed in 1 cm from 4cm to 15 cm and the level of vibration intensity were varied by 0.5V from 3V to 6V. The vertical direction was used because the vertical direction was known to be less sensitive from experiment I.. The participants were asked to answer 'yes' when he/she perceived two spatially separated vibrotactile stimuli, or 'no' when he/she perceived just one vibrotactile stimulus.

Results

We declared that two stimuli sere perceived as spatially separated when 75% of participants agreed. Note that 75% agreement is the generally accepted criterion in the psychological method (Gescheider, 1997). The seat pan showed that it recognizes the difference of the two position of the vibration at over 4V and 8 cm. Also in the case of the back support, it showed to be able to recognize the difference of the two vibrations at the point of over 4V and 9 cm. In other words, it is estimated to need more than 4V of intensity to design the information using the difference between the vibrations, and to be more than 8 to 9 cm in vibration motor's distance (see Figure 2).

In the comparison between the gender groups, first, at the seat pan, it was measured that the male have higher recognition level than female. In the contrary, at the back support, the female group was measured to have higher recognition level than the male group. This kind of result seems to be estimated due to heavier weight that male has than the female. In the comparison between the gender groups, at the seat pan and back support, it was shown that the 40s and 50s group could distinguish slightly narrower gap than the 20s and 30s.

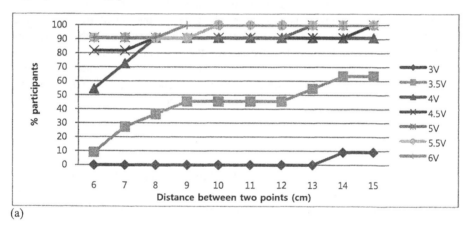

(a)

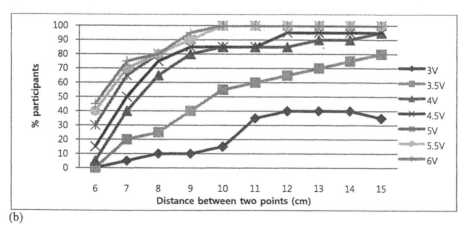

(b)

Figure 2 Rate of participants who recognize the difference of the two vibrations. (a) is result of seat pan and (b) is result of back support.

EXPERIMENT III: FINDING PROPER POSITION AND DIRECTION

The vibration stimulus that is presented to the user differs by the position of the vibration and the direction of it. Therefore, this experiment III is to determine the position and direction that has highest level of satisfaction through measuring the subjective level of satisfaction.

Method

By the experimental method, we divided the position of the vibration into seat pan and back support, the direction of the vibration into vertical, horizontal and indirect to design by the factor of 2x3 and presented the vibration by each factor. The presented vibration intensity is based on 4V through the result of appropriate intensity level measured in the experiment I. The measurement of the subjective satisfaction level was taken after presenting the vibration stimuli. In order to measure the satisfaction level, we referred to the existing studies containing the 4 criteria such as, 'pleasant to vibration', 'easy to perception', 'felt comfortable', 'easy to learn' (Koskine, 2008), which measured in 7 point-scale.

Results

The seat pan was measured in the order of Indirect > Horizontal > Vertical in all 4 criteria. At the back support, the average score was measured in the order of Horizontal > Indirect > Vertical in 3 criteria excluding the felt comfortable criteria. In order to determine the statistical difference of the measured average score, the LSD forensic analysis by the ANOVA analysis. By the result of statistical analysis based on the seat pan, the indirect direction showed significant difference at 0.05 than the vertical at the intimacy and comfort criteria (see Table 1).

Table 1: Satisfaction evaluation of position and direction

Position	Direction	Satisfaction evaluation			
		Pleasant to vibration	Easy to perception	Felt comfortable	Easy to learn
Seat pan	Vertical	4.450	4.550	3.600	4.850
	Horizontal	4.700	4.650	4.000	5.250
	Indirect	**5.550**	**5.300**	**4.500**	**5.650**
Significant difference		Vertical vs. Indirect (p= 0.048)	-	Vertical vs. Indirect (p=0.027)	-
LSD		F(2,57)=2.255, p=0.144	F(2,57)=0.966, p=0.387	F(2,57)=2.581, p=0.084	F(2,57)=0.849, p=0.433
Back support	Vertical	4.950	4.950	3.500	5.300
	Horizontal	**5.900**	**5.800**	4.050	**6.250**
	Indirect	4.750	4.700	**4.150**	5.150
Significant difference		Horizontal vs. Indirect (p=0.025)	Horizontal vs. Indirect (p=0.035)	-	Horizontal vs. Indirect (p=0.038)
LSD		F(2,57)=3.020, p=0.057	F(2,57)=2.555, p=0.087	F(2,57)=0.953, p=0.392	F(2,57)=2.660, p=0.079

DISCUSSION

From the experimental results in this study we can derive the guidelines for the haptic interfaces in the vehicle seat as follows: First, if we want to get high sensitivity of vibrotactile stimuli, the eccentric motors as the vibration actuators should be equipped at the horizontal direction in the seat pan. In addition it is recommended to use a vibration intensity of 3.5V through 4V in the seat pan and a vibration intensity of 4V through 4.5V in the back support, because the drivers perceive those ranges of vibration intensity as moderate or appropriate. Moreover, note that the drivers aged between 20 and 39 perceive the vibration intensity more strongly than the drivers aged between 40 and 59. It means that it would be desirable to apply a lower intensity of vibration to the seat pan than to the back support, to the vehicles for the young drivers than the vehicles for the old drivers. Second, it is recommended for designing 'moving patterns' that the vibration actuators should be equipped at least 8 cm apart from each other in the seat pan, and at least 9cm apart from each other in the back support, when the vibration intensity of 4V is applied. In addition, note that in the seat pan, the minimum distance for spatially distinguishing two vibrations is consistent as 8cm, but not in the back support. Thus, 'moving patterns' could easily be applied to the seat pan rather than to the back support. Third, if we want the drivers to be satisfied with the vibrotactile stimuli in the vehicle seat, the indirect direction of vibration should be used in the seat pan and the horizontal direction of vibration should be used in the back support. More specifically, the indirect vibration from the bolster in the seat pan makes the drivers feel comfortable, and the horizontal direction of vibration in the back support is pleasant, easy to perceive and easy to learn to the drivers.

Table 2: Summary of experimental results

Experiment	Relevant design variables	Experimental variables	Results of experiments
Experiment I	Frequency, Amplitude	Appropriate intensity	Seat pan: 3.5V~4V Back support: 4V~4.5V
Experiment II	Moving Pattern	Minimum distance between vibrations	Seat pan: Minimum 4V, 8cm Back support : Minimum 4V, 9cm
Experiment III	Position, Direction	Satisfaction for position & direction	Seat pan: Indirect>Horizontal>Vertical Back support : Horizontal>Vertical>Indirect

CONCLUSIONS

This paper is suggesting the new way of interaction methodology for the enhancement of the safety and to increase interaction between the user and system. Also it presents the guideline of the attributes of the vibration required by the design elements of the haptic interface drawn through user's experiment. These guidelines would be practical and comprehensive because the experiments were conducted in a real car seat and with vibration actuators for commercial use. Although the experiments in this study were conducted at the driving simulation environments, the above guidelines would provide foundation for designing the haptic seat interfaces for the intelligent vehicle in the future. We expected the guideline from this study complemented by the experiments based on the real driving environments in the further study.

REFERENCES

Iwata, H. (2008). History of haptic interface. In M. Grunwald (Ed.), Human haptic perception: Basics and applications (pp. 355-361). Basel, Switzerland: Birkhäuser Verlag.

Van Erp, J. B. F., & van Veen, H. A. H. C. (2004). Vibrotactile in-vehicle navigation system. Transportation Research Part F: Traffic Psychology and Behaviour, 7, 247-256.

Tijerina, L., Johnston, S., Parmer, E., Winterbottom, M. D., & Goodman, M. (2000). Driver distraction with wireless telecommunications and route guidance systems (NHTSA Pub. No. DOT HS 809-069). Washington, DC: U.S. Department of Transportation. Retrieved on 10 October, 2007, from http://www-nrd.nhtsa.dot. gov/pdf/nrd-13/DDRGS_final0700_1.pdf.

Robles-De-La-Torre, G. (2008). Principles of haptic perception in virtual environments. In M. Grunwald (Ed.), Human haptic perception: Basics and applications (pp. 363-379). Basel, Switzerland: Birkhäuser Verlag.

Morioka, M.,& Griffin, M. J. (2008) Absolute thresholds for the perception of fore-and-aft, lateral, and vertical vibration at the hand, the seat, and the foot, Journal of Sound and Vibration, 314(1-2): pp 357-370

Tan, H. Z., Gray, R., Young, J. J., & Traylor, R. (2003). A haptic display for attentional and directional cueing. Haptics-e, 3(1), 1-20.

Lee, J. D., Hoffman, J. D., & Hayes, E. (2004). Collision warning design to mitigate driver distraction. In: Proceedings of the SIGCHI Conference on Human Factors in Computing Systems (ACM, New York, April 24-29, pp 65-72).

Riener, A. (2008). Age- and gender-related studies on senses of perception for human-vehicle-interaction. In Proceedings of the 2nd Workshop on Automotive User Interfaces and Interactive Applications (Lübeck, Germany, September 7-10).

Self, B. P., van Erp, J. B. F., Eriksson, L., & Elliott, L. R. (2008). Human factors issues of tactile displays for military environments. In J. B. F. van Erp, & B. P. Self (Ed.), Tactile Displays for Orientation, Navigation and Communication in Air, Sea and Land Environments, TR-HFM-122 (pp.3.1-3.17). RTO Technical Report, NATO Research and Technology Organisation.

Gescheider, G. A. (1997). Psychophysics: The fundamentals. 3rd ed. Mahwah, NJ: Lawrence Erlbaum Associates.

Koskinen, E. (2008). Optimizing tactile feedback for virtual buttons in mobile devices. Master's thesis, Helsinki University of Technology.

CHAPTER 42

Cognitive Study on Driver's Behavior by Vehicle Trajectory and Eye Movement in Virtual Environment

*Hirokazu Aoki [1], Hidetoshi Nakayasu[1],
Nobuhiko Kondo[2], Tetuya Miyoshi[3]*

[1]Konan University
8-9-1, Okamoto,Higashinada
Kobe, Hyogo, Japan

[2]Otemae University
2-2-2, Inano-tyou, Itami
Hyogo, Japan

[3]Toyohashi Sozo University
20-1, Mastushita, Ushikawa
Toyohashi, Aichi, Japan

INTRODUCTION

In order to maintain a highly safe and reliable system of driving, it is important that this system be viewed as a man-machine system, made up not only of the movement of vehicles, but also of human perception, cognition, and the responses of the driver. In this paper, we seek to examine and clarify the relationship between eye movements and driving behavior in conventional and unconventional situations. To do so, experiments were conducted using a driving simulator and an eye tracking system. Statistics show that the main cause of traffic accidents is neglect on the part of the driver. Safety, and the reliability of the vehicle, depends not only on physical factors such as vehicle maneuverability or the stability of the vehicle, but also on human factors such as the driver's perception and reflexes. A survey on accidents shows that errors in visual perception are common (ITARDA, 2001). Visual cognition requires the acquisition and judging of information. The more information is omitted, the higher the likelihood of an accident. In order to develop a system which decreases the incidence of accidents, it is necessary to analyze the man-machine system with heed paid to the manner in which a driver operates a vehicle.

The experimental system proposed in this paper is able to realistically mimic the situation of driving and analyses human factors for awareness of hazard and risk. Another aim of this work is to offer relevant information regarding hazards in several different traffic situations, with a focus on human factors in order to forecast human behavior such as visual attention during driving. In the experiments, the visual attention of driver was measured by the use of an eye-tracking system while participants operate a driving simulator. In this paper, eye movements and road scenes are analyzed from a moving driver's perspective. The relationship between eye movement and culture of Japanese road are also investigated.

EXPERIMENT

In order to investigate driver's behavior, including eye movement, vehicle operation and vehicle trajectory, a driving simulator and an eye tracking system were used in this experiment. Fig. 1 shows the experimental set-up and a schematic view of the apparatus. A driving simulator with a 6-axis motion base system (Honda Motor DA-1102) was used to simulate driving an automatic transmission vehicle. Table 1 shows the specifications of the DA-1102. The simulator is controlled by a network of computers which also generate images from the driver's point of view, as well as those in the side-view and rear-view mirrors, car dynamics (such as steering and braking motions), and traffic scenarios. Ethernet-linked computers acquired a log of input and output variables (such as a steering, accelerator, brakes, the position of the car and speed). This data was acquired every 10msec. Several different scenarios were simulated in the driving simulator, such as urban roads and highways and so on. The images and car dynamics were generated in real time based on the participant's maneuvering in the simulator. The instruments on the dashboard, steering wheel, gearshift, side brake, accelerator, and brake pedal were positioned in a manner similar to that in a real car, as shown in Fig. 2.

FIGURE 1 DA-1102.

Table 1 Specifications of the DA-1102.

Elements of DS	Specification
Front view	Wide field (138 deg.) screen projection type
Rear view	3-mirror independent LCD display
CG	Redraw speed: 30 to 60 frame's
Mechanism	Six axis motion base system using G cylinders Control
Frame	Lightweight space frame structure with aluminum extended mechanism
Body	Rear open structure fixed with FRP mold
Operation system	Steering device with reactive force control, accelerator, clutch, brake
Mission	AT/MT switch mechanism
CG computer	Cuamum3D Alchemy (1.5 million polygons)
Control system	6-axis servo amp
Dimensions	2,440mm(D) × 2,280mm(W) × 1,855mm(H)

FIGURE 2 Schematic view of the DS experiment where the driver is equipped with an eye tracking system.

Eye movements were recorded at 250 Hz with a spatial resolution of 0.022degree within a range of ± 30degree in the horizontal axis and ± 20degree in the vertical axis by an eye tracking system (SR Research Ltd. EyeLink II) controlled by a personal computer, which managed the timing of the experiment and collected data. Table 2 shows the specifications of the EyeLink II 4). Head movements were monitored, and the system automatically modified the data on eye position by using an eye camera. The horizontal and vertical eye movements of the participants were recorded from both eyes. At the same time, the scene from the driver's perspective was recorded by a CCD camera equipped on a head mounted cap as shown in Fig. 2. The CCD camera equipped on a head mounted cap was capable of recording the front view of the driver as video clips with a spatial resolution of 720 X 480 pixels and a frame resolution of 30fps. All data of eye movements were analyzed off-line by computer programs that calculated the number of saccades and eye movement distance.

Table 2 Specifications of the EyeLink II.

Sampling rate	500Hz(Pupil One), 250Hz(Pupil and Corneal Reflection)
Error of fixation	less 0.5 °
Accuracy of pupil size	for diameter 0.1%
Range of trace	Horizontal ± 30 ° Vertical ± 20 °
Data file	EDF
Rate of data transfer	3mSec ~ 14mSec
Marker of infrared rays	900nm
Eye camera	925nm 1.2W/cm2
Weight	420g

Six male students (mean age = 23.3; std.dev = 0.84) participated in this experiment as drivers. The participants had standard Japanese automobile licenses and more than one year's worth of driving experience. All participants received more than 5 hours of training on how to use the driving simulator before the experiment. Prior to the experiment, the purpose of the study and the procedures were explained to the participants, and the informed consent of all participants had been obtained. All participants had normal or corrected-to-normal vision. The following two instructions were given to participants in order to ensure the psychological conditions of the experiment were consistent for all. 1) Do not cause an accident. 2) Follow traffic laws during the experiment.

Fig. 3 shows the driving course of an urban road in Japan in which several traffic incidents occur, as shown by the abbreviations ST, RT, LT, MG and STOP which, respectively, mean go straight, right turn, left turn, merge, and stop. The details of the incidents in Fig. 3 are as follows. 1) The course featured a number of dangerous incidents such as a car suddenly swerving into the driver and a child jumping out. 2) The course featured fine weather as clear as day.

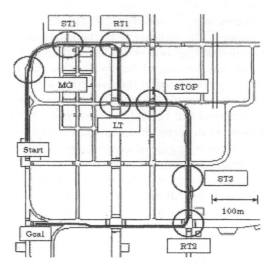

FIGURE 3 Driving course with containing various traffic incidents.

In the analysis, differences in eye position at successive frames were calculated. These data were analyzed on the basis of velocity and position for each traffic incident in order to evaluate the mean distance of eye movement, fixation frequency and duration. This paper defines a saccade of eye movement occurring as when an eye moves at a velocity of more than 90deg/sec. An eye moves after the occurrence of a saccade. This paper also defines a fixation of the eye as when an eye does not move for more than 48msec after an eye movement and a saccade. Therefore, the fixation time refers to the duration time from the saccade after an eye movement to the saccade before an eye movement.

RESULTS

The eye movement data were recorded in the experiments when participants maneuvering through a Japanese road, as shown in Fig. 3. Fig. 4 shows the mean value of average eye movement distance for the six participants in each traffics incident, where the vertical line designates standard errors (SE). The average eye movement distance refers to the mean values of the average number of fixations and the duration of fixations for the six participants, as illustrated in Figs. 5 and 6. The average number of fixations refers specifically to the average number of fixations that occurred during selective traffic incidents. On the other hand, the average duration of fixation refers to the average time required for each of the selective traffic incidents. In Fig. 4, the vertical line shows the standard error (SE). From Fig. 4 it is seen that the mean distance of eye movement at the merging incidents (MG) is much larger than at other traffic incidents. It is seen that in the MG incident that the driver sometimes watched a side mirror and a forward direction in merging point of road However, the same behavior of driver was not found in the straight road (ST1). As a typical example, Fig. 7 and 8 shows the trajectory of eye movement of participants in incidents MG and ST1. An image from one video clip of the incidents MG and ST1 is shown in Fig.7 and 8. It is seen in Fig.7 that the visual lire of the driver moves around to the side mirror, the direction of forward movement, at the MG incident. On the other hand, it is not observed in Fig.8, and there are few movements of visual lines at the ST1 incident. In other cases, it was observed that the visual line moved to the speedometer or the rearview mirror during the ST1, ST2, and STOP incidents. From Fig. 5 and Fig. 6 it is noted that the fixation frequency at STOP is smaller than that at other traffic incidents, while the fixation duration at STOP is longer than that at other traffic incidents. This suggests that the driver recognizes the pedestrian at the crosswalk. Many participants moved their eyes in such a way that they appeared to recognize the pedestrian. In short, it was not necessary to collect very much information at the STOP incident. When there is little fixation frequency, the driver can collect a great deal of information at once. This suggests that the field of view for collecting information becomes larger.

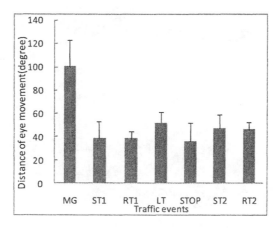

FIGURE 4 Distance of eye movement.

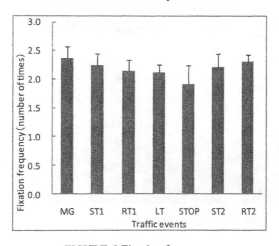

FIGURE 5 Fixation frequency.

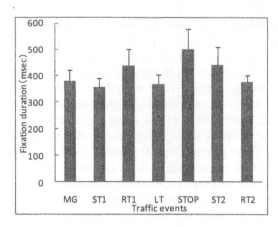

FIGURE 6 Fixation duration.

FIGURE 7 Trajectory of eye movement. At the merging (MG) incident scene.

FIGURE 8 Trajectory of eye movement. At the straight (ST1) incident scene.

Another analysis was also conducted of the vehicle trajectory and the eye movement of the driver. In Fig. 9, the relationship of eye movement data such as the axis of the driver's eye (X-eye, Y-eye) and driver behavior (Steering angle, Throttle, Brake) such as steering angle, degree of throttle and brake are illustrated on the same axis for comparison. Therefore, different units of measurement were used in the vertical axis of Fig. 9. In the figure, X-eye expresses the horizontal eye movement as the eye angle (degree) unit. Y-eye refers to the vertical eye movement of the eye angle (degree) unit. This figure shows the behavior of drivers at the right turn (RT1) incident. Vehicle trajectory during the right turn (RT1) incident is also drawn in Fig. 10 with an eye fixation time stamp, where one of the video clips shows the scanning behavior of a driver reacting to the scene as shown in Fig. 11. It is seen at the right turns (RT1, RT2) and the left turn (LT) that a similar tendency was observed during the right and left turn incidents. It is found in Fig.9, from 0sec to 3.60sec that the driver became aware of a vehicle coming towards him or a person crossing. The driver of this section confirmed that is was safe before making a right turn. That is defined, in this paper, as the introduction stage of turning. In section Fig. 9 from 3.60sec to 6.468sec, the value of X-eye increases. Generally it is known that the human eye repeats a saccade (rapid eye movement) and the stop (The Vision Society of Japan Eds, 2001). It is shown that saccade and "smooth pursuit eye movement (The Vision Society of Japan Eds, 2001) are repeated in turning behavior of driver. For the driver, the forward scenery changes to drift. Therefore, the driver's eye movement changes to drift. This is referred to as smooth pursuit eye movement. Fig. 11 features two video clips which show the observation position of drivers at t = 5.34(sec) and t = 5.804(sec). The filled circle in the figure expresses the fixation point of the driver. The eyes engage in multiple small smooth pursuits' eye movements. The driver's eye movements demonstrate that he/she can only see in front of the vehicle. This behavior is defined in this paper as the practice confirmation stage of turning. In section Fig. 9 from 6.468sec to 9.52sec, the driver acquires information about the road after turning. This behavior can be referred to as the completion stage of turning. The above suggests that there are three phases in the turning behavior of a driver: the introduction, practice confirmation and completion stages. At the introduction stage, the driver performs perception and the cognition. At the practice confirmation stage, the driver judges, practice and confirms his/her actions. At the completion stage, the driver confirms the completion of the operation and then shifts to the next target, once again moving back to the introductory stage.

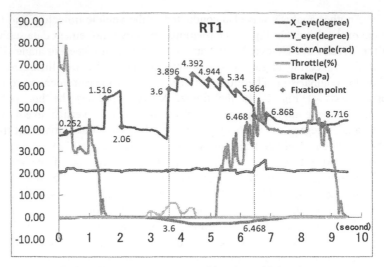

FIGURE 9 Eye movement and log data of DS at RT1.

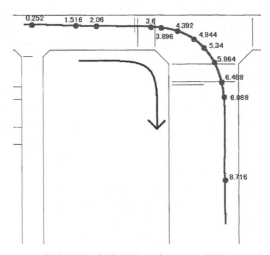

FIGURE 10 Vehicle trajectory at RT1.

CONCLUSIONS

To summarize, a study was conducted in order to provide information useful for reducing the prevalence of automobile accidents caused by human error. Eye movement and the trajectory of vehicles were analyzed to investigate the role of human error in every driving behavior. As a result, this experimental study, using a driving simulator, found that the eye movement of the driver was more pronounced in the merging scene compared to the other scenes.

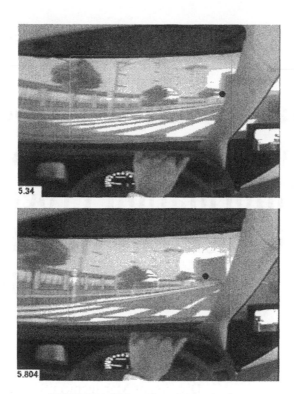

FIGURE 11 Video clips of driver's view.

It is necessary for driver's to collect information, not only regarding what is in front of them, but also in the direction of lanes from which cars may merge. For this reason, since information decreases the probability a dangerous situation might take place, a driver looking away, for example to the side, increases the dangers of driving. Secondly, it was observed that the frequency of fixation becomes small and the eye tends to stop when the driver waits and stops his/her car. Finally, it was found that the act of turning consists of three stages on the part of the driver: introduction, practice confirmation and completion. At the practice confirmation stage, the driver tends to continue looking in the direction in which he wishes to turn and smooth pursuit eye movement typically occurs.

REFERENCES

ITARDA. (2001), Institute for Traffic Accident Research and Data Analysis Information, What Sort of Human Errors Cause Traffic Accidents? -Focus on Mistake Assumption-, No.33. (in Japanese)

The Vision Society of Japan Eds. (2001), Handbook of visual information process, pp.91–92., pp.393–398. (in Japanese)

Chapter 43

A Basic Study on Model of Human Behavior by Visual Information Processing Mechanism for Reliability-Based System Design

Kimihiro Yamanaka , Mi Kyong Park, Mitsuyuki Kawakami

Division of Management Systems Engineering
Tokyo Metropolitan University
Tokyo, 191-0065, Japan

ABSTRACT

The aim of this work is to discuss the relation between reliability of structural system and human performance of perception in man-machine system. It is studied to evaluate the effect of performance deterioration on structural safety from the viewpoint of uncertainties of human behavior. The subjective uncertainties are two kinds of uncertainties by designer and operator. From these points of view, the performance of operator on visual perception was investigated as a case study on visual task.

Keywords: Human Behavior, Visual Information Processing Mechanism, Response Time, Useful Field of View, Man-machine System

INTRODUCTION

The important factor of subjective uncertainties in evaluation on structural safety is derived from human error. Analysis of human errors provides useful information required for the advance of an efficient management of structural system including man-machine system, since structural failures are due mostly to human performance concerned with lack of perception, omissions, misunderstanding, and mistake of recognition or judgment. In order to meet these problems, a lot of models of HRA (Human Reliability Analysis) have been proposed until now from the widespread fields such as psychological observation, natural science tradition, behavior or social science tradition and cognitive science tradition (Reason, 1990; Hollnagel, 1993; Card et al., 1983). However the need for the models can be made clearer if a distinction is made among various instances of the models.

It is well known that there are three kinds of response of human behavior in the operator model (Reason, 1990), when one receives perceptive stimulation. One is a skill-based response, which means the unconscious reaction. In this case, one can play a reactive action within the shortest response time after one receives some information on danger and creates the reactive patterns without the process of recognition, identification and judgment. The second type of response is called a rule-based response, which consists of the information processes such as recognition of situation, matching between situation and task. The third type of response is called a knowledge-based response whose feature is in identification and judging process. Since the latter two types of reaction are conscious reaction, the response time will be longer than unconscious reaction because of the existence of recognition process, identification process, judgment process or forecasting process.

In this paper, we discuss the relation between the human performance and structural safety with objective and subjective uncertainties. Therefore, the effect of human performance on reliability of man-machine system is considered from the viewpoint of cognitive science and psychophysiology in ergonomics. In addition, it is focused on the mechanism of visual information such as response time (RT) and useful field of view (UFOV) where one can receive the useful information for recognition, and response when one can respond to visual stimuli. It is important that the performance deterioration of human visual information processing is relevant to the degradation of the reliability of man-machine system.

RELIABILITY OF MAN-MACHINE SYSTEM

In man-machine system, the reliability or structural safety is strongly dependent on human performance. The most pervasive model of human is the Stimulus-Organism-Response (S-O-R) paradigm. The S-O-R is no longer widely accepted as a proper paradigm (Hollnagel, 1993). A metaphor is a way of conveying a specific meaning, i.e. to understand of human cognition. A more recent metaphor is found in the view of the human as an Information Processing System (IPS) (Card et al.,

1983). The classical information processing view has been extended to Step-Ladder-Model (SLM). In an IPS model the internal mechanism are typically discussed in far greater detail than in an S-O-R model. At same point focusing on the O, the model includes the function of three kinds of IPS model such as skill-based, rule-based, and knowledge-based IPS developed by Rasmussen (Reason, 1990).

Sometimes, we evaluate or judge the response behavior to input information outside by sensory perception typically illustrated by the operator model by Rasmussen. When one can receive outside information on safety by perceptive signal from visual or auditory organ, the response to the input signal was processed by human performance model in Figure 1. As shown in this figure, it is well known that there are various kinds of response pattern when one receives outside information.

The human performance is important factor in man-machine system, when the information was received with objective uncertainty by perceptive stimuli, since the response and safety is dependent on the result of judgment or accuracy of recognition of human. Therefore, if the performance comes down on perceptual information processing such as RT or UFOV, they affect the degradation of system strength in man-machine system. It becomes to increase the failure probability of structural system including man-machine system.

CASE STUDY ON HUMAN PERFORMANCE ON RT & UFOV

In this chapter, in order to discuss the aspect on subjective uncertainty, experimental results by authors are introduced in the following as a case study on visual perception and human behavior. Two kinds of studies on human performance are examined. The former is the study on RT, and the latter is the study on the region of cognition that is called as a term of UFOV.

EXPERIMENTAL STUDY ON RESPONSE TIME

The response time is measured when one can recognize a visual stimulus in the electromagnetic shield room. The 21 inches CRT display (SONY, GDM-F500) was equipped with 600mm distance from the eye of the subject. The output data of NTSC signal from imposer is converted into RGB signal by the up-scan converter.

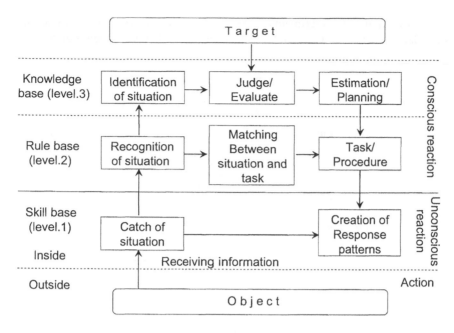

FIGURE 1 Operator model by Rasmussen

In the experiment for RT, a visual stimulus consists of foreground and background scenes. In the foreground scene, the fixation point (S_1) is used to keep the visual attention on the center. A simple white circle as S_1 with the size of 0.36° disappears at duration of 0.88Hz in order to keep the visual attention. The target index (S_2) with the size of 0.36° has a specific shape in Snallen's chart as S_2 and a similar shape as a standard index. The ratio of target mark to standard mark is 0.25. The duration between S_1 and S_2 is 1.7sec. S_2 appears within the zone (0~3.9°, 4.0~7.9°, 8.0~12.0°) at the coordinates randomly distributed. The two kinds of buttons were given to the subject in one's hands. The subject must push the button of right side as soon as possible, when one can recognize S_2. On the other hand, the subject must put S_1 on by one's left side button when S_1 disappears.

3D driving scenes is used for the disturbance effects on the visual attention and recognition. As 3D driving scenes, four kinds of patterns of driving scenes are used by the factors of the velocity (V) and demand of traffic (D). The recorded scenes consist of two kinds of level for factors V (V_1=40km/h, V_2=80km/h) and D (D_1=non-crowded, D_2=crowded).

Two kinds of experiments were carried out in order to examine the sensitive factors. One is to measure the detection region that is the region to detect a

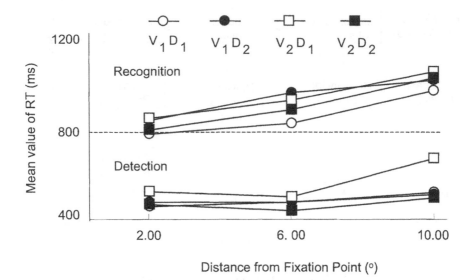

FIGURE 2 RT and Distance from Fixation Point

simple white circle S_2. The other is to measure the recognition region that is the region to be able to recognize a specific shape in Snallen's chart S_2. The former experiment called as "detection experiment", and the latter experiment is called as "recognition experiment". In this experiment, 17 university students from 20 to 23 years old were selected as subjects who have regular class road vehicle licenses with normal vision or corrected normal vision.

In the detection experiment, S_2 appeared 48 times per experiment. In the recognition experiment, S_2 appeared 256 per experiment, the number of standard mark of S_2 is 208 times and the target index is 48 times. In the experiment, the mark S_2 appears the same times per zone.

The behavior on RT is represented in Figure 2 for the appearance region of target mark S_2. From this figure, in the detection experiment it is seen that there are no differences among the appearance region of S_2. On the other hand in the recognition experiment, there is a typical tendency that the wider the appearance region of S_2 becomes, the longer RT is. For three kinds of zones of appearance, RT in recognition is larger than that in detection, whose differences are from 230 to 290ms in average.

REGION OF USEFUL FIELD OF VIEW

UFOV is the region where one can recognize the distinction of the shape, direction and so on. This region is located at middle area in comparison with static/kinetic field or conspicuity field. The field of view is defined by the region from fixation point (FP) (Ohyama et al., 1994). It is problem that there are no standards how to measure UFOV with visual task at fovea in practical work. The term of UFOV is generally used implying that peripheral performance depends not only upon retinal sensitivity but also upon the nature of the perceptual task. However, since such fields vary greatly with respect to shape difference among target and background noise, density of background noise, and so on, it is very difficult to measure the boundaries exactly. In the recent study, Ball measured the region as UFOV that one can detect the target index in the unique and ambiguous background (Ball & Owsley, 1993). Ikeda evaluated the influence of fovea load on UFOV. However, UFOV obtained the above has not been guaranteed the quantitative degree of accuracy or precision (Ikeda & Takeuchi, 1975).

In the following, it is demonstrated to show the new method of measuring UFOV and its results by the authors. The experimental paradigm is also shown in Figure 3. A simple black circle as fixation point (S_1) with 0.86° visual angles is used to keep the visual attention on the center of CRT. The recognition index as (S_2) with the size of 0.86° visual angles appears in the eight kinds of directions (0, 45, 90, 135, 180, 225, 270, 315°). S_2 is a specific object with the three kinds of attributes such as character, arrow and shape as target indices among the three kinds of standards indices with same attribute. The reason why these three kinds of S_2 are

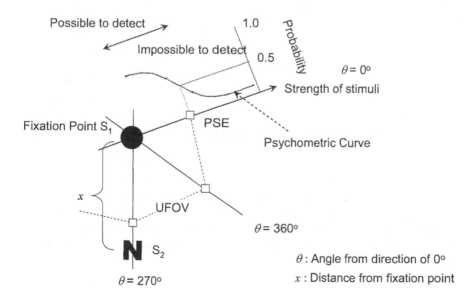

FIGURE 3 Region of UFOV

424

selected is to investigate the difference of human function among the different visual stimuli such as symbol, direction and shape.

In the experiment, the two kinds of response category such as "possible to detect" or "impossible to detect" for distance from S_1 to S_2 are measured. As shown in Figure 3, the distance from S_1 to S_2 means the strength of stimuli that is a

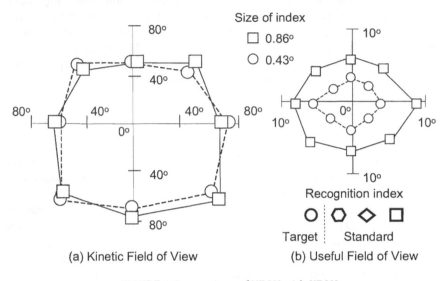

(a) Kinetic Field of View   (b) Useful Field of View

FIGURE 4 Comparison of UFOV with KFOV

psychophysical quantity. It is possible to detect S_2 when S_2 appears near from S_1. It is impossible to detect S_2 when S_2 appear far away from S_1. In other word, the longer the distance from S_1 to S_2 is, the weaker the strength of stimuli is.

This experiment is to measure the threshold of recognition when response category changes from "possible to detect" to "impossible to detect" or from "impossible to detect" to "possible to detect". It is well known that the function between the probability of detection and the strength of stimuli can be obtained by psychometric curve in the experiment (Ohyama et al., 1994) . Since the distance of boundary between "possible to detect" and "impossible to detect" means threshold of recognition, the threshold of recognition can be estimated by psychometric curve. Therefore, it may be possible to obtain the psychometric curves to the any direction from fixation point. It is also known that the stimulus threshold can be obtained as a probabilistic percentile of this psychometric curve. One of the examples of this threshold with 50% probability is the PSE (point of subjective equality) that is equivalent to threshold of recognition. As shown in Figure 3, if it is possible to assume that the region plotted within these stimulus thresholds is defined by UFOV, the outer limit of the region connected with PSE for each angle is defined by UFOV.

Figure 4 (a) shows the kinetic field of view (KFOV). In this figure, the regions shown in solid and dotted lines are the fields for S_2 with 0.86 and 0.43° respectively. On the other hand, the regions shown in Figure 4 (b) are the UFOV for the recognition index with 0.86 and 0.43°. Comparing with Figure 4 (a) and (b), it is clear that there is the difference of the region between KFOV and UFOV. In UFOV as shown in Figure 4 (b), it is seen that there is effect of size of recognition index. This tendency is seen for all kinds of recognition index and subjects. These facts suggest the important differences of the mechanism of IPS for KFOV and UFOV. The former has no differences among the size of S_2, since the mechanism of IPS is simple such as skill-base of operator model in Figure 1. This means that there are little variation at the visual task by KFOV. On the other hand, the latter at UFOV means that the mechanism at UFOV is due to knowledge-based task in operator model. Therefore the deterioration of human performance by KFOV is so small that the failure probability due to human is small rather than that by UFOV. From these results, it is seen that there is a little tendency between RT and UFOV, though the correlation between RT and UFOV is useful tool to estimate the risk or failure probability by human error.

DISCUSSIONS

It is considered in this chapter that there are several relationships between human perceptual performance and safety. Therefore some discussions are introduced concerned with the wide fields on psychology and cognitive science.

EXPERIMENTAL RESULTS AND HYPOTHESES

The phenomena that the more the additional visual task, the narrower the region within the visibility field is, which is called as visual tunneling, perceptual narrowing or tunnel vision. This is strongly concerned with IPS. In order to explain the mechanism, a new hypothesis has been proposed on visual attention from the viewpoints of cognitive and psychophysical science. In this hypothesis, the reason of occurrence of visual tunneling dues to two kinds of rules (Miura, 1985; Miura, 1986):

R1: When the visual demand is more than the ordinal visual task, the distance of moving visual attention is out of the region of UFOV.

R2: When the visual demand is more than the ordinal visual task, there is less amount of overlapping of UFOV.

These rules were derived from the principle of consistency of the amount of the resource for information processing. In other words, the hypothesis proposed by Miura is based on the concept of the depth and width of visual processing and constant amount of resources of information processing (Miura, 1992).

In addition, recently a new concept called cognitive momentum was also proposed which means the active characteristics for information processing (Miura, 1988; Miura, 1992). From this concept, the reason why the visual tunneling occurs when the visual demand includes is not due to the degradation of information processing, but due to adaptation to the change of situation for information processing. In other word, cognitive momentum means adaptation and optimization under the change of situation of information processing because of partition of resources on information processing.

COGNITIVE MODEL WITH PROCESSING MECHANISM

Some typical results of the experiments mentioned the previous chapter; the important questions in the following were pointed out.

Q1: Why are there from 230 to 290ms in average and from 1.5 to 3.15° differences between RT and angle by detection and recognition experiments as shown in this experimental paradigm?

Q2: How can we explain the mechanism of the reason why several kinds of distinctions of RT happens depending on the zone 1, 2 and 3 in recognition experiment?

The most pervasive model of human is probably the Stimulus-Organism-Response (S-O-R) paradigm (Hollnagel, 1993), where the response is a function of the stimulus and the organism. The Rasmussen's Model (Reason, 1990) was one of these kinds of models, which became the well-known Step-Ladder Model (SLM) in cognitive science (Reason, 1990). It is nevertheless difficult to appreciate that the S-O-R and SLM have two fundamental similarities such as the sequential progression through the internal mechanism and the dependency on a stimulus or event to start the processing. Therefore, it seems to be difficult to answer the questions by the cognitive model such as S-O-R or SLM. One of the latest versions of this metaphor is the notion of human cognition as a fallible machine. This idea is most clearly expressed as an indirect definition of the feasible machine that is an information-processing device (Reason, 1990)

It is well known that the channel of visual information processing from brain science has two kinds of channels of information processing. One is the channel of processing on cognition of the motion of a figure and the other is that of character of an object (Amari & Toyama, 2000). From this physiological knowledge, it is expected to construct the model that means the mechanism of information processing corresponding to the experimental paradigm in this paper. Figure 5 is IPS model for visual task proposed by the author (Yamanaka & Nakayasu, 2003), where two kinds of channels correspond to information processing for foreground and background scenes respectively. As shown in this figure, the processes for both experiments such as detection and recognition experiments are also parallel processing. Therefore, it needs more time in "recognition experiment" at the visual task far from fixation point. That is an answer to the question Q1 and Q2 by this IPS model.

CONCLUDING REMARKS

The relation between the structural safety and the performance of operator by visual perception was investigated. It is focused on the response time (RT) and the useful field of view (UFOV), in order to evaluate the effect of the deterioration of human performance in visual task. It is also considered that there are several relationships

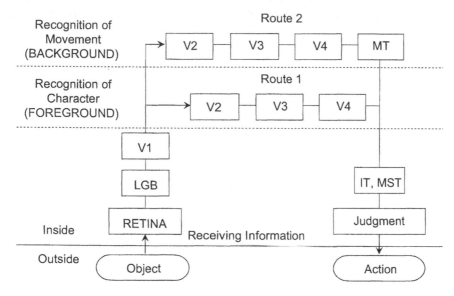

FIGURE 5 IPS Model for Visual Task

between human perceptual performance and structural safety in man-machine system. The causal effects of the deterioration of human performance such as perceptual properties are discussed on IPS model. These issues are remained in order to promote this work in the future. Firstly, it is necessary to formulate the method of measuring and standardizing the amount of human performance. Secondly, in order to evaluate the effect of the change of human performance, the method of structural reliability and safety must be linked with the method for human performance.

REFERENCES

Amari, S, & Toyama, K (2000) *Dictionary of Brain Science*, Asakura Publiser.
Ball, K & Owsley, C (1993) The useful field of view test: A new technique for evaluating age related declines in visual function, *Journal of American Optometric Association*, 16(1), 71-79.

Card, S. K, Noran, T. P & Newell, A (1983) *The Psychology of Human-Computer Iteraction* , Lawrence Erlbaum Associates Publishers.

Hollnagel, E (1993) *Human Reliability Analysis -Context and Contro* , Academic Press.

Ikeda, M & Takeuchi, T (1975) Influence of foveal load on the functional visual field, *Perception & psychophysics*, 18(4), 225-260.

Miura, T (1985) "What is the narrowing of visual field with the license of increase of speed", *Proc. 10th Int. Congress. on Association for Accident and Traffic Medicine*, 2(1), 2-4.

Miura, T (1986) *Behavior and Visual Attention*, Kazama Publisher.

Miura, T (1992) Visual search in Intersections -An Underlying Mechanism-, *IATSS*, 16(1), 42-49.

Miura, T, Shinohara, K & Kanda, K (1988) Visual Attention in Automobile driving: From Eye Movement Study to Depth Attention Study, *Proc 2nd Int. Conf. on Psychophysiology in Ergonomics*, 7-8.

Ohyama, T, Imai, S & Wake, T (1994) *Handbook of Sensory and Perceptive Psychology*, Seishinn Publisher.

Reason, J (1990) *Human Error*, Cambridge University Press.

Yamanaka, K & Nakayasu, H (2003) A model of information processing for visual task at driving, *Proc. 3rd Int. Conf. Artificial Intelligence and Applications*, 545-550.

Chapter 44

The Usability of Data Entry in GPS Navigation Systems

Manuela Quaresma, Anamaria de Moraes

PUC-Rio Pontifical Catholic University of Rio de Janeiro – Brazil
LEUI – Laboratory of Ergonomics and Usability of Interfaces
manu@manuelaquaresma.com

ABSTRACT

This paper presents a study about the usability of destination data entry in three GPS navigation systems with different data entry methods. The study aimed to evaluate which method of entry is easier to use by both experienced and non-experienced users, considering the effectiveness and the efficiency of the methods and the satisfaction with its use. Well-known usability methods and techniques were applied where it was possible to obtain performance metrics such as task success and task efficiency, and user satisfaction.

Keywords: Usability, GPS navigation system, human-computer interaction

INTRODUCTION

Recently, with the development and releasing of new technologies, many new electronic devices have been installed into vehicles, like digital audio systems, wireless communication systems and navigation systems. Due to the time spent in traffic congestion on transport routes and the lack of time, many drivers have been using these equipments with the vehicle in motion. Since they demand a complex interaction between user-equipments, and the way they are positioned on the vehicle dashboard, these devices could cause potential drivers' distractions and thus cause traffic accidents.

Typically, in-vehicle electronic systems are composed by controls and displays. Most of equipments available on the market have very small displays and computer systems with many levels of navigation, due the arrangement of the dashboard, the cost of the displays and the amount of information contained in the systems. From the Ergonomics standpoint, these characteristics are detrimental to effectivesimpler and efficient human-machine interaction, because the use of these equipments requires more visual and cognitive demand from the driver than the use of simpler equipments, such as the conventional audio systems and the onboard computers, which provide simple information of vehicle performance.

The GPS navigation system is a in-vehicle information system that has as primarily objective of guiding the driver to a given destination. Through the GPS antenna, the device locates the position of the vehicle on a map inserted in the database system. To guide to the desired destination, the driver enters the address data in the system and then it calculates its route. The driver enters the data usually by physical buttons or virtual buttons on touch-screens. Once the route is calculated, the system guides the driver with the vehicle moving through maps, voice instructions and indicators (symbols, graphics and messages), throughout the itinerary until the destination.

Although the systems already exist for several decades, they only had a wide spread in Brazil since 2006, when they were authorized for use in automobiles (Contran, 2006, 2007). Before March 2006, the CONTRAN (National Council of Brazilian Traffic) prohibited any device that generates moving images for the driver. Since then, several portable navigation systems have emerged in the Brazilian market. However, as this type of technology is still very new in Brazil and the market is still selling products with foreign translations of content, it is believed that these systems shall be conformed to the Brazilian people and usability tests should be done with this audience.

According to Nowakowski, Green and Tsimhoni (2003), "a well-designed navigation system can prevent wrong turns, reduce travel times, and hopefully, alleviate some of the driver's workload. However, poor usability can misdirect drivers, increase driving workload, and lead drives to make unsafe maneuvers."

The way their interfaces are designed, with content translations of foreign devices, it is believed that this type of system has many usability problems. Therefore, the hypothesis of this research is that the GPS navigation systems sold in Brazil have several usability problems. These problems in human-computer interaction (driver-system) could influence the driving task and, consequently, may cause driver distraction and consequent accidents. Thus, the object of this research is the interaction between drivers and GPS navigation systems available for use in-vehicles.

The research aims to provide GPS navigation systems easy to use and safe during driving. Its overall objective was to define design recommendations for the development of systems interfaces, for use in automobiles.

USABILITY

The usability is defined by ISO 9241-11 (1998) as the "extent to which a product can be used by specified users to achieve specified goals with effectiveness, efficiency and satisfaction in a specified context of use". Where: *Effectiveness* is the accuracy and completeness with which users achieve specific goals, i.e. the degree which a task is performed, if it can finish it or not; *Efficiency* is the resources expended in relation to the accuracy and completeness with which users achieve goals, that is, the level of effort expended by the user to complete a task; and *Satisfaction* is the absence of discomfort and the presence of positive attitudes towards the use of a product.

In order to meet these three requirements in designing a particular product, be it hardware, software or both (an electronic product), some researchers have created a number of usability principles, criteria and heuristics, such as Bastien & Scapin (1993), Nielsen (1994), Shneiderman (1998), Jordan (1998) and Norman (2002, 2006).

The main objective of all these principles is to facilitate user interaction with the product. However, most of these principles is related to user interaction with a computer interface, such as the principles stipulated by the first three authors mentioned above, who are researchers in human-computer interaction (HCI). But the authors Jordan and Norman, defined more general principles and applicable to both physical and computing interfaces. So, this set of principles satisfies the determination of design requirements for electronic products, which are products that contain interfaces both physical and computational, such as GPS navigation systems, digital cameras, cell phones, PDA, MP3 players, etc..

Brangier and Barcenilla (2003) classify these principles into four categories according to their specific purposes:

1) Facilitate learning of the system - principles that deal with issues related to the first use of a system, when the user makes deductions about how to interact, helping the beginner to start interacting with the system. This category includes the principle of compatibility between products and situations, and explicitness of the functions and procedures of the system/product;

2) Facilitate the information search, perception, recognition and understanding in the system - principles related to the presentation of information in the system. This category includes the principles of grouping, visual clarity, readability, user cognitive workload, memorization, consistency and standardization of information.

3) Facilitate the interaction control with the system - principles that address issues related to the conduct of activities. The principles of this category are the feedback, the user control and error Management (error protection, quality of error messages and error correction)

4) Consider the system context of use and the type of user - principles related to issues about the advanced use of the system, like the adaptability and flexibility that the system provides to user.

METHODS AND TECHNIQUES

To verify the validity of the hypothesis usability tests were applied with three GPS navigation systems Brazilians, in order to observe its compliance with the usability principles. According to Dumas and Loring (2008) "usability testing is a systematic way of observing actual and potential users of a product as they work with it under controlled conditions. It differs from other evaluation methods (such as quality assurance testing and product demonstrations) in that user try to complete tasks with a product on their own, with little help."

The usability test plan were organized according to the steps outlined by Rubin and Chisnell (2008).

PARTICIPANTS

Eighteen licensed drivers participated in the test, 9 experienced in GPS navigation systems and 9 non-experienced, with a gender split. The participants' ages ranged from 21 to 60 years. More than half (61%) answered that they drive on average more than four hours per week, while the other participants were balanced between up to one hour per week and up to 4 hours per week.

THE SYSTEMS EVALUATED

Three different systems sold in Brazil were evaluated: A – Nav N Go iGO 8.3; B – Route 66 Navigate 7; C – TomTom Navigator 7. These systems have distinct methods of data entry. In the first, the data is typed through a keyboard that automatically reduces the options (by eliminating some keys) as the user types the names of streets, according to its database (Figure 1). In the second, the user enters the address (fully or partially) through a static keyboard and then searches for it in a list of possibilities, on the next screen (Figure 2). And in the third, as the user enters the name of the street, the system filters the options in the database and presents some possible choices in two lines on the top of the screen (Figure 3).

Figure 1. Screens of the first system tested (system A)

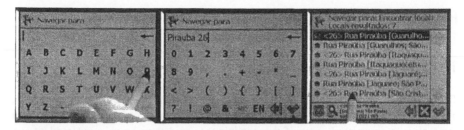

Figure 2. Screens of the second system tested (system B)

Figure 3. Screens of the third system tested (system C)

TASK AND PROCEDURE

To evaluate the different data entry methods it was asked to each participant to define a destination by address, with the following instruction (scenario): *"You have ordered some invitations at a print shop in São Cristóvão* [a far place and not very well known by the participants in Rio de Janeiro, Brazil] *and need to get them. Since you do not know how to get there, you will use your GPS system to guide you. For this, you need to put the address (below) in the system so that it calculates your route and guide you. How would you do that? <ADDRESS>"*

The tests were performed inside a parked vehicle (always the same vehicle) during the morning and afternoon, so that the incidence of light could be similar. The GPS equipment where the systems were installed has always been installed in the same place - stuck at the bottom of the windshield near the center of the dashboard. This is the place recommended to perform secondary tasks with displays in vehicles, according to the European (EC, 2008) and American (AAM, 2003) guidelines. All stages of the test were recorded on digital video through a camera mounted on a tripod in the back seat of the vehicle.

All participants evaluated the three systems, but the order in which the system were tested was counterbalanced, to avoid the results were biased to either system. Also, three different addresses were used with the same amount of letters and numbers. Before performing the task, the participant did some trials to know the system that would be tested. After the trials, the participant received a card with the task and address to be inserted and then performed the task. At the end of the test of all systems, debriefings sessions were conduct with the participants to clarify some issues that were observed and learn more about their beliefs and preferences.

RESULTS

TASK SUCCESS

To measure the task success, all the clicks made by participant during the execution of tasks were observed and tabulated, through the video recordings. To confirm that the task was completed, even if the participants said they have completed, it was used the activities' task flow defined previously as completeness criteria. If the participant had passed through all the activities of the task from beginning to end, the task was considered as completed. But, if the participant did not end them, failing in one or more activities, the task was considered not completed. For those tasks that have been completed with problems, it was verified that the participant passed through all the activities from beginning to end, but he/she also did other unnecessary activities.

The following graph (figure 4) presents the task levels of success in the systems. It is possible to observe that the system A was the most effective system, with 39% of completion without problems. The system C has also presented reasonable results in effectiveness, but the level of success without problems was very low (11%), which shows that there are efficiency problems. But, the system B has presented the most unsatisfactory results; there was no completion of the task without problems.

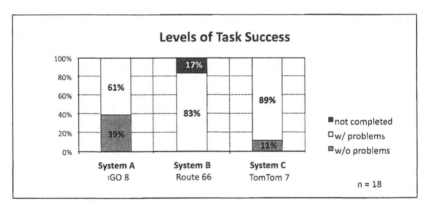

Figure 4. Levels of task success

The reason which led to an incidence of 17% of non-completion of the task, in the system B, was the fact that some participants didn't entered the building number in the system, to calculate the route to the destination. Typically, the entry of an address is done in steps; first the city name is entered, after the street name and at the end building number is entered, in different screens. In this system, the address entry is done differently from both the other systems tested and the other systems in the market. To insert the address in the system B, the user must enter the complete address (including street name, number and city) in one step (screen) and then looks for it and selects it in the list of addresses possibilities, on other screen (figure 2). With this way, some participants didn't know how to enter the number or not realized that they not inserted it, entering only the street name and concluding that the task was completed.

TASK EFFICIENCY

To assess how costly the task may be to the user, or how effective is the task, the numbers of clicks (commands selections or keystrokes) that each participant performed was counted in each task and each system. In addition, the minimum clicks that would be needed to complete the task on each system were counted. Comparing these two values, it was possible to measure the average level of effort that the participants had to complete the task. In this measurement, the tasks that were considered were only the completed ones - tasks completed without problems and completed with problems.

With these comparisons between clicks performed and minimum clicks, it was possible to observe two points: 1st) by the number of minimum clicks, which systems has the task design more efficient, i.e. the system that have lower number of clicks is theoretically the faster and more efficient; 2nd) by the number of clicks performed, which task presents the higher or the lower cost to complete, in each system.

In the following graph (figure 5), it can be seen that the systems A and C have the lowest number of minimum clicks (14 clicks) to complete the task. This means that the task designs in these systems, in principle, are more efficient than in system B. By observing the average number of clicks exceed and/or wrong made by the participants in the graph on the next page (figure 6), it can be concluded that the design of the system A is the most efficient among the three systems, with an average of 70% more clicks than needed to complete the task. To obtain this result, in the graph (figure 6), the amount of exceeded/wrong clicks made by the participants who completed the task with problems was computed. Thus, the 61% of participants who completed the task with problems in system A, they performed an average of 70% more clicks than needed. In system B, 83% of participants clicked on average 165% more clicks, and in system C, 89% of participants exceeded the amount of clicks by 121%. Therefore, it was concluded that the system A is more efficient, even if it is considered their level of success with no problems (39%) compared to the same level from other systems.

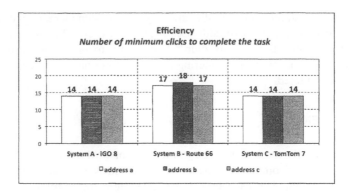

Figure 5. Number of minimum clicks needed to complete de task

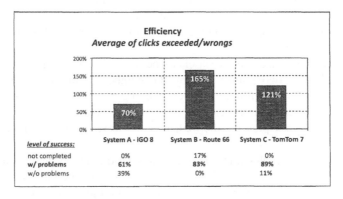

Figure 6. Average of clicks exceeded/wrongs made by the participants

Although the system A has presented more efficiency, it also occurred usability problems. To set a destination, the user first find address and then says he/she wants to use it as a destination. This way of thinking is not compatible with the way users think, because usually when the user wants to go somewhere, first he/she tell the system what is your goal, enters the address and then confirms the address entered. In this system the logic is different, the goal with the address is given only after it was found. Thus, after entering the address, some participants thought that they had already finished or were in doubt whether they had completed the task before making the last click in the command "Set as Destination". This shows an inconsistency of this system with others, that instead of use the label "Find", use terms like "Navigate to," "Go to", "Navigate", "Where to?" etc.

In system B, which presented a greater number of problems, one of the reasons that led to excessive clicks was the same problem occurred with the non-completion of the task, where many participants did not entered the building number with the name of the street. Therefore, the participants who noticed this problem of compatibility restarted the task and, consequently, exceeded the number of minimum clicks needed.

In the system C, one of the reasons that led to excessive clicks was a lack of information/confirmation, since at no time it presents the neighborhood where the address is located. Although this information is not required in the process of data entry, the other two systems (A and B) show the neighborhood at the time of address selection. According to what was discussed with some participants, this feedback is important because in a city may have streets with the same name or parts of name. For some participants this was seen as a problem because they did not trust in the system, so they restarted the task entering the address by ZIP code.

THE USER SATISFACTION

One of the issues most discussed by participants in the debriefing session was the long process required for entry the address data. In general, it is needed to enter the three main components of an address – the city name, the street name and the building number, usually on separate screens. Many participants questioned the need to insert the city name, since the GPS signal has recognized the city where the user is. In this case, most participants preferred the data entry method of system A, which has a screen where all the address data are inserted, and on this screen, it is possible to leave as default the city name, and the country name. This makes the process more efficient, because the user only needs to enter the street name and the building number.

Despite the participants criticized the long process, it was almost unanimous the preference for the data entry in separate screens, such as systems A and C.

The system B was very criticized for the fact of having to enter the building number with the street name and then select the address in long list of all places in the country, because there is no filter for the city.

The method of typing the street name in the system A was seen also as a positive item of the system. The fact that the letters fade, during the typing, was very remarked. However, some participants doubted about the effectiveness of this type of method, because if a user tries to enter an address misspelled (with one letter wrong), the system blocks the entry process. In this case, participants preferred the system C, which gives options to the streets as the name is entered, even if the name contains wrong letters. For example, when someone tries to enter the "*Paissandu*" street, typing with only one "s" or "ç", in the system C, there will come a time that the system will display the correct street and then the user can select it. In the system A this is not possible, because it does not recognize that name in the database.

CONCLUSION

Therefore, it was concluded that for an effective, efficient and satisfactory destination entry, it is necessary a task execution in short and sequences steps, with a navigator aid (with a filter of street names) and with a screen of address data where the city can be left as default. Also, the task must be initiated by a command

438

like "Navigate to". The diagram below (figure 7) shows an ideal sequence of screens for entering address data into GPS navigation systems.

With the research results and conclusions, the hypothesis mentioned above could be proved and the research goal was achieved – set design recommendations to GPS navigation system interfaces. It is emphasized that this paper presents only a few results. Other tasks were also tested, as well as other techniques were applied in Quaresma (2010).

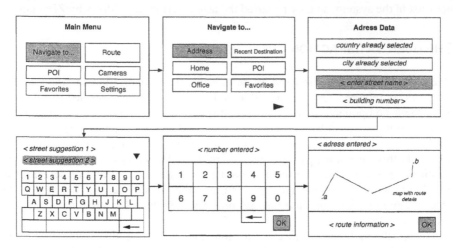

Figure 7. Ideal sequence of screens for address destination data entry

REFERENCES

Alliance of Automotive Manufacturers (2003) Statement of principles, criteria and verification procedures on driver interactions with advanced in-vehicle information and communication systems. Washington, D.C.: Alliance of Automobile Manufacturers.

Bastien, J.M.C. & Scapin, D. (1993). Ergonomic Criteria for the Evaluation of Human Computer Interfaces. Le Chesnay: INRIA.

Brangier, E.; Barcenilla, J. (2003). Concevoir un produit facile à utiliser. Paris: Éditions d'Organisation.

Dumas, J.; Loring, B. (2008). Moderation Usability Tests – Principles & Practices for Interaction. Burlington: Morgan Kaufmann Publishers.

European Communities (2008). Recommendations on safe and efficient in-vehicle information and communication systems: update of european statement of principles on human machine interface. Brussels: European Union (Official Journal 2008/653/EC), 2008.

International Organization Standardization (1998). Ergonomic requirements for office work with visual display terminals (VDTs) -- Part 11: Guidance on usability. Génève: ISO.

Jordan, P. W.. An introduction to usability. London: Taylor & Francis, 1998.

Nielsen, J.; Mack, R. (1994). Usability inspection methods. New York: John Wiley & Sons.

Norman, D. (2002). The design of everyday things. New York: Basic Books, 2002.

Nowakowski, C.; Green, P.; Tsimhoni, O. (2003). Common Automotive Navigation System Usability Problems and a Standard Test Protocol to Identify Them. In: Proceedings of

ITS-America 2003 Annual Meeting. Washington DC: Intelligent Transportation Society of America.

Quaresma, M. (2010). Usability evaluation of in-vehicles information systems: an ergonomic study of GPS navigation systems. Rio de Janeiro, 2010. 340p. DSc. Thesis – Departamento de Artes e Design, Pontifícia Universidade Católica do Rio de Janeiro.

Rubin, J.; Chisnell, D.(2008). Handbook of Usability Testing – how to plan, design and conduct effective tests. Indianapolis: Wiley Publishing. 2ed.

Shneiderman, B. (1998). Designing the User Interface. Massachusetts: Addison-Wesley. 3rd Ed.. 639 p.

Chapter 45

Optical Flow and Road Environments: Implications for Ergonomic Road Design

Paulo Noriega[1], Jorge A. Santos[2]

[1]Ergonomics Laboratory - Technical University of Lisbon
Estrada da Costa, 1499-002 Cruz Quebrada – Dafundo, Portugal

[2]Laboratory of Visualization and Perception - Minho University
Campus de Azúrem, 4800-058 Guimarães, Portugal

ABSTRACT

Applied and fundamental studies show that the optical flow generated by observer self-motion impairs other objects motion discrimination. Data from lab studies quantified this effect in function of the visual environment optical flow. However the preceding studies had left space to question the validity of these results in a more realistic driving situation and in the peripheral vision with a driving like situation. A first study using a driving simulator answered the first question, and a second lab study answered the question of knowing if the effect extended to peripheral vision. There was a clear effect of road environment optical flow in the vehicle motion discrimination during a following task procedure. The higher the environment optical flow density worse was the vehicle motion discrimination task. The effect was sharper in this realistic study with driving simulator than in previous lab studies. The discrimination of approaching objects throughout the visual field is slightly impaired but if is added to the task the effect of global optical flow the performance degrades substantially. In conclusion the environment optical flow is an intervening variable in the process of vehicle's motion discrimination during a following task. Study two, strengthened this information showing that optical flow intervenes in the motion discrimination in the peripheral vision. As vehicles motion discrimination is a vital task for drivers, implications for an ergonomic road design can be derived from this study.

Keywords: Optical Flow, Road Design, Motion Perception, Peripheral Vision

INTRODUCTION

Optical flow, as defined by Gibson (1950) refers to the change in the pattern of light reaching an observer, created when he/she moves or parts of the visual environments move. Several fundamental (Gray, Macuga, & Regan, 2004; Gray & Regan, 1999, 2000b; Probst, Brandt, & Degner, 1986) and applied studies (Gray & Regan, 2000a; Probst, et al., 1986; Santos, Noriega, & Albuquerque, 1999) show an interaction between self-motion and other objects motion.

One kind of interaction shows that when an observer is moving forward and the object is approaching, object motion perception is impaired (Gray, et al., 2004; Probst, et al., 1986; Probst, Krafczyk, & Brandt, 1987; Santos, Berthelon, & Mestre, 2002; Santos, et al., 1999). These findings have implications for road safety in the task of following other vehicles.

When following another vehicle, motion discrimination is fundamental for safe distance keeping and rear-end collisions avoidance. Contrary to general thinking, several reports and studies about rear-end collisions (D.G.V., 2005; Horowitz & Dingus, 1992; Golob et al. 2004, cited by Oh, Park, & Ritchie, 2006; Transportation, 2004) show that they are very costly in human deaths and injuries as well as economically. Besides the safety reasons for making this study, validity issues in previous studies justify it. Several lab studies reveal a relation between optical flow and motion discrimination. Concerning object approaching motion, relevant to avoiding collisions, the greater the optical flow densities (percentage) the lesser the performance in motion discrimination (Santos, et al., 2002; Santos, et al., 1999; Santos, Noriega, Correia, Campilho, & Albuquerque, 2000). However, those laboratory studies used very controlled environments with abstract stimuli or natural stimuli but short time ones. In spite of their high internal validity, we might be tempted to discard such findings, considering arguments such as those related to ecological validity. In a real driving situation, in addition to other vehicle motion discrimination, there are many more motor and cognitive tasks increasing driving demands. Thus, in the first experiment of this study our main objective was to test the optical flow effect on vehicle motion discrimination in a more realistic and ecological valid situation.

Previous studies (Gray, et al., 2004; Probst, et al., 1987; Santos, et al., 2002; Santos, et al., 1999) showed the interference of self-motion, on the perception of moving objects approaching the observer. However, the response to this effect has not been studied in peripheral vision. Regan and Vincent (1995) studied what sources of visual information (e.g. expansion rate, speed) was used for the estimation of TTC in the periphery, showing a degradation of performance along the periphery. Bex and Dakin (2005) described effects of interference in the discrimination of object motion direction in the periphery (Bex & Dakin, 2005).

Altough these studies reveal some degradation in the perception of object motion in the periphery, there are no studies that examine the effect of self-motion on the discrimination of objects approaching the observer along the peripheral vision. This was the aim of the second experiment.

EXPERIMENT 1: ENVIRONMENT OPTICAL FLOW AND VEHICLE MOTION DISCRIMINATION

METHODOLOGY

Participants

Twenty-three drivers, three female and twenty male, with an average age of twenty seven, with a drivers' license for at least two years and more than ten thousand kilometres experience, participated in this study. The sample was selected in order to be homogeneous regarding age and several visual parameters (e.g. visual acuity; and without clinical visual problems). Participants had also to be experimented in the driving simulator, so we used a sample of subjects from our database that had simulator experience.

Driving Simulator and Road Environments

We used a base fixed simulator with participants driving in a real car. Viewing point inside vehicle was at a distance of 3,86m from the screen and completed a visual angle of 45° horizontal and 34° vertical. Stimuli were generated at a rate of 30 images per second, and video projection system had a resolution of 1280*960 pixels at a refreshing rate of 60hz.

Three types of road environments were used, each one with 28,3 km. According to our study objectives, each road type was modelled to have different densities of optical flow. Thus we had road one with trees and shadows on the ground and an optical flow density 13° around the leading vehicle of 37%. Road two with walls and a density of 26% and plain road tree (without trees or walls), with a density of 11%. (Figure 1). The optical flow was measured with the algorithm of Lucas and Kanade implemented by Barron, Fleet and Beauchemin (1994). (For technical details see: Correia, Campilho, Santos, & Nunes, 1996).

Driving Task and Experimental Design

Participants started the engine and the vehicle achieved a speed of 50 km/h, without the need of pressing either accelerator or pedals. While driving, the participant's task was to discriminate desaccelerations (approaching) and accelerations (receding) of a leading grey vehicle that was twenty meters ahead. Stimuli were organized according to the constant stimulus method. The type of answer, right, wrong or omission and discrimination time in milliseconds was gathered through a two-forced choice method. Discrimination was accomplished through two buttons near the steering wheel. Pressing those commands was done without taking the hands from the wheel. Each desacceleration or acceleration was interrupted after six seconds. In that moment the leading vehicle returned to the former distance of twenty meters. If there was no discrimination after six seconds, the response was considered as an omission.

The experimental design involved the three road types, ten leading vehicle conditions (five desaccelerations and five accelerations levels) and five trials for each condition. The experimental model involved a total of one hundred and fifty trials per participant. As there were twenty-three participants, we had a total of 3450 observations.

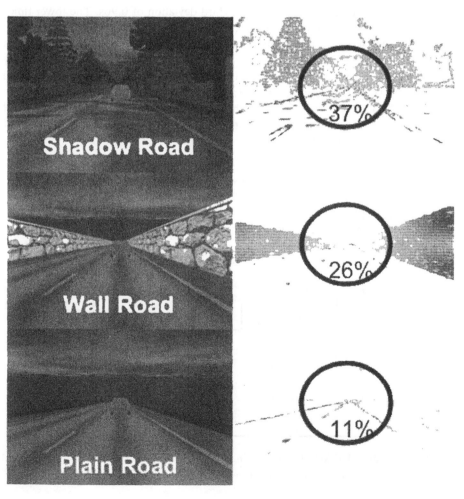

Figure 1. Three roads used in experiment 1. On the left snapshots from the simulator and on the right the vectorial representation of road optical flow. The percentage values are the optical flow densities measures in a circle of 13º around optical expansion focus

Results

Analyzed results are only for approaching situations, therefore we only show the results of 1725 observations. This option was chosen for two reasons. Forced-choice method demands a choice between two responses. Approaching motion of leading vehicle is the relevant and vital situation for: safe vehicle detection, keeping a safe distance and avoiding a collision while following another vehicle.

In the figure 2 we present times of discrimination distributions for each environment, and also the percentile 85. Higher discrimination times were obtained in the shadow environment (Mean=2.86s, SD=0.99s), followed by the environment of wall with an average of 2.6s and a standard deviation of 0.96s. The lower times were obtained in the plain environment (Mean=2.31, SD=0.91). The nonparametric Kruskal-Wallis test revealed significant differences between the environments ($\chi 2$=100.6, df=2, p<0.001). In the post-tests (Mann-Whitney) statistically significant differences were obtained between the environment of shadow and wall (z=-4.83, p<0.001), shadows and plain (z=-9.96, p<0.001) and walls and plain (z=-5.35, p<0.001). The results of the 85th percentile followed the same pattern of results of discrimination time. The 85th percentile value in the environment of shadow has a value of 3.91s, the walls of 3.63s and 3.3s in the plain.

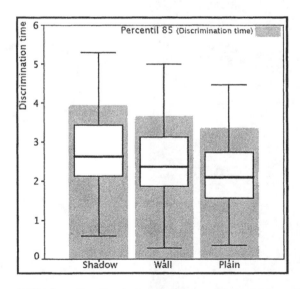

Figure 2. Box and Whiskers for discrimination times and road types. The background gray bars are for discrimination time for percentile 85.

EXPERIMENT 2: OPTICAL FLOW AND OBJECT MOTION DISCRIMINATION ALONG PERIPHERY

METHODOLOGY

Participants

Seven female and six male, with an average age of twenty eight years, participated in this study. The sample was selected in order to be homogeneous regarding age and several visual parameters (e.g. visual acuity; and without clinical visual problems).

Apparatus and Stimuli

We used a three channel retroprojection visualization system with a continuous surface with 7,10m large and 2,10m high. Observer was seated at a distance of 2m from the screen and completed a visual angle of 121,2° horizontal and 55,4° vertical. Stimuli were generated at a rate of 30 images per second, and video projection system had a resolution of 3651*1050 pixels at a refreshing rate of 60hz.

The stimuli represented two spheres vertically symmetrical to the central axis of vision (Figure 3). Spheres approached the observer simultaneously but at different speeds. Duration of each stimulus was 0,8 sec. After that time subjects had more 1,5s to respond. Interval inter-stimuli were of 1,6s.

Observer's task was to discriminate which of the two spheres, the left or right, would reach him/her first.

Experimental Design

Stimuli were organized according to the constant stimulus method. The type of answer, right, wrong or omission and discrimination time in milliseconds was gathered through a two-forced choice method. Discrimination was accomplished through mouse buttons.

The experimental design involved: two eccentricities of the spheres (25° and 55°); a condition with self motion (vection) with the background moving and a condition without self motion (static); the speed of the experimental sphere could assume one of nine speeds, ranging from 60km/h to 100km/h. (The reference sphere had always a speed of 80 km/h); The reference and the experimental sphere appeared randomly in the left and right side; For each condition we had twelve repetitions. Thus, the experimental model involved a total of eight hundred and sixty four trials per participant. Note that both spheres aspect was equal, the name reference and experimental are used to explain the methodology.

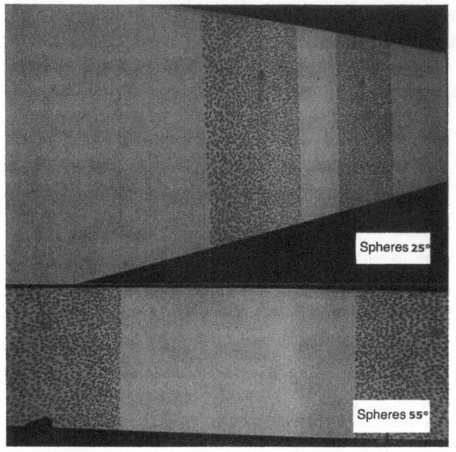

Figure 3. Stimuli of experiment 2. Uper section photo with stimuli at 25° of eccentricity. Down section with photo with stimuli at 55°. The background behind the spheres was used to simulate self-motion (vection). Observer fixed a visual reference (cross), to assure that peripheral vision was used.

Results

We present results for error responses. The maximum error possible was 50%. Even if observers pressed in all trials the same mouse button, they would achieve 50% of correct responses and 50% of wrong responses. In the table 1, we present results for discrimination errors in each eccentricity and for variable of self-motion (vection vs. static).

A Wilcoxon test for two dependent samples revealed that for both 25° ($z=-2.45$, $p < 0.05$), and 55° ($z = -3.11$, $p < 0.01$) the mean error were significantly higher in condition vection than in the static condition. The magnitude of the difference between the conditions of variable self-motion was less pronounced in the 25 ° than in 55 °. The average error increased by 5.2% in 25° (From 18.3% in static to 23.5% in vection). In the 55 ° increased 16.67% (static=23.9%, vection=38,7%).

Table 1. Descriptive statistics for errors in experimental conditions of eccentricity and self-motion

	25° eccentricity		55° eccentricity	
	Static	Vection	Static	Vection
Mean	18,3	23,5	23,9	38,7
s.d.	7	8,2	7,4	5,7
Min.	12	12	14,1	28,1
Max	32,8	42,1	39,4	48,4

CONCLUSIONS

Concerning experiment 1, our results support former studies: on road environments with higher optical flow densities, vehicle motion discrimination times increase. A close observation of previous results (Santos, Noriega & Albuquerque, 1999; Santos, Berthelon and Mestre, 2002), even show that discrimination times increased in this study. We might think that, by integrating more driving task components in our experiment, the optical flow effect would fade out. However, it seems that interference of the global optical flow on the discrimination of vehicle motion is a strong phenomenon.

It's necessary to find a way to add a "non constant reaction" time to formulas of safe distance keeping, and integrate a more "optical flow" dependent reaction time in the design of road environments. Consider the example in this study, where the average difference in times of discrimination between the worst environment (shade) and intermediate (wall) is 0.26 seconds. In a braking situation with the speed used in the study of 50km/h (13.89m/s), the start of braking in the environment of wall would be carried out before 3.6m (13.89 * 0.26) than in the shadow environment. This difference could, in an emergency break, meaning the difference between crashing or not.

Concerning experiment two, is clearly demonstrated that along the periphery, exists an inhibition effect of self-motion on the discrimination of approaching objects.

This study gives a contribution to road design based on human factors. The reaction times used in calculations for acceptable time-headway or inter-vehicle safe distance may benefit from this knowledge. The second experiment on peripheral vision can provide useful indicators for investigating the interaction with in-vehicle information systems (IVIS) or mobile phones. These devices when they are used only leave the peripheral vision available to process the movement of vehicles traveling ahead. The further the devices are from a central line of sight, the worse the performance is in the discrimination of moving objects. Worse results in the eccentricity of 55° gives support to this affirmation.

REFERENCES

Barron, J. L., Fleet, D. J., & Beauchemin, S. S. (1994), "Performance of optical flowtechniques." *International Journal of Computer Vision*, 12, 43-77.

Bex, P. J., & Dakin, S. C. (2005), "Spatial interference among moving targets." *Vision Res.*, Vol. 45(11), 1385-1398.

Correia, M. V., Campilho, A. C., Santos, J. A., & Nunes, L. B. (1996), *Optical flow techniques applied to the calibration of visual perception experiments.* In I. A. f. P. Recognition (Ed.), 13th International Conference on Pattern Recognition (Vol. I-Track A - Computer Vision, pp. 498-502). Vienna, Austria: IEEE Computer Society.

D.G.V. (2005), *Sinistralidade rodoviária 2004: elementos estatísticos. Observatório da Segurança Rodoviária.* Relatório 2005. Lisboa: D.G.V.

Gibson, J. J. (1950), "The perception of the visual world." Boston: Houghton Mifflin.

Gray, R., Macuga, K., & Regan, D. (2004), "Long range interactions between object-motion and self-motion in the perception of movement in depth." *Vision Research*, 44(2), 179-195.

Gray, R., & Regan, D. (1999), "Adapting to expansion increases perceived time-to-collision." *Vision Research*, 39(21), 3602-3607.

Gray, R., & Regan, D. (2000a), "Risky driving behavior: A consequence of motion adaptation for visually guided motor action." *J Exp Psychol Human Percept Perform* 26(6), 1721-1732).

Gray, R., & Regan, D. (2000b), "Simulated self-motion alters perceived time to collision." *Curr Biol*, 10(10), 587-590.

Horowitz, A. D., & Dingus, T. A. (1992, October 12-16), *Warning signal design: a key human factors issue in an in-vehicle front-to-rear-end collision warning system.* Paper presented at the Human Factors and Ergonomics Society 36th Annual Meeting, Atalanta.

Oh, C., Park, S., & Ritchie, S. G. (2006), "A method for identifying rear-end collision risks using inductive loop detectors." *Accident Analysis & Prevention*, 38(2), 295-301.

Probst, T., Brandt, T., & Degner, D. (1986), "Object-motion detection affected by concurrent self-motion perception: psychophysics of a new phenomenon." *Behavioural Brain Research*, 22(1), 1-11.

Probst, T., Krafczyk, S., & Brandt, T. (1987), "Object-motion detection affected by concurrent self-motion perception: applied aspects for vehicle guidance." *Ophthalmic & physiological optics : the journal of the British College of Ophthalmic Opticians (Optometrists)*, 7(3), 309-314.

Regan, D., & Vincent, A. (1995), "Visual processing of looming and time to contact throughout the visual field." *Vision Research*, 35(13), 1845-1857.

Santos, J. A., Berthelon, C., & Mestre, D. R. (2002), "Drivers' perception of other road users." In R. Fuller & J. A. Santos (Eds.), *Human Factors for Highway Engineers* (pp. 115-130). Oxford: UK: Elsevier Science.

Santos, J. A., Noriega, P., & Albuquerque, P. (1999), *Vehicle's motion detection: Influence of road layout and relation with visual drivers assessment.* In A. G. Gale (Ed.), Vision in vehicles VII (pp. 337-352). Amsterdam: Elsevier.

Santos, J. A., Noriega, P., Correia, M., Campilho, A., & Albuquerque, P. (2000), *Object Motion Segmentation and Optical Flow Contrast*: Research Reports - C.E.E.P. Braga: Universidade do Minho.

Transportation, U. S. D. o. (2004), *Traffic safety facts 2003: A compilation of motor vehicle crash data from the Fatality Analysis Reporting System and the General Estimates System* (No. Report DOT HS 809775). Washington, DC: National Highway Transportation Safety Administration.

Chapter 46

Using Mobile Information Sources to Increase Productivity and Quality

Peter Thorvald[1], Anna Brolin[1], Dan Högberg[1], Keith Case[2]

[1]School of Technology and Society
University of Skövde
Skövde, Sweden

[2]Mechanical and Manufacturing Engineering
Loughborough University
Loughborough, Leicestershire, UK

ABSTRACT

This paper presents an experimental study made on the use of different kinds of information sources in manual assembly. The general idea is that only the necessary information should be presented to the worker and it should be presented where and when the worker needs it as this is believed to both save time and unload cognitive strain. To account for the latter two aspects of this thought, where and when, this paper investigates the use of a handheld unit as an information source in manual assembly. Having a mobile information system, such as a Personal Digital Assistant (PDA), that can be carried with you at all times, as opposed to a stationary one, such as a computer terminal, is hypothesized to greatly improve productivity and quality. Experimental results show that the use of a PDA significantly improves quality whereas productivity does not significantly improve.

Keywords: Assembly, information, range of information, PDA, information presentation

INTRODUCTION

Cost, environmental impact and quality are only a few of the challenges that automotive manufacturers need to face in order for their product to stand out against the competition. As a result of this increased competition in the industry, the demands on each individual worker become even greater. Manual assembly workers are given additional and better tools, more tasks, more quality checkpoints and less time to do all this in. Regarding information presentation, there seems to be a huge misconception within many of these organizations that more information leads to better quality. However, this is a belief that can be heavily criticized, especially since studies have shown that many manufacturers seem to have a problem getting their workers to attend to the information in the first place (Bäckstrand, et al., 2005, Thorvald, et al., 2008a, Thorvald, et al., 2008b). Adding even more information in such cases, which at first might seem like a sensible solution, would probably only lead to the workers paying less attention to information or reduced ability to sort out the relevant information. The likely outcome is that they 'will not see the forest for all trees'.

Looking at assembly domains, one can clearly see that information presentation technology is rarely adapted to humans. It's rather the other way around; the workers have to adapt their work strategies to fit the information technology. A clear example is the use of stationary computer terminals as information sources. These are not very mobile, as at best they can be turned or tilted to afford better visibility but can rarely be moved from their position. While working in a work cell which has limited space this might not be much of a problem. The information is never farther away than can be accessed by a turn of the head. However, in automotive assembly, which is where this paper sets out from, work cells can range in length from 1 to 15 meters with this paper focusing on truck chassis assembly which are typically about 11 meters. With this length of the work cell, you would also need a reasonable range of the information. A stationary computer terminal or even a large binder of printed papers, which can be very unpractical to move around and is common practice in many assembly factories, has limited range and depending on font and screen sizes can be difficult to see from a few meters away. When working in such a work cell, the worker has two choices for information gathering:

1. Use precious time and physical effort to move towards the information source and gather information.
2. Rely heavily on memory and experience to make correct choices in assembly.

Obviously, neither of these is desirable. Making the first choice would probably be considered the preferred alternative from an organizational perspective as it would be supposed that the correct assembly is made as long as the assembly instructions are followed. However, assembly workers are often stressed and short of time and this behaviour would eventually increase stress and ultimately result in quality risks. The second alternative is better in terms of work load and stress but would obviously entail a huge quality risk if memory and experience fails.

So, how can this problem be remedied? Aside from the possibility of removing the need for information, logically there is only one way to go about this problem; move the information spatially closer to the task area. However, there are often many tasks spread out on for example a truck chassis, and it would be difficult to have separate information sources for all tasks. Instead, we suggest the use of a mobile information source such as a Personal Digital Assistant (PDA) which the worker can carry with him at all times. This would allow the worker quick and effortless access to information where and when it is needed.

HYPOTHESIS

The hypothesis for the experiment is as follows:

- Having a mobile information system, such as a PDA that can be carried at all times, as opposed to a stationary one, such as a computer terminal, improves productivity and quality.

The time saved due to the reduction of movement at the workstation is argued to benefit productivity in terms of hours per vehicle or similar measures. This decrease in physical strain ultimately leads to benefits in physical ergonomics and productivity. It is also plausible that the time to come to a decision of whether or not to consult the assembly instructions is diminished, as the internal calculation of the costs and benefits of the decision are more distinct.

The argument for better quality in the hypothesis is based on a cost-benefit (expectancy-value) reasoning (Jonides & Mack, 1984, Wigfield & Eccles, 2000) where it is believed that by reducing the cost of gathering the information, through walking distance, cognitive effort etc., subjects are more prone to use the information system than if it had been stationary and far away. However, the challenge is to determine what the actual effects on quality are.

METHOD

The challenge of satisfying testing of the productivity part of the hypothesis mainly consists of having a task with certain spatial properties. To be able to measure productivity the work cell needs to have a length of about 6 to 8 meters. This is simply to ensure that there is a considerable cost to return to the stationary information conveyer and this also corresponds well with the assembly tasks to be emulated. The second part though, quality, needs to be approached more delicately. It is believed that, because of the cognitive and physical strain of gathering information, assembly workers do not always judge it as worthwhile to do. However, this mentality might not occur until the worker is fairly experienced in assembly. Arguably, novices are unsure and tend not to stray from the assembly instructions too much. However, more experienced workers are more confident in their work and can sometimes value their own expertise too highly. Although it is plausible that the experience-factor is not necessary for this quality risk to occur, the situation might also arise where the worker has gathered the necessary information but forgotten it due to a prolonged stimulus-response gap (Dix, et al., 1998). In this case it could possibly be argued that the worker, instead of going back to the information source, puts more faith in their memory and thus might make an error. If a mobile information system is available at this point, the worker is argued to be less prone to trust their memory and consult the information system due to the low cost in cognitive and physical strain of gathering the information. This argument is made with a basis in cost-benefit and expectancy value theories (Jonides & Mack, 1984, Wigfield & Eccles, 2000).

The subjects were divided into two independent groups where each contained 12 to 14 university students/staff. Assembly workers, as experience and observations tells us, are a very heterogeneous group consisting of people of all ages, genders and personalities. In Skövde, Sweden, it is very common that young people spend a year or two in the automotive industry before moving on to University studies. Therefore, no extensive selection of subjects is undergone. This is to secure a mix of subjects that corresponds to the actual people working in assembly.

The task consisted of assembling several building blocks on a wooden chassis about 6 meters long. See figure 1 for a description of the laboratory layout. The test consisted of completing 20 iterations of the main task. Each task consisted of five subtasks where five different blocks were assembled with different bolts. As can be seen in figure 1, the wooden structure, which is supposed to emulate a truck chassis, has two sides. Each main task was performed on one side and as the worker has come full circle around the chassis, she has performed two main tasks, each consisting of the assembly of five blocks to the chassis.

Figure 1. Picture of the work cell. Blocks assembled in the previous task can be seen in the foreground and a laptop as information source in the background.

In the two conditions that are used, one for each independent group, the independent variable is the information conveyer, i.e the stationary information source, a computer laptop (figure 1) and the handheld device which in this case is an iPod Touch. PDA setup is exactly the same as in figure 1, except that the laptop is removed and the worker is fitted with a PDA on their arm. Figure 2 shows a subject fitted and working with the PDA. In figure 2, it can be seen how the assembly worker, while mounting a block, is already focusing on what the next task is. His eyes are already fixated on the PDA.

Figure 2. Example of an actual assembly worker interacting with the mobile device. The PDA is fastened on the subject's arm with a retail holster.

As the test continued the assembly information changed, blocks and bolts were assembled at different places or are replaced by other types of blocks and bolts. Both blocks and bolts were identified through unique four digit article numbers. Handheld power drivers were used to fasten the blocks to the chassis.

The interface for the information conveyers was a newly developed prototype, designed for use in the automotive assembly industry, which is a domain that this test is intended to emulate in terms of the task and laboratory layout. See figure 3 for a screenshot from the software as shown on the PDA. The software looks identical on a computer terminal as it is web based.

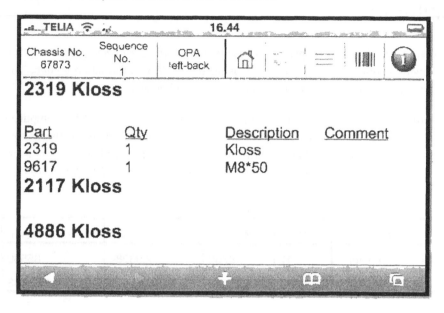

Figure 3. The interface of the software as seen in the PDA. The screen is touch sensitive and a tap on each subtask headline shows its contained parts.

The test consisted of, as mentioned earlier, two independent groups performing the same task which is assembling building blocks to a wooden chassis. One group used a computer terminal as information conveyer and the other used the PDA. The dependant or measured variables were:

1. Productivity – total time.
2. Quality – number of errors. Wrong blocks or wrong bolts.

As the subject worked on one side, a researcher disassembled blocks from the chassis on the other side.

There is an inherent risk in all types of tests where researchers are involved in operations, in this case disassembling parts during the test, that they may disturb or somehow bias the results. However, this is, first of all, balanced between the groups as both conditions have the same possible disturbance from the researcher. Secondly, these kinds of disturbances, and even worse, are expected in the actual assembly domains that we are trying to emulate. Thus there should be no confounding of the data and this fact merely adds to the external validity of the test.

RESULTS

A total of 26 subjects took part in the experiment with 12 subjects performing the tasks with a computer and 14 subjects using the PDA as information source. The groups were evenly balanced with respects to age and gender.

For productivity the groups differed in total time with the means 41,9 (Computer) and 39,5 (PDA) minutes, as can be seen in table 1. Analysis between the groups with respect to time proved not significant in a 2-tailed independent samples t-test (p=0,134). Effect size was 0,59 and post hoc statistical power 0,579 (Cohen, 1992).

Table 1. Descriptive statistics of productivity measured in minutes.

	Group	N	Mean	Std. Deviation	Std. Error Mean
Total	Computer	12	41,8750	2,96284	,85530
	PDA	14	39,4643	4,62628	1,23643

The mean number of errors for both groups was 3,36 as can be seen in the descriptive statistics in table 2. The participants in the computer group made an average of 4,58 errors while the participants in the PDA group made an average of 2,14 errors. For analysis of this dependent variable a non-parametric test was chosen as the distributions are slightly skewed. A 2-tailed Mann-Whitney test showed significant difference between the groups (p=0,045).

Table 2. Descriptive statistics of quality measured in number of errors.

	Group	N	Mean	Std. Deviation	Std. Error Mean
Quality	Computer	12	4,58	3,397	,981
	PDA	14	2,14	2,143	,573

Notable for the analysis of quality is that the effect size was large (ES=0,81). This is probably the main reason why a significant difference between the groups could be shown with only 26 participants in the test. A histogram for both groups showing the difference in distributions can be seen in figure 4. The figure illustrates both differences between means and also the differences in width of distributions which is also evident through the major differences in standard deviations. Although, homogeneity of variances tests did show that variances between the groups were equal.

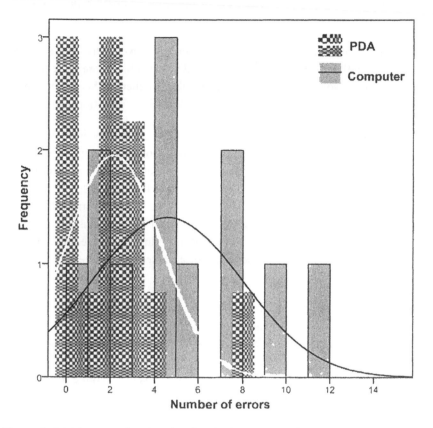

Figure 4. A histogram showing the distribution curves of the two groups.

CONCLUSIONS

Although productivity did not show a significant difference in our analysis this is thought to be because of the poor statistical power of the experiment (Cohen, 1992). The focus in the experimental design was to design quality hazards and avoid floor effects in number of errors. Post hoc statistical power for productivity was 0,579, which gives us about a 57% chance of finding a significant difference, should there be one. More subjects would probably have resulted in this variable also being significant. Also, the reinforced hypothesis that subjects using the PDA are more inclined to use the information source probably results in their measurement for productivity being increased. The time spent on the information source obviously affects productivity if productivity is seen as an isolated factor. Naturally, in production, productivity is also affected by quality from an organizational point of view.

However, quality did show significant differences between the groups. This reinforces the hypothesis and the argument behind it that subjects are more prone to use the information source if it is more accessible to them. This is thought to include both physical effort and time wasted to gather this information.

To conclude, the experiment has shown how quality is greatly improved by a mobile information source. Differences in the means show that the group using the PDA produced less than half the number of errors compared to the computer-group. However generalization and applications to industry of these results must be made with great care. The experimental design was made to generate errors and these built in "traps" might not correspond fully with state of the art industry. However, many of these errors have been observed in modern industry and are a compilation of errors found through personal experience as an assembly worker and through observation.

To sum up, these results are valid for application in industry if handled with care. Since the experiment has been designed to generate as many errors as possible and the task might be considered trivial, analysis within groups may be somewhat distorted. However, the difference between groups, which is what is being investigated, is there and is significant. The main result: 'mobile information sources result in better quality than stationary ones,' is valid and confirms previous, unpublished, observational studies.

ACKNOWLEDGEMENTS

This work has been made possible with great help from the EU integrated project MyCar. Many thanks also go to the subjects who participated in the study.

REFERENCES

Bäckstrand, G., de Vin, L. J., Högberg, D. & Case, K. (2005). Parameters affecting quality in manual assembly of engines. In Proceedings of the International Manufacturing Conference, IMC 22, Institute of Technology, Tallaght, Dublin, August. pp. 165-172.

Cohen, J. (1992). Statistical power analysis. Current Directions in Psychological Science, 98-101.

Dix, A., Ramduny, D. & Wilkinson, J. (1998). Interaction in the large. Interacting with Computers, 11 (1), 9-32.

Jonides, J. & Mack, R. (1984). On the cost and benefit of cost and benefit. Psychological Bulletin, 96 (1), 29-44.

Thorvald, P., Bäckstrand, G., Högberg, D., de Vin, L. J. & Case, K. (2008a). Demands on Technology from a Human Automatism Perspective in Manual Assembly. In Proceedings of FAIM2008 Skövde, Sweden, June-July 2008. pp. 632-638.

Thorvald, P., Bäckstrand, G., Högberg, D., de Vin, L. J. & Case, K. (2008b). Information Presentation in Manual Assembly – A Cognitive Ergonomics Analysis. In Proceedings of NES2008, Reykjavik, Iceland, August, 2008.

Wigfield, A. & Eccles, J. S. (2000). Expectancy-Value Theory of Achievement Motivation. Contemporary Educational Psychology, 25 (1), 68-81.

Chapter 47

Supporting Attention in Manual Assembly and Its Influence on Quality

Gunnar Bäckstrand, Anna Brolin, Dan Högberg, Keith Case

School of Technology and Society
University of Skövde
Skövde, Sweden

Mechanical and Manufacturing Engineering
Loughborough University
Loughborough, Leicestershire, UK

Volvo Powertrain AB
Skövde, Sweden

ABSTRACT

Modern manufacturing information systems allow fast distribution of, and access to, information. One of the main purposes with an information system within manual assembly is to improve product quality, i.e. to ensure that assembly errors are as few as possible. Not only must an information system contain the right information, it must also provide it at the right time and in the right place. The paper highlights some of the concerns related to the design and use of information systems in manual assembly. The paper describes a study that focuses on the correlation between active information seeking behaviour and assembly errors. The results are founded on both quantitative and qualitative methods. The study indicates that by using simplified information carriers, with certain characteristics, the assembly personnel more easily could interpret the information, could to a higher degree be prompted (triggered) about product variants and could also be able to prepare physically and mentally for approaching products arriving along the assembly line. These conditions had positive influence on quality, i.e. gave a reduction of assembly errors.

Keywords: Manual Assembly, Active Attention, Passive Attention, Situation Awareness, Information System, Product Quality

INTRODUCTION

In a modern manufacturing environment the information system is a vital part of the assembly process. The "receiving" of information is an important part of the flow of information between the information system and the receiver. The receiver must interpret the information so that an appropriate action can take place. That is, one has to comprehend the meaning of the information so that an action can take place, and one has to understand how this action will affect the future status of a product. This follows the definition of Situation Awareness (SA).

There are many concerns related to the flow of information in a manual assembly context. One concern is related to active and passive attention. James (1890/1950) describes active attention as a state where humans actively seek information, whereas passive attention is when humans are passively awaiting a situation where active attention is required.

Related to information system use in a manual assembly context, the main focus for the personnel should be to translate the information into actions to reach a specific goal. This requires that the information is available at the right time, in the right place and that the assembly personnel have identified a need for the specific information. During a study at an engine assembly plant an evaluation of some of the workstations indicated that there are problems connected to the distribution of information that can be defined as "Delivery versus Demand" of information (Bäckstrand, et al., 2005). They further argued that *Information Delivery* is the event that occurs when a specific type of information must be accessible in a specific work environment, for example an assembly workstation. *Information Demand* occurs when an "object", in this case one of the assembly personnel, has identified a need for information. This need originates from a need to fulfil a goal (Figure 1), and possibly the satisfaction of fulfilling the goal (Losee, 1990).

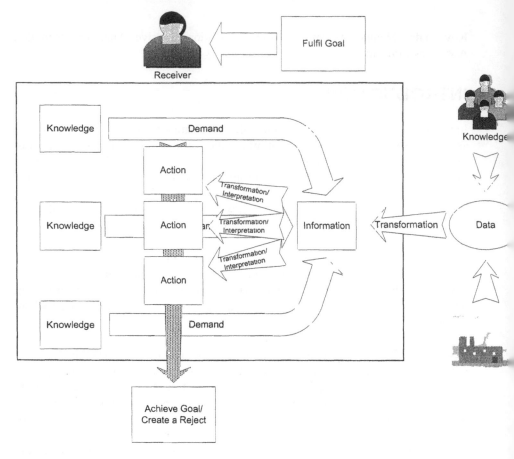

Figure 1. The information flow with respect to demands and goals (Bäckstrand, et al., 2006).

For the goal driven assembly personnel there are two ways to view visual attention: top-down (goal driven) or bottom-up (stimulus driven). Attention is said to be both goal-driven and stimulus-driven, according to Yantis (1998).

- Goal-driven (active attention): The product is known to have a specific marker, and one searches for that marker. This product is likely to be selected by our attention and recognized.
- Stimulus-driven (active or passive attention): All products look similar or the same, except one, that seems to "pop-out of the background and draw attention automatically".

The ability to actively detect information in our environment is of great importance in an assembly context. An interesting aspect is whether there is a connection between attention and assembly errors. According to Reason (1990), attention can be seen as work performed on discrete elements in a restricted work space, our working memory. Hence, the work uses energy from a strict limited "pool of attention resources" (Wickens, 1992; Downing, 2000), and it is important to use this energy in an optimal way. If one presumes that passive attention draws less energy from the "pool of attention", one of the goals with information must be to support the personnel in a way that makes it possible to continue working in a passive attention mode, where the use of skills and sub skills is more or less automatic, (Shiffrin and Schneider, 1977). Norman (1990) describes this kind of automation as a useful advance that can replace tedious or unnecessary tasks and monitoring. This must (if one presumes that the Reason and Norman theories are correct) result in less use of cognitive resources. Less use of cognitive resources by one process, for example the information search process, should make it possible to use the limited resources more effectively (Downing, 2000).

If one focuses on the need of information for the assembler, one realises that there are some issues that the discussion regarding active and passive attention can explain. The move from passive to active attention is triggered by an exogenous event (from outside, in this case the body) that cannot be suppressed or ignored by the assembly personnel. This exogenous event is the trigger that should start the expected information search process. Therefore, an information system that triggers the information search process is essential, and when the information search process has started it should support a passive attention mode. This means that the need for an active (active from an attention perspective) use of the information is to be kept at a minimum.

An interesting finding when one studies the theories of Situation Awareness (SA) is the absence of the important issue concerning how triggers can create change of state, from passive to active attention, from active to passive attention. For example a trigger that indicates when it's time to relax from a mentally point of view. Endsley (1999) states that due to several factors, although the information is directly available, it is not observed or included in the scan pattern, due to for example not looking at the information, attention narrowing etc. This is according to Endsley, the main contributor within SA related errors and is where it is believed that triggers can be a major contributor to preventing this type of human error. The trigger in this case is the event that creates awareness of the presence of important information, and that draws the personnel's attention to the correct information source. It is obvious that if an assembly worker is not aware of a need for information, a correct information search process will not be started. This will result in a quality risk that can influence the assembly errors severely.

Based on this argument it is hypothesised that by using simplified information carriers, in this case based on a colour coded information system, the assembly personnel more easily could interpret the information, could to a higher degree be prompted (triggered) about product variants and could also be able to prepare physically and mentally for approaching products arriving along the assembly line.

Accordingly, it is assumed that these conditions will have a positive influence on quality, i.e. will cause a reduction of assembly errors.

METHOD

The experiment consisted of testing how the colour coded information system could affect the personnel and how that would affect the quality.

The field experiment focused on four assembly stations in an engine assembly plant (two stations S0800 and two stations S1100 on parallel assembly lines) where the main component for S800 was servo pumps assembly and for S1100 cable harness assembly. The gathering of empirical data consisted of recoding assembly errors, complimented by semi-structured interviews. Historic data on assembly errors was used as reference performance indicators. Approximately 33000 historic data was recorded from the reference period: 13th March 2006 to 22nd December 2006. The most relevant data included: date and time, engine family, engine variant, effect number, effect description, part number and a free text field (this field could be used by the quality assurance personnel to describe the cause of rejection).

The production at the assembly line continued during the experiment and the conditions were more or less the same as on regular production days. The personnel involved in the study were all employed by the company and performed their regular duties. The experimental environment and the assembly environment, as it was before and after the case study, differed only in the parts that was included in the study: the *informative triggers*. The triggers were coloured magnetic rubber sheets with a size of approximately 300x300x0.85, 300x50x0.85 and 60x60x0.85 millimetres (length, width and thickness) attached to the carrier (Figure 2) and 400x300x0.85 attached to the material racks (Figure 3 and 4). The triggers had two purposes: firstly, to create awareness that a different engine variant was to be assembled; secondly, to give information regarding what part should be used.

Figure 2. The location of the informative triggers on the engine carriers.

Figure 3. The location of the informative triggers on the material racks in assembly station S0800.

Figure 4. The location of the informative triggers on the material racks in assembly station S1100.

At the beginning of the assembly line a station was built specially for the experiment. This station was manned day and night during the study period: 24th May 2007 to 6th June 2007, and was responsible for engine identification, attachment and detachment of the triggers to the carriers (Figure 2).

The result from the study was obtained by comparing number and type of assembly errors during the reference period and during the experiment. June is traditionally the month where summer employees are hired, therefore it was decided to consider this month specially, since the experiment was conducted during June.

A qualitative approach was added in the later part of the study with a questionnaire answered by 171 workers. The main purpose was to establish if there were connections between the results from the quantitative evaluation and what the assembly personnel had experienced.

RESULTS

In the figure, the results for three periods are presented, the reference period, the case period and the amount of reject during the month of June 2006, a period within the reference period. The purpose of investigating and presenting data for June 2006 was to determine if a major difference could be present during the month of June. However, during the case study period now.summer employees were present when the study was conducted. The quantitative results showed that the assembly errors reject data during the study decreased to a normalized value of 59 (baseline = 100), for station S0800 (Figure 5) and to 0 (baseline = 100) for station S1100 (Figure 6).

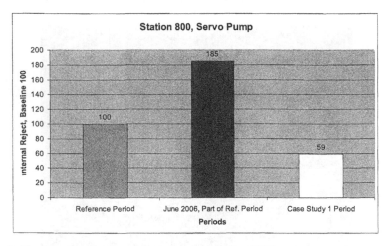

Figure 5. The assembly errors reject data for station S800.

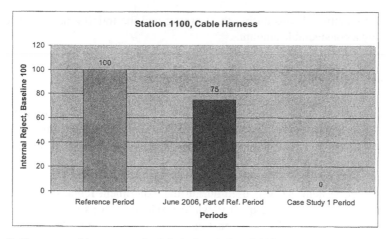

Figure 6. The assembly errors reject data for station S1100.

The decrease during the study for station S800 was approximately 41 points compared to baseline while for station S1100 the decrease was 100 points, i.e. no rejects occurred during the case study period.

An overall evaluation of the data gave:

- Quality: There was a decrease in the number of assembly errors at assembly station S800 and S1100 during the experiment period, compared to the reference period.
- Productivity: No effect could be found regarding the assembly time for each station, i.e. there were no differences in assembly time when comparing the reference and the experiment period

The qualitative survey gave that the colour coded way the information was presented improved the information by:

- Presenting the information so that the assemblers can see the information from their station.
- Making it easier to identify from a distance what to assembly.
- Providing the information so that the information is presented when a need has occurred.
- Eliminating the need to learn by heart where in space the parts that should be assembled are located.
- Creating a less demanding workload due to the elimination of noise (i.e. redundant information).
- Triggering the information search process when it is needed.
- Supporting part changes, for example when a design change note is initiated. When this occurs, it is possible to make the change without changing the colours in the work environment. If a new part is added, a new colour can be added. By using the existing colours or just adding a colour, the need to learn by heart where the parts are located, the

productivity losses during the learning and the training needs are lowered by a considerable amount.

DISCUSSION

The main purpose with the study was to try to find explanations of the quality problems that occur in the assembly environment at the company involved in this study, and to try to understand how information and the use of information can influence humans in both negative and positive ways.

The simplified system that was used during the experiment reduced the number of assembly errors, i.e. it gained quality, while productivity was not influenced. It is believed that the reasons for this effect are that the simplified information system made it easier for the personnel to interpret the information, and they were to a higher degree prompted about product variants. They were also able to prepare physically and mentally for approaching products arriving along the assembly line.

The survey reviled that a common reaction was that the reduction of the amount of information was a main benefit with the colour coding. This made it possible for the personnel to quickly find relevant information without noise from redundant information.

It is obvious that if an assembly worker is not aware of a need for information, a correct information search process will not be started. This will result in a quality risk that can influence the assembly errors severely. Knowledge and how to gain it is an important part of manufacturing, but in this case, attention is the subject of interest. This is because of the strong connection between attention and the investigated assembly errors. An interesting part of Jones and Endsley's causal factors is that 76% of the errors in situation awareness are in their study caused by problems in perception of needed information (Endsley, 2000). If one assumes that there is a connection between "problem in perception" and attention, one can argue that there is a high risk of exclusion due to changes of state in the information flow. This exclusion can be caused by a framing effect that limits the possibilities to choose information (Beach and Connelly, 2005), or a failure to observe provided information. The framing can be connected to a suppression of endogenous signals (signals from the brain) so that the information for some reason is not observed, i.e. a person misses the exogenous signals in the work environment.

To summarize, the results from the study clearly indicate that information and the way it is presented has a strong effect on quality as well as on the ability for the assembler to identify information needs. The study supports the hypothesis that the use of a triggered information system affected the assembly personnel in a positive way, which increased the quality.

ACKNOWLEDGEMENTS

Many thanks to the staff at Volvo Powertrain Skövde, Sweden, that participated in the study.

REFERENCES

Beach, L. R. and T. Connelly (2005). The Psychology of Decision Making. People in Organizations. Thousand Oaks, SAGA Publications.

Bäckstrand, G., De Vin, L. J., Högberg, D. & Case, K. (2005). Parameters affecting quality in manual assembly of engines. Proceedings of the 2nd International Manufacturing Conference: Challenges Facing Manufacturing, Dublin. 395-402.

Bäckstrand, G., De Vin, L. J., Högberg, D. & Case, K. (2006). Attention, interpreting, decision-making and acting in manual assembly. Proceedings of the Innovations in Manufacturing, 23rd International Manufacturing Conference. Belfast: 165-172.

Downing, P. E. (2000). "Interaction between visual working memory and selective attention." Psychological Science 11(6): 467-473.

Endsley, M. R. (1999). Situation Awareness and Human Error: Designing to Support Human Performance. Proceedings of the High Consequence Systems Surety Conference Albuquerque, NM.

Endsley, M. R. (2000). Theretical Underpinnings of Situation Awareness: A critical Review. Situation Awareness Analysis and Measurement. M. R. Endsley and D. J. Garland. New Jersey, Lawrence Erlbaum Associates, Inc.: 3-33.

James, W. (1890/1950). The Principles of Psychology, Vol. 1, Dover Publications.

Losee, R. M. (1990). The Science of Information; Measurement and Applications. San Diego, California, Academic Press, Inc.

Norman, A. D. (1990). The Design of Everyday Things. London, England, The MIT Press

Reason, J. (1990). Human Error. New York, Cambridge University Press.

Shiffrin, R. M. and W. Schneider (1977). "Controlled and automatic human information processing: II. Perceptual learning, automatic attending, and a general theory." Psychological Review 84(2): 127-190.

Wickens, C. D. (1992). Engineering Psychology and Human Performance. , HarperCollins Publishers Inc.

Yantis, S. (1998). Control of Visual Attention. Attention. H. Pashler. Hove, Psychology Press Ltd: 223-252

Chapter 48

Study on Judgment of Visual Similarity of Lumber

Nanae Teshima, Hiromitsu Takasuna, Takamasa Mikami

Graduate School of Information Science and Engineering
Tokyo Institute of Technology
O-okayama, Meguro, Tokyo, Japan

ABSTRACT

Various imitation lumber are used as building materials. It is important to judge the degree of visual similarity between such materials, because building materials are often chosen on the basis of appearance. Therefore, a study was performed in which assessors were asked about the degree of visual similarity between various samples. We found that each type of lumber can be classified into two broad groups, one with a rectilinear grain and the other with a rounded grain. We used image features obtained from a sectional wave approach to predict the judgment of similarity between pieces of lumber with reasonably good results.

Keywords: Lumber, Visual similarity, Image feature

INTRODUCTION

Lumber has recently diversified to include not only natural wood but also imitation materials. Many users and architects choose imitation lumber, because real lumber is expensive and requires special care. Different types of imitation lumber create different impressions on users because of differences in parameters such as their color and grain. For example, some imitation lumber might seem cheap to the users. Furthermore, it is not unusual for building materials to be chosen on the basis of appearance. The people who choose building materials need to know the factors that influence the judgment of visual similarities between building materials. Thus, this study aims to determine the factors affecting the judgment of visual similarity between different types of lumber and quantify the relationship between visual similarity and image features. Although the gloss or roughness of lumber might influence the judgment of visual similarity between different types of lumber, we do not consider these two factors in this report, because this study is in its initial stage.

OUTLINE OF THE TEST

We used digital images of fourteen types of lumber as test samples (Table 1). We chose those types of lumber that have characteristic grains and are often used for floor finishing materials. We used three types of digital images—color, gray-scale, and binary—to study the effects of different levels of image information on the judgment of visual similarity. To produce the images, we placed the samples on a scanner (Canon MP960) and obtained color images (300 dpi). Then, we edited the color images to produce gray-scale and binary images.

Table 1 Color images and details of test samples

1.Aomori hiba arborvitae *	2.Ash **	3.Katsura tree**	4.Japanese larch *	5.Kiso hinoki cypress **
L*: 66.36 a*: -3.10 b*: 32.69	L*: 77.75 a*: -1.11 b*: 23.25	L*: 53.18 a*: 2.00 b*: 27.58	L*: 67.15 a*: 0.51 b*: 31.16	L*: 73.76 a*: 1.28 b*: 36.12
6.Keyaki *	7.Jisugi **	8. sugi **	9. Douglas fir**	10. hinoki **
L*: 68.49 a*: -1.49 b*: 30.25	L*: 65.05 a*: 1.39 b*: 29.23	L*: 58.58 a*: 8.35 b*: 37.09	L*: 66.98 a*: 5.07 b*: 37.28	L*: 73.27 a*: 2.15 b*: 31.68
11.western red cedar **	12.Yellow cedar **	13. white ash *	14.ponderosa pine **	*: acicular tree **: broad-leaf tree
L*: 42.52 a*: 6.10 b*: 32.69	L*: 78.75 a*: -0.22 b*: 33.57	L*: 70.22 a*: -0.22 b*: 27.16	L*: 79.15 a*: -1.12 b*: 30.81	

We asked the assessors to judge the degree of visual similarity between each pair of lumber, showing them sheets on which pairs of images appeared. Examples of the sheets are shown in Figure 2. For pairs of color images, we asked the assessors about overall similarity, similarity of the grains, and that of colors. For gray-scale and binary images, we asked them about the similarity of the grains and the factors that they found to be similar or dissimilar. We chose the factors by referring to the results of a prior test and to a study of Ono, et al.[1] The questions and information about the assessors are given in Table 2. Arikawa, et al.[2],[3],[4] reported that different examination environments can affect one's impressions. Therefore, we performed the test under artificial lighting, which we stabilized by getting rid of natural light. Other environmental factors may also have some effect on the judgment of visual similarity, but stabilizing them would be unreasonable for studying the appearance. Therefore, we did not stabilize other environmental factors. After the test, we used the method of successive categories to scale the judgment of visual similarity between different types of lumber on the basis of the results.

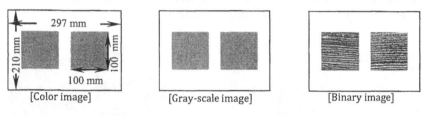

[Color image]   [Gray-scale image]   [Binary image]

Figure 2 Examples of the sheets

Table 2 Outline of the test

Questions	Color images	
	Question1 How similar is one to the other overall?	
	Question2 How similar is one to the other considering only the grain?	
	Question3 How similar is one to the other considering only the color?	
	Gray-scale and binary images	
	Question1 How similar is one to the other considering only the grain?	
	Question2 What factors did you find to be similar?	
	Question3 What factors did you find to be dissimilar?	
Choices of answer	**About similarity**	**Factors**
	1.One has little similarity to the other.	1.Type of the grain (slash grain or straight grain)
	2.	2.Regularity of the grain
	3.	3.Size of the grain
	4.Cannot say as I do.	4.Width of the grain
	5.	5.Contrast intensity of the grain
	6.	6. Fineness of the grain
	7.One has huge similarity to the other.	7.Other
Sample	Images of fourteen types of lumber	
Assessors	Male;5, Female;5, Age:18–25	

VISUAL SIMILARITY IN THE COLOR IMAGES

DIFFERENCE OF VIEWPOINT

Figure 3 shows the relationship between ratings of overall similarity and color similarity. Figure 4 shows the relationship between ratings of overall similarity and grain similarity. The overall similarity is more closely correlated with visual similarity than with grain similarity. Figure 5 shows the relationship between grain similarity and color similarity. There is a weak relationship between the two. We used multiregression analysis to study how well the ratings of overall similarity can be explained by a combination of color similarity and grain similarity ratings. The results are shown in Figure 6. The overall similarity could be expressed reasonably well as by a combination of judgments of grain similarity and color similarity.

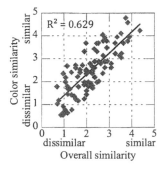

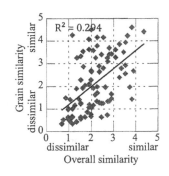

Figure 3 Relationship between overall similarity and color similarity

Figure 4 Relationship between overall similarity and grain similarity

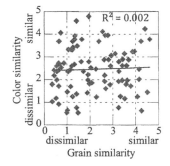

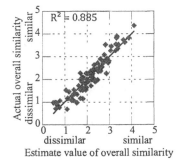

Figure 5 Relationship between color similarity and grain similarity

Figure 6 Relationship between estimate valuᵣ of overall similarity and actual overal similarity

QUANTIFICATION OF COLOR SIMILARITY

We averaged the lightness indexes (L*) and the chromatic indexes (a* and b*) in CIE L*a*b* color spaces for each color image and evaluated the average color difference from the average of L*, a* and b*. We corrected L*, a* and b* by determining L*, a* and b* from a standard color chart (GretagMacbeth 24 color chart) for which the real L*, a* and b* were known with colorimeter (KonicaMinolta CR-300). Figure 7 shows the relationship between the average color difference and the rated color similarity. We found that the pairs of images for which the average color difference is low tend to show strong color similarity. The average color difference does not capture all factors that affect ratings of color similarity, because it does not include information about bias or uniformity of color in the images. Thus, we used a multiregression analysis to study the effect of average color differences and the standard deviation of L*, a* and b* on the ratings of color similarity. The result shown in Figure 8 is not good. We can conclude that the judgment of grain similarity has some effect on the judgment of color similarity.

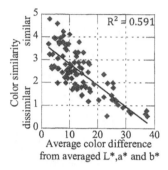

Figure 7 Relationship between average color difference and color similarity

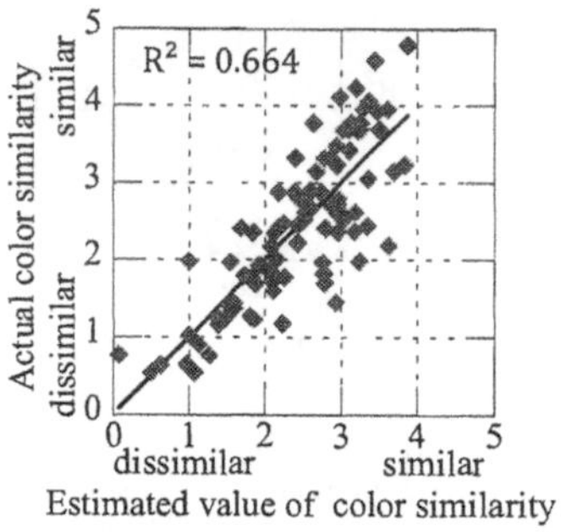

Figure 8 Relationship between estimated value of color similarity and actual color similarity

STUDY OF GRAIN SIMILARITY

STUDY OF DIFFERENT TYPES OF IMAGES

Figure 9 shows the relationship between the rated grain similarity in the color and gray-scale images, while Figure 10 shows the relationship between the grain similarity in the color and binary images. We found strong relationships between the rated grain similarity between the color and gray-scale images and also between the color and binary images. Thus, we concluded that the type of image had little effect on the ratings of grain similarity.

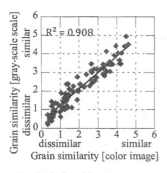
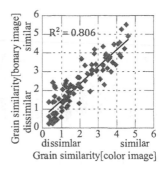

Figure 9 Relationship between grain similarity in color image and in gray-scale image

Figure 10 Relationship between grain similarity in color image and in binary image

FACTORS AFFECTING RATINGS OF GRAIN SIMILARITY

Figure 11 shows the visual similarity scale for binary images composed of pairs of binary images ranging from dissimilar to similar in grain pattern. We might suppose that the assessors judged the degree of grain similarity on the basis of the type of the grain, grain size, and regularity of the grain. Figure 12 shows the results of the question about which factors led to similarity and dissimilarity in the grains. Many assessors selected the type of the grain and the size of the grain as factors in the grain similarity or dissimilarity. This implies that the assessors consciously focused on the type and size of the grain and unconsciously focused on regularity of the grain, when judging the degree of similarity in the grains.

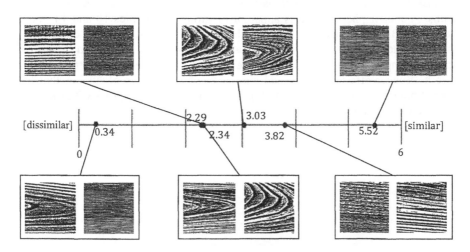

Figure 11 Visual similarity scale and examples of binary images

476

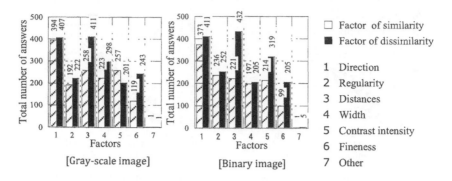

Figure 12 Result of the questions about which factors were similar and which were dissimilar in the grain

STUDY BY PRINCIPAL COORDINATE ANALYSIS

We performed a principal coordinate analysis to study the visual characteristics of each sample. Figure 13 shows the results of the principal coordinate analysis of grain similarity. We found that all samples could be classified into two groups, group A and group B. Figure 14 shows examples of samples of each group. Samples in group A have a straight grain, while samples in group B have a slash grain. Samples 7 and 12 both fit into group A, but the distance between samples 7 and 12 on 2nd principal coordinate is rather large. The results imply that the grain size and contrast intensity of the grain both have a huge effect on the visual similarity of two grains. A similar phenomenon is apparent in group B. Samples 10 and 14 both fit into group B, but the distance between them along the 2nd principal coordinate is not small. Indeed, the distance between samples 7 and 12 is larger than the distance between samples 10 and 14. This indicates that the grain size and contrast intensity have a stronger effect on the visual similarity of two pieces of lumber that have straight grain than on that with a slash grain. This conclusion agrees with the results described in the previous subsection.

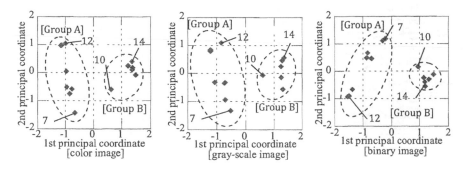

Figure 13 Results of principal coordinate analysis of grain similarity

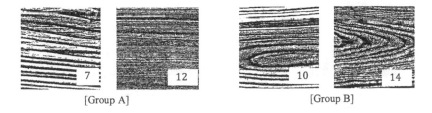

[Group A]   [Group B]

Figure 14 Examples of samples fit into Group A and B

PERCEIVED GRAIN SIMILARITY

We used only one type of image for our study of image features, because, as discussed above, the type of image had little effect on the perceived grain similarity. We chose binary images to study the image features, because they had the simplest grain among the three types of images. Figure 15 illustrates how to extract image features. First, we produced wave patterns from two types of sections on the basis of contrasting density of the grain (step1). One type of section is cut across the grain, and the other is cut on the grain. We divided the images into 16 parts for each of the 32 pixels and produced sectional waves for each part. Next, we created spectral waves from the sectional waves by spectral analysis (step2). We set the step size at 0.01 mm^{-1} and obtained 256 amplitude spectra by spectral analysis. It is hard to use all these amplitude spectra as image features because of the necessary amount of calculation. Therefore, we reduced the number of amplitude spectra. We divided the space frequency into eight parts and called these parts "space frequency area." The range of each space frequency area is shown in the table of Figure 15. We averaged the amplitude spectra for each space frequency area. Finally, we relativized all the amplitude spectra that we averaged (step3).

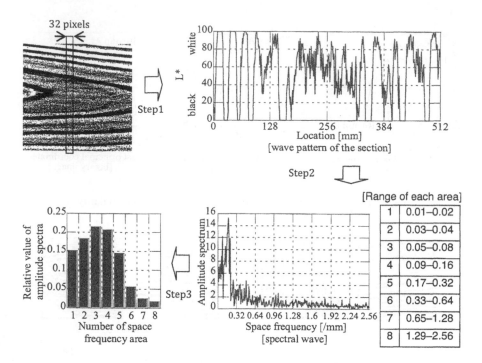

Figure 15 Method of extracting image features

Figure 16 shows examples of distinguishing samples and the results of the above process. The amplitude spectra of high space frequency areas, which are obtained from the along-the-cut sectional wave of fine grain lumber show large values. In addition, the amplitude spectra obtained from the across-the-cut sectional waves of slash grain lumber are different for each location of the section. Therefore, we used the average of the amplitude spectra and standard deviation of the amplitude spectra of across-the-cut sectional waves as the image features.

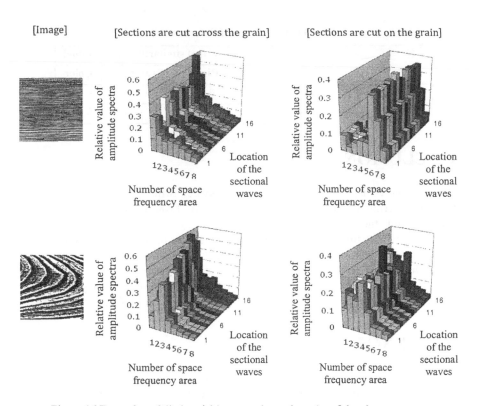

Figure 16 Examples of distinguishing samples and results of the above process

We estimated the similarity of the grains via these image features. With a trial and error process, we chose some image features from the image features evaluated in the previous section. Figure 17 shows the relationships between the estimated similarity and actual similarity. We found that the image features we chose captured the visual similarity partially. Table 3 shows the coefficients of determination. We found that the image features we obtained from binary images performed better than those obtained from the gray-scale images.

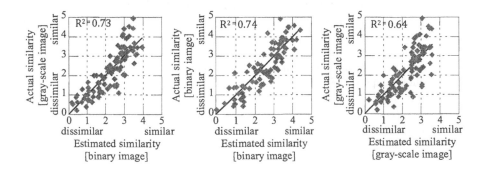

Figure 17 Relationships between estimated similarity and actual similarity

Table 3 Coefficients of determination

		Visual similarity [actual value]		
		Binary	Gray-scale	Color
Visual similarity	Binary image	0.742	0.733	0.705
[estimated value]	Gray-scale image	-	0.640	0.616

CONCLUSION

We studied the visual judgment of similarity between different types of lumber and the factors determining such judgments of similarity. From the results of this study, we can conclude the following:

1) The overall similarity between the color images of different types of lumber is mainly a function of the grain similarity and color similarity.
2) The type of grain has a huge effect on the perceived similarity.
3) People judge the similarity between different types of straight grain lumber more easily than they judge similarity for slash grain lumber.

REFERENCES

1) Ono, H., and Takagi, S. (2000), "Consideration to the factors that influence the imitativeness of the imitation from the visual point: Study of "real" & "imitation" on building finishing materials (Report 1)", *J. Struct. Constr. Eng.*, AIJ, No.533, 21–27.
2) Arikawa, S., Yasuda S., and Mihashi H. (1997), "A note on the concept of "Sincere Material": Part1. Authenticity the perception of the surface of finishing materials", *Summaries of technical papers of annual meeting architectural institute of Japan. Materials and construction*, Chiba, Japan, 511–512.
3) Yasuda S., Arikawa S., and Mihashi H. (1997), "A note on the concept of "Sincere Material": Part2. The senses of sight and touch the perception of the grain", *Summaries of technical papers of annual meeting architectural institute of Japan. Materials and construction*, Chiba, Japan, 513–514.
4) Arikawa S., Yasuda S., and Mihashi H. (1998), "A note on the concept of "Sincere Material": Part3. The Change of Perception of Surface Properties of Materials under Variable Conditions", *Summaries of technical papers of annual meeting architectural institute of Japan. Materials and construction*, Fukuoka, Japan, 175–176.

Chapter 49

Effect of Operator Experience and Automation Strategies on Autonomous Aerial Vehicle Tasks Performance

Christopher Reynolds, Dahai Liu, Shawn Doherty

Human Factors and Systems
Embry-Riddle Aeronautical University
Daytona Beach, FL 32114-3900, USA

ABSTRACT

The demands for Unmanned Vehicles are rapidly increasing in the civilian sector. However, operations will not be carried out in the National Air Space (NAS) until safety concerns are alleviated. Among these concerns is determining the appropriate level of automation in conjunction with a suitable pilot who exhibits the necessary knowledge, skills, and abilities to safely operate these systems. This research examined two levels of automation: Management by Consent (MBC), and Management by Exception (MBE). User experiences were also analyzed in conjunction with both levels of automation while operating an unmanned aircraft simulator. The user experiences encompass three individual groups: Pilots, Air traffic controllers and non-pilot-non-controller group. Results showed that there were no significant differences among the groups. Shortfalls and constraints were examined and discussed to help pave the wave for future research.

Keywords: Unmanned Aerial Vehicles, UAV, Management by Consent, Management by Exception, Automation, Pilot Skills

INTRODUCTION

Unmanned Aircraft Systems (UASs) have demonstrated their true potential through military endeavors, and their wide range of capabilities has inspired civilian agencies to harness the benefits that these systems provide. However, special attention must be made to safety concerns associated with separating the pilot from the aircraft. One of the issues is how the human controller is safely integrated into this highly automated and very complex system.

Of primary importance is the skill-set required on behalf of the pilot to safely and effectively fly the Unmanned Aircraft (UA) remotely. Due to the unique nature of UAV operations, two skill sets are thought very closely related, manned aircraft pilots and air traffic controllers (ATC). Certification requirements for pilots of unmanned aircraft have yet to be developed and little research has been done to evaluate the appropriate knowledge, skills, and abilities (KSAs) that a UA pilot should possess (Williams, 2005). It has been suggested that conventional pilots of manned aircraft are comprised with the fundamental KSAs necessary to develop an accurate mental representation of the UAs current status. However, research that assessed the applicability of pilot KSAs applied to UAS operations are rare and has arrived at conflicting conclusions (McCarley & Wickens, 2005; Tirre, 1998; Flach, 1998). Research that analyzes the transfer of non-pilot KSA's, such as those pertaining to air traffic controller (ATC) and skilled computer gamers, could not be found. It is important to note that UAS applications, scenarios, and designs vary significantly, thus the skill-sets required on behalf of the pilots may be just as diverse. Currently, there is no consistency in UAS pilot selection. The U.S. Air Force tends to select UAS pilot candidates that have received formal military flight training, but have recently graduated their first class of pilots trained specifically for UAS (Brinkerhoff, 2009). The Navy and Marine Corp select UAS pilots that already hold a private pilot license, while the Army selects UAS pilots who have never even flown a manned aircraft (McCarley & Wickens, 2005). Tirre (1998) noted that pilots transitioning from manned aircraft to UAS operations have faced boredom and difficulty maintaining situation awareness. It is also documented that transitioning pilots have difficulty switching flight environments due to the lack of vestibular and "seat-of-the-pants" sensory input obtained in manned flight. Weeks (2000) performed limited research in this area and concluded that there is a wide range of necessary qualifications that exist among UAS pilots, and more research is necessary to identify the KSAs of pilots that would best fit into UAS operations. Currently, the U.S. Air Force and U.S. Navy calls for pilots with manned aircraft training, but this often results in a large amount of negative transfer effects when training them in a UAS environment (Pedersen et al., 2006). The Human Systems Wing in the U.S. Air Force strongly recommends that, "Future work should focus on improving the empirical knowledge base on UAS human factors so evidence-based recommendations can be made when incorporating control migration in UAS design and operations (Tvaryanas, 2006)." The FAA Civil Aerospace Medical Institute furthers this notion, by acknowledging that much research is needed to assess the KSAs for future UAS pilots (Williams, 2005).

On the other hand, it is important in the design of UAS that automation strategies be integrated in a way that allows for the pilot to remain actively involved and aware of the functions taking place within the system. The high-performance nature of the system requires an extensive amount of autonomy in order to operate, but a fully-autonomous system would leave out important human oversight and is deemed unsafe. Therefore, an appropriate level of automation is critical to the safety and performance characteristics of UAS design. It is well accepted that automation offers several advantages over a human operator, but human operation also has advantages over the automation (Endsley, 1996; Billings, 1997; Sheridan, 2002). Particularly in UAS, automation has a significant effect on mission performance, workload and situational awareness (Wiener, 1988; Dixon et al., 2003). Of all the automation levels studied, the most historically used levels in the aviation arena is management by consent (MBC) and management by exception (MBE). MBC is a management style that incorporates lower levels of automation. This management style allows the UAS to perform functions only when given permission by the operator; while MBE is the management style that incorporates higher levels of automation. According to Billings (1996) management by exception is "a management-control situation in which the automation possesses the capability to perform all actions required for mission completion and performs them unless the manager takes exception". Effect of MBC and MBE on pilot performance has been studied for both manned aircraft (Olson & Sarter, 1998; Sarter et al., 1997) and UAS (Ruff, et al, 2002; 2004). Much research is needed to determine which levels of automation are optimal for UAS operations (McCarley & Wickens, 2005), especially its effect on different skill sets which are not very well understood.

The objective of this study is to analyze the pilot's role (in terms of their skill sets) in the UA system, and investigate how the system automation strategies best accommodates this role.

METHOD

Twenty-four participants from Embry-Riddle Aeronautical University were selected to participate in the study. All students were upper-classmen with an average age of 22 years. 16 students were male and 8 were female. The targeted population groups were inclusive of eight flight students, eight ATC students, and eight additional Human Factors students to represent a baseline. Each participant was compensated $15 for their time.

The apparatus consisted of a UAS simulation software called MIIIRO (Multi-modal Immersive Intelligent Interface for Remote Operations) (Nelson et al. 2004; Tso et al. 2003). MIIIRO was developed to support Human Factors research for long-range, high-endurance UASs. The hardware component was comprised of a standard PC with a dual monitor setup. The primary monitor portrayed the Tactical Situation Display (TSD) which encompassed the topographical image of the UA environment, the UA(s), a color-coded assignment of UA routes, critical targets, other intruding aircraft (IA), and the Mission Mode Indicators (MMI). The secondary monitor portrayed the Image Management Display (IMD) which consisted of an image cue and an image display necessary for completing mission assignments and maintaining SA (Ruff et al. 2004). Figure 1 illustrates the MIIIRO interface.

Figure 1. MIIIRO Testbed Display. Left: Tactical Situation Display (TSD); Right: Image Management Display (IMD)

A 2x3 mixed fully-factorial design was used in the study. The within subject independent variable consisted of the two levels of automation: MBC and MBE. The second between subject independent variable consisted of the pilot experience groups: Pilot, Air Traffic Control, and the control group. Dependent measures included workload, situation awareness, accuracy, and response times for both primary and secondary tasks. Workload and SA were subjective measures filled out by the participants, whereas accuracy and response times were objective measures collected by the MIIIRO software.

The MBC and MBE flight mission scenarios were set up similar to a highly-automated UAS, that is, predetermined waypoints made up the flight path in which the UA autonomously followed. Along the flight path, 15 image capture location are also preset and the associated images were automatically displayed to the participant, once the UA approached the preset waypoint. The primary task of the participant was to view the images collected by the UA and verify that the Automatic Target Recognizer (ATR) had selected the correct target(s) present in the image. Each image collected along the flight route contained at least one ground vehicle, but a threat was not always present. The threats and non-threats were depicted as ground vehicles and were visually discernable by color, but were not always selected correctly by the ATR. The ATR attempted to distinguish between the two or more vehicles by placing a red box around the threat(s). The reliability of the ATR is set at 80%, so the participant had to verify that threats were correctly selected. In cases where the ATR had incorrectly dissociated threats from non-threats, the participant needed to manually select and/or deselect the images by directly clicking on the targets. During MBC scenarios, the participant processed the image manually by accepting or rejecting each image in the image cue. During MBE scenarios, the computer automatically processed the images after a 15 second duration, unless the participant overrode the automation by selecting an image hold button prior to it timing out. Primary task performance data was collected automatically by the MIIIRO software. The primary dataset was inclusive of: image response time, image queue time, image processing time, target selection accuracy, manual accepts/rejection, automatic accepts/rejections and image hold times.

There are two secondary tasks associated with the experiment. The first task encompassed Intruder Aircraft (IA) events that mimicked an unexpected aircraft entering within the UAs airspace. This random event occurred twice per trial, and was deemed a highly critical situation that necessitated a quick and attentive response. The event was depicted by a red aircraft-shaped icon instantly appearing on the TSD at random times. To alleviate the threat, the participant needed to click on the aircraft and enter a pre-determined code. The second task encompassed a MMI that mimicked an indicator representing the status or health of the UA system. This indicator was constantly displayed on the TSD and looked similar to a horizontal traffic light. It was made up of three round lights that changed from green to yellow or red, depending on the UAs status. A green status indicated that the UA was in good health. The light would randomly change to yellow or red, indicating that attention was needed from the simulated pilot to correct the situation. To correct the situation, the participant was required to click on the light panel and correctly type in a text string of numbers shown in a pop-up window. Once the text string was entered correctly, the status indicator returned back to green, indicating a healthy status. Secondary task performance data was collected automatically by the MIIIRO software. The secondary dataset is inclusive of: MMI event occurrences, MMI response times, IA occurrences, and IA detection response times.

A NASA-TLX rating scale was used in the study to measure workload experienced by the participants (Hart, S.G., & Staveland, L.E., 1988). The NASA-TLX provided an overall workload score based on a weighted average and rating of six subscales: Mental Demands, Physical Demands, Temporal Demands, Performance, Effort, and Frustration. The participant first responded to a series of pair-wise comparisons to determine the *ranking* order in which each subscale topic contributed to overall workload during the task. These subscales were then weighted in order of its rank, with the top ranking subscale given the most weight. The participant then *rated* each workload subscale individually as it pertained to the mission scenario.

RESULTS

Results were divided into four main areas of interest: accuracy, task processing time, workload, and situation awareness. The data was analyzed using several repeated measure factorial designs to assess the effects of level of automation on user experience resulting in each of the following independent variables: image accuracy, image processing time, MMI processing time, IA processing time, and workload assessment. A significant level of 5% was applied to determine the significance.

ACCURACY

For the primary task, image accuracy was calculated by using the number of correctly processed images divided by the total number of images for each simulated mission. It was found that the mean image accuracy score for MBC was not significantly different from the mean image accuracy score for MBE with $F(1,21)=.015$ and $p>.05$. For the effect of pilot experience on image accuracy, it was found that there are no significant differences among the mean image accuracy scores for the Pilot with $F(1,21)=2.608$ and $p>.05$, which implies differences in image accuracy scores among Pilot, ATC, and control groups are non-significant. The interaction between level of automation (LOA) and pilot experiences were not found significant, with $p>.05$.

PROCESSING TIME

Task processing times are separated into three individual times pertaining to three different tasks. The primary task processing time represents the total time it took the simulated pilot to recognize and process the ground-based images displayed in the primary task. The MMI processing time represents the total time it took the simulated pilot to identify and accurately respond to the multiple mission mode indicator events. The IA processing times indicate the total time it took the simulated pilot to identify an intruder aircraft and resolve the conflict using the Intruder Friend or Foe (IFF) code.

For the primary task processing time, ANOVA did not find any significant differences between the LOA and pilot experiences ($p>.05$), nor did we find any effect between the interaction of these two variables. For the secondary MMI task processing time, it was discovered that the main effects of LOA and pilot experiences as well as the interaction between these two have significant impact on pilot MMI processing time under the 5% significance level. Similar effects were also found for the secondary IA processing time, that is, none of the effects, including the main effects and interaction effects were found significant.

SUBJECTIVE WORKLOAD

Workload was measured subjectively using the NASA-TLX workload rating scale. From the results, it was found that the mean workload scores for MBC (M=37.833,

SD=16.88) was not significantly different from the mean workload scores for MBE (M=38.583, SD=20.70) with $F(1,21)=.047$, $p>.05$. For the experience it was found that there are no significant differences among the mean workload scores for the pilot experiences (M=26.938, SD=19.84), ATC (M=45.875, SD=17.10), and control (M=41.812, SD=11.54) groups with $F(2,21)=3.384$, $p>.05$. For the interaction between the LOA and experience levels on workload scores, it was found the effect was non-significant as well with $F(2,21)=2.512$, and $p>.05$.

DISCUSSIONS

For the primary tasks, we did not found any significant differences, regardless of the level of automation, or the experience level of the participant. Several factors may contribute to a non-significant difference among prior experience and levels of automation. First, the majority of the images were relatively easily deciphered at first glance, while a select few were rather obscure in detail. In other words, it was easy to distinguish the threats from the non-threats in the vast majority of the images. Yet a few of the images left very little evidence to distinguish between the targets, no matter how long the image was observed. For the non-obvious images, it was more of a guessing game, rather than a need to further analyze the image. Therefore, a quick decision and response could be made at first glance. Some participants revealed that even in cases where there was minimal doubt, they would not risk targeting a 'friendly', or non-threat. This type of rationale was never anticipated by the author of this research, but possibly played a significant role in the outcome of the accuracy scores.

The primary task processing time pertains to the average amount of time it took for a participant to respond and fully process an image, by discerning 'threat' vehicles from 'non-threat' vehicles. The results of these image processing times did not indicate any significant differences, regardless of the level of automation, or the experience level of the participant. A further look at the results indicated that participants processed images at an average rate of 3533milliseconds for MBC and 3689milliseconds for MBE. A reasonable explanation for the faster processing times may be that there was a monetary incentive for the top performer in speed and accuracy of the primary task. Additionally, the MBE option was very rarely used among participants. In fact, the most it was ever used by any single participant was once. Ruff et. al. (2004) pointed out that participant's typically responded to images rather than allowing automation to process them. This finding is also supported by Olson & Sarter (1998), specifically among experienced pilots conducting flight tasks under MBC/MBE strategies. This study found that all three levels of experience (pilot, ATC, and control group) chose to process the images on their own, rather than rely on MBE strategies.

There were also two secondary tasks, MMI and IA, which essentially competed for the same mental resources as the primary task. The MMI times reflect the amount of time it took a participant to become aware of an abnormal MMI indication of yellow or red (indicating a need for a response), and respond to it by clicking directly on the indicator and typing in a string of numbers displayed in the resulting pop-up box. The IA times reflect the amount of time it took a participant to become aware of IA, and respond to it by clicking directly on the IA and typing in the preset code in the resulting pop-up box. The non-significant difference found on these secondary tasks may be attributed to the fast responses that were motivated by monetary incentive for the top performer in the primary task (despite this not being part of the primary task). It was noted that response times are indicative of adequate SA, alertness, scanning abilities, and responses times, yet two out of three of the quickest responders were among the control group, having no flight or ATC experience.

The NASA-TLX workload assessment did not find any significant differences, regardless of the level of automation or the experience level of the participant. A further look at the workload scores found that the level of significance between the two levels of automation was .831 with a power level of .055. Whereas, the significance level among the three levels of experience was .052 with a power level of .573. The mean workload score of the pilot group was 27, in comparison to the mean workload scores of ATC and control groups of 46 and 42, respectively. This may suggest that there were very little differences between the MBC and MBE mission scenarios. Since both levels of automation were within-subject treatment, rating of the two largely relied on participants ability to recall and distinguish between the two scenarios. Furthermore, the workload score of pilot group are lower than the ATC and control groups, despite performing the same mission scenarios. It is a possibility that a pilots perception of workload under the tested conditions are in direct comparison to actually flying an aircraft, whereas no other tested group can make this comparison.

CONCLUSIONS

UASs are on the verge of taking flight alongside manned counterparts. However, regulatory constraints will not permit UAS operation in the NAS until technological constraints and human factors concerns have been overcome. Removing the human component from the flying platform poses several advantages, but does not come without an abundance of risk. This study has initiated a much needed area of research pertaining to the user-interface design, as well as understanding the capabilities and KSAs required on behalf of the pilot. Unfortunately, no significant differences were determined among the experience levels of the simulated pilot, nor the level of automation that was implemented into the system design. The possibility of replicating realistic, real-world UAS operations in a laboratory setting should be enough motivation to further this type of research in a simulated environment, rather than allowing shortfalls to be discovered at the expense of human life. The results discovered in this study revealed that humans, regardless of prior training in aviation realms, can perform substantially well under foreseen and expected circumstances. However, pilots are expected to remain in-the-loop of UAS operations for reasons that automation cannot mediate: the unforeseen, unexpected, and·unintended situations. It is for this reason that future research should be carried out in these areas to determine the best approach at aligning adequate UAS pilots with an appropriately level of automation in an effort to promote coordinated team-play.

REFERENCES

Billings, C.E. (1996). *Human-centered aviation automation: Principles and guidelines* (NASA Technical Memorandum 110381). Moffet Field, CA: NASA Ames Research Center.

Billings, C. E. (1997). *Aviation automation: The search for a human centered approach.* Mahwah, NJ: Lawrence Erlbaum Associates, Inc.

Brinkerhoff N. (2009). Air force graduates first class of drone pilots. *AllGov.* Retrieved August30, 2009, from http://www.allgov.com/ViewNews/Air_ Force_Graduates_First_Class_of_Drone_Pilots_90612

Dixon, S.R., Wickens, C.D., and Chang, D. (2003). Comparing quantitative model predictions to experimental data in multiple-UAV flight control. *Proceedings of the 47th human factors and ergonomics society*, pp.104-108.

Endsley, M.R. (1996). "Automation and situation awareness", in: Automation and human performance: Theory and applications, R. Parasuraman and M. Mouloua (Ed.), pp. 163–181). Mahwah, NJ: Erlbaum.

Flach, J. (1998). Uninhabited combat aerial vehicles: Who's driving? *Proceedings of the human factors and ergonomics society 42nd annual meeting,* pp. 113-117. Chicago, IL.

Hart, S. G. and Staveland, L. E. (1988). Development of NASA-TLX (Task Load Index): Results of empirical and theoretical research, In: P. A. Hancock and N. Meshkati (Eds.), *Human Mental Workload,* pp. 139-183. North-Holland: Elsevier Science.

McCarley, J. S. and Wickens, C. D. (2005). *Human Factors Implications of UAVs in the National Airspace*, Technical Report AHFD-05-05/FAA-05-1. Savoy, IL: University of Illinois, Aviation Human Factors Division.

Nelson, J.T., Lefebvre, A.T., and Andre, S.T. (2004). Managing multiple uninhibited aerial vehicles: Changes in number of vehicles and types of target symbology. *Proceeding of the 26th Interservice/Industry Training, Simulation, and Education Conference.* Orlando, FL.

Pederson H.K., Cooke N.J., Pringle H.L., and Connor O. (2006). UAV Human Factors: Operator Perspectives, In P. H. Cooke (Ed.) *Human Factors of Remotely Operated Vehicles*, pp. 21-33. New York: Elsevier.

Olson, W.A., and Sarter, N.B. (1998). As long as I'm in control: Pilot preferences for and experiences with different approaches to automation management. *Proceedings of the 4th Symposium on Human Interaction with Complex Systems,* pp. 63-72. Los Alamitos, CA.

Ruff, H.A., Calhoun, C.L., Draper, M.H., Fontejon, J.V., and Guilfoos, B.J. (2004). Exploring automation issues in supervisory control of multiple UAVs,

Proceedings of the Second Human Performance, Situation Awareness, and Automation Conference (HPSAA II). Daytona Beach, FL, March 22-25.

Ruff, H.A., Narayanan, S., and Draper, M.H. (2002). Human interaction with levels of automation and decision-aid fidelity in the supervisory control of multiple simulated unmanned air vehicles. *Presence, 11*, 335-351.

Sarter, N.D., Woods, D., and Billings, C. (1997). Automation surprises, In G. Salvendy (Ed.), *Handbook of human factors and ergonomics*, pp. 1926-1943. New York: Wiley.

Sheridan, T.B. (2002). *Humans and automation: System design and research issues*. Hoboken, NJ: Wiley.

Tirre, W. (1998). Crew selection for uninhabited air vehicles: preliminary investigation of the air vehicle operator (AVO). *Proceedings of the human factors and ergonomics society 42nd annual meeting*, pp.118-122.

Tso, K.S., Tharp, J.K., Tai, A.T., Draper, M.H., Calhoun, G.L., and Ruff, H.A (2003). A human factors testbed for command and control of unmanned air vehicles. *Proceedings of the Digital Avionics Systems 22nd Conference*. Indianapolis, IN.

Tvaryanas, A.P. (2006). *Human Factors Considerations in Migration of Unmanned Aircraft System (UAS) Operator Control*. Technical Report HSW-PE-BR-TR-206-0002, United States Air Force 311th Human Systems Wing. Brooks City-Base, TX.

Weeks, J.L. (2000). *Unmanned aerial vehicle operator qualifications*. United States Air Force Research Laboratory, Warfighter Training Research Division, AFRL-HE-AZ-TR-2000-0002.

Wiener, E.L. (1988). Cockpit Automation, In E.L. Wiener and D.C. Nagel (Ed.), *Human Factors in Aviation* pp. 433-450. San Diego, CA: Academic Press.

Williams, K.W. (2005). Unmanned Aircraft Pilot Medical and Certification Requirements, In: Krebs, W.K. (ed.), *Unmanned Aerial Vehicles Human Factors*, FAA-AAR-100, FY-05. Washington, D.C.

Chapter 50

Effects of Modes of Cockpit Automation on Pilot Performance and Workload in a Next Generation Flight Concept of Operation

Guk-Ho Gil[1], Karl Kaufmann[1], Sang-Hwan Kim[2], David Kaber[1]

[1]Department of Industrial & Systems Engineering
North Carolina State University
Raleigh, NC, 27695-7906

[2]Department of Industrial and Manufacturing Systems Engineering
University of Michigan-Dearborn
Dearborn, MI 48128-1491

ABSTRACT

Objective: The objective of this study was to compare the effects of various forms of advanced cockpit automation for flight planning on pilot performance and workload under a futuristic concept of operation. **Method**: An enhanced flight simulator was developed to present a Boeing (B) 767-300ER cockpit with interfaces and functions of both existing and futuristic forms of automation, including a control-display unit (CDU) to the aircraft flight management system, an enhanced CDU (CDU+), and a continuous descent approach (CDA) tool. A lab experiment was conducted in which pilots flew tailored arrivals (TAs) to an airport using each mode of automation (MOA). The arrival scenario required replanning to avoid convective activity and was constrained by a minimum fuel requirement at the initial approach fix (IAF). The three different MOAs ranged from minimal assistance to a fully automated system. Flight task workload was also manipulated in the test trials by adjusting the starting position of the aircraft relative to the airport, depending upon the MOA, in order to allow more or less time for the route replanning task. Responses measures included time-to-task completion (TTC) and success/failure in completing the replanning task before the first waypoint on the arrival. The Modified Cooper-Harper (MCH) cognitive workload rating scale and pilot heart-rate (HR) were used as measures of workload in the test trials. **Results**: The use of low-level automation (CDU mode) led to significantly higher pilot workload (HR and MCH) and longer TTC compared with the CDU+ and CDA modes. Contrary to our expectation, the use of high-level automation (the CDA tool) led to the lowest success rate (correct route selections by the first point on the arrival); whereas, the use of intermediate-level automation (CDU+) led to the highest success rates. This was possibly due to the definition of the success criterion under the CDA mode. The CDU and CDU+ modes of automation revealed learning effects during the earliest test trials, in terms of TTC. In general, results indicated that the MOAs influenced pilot performance and workload responses according to hypotheses. **Conclusion**: This study provides new knowledge on the relationship of cockpit automation and interface features with pilot performance and workload in a novel next generation-style flight concept of operation.

Keywords: level of automation, modes of automation, cockpit automation, flight simulation, NextGen, cognitive workload

INTRODUCTION

In a recent report on fatalities in aviation accidents, loss of control in flight was found to explain 65% of all accidents (Boeing, 2008). When a pilot must replan a route due to either non-nominal (e.g., runway closure or change, weather disturbance, etc.) or emergency (e.g., cargo fire, medical emergency, etc.) situations (Kalambi et al., 2007), there are required functions, including systems monitoring, generating decision alternatives, selecting and implementing options. The functions actually performed by the pilot depend on the level of aircraft automation. Levels of automation in human-in-the-loop systems define function allocation schemes for the human operator and/or machine (Endsley & Kaber, 1999). Pilots using low level cockpit automation may need to perform selection or implementation functions as compared to using high level automation, where they may only need to perform monitoring or generating functions. In general, lower level automation may increase pilot workload but high-level automation may result in a loss of pilot awareness of the states of the aircraft or flight environment, particularly during low workload phases of flight. In a reroute scenario, the latter human performance consequence may be critical. The route is typically extended in this process and pilots must effectively monitor fuel consumption to verify sufficient levels before landing. A loss of situation awareness can also increase pilot workload with the occurrence of non-nominal or emergency conditions and lead to serious performance problems or flight control errors due to a need for pilots to effectively reorient to the state of the situation and then begin to process the current conditions (Sarter & Woods, 1995). Previous research (Wright et al., 2003; Gil et al., 2009) has investigated the effect of various forms of cockpit automation on pilot performance and workload in replanning scenarios. This study investigates these issues in the context of an automated next-generation (NextGen)-style tailored arrival (TA) following a continuous descent approach (CDA). (NextGen is the FAA and NASA acronym to refer to technologies and procedure to be developed as part of the Next Generation Air Transportation System.)

Automated tools have been developed for CDA, providing pilots with the capability to define arrivals to an airport under low power settings and without the typical requirement of altitude "step-downs" (Coppenbarger et al., 2007). This type of arrival is referred to as a Tailored Arrival (TA). It is "tailored" in the sense that the arrival is constructed to take into account ATC constraints, such as traffic patterns, airspace restrictions and minimum altitudes, as well as the performance of the specific aircraft type. TA routes are pre-determined and coordinated across ATC facilities so that a clearance for an arrival can be given well in advance of the point at which an aircraft begins its descent. Furthermore, TA clearances are generally given by data link (electronic commutation between a control center and the aircraft) to reduce flight crew and ATC workload. In general, the TA scenario requires a high-level of aircraft and ATC automation for effective performance.

The objective of this study was to compare the effects of various types and levels of cockpit automation for flight planning on pilot workload, time-to-task completion, and decision making in a TA following a CDA.

METHODS

FLIGHT SIMULATOR SETUP

An enhanced flight simulator (see Figure 1), based on the X-Plane simulation software, was setup to present a B767-300ER cockpit with interfaces and functions of existing and futuristic forms of cockpit automation, including a control-display unit (CDU) for the aircraft flight management system (FMS), an enhanced CDU (or CDU+), and a continuous descent approach (CDA) tool. Each type of automation was integrated with an ATC datalink system for electronic relay of clearance s between the simulated cockpit and control center. The simulator also presented pilots with an out-of-cockpit view of the world. A touch screen monitor was used to present the CDA planning tool (see right side of Figure 1). Pilots used the screen to directly manipulate graphical buttons and select waypoints. The display content of the monitors was synchronized using a TCP/IP network supported by the X-Plane software. The simulator was used to present dynamic flight situations.

SCENARIO DEVELOPMENT

Reno-Tahoe International Airport (KRNO) was selected as the site for the flight scenario because it is located in a valley with terrain rising to approximately 6,000 feet above the airport. This makes maintaining the published flight path more critical than at other airports and increases the utility of automation for pilots.

The scenario began with the simulated aircraft flying westbound toward the Wilson Creek VOR (ILC). At the beginning of each trial, the pilot received a datalink clearance for the TA from ATC, with the route appearing automatically on the ATC uplink display. The pilot then reviewed the route focusing on whether it was flyable under the constraints of weather and fuel consumption limits. Once an acceptable clearance was received, the pilot would accept the clearance via datalink using a button on the planning tool touchscreen. The TA route was then loaded into the FMS, either by data connection between the DL system and FMS (for the CDA tool or CDU+ mode) or manually (for CDU mode).

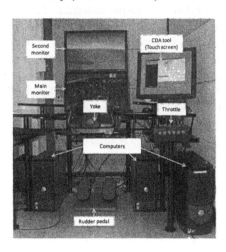

FIGURE 1. Flight simulator setup.

INDEPENDENT VARIABLES

Modes of Automation: The three MOAs used in the experiment were CDA, CDU+ and CDU. The CDU mode provided minimal automation to pilots for replanning. The interface displayed all possible waypoints from which a pilot could construct a arrival route and (s)he selected each waypoint individually. The mode also required pilots to manually perform fuel-burn calculations and determine if the minimum fuel requirement would be met. The intermediate level of automation, CDU+, displayed a set of alternative routes for pilots to select from, some of which failed to avoid the weather or meet the minimum fuel criteria. The CDA mode was a fully automated system that generated a single acceptable route meeting both flight constraints. This mode required less use of pilot working memory in making a reroute decision.

Workload: Flight Workload was manipulated in terms of the time available for route choice or the distance from the starting position of the aircraft to the decision point. Pilots were required to complete the reroute before ILC, which the first fix on the arrival. For the CDU MOA, the low and high workload conditions provided 80 and 60nm of travel, respectively. For the CDU+ MOA, the low and high workload conditions provided 27 and 20nm of travel, respectively. For the CDA MOA, the low and high workload conditions provided 13 and 10nm of travel, respectively.

Sequential Blocks of Trials: There were three sequential blocks (SET1, SET2 and SET3) of trials to allow for assessment of any pilot learning effect. Each block included all three modes of automations. However the order of appearance of the modes differed from block-to-block (e.g., SET1: CDU → CDU+ → CDA, SET2: CDA → CDU+ → CDU).

PHASES OF FLIGHTS, DEPENDENT VARIABLES AND HYPOTHESES

Each trial had two phases, a decision phase involving planning tool use and an implementation phase involving FMS programming. The decision phase ran from the start of the trial until the pilot accepted an arrival clearance. The implementation phase ran from the start of pilot use of the CDU/FMS they pressed the "EXEC" button to activate a route. Finally, total time to completion spanned from the beginning of a trial until the pilot pressed the "EXEC" button on the CDU/FMS.

Time-to-task Completion (TTC): Flight time was recorded by a "Data Manager" module developed for the X-Plane simulator in each trial phase. This yielded three data sets including total TTC, decision phase TTC and implementation phase TTC.

Modified Cooper-Harper (TTC) Workload Rating: In order to measure pilot subjective impressions of workload, the Modified Cooper Harper (MCH) rating scale was used. The "Cooper Harper manned aircraft assessment tool" was originally used to evaluate pilot physical workload associated with use of controls in aircraft (Harper & Cooper, 1986). In this experiment, a modified version of the Cooper Harper scale was used to evaluate how well the displays supported basic operator information processing and the general cognitive load pilots experienced in various segments of the flight (Cummings et al., 2006).

Heart-Rate (HR): In order to obtain an objective measure of pilot workload, cardiovascular activity, specifically pilot heart-rate (HR), was also recorded during each phase of the test trials. A Polar S810i heart-rate monitor system was used for this purpose. HR data was recorded every 2 seconds during the test trials. Pilot baseline HR was also collected to calculate the percent change in HR for test conditions as compared to the baseline using the following formula: ((TestHR-BaselineHR) / BaselineHR) * 100%

Trial Success/Failure: At the end of each trial, the experimenters checked whether the pilot completed replanning by the first waypoint on the arrival.

With respect to hypotheses on the response measures, the CDU trials were expected to produce the highest workload response in terms of MCH and HR

(Hypothesis (**H**)**1**). The CDU MOA required manual decisions (selecting individual waypoints) and action implementation (programming the FMS). With respect to TTC, the CDU MOA was expected to produce the worst performance (greatest time) since it provided minimal automation (**H2**). Lastly, the CDU MOA was expected to produce the worst performance in terms task success rate (**H3**) since it provided little assistance with information processing functions.

DESIGN OF EXPERIMENT, PARTICIPANTS AND PROCEDURE

The experiment design was mixed with nested and crossed factors, including: (1) data collection block; (2) trial set; (3) MOA; (4) order of presentation of MOAs within a trial set; and (4) workload condition. The block represented a complete crossing of all the settings of the main effects (MOA and workload) for four subjects. The design included three blocks in total. Each subject performed three sets of trials. The trial set variable represented the time period within a block. Each trial set included pilot exposure to all three MOAs, CDU, CDU+ and CDA. In an attempt to minimize condition carryover effects, there were two different orders of presentation of MOAs among the trial sets. The first order was from the lowest to the highest level of automation (CDU→CDU+→CDA), and the second was from the highest to the lowest level of automation (CDA→CDU+→CDU).

Thirteen pilots were recruited for the experiment from several sources in the Raleigh, NC area. An instrument rating and flight experience in aircraft equipped with an FMS were the minimum requirements for participation in the study. Since the simulation presented a B767-300ER cockpit with a PFD and NAV display, pilots with "glass" cockpit experience were recruited. Eight subjects had FAR Part 121 (scheduled airline) experience. Five subjects had FAR Part 135 (charter) or Part 91 (corporate) experience. The pilot's ages ranged from 25 to 72 years with an average age of 51. Total instrument time ranged from 175 hours to 20000 hours with an average of 3706 hours. One subject's data was excluded, as he did not follow the Modified Cooper-Harper rating instructions.

After a brief survey session, the pilots were familiarized with the simulator and the experimental setup. Pilots were trained on all three MOAs (CDA, CDU+, CDU) following a TA to Raleigh-Durham Airport (KRDU). Subsequently, they were instructed on how to complete the MCH rating form and they donned the HR monitor. During test trials, each subject followed the flight scenario under each of the three different MOAs in three sequential blocks. At the close of each trial, the pilots were asked to complete the MCH rating form. At the end of the experiment, pilots were instructed to relax without performing any tasks for approximately 10 minutes in order to measure their baseline HR. Finally, pilots were debriefed and received a $100 honorarium for participating.

RESULTS

TIME-TO–TASK COMPLETION (TTC)

Total TTC: Analyses of variance (ANOVA) results revealed a significant effect of MOA (F(2,64)=349.05, p<0.0001). Post-hoc tests using Tukey's tests indicated the CDU MOA produced the longest TTC, while the CDA MOA required the shortest TTC. There was also a significant effect of SET (F(2,64)=15.79, p<0.0001) or pilot learning. Post-hoc tests indicated the TTC for SET1 to be different from the TTCs for SET2 and SET3, with SET1 requiring greater time to complete. Figure 2 summarizes the TTC results.

Decision phase TTC: There was a significant effect of MOA (F(2,64)=44.50, p<0.0001). Tukey's tests indicated the CDU MOA required more decision time than the CDU+ and CDA MOAs. There was a significant effect of SET (F(2,64)=13.52, p<0.0001). Tukey's tests indicated there was pilot learning in decision making from SET1 to SET2 and SET3, which did not differ in decision time.

Implementation phase TTC: There was a highly significant effect of MOA (F(2,64)=1152.22, p<0.0001). Tukey's tests indicated that the CDU MOA required the longest time, while the CDA MOA took the shortest time to implement. There was a significant effect of SET (F(2,64)=9.34, p<0.0001). Tukey's tests indicated SET1 took longer to implement than SET2 and SET3.

Analysis of Covariance (ANCOVA): There was a significant effect of age (F(1,63)=16.04, p<0.0002) on pilot performance, which was consistent across MOAs. The results of the ANCOVA were consistent the ANOVA results.

Workload (WL): There was no significant effect of WL across all trial sets.

Learning Effect: With respect to the CDU MOA, there was a significant effect (F(2,33)=5.808, p=0.007) of trial SET. Tukey's tests indicated that SET1 took longer than SET2 and SET3, while there was no difference between SET2 and SET3. With respect to the CDU+ MOA, there was also a significant effect (F(2,33)=4.124, p=0.025) of SET. Tukey's tests indicated SET1 took longer than SET3, while there was no difference between SET1 and SET2 or SET2 and SET3. With respect to the CDA MOA, there was no significant effect of SET.

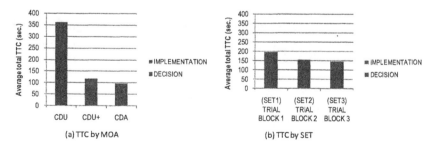

FIGURE 2. Average total TTC by MOA and SET.

HEART RATE (HR)

Total HR: ANOVA results revealed a significant effect of MOA (F(2,63)=3.21, p<0.0472) (see Figure 4 (a)). Post-hoc tests using Duncan's method indicated that the CDU MOA produced a higher workload response (increase in HR over baseline) than the CDA MOA. There was also a significant effect of SET (F(2,63)=3.18, p<0.0482). Duncan's tests indicated pilots experienced greater arousal during SET1 than SET2 and SET3 (see Figure 4 (a)). Pilots might have experienced some anxiety in adapting to experiment conditions.

Decision phase HR: There was a significant effect of MOA (F(2,63)=3.45, p<0.0377). Duncan's tests indicated that the CDU MOA produced a greater increase in HR compared to the CDA MOA. There was a significant effect of SET (F(2,63)=3.58, p<0.0336). Duncan's tests indicated that the decision phase HR for SET1 was different from that for SET2 and SET3.

Implementation phase HR: There was no significant effect of MOA or SET.

ANCOVA: There was a significant and consistent effect of age (F(1,63)=16.04, p<0.0002) on pilot performance across MOAs. The results of the ANCOVA were consistent the ANOVA results.

WL: There was no significant effect of the WL conditions across all trial sets.

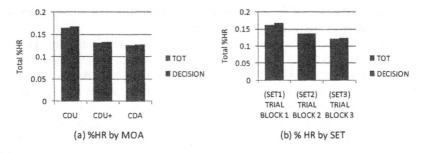

(a) %HR by MOA   (b) % HR by SET

FIGURE 4. Total % HR by MOA and SET.

MODIFIED COOPER-HARPER (MCH) RATINGS

Fisher's exact test revealed mean pilot perceived workload to significantly differ among all MOAs (i.e., CDU vs. CDU+ (p<0.001), CDU vs. CDA (p<0.001) and CDU+ vs. CDA (p=0.001)). Figure 5 presents the frequency of MCH ratings across MCH scores (1-9).

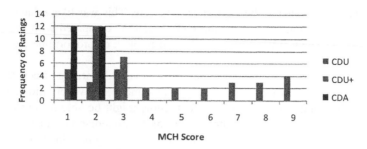

FIGURE 5. MCH frequency of ratings.

SUCCESS/FAILURE OF TRIALS

Based on a contingency table analysis, there was a significant effect of the WL manipulation on pilot success/failure (χ^2=7.368, p=0.007) under the CDA MOA condition. There was also a significant effect of the MOA on success/failure rates (χ^2=21.70, p<0.0001) under the high task workload condition (see Figure 6 (b)).This was not the case for the low workload condition. As shown in Figure 6 (a), the low workload condition produced higher success rates than high workload. As can be seen in Figure 6 (b), there was no difference in success rates between the CDU and CDU+ MOAs; however, both of these MOAs produced more successes than the CDA MOA.

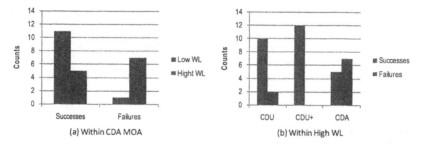

FIGURE 6. Number of successes/failures within MOA and among WL settings.

DISCUSSION AND CONCLUSIONS

With respect to the hypotheses on pilot workload and performance under the three different forms of cockpit automation, the use of low-level automation (the CDU) led to significantly higher workload (HR and MCH) and longer TTCs when compared with the CDU+ and CDA modes. These results supported our first and second hypothesis (**H1** and **H2**). Contrary to our third hypothesis (**H3**), the use of high-level automation (the CDA tool) led to the lowest success rate (correct route selections before the start of the arrival); whereas, the use of intermediate-level automation (CDU+) led to the highest success rate. This was likely due to an overly rigid success criterion defined for the CDA mode as compared with the CDU and CDU+ modes. The CDA mode trials allotted pilots a disproportionately short time to accomplish the decision task.

The CDU and CDU+ modes of automation revealed learning effects through the early test trials, in terms of TTC. The CDA mode, however, did not reveal a significant learning effect, suggesting that pilots might be able to quickly adapt to this futuristic form of cockpit automation for flight replanning tasks in NextGen operations. There was no significant effect of the starting position (flight workload) manipulation on pilot TTC or workload responses (HR and MCH) across all modes of automation. In general, results indicated that the modes of cockpit automation influenced pilot performance and workload responses according to the hypotheses.

This provides new knowledge on the relationship of cockpit automation and interface features with pilot performance and workload experiences in a novel NextGen-style flight concept of operation. It is expected that the results may be used as a basis for developing cockpit automation interface design guidelines and models of pilot cognitive behavior to predict performance and workload with other forms of futuristic automation as part of the aircraft systems design process.

ACKNOWLEDGEMENTS

This research was supported by NASA Ames Research Center under Grant No. NNH06ZNH001. Mike Feary was the technical monitor. A team from APTIMA Corporation, led by Paul Picciano, developed the CDA tool prototypes used for this experiment. The opinions and conclusions expressed in this paper are those of the authors and do not necessarily reflect the views of NASA. This research was completed while the third author, Sang-Hwan Kim, worked as research assistant at North Carolina State University.

REFERENCES

Boeing (2008), "Statistical summary of commercial jet airplane accidents." *http://www.boeing.com*.

Cummings, M.L., Myers, K., and Scott, S.D. (2006), "Modified Cooper Harper evaluation tool for unmanned vehicle displays." *Fourth Annual Conference of UVS Canada*, Montebello, Quebec.

Coppenbarger, R., Mead, R., and Sweet, D. (2007), "Field evaluation of the tailored arrivals concept for datalink-enabled continuous descent approach." *7th AIAA Conference (ATIO)*, Belfast, Northern Ireland.

Endsley, M.R., and Kaber, D.B. (1999), "Level of automation effects on performance, situation awareness and workload in a dynamic control task." *Ergonomics*, 42, 462-492.

Gil. G.H., Kaber. D.B., Kim. S.H., Kaufmann. K., Veil. T., and Picciano. P.

(2009), "Modeling pilot cognitive behavior for predicting performance and workload effects of cockpit automation", 15th *Int. Symp. on Avia. Psych.*, Dayton, Ohio.

Harper, R.P., and Cooper, G.E. (1986), "Handling qualities and pilot evaluation." *Journal of Guidance, Control, and Dynamics*, 9(5), 515-529.

Kalambi, V.V., Pritchett, A.R., Bruneau, D.P.K., Endsley, M.R., and Kaber D.B. (2007), "In-flight planning and intelligent pilot aids for emergencies and non-nominal flight conditions using automatically generated flight plans." *Proceedings of the Human Factors and Ergonomics Society 51st Annual Meeting, Baltimore*, Maryland.

Sarter, N.B., & Woods, D.D. (1995), "How in the world did we ever get into that mode? Mode error and awareness in supervisory control." *Human Factors*, 37(1), 5-19.

Wright, M.C., Kaber, D.R. and Endsley, M. R. (2004), "Performance and situation awareness effects of levels of automation in an advanced commercial aircraft flight simulation." *In Proceedings of the 12th Int. Symp. on Avia. Psych.* (pp. 1277 - 1282), Dayton, OH, Wright University.

CHAPTER 51

Potential Human Factors Issues in NextGen ATM Concepts

Lynne Martin, Christopher Cabrall, Paul Lee, Kimberly Jobe

San José State University Foundation
at the NASA Ames Research Center

ABSTRACT

As part of a recent project to identify and prioritize human performance issues (Lee, Sheridan, Poage, Martin, Cabrall & Jobe, 2009) related to NASA's Next Generation Air Transportation System ATM-Airspace Project (NextGen; SLDAST, 2007), a walkthrough of a future Separation Assurance (SA) concept was conducted. The walkthrough revealed that the Human Factors (HF) issues initially identified as problematic from a theoretical standpoint were not necessarily those that experienced Air Traffic Controllers (ATC), i.e., expert users, saw as stumbling points in the concept. These findings had particular implications for the example concept but more broadly emphasized that a range of human factors issues should be considered from the earliest points of future airspace concept designs.

The initial driver for a detailed study, albeit a snapshot, of a NextGen operator was to take a "bottom-up" approach to complement our primarily theoretical, "top-down", analysis of human factors issues across NextGen Research Focus Areas (RFA) in the NASA Airspace Project. One concept instantiation from the Separation Assurance RFA, namely ground-based Automated Separation Assurance (ASA) (Erzberger, 2001), was chosen for a walkthrough. ASA was selected because prototype automation for the concept had been built in the NASA Airspace Operations Laboratory (AOL, NASA Ames Research Center). Therefore, one possible instantiation was available for review, which assisted with scoping a more general ground-based SA concept for the purposes of the walkthrough.

This study was based on the Cognitive Walkthrough method developed by Polson, Lewis, Reiman and Wharton (1992), and aimed to explore the potential human performance issues that could arise for an operator in the future National Airspace System (NAS). The walkthrough was focused on one human operator, namely an Air Navigation Service Provider (ANSP), who would be the operator interacting most closely with the SA automation and who would be on-position in control of a future ATC sector.

IDENTIFYING HUMAN FACTORS ISSUES

Our interest was to explore the human factors issues related to future SA concepts. As the ASA concept is a prototype of a far-term NextGen concept, there have been no field reports from which we could determine HF issues. Our approach was to develop some hypotheses from experience of current systems and general HF expertise and to corroborate these with a group of expert operators.

First we conducted a review of the existing human factors literature related to aviation (which can be found in Lee, et al., 2009) and of future aviation concepts. To identify the issues specifically relevant to the NASA NextGen ASA concept, the concept description, its benefit mechanisms, the enabling technologies and the general technology assumptions were researched. From these sources, we brainstormed important HF themes based on our expertise and knowledge of the RFAs. We developed eight general human factors themes (Table 1) that we believe will play a critical role in future airspace concepts. It has been well-documented that these areas of human factors research apply in a variety of aviation and other human-systems contexts. In the NextGen concepts, many of these factors play a critical role, but it was judged that their importance will differ based on the individual concept.

GENERAL FUTURE SEPARATION ASSURANCE CONCEPT

For this analysis, a general ground-based separation assurance concept was modeled closely after the ASA concept (Erzberger, 2001), in which the ground-based automation assumes separation responsibilities and an Air Navigation Service Provider plays a supervisory role to the automation. A subset of future-ANSP responsibilities were the focus in our events, those where, in conjunction with short-term conflict resolution automation, the ANSP acted in a back-up safety position to the overall system for off-nominal situations that created unresolved short-term conflicts. In the walkthrough the ANSP was not presented with additional complexities, like user preferences or pop-up weather.

Table 1. Human Performance Themes and SA Subtopics

Future concept Human Performance Themes	SA HF Topics
Attention	Attention
	Monitoring
	Situation Awareness
Decision Making	Decision making
	Time Pressure
Workload	Workload
	Task Management
Communication	Communication
	Coordination
Memory	Memory
Organizational	Roles
	Responsibilities
Interaction with Automation	Automation interaction
	Trust
Qualification & certification	Training
	Procedures

Our separation assurance environment had two automation tools operating to maintain separation. The first tool probed from three to twelve minutes into the future for conflicting aircraft trajectories, identifying conflicts, and then solving them by sending a trajectory change to one of the aircraft. These avoidance maneuver solutions also included a trajectory to return the aircraft to its original route. The second tool acted as a redundancy to the first tool, only looked ahead in the simulation from zero to three minutes. It functioned by projecting the aircrafts' current heading into the future and resolving conflicts by vectoring an aircraft away from the conflict without including a trajectory path that would return this aircraft to its original route. In our activity we called these tools a Mid-term Separation Assurance Tool (MSAT) and a Short-term Separation Assurance tool (SSAT) respectively. Both of these tools operated by looking only at pairs of aircraft and assigning either a vertical or lateral solution to one of the aircraft via an uplink message.

Focusing on this Separation Assurance concept and using the human performance themes, we generated a number of issues that were anticipated to be salient to controllers (these are listed in the right column of Table x.1). From these issues, a set of working hypotheses were generated, five of which are listed below:

- ANSP workload could be high when traffic levels are high.
- As the ANSP is out of the loop (due to the conflict detection and resolution tools acting without human review), s/he may have trouble forming an understanding of a situation s/he is required to step into,

including requiring longer to formulate resolutions and take action.

• As the automation dictates the timeframe in which the ANSP sees a problem, s/he may object that his/her decision time is constrained.

• Now that ANSPs are part of a human-automation team, the division of responsibilities within their sectors is unclear.

• An ANSP will favor using voice for communication with aircraft over data link because voice is familiar.

These hypotheses refer to five of the eight human performance themes from Table 1 (left column), covering six of the HF SA topics we identified (right column). These are workload, situation awareness, decision making, time pressure, responsibilities, and communication. There were not expected to be any issues arising with the areas of memory, and the concept was not considered to be developed enough to be able to query certification and training.

Using the list of SA HF topics and our hypotheses as the framework, a question or two was generated for every step in our scenarios (see below for scenario development). Initially as many relevant questions as possible were constructed for each step of each event. In a second step, these were pared down to the most relevant or fitting questions in order to keep the size of the walkthrough manageable at the desired granularity. Efforts were then taken to balance the questions both within and across events so that each category was equally represented. Overall 136 human performance themed questions were generated across the 16 HF topics; 103 around our hypotheses and the remainder (33) covered the other topics (memory, training, monitoring and coordination).

WALKTHROUGH DEVELOPMENT AND EXECUTION

The scenarios that were the stimulus material for our question set came from a human-in-the-loop (HITL) study that the Airspace Operations Laboratory ran in 2008 (Prevot, Homola, Mercer, Mainini, & Cabrall, 2009). This simulation included a series of nominal and off-nominal events under NextGen forecast levels of traffic and automation. These traffic levels were representative of both two and three times the current day traffic – at about 30 and 45 aircraft per sector. Throughout this simulation, recordings were taken of the controllers' scopes and these were made available as aids to our walkthrough activity. The silent-screen recordings were reviewed by the research team for situations where something went awry for the controller (whether scripted or otherwise) and aircraft on the scope were detected and displayed as being in conflict for potential loss of separation minima (LoS). Such situations were judged critical for exploring an operator's point of view in a future separation assurance operational environment and ripe for human factors investigation.

An exhaustive list of such conflict prediction events was anticipated to be too cumbersome for fruitful discussion. Some conflicts were too complicated to conversationally tease apart or did not make sense for our purposes. For example, when three aircraft or more were in conflict at one time the automation rapidly "changed its mind" over how it identified the combination of conflicting pairs in the situation. Four events that could reasonably drive sensible dissection by our walkthrough participants were down-selected from all the available data.

Each of the four events was analyzed to identify its progression as a series of steps. Steps were defined as turning points in the events where an action or key decision was made by either the controller, the aircraft, or the automation. For example, when an aircraft flight-deck was sent a clearance, this was counted as a step. This example is shown in Figure x.1 below for Event 2, when the controller sent a message to an American aircraft to turn left. On average, the events were broken down into twelve steps, although Event 4 had an additional seven steps where a data link (DL) malfunction was reported. Because the audio channels from the screen recordings were not available due to data privacy policies, breaking down each event into such detailed steps proved an invaluable process for understanding what occurred. The absence of an audio track also afforded a desirable level of flexibility in reconstructing what we wanted to portray to our participants, thus un-constraining ourselves from minor variances in what might have actually happened in the previous study's simulation. The steps were diagrammed and compiled to form a "storyboard" for each event. For example, Event 2 involves a Southwest aircraft 1874 entering the sector on a north-westerly climb that will take it through the trajectory of American aircraft 140, which is in level flight traveling north-east. The conflict is first flagged by SSAT at one minute. The ANSP asks the American to turn left and, when this is possibly not enough for separation, asks the Southwest to also turn left. All these events were depicted as fourteen steps for the Event 2 storyboard.

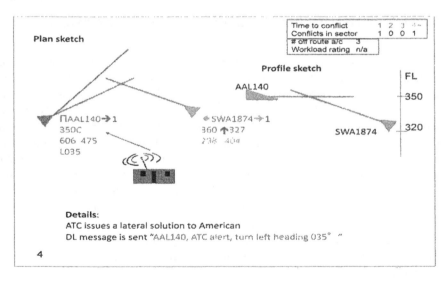

Figure 1: Storyboard of Step 4 from Event 2

The other three scenarios depicted similarly off-nominal events: an aircraft climbing through the trajectory of another or a level conflict. In each case, the SSAT flagged the conflict, usually at three minutes, but may or may not have sent a resolution. The aircraft receiving the resolution may or may not have responded for a variety of reasons, including the loss of data link capability in Event 4. In each case, the ANSP had to step in to prevent a loss of separation and to issue a resolution, sometimes with only seconds to make an assessment and take an action.

The action, or occurrence, in each event step was reviewed and assigned an HF theme then, as described above, the most appropriate HF question from the set written for that step was selected. Taking the example in Figure 1, showing Step 4 from Event 2, one possibly significant human factors topic was identified as time pressure and the question assigned at this point was "Is DL fast enough for this resolution?" Once all four events had storyboards prepared in this way, they were presented to our expert participants.

METHOD: PARTICIPANTS, APPARATUS AND PROCEDURE

Six recently retired controllers, all with more than 25 years of experience with the current air traffic control system, took part in our study as expert participants. In addition to their field experience, this group had a reasonable amount of familiarity with the initial operationalization of some of the NextGen concepts through their participation with tool prototypes in prior human-in-the-loop studies.

Digital screen capture files were originally recorded in the 2008 HITL study through the Camtasia Studio 6 screen recorder software. For our walkthrough, the videos were displayed to participants on a Barco 28" LCD monitor with a resolution of 2000 x 2000 pixels. Additionally, the event storyboard frames (see Figure x.1) were constructed in MS Powerpoint and printed as talking point references. Participants' answers were recorded on an Olympus digital voice recorder.

Participants received a short orientation, which explained that all the events took place in one high altitude en-route sector (above Indiana) in the Kansas City Center. This sector is a busy area, taking arrivals and departures from St Louis to the west and Louisville to the east. The automation notation on the display was briefly described but, as all participants had worked with these tools in the AOL before, it was assumed that they understood the icons and display elements. Beyond a list of tasks that they would be required to assume the responsibility of performing (e.g., put aircraft separated by SSAT back onto their original routes, deal with off-nominal or emergency circumstances), ANSP procedures, operating rules and jurisdiction were not formally defined for our participants.

Each of the four events was shown three times to each participant. First, the event was played in real time. Participants were free to ask questions or make comments as they watched. Second, the event was stepped through. The recording was forwarded to match each storyboard frame and paused at a point just as the action for that step had happened. Figure x.2 below shows an example of the way the problem area on the display for Event 2 looked at as it was paused at the step described in Figure 1 above. One member of the team briefly described the occurrence in the step and asked the associated HF question. In the third playback, a subset of the storyboard steps were selected and an additional organization question was asked at each one of these. The organizational theme questions were asked in a separate round due to their wider concern with aspects prior to and off the scope. The event was reviewed with fewer steps because asking an organizational question at every step a third time around proved too repetitive in practice runs of the walkthrough.

Participants' answers were digitally recorded and a second researcher took notes during the participant interviews. Sixty-seven questions were transcribed from the recording – approximately two thirds of the 103 hypothesis-focused questions. Recordings of the remaining 61 questions were used to augment the notes but were not transcribed. Qualitative analyses were planned.

ANALYSIS AND DISCUSSION

The questions and answers exchanged during our walkthrough are too numerous to list in this paper. However, to illustrate the method one question and the issues it addresses are discussed below. This question was asked in the example above (see Figures 1 and 2). The questions was "Is data link fast enough [to issue] this resolution?" and is one in the set of five questions constructed to ask participants about the pressures of time in decision making. For this question, participants replied that as time to put a resolution in place gets shorter, they favor using voice over data link to issue clearances. They thought voice was faster (in 2008), although, most of the participants conceded that data link may become faster in the future and, if it became as fast as voice is currently, it would be a better medium because it is "cleaner" (no static, no step-ons, no mis-hears).

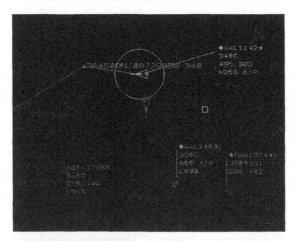

Figure 2: Screen snapshot of Step 4 from Event 2

Other questions in the set indicated that participants were able to estimate whether they had enough time to solve a conflict, when their ability to do that would become marginal and when they needed to act quickly for any solution to be put in place. "It's getting to the point of no return," commented P5 as he watched one of the events. Participants were also able to identify which types of solution maneuvers could still be executed in the remaining time and which were no longer possible. When challenged with a situation of heightened severity, most participants indicated that less than three minutes was too short a time horizon to reasonably be expected to take action if their workload was anything above minimal.

Using the process described above, participant responses were reviewed and catalogued to address as many of the SA and human factors issues identified as possible. They were also used to support or oppose our five hypotheses. Three of these five hypotheses were not supported by participants' answers to our questions.

Hypothesis 1 was refuted: ANSP *did not* think workload would be high when traffic levels are high. At least, this is not necessarily the case. Based on the assumption that the automation would be separating all equipped aircraft (flying using a 4-dimensional trajectory), participants judged that they could handle ten to twelve aircraft that needed controller assistance, e.g., a reroute, or two to three pair-wise conflicts at a time (with a three or more minute warning).

Although the ANSP was out of the loop due to the tools acting without human review, participants *did not* have trouble forming an understanding of the situation, refuting our second hypothesis. Even without specific sector training, participants were able to follow the traffic and often predicted future conflicts before the automation flagged them, suggesting they had an awareness of the presented situations. However, our participants were long-time controllers who had been trained in manual methods. When training is updated to encompass NextGen tools, incoming ANSPs may lack this kind of pattern recognition due to lack of practice, since NextGen tools may supplant controllers in performing routine separation tasks. Participants did not seem to require a longer time to take action. During their first viewing of an event, a participant would often comment on what their strategy would have been. From this, we observed that participants would have developed action plans in time to issue clearances had they been actually working the sector.

As explored above, hypothesis five was refuted: ANSP *did not* favor using voice for communication with aircraft. Under normal circumstances, participants preferred data link, as it is "cleaner". At the moment, they favor using voice over data link to issue critical clearances because it is faster.

Our remaining hypotheses were modified by the answers received. The prediction that as the automation dictates the timeframe in which the ANSP sees a problem, s/he may object that his/her decision time is constrained, was supported. If the automation flagged a conflict at or before three minutes, participants estimated that they would have enough time to solve a conflict. However, in three of the off-nominal events presented, the automation did not flag the conflict until there were two minutes or less to LoS. In these cases, participants felt that they had not been given enough time to create an elegant solution to the problem, and perhaps not enough time to solve the LoS at all.

The prediction that "now the ASA Manager is part of a human-automation team, the division of responsibilities within their sector is unclear" was supported. In contrast to today's responsibilities, which at their most basic level are very clear cut and straightforward (if an aircraft is in your sector it's your responsibility), it was not clear when aircraft in our SA examples were the ANSP's responsibility and when this fell to the automation. Participants, in the absence of other guidance, had to assume today's rules still applied. That these rules would be unworkable and unacceptable under a future concept became very clear when imminent LoS problems popped up in our examples. Also unclear was when (or if) responsibility shifted as an ANSP began to work on a problem. Our participants usually assumed that as soon as they took an action they were now responsible for separating the aircraft, but the display did not indicate that responsibility had been transferred.

DISCUSSION – BROADER HF INTERPRETATION OF RESULTS

Although some of our conclusions were specific to ground-based separation assurance concepts, many relate to future concepts in general. From this review, it would seem concerns that usually spring to mind when designing automation, such as workload, communication and situation awareness should be addressed by functional studies using future automation prototypes. However, a raft of issues that tie time constraints with responsibilities, procedures, and training need to be addressed if the prototype tools are to be successfully developed further into an operational realm. As well as the specific expert replies that qualified our hypotheses we learned three major human factors lessons from our study and from our wider project analyses:

1. Procedures are the gaps in future airspace concepts. Although not always true, NextGen concepts often lack necessary details in operational procedures – how the concepts will work in real world situations with people in-the-loop sharing responsibility with automation. The focus is often biased toward technology development and validation, instead of technology design and integration with humans as a human-system environment. The problem may lie in that concepts are designed by developers who are caught in a "catch 22" – i.e., it is difficult to generalize on human capabilities to perform tasks until and unless there is specification of what these tasks are, but if the hardware/software is "cast in concrete" without some simulation to check out the human operator interactions with that automation, then it is too late for substantial change. This suggests that developers should be encouraged to develop procedures as soon as possible within any concept, because consideration of operator (human) behavior will be a factor in system performance.

2. The key HF concerns of current systems may not be those of future systems. Design of future systems is largely based from current systems and specifically in correcting what are observed to be the pitfalls of the current system. For example, for air traffic control, it is claimed that

controller workload limits the number of aircraft that can be in any one sector and hence the throughput and overall amount of traffic that can be supported in the NAS. Developers of future concepts have worked hard to remove this constraint. However, HF issues are still present in the concepts, albeit not the same HF issues that plague current systems. This suggests that HF researchers may need to refocus their approaches to tackle the new issues that are arising.

3. HF issues do not only concern one type of HF theme or another but tend to evolve from the interaction between two or more HF areas. For example, decision making per se was not an issue for our participants but decision making under time pressure was. The idea of trusting the automation was not an issue for our participants but they described their trust varying by the event and circumstances – it was the interaction, or complexity, of circumstances that seemed to create issues for our participant controllers. Again, HF enquiry should help to identify and develop solutions to these multi-faceted issues.

CONCLUSION

The walkthrough provided important unique insights into human factors issues identified to be problems for the NextGen Separation Assurance concept from a bottom-up perspective at the eye-level of a "front-line" operator. Some factors, such as workload and trust were found to be less problematic than expected. For other HF topics, such as division of roles and responsibilities, our participants confirmed and provided concrete examples of the issues.

From this review we found that concerns that usually spring to mind when designing ATC automation, such as workload and situation awareness, are being addressed by functional studies with future automation prototypes. However, a raft of human factors issues that tie time constraints with responsibilities and procedures were flagged as highly important topics that need to be addressed.

REFERENCES

Erzberger, H. (2001). The Automated Airspace Concept, *Proceedings of the 4th USA/Europe Air Traffic Management R&D Seminar*, Santa Fe, New Mexico.

Lee, P., Sheridan, T., Poage, J., Martin, L., Cabrall, C. & Jobe, K. (2009). *Identification, characterization, and prioritization of human performance issues and research in the Next Generation Air Transportation System (NextGen)*. NASA NRA NGATS Subtopic 9, System Level Design Analysis and Simulation Tools element, NASA Next Generation Air Transportation System ATM-Airspace Project. San Jose, CA: San José State University Foundation.

Polson, P.G., Lewis, C.H., Rieman, J. & Wharton, C. (1992). Cognitive walkthroughs: A method for theory-based evaluation of user interfaces. *International Journal of Man-Machine Studies, 36*, 741-773.

Prevot, T., Homola, J., Mercer, J., Mainini, M., & Cabrall, C. (2009). Initial evaluation of NextGen Air/Ground Operations with ground-based Automated Separation Assurance. *Eighth USA/Europe Air Traffic Management Research & Development Seminar ATM 2009*, June, Napa, California, USA.

SLDAST: System-Level Design, Analysis, and Simulation Tools (2007). *NASA NextGen Air Transportation System ATM-Airspace Project SLDAST Technical Integration Report*, November, 26. Moffett Field, CA: NASA ARC.

CHAPTER 52

Preliminary Guidelines on Controller's Procedures for Pairing Aircraft for Simultaneous Approaches under Different Levels of Automation

Savita Verma, Thomas Kozon, Deborah Ballinger

NASA Ames Research Center
Moffett Field, CA 94035

ABSTRACT

The capacity of an airport can be halved during poor visibility conditions if the airport allows simultaneous approaches only during visual conditions. Several concepts are defining new standards, procedures and operations to allow simultaneous operations during poor weather conditions. However, all the concepts assume that the controllers pair the aircraft and align them for simultaneous approaches, but there are no decision support tools that aid them in this process. This study investigates different levels of automation for pairing and aligning aircraft and evaluates the role of the air traffic controller while interfacing with the tool. In all the conditions the goal was to deliver a pair of aircraft with a temporal separation of 15 s (+/- 10s error) at a "coupling" point that is about 12 nmi from the runway threshold. The logical pairing of aircraft is completed much earlier than the physical pairing of the aircraft that occurs at the coupling point. Four levels of automation were selected that ranged from no automation, to full automation suggesting optimal pair assignments. The metrics in this paper describe the highlights of what has been analyzed and include number of pairs made under different conditions, number of pairs broken and controlled as a single aircraft to prevent potential loss of separation, and excessive workload. It was found that the controllers pair aircraft differently from the pairing algorithm. Also the area coordinator responsible for creating aircraft pairings experienced higher workload than the sector controllers, suggesting that the roles of the controllers, when using this automation need further refinement.

INTRODUCTION

Operations on closely spaced parallel runways have been prevalent in the National Airspace (NAS) for about 40 years. There are several concepts in development and in operational use that define procedures for operations on parallel runways. One concept under development has been Airborne Information for Lateral Spacing (AILS) [Abbott et al., 2001]. Simultaneous Offset Instrument Approach (SOIA) [Magyratis et al., 2001] is currently used at airports like San Francisco International airport. Both concepts support arrivals on runways that are only 750 ft apart and assume that air traffic control will pair the appropriate aircraft for simultaneous landings. However, no tool or formal process exists to facilitate the pairing of aircraft. This paper will evaluate the role of the controller for pairing aircraft under different levels of automation using another pairing concept. The levels of automation define how much functionality the tool and interface provide to facilitate pairing aircraft for simultaneous approaches on parallel runways 750 ft apart.

BACKGROUND

The FAA recognizes that significant capacity is lost when simultaneous operations performed under visual conditions are not operational under poor weather conditions. The FAA, as a part of its NextGen plan [FAA, 2008], aims to reduce the allowable spacing between runways used for simultaneous operations in poor visibility, currently 4300 ft., by revising standards and improving technologies. Several concepts that address the revision of separation standards and new technologies include SOIA, AILS and Terminal Area Capacity Enhancing Concept.

The role of the air traffic controller during simultaneous approaches is different for each of the above mentioned concepts. Under SOIA, the controller has positive control over the aircraft until they break through the clouds and the follower aircraft has visual contact with the leading aircraft, at which time separation authority is delegated to the flight deck. Under the AILS concept, the final approach controller has positive control over the aircraft pair until the flight deck of the trailing aircraft is given a clearance for AILS approach. This clearance is given by the final approach controller just prior to transfer of communications to the tower. Once the AILS clearance is given the trailing aircraft crew is responsible for maintaining separation from traffic on the adjacent parallel approach, while ATC remains responsible for longitudinal separation of in-trail traffic operating in the same stream [Waller, et al., 2000].

The concept investigated in the current study, called Terminal Area Capacity Enhancing Concept (TACEC) [Miller, et. al., 2005], was collaboratively developed by Raytheon and NASA Ames Research Center. TACEC is a technique that can be used for conducting simultaneous instrument approaches to two or even three closely-spaced parallel runways that are 750 ft apart. TACEC operations could double the landing capacity of airports with closely-spaced parallel runways (closer than 2500 ft) during low visibility conditions, approaching arrival rates comparable to visual approach operations. The concept defines a safe zone behind the leader (range 5s to 25s) where the trailing aircraft is protected from the wake of the leader. The trailing aircraft flies an approach with a 6 degree slew, and at a coupling point, which is about 12 nmi from the runway threshold, the two aircraft are linked, with the trailing plane using flight deck automation to control speed and maintain precise spacing of 15 sec in trail behind the leader (Figure 1). The concept assumes Differential Global Positioning System (DGPS), augmented ADS-B, 4 dimensional flight management system, wind detection sensors onboard the aircraft, and cockpit automation that are not extant in today's NAS.

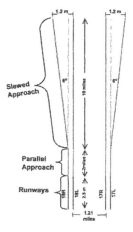

FIGURE 1 Example of aircraft geometries for the concept under investigation.

All the concepts discussed assume that the air traffic controller will assign aircraft to pairs with the knowledge that they are properly equipped. Given this, the TACEC research explores the role of the air traffic controller in assigning aircraft to pairs so they can perform simultaneous approaches. The pairing of aircraft was done under different levels of automation in order to investigate the appropriate human/automation mix for the given task.

Previous research that explored the role of the controller under different levels of automation included a simulation study by Slattery et al., [1995] who examined the effects of the Final Approach Spacing Tool (FAST) with aircraft landing simultaneously on parallel runways. The simulation contained various combinations of aircraft, equipped and unequipped with advanced navigation systems. Similarly, another study [McAnnulty and Zingale, 2005] investigated the effect of using a

Cockpit Display Traffic Information (CDTI), for enhanced visual operations, on controller workload and situation awareness. They found that advanced concepts involving the use of more sophisticated CDTI functions require modifications to current procedures and additional controller workstation tools. Verma et.al. (2009) also investigated the pilot procedures for breakout maneuvers for simultaneous arrivals that were flown under the manual and auto pilot flight control, but did not explore procedures for controllers.

The work described in this paper involves a simulation experiment to examine the role of controllers using a pairing tool, under four different levels of automation, to assign pairs for simultaneous approaches to runways 750 ft apart. A new ATC position, the area coordinator, was added and given responsibility for pairing the aircraft. The simulation also included flight deck automation on the following aircraft that enabled pilots to achieve precise 15 s in trail spacing between the leader and the follower at the coupling point (Figure 1). Results from the different levels of the human/automation mix are presented with the dependent variables of (1) the number of aircraft pairs made by the controllers, (2) the number of aircraft pairs that had to be broken and brought in as single aircraft to prevent potential loss of separation, and (3) controller workload. The experimental approach section defines the airspace used, the scenarios, and the experimental setup. The results and discussion section focuses on the description of the metrics collected and analyzed.

EXPERIMENTAL APPROACH

AIRSPACE ORGANIZATION

San Francisco International (SFO) airport was used as the test bed. SFO has parallel runways, 28L and 28R that are used for all arrival streams. The traffic scenarios consisted of four arrivals streams – Yosem and Modesto from the east, Point Reyes from the north, and Big Sur from the south. The airspace was modified to split the route to the two coupling points (CP28L and CP28R) on each of the four streams. This would allow for runway changes and for aircraft from the same stream to be paired. The routes were modified so they were de-conflicted and were set up with a Required Navigation Performance (RNP) of 1.14 nmi meaning that standard separation was not applied. Instead, the closest distance between the routes before the coupling point was 1.14. The RNP level after the coupling point was 0.01.

For this study, the two approach sectors, Niles and Boulder, were configured such that the controller was responsible for the airspace from the TRACON boundary up to the coupling point which is at 4000 ft. AGL. The Niles Sector managed traffic from the two east-side routes- Yosem and Modesto. The Boulder sector managed the routes from the north and south - Point Reyes, and Big Sur respectively.

Traffic Scenarios

Two different traffic scenarios were used for the simulation. Both the scenarios were equivalent in the rate of arrivals, (approximately 60 arrivals per hour), to the rate of arrivals under visual flight rules (VFR) conditions at SFO. The scenarios also approximated the current distribution of traffic across the four arrival routes simulated for the study.

TEST CONDITIONS

The study included a pairing interface on the Standard Terminal Automation Replacement System (STARS) display, and an algorithm that created pairs. To manipulate the level of automation used for the study, the capabilities of the pairing algorithm and the pairing interface were varied. The role of creating aircraft pairs for simultaneous approaches was assigned to the area coordinator who looked beyond the TRACON boundary, with the sector controllers managing the pair inside the TRACON boundary such that the follower arrived 15 sec behind the lead aircraft at the coupling point. The controllers were also responsible for standard separation between the pairs. They were allowed to use speed only to manage the flow and create adequate separation; vectoring of aircraft was not allowed in any of the conditions.

In the no-automation condition, the area coordinator used current day technologies and flight strips to make pairs and communicate them to the two sector controllers. There was no pairing algorithm or controller interface to assist the area coordinator with creating pairs for simultaneous approaches. The goal for the sector controllers (Niles and Boulder) was to bring the trailing aircraft slightly behind the lead aircraft at the coupling point sans automation.

In the next level of automation (Mixed-1) the area coordinator was responsible for creating pairs, using an interface provided on the STARS display. The area coordinator was able to mouse over the data tag and click on a lead aircraft and a following aircraft to create pairs in the "pairs table" - a new feature added to the STARS display. The area coordinator sent a data link message with pairing information to the two flight decks and waited for an acknowledgement from the pilots. After both acknowledgements were received, a finalized pair was displayed on the area coordinator's and both sector controllers' displays. Under all automated conditions, merging and spacing flight deck automation was used on the simulated flights to achieve 15s temporal distance between leader and follower at the coupling point without the intervention of the controller.

The Mixed-2 condition increased automation. In this condition, the area coordinator selected a leader and a pairing algorithm provided up to three options for trailing aircraft in the "pairs table." The area coordinator evaluated each option offered by the automation against the timeline and finalized the best option by sending datalink messages to the aircraft as in Mixed 1.

The Full Automation condition further increased the role of automation and reduced the role of the human for the pairing task. The pairing algorithm offered one best option for aircraft pairs to the area coordinator, who finalized the pairs by sending the datalink message after evaluating the pair against the timeline.

Methodology

The experiment was a 3x2 within subjects design, with three controller positions and two scenarios. The three controller positions consisted of one area coordinator, and Niles and Boulder sector controllers. The three participant controllers on each team rotated between the three positions. A total of 24 runs (4 conditions x 6 runs each) were conducted per week for two weeks, with a different team of recently retired controllers each week. The four experimental conditions were not randomly distributed. All six runs for every condition were conducted before the participants were trained on the procedures for the next condition and training always preceded actual data collection runs. This was done to avoid confusion between the different procedures and displays used in the four conditions.

EQUIPMENT/ DISPLAYS

The simulation used the Multi Aircraft Control Systems (MACS) simulation environment including a STARS display that could be used in the Terminal Radar Approach Control (TRACON). MACS is an aircraft target generator system [Prevot et. al, 2004] that provides current controller displays and can be used for rapid prototyping of new displays.

The Airborne Spacing for Terminal Area Routes (ASTAR) modeled flight deck merging and spacing to achieve the 15 sec in trail interval between the lead and following aircraft at the coupling point. ASTAR builds 4D trajectories for both the ownship and the lead aircraft approaching the adjacent runway [Barmore et al., 2008], then provides target speed inputs to the follower's FMS, to achieve the assigned temporal spacing between the leader and follower.

A pairing algorithm was integrated with MACS to identify overlapping Estimated Time of Arrivals (ETAs) between aircraft and chose possible pairs (in Mixed 2) or best pairs (in Full Automation). The window of opportunity for pairing was reduced as the aircraft moved closer to the airport, and the distance that could be made up or lost by speed adjustment shortened. The pairing algorithm assessed and offered pairs that could be made by changing the arrival runway for one aircraft [Kupfer, 2009].

TOOLS FOR DATA COLLECTION

All participants completed a demographic survey that included information such as age, experience at different facilities etc. before the simulation started. Controller workload data was also collected using the Workload Assessment

Keypad (WAK). Metrics such as situation awareness, intra pair spacing and others were analyzed but not presented in this paper.

RESULTS AND DISCUSSION

The data analysis paradigm used two independent variables, consistent with experiment procedures described earlier. The independent variable of automation condition had four levels: no automation, mixed automation1 (mixed1), mixed automation2 (mixed2), and full automation. The independent variable of controller position had three levels: Boulder, Niles, and area coordinator. The effects of these two independent variables on the three dependent variables are described in this section. The three dependent variables include controller workload, number of aircraft pairs, and number of deleted aircraft pairs.

CONTROLLER WORKLOAD

Participants recorded their current level of workload by pressing a key on the electronic Workload Assessment Keypad (WAK) [Stein, 1985] at 5 minute intervals throughout the simulation runs. Workload assessments are subjective and could range from 1 (very low workload) to 7 (very high workload).

Workload By Controller Position and Automation Level

Analysis of variance results indicated a significant main effect of position on controller workload, $F(2,4)=11.56$, $p<0.05$ (Figure 2).

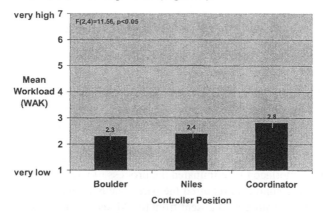

FIGURE 2 Controller Workload by Position

While overall workload across all positions and conditions was low (mean=2.5, SD=1.0), Figure 2 shows that the area coordinator had a higher level of workload relative to the other two positions. Post-hoc analysis yielded a statistically significant difference between the area coordinator and Boulder, $F(1,5)=25.27$, $p<0.01$, and the area coordinator and Niles, $F(1,5)=25.55$, $p<0.01$. This finding is not surprising since the area coordinator is responsible for the area covered by multiple sectors, pairing the aircraft under different positions, and monitoring the aircraft pairs and the flow. In this sense, the area coordinator is required to perform a higher level of multi-tasking relative to the other two positions. Also, the experiment procedures did not allow the sector controllers to form pairs if the area coordinator was too busy. Similarly, the procedures did not allow the area coordinator position to break pairs or directly swap runways for any aircraft - this had to be done through the sector controllers who had ownership of the aircraft. Again, this increased workload suggests the need for additional fine-tuning of the area coordinator's responsibilities.

While statistically significant, the mean difference of less than 1 scale point between positions might also serve to reinforce the main finding, which is that workload was found to be consistently low across all positions. To further illuminate position differences, analysis of the current study factors is currently underway which explores the sub-components that contribute to overall workload.

Workload was also found to be low across the four automation conditions, with the Mixed-2 condition showing the highest workload (mean=2.8, stdev=1.2) and the Full automation condition showing the lowest workload (mean=2.3, stdev=0.7). The higher workload level in Mixed-2 may reflect the excessive task load, which was substantiated with participants' open-ended feedback. However, this result should be viewed as preliminary, since the range was less than 1 scale point. Again, further analysis on workload sub-components might help to illuminate this finding, which may provide an excellent input for the heuristics used by the pairing algorithm.

CREATING AIRCRAFT PAIRS

Analysis of variance results yielded a significant main effect of automation on the number of aircraft pairs made by the study participants, $F(3,44)=4.69$, $p<0.01$. A Tukey HSD post-hoc analysis yielded a statistically significant difference between the Mixed-1 and Mixed-2 conditions (mean difference=4.7, std error=1.266, $p<0.01$). Clearly, the controller-participants were more productive in making aircraft pairs under the Mixed-1 condition (mean=18.1, sd=3.3), as compared to the Mixed-2 (mean=13.4, sd=3.9) and the full automation (mean=16.0, sd=2.7) conditions. The controllers used their own judgment in creating pairs under the Mixed-1 and No-automation conditions. However, Mixed-1 provided the option of an alternative interface that eliminated the process of writing pairs on flight strips.

The Mixed-2 condition required the participants to evaluate several pairs before choosing one – a requirement absent from the Mixed-1 condition. Also, the Mixed-1 condition allowed the controller-participants the greatest level of flexibility in aircraft pairing procedures. Controller-participant feedback on the Mixed-1 condition indicated that the display and flight deck automation were very helpful in making aircraft pairs, while being allowed to use their own judgment to create pairs meant they were not constrained by the automation. The Mixed-2 and full automation conditions sometimes frustrated the controllers if they did not agree with the pairs suggested by the automation, which may have contributed to the relatively low number of aircraft pairs made under those conditions. During discussions, controllers indicated preference for the Mixed-1 condition and expressed a desire for another condition where automation would suggest one good pair while a manual override was allowed. Another heuristic for the algorithm would be to not show pairs that would likely be unacceptable to the controller.

BREAKING AIRCRAFT PAIRS BY AUTOMATION LEVEL

Some aircraft pairs were broken by the controller-participants because flight deck automation could not achieve 15 s temporal separation at the coupling point and standard separation between the aircraft was not possible (Table 1).

Table 1 Percentage of Aircraft Pairs Deleted by Automation Level

condition	percentage of aircraft pairs broken
No automation	7.6
Mixed-1 automation	15.7
Mixed-2 automation	15.0
Full automation	11.5

It is interesting to note the relatively small percentage aircraft pairs broken under the No automation condition, possibly because the controllers had the goal to bring the aircraft slightly behind each other and they achieved it through speed intervention. In the Mixed-1 condition, the area coordinator created the pairs using the display tools. The higher number of pairs broken under Mixed-1 may have been caused by the flight deck automation's speed manipulation constraints, making it impossible to drive the following aircraft to meet the temporal separation of 15s at the coupling point for some pairs created by the area coordinator.

CONCLUSIONS

This simulation study examined the human–automation mix for pairing aircraft for simultaneous approaches to closely spaced parallel runways under different levels of automation. Four levels of automation and three controller positions were examined, and results include analyses of controller-participant workload, the number of pairs made by the controller-participants, and the number of pairs that were broken before the aircraft landed.

Results show that the controller-participants were most productive in forming pairs under the Mixed-1 condition where they used their own judgment to create pairs and used the automation as an interface and for communicating the pairs information to the controllers. Under the Mixed-2 and Full conditions, the study participants did not perform as well on the number of pairs created because the pairing algorithm suggested pairs that were not acceptable to the controller. The heuristics for the pairing algorithm need further refinement. Allowing the controller to have the final say and override any pairing suggestion made by algorithm will be the key for maintaining flexibility for the controller. Finally, while controller workload remained at a manageable level across all automation levels and controller positions, there was higher workload under the Mixed-2 condition and for the area coordinator position, which may suggest the need for additional fine-tuning of the pairing procedures and the area coordinator's responsibilities.

REFERENCES

Abbott, T. & Elliot, D. (2001). Simulator Evaluation of Airborne Information for lateral Spacing (AILS) Concept. NASA/TP-2001-210665.

Barmore, B.E., Abbott, T.S.,Capron, W.R., Baxley, B.T. (2008) "Simulation Results for Airborne precision Spacing along with Continuous Descent arrivals." ATIO 2008, AIAA 2008-8931.

Federal Aviation Administration (2008). National Aviation Research Plan, February 2008. http://nas-architecture.faa.gov/nas/downloads/

Kupfer, M. (2009). Scheduling Aircraft Landings to Closely Spaced Parallel Runways, Eighth USA/Europe Air Traffic management Research and Development Seminar (ATM2009). Napa, CA.

Magyratis, S., Kopardekar, P., Sacco, N., & Carmen, K. (2001). Simultaneous Offset Instrument Approaches – Newark International Airport: An Airport Feasibility Study, DOT/FAA/CT-TN02/01.

McAnulty, D., & Zingale, C. (2005) "Pilot-Based Separation and Spacing on Approach to Final: The Effect on Air Traffic Controller Workload and Performance," DOT/FAA/CT-05/14.

Miller, M., Dougherty, S., Stella, J., and Reddy, P. (2005) "CNS Requirements for Precision Flight in Advanced Terminal Airspace," Aerospace, 2005 IEEE Conference, pp. 1-10.

Prevot T., Callantine T., Lee P., Mercer J., Palmer E., & Smith, N. (2004). Rapid Prototyping and Exploration of Advanced Air Traffic Concepts. *International Conference on Computational and Engineering Science,* Madeira, Portugal.

Slattery, R., Lee, K., and Sanford,B. (1995). Effects of ATC automation on precision approaches to closely spaced parallel runaways. In *Proceedings of the 1995 AIAA Guidance, Navigation, and Control Conference.*

Stein, E.S. (1985) "Air Traffic Controller Workload: An Examination of Workload Probe," *Report No. DOT/FAA/CT-TN84/24*, Federal Aviation Administration, Atlantic City, NJ.

Verma, S.A., Lozito, S., Ballinger, D., Kozon, T., Hardy, G. & Resnick, H. (2009) "Comparison of Manual and Autopilot Breakout Maneuvers for Three Closely Spaced Parallel Runways approaches." Digital Avionics Systems Conference, Orlando, FL.

Waller, M., Doyle,T., & McGee,F. (2000). *Analysis of the Role of ATC in the AILS Concept*, NASA/TM-2000-210091.

Chapter 53

An Approach to Function Allocation for the Next Generation Air Transportation System

Steven J. Landry

School of Industrial Engineering
Purdue University
West Lafayette, IN 47907 USA

ABSTRACT

A method for function allocation was developed and applied to the conflict detection and resolution problem in the next generation air transportation system. The method identifies key capabilities required of agents, automated or human, in order to achieve the goals of the system through state-based modeling. For conflict detection and resolution in the next generation air transportation system, those capabilities include the ability to accurately identify current separation, predict future losses of separation and collisions, and the ability to identify four-dimensional trajectory changes that, when implemented, will result in proper future separation. The modeling method feeds into an analysis of potential levels of automation across four stages of information processing, yielding a taxonomy of possible function allocation schemes, which can then be analyzed with respect to system and operator performance needs. The taxonomy is identified and analyzed for the current and future conflict detection and resolution system in air traffic control.

Keywords: function allocation, air transportation, air traffic control, human systems integration, statecharts

INTRODUCTION

In order to dramatically increase the capacity of the airspace system, it will be necessary to automate the primary safety-critical function of air traffic controllers – that of conflict detection and resolution. In doing so, controllers will be no longer able to supervise the automation, and their role in the system is therefore in question.

In order to guide function allocation decisions, a framework was recommended for analysis of the levels of automation of a system (Parasuraman, 2000), and it was recommended that more detailed system models drive the application of this framework. A state model-based method has been developed to identify key agent capabilities required for achieving system objectives. A mapping of these capabilities to levels of automation can then be accomplished and analyzed, resulting in recommendations for function allocation.

In this paper, we will first briefly describe the application in question, and then review current guidance on function allocation. That guidance indicates the need for more detailed modeling methods to properly allocate function. The application of a recent state-based modeling method will be described that provides such guidance, and used to guide the levels-of-automation framework. Analysis then leads to recommendations regarding system designs.

BACKGROUND

AIR TRAFFIC AUTOMATED SEPARATION ASSURANCE

Programs in both the U.S. and Europe aim to increase the capacity of the air transportation system by 2-3 times over the next few decades (EUROCONTROL, 2007; Joint Planning & Development Office, 2004). One major bottleneck to accomplishing this is the capacity of enroute air traffic control sectors. Controllers managing aircraft within these sectors can accommodate only about 15 aircraft in their sector at any one time (Erzberger, 2004). Moreover, empirical research has shown that beyond about 1.5 times that level, the problem is unmanageable (Prevot, Homola, & Mercer, 2008b).

Sector capacity is primarily limited by controller workload, and solutions other than removing the separation assurance task from manual control do not appear to substantially reduce a controller's workload, allowing for substantial increases in traffic capacity (Erzberger, 2004). Based on this, a major shift in responsibility to achieve the goals of 2 – 3 times capacity appears necessary (Prevot, Homola, & Mercer, 2008a).

Controllers are required to maintain a minimum horizontal and vertical separation between aircraft, which is typically 5 nautical miles horizontally and 1,000 feet vertically in enroute airspace. The function of monitoring and ensuring this separation is referred to as "separation assurance," and a failure to maintain this separation is referred to as a "loss of separation" (LOS).

Such separation standards are regulatory in nature, and are designed to keep sufficient distance between aircraft that, if a LOS occurs, there is sufficient time to detect the LOS and intervene to prevent a collision. The pilots, who are responsible for the safe conduct of the aircraft, are tasked with ensuring that collisions are avoided. This function is referred to as "collision avoidance." For commercial aircraft, automation has been developed to aid the pilot in this task. That automation is referred to as an automated collision avoidance system (ACAS). The combination of separation assurance and collision avoidance is referred to collectively as "conflict detection and resolution."

In today's system, there are few automated aids to conflict detection and resolution. In addition to ACAS, which aids pilots in this task, controllers have an aid called "conflict alert," which identifies predicted LOS within about 2 minutes. However, this system is highly prone to false alarms since it uses trajectory predictions based on closure rate rather than on knowledge of pilot intent (Wickens et al., 2009). Other than these systems, conflict detection and resolution is a manual task, particularly for air traffic controllers.

As mentioned, to obtain 2-3x traffic, it is likely that this function will have to be automated. Various systems for this purpose are being studied (Consiglio, Carreño, Williams, & Muñoz, 2008; Erzberger & Paielli, 2002; Everdij, Blom, & Bakker, 2007). The role of controllers in such systems is not clearly defined. To date, no clear methodology has been applied to try to define a controller's role in the next generation air traffic system.

ATTEMPTING TO APPLY FUNCTION ALLOCATION METHODS TO AUTOMATED SEPARATION ASSURANCE

There are two general classes of function allocation methods: static and dynamic. In static methods, functions are predefined as being performed by human or by automation. In dynamic methods, functions can be shifted between human and automation, either by the human (Miller & Parasuraman, 2007), or by automation (Hancock, 2007). The former is referred to as "adaptable" automation, whereas the latter is referred to as "adaptive."

Static function allocation methods assign task functions to humans or automation based on assumptions about the relative capabilities of humans and automation. However, static allocations are inflexible, relying on a system designer's understanding of the functions. Static allocations can "(foster) a division of tasks rather than (cooperation)." In addition, static allocations permanently separate the human operator from particular functions, creating over-reliance, loss of skill at the allocated function, and losses of situation awareness (Parasuraman, 2000). These considerations are central to human-centered design.

Dynamic function allocation methods attempt to foster better compliance, situation awareness, and overall performance by reallocating functions from automation to human when the human is under-loaded, and vice-versa when the human is over-loaded. The resulting allocation can be viewed as corresponding to a particular "level of automation" (LOA). A recent review is given in Lagu & Landry (In press). Dynamic function allocation implies that functions in the task can be assigned along a range of LOA in that the system design can allow for various levels of involvement by the human operator. This involvement can be either supervisory, at the level of decision making, or even execution.

The concept of automation acting on a particular LOA was expanded to include a second dimension, that of information processing level (Parasuraman, 2000). Specifically, LOA had to be considered not at the level of an entire function, but at the level of "information acquisition," "information analysis," "decision selection," and "action implementation." A more recent paper argued that this decomposition was an "abstraction," and that other, more detailed modeling methods might be better at identifying subtasks (Miller & Parasuraman, 2007).

The methods mentioned include the GOMS method, operator functional modeling, and Petri nets. However, these models are extremely detailed and rely on an existing form for the system. As such they require an extraordinary amount of work to develop, and would have to be redeveloped for each possible design of the system.

What is needed is a higher-level model that is easy to produce and is largely independent of the particular form of the system. This modeling method should allow for identification of key capabilities to achieve the overall goal of the system without the need to identify the specific form of the system or the procedures used in one existing or planned instantiation of the system.

The method proposed in this paper does precisely that – constructs a model of a system to identify the range of opportunities for function allocation, without the need to identify *a priori* the actual allocation of function in the system. Moreover, the model is constructed at a high level, simplifying its use. The model identifies critical capabilities, which then form a third dimension along which functions can be allocated according to LOA.

Furthermore, we specify constraints on LOA, consistent with past work that has indicated the utility of recommendations about lower and upper bounds to satisfy performance requirements. This introduction is made possible by the modeling method, which deconstructs the task into functions for which human and automation capabilities can clearly be mapped.

MODEL AND APPLICATION

A model of the conflict detection and resolution function in air traffic control is shown in Figure 1. The goal of the system is to prevent a collision, which is approximated by the system reaching state 2, where a "near midair collision" (NMAC) occurs. An NMAC is defined as the separation between a pair of aircraft being less than 500 feet, or $s(t_{now}) < 500$ for the current separation or $\min(s(t)) < 500$ for future separation. A description of the model and its generation

528

is given in Landry et al. (2010).

Briefly, the model is constructed by identifying a goal state, which in this case is the avoidance of collision. Nominal states are identified, and each state is hierarchically decomposed such that the resulting model is a complete description of the behavior of the system, albeit at a high level. Transitions between states are identified, including the control agents, human or automated, must apply.

Generally, agents must be capable of identifying current and future system states and controlling transitions between states. In addition, if off-nominal behavior is possible, agents must be capable of controlling the intensity of transitions so that transitions cannot happen in less time than the agents need to detect and intervene. These capabilities can be broadly mapped into two functions: (1) detect and control system state and (2) detect and control system intensity.

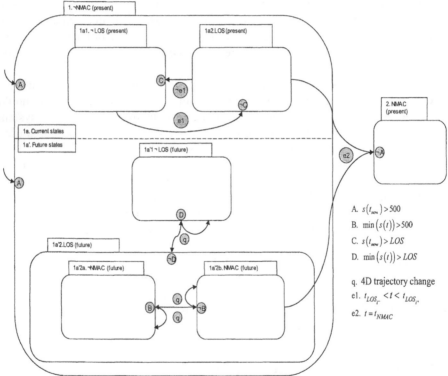

Figure 1. State model of conflict detection and resolution. Originally published in Landry et al. (2010).

For this application, the model is a complete description of the conflict detection and resolution problem and maps these critical capabilities for agents in the system to particular state variables. Mapping these to the four stages of human information processing produces Table 1.

Table 1. Mapping of critical capabilities to stages of human information processing.

Function	Information acquisition	Information analysis	Decision selection	Action implementation
1	Aircraft positions identified by primary and secondary radar	Conflict alert predicts LOS		
		Controller predicts LOS	Controller determines resolution	Controller instructs aircraft, pilot complies
	Aircraft positions inferred by radio position reports	Controller predicts LOS		
2	Controller estimates intensity	Controller predicts intensity resolution needed	Controller determines resolution	Controller instructs aircraft, pilot complies

For function 1, there are two channels for information acquisition, one manual (level 1) and one fully automated (level 10). There are also two channels for information analysis, one at level 1 and one at level 10. The function allocation in the current system is static, although there is a manual backup for each automated function.

APPLICATION TO NEXT GENERATION CONFLICT DETECTION AND RESOLUTION

In applying the model to the next generation problem, there is no actual implementation defined. The purpose here is to identify ranges of possible levels of automation available and, roughly, how they could be implemented. Analysis can then be conducted on each possible implementation.

Such an application results in Table 2. In Table 2, options for information acquisition are expanded by new surveillance systems such as ADS-B and Mode S, and new automation will replace or augment conflict alert to identify and resolve conflicts. However, the choice still remains between a level 10 and level 1 system, or a combination of the two, rather than a range of possible levels of automation for those stages of information processing.

Action implementation can be automated across two dimensions – controller feedback and execution. Decision selection automation can provide a wide range of feedback to the controller, from no feedback to identifying all possible resolutions. This allows for an almost full range of levels of automation for action implementation. (The items in italics represent a range and not the only choices.) With respect to execution, data communications (datacomm) can allow the automation to close the loop to action implementation using the flight management system (FMS) on the flight deck, which represents a level 10 alternative to manual radio calls by the controller. A discussion of various implementation options is provided in Dwyer & Landry (2009).

Table 2. Function allocation options for next generation conflict detection and resolution.

Function	Information acquisition	LOA	Information analysis	LOA	Decision selection	LOA	Action implementation (Feedback)	LOA	Action implementation (Execution)	LOA
1	Aircraft positions identified by primary and secondary radar/Mode S/ADS-B	10	Automation identifies conflicts	10	Automation determines resolutions	10	None	10	Datacomm, FMS executes	10
							Identify all resolutions	2	Controller radios resolution to pilot, pilot initiates	1
			Controller identifies conflicts	1	Controller determines resolutions	1	N/A	-1	Controller radios resolution to pilot, pilot initiates	1
	Aircraft positions inferred by radio position reports	1	Controller identifies conflicts	1	Controller determines resolutions	1	N/A	1	Controller radios resolution to pilot, pilot initiates	1
2	Automation calculates intensity	10	Automation identifies intensity resolution needed	10	Automation determines resolutions	10	None	10	Datacomm, FMS executes	10
							Identify all resolutions	2	Controller radios resolution to pilot, pilot initiates	1
			Controller identifies intensity resolution needed	1	Controller determines resolutions	1	N/A	1	Controller radios resolution to pilot, pilot initiates	1
	Controller estimates intensity	1	Controller estimates intensity resolution needed	1	Controller determines resolutions	1	NA	1	Controller radios resolution to pilot, pilot initiates	1

RECOMMENDED LIMITS TO LEVELS OF AUTOMATION FOR ACTION IMPLEMENTATION

Although an almost full range of options with respect to levels of automation are available for action implementation (feedback), many are not practical, technically feasible, or have significant drawbacks. Eliminating these options is a critical aspect of limiting the range of potential design options for consideration.

Recall that the primary human-centered design considerations for function allocation are that it foster proper reliance on the automation, it not lead to degradations of needed skills, and that it allow proper situation awareness. The range of possibilities for action implementation should be analyzed with respect to these considerations.

Another important consideration is the range of traffic likely encountered by controllers. At times of low demand, or when the automated system is not functioning, the system should be able to run as it does today. However, to accommodate 2x – 3x demand, the system will need to be largely automated since the controller will be unable to perform the task.

Currently conceived resolution determination automation can at best identify a few resolutions, since it must do so in real time. Because of this, the lower limit of levels of automation is likely at LOA 3. At the high end, not providing feedback would result in poor situation awareness. Moreover, reliance on the automation would be affected by the potential confusion induced by the automation deciding on whether to provide feedback. For these reasons, a recommendation for the upper limit on LOA is 8.

For function 1, because the range of traffic includes current levels and the controller will still need to be able to identify and resolve conflicts should the automation fail, the controller will still require that skill and should therefore not be fully divorced from that function. However, there will be conditions under which the controller cannot perform that function. Such a situation strongly suggests that adaptive or adaptable function allocation will be necessary.

However, the transitions into and out of different LOAs is problematic. Current systems for adaptive or adaptable function allocation are at early stages of development, and a great deal more research is needed to identify a practical system for this purpose.

During the likely extended times when function 1 is automated, controller situation awareness will be significantly reduced, because they will not be able to monitor the automation, as suggested earlier. In order to keep the operator "in the loop," a substantive task must be identified.

Current estimates of traffic intensity for 2x and 3x traffic suggest that, at least under some conditions controllers may be able to perform the action implementation of function 2, even at LOA 1. Performing this task, which is to some degree similar to function 1, may allow for controllers to maintain situation awareness, but this option has not been considered to date, as the automated algorithms for detecting and resolving intensity problems has only just been developed.

CONCLUSIONS AND FUTURE WORK

The most widely-accepted framework for function allocation was applied, as supplemented by a novel state-model based method for identifying key capabilities required of agents for conflict detection and resolution. This effort represents a structured approach to identifying system design options for next generation conflict detection and resolution.

The method yielded important structural insights into design options for this system, and demonstrated that the framework could be applied to identify these insights. These options allow for a broad range of system designs, but additional analyses regarding human-centered design principles narrowed this range.

The results generally point to a system where the feedback portion of the action implementation stage of information processing is performed by an adaptive or adaptable automation system. In addition, it is suggested that the intensity-related function could be allocated primarily to the controller during all phases, thereby leaving it at a low LOA. Such an option may keep controller situation awareness high.

The results also suggest that fully automating conflict detection and resolution is undesirable. Such a choice would likely degrade controller situation awareness and skill.

Considerable additional work must be accomplished to investigate each option in detail. In addition, adaptive or adaptable function allocation methods must be improved to the point where they could be introduced into the system.

REFERENCES

Consiglio, M. C., Carreño, V., Williams, D. M., & Muñoz, C. (2008). Conflict prevention and separation assurance in small aircraft transportation systems. *Journal of Aircraft, 45*, 353-358.

Dwyer, J., & Landry, S. J. (2009, July). *Separation assurance and collision avoidance concepts for the next generation air transportation system.* Paper presented at the Human-Computer Interaction International Conference, San Diego, CA.

Erzberger, H. (2004). *Transforming the NAS: The next generation air traffic control system.* Paper presented at the 24th International Congress of the Aeronautical Sciences, Yokohama, Japan.

Erzberger, H., & Paielli, R. A. (2002). Concept for next generation air traffic control system. *Air Traffic Control Quarterly, 10*(4), 355-378.

EUROCONTROL. (2007). *Single European Sky ATM Research.*

Everdij, M. H. C., Blom, H. A. P., & Bakker, B. G. J. (2007). Modelling lateral spacing and separation for airborne separation assurance using Petri nets. *Simulation, 83*, 401-414.

Hancock, P. A. (2007). On the process of automation transition in multitask human-machine systems. *IEEE Transactions on Systems, Man and Cybernetics - Part A: Systems and Humans, 37*, 586-598.

Joint Planning & Development Office. (2004). *Next Generation Air Traffic Control System Integrated Plan.* Washington, DC: Joint Planning & Development Office.

Lagu, A., & Landry, S. J. (In press). Roadmap for the next generation of dynamic function allocation theories and strategies. *Human Factors and Ergonomics in Manufacturing.*

Landry, S. J., Lagu, A., & Kinnari, J. (2010). State-based modeling of continuous human-integrated systems: An application to air traffic separation assurance. *Reliability Engineering and Safety Science, 95*, 345-353.

Miller, C. A., & Parasuraman, R. (2007). Designing for flexible interaction between humans and automation: Delegation interfaces for supervisory control. *Human Factors, 49*, 57-75.

Parasuraman, R., Sheridan, T. B., and Wickens, C. D. (2000). A model for types and levels of human interaction with automation. *IEEE Transactions on Systems, Man, and Cybernetics, 30*(3), 286-297.

Prevot, T., Homola, J., & Mercer, J. (2008a). *Human-in-the-loop evaluation of ground-based automated separation assurance for NEXTGEN.* Paper presented at the 8th AIAA AIrcraft Technology, Integration, and Operations Conference, Anchorage, Alaska.

Prevot, T., Homola, J., & Mercer, J. (2008b, 18 - 21 August). *Initial study of controller/automation integration for NextGen separation assurance.*

Paper presented at the Guidance, Navigation, and Control Conference, Honolulu, HI.

Wickens, C. D., Rice, S., Keller, D., Hutchins, S., Hughes, J., & Clayton, K. (2009). False alerts in air traffic control conflict alerting system: Is there a "cry wolf" effect? *Human Factors, 51*, 446-462.

CHAPTER 54

Impact of Airspace Reconfiguration on Controller Workload and Task Performance

Paul U. Lee[1], Nancy Smith[2], Thomas Prevot[1], Jeffrey Homola[1], Hwasoo Lee[1], Angela Kessell[1], Connie Brasil[1]

[1]San Jose State University / NASA Ames Research Center
Moffett Field, CA, USA

[2]NASA Ames Research Center
Moffett Field, CA, USA

INTRODUCTION

In the National Airspace System, a key aspect of air traffic management is to adapt to changing traffic demand, traffic flow, and airspace/system constraints while maintaining safe and efficient operations. In the Next Generation Air Transportation System (NextGen), the traffic is predicted to increase substantially, creating an environment in which effective balancing of demand and capacity becomes a high priority. When a particular airspace cannot meet the traffic demand due to factors such as aircraft density, traffic complexity, or weather, the Air Navigation Service Provider (ANSP) manages the problem by various flow contingency management techniques, such as rerouting traffic flow away from constrained areas, issuing miles-in-trail requirements and/or ground stops.

In Flexible Airspace operations, we expect that the demand-capacity balance can be achieved by selectively managing the airspace capacity in conjunction with managing the traffic demand. Instead of reducing the traffic demand to address the demand-capacity imbalance, sector boundaries can be flexibly reconfigured to redistribute the traffic volume and demand across sectors (Kopardekar, Bilimoria, & Sridhar, 2007; Lee, et al., 2008). In such operations, the demand and capacity can be calculated for one to two hours into the future to identify sectors that could exceed the traffic threshold as well as sectors that are under-utilized. Using various airspace optimization algorithms, airspace can be reconfigured to manage the existing traffic demand without moving aircraft away from existing routes. A number of airspace optimization algorithms are currently being explored to find the best ways to reconfigure the airspace (e.g., Yousefi, Khorrami, Hoffman, & Hackney, 2007; Klein, Rodgers, & Kaing, 2008; Brinton & Pledgie, 2008; Zelinski, 2009).

EXPLORING FEASIBILITY OF FLEXIBLE AIRSPACE RECONFIGURATION

Flexible airspace management already exists today to a limited extent. For example, sectors are combined daily whenever traffic flow significantly decreases through an airspace and reopened as traffic increases. For a wider implementation of flexible airspace management, general questions related to where, how often, and how fast the sector boundary changes can occur need to be examined, because there may be an adverse impact of flexible sector boundary changes on the ANSPs. Better understanding of the ANSPs' abilities to handle the transition is needed. Some of the fundamental questions related to airspace changes and their impact on the ANSPs are as follows:

- Which airspace-related factors (e.g., airspace volume change, number of aircraft affected by the boundary change, etc.) significantly impact controllers during a boundary change?
- How often can airspace be changed?
- When is airspace change feasible?

A human-in-the-loop simulation was conducted in 2009 to address some of the questions posed above. Traffic scenarios with varying types and severity of boundary changes (BCs) were used to test their impact on the controllers. Per each boundary change, metrics such as airspace volume change, number of aircraft, and various task loads (e.g., handoffs, pointouts, etc.) were compared against subjective metrics such as workload and acceptability, as well as the safety implications in terms of separation losses and other operational errors.

PARTICIPANTS

There were four test participants. Three were operations supervisors from Washington Center (ZDC), Atlanta Center (ZTL), and Indianapolis Center (ZOA), and one a recently retired controller for Oakland Center (ZOA) who had actively controlled traffic within the last four months prior to the start of the simulation. Their air traffic control (ATC) experience spanned from 20 to 25 years with an average of 22.5 years of ATC experience.

In addition to the test participants, retired controllers from ZOA performed the duties of Area Supervisor, two Radar Associates (RAs), and "ghost" controllers responsible for all aircraft outside of the test airspace. The Area Supervisor and the two RAs played an integral role in the study. The RAs had recently retired within 2.5 and 2 years, respectively, and the Area Supervisor had retired within 6 years. All of the simulated aircraft were flown by pseudo-pilots, who were active commercial pilots or students from the San Jose State University aviation department.

AIRSPACE

The test sectors were adapted from four high altitude sectors in Kansas City Center (ZKC). The four test sectors, ZKC sectors 94, 98, 29 and 90, were surrounded by the "ghost" sectors that handled the traffic that entered and exited the test sectors.

The flows in the test scenarios consisted mostly of aircraft in level flight, with a small mix of arrivals and departures to and from the area airports. The minimum altitude of these over-flights was FL 290 with maximums being dependent upon aircraft characteristics. In general the East-West flows in these scenarios were slightly heavier than the flows running North-South. Two main traffic scenarios were created for the study. Both scenarios created traffic overload for sectors 94 and 90 while sectors 98 and 29 had capacity to absorb the excess demand.

EXPERIMENT DESIGN

The experiment consisted of four test conditions. A **Baseline** condition with no boundary changes was used to establish the baseline workload and other performance metrics. Three additional conditions consisted of **Low, Medium,** and **High** severity of BCs (see Figure 1). Three airspace resectorization algorithms were selected based upon their approach and aggressiveness related to the magnitude of the sector boundary change and they were labeled as Low, Medium, and High according to the severity of the BCs. The algorithms that were leveraged for this study are a part of an ongoing research effort at NASA to explore different ways to create dynamic sectorizations.

FIGURE 1: Example of Low, Medium, and High Magnitudes of BC Severity

For each BC, we measured various airspace-related factors, BC frequency (i.e., time elapsed since the last BC) and the total number of aircraft. We expect to correlate these factors with subjective metrics such as workload and acceptability ratings and objective metrics, like operational deviations and errors.

The simulation was conducted over eight days in 2009. There were three days of training followed by 4.5 days of data collection runs that concluded with debrief discussions and questionnaires. There were 16 full data simulation runs in total.

TECHNOLOGY ASSUMPTIONS AND TOOL CAPABILITIES

The technology assumptions for the study were modeled after the assumptions in High Altitude Airspace (HAA). For the study, all aircraft were flying under Trajectory-Based Operations, flying along 4-D trajectories. They were assumed to be equipped with air-ground Data Communication (Data Comm) with automated transfer-of-communication (Auto-TOC) as they were handed off between sectors. All positions still had ground-ground and air-ground voice communication channels as they do today. The radar controllers (R-side) had integrated conflict detection and resolution (CD&R) capabilities integrated into their displays.

The simulation platform used for the study was Multi-Aircraft Control System which provided a high fidelity emulation of the Display System Replacement (DSR) controller workstation. This DSR emulator was highly configurable to mimic both DSR workstations in the field today and future DSRs with advanced decision support tools (DSTs). For the study, air-ground Data Comm and CD&R were integrated with a route planning tool. A conflict detection tool probed for conflicts along the 4-D trajectories and alerted the controllers in case of conflicts. Controllers then used an interactive trial planning tool to plan either a lateral or vertical maneuver to resolve the conflict. A trial plan was constructed and manipulated using a trackball and the route information was displayed graphically. The conflict resolution could also be constructed using an automated conflict resolver, which could be invoked by the controller and used as another DST. Once a resolution was completed the resultant trial plans were uplinked to the aircraft via Data Comm. The advanced air and ground-side DSTs were integrated with Data Comm and the Flight Management System to allow controllers and pilots to exchange and implement 4-D trajectory information quickly. Sector handoff was manually initiated by the transferring controller. When the handoff was accepted, a frequency change uplink message was automatically sent to the aircraft.

In addition to assumptions related to HAA, we assumed that both radar and radar associates had the same displays and tools to monitor and issue clearances to the aircraft. We also assumed that sector boundary changes can be constructed from one position and propagated flexibly both at the sector positions and other locations throughout the Center and beyond (e.g., Command Center). The sector boundary changes were assumed to be accessible to the controllers via a "preview" function which displayed both the current and future sector boundaries on their displays along with the impacted traffic.

538

PROCEDURE

Upon arrival, participants were given a brief introductory briefing on the Flexible Airspace concept, followed by hands-on training on the airspace, tools, and the traffic scenarios. For the training and the data collection, the participants used radar controller stations that were similar to the ones that they use in actual operations. Stations were used as four radar displays in the corners and two RA displays were located between them. In addition to these six stations, an Area Supervisor's station was configured similarly to the controller stations but with extra displays for load awareness. Two side-by-side projectors were connected to the Area Supervisor's station and projected a Traffic Situation Display with a real-time display of traffic.

When the participants felt sufficiently comfortable with the new tools and NextGen operations in general, they continued their hands-on training with the BCs. In a BC, the Area Supervisor previewed the BCs and monitored when the next BC would occur. In the simulation, the sector boundary changes were pre-calculated by various algorithms and scripted to occur at a pre-designated time.

Once the Area Supervisor reviewed the BC and the predicted traffic levels for all sectors before and after each BC, he assigned RA controllers to the sectors that needed the most help. The supervisor first coordinated the plan with the RA controllers (5 to 10 minutes in advance) and then they coordinated with the R-side controller at three minutes prior to the BC. At three minutes, the R-side saw an upcoming BC preview displayed on their DSR screen. Figure 2 shows what a boundary preview might look like during the simulation.

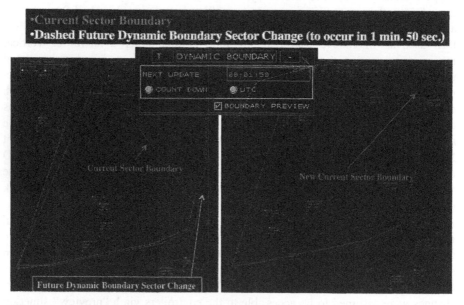

FIGURE 2: Controller Display during a Boundary Change

The R-side controllers previewed the BCs on their displays. By examining their own sector before and after a BC, controller participants calculated which aircraft should have their control transferred. The sector controller, who no longer owned the airspace but still owned the aircraft, initiated handoffs for the impacted aircraft to the appropriate receiving controller. Radar Associate's help was available to handle workload during this transition. The receiving controller accepted the handoff. The Auto-TOC was then executed via Data Comm. Either the initiating R-side or RA briefed the receiving controller about the traffic situation if necessary. Finally, pilots checked in to the new sector once TOC was completed.

EXPERIMENTAL METRICS

We have collected and analyzed a number of airspace-related factors in this NextGen environment which may impact controller workload and the feasibility of the operations during BCs. We list below the independent and dependent variables collected for the study.

Independent Variables

The Table 1 describes the airspace-related factors that may impact controller workload and operational feasibility during boundary changes. These metrics are calculated for each BC by averaging the values across the four test sectors. For some metrics such as aircraft count, handoffs, and pointouts, a period of +/- 3 minutes around BCs was used to average the values for the corresponding BC. This duration was chosen because the controllers started to preview the new sector boundaries starting at -3 minutes to the BC and it took approximately 3 minutes after the BC for the controllers to become accustomed to the new sector boundaries.

Dependent Variables

To assess the impact of airspace reconfiguration on controller workload and operational feasibility, the following four metrics were measured and calculated during the BCs (see Table 2).

Table 1 Potential Factors that Impact Controllers

Variable Name	Description
BC_Frequency	Frequency of boundary changes = time duration (in min) from the last BC to the current BC.
AC_Gained_Lost	Average number of aircraft (AC) that changed geographic sector at the BC. For each BC, average AC gained across four test sectors are the same as average AC lost.
SectorCnt_AC_Gained_Lost	Number of sector pairs (12 possible pairs = 4 sectors X 3 neighboring sectors) that have aircraft that change sectors at the BC. This metric is calculated in case there is an adverse impact due to the number of sector pairs that need to coordinate rather than the number of aircraft that need to change ownership.
P_Vol_Gained	Average percentage of volume gained per sector.
P_Vol_Lost	Average percentage of volume lost per sector.
SectorCnt_Vol_Gained_Lost	Number of sector pairs (12 possible pairs) that have sectors that gained or lost volume at the BC.
AC_Count	Average number of aircraft in the sector at +/- 3 minutes around the boundary change.
Conflict_Count	Average number of conflicts in the sector at +/- 3 minutes around the boundary change.
HO_Init	Average number of handoffs initiated per sector during +/- 3 minutes around the boundary change.
HO_Accept	Average number of handoffs accepted per sector during +/- 3 minutes around the boundary change.
HO_Cancel	Average number of handoffs initiated but subsequently cancelled per sector during +/- 3 minutes around the boundary change.
Pointout	Average number of pointouts per sector during +/- 3 minutes around the boundary change.
Sector_Dir_Change	Change in the long axis of the sector (in degrees).
Hausdorff	Calculates "similarity" of sectors before and after the BC.

Table 2 Controller Ratings on Workload and Acceptability

Variable Name	Description
PostRun_WL	Workload ratings (1 – 7 scale) for each boundary change taken after each simulation in a post-run questionnaire.
RealTime_WL	Average workload ratings (1 – 7 scale) at +/- 3 minutes of the boundary change, taken real-time during the simulation run.
BC_Workload	Workload ratings at the boundary change – workload ratings in the Baseline condition at the corresponding traffic scenario.
Acceptability	Acceptability ratings (1 – 7 scale) for each boundary change taken during post-run questionnaire.

In addition to the metrics above, metrics related to operational deviation were measured (see Table 3). These metrics are expected to occur whenever the BCs and/or traffic situation become severe and infeasible, which in turn are likely to impact controller workload and result ultimately in operational deviations.

Table 3 Metrics Related to Operational Deviation

Variable Name	Description
Late_HO_Init	Number of handoffs that were initiated after the aircraft already entered the downstream sector.
Late_HO_Accept	Number of handoffs that were accepted after the aircraft already entered one's own sector (Late handoffs due to late handoff initiations were excluded).
Sector_Bypassed	Number of aircraft that are handed off to the next sector prior to entering one's own sector. This often happens when aircraft is handed off to a sector with short transit time. Short transit time can be due to an inappropriate handoff (instead of pointout) or a bad sector design.

RESULTS

Our initial hypothesis on the impact of sector boundary changes on the controllers was that the BCs would cause high workload whenever a large number of aircraft changed ownership from one sector to another via handoff. The prediction was that large airspace volume changes would require greater numbers of aircraft to change sectors, which would result in excessive workload thereby making the transition infeasible. We also identified and assessed other factors that may be correlated with operational feasibility.

Results were initially analyzed by BC severity conditions and summarized by Homola and colleagues (submitted). The summary is given in the following section but will quickly move onto the main focus of this paper, which is to link the airspace-related factors directly to the workload, acceptability, and performance metrics to examine their relationships.

SUMMARY OF RESULTS BY CONDITIONS

Table 4 shows the controller workload/acceptability ratings of the BCs by BC condition. The results suggest that both the overall workload and BC workload at the boundary change increased (1 to 7 scale; 1=low; 7=high) and acceptability decreased with increasing BC severity as expected. The absent data in the Baseline condition are due to the questions not being asked of the participants.

Table 4 Results from Controller Workload and Acceptability Ratings

Metrics	Baseline (No BC)	Low	Medium	High
PostRun_WL		4.75	5.06	5.63
RealTime_WL	4.38	4.88	5.05	5.49
BC_Workload		0.50	0.67	1.11
Acceptability		6.65	6.04	4.46

Table 5 presents the per condition means for the airspace-related factors identified as potentially impacting controller workload and performance. Given that controller workload and acceptability ratings were impacted by BC severity conditions, we examined whether these airspace factors also correlated with the experimental conditions. The results suggest that most of the factors that we identified, such as aircraft gained/lost, airspace volume change (e.g., P_Vol_Gained), and handoff related events (e.g., HO_Init), increased with increasing BC severity as we hypothesized. Factors such as overall aircraft count, conflict count, and aircraft density did not significantly differ as expected since the aircraft in the simulation were left on their original path as much as possible, leaving the demand set and the traffic situation similar across conditions.

Table 6 shows the results related to late handoff initiation/acceptance and handoffs to the downstream sector prior to entering one's own sector (Sector_Bypassed). The results show a general increase as BC severity increased, with an exception of relatively high counts for late handoffs in the Low severity condition. A more detailed look at the data suggested that the high values come mainly from one BC in particular which had numerous late handoffs in two of the sectors. The actual explanation for the deviation is yet undetermined and needs further investigation.

Table 5 Results from Potential Factors that Impact Controllers

Metrics	Baseline (No BC)	Low	Medium	High
AC_Gained_Lost		1.77	2.48	4.48
SectorCnt_AC_Gained_Lost		3.00	3.83	4.75
P_Vol_Gained		9.56	15.99	23.66
P_Vol_Lost		7.69	13.78	21.75
SectorCnt_Vol_Gained_Lost		3.67	5.50	6.17
AC_Count	18.23	18.09	17.94	17.88
Conflict_Count	0.77	0.88	0.86	0.87
HO_Init	11.29	12.58	13.38	14.10
HO_Accept	10.58	10.98	11.85	12.29
HO_Cancel	0.17	0.54	0.46	0.75
Pointout	1.06	1.75	2.00	2.50
Sector_Dir_Change		23.33	19.14	32.10
Hausdorff		34.35	38.88	48.81

Table 6 Results from Metrics Related to Operational Deviation

Metrics	Baseline (No BC)	Low	Medium	High
Late_HO_Init	0.04	0.31	0.15	0.77
Late_HO_Accept	0.23	0.38	0.25	0.25
Sector_Bypassed	1.00	1.83	1.75	2.65

Overall, the results from this analysis suggest that BC severity has an impact on both the controller workload and operational feasibility and is correlated with our proposed airspace-related factors. In the following section, this relationship will be more directly examined via correlation and regression analyses.

ANALYSIS OF BOUNDARY CHANGE FACTORS

The examination of the BC factors was done by taking each BC as a sample. The data from the four test sectors were averaged into a single value for the analysis. Some metrics, such as workload and conflict count, were taken over +/- 3 minutes of the BC as an acceptable time duration that was impacted by the BC.

Correlation of Dependent Variables

PostRun_WL and RealTime_WL metrics had high correlation with each other, as expected since they both evaluated workload (see Table 7). Since these two metrics were highly correlated, subsequent analyses will focus on only one of the variables, namely RealTime_WL. Correlation between the workload ratings and the other two metrics, namely BC_Workload and Acceptability, also had high correlation. Operational deviation factors (e.g., late handoffs) were also correlated with each other and the results show that Late_HO_Init correlated well with Late_HO_Accept but Sector_Bypassed did not correlate well with the other factors.

Table 7 Correlation between Workload and Acceptability Ratings

Pearson Correlation	PostRun_WL	RealTime_WL	BC_Workload	Acceptability
PostRun_WL	1			
RealTime_WL	0.768**	1		
BC_Workload	0.453**	0.608**	1	
Acceptability	-0.599**	-0.513**	-0.412*	1

** Correlation is significant at the 0.01 level (2-tailed).
* Correlation is significant at the 0.05 level (2-tailed).

Correlation of Independent Variables

A number of variants on the airspace volume change and aircraft with ownership change were defined to see which of these factor variants would be most relevant. Unfortunately, five of the factors correlated strongly with each other (see Table 8). This implies that the later regression analyses will not be able to identify the individual contribution from these factors. When one of the factors is chosen in the model, the others are likely to be excluded since they capture the same variance in the model. These factors also correlated with SectorCnt_Vol_Gained_Lost, HO_Init, Pointouts, Sector_DirChange, and, to a lesser extent, with HO_Accept.

Table 8 Correlation of Factors related to Airspace Volume Change

Pearson Correlation	AC_Gained_Lost	SectorCnt_AC_Gained_Lost	P_Vol_Gained	P_Vol_Lost	Hausdorff
AC_Gained_Lost	1				
SectorCnt_AC_Gained_Lost	0.806**	1			
P_Vol_Gained	0.871**	0.777**	1		
P_Vol_Lost	0.911**	0.803**	0.967**	1	
Hausdorff	0.709**	0.733**	0.856**	0.827**	1

** Correlation is significant at the 0.01 level (2-tailed).

Factors that Impact Overall Workload Ratings

After each simulation run, the controller participants were asked to rate the overall workload at the BC and identify factors that had caused high workload. They listed the following factors (accompanied by its simulation metric names) as high

workload contributors:
- *Heavy traffic volume* – AC_Count
- *Large number of aircraft that change ownership* – AC_Gained_Lost and SectorCnt_AC_Gained_Lost
- *Tasks initiated by a controller* –Pointout
- *Too many overlapping data blocks*

Overall workload ratings (RealTime_WL) were correlated with dependent variables using Pearson correlation with α level, $p < 0.05$. The correlated factors are:
- *Number of aircraft* – AC_Count
- *Aircraft that change ownership* – AC_Gained_Lost and SectorCnt_AC_Gained_Lost
- *Airspace volume change* – P_Vol_Gained, P_Vol_Lost, and SectorCnt_Vol_Gained_Lost
- *Sector similarity* – Hausdorff
- *Operational deviation* – Late_HO_Init, Late_HO_Accept, and Sector_Bypassed

Hierarchical stepwise regression was used to narrow which of these factors contributed most to controller workload. The stepwise regression was set up in two levels – the first level contained all of the correlated factors described above and the second level contained the rest of the factors. The results from this analysis suggested that three factors, namely, airspace volume change (P_Vol_Gained), overall aircraft count (AC_Count), and late handoff acceptance (Late_HO_Accept), provided good fit to the workload data, resulting in R^2 of 0.683 (see Table 9).

Table 9 Factors related to Overall Workload

		b	SE b	Beta	R^2
Step 1					0.367
	Constant	4.44	0.18		
	P_Vol_Gained	0.04	0.01	0.61***	
Step 2					0.588
	Constant	0.64	0.92		
	P_Vol_Gained	0.04	0.01	0.58***	
	AC_Count	0.21	0.05	0.47***	
Step 3					**0.683**
	Constant	0.97	0.82		
	P_Vol_Gained	0.04	0.01	0.59***	
	AC_Count	0.19	0.05	0.41***	
	Late_HO_Accept	0.45	0.14	0.31**	
$*p < 0.05, **p < 0.01, ***p< 0.001$					

The workload ratings at the BC seem to be driven by airspace volume change (and the associated highly correlated factors such as airspace gained/lost) and the number of aircraft at the BC. The late handoff acceptance may be driven by high workload, but is also likely to create high workload once it occurs.

Factors that Impact BC Workload Ratings

The factors that correlate with workload change from Baseline (BC_Workload) were identified using Pearson correlation with α level of $p < 0.05$. Unlike the overall workload ratings, BC_Workload did not correlate with the overall aircraft count. It was correlated with following variables:

- *Aircraft that change ownership* – AC_Gained_Lost, and SectorCnt_AC_Gained_Lost
- *Airspace volume change* – P_Vol_Gained and P_Vol_Lost,
- *Tasks initiated by a controller* – HO_Init and Pointouts,
- *Sector direction change*
- *Sector similarity* – Hausdorff
- *Operational deviation* – Late_HO_Init and Sector_Bypassed

Both aircraft gained/lost and increased pointouts were mentioned in the subjective feedback as workload contributors. Hierarchical stepwise regression on the above factors identified only a single factor, namely, handoff initiation (HO_Init) to explain BC_Workload. The model provided good fit to the data, resulting in R^2 of 0.501 (see Table 10).

Table 10 Factors related to the BC Component of Workload

	b	SE b	Beta	R^2
Step 1				**0.501**
Constant	-2.56	0.76		
HO_Init	0.26	0.05	0.71***	
*p < 0.05, **p < 0.01, ***p< 0.001				

Factors that Impact Acceptability Ratings

After each simulation run, the controller participants rated the overall acceptability of each BC and identified the following factors as causing low acceptability:

- *Large changes in sector size or geometry* – P_Vol_Gained, P_Vol_Lost, SectorCnt_Vol_Gained_Lost, and Hausdorff
- *Too many coordinations due to pointouts and handoffs* – HO_Init, HO_Accept, and Pointout

- *Too much or not enough airspace* – overall sector volume
- *Sector shape not aligned with the traffic flow*

The factors that correlate with overall acceptability ratings capture some of the same factors but included many others not mentioned explicitly by the controller participants. The factors were identified using Pearson correlation with α level of p < 0.05 and are shown below:

- *Aircraft that change ownership* – AC_Gained_Lost and SectorCnt_AC_Gained_Lost
- *Airspace volume change* – P_Vol_Gained, P_Vol_Lost, and SectorCnt_Vol_Gained_Lost
- *Tasks initiated by a controller* – HO_Init , HO_Cancel, and Pointout
- *Sector Direction Change*
- *Sector similarity* – Hausdorff
- *Operational deviation* – Late_HO_Init and Sector_Bypassed

Despite numerous factors that correlated with acceptability ratings, Hierarchical stepwise regression on the above factors identified a single factor, namely the number of aircraft gained/lost. The model fit the data very well, resulting in R^2 of 0.794 (see Table 11). The acceptability of the BC seems to be driven by the number of aircraft that changed ownership as a result of the BC.

Table 11 Factors related to the Acceptability Ratings

	b	SE b	Beta	R^2
Step 1				**0.794**
Constant	7.27	0.23		
AC_Gained_Lost	-0.54	0.06	-0.90***	
*p < 0.05, **p < 0.01, ***p< 0.001				

It is somewhat interesting that the participants considered the airspace volume change to be a contributor to the acceptability and the aircraft ownership change to the workload, while the regression analyses suggest the inverse – i.e., airspace volume change is the main workload predictor and the aircraft ownership change is the acceptability predictor. Since these two metrics were highly correlated, however, these factors can probably be used interchangeably for the regression analyses.

Factors that Impact Operational Deviations

Due to space limitations, only the regression results are reported for the operational deviations (e.g. late handoff initiation/acceptance). For the late handoff initiations, the regression identified the acceptability rating to be the only contributor with R^2 of

0.487 – i.e., low acceptability BCs are likely to result in high number of late handoffs. The late handoff acceptance identified the frequency of the BC with R^2 of 0.251. The relationship between these two factors is not clear and requires further examination.

Finally, the regression analysis for the bypassed sectors identified three factors, namely, aircraft gained/lost, overall aircraft count, and the number of pointouts with combined R^2 of 0.670. The frequency of bypassing a sector seems to be correlated with increased number of aircraft and aircraft that switch sectors, as well as the number of pointouts that are needed possibly due to bad sector design and/or routes that clip a corner of a sector. Other sector design related factors, such as sector transit time, may provide an additional insight into the traffic situation that causes the bypassed sectors.

SUMMARY AND CONCLUSION

This study explored the controllers' ability to handle the sector boundary changes (BCs) in various conditions from relatively easy (e.g., small volume changes, few aircraft that change sector ownership, etc.) to very difficult (e.g., large volume and aircraft changes, rapid frequency of the change, etc.). The main questions of the study were identification of the airspace-related factors that predict controller workload and operational feasibility during the BCs and whether the frequency and/or the timing of the BCs adversely impact operational feasibility.

In the study, controllers managed high traffic load during each BC, gave workload/acceptability ratings for each BC, and commented on the BCs that were considered problematic. Subjective feedback suggested that overall traffic volume and the task load related to aircraft gained/lost during BCs were the main workload contributors. They also suggested that a BC was less acceptable for severe volume changes and when it required excessive coordination (e.g., pointouts) either due to bad sector design or short transit time. Additionally, workload/acceptability ratings and airspace-related factors were significantly correlated with many of the variables identified by the participants.

Hierarchical stepwise regression narrowed the explanatory variables for workload down to airspace volume change, aircraft count, and number of late handoff acceptances. Since prior research showed aircraft count to be the main predictor of workload, it is notable that airspace volume change was a better predictor than the aircraft count during BCs. Hierarchical stepwise regression of the acceptability ratings identified aircraft gained/lost as the single predictor of the ratings.

The BC transition component of the workload was isolated by subtracting Baseline workload from the BC condition workload for the same traffic scenario at the same elapsed time into a simulation run. Hierarchical stepwise regression of the BC workload component suggested that the number of handoffs initiated was the single predictor of the BC workload component.

Subjective feedback on workload and acceptability identified aircraft gained/lost during BC and airspace volume change as their main predictors, respectively, while the regression analysis swapped the predictors, which suggests the high correlation between these two predictors might make them interchangeable in this analysis. Unless a BC is pre-selected to be at a time when the aircraft count is low, a larger volume change will naturally result in an increased number of aircraft that need to change sector ownership. Further studies that control for the aircraft in transition while varying the airspace volume change are needed to tease apart the individual impacts of these two predictors.

In addition, the two predictors, aircraft and volume change, may have had weak correlations if the operational procedure and tools allowed the handoffs to be automated during the BC. In such situations, a large volume change may still cause high cognitive workload to monitor the changes but the number of aircraft that change sectors may no longer matter as much. Automated handoff will also likely eliminate handoff initiation as the main predictor of the BC component of the workload.

In the overall analysis, BC frequency was not correlated with either workload or acceptability. Observations also supported that as long as controllers had enough time to prepare for each BC (three minutes in this study), high BC frequency did not pose a major problem. In terms of the timing of the BC, finding and/or creating an appropriate time when fewer aircraft are present would help reduce the BC workload. Participants commented that they would be able to handle large volume changes if they had sufficient transition time to monitor the traffic and prepare for the BC. In actual operations, the BC should not have a fixed preparation/preview time (three minutes in this study); instead, it should be done when the controllers are ready for the change. An important caveat to the concept feasibility is that participants needed a reliable conflict probe to manage the BCs. They reported that they did not have adequate situation awareness of the incoming traffic for separation management without the help of the decision support tools.

Overall, the results and feedback from the study showed that Flexible Airspace is a promising concept worth further development and refinement. A number of tradeoffs may be required in finding the most effective way to address the demand-capacity imbalance while keeping the human controller integrated and functioning meaningfully within the system. Based on the results from this study, further research can begin in addressing these issues.

REFERENCES

Brinton, C. & Pledgie, S. (2008). Airspace Partitioning using Flight Clustering and Computational Geometry. In Proceedings of the *27th Digital Avionics Systems Conference (DASC),* St. Paul, MN.

Homola, J., Lee, P. U., Smith, N., Prevot, T., Lee, H., Kessell, A., & Brasil, C. (submitted). A Human-in-the Loop Exploration of the Dynamic Airspace Configuration Concept. *AIAA Guidance, Navigation, and Control (GNC) Conference and Exhibit,* Toronto, Canada: American Institute of Aeronautics and Astronautics.

Klein, A., Rogers, M., & Kaing, H. (2008). Dynamic FPAs: A New Method for Dynamic Airspace Configuration. *Integrated Communications Navigation and Surveillance (ICNS) Conference.* Bethesda, MD.

Kopardekar, P., Bilimoria, K., & Sridhar, B. (2007). Initial concepts for Dynamic Airspace Configuration, *7th Aviation Technology, Integration and Operations (ATIO) Seminar.AIAA*, Belfast, Northern Ireland.

Kopardekar, P., & Magyarits, S. (2003). Measurement and prediction of dynamic density. *5th USA/Europe Air Traffic Management R&D Seminar*, Budapest, Hungary, June, 2003.

Lee, P.U., Mercer, J., Gore, B., Smith, N., Lee, K., & Hoffman, R. (2008). Examining Airspace Structural Components and Configuration Practices for Dynamic Airspace Configuration, *AIAA Guidance, Navigation, and Control Conference and Exhibit* 18 - 21 August 2008, Honolulu, HI.

Yousefi, A., Khorrami, B., Hoffman, R., & Hackney, B. (2007). Enhanced Dynamic Airspace Configuration Algorithms and Concepts, Metron Aviation Inc., Technical Report No. 34N1207-001-R0, December 2007.

Zelinsky, S. (2009). A Comparison of Algorithm Generated Sectorizations. *Eighth USA/Europe Air Traffic Management Research and Development Seminar (ATM 2009)*, Napa, CA.

CHAPTER 55

The Virtual Camera Concept: A Third Person View

Guy A. Boy[1], Rebecca Mazzone[2], Michael Conroy[2]

[1]Florida Institute for Human and Machine Cognition (IHMC)
40 South Alcaniz Street
Pensacola, Florida 32502, U.S.A.

[2]NASA Kennedy Space Center
Information Technology Division, Mail Code IT
Florida 32899, U.S.A.

ABSTRACT

A virtual camera (VC) encapsulates the "third person view" concept. This paper presents the VC perspective in planetary exploration. The concept was initially designed to support astronauts driving a Lunar rover. We extended the VC concept to robotic support to planetary exploration from the Earth. Human-computer interaction properties are presented together with a use case. Furthermore, the VC concept can be extended to a collaborative sensing-acting multi-agent system that is introduced in the balance of the paper.

Keywords: Virtual camera, planetary exploration, HCI, multi-agent systems.

INTRODUCTION

The virtual camera (VC) concept emerged from the early test of the Lunar Electric Rover (LER) developed by NASA for the exploration of the Moon. Indeed, driving a vehicle in a seldom known environment is a difficult task. Even in a well-known

---

[1] Dr. Boy is also affiliated with the Florida Institute of Technology (FIT), 150 West University Boulevard, Melbourne, Florida 32901, U.S.A.

environment such as reconstructed scenery of the moon at Johnson Space Center, we realized that the astronaut driving the LER needed external advice to move safely. The idea of a virtual camera came up as a "third person view", as if someone outside the vehicle was able to see the scene and help the driver to move safely and efficiently.

We extended the VC concept to more general planetary exploration. In particular, when the exploration is performed from the Earth. A virtual camera basically provides data that experts will further analyze to produce knowledge, e.g., the rock is brittle, round, or sparkly. In addition, current knowledge can be visually represented as complete or incomplete. For example, a planetary surface could be displayed with shaded areas representing unexplored regions, i.e., knowledge holes. These knowledge holes could be assessed by domain experts through possible interpolations or extrapolations, and eventually decide to send scout rovers to explore and appropriately sense terrain. VC is being designed to support risk mitigation by providing the ability to investigate possible futures based on the best possible data as well as choosing the right tools and systems to achieve the mission.

A virtual camera includes geometrical information, in the form of a database, which evolves with time. Typically, consecutive versions are incrementally published. They should also be traceable and easily retrieved, possibly over long periods of time. Each version should be published with associated meta-data. New data recently observed should also be easily incorporated into the existing intrinsic VC information, and be a departure for new investigations. This paper presents the implementation of such a VC concept as well as the potential use of it

DEFINITIONS

WHAT IS A VIRTUAL CAMERA?

Let's assume that we have a very well known environment that is watchable via appropriate numerical sensing devices, i.e., providing a set of 3D pixels, which we will call a 3D scene. This 3D scene can be approximated by various kinds of finite element representations. A purposeful 3D scene needs to have useful and usable attributes to support the task it is needed for. In the case of planetary exploration, such attributes are typically related to geometrical and/or geological dimensions. A virtual camera (VC) is a piece of software that provides such a 3D scene resulting from various kinds of available data.

How can we generate purposeful 3D scenes? First, a VC is equipped with a data/knowledge base and a processor that controls what to present to its user with respect to the situation (context) and user's demand. There might be areas of the scene that are not very well known, e.g., either some attributes of the scene are not sufficiently known or the resolution is poor. In this case, VC may adapt the scene using extrapolations. A virtual camera includes augmented reality features that either compensate 3D data-poor scenes or provide useful interpretations and advice to the user.

VC requirements are the followings: (x, y, z) position handling: these coordinates will be provided through an electro-magnetic sensor that should be initialized and calibrated (this should be studied in more detail to provide a good user experience; (roll, pitch, yaw) attitude handling: the VC control and display unit (CDU) can be equipped with an accelerometer that will provide attitude data in real-time; focus and zoom handling: we will choose an automated focus, and the zoom function will be handled by a cursor on the VC CDU; a precision mode button: this control device will enable the astronaut to select an appropriate level of precision. Several kinds of inputs will be available including direct camera view, re-construction of the relative position of the LER with respect to the planetary surface, and laser and mass spectrometry data. Usability of these options still needs to be tested, i.e., ease of learning, efficiency and precision, recovery from errors and subjective feeling in the manipulation. Note that VC control can be done either directly in the vehicle itself or from a remote station.

Consequently, a virtual camera is a mathematical object that is able to move in 3D space around physical objects, a rover for example, to provide the view of these objects in their environment from the point where the camera is located (Figure 1). Obviously, a VC should be easy to manipulate and visualization should be clearly understandable and affordable. The VC CDU should enable its user to get an appropriate mental representation of the actual situation.

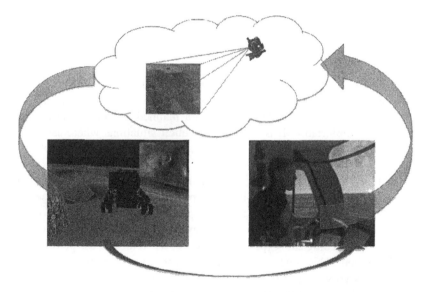

FIGURE 1. VC providing an appropriate situational representation to the LER driver.

WHAT DO WE MEAN BY SITUATION AWARENESS?

Situation awareness (SA) can be decomposed into three processes: perception, comprehension and projection (Endsley, 1995). (1) The perception process provides images and raw data of observed scenes. There are several levels of fidelity of SA according to the available data (from far to immediate area) that can be easily and incrementally identified, structured and refined. (2) The comprehension process provides interpretations of these images and raw data of observed scenes. Comprehension involves reasoning. (3) The projection process provides what to do next according to the interpretations.

A virtual camera has these kinds of processes using human-computer interaction and artificial intelligence techniques. It is designed to bring accurate representations of the observed world, i.e., what is actually generated from real cameras and/or other sensors (machine vision systems or mass spectrometry for example). NASA Kennedy Space Center has already developed a 3D graphical environment that enables management of large sets of information over long periods of time (Conroy & Mazzone, 2008). Within such an environment, VC is intended to be used to elicit richer geographical and geological information and knowledge for example. Such an environment enables data-rich manipulations involving high-fidelity models.

Note that augmented reality features of the VC provide users with enhanced situation awareness. Consequently, reliability and robustness of such augmented reality features are crucial.

USE CASE

LUNAR SURFACE EXPLORATION

Driving the Lunar Electric Rover (LER) on a lunar surface is totally different from driving a car on a road. There are lots of unknown aspects of the planetary surface that need to be identified in order to perform a manned or robotic mission safely, efficiently and comfortably. It is important to elicit situations where astronauts or robots could have difficulty to assess relevant attributes of a planetary surface, e.g., whether they will bump into a rock, fall into a crater, experience an excess of dust, have a stiff hill to climb and so on. In such situations, they might need extra assistance to find appropriate solutions to the problems encountered. Consequently, timely and useful information for the anticipation of these situations will help astronauts or robot "drivers" to choose appropriate trajectories. The representation provided to the geological experts should be as realistic as possible to be meaningful to them. This is why a high-resolution solution is crucial.

When things go wrong, it would be useful to access appropriate views of what the situation is. In order to improve situation awareness, it would be useful to have views of the position and attitude of the rover with respect to the terrain, in the same way as if someone would be outside the vehicle and was watching the scene and would be able to provide appropriate information to the driver. For example during a demonstration at NASA JSC it was observed that the driver of the LER needed a third person view to decide what to do in some specific situations. A typical situation is when the rover is standing on top of a hill and the astronaut wants to turn around and backtrack (Figure 2). He or she needs to know if this at all possible in terms of safety because he or she needs to move a little bit on the right before turning for example, and he or she may not be able to see the terrain underneath. There are other situations such as the one in which the astronaut or robot wants to explore rocks, but with no knowledge of what is underneath.

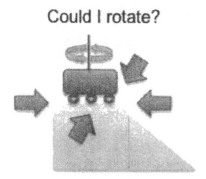

FIGURE 2. LER on top of a hill (observed and represented 2D scenes)

DATA PROCESSES

DATA SOURCES

There are several complementary ways to compute a digital image of the scene that the VC is supposed to show to its user. VC data sources can be multiple.

- A 3D digital model of the environment and various artifacts is necessary, as well as a model of the LER dynamics.
- A GPS view of the scene.
- A few real cameras can provide useful additions. They need to be well located to enable the reconstruction of scene elements from image analyses. Early use of the VC is likely to help fine-tuning such locations.
- As much as possible, cartography of the terrain is also likely to help. This cartography could be updated incrementally from previous missions. In any case, observations made on the way in should be stored for use on the way back.
- Other means such as a laser scanner and mass spectrometer could be used to increase the precision of the scenes

DATA CATEGORIZATION AND REPRESENTATION

There are various categories of situation patterns that need to be identified in order to anticipate necessary sources of information. For example, in geographical situations where LER safety and efficiency could be issues, exploration efficiency could be improved, trip planning could be facilitated, and so on.

The categorization effort will be carried out with NASA lunar experts and cognitive research scientists in the form of brainstorming sessions and interviews. Situation pattern categories will be incrementally created and refined using conceptual maps (CMaps) (Novak & Cañas, 2008). The development of CMaps will be performed with a main objective to incrementally discover generic situation patterns.

Each situation pattern will be associated to one or several possible information sources. Information sources will be described in order to define the type of data that they will be able to generate. Data types could be video images, laser images, or reconstructed geometrical data.

DATA FUSION

Since the Apollo 17 expedition, there have been no opportunities for geologists to get pictures and direct sensing of the Moon's surface. The only possibility is remote sensing of data that can be incrementally added to existing topographical data. This is a first example of data fusion for the virtual camera.

The field of data fusion is very rich in methods, techniques and tools today. It will be necessary to work cooperatively with NASA experts in that field in order to choose the appropriate solutions.

Data fusion will be performed in real-time using appropriate information sources that are relevant for a given situation pattern. The various information attributes will not be uniform and interpolations will need to be performed. Since we will have heterogeneous sources of information, data integration will be necessary. Of course, data fusion depends on the type of mission, and models will differ from one mission to another. This has to be further discussed with NASA and potential commercial space transportation partners.

DATA AUGMENTATION

Since the visual images provided by the virtual camera will be constructed from various sorts of information, it will be easy to incorporate additional information that would help the LER driver in improving his/her understanding of the current situation. For example, relevant geographical data such as the angle of the slope of the terrain improves comprehension of the scene, or appropriate suggestions of actions to take in the current situation (Figure 3) could be inserted in the initial picture, i.e., provide projection.

FIGURE 3. VC providing additional appropriate information and advice.

INTERACTION WITH DATA

As already described above in this paper, one of the most important requirements for a successful virtual camera is the ease of manipulation by its user. There are two main types of objects that will need to be analyzed and experimented: control devices and displays.

Note that a smart combination of VC position and attitude control and zoom control could help in the navigation aspect of the mission also. This leads to the concept of a *navigation display* (ND), in the same way as an aircraft ND, but augmented with appropriate controls. This CDU-ND displays two windows: (1) visualization of the location of the VC; and (2) visualization of the field of view of the VC.

From a purely functional point of view, instead of targeting 6 degrees of freedom for the VC, we could easily reduce VC degrees of freedom by considering that the camera moves on a sphere around the LER. Therefore, in addition to the attitude parameters and the location of the VC on the sphere, the radius of the sphere could be an additional parameter of adjustment. From a user interaction viewpoint, the usability of VC manipulation in polar coordinates instead of (x, y, z) Cartesian coordinates should be tested.

RELATION TO OTHER WORK EFFORTS

The concept of virtual camera in a 3D world, and a third person view, is not new. People involved in videogames have used it for a long time in order to show the action in the best possible angle, e.g., Resident Evil, Sonic Adventure, Tomb Raider, Super Mario and The Legend of Zelda, to cite a few commercial videogame contributions where virtual cameras were implemented. In the scientific world, contributions are generally coming from artificial intelligence, autonomous agents and multimedia communities (He, Cohen & Salesin, 1996; Tomlinson, Blumberg & Nain, 2000; Bares, McDermott, Bourdeaux & Thainimit, 2000).

The concept of virtual camera in videogames can be very different whether the camera is fixed displaying snapshots, tracking movements in the environment, or interactive enabling the user (player) to move the camera itself with respect to his or her intentions. The current concept that we are developing belongs to the third class. A major difference in our case is that the link between the scene being "filmed" and the camera is incrementally built. There is no a priori fixed and permanent link between the scene and the orientation of the camera. Users need to build the necessary constraints that will improve the use of the VC. Another difference comes from the fact that we want to build an image of a real-world 3D scene, and not film an already constructed virtual environment. In other words, the story is not invented, but discovered incrementally.

In the same domain of expertise, Terry Fong and his team at NASA Ames work on planetary exploration. They developed a robotic approach for the production of high quality maps of lunar terrain. They recognized that "although orbital images can provide much information, many features (local topography, resources, etc) will have to be characterized directly on the surface" (Fong et al., 2006). Consequently, they developed a system to perform site survey and sampling. The system includes multiple robots and humans operating in a variety of team configurations, coordinated via peer-to-peer human-robot interaction.

FUTURE PLANS

One VC major purpose is to incrementally design and refine a map of the explored terrain, but it could be any kind of environment. Let's imagine a set of robots, and eventually manned rovers, cooperating in exploring a planet. We could start with a low precision map and improve its definition and resolution by integrating information as it is acquired from cameras and sensors installed on exploring robots.

We need to extend the concept of a scout to an interactive scout that is able to send information, to a ground station for example, and receive orders to go and visit a specific site. Data are immediately integrated within the evolutionary map. Consequently, a dialogue could be established between scouts and ground controllers (and mission managers). The current design leads to the definition of a finite element approach; finding the most appropriate level of grain, and going from low to high resolution. We will explore automated construction and human-driven construction, and a combination of both.

In addition, a robot team could evolve on the explored planet, self-protecting itself by observing the evolution of each other and providing appropriate recommendations to each other. We could define the following functions: (1) Each agent watches other visible agents and their environments; (2) Each agent sends images to other agents; (3) Each agent receives images from a set of other agents; (4) Each agent has a system that integrates received images in real-time. This way we create a collaborative sensing-acting multi-agent system.

REFERENCES

Bares, W., McDermott, S., Bourdeaux, C. & Thainimit, S. (2000). Virtual #D camera composition from frame constraints. *International Multimedia Conference*, Marina del Rey, California, USA, pp. 177–186.

Boy, G.A. (2005). Decision Making: A Cognitive Function Approach. *Proceedings of the Seventh International on Naturalistic Decision Making (NDM7) Conference* (Ed. J.M.C Schraagen), Amsterdam, The Netherlands, June.

Conroy, M.P. & Mazzone, R.A. (2008). Multi-Dimensional Publication of Analysis Results for Future Generations. *International Simulation Multi-Conference*, Edinburgh, Scotland.

Endsley, M.R. (1995). Toward a theory of situation awareness in dynamic systems. *Human Factors, 37 (1)*, p. 32-64.

Dendrinos, D.S. (1994). "Traffic-flow dynamics: a search for chaos." *Chaos, Solitons and Fractals*, 4(4), 605–617.

Dennis, J.E., and Schnable, R.B. (1983), *Numerical Methods for Unconstrained Optimization and Nonlinear Equations*. Prentice Hall, New Jersey.

Fong, T.W., Bualat, M., Edwards, L., Flueckiger, L., Kunz, C., Lee, S.Y., Park, E., To, V., Utz, H., Ackner, N., Armstrong-Crews, N. & Gannon, J. (2006). Human-Robot Site Survey and Sampling for Space Exploration. *AIAA Space 2006*, September.

He, L., Cohen, M.F. & Salcsin, D.H. (1996). The virtual cinematographer: A paradigm for automatic real-time camera control and directing. *International Conference on Computer Graphics and Interactive Techniques*. New York, pp. 217–224.

Novak, J. D. & A. J. Cañas (2008). The Theory Underlying Concept Maps and How to Construct Them, Technical Report IHMC CmapTools 2006-01 Rev 01-2008, Florida Institute for Human and Machine Cognition, 2008", available at: http://cmap.ihmc.us/Publications/ResearchPapers/TheoryUnderlyingConcept Maps.pdf.

Tomlinson, B., Blumberg, B. & Nain, D. (2000). Expressive autonomus cinematography for interactive virtual environments. *4th international conference on Autonomous Agents (Agens 2000)*, Barcelona, Spain.

Chapter 56

Enhancing Situation Awareness with Visual Aids on Cognitively-inspired Agent Systems

Hyun-Woo Kim[1], Sooyoung Oh[1], Dev Minotra[1], Michael McNeese[1], John Yen[1]
Timothy Hanratty[2], Laura Strater[3], Haydee Cuevas[3], Daniel Colombo[3]

[1] Pennsylvania State University
University Park, PA, USA

[2] U.S. Army Research Laboratory
Aberdeen Proving Ground, MD, USA

[3] SA Technologies, Inc.
Marietta, GA, USA

ABSTRACT

This paper describes the strengths of two types of visual aids: VADS (Visualization of Agent Decision Space) and ADT (Agent Decision Table). The VADS is a newly developed visual aid that intuitively represents complicated target attributes in a graphical form, whereas the ADT shows such things in a traditional tabular form. The strengths of each visual aid have been obtained through the analysis of real-time situational awareness queries. Graphical icons expressing multiple attributes are useful in achieving overall situation awareness in most cases. However, there are cases where presentation by a simple text is better in recognition than that by a graphical notation. We discuss the recognition of the ethnic affiliation of crowds as an example at the end.

Keywords: Decision aid, Situation awareness, Knowledge visualization, Decision Support, Cognitively-inspired agent

INTRODUCTION

The information age has brought enormous changes to modern combat operations. In an information technology perspective, network-centric warfare (Cebrowski and Garstka 1998), a military doctrine of war, can be viewed as a concept of operations for increasing combat power through the use of networking sensors, decision-makers, strategies to achieve a shared situation awareness and so on (Fan, Sun et al. 2005). As effective teamwork plays a key role in the doctrine, and the teamwork can be accomplished by the use of intelligent agents, *Three-Block Challenge* has been proposed to study and evaluate agent-aided teamwork in C2CUT (command and control in complex urban terrain) environment where operations officers must react to a constant flow of event reports and make timely decisions for many tasks including combat, peacekeeping and humanitarian missions (Fan, Sun et al. 2006).

We previously developed a human-centered agent architecture for supporting decision making of the armed forces especially when they need to work together in teams with the requirement to analyze a large amount of dynamic information in order to eliminate potential threats from targets. The goal of the past research on this topic was to find a way of enhancing team collaboration and performance; it could be done by modeling a cognitively-inspired agent architecture. The architecture is suitable for proactive seeking, linking, and sharing information through the use of knowledge databases. The collaborative RPD (Recognition-Primed Decision) model has been adopted into the agent architecture for team-based decision making (Airy, Chen et al. 2006).

However, when there are too many targets to deal with and when decisions are made under time pressure, it is hard for the commander to keep track of them and predict what they will likely to behave. The Visualization of Agent Decision Space (VADS) has been introduced for ensuring better situation awareness of the subject (Yen, Strater et al. 2009). The VADS shows prior knowledge for making right decisions. The VADS maps a collection of the past experiences called common historical cases (CHCs) and current targets as icons on the display based on their relative similarities. If a new event occurs, the VADS finds several CHCs that are similar to the current event, and shows it as small bubbles on the left-hand side of the display. Each small bubble indicates one common historical case. Also, the VADS places another bubble, which indicates the current event, among the CHC bubbles. The nearest CHC bubble from the current event bubble in distance is much likely to contain the most appropriate information for dealing with the current target.

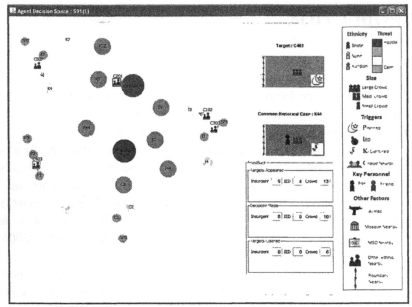

Figure 1: Visualization of Agent Decision Space (VADS)

TARGE...	THREAT LEVEL	WARNING	TASKING	ETHNICITY	TRIGGER	CROWD SIZE	KEY PERSONNEL	ARMED	MOSQUE	BOUNDARY	MSO	OTHE...
C201	HIG~ (Insurgency)			Kurdish	Key Insurgent	Large	Yes	Armed		Near	Near	
C302	LOW			Sunni	Key Insurgent	Small						
C102	MEDIUM			Shate	IED	Medium	Yes				Near	
C403	LOW			mixed	Planned	Small				Near		
C502	MEDIUM			Shate	Key Insurgent	Small	Yes			Near		

Figure 2: Agent Decision Table (ADT)

METHOD

In this experiment, participants interacted as command officer with an urban combat simulation. There were three types of targets with limited number of units on the simulation. The participants were supposed to assign their units to remove the targets and give an answer to a situation awareness (SA) query. The subject's goal was to achieve high scores in both the simulation and SA queries. Each participant was randomly assigned to either the control group or the experimental group. Each group had a different kind of secondary visual-aid; the control group was assisted by a tabular form situation display whereas the experimental group was assisted by visualized situation display. Their performance was evaluated in terms of their game scores as well as SA scores.

PARTICIPANTS

32 U.S. Army ROTC (Reserve Officer Training Corps) students from Pennsylvania State University, 3 female students and 29 male students, were recruited for this experiment. The participants ranged in age from 18 to 22 (Mean: 20.15, Standard Deviation: 1.48) years. Most of them described themselves as ones familiar with video games. On average, at the time of the experiment, they had been playing video games for 10 years, and 4 hours a week. Each participant was compensated with $20.

APPARATUS

The experiment took place in the Laboratory for Intelligent Agents at Penn State. Each human subject was seated in front of two 21 inch (1680x1050 pixels) computer monitors. The urban combat simulation environment on R-CAST, a multi-agent framework, was used with four different scenarios based on the *Three-Block Challenge* (Fan, Sun et al. 2006) where three types of operations are carried out: peacekeeping, combat, and humanitarian operations.

For the apparatus in the simulation, a limited number of units (5 police units, 6 combat units and 1 explosive ordnance disposal unit) are given to the human subject to conduct operations. The peacekeeping operation is to disperse violent crowds with a combination of police and combat units according to the threat level of the crowds. A crowd target is a moving target representing a group of people that may contain activists who are either friends or foes, and has an ethnic affiliation. Also, its threat level may change over time from low to very high. The combat operation is to capture a key insurgent, also a moving target, with two combat units. The humanitarian operation is to remove IEDs, or improvised explosive devices, for the protection of logistic routes; they are stationary targets that cause damage to nearby objects if exploded. Static objects of interest include MSRs, or main supply routes, checkpoints and key buildings such as religious buildings, schools, and hospitals. These objects may affect the movement of crowd targets.

Target		Police	Combat	EOD	Total
IED		0	1	1	2
Key Insurgent		0	2	0	2
Crowd	Low Threat Level	1	0	0	1
	Medium Threat Level	1	1	0	2
	High (Insurgency)	1	2	0	3
	High (Sectarian)	2	1	0	3
	Very High (Insurgency)	1	3	0	4
	Very High (Sectarian)	3	1	0	4

Table 1: Required units for each type of targets

EXPERIMENTAL TASKS

The subjects were asked to assign appropriate types and numbers of units to a particular location in response to terroristic event information received from other agents, maximizing the utility of the human resource. Decision making in target selection and resource allocation requires the officer to consider tradeoffs among multiple factors such as target type, threat level, the unit-target distance, etc. The type and the threat level of a target determine how many units will be necessary to deal with the target. For the purpose of a situation awareness measurement, subjects were also asked to answer a situation-related question, or a Real-time Situation Awareness Questions (RTSA), each time when they assigned their units to a target.

EXPERIMENTAL DESIGN

The experiment was a 2x2x2 mixed factorial design with one between-group variable (visual-aid) and two within-group variables, crowd size and fast burner ratio. The size of crowd was included to determine whether the usage of knowledge visualization was differently influenced by the workload.

Participants first completed an informed consent form and took a demographic survey, and were randomly assigned to one of two conditions (experimental and control). Each participant then watched a 20-minute training video. After the training, the participants had played with a 5-min practice trial until they became familiar with the whole rules in the simulation. And then they proceeded to four 10-minute simulation trials that were given in a random order.

RESULTS AND CONCLUSION

A previous study showed that the group with the VADS achieved a 20 percent improvement over the group with the ADT in terms of task performance, the product of game score and the degree of RTSA (Yen et al. 2009, Hanratty et al. 2009). Even though the task performance implicates overall situation awareness and the VADS helps commanders make right decisions as a result, it is important to understand what type of visual aids is good for recognizing a certain attribute of events (e.g. ethnicity of crowds) when designing a visual aid for decision support system.

This study analyzes the advantages of the two types of visual-aid: the VADS and the ADT. The VADS is good for representing multiple objects with many attributes on a limited space in an intuitive way, and for showing the likely future behaviors of crowd targets predicted by the agent. However tabular forms are also good in some cases. The following table shows the RTSA queries that the control group got better scores than the experimental group did (left-hand side), and vice versa (right-hand side).

Control group with ADT	Experimental group with VADS
(Q1) What was the affiliation of the last crowd target that you assigned units to? { Shiite, Sunni, Kurdish, Mixed }	(Q3) How many active crowd targets contain escalating elements? { 0-1, 2-3, 4-5, 6-7, 8-10 }
(Q2) Did the last crowd target you assigned units to require more police units or more combat units? { more police units, more combat units, same number of police and combat units }	(Q4) In the last crowd target you assigned units to more likely to become an insurgency event or a sectarian event? { Insurgency, Sectarian }
	(Q5) What has been most frequent trigger event for crowd targets in this scenario? { IED exploded, Key insurgent captured, Nearby crowd, Planned Event }

Table 2: Selected RTSA queries (Q1 – Q5 mean RTSA query numbers)

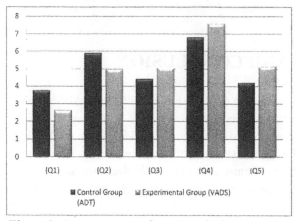

Figure 3: Average scores for selected RTSA queries
(Y-axis: Average score, X-axis: RTSA query numbers)

The ethnic affiliation of crowd targets was shown as texts in the ADT, and as a result, the subjects could easily recognize it and recall it after unit assignments. The VADS also graphically represented it in its display, but it was not as good as the ADT in terms of the average RTSA score. The ADT was shown to be good for recognizing how many police and combat units were required for the previous target. On the other hand, the VADS turned out to be useful in showing whether a crowd event would be more likely to become a sectarian event or an insurgency event, and it is important to assign the appropriate number of units to the target in advance before the threat becomes more dangerous; it may be too late if the subject tries to deal with targets after they reach to such states.

The following is the table of the RTSA queries that both groups got similar degree.

Both Groups
• What was the trigger event for the last crowd target you assigned units to?
• How many high or very high threat crowd targets are currently on your display?
• Will you need to assign additional units to any crowd target that you've already assigned units to?
• What is the intent of the last crowd you assigned units to?
• How many of the active key insurgent targets will your units capture before they appear?

Table 3: RTSA queries that did not distinguish the groups

The VADS and the ADT have different strengths in ensuring situation awareness. In particular, a simple text can be better than a graphical icon if there is something to be memorized for precise operation planning. However, an intuitive graphical notation implying many attributes can be useful in achieving better task performance.

ACKNOWLEDGEMENT

This research was supported by the U.S. Army Research Laboratory (ARL) through the Advanced Decision Architectures (ADA) Collaborative Technology Alliance (CTA) under Cooperative Agreement DAAD19-01-2-0009. The opinions, views, and conclusions contained herein, however, are those of the authors and should not be interpreted as representing the official policies, either expressed or implied, of the U.S. Army Research Laboratory.

REFERENCES

Airy, G., P. Chen, et al. (2006). Collaborative RPD Agents Assisting Decision Making in Active Decision Spaces, IEEE Computer Society Washington, DC, USA.

Cebrowski, A., J. Garstka (1998). Network-centric warfare: Its origin and future, Proceedings in US Naval Institute 124(1): 28-35.

Endsley, M., Garland, D. (2000). Situation awareness: analysis and measurement, Lawrence Erlbaum Associates.

Fan, X., S. Oh, et al. (2008). The influence of agent reliability on trust in human-agent collaboration, ACM New York, NY, USA.

Fan, X., Sun, B., Sun, S., McNeese, M., Yen, J. (2006). RPD-enabled agents teaming with humans for multi-context decision making, The 5th International Joint Conference on Autonomous Agents and Multi-Agent Systems (AAMAS).

Fan, X., S. Sun, et al. (2005). Collaborative RPD-enabled agents assisting the three-block challenge in command and control in complex and urban terrain, Conference on Behavior Representation in Modeling and Simulation (BRIMS)

Hanratty, T., Hammell II, et al. (2009). Knowledge Visualization to Enhance Human-Agent Situation Awareness Within a Computational Recognition-Primed Decision System, The 5th IEEE Workshop on Situation Management (SIMA).

Yen, J., L. Strater, et al. (2009). "Cognitively-inspired Agents as Teammates and Decision Aids", in: Advanced Decision Architectures for the Warfighter: Foundations and Technology, McDermott, P., Allender, L (Ed.). pp. 219-236.

Chapter 57

Establishing Trust in Decision-Aiding: The Role of a Visualization of Agent Decision-Space in Improving Operator Engagement with Reduced Cognitive Load

Dev Minotra[1], Sooyoung Oh[1], Hyun-Woo Kim[1], Michael McNeese[1], John Yen[1], Timothy Hanratty[2], Laura Strater[3], Haydee Cuevas[3] and Dan Colombo[3]

[1]The Pennsylvania State University

[2]U.S. Army Research Laboratory

[3]SA Technologies, Inc.

ABSTRACT

Decision recommendation systems relieve operators from high cognitive-load during stressful situations. However, automation over-trust can disengage complacent operators from the task leading to lower situation awareness and inability to intervene and override incorrect recommendations. Our recent research effort was focused on a visualization of agent decision-space to improve automation transparency and aid the operator's perception of the environment. We describe specific properties of the interface and their anticipated benefits such as improved situation-awareness and expectancy. The visualization is compared with an alternative static representation with an emphasis on how the visualization improves expectancy. An experiment was conducted with a command and control simulation environment to compare the two representations. The results of the experiment have been encouraging. Observed performance improvements in specific scenario conditions, are in accordance to anticipated benefits of the visualization.

Keywords: Decision-aids,   Human-automation-interaction,   Automation-transparency, Multi-dimensional scaling, Cognitive-load.

INTRODUCTION

Decision-aids for assisting commanders in uncertain and time-constrained environments are becoming ubiquitous in the military. They relieve operators from the mental burden of handling large amounts of information. However, decision-aids are often inaccurate, and can sometimes provide operators with incorrect recommendations. Our problem context is the scenario of a military commander dealing with large amounts of incoming reports in a socially unstable, urban environment. The operator's task in our study involves receiving information about active-crowds in the environment, and making decisions to deploy resources to mitigate or combat hostile crowds. We particularly discuss the design features of a visualization and how it augments the operators situation-awareness in order to better interact with a recommending system. As crowds can have a large number of attributes, identifying their qualitative nature can be mentally challenging in a time-constrained and noisy environment. Decision-recommendations are made using R-CAST, an intelligent-agent based on the Recognition Primed Decision model (Fan, Sun, Sun, McNeese, & Yen, 2006). However, decision recommendation systems can throw operators out of the loop as a result of over-trust and vigilance decrement (Endsley & Kiris, 1995). The visualization of agent decision-space which we hereafter refer to as the VADS, was designed to make operators perceive the link between the experience-space of the intelligent-agent and, the nature of the crowd target (Hanratty, et al., 2009). This strategy is intended to keep operators engaged and maintain situation awareness.

The next section provides the reader with the description of the VADS. This is followed by a section that describes properties of the interface and the potential advantages they may have in augmenting situation awareness. The experimental-design and results are discussed on the 4th and 5th sections respectively. Finally, we conclude with a discussion of the implications of the study and potential directions for future research.

VADS: THE ESSENTIAL IDEA

The VADS displays a fixed number of prototypical events indicated by solid circles as shown in figure 1. Their positions on the two dimensional space are based on their attributes, and these positions are calculated by multi-dimensional scaling (Cox, 2000). When an active crowd-target appears, its qualitative-nature can be interpreted by its position on the display. The 'nearness' to prototypical crowds makes the operator perceive the qualitative-nature of the target. When a large number of targets appear, that awareness is critical in prioritizing targets in order to optimally and appropriately allocate resources. Green prototypes as shown in Figure 1 represent crowds that are typically not hostile, and they are perceived as non-events. The yellow, orange and red prototypes represent crowd-targets in increasing order of threat-level. An active crowd-target can undergo transitions across these threat levels. In our simulation environment, crowd-targets can either be fast-burners or slow-burners. Fast-burners are quicker in progressing to higher threat-levels than slow-burners. Representing an active crowd-target on the VADS is an alternative to listing its attributes on a table along with its threat-level. Attributes to a crowd-target include details such as size, proximity to a military-significant-object, presence of a key-insurgent etc. We refer to this tabular representation of crowd target attributes as the Agent Decision Table (ADT) (Hanratty, et al., 2009). The next section discusses potential merits of the VADS over the ADT.

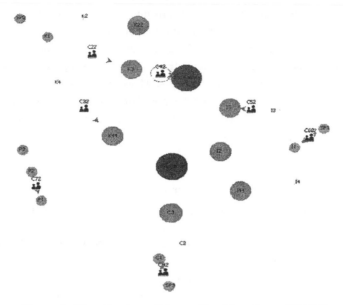

Figure 1. Visualization of Agent Decision-Space (VADS)

INTERFACE PROPERTIES FOR IMPROVED AWARENESS

R-CAST in our simulation environment only recommends the number of resources that need to be allocated to handle an active crowd-target. However, the prioritization of targets, or, the order in which they receive attention is solely decided by the operator. Additionally, R-CAST does not consider the travel time a dispatched resource would take until it reaches its crowd-target. Sometimes, this causes R-CAST to recommend an insufficient number of resources. This is a problem especially when a crowd-target is progressing in threat-level over time, resulting in an increased resource demand. Therefore, the operator is left with the onus to change resource allocation decisions when necessary. These two issues of prioritization, and the anticipation of resource requirements for a progressing target, are both linked to the ability to project a targets future status. Acquiring the ability to project the outcome of a resource allocation decision, or to discriminate between slowly-progressing targets and quickly-progressing targets, is dependent on sufficient comprehension of crowd-relevant data at an early stage. This ability to make predictions is better known as level-3 SA (Endsley & Garland, 2000). Therefore, an important consideration of interface design is providing more information about the dynamics of the environment. Crowd-targets progressing to higher threat-level should be identified by the operator at an early stage. However, on the ADT, information about threat-level is static and is abruptly replaced by a new threat-level when a target completes transition. This is a problem when a large number of active targets are present on the ADT as more targets would fall outside the immediate visual focal attention of the operator. Peripheral vision cannot detect color (Wyszecki & Stiles, 1982), and abrupt transitions in the periphery of vision would go unnoticed during saccadic shifts or rapid eye movements (Burr & Ross, 1982; Ware, 2004).

However, on the VADS, transition information is shown by motion, which should help in attracting attention, and reducing the reaction-time in responding to a crowd-target (Peterson & Dugas, 1972). Crowd-target motion can also help in the perception of transition-rate which would eventually be helpful in discriminating between slow-burners and fast-burners. Although the ADT can facilitate much of this information as the user gets accustomed to using it, affordances on the VADS facilitate this with less mental-workload resulting from the vigilance aspect of the task.

On a real-world task, the selection of targets and resource allocation can depend on the qualitative-nature of the targets as well. External representations should match internal representations of how crowd-targets are perceived, for more efficient perceptual processing and reduced cognitive-load (Zhang, 2001). Crowds of the same threat-level that indeed differ in terms of how they are qualitatively perceived are associated to different types of resource requirements in a real-world setting.

The VADS facilitates this perception of qualitative-type as its representational layout is designed on the basis of prior experience. On the ADT, crowd attributes can indirectly facilitate this perceptual process in a similar way, however, interpreting the qualitative-nature can be mentally challenging or may require significant prior experience in the environment, due to a large number of attributes.

The ability to predict future status can also depend on the qualitative-nature of the target, as some prototypical crowds have a higher probability to progress to a higher threat-level than other ones. It is also important to consider that changes in the environment during a period of long term usage of the agent, can lead to predictable errors in the agents recommendation. Such errors, that have fixed patterns, can be linked to the qualitative-nature of crowd-targets. Thus, displaying this qualitative information can improve the ability to change agent-recommendations about resource-type when necessary. Predicting future target status based on the dynamics of progressing targets and developing expectancies towards errors in recommendation are also potential advantages of the VADS.

These design decisions are intended to make the operator perceive the environment differently, with better situation awareness. Promoting transparency in automation is intended to engage the operator into the loop without increasing cognitive-load.

EXPERIMENTAL DESIGN

We will only provide an overview of the experimental design in this paper. Hanratty et al (2009), describe the experiment in greater detail. Our simulation of an urban command and control environment based on the three-block-challenge problem (Fan, et al., 2005) was the test-bed for the experiment. The experiment was a between subjects design with a total of 32 participants recruited from the Army ROTC at the Pennsylvania State University. Both groups were presented with the display of the battlefield environment. However, the presentation of crowd-target information differed between groups where the experimental group received the VADS, and the control group received the ADT. Four scenarios were developed that were presented in random order to each participant. Scenarios differed in workload and the proportion fast-burners to slow-burners.

RESULTS

Hanratty et al(2009), reported the initial results of the experiment. For each scenario, a one-way ANOVA was conducted to compare scores between the two groups. Score differences were insignificant with the exception of scenario 3.

In the case of scenario 3, the experimental group scored significantly higher than the control group ($F(1,30) = 6.370$, $p = .017$). A possible explanation is that scenario 3, that was designed for higher workload had an equal proportion of slow-burners and fast-burners. Therefore, the ability to discriminate between fast and slow burners became more critical in order to predict future threat level of targets. The VADS possibly facilitated participants for meeting this requirement. Our analyses revealed that SAGAT scores did not significantly differ between the two groups. RT-SA scores was higher for the experimental group although this difference was not statistically significant ($p = .1424$).

DISCUSSION AND FUTURE WORK

Design decisions for the VADS were made with the objective of improving operator situation awareness relevant to the task environment. We discussed how various properties in the design are aimed at reducing the mental effort required to identify crowd-targets making transitions, and in projecting their future threat-level. Our experimental results are encouraging. Even though the VADS did not improve overall performance, it appears to have facilitated the scenario in which the ability to predict future threat-level was a necessity. Our analyses on situation awareness measures did not reveal statistically significant differences, however, an experiment with a larger sample size may lead to more convincing results.

The inherent nature of VADS that abstracts crowd information, aids the operators interpretation of the qualitative-nature of targets. Although experimental results may not reflect on the direct advantages of this property, it has the potential to provide the necessary awareness to an operator interacting with automation. We hope that future experimental studies on this concept may throw more light on its effects in naturalistic decision-making.

ACKNOWLEDGEMENTS

This research was supported by the Army Research Laboratory through the Advanced Decision Architecture (ADA) Collaborative Technology Alliance (CTA) under Cooperative Agreement DAAD19-01-2-0009. The views contained in this document are those of the authors and should not be interpreted as the official policies of Army Research Laboratory.

REFERENCES

Burr, D., & Ross, J. (1982). Contrast sensitivity at high velocities (measurement of visual sensitivity to rapidly moving stimuli). *Vision Research, 22*(4), 479-484.

Cox, M. (2000). *Multidimensional scaling* (Vol. 88): Chapman and Hall.

Endsley, M., & Garland, D. (2000). *Situation awareness: analysis and measurement*: Lawrence Erlbaum Associates.

Endsley, M., & Kiris, E. (1995). The Out-of-the-Loop Performance Problem and Level of Control in Automation. *Human Factors, 37*(2).

Fan, X., Sun, B., Sun, S., McNeese, M., & Yen, J. (2006). *RPD-enabled agents teaming with humans for multi-context decision making*. Paper presented at the Fifth International Joint Conference on Autonomous Agents and Multi-Agent Systems (AAMAS'06)

Fan, X., Sun, S., Sun, B., Airy, G., McNeese, M., Yen, J., et al. (2005). *Collaborative RPD-Enabled Agents Assisting the Three-Block Challenge in Command and Control in Complex and Urban Terrain*. Paper presented at the Conference on Behavior Representation in Modeling and Simulation (BRIMS)

Hanratty, T., Hammell II, R., Yen, J., McNeese, M., Oh, S., Kim, H., et al. (2009). *Knowledge Visualization to Enhance Human-Agent Situation Awareness*

Within a Computational Recognition-Primed Decision System. Paper presented at the 5th IEEE Workshop on Situation Management (SIMA 2009).

Peterson, H., & Dugas, D. (1972). The relative importance of contrast and motion in visual perception. *Human Factors, 14*, 207–216.

Ware, C. (2004). *Information Visualization: Perception for Design* (2 ed.): Morgan Kaufmann.

Wyszecki, G., & Stiles, W. (1982). *Color science: concepts and methods, quantitative data and formulae*: Wiley Interscience, New York.

Zhang, J. (2001). External representations in complex information processing tasks. *Encyclopedia of library and information science, 68*(31), 164-180.

Chapter 58

Are People Able to Develop Cognitive Maps of Virtual Environments While Performing a Wayfinding Task?

Elisângela Vilar[a], Francisco Rebelo[a], Luís Teixeira[a] and Júlia Teles[b]

[a] Ergonomics Laboratory. FMH - Technical University of Lisbon
[b] Mathematics Unit. FMH /Technical University of Lisbon
Estrada da Costa 1499-002 Cruz Quebrada, Portugal

ABSTRACT

The main goal of this pilot study is to investigate the ability in developing a cognitive map after an interaction with a simple immersive VR system. The influence of environmental cues in cognitive mapping ability was also investigated. Two experimental conditions were considered, one Neutral (without signs which facilitate the task execution) and the other Dynamic (with signs to facilitate the task execution). 28 participants had to perform 4 tasks in the environment. A map-identification method was used. The escape route, contrary to expectations, were identified as fundamental for plant identification instead the number of the rooms or the corridors intersection. We did not find correlations between the distance travelled and experimental conditions with the plant choice. However, findings suggest that users are able to acquire basic spatial knowledge after interacting with simple immersive VR system.

Keywords: Cognitive map, Wayfinding, Virtual Reality

INTRODUCTION

The urban areas and buildings in which we live are rich in spatial structure with which we interact every day, and the ability to learn this environment and to structure it in a mental representation is an important factor to improve wayfinding.

According to Golledge (1999) there are two most common ways of learning the environment or to acquire spatial knowledge: i) experiencing the environment through a travel process, and ii) learning the layout either from an overlooking vantage point or via some symbolic, analog modeling (e.g., maps or photographs). In this paper we will focus in the first one, when a user interacts with an environment to perform a wayfinding task. The opportunity to move through an environment allows people to integrate various routes into a *cognitive map* – "internal representation of perceived environmental features or objects and the spatial relations among them" (Golledge, 1999).

It is commonly held that the longest standing model of spatial knowledge representation is the Landmark-Route-Survey (LRS) model, described by Siegel and White (1975). In this model they assumed that spatial knowledge develops from a initial stage of landmark knowledge (knowledge of salient cues/objects, which are static, orientation dependent and disconnected from one another) to an intermediate stage of route knowledge (knowledge of the orders of these landmarks which are connected by paths) to the final stage of survey knowledge (a representation of the layout, much like a map). However the systematic development towards survey maps is not universally accepted. It has been shown that survey maps are not necessarily created or derived from route maps in a rigid progression (Lindberg & Gärling, 1982)

Golledge (1999) and Carassa, Geminiani, Morganti and Varotto (2002) agree that the spatial knowledge of large-scale environments is organized in two types of mental representation or *cognitive maps*: route and survey maps. In route maps the environment is represented in a viewer-centered frame of reference that reflects the person's navigational experiences, while in survey maps distant places are linked together to form a coherent global overview of the entire environment.

Apart different points of view, authors agree that survey maps are the most difficult to acquire and in many situations they are never constructed (Moeser, 1988) or take a long time to develop (Thorndyke & Hayes-Roth, 1982). According to Carassa *et al.* (2002), the reasons for this are either that the environment is too complex or that simpler representations are perfectly adequate for the needs of the individual concerned.

Cognitive maps are used for a wide variety of purposes, but in wayfinding studies they are fundamental. According to Morganti, Carassa and Geminiani (2007), in wayfinding, these representations serve to aid navigation within the mapped environment in order to reach a target. Thus, it is important to study if, when needed, people are able to organize the information about an environment in a cognitive map, in a way to predict the ability of an individual in performing a spatial task.

In the last years, many studies in several areas, namely in wayfinding (Cubukcu & Nasar, 2005; Omer & Goldblatt, 2007), have been conducted using virtual environments (VE). VEs shows a great potential to create experimental contexts in Virtual Reality (VR), where the study variables can be diversified and systematically manipulated is a controlled way. In these studies, VEs are used as interaction environments assuming that the users are able to acquire the spatial knowledge in an active way as they were interacting with real environments.

However, ecological validity of VR-based studies is a matter of debate. According to Cánovas, Espínola, Iribarne, & Cimadevilla (2008), in VR locomotor-based proprioceptive and vestibular cues are largely absent and this lack can difficult the formation of cognitive maps. Billinghurst and Weghorst (1995) states that in VR there is a typically sensory degradation and a lack of many of the perceptual cues used in the real world, for example, the visual modality may suffer from low image resolution, poor image quality or reduction of the peripheral field.

This is a pilot study which takes part of an ongoing PhD project being carried out in the Ergonomics Laboratory, of FMH -Technical University of Lisbon, Portugal, that uses VR as a simulation tool to display indoor environments with different environmental features, with the purpose to evaluate the effect of environmental features in the wayfinding process. For this paper, othe objectives are: i) to know if people are able to acquire basic spatial knowledge from an interaction with a environment in an immersive VR; and, ii) to investigate the influence of environmental cues (signs which facilitates the tasks execution) in cognitive mapping ability.

THE EXPERIMENT

METHOD

One of the difficulties in studying cognitive mapping is the issue of the external representation of an individual's internal map. Cognitive maps are highly subject-specific. Golledge (1999) identified the analysis of external representations as one of the four methods which could be useful for extracting environmental cognition information.

In this study, it is of particular interest the subject's topological understanding of the VE, mainly if the subject is able to identify the VE spatial morphology. Thus, it was used the technique of map-identification which is to ask a subject to choose between correct and incorrect map or between views of an area (Hunt & Waller, 1999).

For this pilot study, it is predicted that when interacting with VEs, people are able to acquire basic spatial knowledge and the number of visited rooms (related to the success in accomplish a task) will influence the users' opinion about the correct environment's plant. It is also considered that in the condition without signs

(Neutral), people will spend much time and will travel longer distances than with signs. It also may influence in the correct choice of the environment's plant.

Participants

Twenty eight participants (17 males and 11 females), aged between 18 and 35 years old (mean= 21.18; SD=3,632) participated in the study. All participants interacted individually with the VR and were randomly assigned to one of the two experimental conditions (with signs or without signs). All participants were asked to sign a form of consent and were advised they could stop the simulation at any time they wanted to.

Apparatus

The VR system used for the experimental sessions consists of the following equipments: i) two magnetic motion trackers from Ascension-Tech, model Flock of Birds. One of them, positioned on the top of the user's head was used to control the head motion and another, attached to the user's left wrist, was used to control a hand icon (like a mouse icon) used to select buttons in the environment; ii) a joystick from Thrustmaster which was used as the locomotion device; iii) a Head-Mounted-Display (HMD) from Sony, model PLMS700E; and, iv) headphones. The body and head movements were controlled separately to give the users a higher level of autonomy. Participants remained seated during the entire experimental session.

Virtual Environment (VE)

The VE was developed in a way to accomplish the study's objectives. It was created from a requirements list generated in brainstorming meetings with experts in Ergonomics, Architecture and Design.

A 2D project was firstly developed and it was the base of the VE. It was designed using the software Autocad® and exported to 3D StudioMax®, both from Autodesk, Inc. in order to model the 3D environment. Several objects, such as furniture, and general properties (colors, texture and light) were created in order to generate a realistic scenario. A collision detection system which does not allow people passing through the walls and objects was considered in order to make the interaction closer to the real world. The scenario was exported using a free plug-in OgreMax, to be used by the ErgoVR system, developed by the Ergonomics Laboratory of FMH - Technical University of Lisbon.

The VE consists of 2 distinct areas: Area 1 (start area) is a square plant with four rooms (12x12 meters each) symmetrically distributed. The rooms are interconnected by two symmetrical and perpendiculars axes of corridors (2 meters wide) and circumvented by another corridor; Area 2 (escape area/decision points) consists of several consecutive "T" corridors intersections. A top down view of the VE can be seen in Figure 1.

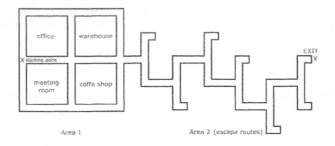

FIGURE 1 - A top down view of the VE plan, with the 2 areas

Design of the Experiment

The experiment used a between-subject design. It consisted of one preliminary practice and two experimental conditions, one Neutral (with no signs for helping in the task execution) and the other Dynamic (with signs that help in the task execution). The participants were randomly assigned to one experimental condition and were unaware of the real objective of the experiment. They were told that they had to perform some tasks as accurate as possible.

Procedure

Participants were introduced to the equipment and started in a practice trial, with the objective to get them acquainted with the equipment and their movement inside the simulation and to check for eventual simulator sickness cases. In the practice trial, participants were encouraged to freely explore and navigate through a virtual room as quickly and efficiently as they could. When they declared that they were able to control the navigation devices they were assigned to one of two experimental conditions (Neutral and Dynamic).

In the Neutral condition there were no signs which could facilitate the task execution. The environment had a small amount of concurrent visual information, it was organized and with good visibility (lighting). In the Dynamic condition there were signs which could facilitate the task execution. The signs were self-illuminated and had intermittent lights.

The same series of four tasks were given for both experimental conditions. Firstly, participants had to go to the meeting room where they received the instruction to go to the Laboratory and to turn on the security system. After this, in the laboratory, they received an instruction to go to the coffee shop where they must to turn the gas off. At the coffee shop, participants received the last instruction which was to go to the warehouse in order to cut the energy to the machines room. An explosion occurs and some fire and smoke are displayed if the participant enters the warehouse or after 5 minutes after entering the corridor that leads to the warehouse. Thus participants should leave the building as fast as possible through

Area 2. Participants in the Dynamic condition had emergency signs which should guide them to the exit. There were no emergency signs in the Neutral condition.

Map - Identification Technique

After the experimental session participants were asked to choose among 6 options of top down plants, only one represented the VE they interacted with. The number of options was chosen in order to reduce the possibility of a random choice. The options were designed based in the following criteria:
1. Area 1 (start area) with a pair of corridor's intersection (4 rooms);
2. Area 1 (start area) with two pairs of corridor's intersection (6 rooms);
3. Area 2 (escape area) with alternative choice (right);
4. Area 2 with alternating direction (zig-zag path);
5. Area 2 with redundant route (turns).

This way, two morphologies were defined: one where Area 2 was highlighted from Area 1 (include in the morphology all plants that meet the criteria 3 and 4) and another one where Area 1 and Area 2 were together (criterion 5). The Area 1 variations (criteria 1 and 2) were considered for both morphologies. The 6 options that were presented to the participants are shown in Figure 2.

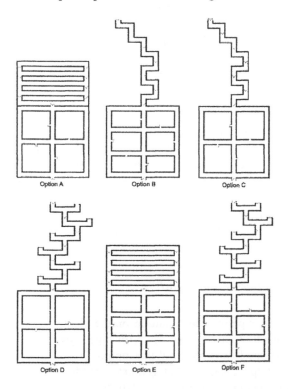

FIGURE 2- Six plants options presented to the participants

RESULTS

All statistical tests were made using the PASW software (v.18; SPSS Inc, Chicago, IL) and the significance level (α) of 5% was considered for all tests.

Plant choice

The total plant choice distribution and the distribution for each experimental condition are displayed in Table 1. Option D represents the right choice (the correct VE plant). In an overview, 50% of the participants chose the right option (D). In a partial analysis of the VE plant, it is verified that Area 2 (escape route) was important in the plants choice (the most chosen were the plants which had the Area 2 displayed correctly - D and F - 86%). However, opposite to the predictions, regarding Area 1, there were no difference among participants that chose between the options with the correct number of the rooms (50%) and the options with higher number of the rooms (50%).

Table 1.0 Plant choice distribution

Experimental Conditions	Plant's options						Total
	A	B	C	D	E	F	
Dynamic (D)	0	1	0	5	3	5	14
Neutral (N)	0	0	0	9	0	5	14
Total	0	1	0	14	3	10	28

The Binomial Test was carried out in order to verify the statistical significance of the percentual incidence of the plant choice. Option D (the correct one) was chosen by 14 participants (50%) while 10 participants chose option F (35.7%). However, inferential statistical analysis indicates no statistically significant differences between the choice of the participants regarding to the options D and F (p=0.541; N=24). The same test revealed statistically significant difference between the options D and B (p=0.001; N=15) and between option D and E (p=0.013; N=17).

Plant choice x Experimental condition

The influence of the experimental condition on the plant choice was verified using the Chi-Square Test for independence. An error type I (α) of 0.05 was considered in all inferential analysis. It was observed that option D was the most chosen for the Neutral condition (Np=9; 64.3%). For Dynamic condition, options D and F were chosen by the same number of participants (Np=5; 35.7%). However, the inferential statistical analysis allows to consider that the plant choice does not depend on the experimental condition ($\xi 2(3) = 5.143$; $p = 0.16$; N=28).

Plant choice x Number of visited rooms

The number of rooms visited by the participant was also considered in order to verify the influence of the VE exploration in the plant's recognition. Participants who accomplished all task had to visit the entire VE.

Only 2 participants (7.1%) did not visited the 4 rooms (2 and 3 rooms respectively) and their choices were options with more rooms than the correct number (6 rooms instead of 4, options E and F respectively). Among those who visited four rooms (N=26, 92.9%), 14 of them chose the right option (D) while 11 chose options E and F, both with 6 rooms.

The Chi-Square Test of independence was carried out in order to verify if participants who visited four rooms had choices statistically different from those who visited less than four rooms. The inferential statistic analysis allows to consider that the plant choice does not depend on the number of rooms visited by the participant ($\xi^2(3) = 4.379$; $p= 0.159$; N = 28).

Plant choice x Distance Travelled

The non-parametric Kruskall-Wallis Test was carried out in order to verify if the distance travelled by the participants in the VE had significant influence on the plant choice. An error probability type I (α) of 0.05 was considered. According to the test results, the distance travelled did not have a statistically significant effect on the plant choice ($\xi^2 KW(3) = 7.177$; $p=0.06$; N= 28).

CONCLUSION

According to Aguirre and D'Esposito (1999), the individual's mental representation of an environment is dependent of: i) age or development stage; ii) the duration of the interaction with the environment; iii) the way that the subject was introduced to the environment (e.g. self-exploration, guided-exploration, through a map); iv) the level of differentiation (details) of the environment; and v) the tasks that the subject has to accomplish into the environment. In this way, it was predicted that a key factor on plant choice would be the number of visited rooms. In this pilot study participants had to perform four tasks and their existence should act as an aid in the memorization of the number of rooms. It was also hypothesized that the experimental condition and the distance travelled in the VE could influence the spatial knowledge acquisition.

The findings from this pilot study suggest that during the interaction with a VE, participants were able to identify the basic morphology of the environment as they were able to identify the independent areas (Areas 1 and 2). The influence of Area 2 (intersections/decision points) was clear in participants' choices due to the fact that this area was identified correctly for most of the participants (86%). However, it was an unexpected finding as it was predicted that the key factor on plant choice would be the number of visited rooms. In this way, the existence of tasks that were related to the number of rooms visited by the participants seems that did not influence the spatial knowledge acquisition. The findings do not confirm the existence of clear influence of neither the experimental condition nor the distance travelled into the VE over the plant choice, thus the hypothesis that longer distance travelled might result in a better understanding of the environment cannot be assumed as true.

This study demonstrated that participants seem to be able to acquire spatial knowledge interacting with a VE, even using a non locomotor-based motion controller, such as a joystick, a HMD with a small field-of-view (30°H and 18°V) and without stereoscopic view) contrary to what was predicted by Cánovas, Espínola, Iribarne, & Cimadevilla (2008) and Billinghurst and Weghorst (1995). However more studies are needed in this area.

The findings from this pilot study suggest that decision points may figure as powerful landmarks, in this way future work with more complex environments and considering others methods to assess cognitive map may reveal different aspects which were not considered as the influence of the decision points in spatial knowledge acquisition. The analysis of participants' path into the VE compared to the analysis of method for acquiring participants' mental representation may be an useful way to understand the spatial knowledge acquisition.

ACKNOWLEDGMENTS

This research was supported by the Portuguese Science and Technology Foundation (FCT) grants (PTDC-PSI-69462-2006 and SFRH/BD/38927/2007).

REFERENCES

Aguirre, G., & D'Esposito, M. (1999). Topographical disorientation: a synthesis and taxonomy. [Review Article]. *Brain, 122*, 1613-1628.

Billinghurst, M., & Weghorst, S. (1995). *The use of sketch maps to measure cognitive maps of virtual environments*. Paper presented at the Proceedings of the Virtual Reality Annual International Symposium (VRAIS'95).

Cánovas, R., Espínola, M., Iribarne, L., & Cimadevilla, J. M. (2008). A new virtual task to evaluate human place learning. *Behavioural Brain Research, 190*(1), 112-118.

Carassa, A., Geminiani, G., Morganti, F., & Varotto, D. (2002). Active and passive spatial learning in a complex virtual environment: The effect of effcient exploration. *Cognitive Processing, 3*(4), 65-81.

Cubukcu, E., & Nasar, J. L. (2005). Relation of Physical Form to Spatial Knowledge in Largescale Virtual Environments. *Environment and Behavior, 37*(3), 397-417.

Golledge, R. G. (1999). Human Wayfinding and Cognitive Maps. In R. G. Golledge (Ed.), *Wayfinding Behavior: Cognitive mapping and other spatial processes* (pp. 5 - 45). Baltimore: Johns Hopkings University Press.

Hunt, E., & Waller, D. (1999). Orientation and Wayfinding: a review. University of Washington.

Lindberg, E., & Gärling, T. (1982). Acquisition of locational information about reference points during locomotion: The role of central information processing. *Scandinavian Journal of Psychology, 23*(1), 207-218.

Moeser, S. D. (1988). Cognitive mapping in a complex building. *Environment and Behavior, 20*(1), 21-49.

Morganti, F., Carassa, A., & Geminiani, G. (2007). Planning optimal paths: A simple assessment of survey spatial knowledge in virtual environments. *Computers in Human Behavior, 23*(4), 1982-1996.

Omer, I., & Goldblatt, R. (2007). The implications of inter-visibility between landmarks on wayfinding performance: An investigation using a virtual urban environment. *Computers, Environment and Urban Systems, 31*(5), 520-534.

Siegel, A. W., & White, S. H. (1975). The development of spatial representations of large-scale environments. In H. W. Reese (Ed.), *Advances in Child Development and Behavior* (Vol. 10, pp. 9-55). New York: Academic Press.

Thorndyke, P. W., & Hayes-Roth, B. (1982). Differences in spatial knowledge acquired from maps and navigation. *Cognitive Psychology, 14*(4), 560-589.

Chapter 59

Characterization of Mental Models in an Inductive Reasoning Task Using Measures of Situation Awareness

Tao Zhang[1,] David B. Kaber[2]

[1]Department of Electrical Engineering and Computer Science
Vanderbilt University
Nashville, TN 37212, USA

[2]Edward P. Fitts Department of Industrial and Systems Engineering
North Carolina State University
Raleigh, NC 27695-7906, USA

ABSTRACT

The objective of this research was to develop and validate an empirical method of situation awareness (SA) assessment as a basis for characterizing mental models in an inductive reasoning task with complex context. An experiment was conducted in which participants watched videos of two structured detective stories while simultaneously playing a simple shooter game. Participant SA was measured during the trial and post-trial measures included concept mapping and a knowledge test on the stories. Three different types of mental models were hypothesized to occur, including a simple list of story elements, elements grouped based on importance, and an organized network of elements. A fuzzy inference model was developed to classify measures of participant SA into patterns that matched with the three hypothesized mental model types. Based on fuzzy inference, the mental model type showed utility for predicting knowledge test performance. The fuzzy model outputs and measures of SA were found to be independent from concept map measures (no significant correlations). Regression analyses showed that for both stories, levels of SA were predictive of knowledge test performance. The experiment results provided empirical evidence of the utility of SA measures for assessing mental models and predicting task performance in a complex inductive reasoning task (i.e., understanding detective scenarios), independent of an existing assessment approach (i.e. concept mapping).

Keywords: Situation awareness, Mental models, Inductive Reasoning, Concept Mapping

INTRODUCTION

The concept of mental models has been widely accepted in various domains, particularly human factors, in attempting to explain how human performance with complex technical or physical systems occurs. From a functional perspective, Rouse and Morris (1986) define mental models as, "the mechanisms whereby humans are able to generate: descriptions of a system's purpose and form; explanations of system functioning and observed system states; and predictions of future system states." This definition implies that the integration of different levels of information is the key to forming a complex representation of a task and determining how to interact with a system (Sanderson, 1990).

Although the concept of mental models as a coherent form of information organization and processing is appealing, in practice the construct may be limited due to abstract description of the structure or functional characteristics of models. In order to generate close approximations of the forms of mental models as a basis for design and training, a variety of approaches have been proposed, including verbal protocol analysis (e.g., Sanderson, Verhage, & Fuld, 1989), analytical and empirical modeling (e.g., Jagacinski & Miller, 1978), concept-based analysis (card sorting, repertory grid technique, causal mapping and pair-wise rating; see Langan-Fox, et al., 2000), and concept mapping (Novak & Gowin, 1984).

In addition to these methods, the close relationship between situation awareness (SA) and mental models may also be used for the assessment of mental models, as suggested by Endsley's (1995, 2000a) SA theory. In general, SA refers to the level of awareness and dynamic understanding that an individual has of a situation. Endsley (1995) describes SA as a state of knowledge resulting from a dedicated situation assessment process with three levels: "the perception of elements in the environment within a volume of time and space, the comprehension of their meaning, and the projection of their status in the near future." As mental models have been established as fundamental for SA (Endsley, 2000a) in tasks that involve

SA as part of performance, mental models may be formed and updated through the same course as SA. Related to this, Endsley (2000b) has attempted to demonstrate the utility of SA for assessing mental models, which may be descriptive of pilot performance.

In inductive reasoning tasks, mental models develop by integrating information in novel ways in order to achieve task goals. The conceptual organization of information is critical to the reasoning process (Holland, et al., 1989). Model construction also serves as the major source of inductive changes in knowledge structures, such as SA, and ultimately reasoning and decisions. It is thus reasonable to suggest that changes in SA can reflect content changes in the underlying mental models. Following this logic, the present study used detective stories as the stimuli in inductive reasoning tasks. The goal for participants was to determine who committed the crime based on their understanding of each story. The understanding process allowed participants to discover related information in the environment, comprehend the situation, and establish hypotheses. This process may also be interpreted as one of forming and updating SA in an attempt to construct an appropriate mental model of the crime. Furthermore, as understanding the detective story is a process of uncovering the hidden truth (i.e., how the crime was committed), it allows for experimental testing of participant performance in the inductive reasoning task in terms of knowledge of the ground truth (i.e., accuracy of their mental model).

METHOD

PARTICIPANTS

Twenty-four undergraduate and graduate students (9 females and 15 males) at North Carolina State University were recruited for the experiment. Their ages ranged from 19 to 36 years (M=23.3, SD=3.24). All participants reported none or little experience in reading or watching detective stories or television in a pre-experiment screening questionnaire.

STIMULI AND APPARATUS

Two detective stories extracted from a past novel-based television show were presented to participants on video, with an average length of 30 minutes. The detective stories were chosen as task stimuli because they were expected to be more motivating to participants, in contrast to simple induction tasks. The videos were edited to be intense and there were clear differences among cues in terms of the degree to which they reduced viewer uncertainty on the identity of the criminal.

During trials, participants were presented with a large projection screen divided into two halves. The left half of the screen presented the detective story video, while the right half of the screen was used to display a Flash®-based first person shooter game. The shooter game was used as a means of promoting a participant's sense of involvement in the detective investigation and a more realistic task loading by requiring some use of divided attention. Participants were instructed to attend to and comprehend as much information as possible on the detective video and to address the shooter task as they could.

HYPOTHESIZED MENTAL MODELS

A typical inductive reasoning task can be conceived as involving three steps with different levels of information processing. First, elements relevant to the task goal need to be identified and extracted (or perceived) from the environment. Second, these elements are assigned different levels of significance and connected according to certain relations, such as categorical, causal and temporal, as part of comprehension. Third, the connected elements are further organized and adjusted into a more meaningful network to allow in depth examinations (or projections to be made). Based on this, it was hypothesized that three specific types of mental models might be formed by participants in the experiment trials, as described below.

Mental Model Type 1: It was expected that some participants would form a linear list of relevant elements (or clues) based on their sequence of appearance in the story. This is a model formed based on perceived information, involving limited voluntary processing employing knowledge from long-term memory.

Mental Model Type 2: It was also expected that some participants would form a model representing grouping or categorization of elements in a story. This model type corresponds to the association of different weights with elements and formation of information groups for simplification of the story.

Mental Model Type 3: Finally, some other participants might form a plausible network model based on the identification of related elements and their causal relations, as well as thorough understanding of the story. This type of model reflects an informative organization of elements from the story and participants who form this type of model should be able to identify the root cause of the crime.

EXPERIMENT DESIGN AND VARIABLES

A randomized complete block design was used for the experiment, in which participants served as the blocking factor. The presentation sequence of the two detective stories was balanced across participants. Predictor variables included measures of participant correct responses to SA queries, targeting three levels (Level 1, 2 and 3), and measures of participant concept mapping in terms of content similarity (CS) and structural similarity (SS), as compared to the "perfect" concept maps of two independent raters. Level 1 SA queries were to test the accuracy of participant perception of information elements in the story ("what", "when" and "where"); Level 2 SA queries were structured to assess participant comprehension of elements of the story, including questions regarding the importance relative to the goal of identifying criminals; Level 3 SA queries were designed to determine participant predictions or expectations in viewing the story, including questions regarding possible causal relationships among elements and estimation of the course of the crime. Participant answers to SA queries were graded in terms of percentages of correct responses according to the original story. Regarding the concept map measures, CS only involved comparison of the elements in participant maps to a rater's map while disregarding the links between elements; SS involved comparing all links in relation to elements in participant maps to the rater's map. These two indices are essentially the percentage of match between elements and links among participant maps and rater "perfect" maps (Gwee, 2005). The correlation of concept map ratings between the two raters (i.e., inter-rater reliability) was calculated and the average values of the two raters' ratings were used. The dependent variable was knowledge test performance in terms of total percent correct responses to questions. The questions on the knowledge test were targeted at participant identification of essential elements in a story, the importance of certain elements, and causal relationships among elements. These questions were different from the SA queries posed to participants during trials, in terms of both content and depth (i.e., the questions were based on the story as a whole). Participant answers were graded according to the ground truth of the story, as recorded by the experimenter.

PROCEDURE

At the beginning of the experiment participants read and signed an informed consent form and completed a short survey addressing their prior experiences in reading or watching detective stories. A practice trial was provided before formal testing, during which participants were familiarized with the experiment setup and SA queries. Participants were then introduced to the concept mapping technique as described by Novak and Gowin (1984). As previously mentioned, in the formal test trials participants watched the detective story videos presented on the left half of the projection screen, while continuously playing the shooter game on the right half of the screen. At pre-selected points in time (relatively inactive periods of the detective story), at least 5 minutes after a story began and 2 minutes before the end of a story, the video and shooter game were paused and six SA questions (two targeting each level) were presented on the display screen. Participants were asked to give their answers verbally within 1 minute. There were four freezes for SA queries in each test trial and the average inter-freeze time was about 5 minutes.

Immediately after the end of each video, participants created a concept map of the specific story based on their understanding. They were asked to identify any concepts or elements they felt were important to illustrate the story and/or identify the criminal(s), in an organized manner. There was no time limit for participants to draw the concept map; however, they generally completed a map within 15 to 20 minutes. After drawing the concept map, participants completed anagram puzzles (making words from scrabbled spellings) for 3 minutes. The anagram task was intended to wipe-out working memory representations of information (i.e., the concept map). Subsequently, participants were presented with the knowledge test questions. Participants took a break after finishing the first test trial and whenever they felt ready, they began the second trial. The experiment lasted about 2 hours for each participant.

RESULTS AND DISCUSSION

DESCRIPTIVE STATISTICS

The descriptive statistics of the predictor and dependent variables are shown in Table 1, including measures of participant SA at the three levels (SA1, SA2 and SA3), concept mapping (CS and SS), and knowledge test performance (KTEST). The range for all variables are in percentages (0 – 1). The inter-rater reliabilities for the CS and SS were 0.81 and 0.87 for the concept map for Story 1, and 0.86 and 0.90 for Story 2. There was no significant correlation between measures of SA and concept map measures. One possible reason for this is that measuring SA for mental model elicitation represents a "bottom-up" approach (i.e., higher levels of SA are generally built on lower levels of SA) focusing on the state of working memory at freezes, while concept mapping is more of a "top-down" approach based on the information organized in long-term memory over time. Therefore, SA measures and concept mapping might not assess the same characteristics of mental models or provide complementary information on models.

Table 1 Descriptive statistics of response variables

STORY	VARIABLE	MEAN	STANDARD DEVIATION	MIN.	MAX.
	SA1	0.73	0.20	0.38	1.00
	SA2	0.81	0.16	0.50	1.00
Story 1	SA3	0.70	0.18	0.25	1.00
	CS	0.55	0.08	0.36	0.67
	SS	0.49	0.08	0.40	0.68
	KTEST	0.60	0.16	0.23	0.85
	SA1	0.74	0.13	0.50	1.00
	SA2	0.84	0.12	0.63	1.00
Story 2	SA3	0.64	0.16	0.38	1.00
	CS	0.51	0.11	0.36	0.79
	SS	0.46	0.12	0.20	0.71
	KTEST	0.71	0.13	0.50	1.00

SA MEASURES AND FUZZY LOGIC MODELING

In general, it was expected that participant SA across levels would increase as mental model complexity increased from Type 1 to Type 3. Specifically, the Type 1 mental model was hypothesized to be essential to all levels of SA; the Type 2 model was expected to provide participants with the ability to respond to Level 2 SA questions in addition to Level 1 questions; and participants achieving the Type 3 model were expected to be able to answer all levels of SA questions. Based on these assumptions, a fuzzy logical model was developed to transform the patterns of participant responses to SA queries into the three hypothesized types of mental models. The fuzzy logic model consisted of three input variables (the three levels of SA), one output (Mental Model Type), membership functions associated with the input and output variables, and fuzzy inference rules. The membership functions of the input variables were determined from the histograms of the raw SA data (i.e., empirical distributions). The dividing points for "low", "medium" and "high" groups were defined based on the sample mean ± standard deviation. The membership function of the output, as the degree of belonging to a certain model type, was triangular for each category and the categories were evenly distributed between 0 and 1 (i.e., 0 to 0.33 for "Type 1", 0.33 to 0.67 for "Type 2" and 0.67 to 1 for "Type 3"). The fuzzy output was indicative of which type of mental model a participant might have formed. The fuzzy inference rules are listed in Table 2.

Table 2 Rules of the fuzzy logic model

IF LEVEL 1 SA IS	LEVEL 2 SA IS	LEVEL 3 SA IS	THEN MODEL IS
High	Not high	Not high	Type 1
Not high	Low	Low	Type 1
Medium	Low	Low	Type 1
Low	Low	Low	Type 1
High	Not high	Not high	Type 2
High	High	Not high	Type 2
Medium	Medium	Not high	Type 2
Low	Not low	Low	Type 2
High	High	High	Type 3
High	High	Not low	Type 3
Medium	High	High	Type 3
Medium	Medium	Not low	Type 3
Low	Low	Not low	Type 3

The descriptive statistics on the fuzzy inference model output are shown in Table 3. For Story 1, there was a significant positive correlation between the inferred model membership and the percentage of participant correct answers to questions on the knowledge test ($r = 0.4840$, $p = 0.0166$); however, this correlation was not present for Story 2. Based on the data for both stories, the model membership values were not correlated with any concept map measures. The most likely model type for each participant was determined by locating the fuzzy logic output in the membership functions (i.e., defuzzification). The results are summarized in Table 4. The majority of participants in Story 1 tended to form mental models similar to the Type 2 and Type 3 hypothesized models. They were able to list elements from the story and group them according to relative importance, but some were not be able to internally formulate all possible causal links among the elements. For Story 2, the majority of participants appeared to develop mental models in-line with Type 2. Furthermore, a fuzzy cluster analysis of the SA measures showed similar groupings to the fuzzy inference output, which suggests that the fuzzy inference rules were representative of the SA patterns associated with the different types of mental models.

Table 3 Descriptive statistics on fuzzy logic model output

STORY	MEAN	SD	MIN.	MAX.	CORR. WITH KTEST
Story 1	0.568	0.212	0.107	0.893	$r = 0.4840$ ($p = 0.0166$)
Story 2	0.596	0.216	0.107	0.893	$r = 0.3342$ ($p = 0.1105$)

Table 4 Counts of mental model types for participants from the fuzzy inference

STORY	MODEL TYPE 1	MODEL TYPE 2	MODEL TYPE 3
Story 1	5 (20.8%)	9 (37.5%)	10 (41.7%)
Story 2	6 (25%)	13 (54.2%)	5 (20.8%)

To further validate the fuzzy inference result, the mental model types indicated by the output memberships were used to classify participants into three groups and a one-way ANOVA model was conducted to determine the extent to which participant mental model type predicted knowledge test performance after exposure to each story. The ANOVA model for Story 1 proved significant ($F(2, 21) = 4.97, p = 0.0171$). Post-hoc analysis using Duncan's multiple range test revealed that participants achieving the Type 3 (most complex) mental model produced significantly higher ($p < 0.05$) performance on the knowledge test than those achieving Type 1 and Type 2 mental models. There was no significant difference of knowledge test performance among the three classified groups for Story 2.

The results showed the fuzzy output values for Story 1 had utility for predicting participant performance and the underlying mental models. The non-significant result for Story 2 suggests the pattern of participant SA was different from that for Story 1. One possible reason for this is that the nature of the second detective story was different from the first, although the difficulty of the two stories was roughly the same, according to the comparison of concept map measures (CS and SS). In

Story 1 (murder of a college girl), there are multiple suspects including the victim's boyfriend, professor, and stepmother; in Story 2 (murder of a famous cricket player), initial evidences all point to a single suspect (the victim's teenage son). A further examination of Story 2 data showed that most of the participants had similar Level 1 and 2 SA, which was different from the Story 1 data. The second detective story may have better facilitated the identification and organization of important information related to the initial suspect, making the measures of SA less consistent with the expected pattern for the hypothesized mental model types. In addition, since the fuzzy inference rules were based on the hypothesized mental model types, the fuzzy logic model did not produce a "good" conversion of the SA measures with different patterns, leading to fewer unique model classifications.

REGRESSION ANALYSIS

Stepwise regression models, including SA (SA1, SA2 and SA3) and concept measures (CS and SS) as predictors of knowledge test performance (KTEST), were used to assess the extent to which performance of inductive reasoning could be predicted by the two different mental model assessment techniques. Results revealed that for Story 1, a regression model included SA2 and SA3 as independent variables (R^2 = 0.358, model p = 0.0095, regression coefficients for SA2 and SA3 were 0.407 (SD = 0.172) and 0.368 (SD = 0.154)). For Story 2, the regression model included SA1 and SA2 (R^2 = 0.398, model p = 0.0156, regression coefficients for SA1 and SA2 were 0.329 (SD = 0.193) and 0.308 (SD = 0.203)). Both models excluded CS and SS as predictors of participant knowledge test performance. The regression coefficients of the levels of SA in both models were relatively close, suggesting that both levels made almost equivalent contributions to mental model-driven task performance.

CONCLUSIONS

The goal of this research was to develop and validate an empirical method of SA and task performance assessment for characterizing mental models in an inductive reasoning task requiring SA and mental model formation. A fuzzy inference model was formed to relate patterns of SA to hypothesized mental model types. The model output was positively correlated with performance on a knowledge test for a first detective story (Story 1); no significant correlation was found for a second story (Story 2), possibly due to its different structure and information presentation. This result suggests that the fuzzy inference model may be sensitive to the patterns of SA expected from hypothesized mental model types. SA was not correlated with concept map measures for both stories. Therefore, SA may be independent of (or complementary to) concept map measures for assessing mental models. This is a contention that needs further investigation. The regression analysis indicated that certain levels of SA were predictive of participant performance on the knowledge test for the detective stories. SA measures may be unique in characterizing task performance requiring the use of mental models. One limitation of the present study was a lack of control over the structure of the detective stories. As the stimuli used in the experiment were edits from an existing television series, it was challenging to ensure the two stories had a similar course of information presentation. This may explain why participant SA responses on Story 2 differed from those on Story 1.

The present study can be fit into existing research efforts (e.g., Moray, 1999; Payne, 2003) that have attempted to obtain a full understanding of mental models in practical contexts. As demonstrated in the present study, one way to tackle the mental model characterization problem may be to start from analyzing the characteristics of tasks involving mental models, identify a task-related concept (e.g., SA in inductive reasoning tasks) that has an essential linkage with mental models in the course of information processing, and then empirically testing the linkage with hypothesized forms of mental models. From a practical perspective, this methodology may be applied in system design and development of operator training programs. Measures of SA may be used to assess types of mental models trainees use, in effect providing trainers the ability to interpret training results in terms of mental models.

REFERENCES

Endsley, M. R. (1995). Toward a theory of situation awareness in dynamic systems. Human Factors, 37(1), 32-64.

Endsley, M. R. (2000a). Theoretical underpinnings of situation awareness: A critical review. In M. R. Endsley & D. J. Garland (Eds.), Situation Awareness Analysis and Measurement. Mahwah, NJ: Lawrence Erlbaum Associates.

Endsley, M. R. (2000b). Situation models: An avenue to the modeling of mental models. Proceedings of the 44th Annual Meeting of the Human Factors and Ergonomics Society.

Holland, J. H., Holyoak, K. J., Nisbett, R. E., & Thagard, P. R. (1989). Induction: Processes of Inference, Learning, and Discovery MIT Press.

Jagacinski, R. J., & Miller, R. A. (1978). Describing the human operator's internal model of a dynamic system. Human Factors, 20(4), 425-433.

Langan-Fox, J., Code, S., & Langfield-Smith, K. (2000). Team mental models: Techniques, methods, and analytic approaches. Human Factors, 42(2), 242-271.

Moray, N. (1999). Mental models in theory and practice. In D. Gopher & A. Koriat (Eds.), Attention and Performance XVII: Cognitive Regulation of Performance, Interaction of Theory and Application (pp. 223-258). Cambridge, MA: MIT Press.

Novak, J. D., & Gowin, D. B. (1984). Learning How to Learn. New York: Cambridge University Press.

Payne, S. J. (2003). Users' mental models: The very ideas. In J. M. Carroll (Ed.), HCI Models, Theories and Frameworks: Toward a multidisciplinary science. San Francisco: Morgan Kaufman.

Sanderson, P. M. (1990). Knowledge acquisition and fault diagnosis: experiments with PLAULT. Systems, Man and Cybernetics, IEEE Transactions on, 20(1), 225-242.

Sanderson, P. M., Verhage, A. G., & Fuld, R. B. (1989). State-space and verbal protocol methods for studying the human operator in process control. Ergonomics, 32(11), 1343-1372.

Chapter 60

Acquisition of Skill Sets and Mental Models over Time

Thomas Fincannon, Scott Ososky, Joseph Keebler, Florian Jentsch

Team Performance Laboratory
University of Central Florida
Orlando, FL 32826, USA

ABSTRACT

This paper is intended discuss issues associated with different measures of ability and performance with respect to the operation of multiple unmanned systems. In describing these differences, we are interested in illustrating general trends of skill acquisition and factors that may influence the rate at which skills are acquired over time. Based on the results across two experiments, we argue that declarative knowledge (e.g. target familiarity) represents a dimension performance that can improve over a short period of time (~2 hours), but other dimensions of performance (e.g. correct localization using ground images) represent difficult dimensions of performance that require long periods of time for significant improvement (+9 hours). Furthermore, individual differences, such as spatial ability, and reconnaissance performance appear to be associated with the rate at which operators improved at performing different types of localization tasks. Implications of these findings are discussed.

Keywords: UAV, UGV, Target Identification, Localization, Familiarity, Mental Model, Skill Acquisition, Time, Learning, Training

INTRODUCTION

Research involving unmanned systems revolves around the study of complex tasks requiring a specific set of interdependent skills and complicated team dynamics between men and machines. Much of the current literature in unmanned systems research has examined the operator to asset ratio, the design of systems to support multiple operators working with multiple vehicles, and the influence of teaming with respect to autonomous and semi-autonomous systems. Little consideration, however, has been given to the process of performance improvement over multiple iterations with the same study. Attending to the improvement of overall performance with unmanned systems requires an examination of an operator's skill acquisition rates. We are interested in understanding the development of different unmanned systems skill sets, their relationship to one another, and their effects on the overall impact of system performance over multiple trials. Understanding these relationships will provide for the improvement of system design and training programs which support the human component of the man-machine system.

There are many measures that can describe different aspects of an operator's ability to perform a given unmanned systems task. One method of categorizing these measures focuses on constructs. Target identification represents the ability of an individual or team to recognize and label objects in an environment. Target identification is typically represented as a measure of accuracy and represents a dimension of performance that is influenced by measures of target and equipment familiarity. Fincannon, Curtis, and Jentsch (2006) demonstrated that military vehicle familiarity was predictive of the amount of correct military target identifications when participants were presented with an unmanned ground vehicle video recording.

Complimentary to the construct of target identification is object localization, which refers to the ability to observe images of objects from a remote vehicle and then use that information to localize. The use of heterogeneous vehicles, consisting of both an unmanned ground vehicle (UGV) and unmanned aerial vehicle (UAV) assets has been found to improve localization performance (Chadwick 2005, 2008). The most likely explanation for this is the presence of multiple viewpoints of the same area, with the UAV providing a wider frame of reference for the operator(s). It has also been demonstrated that the reported workload of a UGV operator can be influenced by the amount of (navigation) support provided by a UAV operator teammate, dependent upon the UAV operators own spatial abilities (Fincannon, Evans, Jentsch & Keebler, 2008). This suggests that skill acquisition rates may have a unique impact at the individual level as well as the team level.

The time required to complete a mission is expected to be influenced by skill acquisition rates. By definition, the learning curve effect describes an inverse relationship between the time to complete a task and the number of times that task has been performed (Wright, 1936). However, it is important to accurately capture the effects of skill acquisition on mission time, as other factors can heavily influence this measure. Chen (2008) demonstrated that proving a single operator with control of multiple assets negatively influences the time to complete a reconnaissance mission. Co-located teams, as well as communication modalities have also been demonstrated to influence mission completion times (McDermott, Luck, Allender & Fisher, 2005).

The development of mental models is also expected to be impacted by the development of skills over time. A mental model is expressed as a representation formed by a user of a system and/or task, based on previous experience as well as current observation, which provides most of their subsequent system understanding (Wilson & Rutherford, 1989, p. 619). Mental models, in turn, shape the strategies used by unmanned system operators. Delaney and colleagues (1998) concluded that skill acquisition manifests as strategy shifts that create disconnected segments in the learning curve. Therefore, we should be able identify changes in mental models and strategy shifts by examining the interactions between the previously mentioned constructs (specifically time) and overall performance metrics.

There is also a need to attend to how these constructs are measured. Many studies of reconnaissance use accuracy to measure a dimension of performance, but few have discussed operator error. If error has been discussed, it is typically examined on its own, and the differences between measures of accuracy and error are not addressed. Localization, which has been examined in the context of accuracy (Fincannon, Evans, Phillips, Jentsch, & Keebler, 2009) and error (Rehfeld, 2006), can be used to illustrate some useful relationships, and while these measures may appear to be exact, mirror opposites that would be expected to provide different types of information, they are assessed in different ways that can have different impacts on performance. First, consider the impact of "no-response". While this can be captured as a zero in a measure of accuracy, it may not be an outright error that can be recorded as a negative response. Second, if there is a large amount of error that is associated with a particular dimension of performance, how do these different measures address degrees of error? If error of localization is assessed as a measure of distance between an actual and reported location, there is more variance, and potentially more power, with a measure of error than a simple correct/incorrect response that is captured with accuracy. An exploration into differences associated with different assessments of a single dimension of performance (i.e. localization accuracy/error) will be considered in this paper.

A final consideration should involve the potential for the influence of individual differences. One construct of interest in the operation of unmanned systems is spatial ability, which is generally described as an individual's ability to perceive a visual field, form representations of objects within that field, and mentally manipulates those objects (Carrol, 1993). As described by Fincannon, Evans, Jentsch and Keebler (2010), spatial ability has the potential to influence the ability of an operator to identify targets, localize objects in a remote environment, and navigate through that environment. While the positive impact of spatial ability has been discussed with these dimensions of performance with unmanned systems, there has been a lack of research that examines spatial ability in the context of skill acquisition. This paper intended to explore this component of individual differences and how they impact usage of unmanned systems.

In summation, this paper intends to assess skill acquisition over time. First, it will consider multiple measures of performance. Second, it will consider multiple-assessments approaches for measuring a single dimension of performance. Third, it will consider the impact of spatial ability on skill acquisition over time. This is intended to provide a stronger understanding of how operators use unmanned vehicles in the context of military reconnaissance missions.

EXPERIMENT 1

The purpose of this first study was to observe long term skill acquisition. Due to the difficult nature of obtaining participants for this period, the sample size is small, and the results are largely qualitative in their description.

METHOD

Participants

There was one participant in this study. This participant used a UAV and UGV to complete reconnaissance missions over the course of 12 sessions.

Testbed

As detailed by (Ososky, Evans, Keebler, & Jentsch 2007), the Military Operations in Urban Terrain (MOUT) facility is a 1:35 scale replica of Al-Najeef, Iraq. There were four quadrants to the city that were intended to represent desert terrain, low income housing, a market place, and a high-rise downtown area. The primary base of operations for all of these missions was located in a virtual foxhole that was intended to simulate a dismounted environment (Ososky et al. 2007). For each mission, the operator had access to an electronic/virtual whiteboard that presented an overhead map of the MOUT facility and allowed for planning of missions. Prior to the planning of vehicle routes, the operator received a Fragmented Order (FRAGO) and mission map that illustrated the location of and provided important information about each of the mission objectives.

Design & Measures

In order to create the missions, six classes of objectives were identified, and each of these classifications varied according to a type of obstacle (aerial obstruction, ground obstruction) and the type(s) of vehicles that should be sent to identify all of the objectives (UAV, UGV, both vehicles). These objectives were evenly split across two missions to create a session, and there were twelve sessions to complete over the course of the study. By using different combinations of objectives and targets to identify each mission was unique.

A variety of assessments were used in the present study. Pretests were used to assess spatial ability (Guilford-Zimmerman Test of Spatial Visualization), familiarity with targets, and familiarity with the equipment that was used to complete the missions. During the mission, performance on reconnaissance and mission time was assessed. Localization was assessed by presenting a series of three ground images; the location of these images was to be written on an overhead map. A score was considered correct if it was within a predetermined radius of the actual location, and error was assessed as the absolute difference between the actual location and the location that was reported. In order to assess mental models, a participant was presented with one of the six classes of objectives mentioned above and then asked to rate the appropriateness of sending a UAV, UGV, or both vehicles to that objective on a one to six scale. In order to measure development, the final assessment was compared to all earlier assessments (e.g. first to final, second to final, etc.) using measures of consistency and agreement. Learning curves were then created for each of these measures.

Procedure

Prior to any training or practice, spatial ability, target familiarity, equipment familiarity, localization, and mental models were assessed as initial intake measures. For each session, there were three segments. First, the participant was allowed to review training slides and a booklet of targets. Second, there was an assessment of target familiarity, equipment familiarity, localization, and mental model. Third, reconnaissance missions were performed. This format was repeated to create a total of twelve sessions.

RESULTS

Not counting the time taken to complete tests, approximately 537 minutes (~9 hours) were spent on the review of the training materials (107 minutes) and executing missions (430 minutes). At the ninth session (429 minutes of training & mission time), plateaus were reached for time spent on review (0 minutes) and time spent to complete the missions (23 minutes).

Measures of declarative knowledge developed at similar rates. At the second session (121 minutes of training & mission time), performance reached a plateau for target familiarity, equipment familiarity, and target identification on a mission. While the score on the target and equipment familiarity test remained constant, the ability of the participant to apply that knowledge on a reconnaissance task oscillated, such that there were drops in performance at sessions three (194 minutes) and seven (371 minutes). As time progressed, the degrees of change between the high and low scores decreased.

Measures of localization using ground images illustrated that this was a difficult task. After twelve sessions, only three of the eight possible locations had been reported correctly, and while this did represent some improvement over initial performance, there was not much improvement. Error on the other hand, appeared to be more sensitive to changes over time. As with training and mission time, error in localization reached a plateau at the ninth session.

While the mental model assessments reached perfect agreement with the final assessment at the tenth session (452 minutes of training & mission time), agreement and consistency showed different patterns of acquisition. Agreement reached its first plateau after the first training session and remained constant until the eighth session, where it gradually increased to perfect agreement. Consistency appeared to oscillate unstably until the seventh session and remained relatively constant until the ninth session, where it rapidly increased to perfect consistency.

DISCUSSION

While this represents a small sample size, this does illustrate some differences in the skill acquisition of different measures of performance. First, it appears as through declarative knowledge (i.e. target & equipment familiarity) and applications of that knowledge (e.g. target identification during a reconnaissance mission) can improve over approximately 2 hours of training & practice. Second, it appears as though knowledge regarding how to use vehicles (i.e. cue-strategy mental model) and applications of that knowledge (i.e. time to complete a reconnaissance mission) took approximately 7 hours. Third, some tasks, such as using UGV images to determine a correct location, are so difficult that they require an extreme degree of proficiency, and a time estimate to obtain this could not be achieved in the current observation.

EXPERIMENT 2

The purpose of this second experiment was to apply statistics to the qualitative data that was described above. As the methodology was unchanged across the two experiments, some data from the previous experiment is incorporated into this study.

METHOD

Participants

Data from six participants were used in this study. Data from one of the participants were used in the first experiment, but data from the remaining five participants were only used in this study.

Testbed, Measures, & Procedure

The testbed, measures, and procedures remained mostly unchanged across experiments one and two. Instead twelve sessions, data was examined across two sessions. For every measure except the mental model assessment, there were six participants used in the data analysis. As the final mental model assessment of experiment one served as an expert reference for experiment two, only data from the five new participants were used in this analysis.

Design

This analysis focuses on several different relationships. First, a paired sample t-test was used to test for significant improvement over time. Second, correlations were used to examine whether spatial ability was associated with the rate of change in measures, which was measured as a difference score. Third, correlations are used to discuss association between different measures in the analysis.

RESULTS

When comparing the first and final measures, only target familiarity $t(5)=16.28$, $p<.05$, equipment familiarity $t(5)=11.49$, $p<.05$, and consistency with the expert mental model $t(4)=3.87$, $p<.05$ increased significantly over time. As with experiment one, improvement in target and equipment was both fast and near perfect, 91.67% and 94.15% respectively, after a relatively short period of practice. While all of the other measures did show some degree of improvement, the differences over time were not significant.

The second pattern of results centered on spatial ability and how it influenced different dimensions of performance. When considering measures of localization, an operator's spatial ability test was predictive of the number of locations that an operator could correctly place on a map after training, $r = .82$, $p<.05$, the first session $r = .75$, $p<.10$, and the second session $r = .85$, $p<.05$. As with experiment one, there was little improvement over time, and this small difference was not correlated with spatial ability. In contrast however, the error score for localization showed more improvement, and rate of improvement for this measure of localization was highly correlated with participant spatial ability $r = .98$, $p<.05$. With respect to target identification, spatial ability showed marginally significant correlations with the number of targets reported by the final familiarity assessment, $r = .85$, $p<.10$, and the number of targets that were identified across, $r = .77$, $p<.10$.

With respect to skill acquisition, some the remaining correlations illustrated "common sense" relationships. For example, the amount of time spent with the execution of the first session was positively correlated with the number of targets that participants were able to report on the subsequent familiarity assessment, $r = .86$, $p<.05$. As illustrated by Fincannon et al. (2006), target familiarity before the start of training was associated with reconnaissance during the first session, $r = .91$, $p<.05$, but this was not significant during the second session, which indicates a leveling of differences across participants over time. If participants that identified more targets across all sessions were better at correct, $r = .90$, $p<.05$, and error, $r = .88$, $p<.05$, measures of localization. If an operator had high consistency with the expert before starting the second session, that operator was faster at completing that session, $r = .94$, $p<.05$.

DISCUSSION

Results from this second experiment illustrate several relationships. First, results with target and equipment familiarity indicate that this type of declarative knowledge can improve to near perfect levels after a relatively short period of time (2 hours). Second, localization was still an incredibly difficult dimension of performance. By the second session, the mean for the harshest measure of localization (accuracy) was ·1.5, and the highest score was four out of eight (50%).

Third, results illustrated that all participants in this second experiment became more like the mental model of the "expert" undergraduate from the first experiment. Fourth, constructs, such as spatial ability, are not only predictive of certain dimensions of performance, but they are also predictive of the rate at which participants can improve a certain dimension of performance. Fifth, the predictive ability of some measures, such as the first assessment of target familiarity, can decrease as participants learn more about the task.

GENERAL DISCUSSION

This study brought light to several points for discussion. First, different dimensions improve at different rates. While a measure of target identification can improve quickly, measures of mental models, mission times, and localization from UGVs take longer periods of time to develop. As a result, researchers need to attend to the relationship between the choice of a construct and whether they can say that results generalize to "expert" operators. If the main focus of a study is target identification, it is likely that participants can be trained to adequate levels of proficiency in a relatively short period of time. If the main focus of a study is localization, neither training nor selection is a useful tool for obtaining participants that are proficient at this task. If a 50% accuracy mark was used for the selection of participants, only one of the six participants, or 16.67% of the available subject pool, would have achieved this level of proficiency, and while the available subject pool might perform the task well, attrition is a new issue that needs to be dealt with.

A second item of interest involves the use of different approaches for assessing the same dimension of performance. As described in the introduction, accuracy and error of localization provided different descriptions of localization. While little improvement was observed with accuracy, the use of error provided a meaningful illustration of skill acquisition with this construct. The differences in these measures can probably be attributed to the difficulty of the task, where degrees of error improved over time, and while a measure of error could capture this, the change in the degree of error was not sufficient to produce meaningful improvements with accuracy.

Third, this paper illustrated a continued effect of spatial ability on the use of unmanned systems. As spatial ability refers to the ability to mentally manipulated complex imagery (Carrol, 1993), it should not be surprising that it improves an operator's ability to identify targets and interpret visual imagery to comprehend vehicle location. The new relationship with spatial ability involved its influence on the rate at which participant improved performance on a localization task, which represented one of the more difficult dimensions of performance.

A fourth item of discussion should address the relative dependence of one dimension of performance on another dimension of performance. For example, consider performance on a reconnaissance mission, where several measures can be tied together. All of the targets could be identified by the second session, but the operator required approximately 25 to 30 minutes to successfully complete this mission. If operators with this level of training are only allowed a certain period of time to complete a mission (e.g. 15 minutes), the dependence of reconnaissance on mission time could result in lower performance on the reconnaissance task. Conversely, operator training that is designed to improve mission times could improve performance on the same reconnaissance task. Understanding these interdependencies across measures can improve our conceptualization of constructs.

Understanding skill acquisition is important for a couple of reasons. First, by illustrating how operator skills sets develop, needs for training and design can be identified. Training interventions may not need to focus heavily on items that improve quickly (i.e. vehicle operation), but interventions that focus on the strategic use of remote vehicles may improve the rate of development for that mental model and relevant measures of performance. If certain dimensions of performance are difficult (e.g. localization from a UGV), those difficulties may identify a dimension of performance that is in need of design interventions (e.g. incorporation of GPS systems) to support operator performance. Second, it is useful for assessing periods of training. This is not only useful in applied domains, where training can be expensive, but it is also beneficial for the design of experimental studies, where a learning curve can introduce variance to a sample of participants. Attending to a wide base of cognitive measures and how they develop is important to develop an understanding of operator performance with unmanned systems.

REFERENCES

Chadwick, R. (2005). The impacts of multiple robots and displays views: An urban search and rescue simulation. *Proceedings of the 49th Annual Meeting of the Human Factors and Ergonomics Society.* 387-391.

Chadwick, R., & Gillan, D. (2008) Cognitive integration of aerial and ground views in remote vehicle operation. *Proceedings of SPIE, 6962,* 1-9.

Carrol, J.B. (1993). *Human cognitive abilities: A survey of factor-analytic studies.* Cambridge England: Cambridge University Press.

Chen, J., Durlach, P., Sloan, J., Bowens, L. (2008). Human-robot interaction in the context of simulated route reconnaissance missions. *Military psychology, 20(3),* 135-149.

Delaney, P. F., Reder, L. M. Stazewski, J. J., & Ritter, F. E. (1998). The strategy-specific nature of improvement: The power law applies by strategy within task. *Psychological Science, 9,* 1-7.

Fincannon, T., Curtis, M., & Jentsch, F. (2006). Familiarity and Expertise in the Recognition of Vehicles from an Unmanned Ground Vehicle. *Proceedings of the 50th Annual Meeting of the Human Factors and Ergonomics Society,*

1218-1222.

Fincannon, T., Evans, A. W., Jentsch, F., & Keebler, J. (2008). Interactive effects of backup behavior and spatial abilities in the prediction of teammate workload using multiple unmanned vehicles. *Proceedings of the 52nd Annual Meeting of the Human Factors and Ergonomics Society*, 995-999.

Fincannon, T, Evans, III, A.W., Jentsch, F, & Keebler, J. (2010). Dimensions of spatial ability and their influence on performance with unmanned systems. In D. H. AndrewsR. P. Herz, & M. B. Wolf (Eds.). *Human factors in defense: Human factors in combat identification* (pp. 67-81). Burlington, VT: Ashgate Publishing.

Fincannon, T., Evans, A. W., Phillips, E., Jentsch, F., & Keebler, J. (2009). The influence of team size and communication modality on team effectiveness with unmanned systems. *Proceedings of the 53rd Annual Meeting of the human Factors and Ergonomics Society*, 419-423.

Ososky, S., Evans, A. W., III, Keebler, J. R., & Jentsch, F. (2007). Using Scale Simulation and Virtual Environments to Study Human-Robot Teams. *Proceedings of the 4th Annual Conference of Augmented Cognition International* (pp. 183-189). Baltimore, Maryland.Daganzo, C.F. (1997), "A continuum theory of traffic dynamics for freeways with special lanes." *Transportation Research B*, 31(2), 83–102.

McDermott, P. L., Luck, J., Allender, L., & Fisher, A. (2005). Effective Human to Human Communication of Information Provided by an Unmanned Vehicle. *Human Factors and Ergonomics Society Annual Meeting Proceedings, 49*, 402-406.

Rehfeld, S. (2006). *The impact of mental transformation training across levels of automation on spatial awareness in human-robot interaction*. Unpublished Dissertation, University of Central Florida, Orlando, FL.

Wilson, J. R., & Rutherford, A. (1989). Mental models: Theory and application in human factors. Human Factors, 31(6), 617-634.

Wright, T.P. "Factors Affecting the Cost of Airplanes." *Journal of Aeronautical Sciences, 3(4)*, 122-128.

Chapter 61

Pre-Motor Response Time Benefits in Multi-Modal Displays

James L. Merlo[1], P. A. Hancock[2]

[1]United States Military Academy
West Point, NY, USA

[2]University of Central Florida
Orlando, FL, USA

ABSTRACT

The present series of experiments tested the assimilation and efficacy of purpose-created tactile messages based on five common military arm and hand signals. We compared the response times and accuracy rates to these tactile representations against the comparable responses to equivalent visual representations of these same messages. Results indicated that there was a performance benefit for concurrent message presentations which showed superior response times and improved accuracy rates when compared to individual presentations in either modality. Such improvement was identified as being due largely to a reduction in pre-motor response time and these improvements occurred equally in a military and non-military population. Results were not contingent upon the gender of the participant. Potential reasons for this multi-modal facilitation are discussed. The novel techniques employed to measure pre-motor response inform computational neuro-ergonomic models for multi-modal advantages in dynamic signaling. On a practical

608

level, these results confirm the utility of tactile messaging to augment visual messaging, especially in challenging and stressful environments where visual messaging may not always be feasible or effective.

Keywords: Visual Signaling, Tactile Signaling, Multi-Modal Advantage.

INTRODUCTION

Humans rely on their multiple sensory systems to continually integrate the environmental stimuli around them in order to build their perception of the world in which they live. While each sense is, in itself, remarkably adept at detection it is the combination and integration of these disparate sensory inputs which provide the rich tapestry of spatial, temporal, and object information on which humans rely to survive and thrive. The cross-modal fusion of these information sources is often more beneficial than simply increasing information from only one sensory modality. For example, Hillis, Ernst, Banks and Landy (2002) found that when combined, the value of multiple visual cues (e.g., disparity and texture gradients) did not produce as accurate performance as when visual and tactile cues were provided in an object property discrimination task. Comparing performance within the same modality versus combinations of two or more different modalities illustrates that information loss can occur during intra-modal presentations that does not occur with the fusion across different modalities. In the specific case of tactile and visual information there seems to be a highly efficient integration of the two sources (Ernst & Banks, 2002). This integration is especially beneficial when the cross-modal cues are congruent and match the top down expectancies generated by past experience.

Humans not only rely on their multiple sensory capacities to integrate different forms of stimuli, they also use these multiple sources to aid them in the initial process of orientation and the subsequent focus of their attention in space and time. When an individual directs their attention, regardless of the primary modality used in the process of detection, the other modalities are also frequently directed toward that same location. Indeed, it is the subject of an on-going debate as to the degree to which such orientation of attention is a multi-sensory construction (Spence & Driver, 2004) versus an over-dominantly visual process (Posner, Nissen, & Klein, 1976). In part, this issue can be approached from a neuro-physiological perspective. For example, Stein and Meredith (1993) have shown that bimodal and tri-modal neurons have a stronger cellular response when animals are presented with stimuli from two sensory modalities as compared with stimulation from only one modality. The combinations of two different sensory stimuli have been shown to significantly enhance the responses of neurons in the superior colliculus (SC) above those evoked by either uni-modal stimulus alone. Such an observation supports the conclusion that there is a multi-sensory link among individual SC neurons for cross-modality attention and orientation behaviors (see also Wallace, Meredith, & Stein,

1998). Multi-modal stimulation in the world is not always presented or received in a congruent spatial and temporal manner. This problem can be resolved in the brain by an over reliance on the one single dominant system which in humans is expressed in the visual modality (see Hancock, 2005).

To date, the exploration into the cross-modal attentional phenomenon has relied mainly on simple stimuli to elicit response (Spence & Walton, 2005). Gray and Tan (2002) used a number of tactors (vibro-tactile actuators) spanning the length of the participant's arm with lights mounted on the individual tactors. Using an appropriate inter-stimulus interval (ISI) and tactor spacing (see Geldard, 1982) to create the illusion of movement, either up or down the arm, they found that response times were faster when the visual target was offset in the same direction as the tactile motion (similar to the predictive abilities one has to know the location of an insect when it runs up or down the arm). Reaction times were slower when the target was offset in the direction opposite to the tactile motion. Such a finding supports the idea that the cross-modal links between vision and touch are updated dynamically for moving objects and are best supported perceptually when the stimuli are congruent.

In another study, Craig (2006) had participants judge the direction of apparent motion by stimulating two locations sequentially on a participant's finger pad using vibro-tactors. Visual trials included apparent motion induced by the activation of two lights sequentially. Some trials also were recorded with both visual and tactile stimuli presented together either congruently or incongruently. When visual motion was presented at the same time as, but in a direction opposite to tactile motion, accuracy in judging the direction of tactile apparent motion was substantially reduced. This superior performance during congruent presentation was referred to as 'the congruency effect'. A similar experiment conducted by Strybel and Vatakis (2004) who used visual apparent motion and found similar effects for judgments of auditory apparent motion. Auditory stimuli have also been shown to affect the perceived direction of tactile apparent motion (see Soto-Faraco, Spence, & Kingstone, 2004).

While all of these experiments with simple tasks are essential for understanding the psychological phenomena being studied, the extension of these findings into real-world conditions to embrace more applied stimuli is as yet largely unexplored. However, with advancements in tactile display technology and innovative signaling techniques, the importance of testing systems capable of assisting actual field communications is now both feasible and pragmatically important. Thus, the purpose of the present experiment was to examine combinations of visual and tactile communications of real-world operational signals in order to evaluate their efficacy for real-world applications. We also sought to distinguish whether multi-modal signal presentation led to performance advantages under such circumstances.

EXPERIMENTAL METHOD

EXPERIMENTAL PARTICIPANTS

To investigate the foregoing propositions, 72 participants (47 males and 25 females) ranging in age from 18 to 21, with an average age of 18.5 years, volunteered to participate. Of these individuals, 31 were from a large public southern metropolitan university and the remaining 41 were from a United States Military Academy. The latter group had prior experience with the visual form of the presented military visual signals, with the tactile form of the signals new to all.

EXPERIMENTAL MATERIALS AND APPARATUS

The vibro-tactile actuators (tactors) used in the present system were the model C2, manufactured by Engineering Acoustics, Inc (EAI). They are acoustic transducers that displace 200-300 Hz sinusoidal vibrations onto the skin. Their 17 gm mass is sufficient for activating the skin's tactile receptors. The tactile display itself is a belt like device with eight vibro-tactile actuators. Examples of the present belt system are shown in Figure 1. When stretched around the body and fastened, the wearer has an actuator over the umbilicus and one centred over the spine in the back. The other six actuators are equally spaced around the body; three on each side, for a total of eight (see also Cholewiak, Brill, & Schwab, 2004).

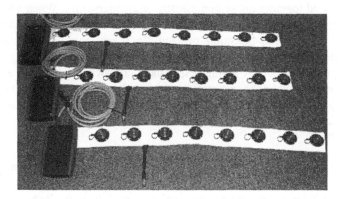

Figure 1. Three tactile displays belt assemblies are shown above along with their controller box.

The tactors are operated using a Tactor Control Unit (TCU) that is a computer-controlled driver/amplifier system that switches each tactor on and off as required. This device is shown on the left side of the tactile displays belts in Figure 1. The TCU weighs 1.2 lbs independent of its power source and is approximately one inch thick. This device connects to a power source with one cable and to the display belt with the other and uses Bluetooth technology to communicate with the computer driven interface. Tactile messages were created using five standard Army and Marine Corps arm and hand signals (Department of the Army, 1987). The five signals chosen for the present experiment were, "Attention", "Halt", "Rally", "Move Out", and "Nuclear Biological Chemical Event (NBC)". The tactile representations of these signals were designed in a collaborative effort involving a consultant group of subject matter experts (SMEs) consisting of former US Soldiers and Marines.

Short video clips of a soldier in uniform performing these five arm and hand signals were edited to create the visual stimuli. Careful editing ensured the timing of the arm and hand signals closely matched that of the tactile presentations (see Figure 2). A Samsung Q1 Ultra Mobile computer using an Intel Celeron M ULV (900 MHz) processor with a 7" WVGA (800 x 480) liquid crystal display was used to present videos of the soldier performing the arm and hand signals. This computer ran a custom LabVIEW (8.2; National Instruments) application that presented the tactile signals via Bluetooth to the tactor controller board and captured all of the participant's responses via mouse input. Participants wore sound dampening headphones with a reduction rating of 11.3 dB at 250 Hz. This precaution was designed to mask any possible effects which could have accrued due to extraneous auditory stimuli produced by tactor actuation. As this is an issue which has caused some degree of controversy in the past, we were careful to control for this potential artifact in our own work (cf., Broadbent, 1978; Poulton, 1977).

Figure 2. A computer screen shot showing what the participant viewed as the signals were presented. The participant mouse clicked on the appropriate signal name after each presentation.

EXPERIMENTAL DESIGN AND PROCEDURE

Participants first completed an informed consent document in accordance with the strictures of the American Psychological Association (APA). Participants then viewed a computer-based tutorial that described each arm and hand signal individually. For each signal, a short description was presented. Participants then viewed a video of a soldier in uniform performing the signal followed by a direct experience of its tactile equivalent. Finally, the participants were able to play the signals concurrently (both visual and tactile representation) together. Participants were allowed to repeat this presentation (i.e., visual, tactile, visual-tactile combined) as many times as they desired. Once the participant reviewed the five signals in the two presentation styles, a validation exercise was performed. Participants had to correctly identify each signal twice before the computer would prompt the experimenter that the participant was ready to begin.

The display of each signal was presented in one of three ways; i) a visual only (video presentation of the arm and hand signal), ii) a tactile only (tactile representation of the arm and hand signal), and iii) both visual and tactile simultaneously and congruent (i.e. exactly the same signal was presented both through the video and through the tactile system at the same time for all of these trials). The participants were presented each signal visually 8 times (8 trials x 5 different signals = 40 total trials to be visual only, tactile only, and combined visual and tactile presentations). This gave a grand total of 120 trials. The order that each participant performed the 120 trials was completely randomized. The entire experiment took less than an hour to complete.

Before each trial began, the mouse cursor had to be placed inside a small square in the center of the screen by the participant. The presentation of the signal, regardless of its modality, started the timer and the following performance responses were collected: i) the initial movement of the mouse, ii) the latency to name the received signal, iii) the signal named and accuracy of that choice. This formatting permitted us to parse the response into pre-motor time (the first movement of the mouse) and motor time (the time to place the cursor in the appropriate response box). It was these responses that were subjected to analysis.

RESULTS

Results were analyzed in terms of the speed of the response and the accuracy of the response under the respective conditions. We did conduct an initial analysis for any potential sex differences but found no significant influence upon any of the measures recorded. The subsequent analysis was therefore collapsed across sex. A one-way Analysis of Variance (ANOVA) was performed on the mean response times across the three experimental conditions of visual presentation, tactile presentation or visual-tactile concurrent and congruent presentation, with the following results: $F(2, 213)=9.37$, $p<.01$, ($\eta^2_p = .961$, $\beta= 1.00$). Post hoc analysis subsequently showed that simultaneously presented congruent signals resulted in significantly faster response times than visual signals presented alone $t(71)=3.15$, $p<.01$, see Figure 3. Also, as is evident from this illustration, responses to the congruent signals were also faster than tactile responses alone $t(71)=10.29$, $p<.01$. Additionally, the visual only presentation of the signal was significantly faster than the tactile only presentation of the signal $t(71)=-4.15$, $p<.01$.

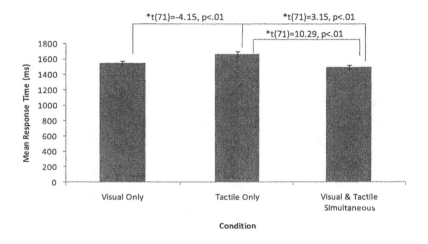

Figure 3. Response Time in milliseconds by signal presentation condition.

Analysis of the response accuracy data showed that there was a significant difference in the accuracy rate between the visual and tactile signals when presented alone $t(71)=-7.10$, $p<.01$. This difference was most likely due to the extraordinarily high accuracy visual performance rate since the military participants were already familiar with and already had some previous level of training for the visual presentation of the signals and no prior experience for the tactile presentations. There was also a significant difference in the accuracy rate when responses using the tactile modality were compared to the concurrent congruent presentation of the signals, $t(71)=7.47$, $p<.01$. Here, response to tactile signals proved less accurate than to the combined visual-tactile presentation. The overall lower accuracy rate for the tactile signaling is again attributed to the confusion between the tactile signal for 'NBC" and 'Halt". Analysis without the "NBC" tactile signal data again removed these significant differences in response accuracy. There was no significant difference between responses for the visual only condition and the combined condition.

A one-way Analysis of Variance (ANOVA) was performed on the mean response times for the pre-motor element (the time that elapsed from presentation of the signal to the first movement of the mouse) across the three experimental conditions of visual presentation, tactile presentation or visual-tactile concurrent and congruent presentation. This analysis produced a significant effect: $F(2, 213)=5.48$, $p<.01$, ($\eta^2_p = .961$, $\beta= 1.00$). Subsequent pair-wise comparisons showed that simultaneously presented congruent signals resulted in significantly faster pre-motor response times than visual signals presented alone $t(71)=4.30$, $p<.01$, see Figure 4. Also, as is evident from the illustration, the congruent signals were faster than those pre-motor times for tactile alone $t(71)=-2.9$, $p<.01$. Additionally, the visual only presentation of the signal was significantly faster for pre-motor response than the tactile only presentation of the signal $t(71)=-2.89$, $p<.01$.

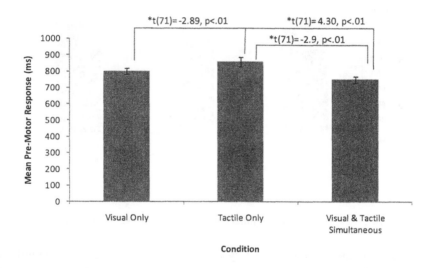

Figure 4. Pre-motor response time in milliseconds by signal presentation condition.

As previously stated, the presentation of the signal, regardless of its modality, started the experimental timer, allowing the capture of the latency from signal elicitation to the initial movement of the mouse, or pre-motor response. The latency to name the received signal, or in other words, the motor time, the time that it takes from the initial mouse movement to the time that the mouse resides in the appropriate response box was regarded to as the motor response time. There were no differences found across any of the experimental conditions for motor response latency.

It was further hypothesized that there could be some differences between the two respective groups of student and cadet participants due to their differential experience with the hand signals communicated. The participants from the military academy had some prior experience with the visual form of message while the university students were encountering them for the first time. To a degree, any such difference should have been mitigated by the practice given. However, we chose to examine this eventuality analytically. A simple t-test did distinguish such a difference which was evident in the pre-motor response time to the tactile signals only (i.e., $t(70)=1.99$, p<.01 [military cadets = 785 ms vs. university students = 956 ms]). Potential reasons for this interesting outcome and an evaluation of all of the present results are discussed below.

DISCUSSION

From a simple 'horse-race' model of combinational processing, one would initially expect that the combined visual and tactile presentation of consistent signals would be equivalent to the faster of the two modalities (i.e., visual or tactile when presented alone). However, this simplistic conception was not supported by the data. Rather, the combinatorial condition was faster than either the visual alone or the tactile alone condition. Neither could enhanced processing speed be attributed to a tradeoff of speed for accuracy since the combined condition was significantly more accurate than the tactile alone presentation, although this latter result might have been affected by a confusion between two specific forms of tactile signal. However, in general, what emerges is a genuine advantage in performance for the multi-modal signal presentation. There are a number of potential reasons why this may occur. At the present, we must postulate some form of multi-signal reinforcement effect that derives from the facilitation due to cross-reinforcement of sensory signals. A more realistic source for the enhancement may lie in the neurophysiologic architecture linkages discussed at the start of this paper. It appears that cross-modal reinforcement has a direct effect on strength of synaptic transmission that is experienced early in the stimulus processing sequence. It was to explore this possibility that the experiment was conducted which parsed the response in order to isolate motor output components of the response sequence. Here, we found a strong confirmation first of the multi-modal presentation advantage and second of the isolation of that advantage into the early, pre-motor stages of response. At present, it is uncertain whether the primary advantage is to be found in the perceptual recognition phase of the response sequence of in the decision-making and response formulation element of that sequence. However, the distinction of such a difference is amenable to further empirical identification. From the assembly of present results it appears that a neuro-physiological argument underlying cross-modal stimulation provides the best candidate account for the early advantage offered by consistent multi-modal signaling.

616

ACKNOWLEDGMENT

This work was supported by Government Contract No.: W911NF-08-1-0196: *Adaptation of Physiological and Cognitive Workload via Interactive Multi-Modal Displays*. P.A. Hancock, *Principal Investigator*, from the Army Research Office. Dr. Elmar Schmeisser was the Technical Director and we are very grateful for his oversight and direction concerning the present effort. The work reported represents the opinions of the authors and should not be construed as representing any official position of the U.S. Army or any other named agency.

REFERENCES

Broadbent, D.E. (1978). The current state of noise research: Reply to Poulton. *Psychological Bulletin, 85,* 1050-1067.

Cholewiak, R.W., Brill, J.C., & Schwab, A. (2004). Vibro-tactile localization on the abdomen: Effects of place and space. *Perception & Psychophysics, 66,* 970–987.

Craig, J.C. (2006). Visual motion interferes with tactile motion perception. *Perception, 35*(3), 351-367.

Department of the Army. (1987). *Visual Signals.* (Field Manual No. 21-60). Washington, DC: Government Printing Office.

Ernst, M.O., & Banks, M.S. (2002). Humans integrate visual and haptic information in a statistically optimal fashion. *Nature, 415*(6870), 429-433.

Geldard, F.A. (1982). Saltation in somesthesis. *Psychological Bulletin, 92*(1), 136-175.

Gray, R., & Tan, H.Z. (2002). Dynamic and predictive links between touch and vision. *Experimental Brain Research, 145*(1), 50-55.

Hancock, P.A. (2005). Time and the privileged observer. *Kronoscope, 5(2),* 177-191.

Hillis, J.M., Ernst, M.O., Banks, M.S., & Landy, M.S. (2002). Combining sensory information: Mandatory fusion within, but not between, senses. *Science, 298*(5598), 1627-1630.

Posner, M.I., Nissen, M.J., & Klein, R.M. (1976). Visual dominance: An information-processing account of its origins and significance. *Psychological Review, 83*(2), 157-171.

Poulton, E.C. (1977). Continuous intense noise masks auditory feedback and inner speech. *Psychological Bulletin, 84,* 977-1001.

Soto-Faraco, S., Spence, C., & Kingstone, A. (2004a). Congruency effects between auditory and tactile motion: Extending the phenomenon of cross-modal dynamic capture. *Cognitive, Affective & Behavioral Neuroscience, 4*(2), 208-217.

Spence, C., & Driver, J. (Eds.). (2004). *Crossmodal Space and Crossmodal Attention.* Oxford; New York: Oxford University Press.

Spence, C., & Walton, M. (2005). On the inability to ignore touch when responding to vision in the crossmodal congruency task. *Acta Psychologica, 118*(1), 47-70.

Stein, B.E., & Meredith, M.A. (1993). *The merging of the senses.* Cambridge, MA: MIT Press.

Strybel, T.Z., & Vatakis, A. (2004). A comparison of auditory and visual apparent motion presented individually and with crossmodal moving distractors. *Perception, 33*(9), 1033-1048.

Wallace, M.T., Meredith, M.A., & Stein, B.E. (1998). Multisensory integration in the superior colliculus of the alert cat. *Journal of Neurophysiology, 80*(2), 1006-1010.

Chapter 62

The Decision-Making Process for Wayfinding in Close Environments by Blind People

Laura Bezerra Martins, Francisco José de Lima,
Maria de Fátima Xavier do Monte Almeida

Programa de Pós-Graduação em Design
Universidade Federal de Pernambuco
Recife, Pernambuco, Brasil

ABSTRACT

This article discusses methodologic procedures for obtaining information which are perceived and reported by blind people during the decision process on orientation and wayfinding. Wayfinding is understood as finding one's ideal pathway by means of planning, execution and descriptive strategies grounded on tactile maps. The experiment has proved the importance of identifying those elements used as references so that users may be aware of them and make decisions for traveling with higher levels of security and comfort.

Keywords: Ergonomy of the constructed environment, accessibility, Universal Design, people with visual disability, wayfinding

INTRODUCTION

Public places such as airports, shopping centers, hospitals, schools, can represent a risk of falls, accidents, spatial disorientation or even stress for blind travelers. Such reality has a negative impact on people's social and professional lives, and autonomy.

The present work reports blind people's perception of environmental elements encountered in indoor places during their process of decision making in orientation and traveling (wayfinding), that is, planning, executing and describing a pathway aided by a tactile map.

According to Passini e Proulx (1988), *wayfinding* is composed of three interrelated processes: decision making, execution, and information processing. Golledge (1999) and May et al (2003), state that 'landmarks' are the best clues for traveling. May et al (2003) in their experiment, present some alternatives in traveling for sighted pedestrians and identify their reference elements.

The data here described was classified according to categories used in the experiment of Passini e Proulx (op. cit.). The reference elements for traveling were identified and hierachized based on an adaptation of the methodology employed by May et al (op. cit.) in their experiment.

Based on the three categories suggested by Passini and Proulx of information perceived by blind people: *"what are reference elements to assist them while traveling?"*

SEVEN TASKS IN WAYFINDING

Arthur and Passini (2002) report seven basic tasks in wayfinding: 01: The recording of a decision plan and/or the construction of a cognitive map; 02: Learning a pathway from a small tactile map; 03: Learning a pathway from a non aligned display, which implies doing a mental rotation; 04: Understand the whole layout of a visited place; 05: Returning to the start point, which means inverting a decision plan; 06: Linking familiar pathways to new ones; 07: Pointing the location of spots visited along the travel, which means making a triangulation.

SPATIAL COGNITION AND COGNITIVE MAPS

Golledge (1999) considers as consensus the fact that man acquires, codifies, stores, decodifies, and uses cognitive information as part of their traveling and wayfinding activities. There is an evidence that this inner representation doesn't necessarily corresponds to the external reality and, for that reason, fragmented, distorted, and incomplete inner representations are constructed.

Passini and Proulx (1988) divide wayfinding into three interrelated steps: **Decision making**: an action plan or decision to reach the destination; **Execution:** Implementation of a plan on the form of environment behavior and movement; **Information Processing** and environmental cognition allow the first two steps to take place. According to Passini e Proulx (1988) cognitive map is the mental image of the environment spaces; whereas cognitive mapping is the mental process which possibilitates the creation of a cognitive map. It is known that the cognitive map created by the blind person is different from the sighted ones, since the latter has visual images of the spaces. The spacial experience of the one who doesn't see comes from hearing, touch and movement (UNGAR, 2000).

LIMA (1998) states that mental representation is produced by totally congenitally blind people, and mental image is produced upon visual experience

THE IMPORTANCE OF LANDMARKS

Lynch (1999) defines as 'landmarks', any noticeable element along the way as the person looks for orientation and mobility. Golledge (1999) says that it is necessary to attend to some major goals, such as: 1] defining the limits of a partial or whole area, 2] Integrate separated information of the pathway into a network 3] Observe from a panoramic point or as a *bird's eye view.*

TRAVELING WITHOUT VISION AND PROBLEMS

The difference between blind and sighted people lays in the use of different clues for traveling. The first, not having visual information about the environment, obstacles, search for orientation and mobility from an egocentric standpoint, and find greater difficulties in open areas when they cannot find reference elements for guidance.

Harper (1998) breaks the task into a series of minor steps, setting forth a model for traveling for blind people. In such description, *waypoints, orientation points* e *information points* are planned in order to represent information related to certain object. He states the importance of the way-edge concept in order to best orient traveling of those who move guided by large scale and uninterrupted elements that limit the way, for example, walls and curbs. Harper also shows the importance of the user to be well informed along the pathway between a reference point and the next. He describes five stages that have been observed in the act of planning and following a certain pathway:

1. Pathway planning: based on maps and/or previous knowledge; *2. Identifying and bypassing obstacles*: stand-alone objects (walls, columns) or mobile (people, chairs); *3. Orientation and reference points:* the pathway is divided by reference points and by pathway limits; *4. Information points:* Points along the travel where information about the pathway is made available on any equipment of environment information; *5. Road guide, combination of a more sophisticated Road planning and provision of orientation landmark such as encountered on portable maps.*

EXPERIMENTAL STUDY

This is an exploratory study, where a learning session and an experimental one were carried out. Each session is divided in three stages containing tasks for each candidate: i] planning a pathway based on a tactile map ii] follow the pathway, and iii] describe the pathway after its reproduction with magnetized bands on a sheet metal. Four totally congenitally blind subjects, four subjects who have been blind for more than 10 years, and four subjects with low vision participated in this experiment.

▪ EXPERIMENT PREPARATION

The study focuses on orientational decisions made by four totally congenitally blind subjects, cb1, cb2, cb3, cb4, four adventitiously blind subjects, ab1, ab2, ab3, ab4, and four with low vision, lv1, lv2, lv3, lv4, in the process of wayfinding. A comparision of orientational decisions between the three groups in the two phases of the experiment (learning session and experiment session) was done and involved the following steps: Identifying orientational decisions spoken by users; Identifying decisions used "at the points" and "between" decision points; Identifying which piece of information is used during this process.

The identification of orientational decisions spoken by users was transcribed into a table. The identification of the orientational decisions used at the points and between the key decision points was coded from the two denominations for nodules and ways. The types of information perceived and reported by users were transcribed and cathegorized in a table, where the frequency of words related to reference elements for helping traveling and orientation. Data studied according to Bardin's (1977) content analysis method. The word frequency was the explicit mention of words related to decisions and orientational iformation spoken by the user. The corresponding indicator was the frequency of those words. The highest indicator was considered the top reference element used by blind people for orientation.

▪ STRATEGY FOR ANALYSIS OF DECISIONS

Subjects with some residual vision were blindfolded for the purpose of comparison with totally blind subjects. The effect of visual experience in performing tasks was evaluated. If the group of totally congenitally blind subjects had a similar or superior performance compared to low vision subjects or those adventitiously blind, it would indicate that totally congenitally blind subjects made use of different strategies for problem solving, therefore not using experience or visual memory.

▪ CATHEGORIZING AND CODING

The scheme of categorizing and coding was used to explore the questions related to the objectives of the research and as a facilitator in the analysis of wayfinding of each participant.

Categorising decisions – the classification of decisions, by Passini e Proulx (1988) was used in this study for establishing orientational decisions, reported by users. Two criteria were used: physical and behavioral. Based on the behavioral criterion, the following decisions are mentioned: 1] Changing direction (for example, turning right), 2] Changing level (stepping up or down the stairs) 3] To maintain the travel direction (such as walking alongside the hallway) and 4] finding architectural elements (such as doors, columns). When following the physical criterion, there are decisions related to the physical characteristics of 1] hallways, 2] intersections, 3] stairways, 4] open spaces and 5] door stills.

Categorising information –. The classification for the type of information was based on three units of information: 1] the building (permanent physical elements), 2] From the indoor environment (mobile physical elements) and 3] the context, which may by the interior, including temperature variation, noise; or outdoors, including the surrounding environment, for example the climate and wind.

Use of information "at" and "between" decision points – two denominations were used for identifying where and how orientational decisions were made. The decision was coded according to the place where it was orally given. For doing so, a map of the pathway was created, adopting a code system similar to the one employed by May et al (2003).

A table for cathegorizing statements was built. At the end of each statement, a notation used taking the letter "n" to represent nodes; the letter "c" to represent pathways segments. Each letter is followed by a number according to the direction of the steps to be taken, from the start point until the end point, along the pathway. For example: n1, c1, n2, c2, n3, c3 and so on.

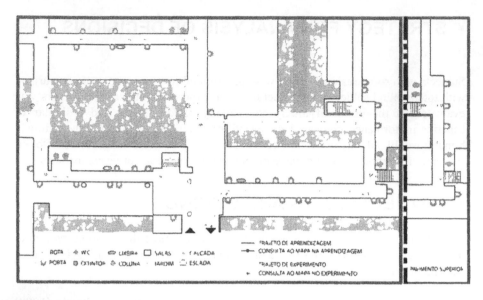

Figure 1. Floor plan of the pathway

Whenever the volunteer verbalizes a decision at the node without a physical critereon, it becomes a behavioural decision. For example, in the phrase "I turned right", it is possible to notice that there is no physical characteristic related to such decision. But, if he says: "I turn right on the corner of the last wall", such decision implies a physical criterion, which is called intersection. The "last wall" becomes a reference point for him/her to make a decision.

If the volunteer verbalizes a decision he was going to take while walking, but saying that He intends to do it further ahead, the notation for identifying the place he has in mind will be inserted beside the phrase. The following example shows a hypothetical decision of a volunteer: "I'm going to turn at the next hallway (n6) on the left, c5". The notation of that spot means that the subject verbalized a decision walking on way 5, which will be carried out further ahead at the node number 6.

▪ DATA COLLECTION

A content analysis of what was expressed by the subjects during planning, executing, and describing a pathway was made. Errors and hesitations were taken into consideration when walking the pathway. Data collection was done by groups. The group of totally blind users cb1, cb2, cb3, cb4; another of adventitiously blind: users ab1, ab2, ab3, ab4; and a group made of low vision subjects: lv1, lv2, lv3, lv4

Each wayfinding register is composed of:

- *Six decision tables* (For analyzing and collecting data). The first three tables present the decisions expressed by the user in the act of planning, following and describing the pathway during the learning session. The last three tables display the decisions taken made by the user in the experimental session. The decisions have been cathegorized according to Passini e Proulx's (1988) decision classification.
- *Tables for recording errors and hesitations*.
- *A map for dislocation*. It shows the two pathways followed by the user during the learning and experimental session. The red line represents the pathway which was followed during the learning session. The blue line describes the pathway followed during the experimental session. The black dotted line corresponds to the way which should be followed by each user, described in terms of ways, c, and nodes, n, a notation based on the two denominations used by May et al (2003) in their experiment. During moments of error or hesitation, the user could be assisted by the tactile map. Such fact was recorded with the use of a circle filled in blue or red on the pathway line: red, if the user would have asked for assistance from the tactile map during the learning session, blue if he/she had asked for information from the tactile map during the experimental session.
- *Two figures reproducing the pathway which was followed by the user*. The wayfinding register has not analyzed the representation error, but, rather, the orientational decisions.

The process of wayfinding in a non familiar pathway was analyzed from the decisions made during planning, following and describing the pathway. The decisions were analyzed and quantified.

▪ FILLING THE WAYFINDING REGISTER

The activities developed for filling the wayfinding register of each user were:

Transcribing - everything verbalized was video recorded; **Selecting**, cathegorizing and tabulate the word, phrase or text which match Passini and Proulx's (1988) classification criterion for decisions ▪ **Coding** at the end of each word, phrase or text, the spot at which those decisions were verbalized, according to May et al's classification (2003);▪ **Selecting**, cathegorizing, tabulating and quantifying information according to the Passini and Proulx's (1988) classification of the type of information. **Grouping**, quantifying and tabulate the denominations and the types of information verbalized by totally blind, adventitiously blind and low vision subjects. **Quantify** the decisions of each user on a table for the use of the reseacher for collecting and analyzing data. **Presenting** the dislocation map based on the pathway register, during execution of the pathway, and the registration of errors and hesitations; **Presenting** the reproductions of the pathway that was recorded during the description of the pathway.

	Learning Session	
Stage	Objective	Description
1	*Planning the pathway*	Each user planned the pathway from a tactile map. Since it was an Unknown instrument for participants, its objective was explained: to identify the indoor pathway they should follow in a UFPE building. They were told of the existence of a legend in Braille, with the symbols for the map, representing the permanent physical elements of the building, such as doors and walls. The beginning and the end of the pathway were indicated. As it was a complex pathway, they were asked, after understanding the map, to read it over, without without timing, with the objective of planning their orientational decisions in order to later describe what they have planned to find their way. The reading register and orientational decisions which were expressed were video recorded
2	*Following the pathway*	Participants were asked they would do to find their way?" Each subject was asked what they found in their pathway and what led them to a decision. Each subject walked the pathway observed closely by the researcher. If the subject veered away, the reseacher would mark it as hesitation of misperception. Veering was considered when the subject was sure to following the right path but was. It was considered hesitation when the subject showed uncertainty what to do next and after having walked some he or she indicated that uncertainty. Participants could choose whether or not making use of the tactile map during task (figure 1).
3	*Describing the pathway*	Subjects reproduced on a 90cmx50cm metallic sheet the pathway they walked as well as what they found on their way. After reproducing the pathway, they were asked: "how did you do your way?". And they were asked to tell what decisions they made. Representation errors were not evaluated, neither orientational errors.

Figure 2. Phase 01 of the study

The parallel activities carried out for a better data reckoning are, as well, recorded below: **Checking** if the representation of the dislocation map corresponds to the reality of the pathway followed by the user, by comparing the representation of the map pathway with the vídeo recording; Cheking if the transcription corresponds to the reality of what was verbalized in the experiment comparing it with the vídeo recording.

▪ APPLICATION OF THE EXPERIMENTAL STUDY

The two phases of the study, learning session and experimental session, are presented in Figures 2 and 3. The phases and objectives of the tasks for each user, the description of how the experiment has been applied are shown as follows:

Experimental session		
Phase	Objective	Description
1	*Planning the pathway*	After a 15 minute interval, the user was asked to plan the same route from the reading of the tactile map. Each one verbalized his/her orientation decisions, with the exploration of the tactile map, and the experience walking the pathway, during the learning session. The same procedures used in the first session were repeated in this one.
2	*Following the pathway*	After planning the pathway, each participant was asked to walk the same route verbalizing his/her orientational decisions as he/she was going.
3	*Describing the pathway*	After following the route, each participant was asked to reproduce the pathway they went through, with the same materials used in session 01 phase 01. After its reproduction, they were asked the following question: "How did you do your way?" The pathway reproduction and the orientational decisions were recorded on a video.

Figure 3. Phase 02 of the experiment

Phases of the experimental study:

Fig 4. Planning the pathway   Fig 5. Following the pathway   Fig. 6 Describing the pathway

CONCLUSION

The present study indicates that the perception of blind people varies as a function of the amount of decisions and by the type of information that is perceived, not due to the lack of vision, visual memory or visual experience. Assistive elements as landmarks were shown to be important for traveling, so that users may be aware of them and make orientational decisions with more safety and comfort. The strategies users employ in following a pathway through a tactile map, errors and hesitations have allowed us to: understand how a blind person elaborates an orientational plan in a non familiar pathway – following and describing the pathway. The study also helps in establishing sallows classifying the orientational decisions for establishing guidelines for projects of environments, based upon decision strategies. A good strategy of haptic reading of the tactile map allows blind users to have a clearer idea about the environment and consequently a pathway planning which facilitates his/her performance. In order to do that, the necessary information should be made available in the right time and place, aiming at providing more elements for the process of orientational decision and traveling (wayfinding) of a pathway prior to its execution. Therefore, it is essential to equip public buildings with information systems which integrate visual graphic, tactile graphic, olphactory, and sonorous information, taking into consideration the abilities and needs of all users.

REFERENCES

Bardin, L. (1977). Análise de conteúdo. São Paulo: Edições 70.

Golledge, R, G. (1999). Human wayfinding and cognitive maps. In: Wayfinding behavior: cognitive mapping and other spatial processes. The Johns Hopkins University press. Baltimore.

Harper, S. (1998). Standardising electronic travel aid interaction for visually impaired people. Thesis (Master of Philosophy) – Institute of Science and Technology, University of Manchester, Manchester.

Lima, F.J. (2004). Prevendo barreiras, antecipando soluções, evitando acidentes. Descrição do Projeto de pesquisa Educação -Programa 25001019001P-7 Educação, UFPE.

Lima, F.J. (1998).. Representação Mental de Estímulos Táteis. Ribeirão Preto, 1998. 166p. Dissertação (Mestrado). Faculdade de Filosofia Ciências e Letras de Ribeirão Preto, Universidade de São Paulo.

Lynch, K. (1999). A Imagem da Cidade. São Paulo: Martins Fontes.

May, A.J, Ross, T., Bayer S.H.; Tarkiainen, M.J. (2003). Pedestrian navigation aids: information requirements and design implications. Personal & Ubiquitous Computing, vol 7, n 6, pp. 331-338.

Passini, R.; Proulx, G. (1988). Wayfinding without vision: an experiment with congenitally totally blind people. Environment and Behavior: http://eab.sagepub.com/cgi/content/abstract/20/2/277

Ungar, S. (2000). Cognitive mapping without visual experience. In: KITCHIN, R.; FREUNDSCHUH, S. (Ed.) Cogntive mapping: past, present and future. London: Routledge: http://www.psy.surrey.ac.uk/staff/SUngar.htm

Ungar, S.; Blades, M.; Spencer, C. (1996). The construction of cognitive maps by children with visual impairments. In: PORTUGALI, J. (Ed.) The construction of cognitive maps. Dordrecht: Kluwer Academic Pub. pp. 247-273.

CHAPTER 63

An Approach of Classifying Categorical Level of Expert of Inspection Panel by Sensoring Yes-No Test

Masao Nakagawa[1], Hidetoshi Nakayasu[2], Tetsuya Miyoshi[3]

[1]Faculty of Economics, Shiga University
1-1-1 Bamba, Hikone
Shiga 522-8522, Japan

[2]Faculty of Intelligence and Informatics
Konan University, 8-9-1 Okamoto, Higashinada
Kobe 658-8501, Japan

[3]Faculty of Business and Informatics
Toyohashi Sozo University, 20-1 Matsushita, Ushikawacho
Toyohashi, Aichi 440-8511, Japan

ABSTRACT

In order to classify the categorical level of experience and skill of inspection panel, the relation of the defect detection probability and non-defect correct-rejection probability was defined by the theory of signal detection (TSD). In this paper, it is shown from the viewpoint of the international standardization in recent years how to measure and classify the degree of skill and experiment. The proposed method is based on the relationship between ROC curve and sensoring Yes-No test. It was demonstrated to show the perception measuring system of sensoring Yes-No test by the test specimen for fracture image of metal. It was found from the results that the value of indicator that evaluated the degree of experienced skill enabled one to determine the threshold level of inspection.

Keywords: Theory of signal detection, Yes-No test, Experienced panel, Detection probability, Stimulus.

INTRODUCTION

The quality of the product of various materials and shape is inspected, to maintain the reliability of the product. The inspection process such as scanning and judgment is the most important for securing the safety of the product. Especially, the nondestructive testing method with the supersonic wave and X rays, etc. has been established.

The psychophysical method by authors enables one to measure the perceptual skill related to the defect inspection (Nakagawa et al., 2009). It seems to be possible to add the knowledge and the experience by requesting information obtained from the visual performance of the experienced panel.

The primary aim of this paper is to prepare the effective tool how to classify the categories of skill level depending on human's perception. It is important in the practical inspection process to evaluate not only the ability to reject the defective product when the product is poor, but also accept the non-defective product when the product is not poor. It is very difficult in practical inspection to keep the high probability in both situation of the former and the latter cases. From these points of view, a new method to classify the category level of panel in proposed in this paper.

In this paper, it is assumed that the human response can be formulated by the theory of signal detection (TSD). In this formulation, there are two kinds of response and stimulus. The former consists of "yes" and "no", the latter consists of "signal" and "noise". Even if the detection subject is a machine or human, the signal is detected from among the noise that always exists in the background. The level of the noise in the background is assumed to be random; it is generated from the outside or inside of the detection subject.

INTERNATIONAL STANDARD OF PANEL'S SKILL

ISO9000 is international standard of the quality, reliability in the world wide sense is defined by the term named dependability, and management on both sides of a equipment failure and a human failure is requested. ISO12100 formally issued in November 2003 is a standard of the reliability field in relation to safety, and the flow of the globalization has accelerated in recent years (Kumamoto et al., 2000). Especially, International Organization for Standardization enacted ISO9712 as a capability recognition standard to the nondestructive inspector in the quality control field of the nondestructive inspection is greatly depended on human's skill and ability.

As shown in Figure 1, the panel's skill level is divided into three categories of level. Level I is a worker, and Level II can execute the evaluation as a expert, and Level III is a super expert. It assumes that all the responsibilities is concerned with the inspection method, equipment, and staff assignment. The panel at a high level should have plenty of experience in, and the accumulation of skill and knowledge of the technology by the education and training is necessary (Japanese Standards Association, 2009).

The skill level is shown by the layered structure like Figure 1. Therefore, the panel at a high level can do a highly accurate inspection, and will become a high defect detection probability. The condition necessary for the skilled person of the panel is not enough only by nimble fingers. The panel's sense acknowledgment ability is important. It is a skill that can adjust in the state of the ability and the thing to observe the state of the object. Therefore, the method of quantitatively evaluating human's sensory property by using the defect detection probability is examined. Therefore, it is assumed in this paper that the evaluation of categories level of experience and skill due to the international standard corresponding to panel's technological level.

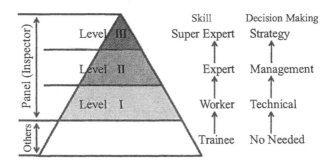

FIGURE 1 Layered structure of the skill level for inspection by human.

MEASUREMENT OF THE DEGREE OF SKILLS

STRENGTH OF STIMULUS AND THE HUMAN RESPONSE

The psychophysical method advanced by authors enables one to measure the perceptual skill related to the defect inspection. It seems to be possible to add the knowledge and the experience by requesting information obtained from the visual performance of the experienced panel. It is assumed in this paper that the difference of the figure attribute is detected in the defect detection work as shown in Figure 2. As for this method, the discrimination capacity of panel who inspects it by the

630

paired comparison is based on the method of constant stimuli. Therefore, it is assumed in this paper that the human response can be identified by the mathematical function of the strength of stimulus (Nakagawa et al., 2009).

THEORY OF SIGNAL DETECTION

The human response is identified by the mathematical function of the strength of stimulus. From this point of view, the model of human response is formulated by the theory of TSD. In this formulation, there are two kinds of response and stimulus. The former consists of "yes" and "no", the latter consists of "signal" and "noise" as shown in Table 1. Even if the detection subject is a machine or human, the signal is always detected with the noise that is imposed on the signal. The degree of the noise is random.

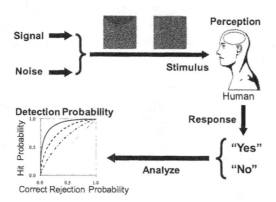

FIGURE 2 Outline of perception measuring and decision by the signal detection theory.

Table 1 Response of human by stimulus.

		Response of Human	
		Yes (SN)	No (N)
Stimulus	Signal + Noise (sn)	A: hit P(SN|sn) **OK**	C: miss P(N|sn) **NG**
	Noise (n)	B: false-alarm P(SN|n) **NG**	D: correct-rejection P(N|n) **OK**

Table 1 shows the combination of the responses of panel for sensory test of signal with noise such as Yes and No to the stimulus such as signal with noise. The response to the stimulation of the SN distribution is said, "hit" for "Yes" and "miss" for "No". Where SN means signal with noise. When each probability is $P(SN \mid sn)$ and $P(N \mid sn)$ the relationship is as follow:

$$P(SN \mid sn) + P(N \mid sn) = 1 \tag{1}$$

If SN is "signal with noise", N is only "noise", the response to the stimulation N (noise) is called "false-alarm" when the response is "yes" and "correct-rejection" when the response is "no". If each probability is $P(SN \mid n)$ and $P(N \mid n)$, the relationship is

$$P(SN \mid n) + P(N \mid n) - 1 \tag{2}$$

Therefore, it is enough only by the calculation of $P(SN \mid sn)$ of "hit" and $P(SN \mid n)$ of "false-alarm".

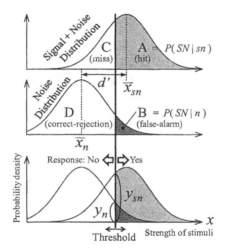

FIGURE 3 Distribution of stimulus and human response.

The participant first receives stimulation, and the judgment from perception. They should judge whether stimulation is a signal or a noise in each trial. However, the judgment is too difficult. One of the reasons is a random change of the noise that it strengthens and weakens. This situation can be shown in the Figure 3 as two probability distributions that show a random change of the noise (N) and the signal-plus-noise (SN) distributions. As for the perception, the SN distribution is always larger than N distribution, because the signal is added to the noise. When the strength of the signal weakens, the difference of the mean value of two distributions becomes small. If SN distribution corresponds to N distribution then the judgment is very difficult. A relative distance of the mean value of the SN distribution and N distribution becomes the index of the sensitivity of the detection that is called d'. Let us designate by $\overline{x}_{sn}, \overline{x}_n$ the mean of the SN and N distributions. The distance d' is designated by the next equation.

$$d' = \overline{x}_{sn} - \overline{x}_n \tag{3}$$

When d' is small, the detection by human becomes difficult. The participant perceives when stimulation is strong or weaker than the standard values by method of constant stimuli, and responds with "Yes" or "No". They should respond yes-no for the stimulation that mixed the signal and the noise was received at random. In the physical value x of stimulation, the probability density of N distribution is defined $y_n(x)$, the probability density of SN distribution is defined $y_{sn}(x)$. The ratio $l(x)$ is shown by the next equation as shown in Figure 3.

$$l(x) = \frac{y_{sn}(x)}{y_n(x)} \tag{4}$$

This $l(x)$ is called a likelihood ratio. In the case where $l(x)$ is greater than 1.0, the probability of responding for the SN distribution is larger than that of N.

CLASSIFYING CATEGORICAL LEVEL OF EXPERT OF INSPECTION PANEL

DETECTION PROBABILITY AND HUMAN ABILITY TO DETECT

The larger the probability A and D of Figure 3 is, the higher the detection probability is. The relation between a technological level and SN, S and N is examined by dividing SN into S and N. If the parameter of probability distributions corresponding to each skill level is determined in Table 2, the probabilities of $P(SN \mid sn)$ and $P(N \mid n)$, and the likelihood ratio $l(x)$ can be calculated as Figure 4(a) and (b) by the TSD and the noise correct-rejection probability. In Figure 4, it is seen that the panel's ability is sensitive to stimulation strength. When the SN distribution at each skill level was set to a constant amount, the parameter of the distribution of S and N had been changed.

First of all, it is noted that the combination of the means of the SN and N distribution is sensitive. When Level I and Level II in Figure 4 are compared, Level II of the detection ability is higher than that of Level I. When the mean of SN is similar to that of N, it is too difficult to find the distinction between the signal and the noise. If the difference of the mean of SN and N is small, it means the probability of success is large, where these are big differences between the signal and the noise. Therefore, it is seen that the difference from Level I to II is depend on the detection ability. Because the detection ability is large, Level III is robust to detect the signal and noise.

RELATION BETWEEN PROBABILITY PARAMETER AND HUMAN SKILL

Table 1 shows the one that the sensory stimulation and the combination of response of panel. When the trial to signal with noise is defined as the SN trial and the trial only to the noise is defined as N trial, the number of the SN trial and N trial is written by number of m_{sn} and m_n. The order of image displays two kinds of trials at random. When the "Yes" in SN trial and N trial is defined as r_{sn} and r_n , the probabilities are

$$P(SN \mid sn) = r_{sn} / m_{sn} \tag{5}$$
$$P(SN \mid n) = r_n / m_n \tag{6}$$
$$P(N \mid sn) = (m_{sn} - r_{sn}) / m_{sn} \tag{7}$$
$$P(N \mid n) = (m_n - r_n) / m_n \tag{8}$$

Figure 3 is drawing of these probabilities by probability distributions. This probability A in this figure is $P(SN \mid sn)$, and the probability D is $P(N \mid n)$. It is seen that the possibility of giving the true judgment is high if the value $P(SN \mid sn) + P(N \mid n)$ is large. The detection ability is defined as this value.

The relations between S, N, SN distributions and the panel's technological levels are shown in Figure 3. The parameter of probability distributions corresponding to each skill level is assumed like Table 2.

Table 2 Probability parameter definition in each level.

Level I	$\bar{x}_n \gg \bar{x}_s$, $\sigma_n > \sigma_s$
Level II	$\bar{x}_n > \bar{x}_s$, $\sigma_s \to 0$
Level III	$\bar{x}_n \leq \bar{x}_s$, $\sigma_s \to 0$

$$\bar{x}_{sn} = \bar{x}_s + \bar{x}_n , \quad \sigma_{sn} = \sqrt{\sigma_s^2 + \sigma_n^2}$$

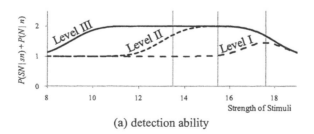

(a) detection ability

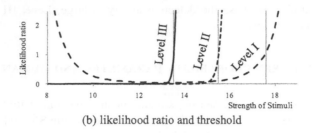

(b) likelihood ratio and threshold

FIGURE 4 Detection ability each level of experienced panel.

Figure 4(a) shows the detection ability, and Figure 4(b) shows likelihood ratio. The threshold of the judgment is a point that the detection ability higher, that is likelihood ratio $l(x) = 1$. The SN distribution is set to a constant amount, and the probability parameter of S distribution and N distribution has been changed in this figure. The difference of the mean value of the distribution of S and N is considered. As for the detection ability is shown, the probability of Level I in Figure 4 is lower than that of Level II. Thus, when the mean of S is very small than that of N, it is difficult to detect the signal and the noise. It is seen that the detection probability of Level I is different from that of Level II.

EVALUATION BY ROC CURVE

Figure 5 is ROC curve (Receiver Operating Characteristic curve) as figure where the relation of the defect detection probability and non-defect correct-rejection probability each experienced panel is expressed. The example of ROC curve is shown in Figure 5 (b). Figure 5 (a) is ROC curve converted to z-coordinates. In Figure 5 (b), the vertical axis is $P(SN \mid sn)$ and horizontal axis is $P(N \mid n)$. Because $Z(SN \mid sn)$ and $Z(N \mid n)$ are the normal variates of $P(SN \mid sn)$ and $P(N \mid n)$, the relation show in Figure 5 (b) can be illustrated as relationship in Figure 5 (a).

The Panel who has ROC curve up can distinguish the signal and the noise accurately. It is guessed that panel's technological level and the relation of ROC curve show the tendency shown in Figure 5.

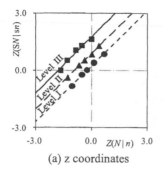
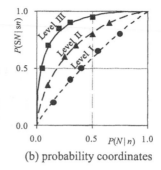

(a) z coordinates   (b) probability coordinates

FIGURE 5 ROC curve each level of experienced panel.

PERCEPTION MEASURING SYSTEM OF SENSORING YES-NO TEST

SPECIMEN

Figure 6 (a) shows image of fractography of metal fracture with crack generated. It was assumed the the image with defective area (signal) and without defective area are generated at randomly by the white noise image. The examples of the image generated are shown Figure 6 (b), (c). These image sizes are 256×256 pixel. The grayscale density synthesizes the image for the signal of the hemicycle (black) by the permeability 50% in the background of the white noise image. The place of the hemicycle is selected at random of four directions.

(a) Sample picture of fatigue crack   (b) Artificial image of SN   (c) Artificial image of N
    (Rummel et al., 1974)

FIGURE 6 Sample pictures for experimental system.

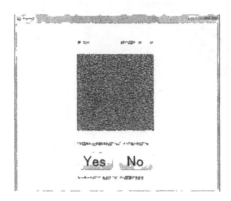

FIGURE 7 Screen image of perception measuring system.

PERCEPTION MEASURING SYSTEM

The defect detection probability and non-defect correct-rejection probability of the panel in the defect inspection are measured with the images such as Figure 6 (b) and (c). The ROC curve obtained by the experiment is shown in Figure 5. Such a tendency is assumed to be a clue, and the experimental methodology that presumes the category at panel's technological level is described in the following.

The defect inspection by the diagnostic imaging was assumed. The artificial image based on the image of the defect (crack) of the structure material shown in Fig.6 as an example was generated. The image that the defective area (signal) synthesized to a random position of the white noise image and the image that doesn't synthesize it are presented. The difference of the two images is small in an actual image though this figure is displayed emphasizing the defective area.

The number of trial of 1 set is 100. The ratio of the image where the signal is included are set to 10%, 30%, 50%, 70% and 90% among those, and five kinds of combinations of the trial is 500. If it is a defect image, the answer is "Yes". If it is non-defect image, the answer is "No". Figure 7 shows the experimental computer screen for the subject to see. When the result of the trial is totaled, the observed value probability of $P(SN \mid sn)$ and $P(N \mid n)$ is calculated. And, ROC curve like Figure 5 is drawn. The technological category is classified as a result.

CONCLUDING REMARKS

The theory of signal detection can be applied to various kinds of fields. Because the sensitivity of the detection and the response to stimulation are individually appreciable, it was impossible in traditional psychophysics. It should be noted that the method of determining the level of threshold proposed in this paper is very efficient and improves the accuracy of the measurement of perceptive sensitivities of experienced panel for visual inspection.

REFERENCES

Japanese Standards Association (2009) JIS Handbook 43 - non-destructive testing, Japanese
    Standards Association.
Kumamoto, H., Henley, E. J. (2000) Probablistic Risk Assesment and Management for
    Engineer and Scientists, 2nd Edition, WILEY-IEEE press
Nakagawa M., Nakayasu H., Miyoshi T. (2009) Determination of Inspection Threshold
    Using Perceptive Sensitivities of Experienced Panel, Human Interface and the
    Management of Information. Designing Information Environments, Springer Berlin /
    Heidelberg, pp. 279-286
Rummel, W., Todd. P. Jr., Frecska, S. and Rathke, R. (1974) The Detection of Fatigue
    Cracks by Nondestractive Testing Methods, NASA Contractor Report (CR-2369).

Chapter 64

System Approach to Integral Valuation of Psychophysiological Condition of Individuals Involved in Professional Activities

Aza Karpoukhina

Ukrainian Academy of Pedagogical Sciences
205 East Ruth Avenue, Suite 330
Phoenix, AZ 85020, USA

ABSTRACT

This article comprises the analysis that is based on Systemic-Structural Theory of Activity (Bedny, G.; Karwowski,W.) and General Theory of Functional Systems (Anokhin, P.K.) and is focused on weighting the possibility of obtaining integral assessment of psychophysiological support for professional activities. Such integral assessment is performed by matching the specifics of the systematic structure of elements designed to achieve psychophysiological support for professional activities with the expert evaluation of the activity attributes. To this effect, the mathematical stepwise linear discriminative analysis was applied. The formula of discriminative function was studied as the tool of integral assessment and classification of psychophysiological support for professional activities. The article also includes the research into and evaluation of methodology used to obtain information about elements of operator's condition systematic structure based on measurements of bio-potentials in representative biologically active skin points. It includes results of research that was conducted to obtain integral assessment of psychophysiological support for professional activities performed by ship operators and ranking navy officers as well as integral assessment of professionally significant personal qualities. The author analyzed advantages and drawbacks of the new methodology and looked into possible ways of its improvement. The prospective development and application of the methodology was also reviewed.

Keywords: Functional System, Psychophysiological Support for Professional activities, Integral Assessment, Formula of Discriminative Function, Biologically Active Points

INTRODUCTION

The importance of developing integral assessment (IA) of the psychophysiological condition (PPC) is explained by the fact that success of professional activities: its effectiveness, efficiency, reliability are determined mostly by man's condition. The earlier approaches to the search for PPC integral index, the quest for the crucial, "pivotal" PPC index or set of indices were either of moderate efficiency or totally inefficient due to the evident lack of appreciation for and understanding of the events' systematic organization.

The paradigm of systematic approach that relies on definition of a system as a "set", "complexity", "interface" of its elements was supplemented in the middle of the last century by the theory of "Functional System" (FS) elaborated by the famous Russian scientist P.K Anokhin. He worked out the universal theoretical model of Functional System, which is currently used by modern researchers to review and analyze different phenomena, which helps to successfully solve crucial problems, including that of optimizing the functioning of the "human-equipment" system.

PPC AND PPCPA SYSTEM ANALYSIS

To achieve the objective of obtaining PPC IA we have used by FS theory and models to analyze the extent to which condition of man is affected by his professional activities. In terms of the system's hierarchy the human activities and condition are interrelated as super- and sub-functional systems. PPC subsystem is part of the supersystem by way of its integral result that is defined as the *psychophysiological support for professional activities* (PPCPA).

Since the model of *Functional System* according to its author P.K. Anokhin, can be and in fact is the analysis and planning tool for a range of various experimental research we deem it necessary to start with a detailed description of architectonics (blocks) and links inside the FS virtual model and then flesh it out with real attributes of human condition. See Picture 1.

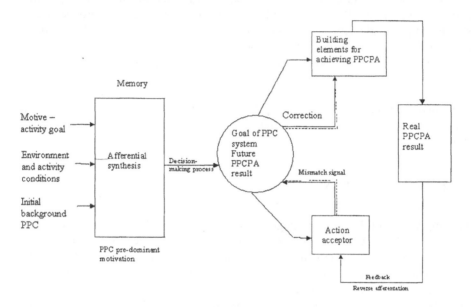

Figure 1. Model of PPC functional system of psychophysiological support

The human activity triggers the function of the *afferentation synthesis* block. Its main components for the PPC system are: the motive/ goal of a concrete activity that requires PPCPA; information about conditions of environment and professional activity and, finally, information about the background or initial human condition. The analysis and synthesis of this information relies on memory; personal life experience (training results) and motivation predominant in PPC. This enables to measure the ratio between motivation and psychophysiological support of professional activities and between motivation and support of life activities, i.e., life preservation. Such measurements are then used to substantiate decision of what the needed (derived from the term *need*) future result of PPC system, i.e., necessary PPCPA should be. This image block of the needed *future result*, in its essence, is the *PPC system goal* block and, according to Anokhin, represents the backbone system factor. The *future result* determines and molds the elements' *systematic structure* that is necessary and sufficient for obtaining real result, real PPCPA. This is the block of the *program designed to achieve the consistent system result*.

It should be stressed that the reason why the author has called functional system functional is that its structure is not a rigid, anatomical and morphological formation. It incorporates only the elements that are capable to ensure that the system results are successfully achieved. In the light of our notions about hierarchic organization of various systems structural elements of a concrete PPC system (in this case, - supersystem) are generated as results of various subsystems. Their organization affects all components of human functioning – from energoinformational, biochemical, biophysical, physiological to psychological and social-psychological. It is extremely important to remember that the elements' structure is built on the basis of the systematic principle that is underpinned by *interaction* and not just by interrelation. It means that if one of the PPC structure elements loses its effect or if its function (input) decreases it is either replaced by another element or its input into achieving the programmed system result is increased. It renders the structure of the PPC system mobile and *dynamic* entity.

Coming back to the block where decision is taken about the necessary future result it should be emphasized that it sends not only the command to the block where the system structure is generated but also the command to the *action acceptance* block that has been discovered by Anokhin. The latter is designed to save the backup copy of the *future necessary result* and subsequently compare it with the *actually obtained system result*. Data about actual result that was actually achieved by the PPCPA system structure is transferred to the action receptor by way of a *feedback* or as Anokhin baptized this phenomenon before Viner, by way of *reverse afferentation*. As soon as the actual result is matched with the backup copy of the programmed result that is stored in the action receptor the system generates the *discrepancy signal* that initiates *correction* inside the PPCPA system achievement structure.

The reason of the critical importance attributed to the chain *actual result – feedback – action receptor – discrepancy signal – correction of the system structure* in FS model is that it explains and constitutes the basis of internal system *self-adjustment* mechanism. Without self-adjustment the existence of FS is impossible. Similarly impossible is implementation of the self-adjustment mechanism if there is no activity receptor in the system, without which feedback having no address would have been directed "in the middle of nowhere".

It was strange to say the least that the revolutionary discovery of the action receptor system block made by Anokhin provoked in the 60s and 70s of last century the wave of a fierce opposition and rabid criticism from the side of scientific community that specialized in biology, physiology and psychology. In the heat of discussion the opponents of the FS theory and model insisted on substantiation of anatomomorphological acceptance receptor and went as far as to accuse Anokhin in lapsing back into medieval concepts of some homunculus located inside the human brain and holding the reins on human behavior. These scholars could not grasp and accept the functional importance of the action receptor and various forms of its implementation in the hierarchy of functional systems.

In 1964-1967 at P.K Anokhin's request we conducted the research of the respiratory functional system at the neuronal brain level in test animals by using microelectrodes. The analysis of the neuron electric activity related to respiratory motility was difficult since the point of the 1 micron (in diameter) monitoring microelectrode had to receive mixed impulses from neurons located within the same range which, consequently, led to different amplitude of logged measurements. Since generation of neuron electrical impulses is governed by the law that in neurophysiology is called the "all or nothing" rule each neuron generates impulses of equal amplitude. Information is coded not by impulse amplitude but the temporary spacing between inter-impulse intervals. We used this concept to propose and develop the amplitude discriminator that was capable of identifying individual activity performed by respiratory center neurons. By matching impulse generation periods with the respiratory motility phases we could register in the brain's respiratory center "inhaling" and "exhaling" neurons and the neutrons that we named "boundary" neurons that generated a short batch of impulses either only at the borderline between "inhale and exhale" or at the border between "exhale and inhale". However what was most interesting for us was the presence of neurons with constant impulsation neurons that differed by their frequency and inter-pulse intervals at inhale and exhale phases. This fact as well as discovery by histologists of the so called Ranshow cells with short circular neuron links ensuring reverberation of impulse activity led us to conclude that such neurons could represent material embodiment of action receptor in the respiratory functional system. It should be noted though that such phenomenon most probably is the product of the long multi-million year evolution. In all likelihood the genetic code that the evolution realized in form of DNA also plays the action receptor in the functional system of entire life cycle.

However, the functional systems that are built over the course of entire human life such as FS to achieve results in concrete professional activity or FS of its psychophysiological support are characterized by a relatively transient and mobile nature and rely predominantly on functional information mechanisms. In this case the role of action receptor can be explained by the "image" of neuron activity, holographic "language" of the brain, mechanisms of short- and long-term memory etc.

The PPC analysis based on "functional systems" methodology showed that PPC is the dynamic self-adjusted functional system where the final result, PPCPA is achieved by the system structure of interaction between member elements. The fairly full description of structures that determine sufficiency or insufficiency of PPCPA is possible only if the following principal characteristics of the system structures are taken into consideration:

- composition (list) of elements included into the system structure;
- "weight" of the element's contribution into the efforts to achieve the system result;

- density (closeness) and dynamics of structure elements' interrelation coefficients.

The PPCPA resulting from functioning of the PPC system structure comprises integral information. However its measurement, i.e., assessment of PPCPA sufficiency or insufficiency is determined by its compliance with direct activity attribute indices such as success-failure, efficiency-inefficiency, high results-low results. The integral assessment of the human activity PPC can be measured by matching, correlating system specifics of the PPC structure to achieve PPCPA with activity attributes.

RESEARCH METHODS AND OBJECTIVES

METHODS USED TO OBTAIN DISCRIMINATIVE FUNCTION FORMULAS FOR PPCPA INTEGRAL ASSESSMENT

From among all possible mathematical tools that can be applied to calculate similar match we decided to select the stepwise linear discriminative analysis. It allows to obtain the formula of discriminative function equation describing dividing plane between n-dimensional (equal to the number of registered element attributes) virtual images of PPC structures and match them with the activity attribute indices, i.e., provide integral assessment of PPCPA. It should be kept in mind that groups of subjects tested to derive the formula of PPCPA integral assessment were initially selected in pairs by criterion of being "good-bad" or "successful-unsuccessful". Such way of selection however does not preclude any eventual variances in assessment of pair groups within their objective differences.

The process of discriminative analysis started by selecting pairs of subjects whose professional activity indices through expert evaluation had been found either high or low. Inside each group individual information about PPC structure elements was collected. To this effect, different PPC indices were registered though a set of commonly accessible methods and equipment.

At this stage it was not yet clear whether these indices were sufficient to describe structures of obtaining PPCPA for a concrete professional activity. For these indices to be combined into a single structure all three above main characteristics had to be taken into account.

At the first stage two virtual images were constructed in the study groups based on average group PPC structure values in n-dimensional space where n was the number of registered indices.

At subsequent stage of discriminative analysis the virtual special images of study groups were "rotated" against each in a fashion to obtain minimal intersection.

Their dividing plane can be described by the following discriminative function equation:

$F = (k1\ A1 + k2\ A2 + k3\ A3 + \ldots + kn\ An) + K$, where:

A means the index value and digit means its number;
k means coefficient that is assigned to this index in this formula;
K means a constant value;
F means discriminative function.

At the final stage the procedure of "index minimization" was performed to eliminate from the formula those indices that did not produce any considerable effect on its information value.

When individual index values that had been collected during the subjects' selection process were inserted in the formula we could refer individuals to the first group if $F > 0$ or to the second group if $F < 0$. Since integral expert evaluation was not applied to individual test subjects as measurement of their professional activity inside the group the integral assessment of their PPCPA was used instead. Differentiation, precision and objectiveness of assessment are considerably improved by virtue of the fact that differential analysis evaluates also the value of the probability that an individual can be categorized by activity attributes and PPCPA.

If the set of indices "A" in the formula indicates that there is a systematic PPCPA structure the coefficients k that apply to these indices can be construed as the "extent" of functional contribution made by each structure element into achieving PPCPA. The indices that remain into a minimized formula signify the group of elements with the highest density of structural links, i.e., those that determine whether PPCPA will or not be achieved. The analysis of functional importance attached to these indices can facilitate understanding of the entire functional mechanism that is designed to achieve system result. Therefore, mathematical apparatus that is applied for analysis can be considered as adequate for obtaining information about major characteristics of the PPC system structure to achieve PPCPA.

METHODS OF PPCPA STRUCTURE INTEGRAL ASSESSMENT BY ELECTRODERMAL BIOPOTENTIAL VALUES IN REPRESENTATIVE BAP AREAS

It is a well established fact that assessment and forecast of professional activity and its psychophysiological support necessitates a battery of expensive and lengthy tests. Besides, integral assessment of results demonstrated in tests requires that the test subject represent various specialists/ experts (psychologists, physiologists, doctors, specialists of activity specifics etc.).

That is why we proposed to use the new method of recoding and analyzing bio-physical electrodermal processes in skin's biologically active points (BAP) (10

BAPs at the finger tips or a set of 21 auricular BAPs), which is also a new way of identifying and assessing man's condition. We found out that the values of BAP electrical characteristics (such as electrical conductivity, resistance, bio-potential) are related with energy support of information processes at all levels of human activity, mostly in man's nervous system. Therefore, parameters of the BAP representative complex carry sufficiently integrated information about PPC. Another advantage is objectiveness of its quantity measurements, its reduced time (entire measurement procedure takes 1 to 2 minutes) and the possibility of applying the method under real-life conditions. The most important feature of this method is its future applicability as a uniform closed circuit BAP-based PPC assessment and adjustment (correction) method.

Over the course of this research we have developed a special device (M.Chepa) to take 1 microampere DC measurements of BAP bio-potentials. It is critical to maintain this current strength since the greater current will provoke purely chemical iontophoresis processes interfering with objective measurement of bio-electric potential.

BAP bio-potentials were measured when "plus" and "minus" DC current was applied, after which the values of their pole and left–right asymmetry were calculated. With 10 BAPs at the fingertips the whole amount of controlled indices amounted to 35. All indices that have been pre-selected according to the integral expert evaluation and categorized by activity attribute and PPCPA classes were then used to calculate the formula of discriminative function that relies solely on organization of BAP bio-potentials. The precision of the new method was determined by the number of coincidence between forecasted BAPs and IEE (integrated expert evaluation) and expressed as percentage. To calculate the formula of discriminative function in the "test" sample the average group indices of this sample were used and its precision was assessed by inserting into formula individual *An* values. In the "validation" sample individual *An* values were inserted into formula obtained from the training sample. Both samples were kept equal by their major characteristics: activity type and motivation, life conditions and life style; age (between 18 and 19) and sex (males).

Below we include results of three test series that have been performed to assess the probability of obtaining PPCPA integral assessment according to the following test assignments:

- professional aptitude test to define capability of being the ranking navy officer. The *target group* for this research was 160 subjects;

- define extent of success achieved by professional ship operators. The *target group* for this research was 300 subjects;

- define human adaptation capacity to withstand antiorthostasis. The *target group* for this research was 154 subjects.

BAP –BASED PPCPA INTEGRAL ASSESSMENT

PPCPA IA OF RANKING NAVY OFFICERS

The study has been conducted for 2 years as observation of the Navy Academy candidates who upon graduation were to become ranking navy officers and assembled in the annual 2-month training camps. The candidates had to report to a special acceptance committee that studied their files and passed final decision of whether these candidates were fit or not to exercise this profession. The final decision was based on the average grade the candidates had acquired in the college, average grade of their entry tests and the results of their professional aptitude tests. The following methods have been used:

Methods of Social and Psychological Research

1. Curriculum Vitae. 2. Professional aptitude test. 3. Professional interest questionnaire. 4. Findings of psychological/ pedagogical observation. 5. Individual interview.

Methods of Personal Psychological Observation

1. Method of defining the subjective control level. 2. Method of defining thinking type. 3. Method of "Three-factor personal questionnaire". 4. Method of defining social behavior type. 5. Method SMRP (variety of MMPI). 6. Method of diagnosing character and temperament accentuation (Leonhard, Smishek). 7. "Prognosis" method.

Methods of Psychophysiological Observation

1. Immediate (short-term) visual memory for numbers. 2. Long-term (delayed) memory for numbers. 3. Short-term aural memory. 4. Long-term aural memory. 5. Recent (short-term) memory. 6 ."Number series". 7. "Analogy". 8. Quantitative relations. 9. Landolt's rings visual acuity test. 10. Subjective state study method.

Supplementary methods depending on profession

1. Reaction to moving object. 2. Step-up exercise. 3. Altitude chamber tests for capability of supporting high pressure and pressure difference. 4. Time and result of diving assignments at 5 meter depth. 5. Time of reaction to the light signal under water (the signal source and response button are inserted into the diving suit and connected through the life support umbilical with the surface instrumentation).

The expert committee has used all above methods to perform the studies; make an IEE and form classes (groups) of successful - unsuccessful, professionally fit – unfit students. In parallel with other methods we measured for bio-potential values in auricular BAPs. These measurements were used to construct the discriminative function formula for integral assessment and discrimination of the above activity attributes and PPCPA. Calculation of new methods' precision led to results shown in Table 1.

A separate section of the table includes the precision values of BAP method that was applied to categorize students according to the level of subjective control (LSC) into internals and externals. Such classification is all the more important since these PPCPA characteristics are critical for the navy ship and submarine officers who must be able to take valid decision and shoulder high level of responsibility. LSC characteristic is determined by the localization of control over significant events and is measured by the Locus of Control Scale: internality - externality. Internal person (an internal) interprets the event as something that results from his activity. He thinks he can be on top of events and, consequently, feels responsibility for what is going on. External person (an external) believes that the events that affect his life are produced through the impact of external forces: an accident, actions of other people etc., therefore he does not think he is capable to control events and their development and be responsible fort their outcome.

Table 1. shows that the precision of discriminative function formula calculated through BAP bio-potentials was quite high and averaged 70 %. On the other hand the new method has incontestable advantages in terms of time consumption. The LSC test lasts about 20 minutes and so does the test interpretation by an expert. All other tests and examinations take pretty much the entire time of two-month camp. For comparison, measurement of BAP bio-potentials lasts only 1 or 2 minutes. We can say with a great degree of certainty that precision of the forecast regarding classification of test subjects was in most cases higher if the formula was calculated for the "native" test group sample as opposed to the "other" test group. It should be emphasized that each assignment requires that a specially tailored formula of discriminative function be elaborated. Usually such formula designed to differentiate groups by criterion of "success in test situation" is virtually useless if applied to categorize groups by their potential professional success. This thesis was corroborated by a series of "cross tests" where all test subjects were equal by age, mode of life and professional activity.

Table 1. Precision of PPCPA integral assessment made by stepwise linear discriminative analysis of the auricular BAP bio-potentials

Basis for integral expert evaluation (IEE) of professional activity attributes	Classes (groups)	Test sample			
		Precision of measurements in test sample (%)	Number of subjects in a group	Number of matches with IEE	Precision (%)
Successful by criterion of average grade in high school	Successful (4.5 - 5.0)	71	17	11	65
	Average (3.5 - 4.4)	72	36	22	61
	Unsuccessful (3.0 - 3.5)	75	24	17	71
Successful by criterion of	Successful (4.0 - 5.0)	73	10	4	40
	Average (3.0 - 3.9)	73	30	20	67

average grade at entry tests in the Navy Academy	Unsuccessful (2.0 and "very poor")	64	40	28	70
Results obtained from battery of tests to define professional aptitude of ranking navy officers	"Professionally fit" (mostly comply with vocational requirements)	74	63	58	92
	"Conditionally fit" (partially comply with occupational requirements; can be allowed to stay if there is lack of candidates).	83	16	9	56
Define level of subjective control (LSC)	Internals (indices exceed 35)	79	28	21	75
	Mixed group (26 - 35)	64	45	42	91
	Externals (less than 25)	72	7	2	29

PPCPA IA IN PROFESSION OF SHIP OPERATORS

The study was conducted at the Ship Operators Postgraduate School for ship operators of various professions (motorists, electricians, radio operators, life support system operators) located in St. Petersburg, Russia. The target audience included more than 300 operators. The study was conducted in conformity with the algorithm described above. At the first stage, the students were broken down into groups of "successful" and "unsuccessful" operators, separately by their professions. The criterion that determined whether the subject had to be referred to a certain group was his/her success measured by 4 professional qualities. Performance of these qualities was then evaluated by experts according to the 5-point scale where:

1 point was average grade for theoretical training (knowledge of equipment, facilities, operation rules, operation manuals);

2 was the average grade for practical training (development of practical skills to operate equipment under different conditions);

3 was the evaluation of practical skills to search for and eliminate various disrepairs and defects;

4 was evaluation according to the criterion of whether or not the subject is likely to commit grave mistakes when operating the equipment.

Within each group the measurements of bio-potential at 10 fingertip BAPs (the total of 35 indices) were taken. The resulting values were used as inputs for the formula of discriminative function to perform integral assessment of PPCPA sufficiency or insufficiency for each study group that included operators of different professions.

The precision of forecast formulas was assessed at 75-80 %.

IA OF MAN'S ADAPTATION CAPACITY TO WITHSTAND ANTIORTHOSTASIS

The BAP bio-potential measurements were also used to develop the IA of the man's capacity to withstand antiorthostasisa (AO). The resistance to AO plays predominant and sometimes decisive role for the man's aptitude to exercise certain professions such as cosmonauts, miners, submariners.

AO was created in laboratory conditions by lying subjects on their back with their head facing downwards at an angle of 30 degrees for duration of 10 minutes.

For the purpose of the subjects' primary classification into groups of those with "bad" and "good" adaptation capacities we used direct values of resistance to AO that was determined by taking measurements of cardiovascular system and haemodynamics, neurodynamics (sensitivity of the nervous system), muscle strength and stamina, short-term memory. The summarized reaction index was calculated by using formula by E. Derevianko:

$$K = \frac{A - B}{A + B + C}, \text{ where}$$

A means the number of cases when background health conditions improved as soon as subject transferred from vertical position to AO position;
B means the number of cases when such health conditions deteriorated;
C means the number of cases when no changes were observed as compared to initial data.

Immediately before and after being exposed to AO test the subjects' auricular BAPs were measured to calculate the discriminative function formula and determine their adaptation capacity of withstanding AO. If classification by direct indices corresponded with that by BAPs we could determine precision of forecast formula that was calculated through study results. It was used as the basis to perform IA of subjects' psychophysiological support for AO adaptation. By the end of tests we came up with precision of formula applied to training sample (75 subjects) equal to 60-72 % and to validation sample (79 subjects) equal to 80-94 %.

Noteworthy most informative and correct were BAP indices that reflected no background values but shifts in bilateral and bilateral / polar asymmetry of BAP bio-potential when subjects were exposed to AO. It should be remembered that brain processes are also characterized by bilateral asymmetry, which may be interpreted as the evidence of BAP-based formula actually reflecting the organization and dynamics of the PPC activation energoinformational component. This finding is corroborated also by analysis of functional importance in BAPs (acupuncture points) with tightest linkage coefficients, i.e., those that constituted the bulk (backbone) of the final formula after minimization of indices. The modern terminology and ancient acupuncture texts define the functional meaning of these auricular BAPs in the following fashion:

A13 or "zero" reflects general physical status;
A7 or "subcortex" reflects emotional and psychic tension;
A8 or "endocrine glands" reflects the level of endocrine control.

CONCLUSION

The analysis of PPC system from perspective of the P.K Anokhin's *Functional System* theory proved to be helpful and instrumental in defining new approaches that can be used to develop a method constructing the IA of PPCPA. Correlation between PPC functional system and "Activity" functional system hinge upon subordination links between sub- and super-functional systems. Results of the PPC – PPCPA system have been conditioned by specifics of PPC system structure and to a considerable extent determine professional activity outcomes and efficiency. Therefore, sufficiency or insufficiency of PPCPA can be indirectly evaluated by comparing specifics of PPC system structure organization that determines PPCPA with the professional activity attributes: success, productivity, i.e., with integral expert evaluation (IEE). To perform this comparison, the mathematical tool of stepwise linear discriminative analysis was selected. As the result of research into PPCPA of ship operators engaged in diverse professional activities we have deducted formulas of discriminative function that were quite informative and found capable of identifying PPCPA level classes and substantiating decisions on integral assessment of PPCPA sufficiency for concrete activities. We think that the information value of formulas is validated by the fact that they include information not only about values of element *set* incorporated into PPC structure but also reveal *specifics of element organization* in this structure. This is achieved by including into the formula main structure characteristics: elements' composition, input of elements into achieving the system results (PPCPA) and density of element interrelation that is conditioned by the system principle of **interaction** between elements for the purpose of obtaining a certain result.

In parallel, we studied informational value of new PPCPA IA method that is based on analysis of representative BAP bio-potentials. The diagnosis sensitivity of formulas for activities of different types and with different tasks averaged 75 % and sometimes reached 90%. Based on research results the following advantages of this method can be emphasized:

- an integral nature of each BAP index since dynamics of their bio-physical characteristics, including that of their bio-potential is interrelated and reflects the implementation of information processes in certain subsystems of the human body represented on human skin by BAPs;
- easy access to BAPs;

- reduced time of registration: the time of measuring BAP bio-potentials in sequential mode that equals to 1 - 2 minutes and in parallel mode (with necessary equipment) can be reduced to 1 - 2 seconds;

- BAP method can be applied both for PPC assessment and for PPC adjustment/ correction purposes. In future, it is possible to create automated close-circuit BAP-based PPC assessment and adjustment systems and enhance PPCPA reliability. Adjusting effect at all levels of PPC produced by low-intensity laser light in red and infrared band was proven by many researchers, including those who participated in 35-year research under auspices of Kostyuk Psychology Institute Psychophysiology Department at the Academy of Pedagogical Science of Ukraine. Most important is the established lack of any unfavorable side effects, including doping effect.

We believe that future work to increase precision of PPCPA IA discriminative function formulas must be performed in two major directions. On one hand, the efforts should be taken to improve sensitivity and objectiveness of expert evaluation for primary classification of operator's professional activities. On the other hand, researchers must specify, refine type, objectives and conditions of activities, for which this formula is designed. The fact of sharp loss in formula informative value once it is applied to a different or to extended variety of activity was interpreted at first as its disadvantage. However the analysis showed that this in fact is quite on the contrary the proof of its advantage and confirms the formula's objective reflection of PPC structure system that allows to achieve PPCPA of a concrete activity content.

At present, the developed method is applied as one of many that are included in the professional tool kit to assess aptitude for certain professional activities, and to measure for man's professionally significant qualities and capacity of adapting to major factors of a given profession. We envision another promising direction of research: to develop methods and formulas of discriminative function for IA psychophysiological support for an *individual operator* in a real time mode and under real life conditions over different professional activity periods. The expert evaluation of successful and unsuccessful activity periods of a concrete operator can be performed and statistically significant number of evaluations can be collected if the subjects operate simulators. At the same time during these periods the BAP bio-potential can be monitored automatically to derive the formula of discriminative function for subsequent determination of PPCPA IA and activity attributes in different periods of real work. Scientifically substantiated conditions and technical capability for implementation of such operator's PPCPA discrete-continuous control automated system already exist. The complexity and cost associated with such research might be high. But is it commensurable with the price of uncontrolled decrease in PPCPA and errors operators make under stress that could result not only in equipment damages but also in human losses?

REFERENCES

Anokhin, P.K. (1962), The Theory of the Functional System as a Prerequisite for the Construction of Physiological Cybernetics. Academy of Sciences of USSR, Moscow.

Anokhin, P.K. (1969), "Cybernetics and the Integrated Activity of the Brain." A Handbook of Contemporary Soviet Psychology. Basic Books Inc. Publishers, New York.

Anokhin, P.K. (1975), Essays on Physiology of Functional Systems. Medicine, Moscow.

Bedney, G., Mester, D. (1997), The Russian Theory of Activity: Current Application to Design and Learning. Lawrence Erlbaum Associates, Mahwah.

Bedney, G., Seglin, S., Mester, D. (2000), "Activity Theory: history, research and application." Theoretical Issues in Ergonomics Science, 1, 2, 168-206.

Bedny, G., Karwowski, W. (2007), A Systemic-Structural Theory of Activity: Applications to Human Performance and Work Design. CRC Press, Boca Raton.

Bedny, G., Harris, S. (2008), "Working sphere/engagement" and the concept of task in activity theory." Interacting with Computers, 20, 2, 251-255.

Karpoukhina, A., Isaevsky, N., Chepa, M. (1982), "The Use of Laser Acupuncture Procedures as Factors of Optimizing Neuro Dynamic and Cognitive Processes." Method Diagnostic Therapy and Rehabilitation in Work Environment of Minors. ZNII Energy in Minor Industry, Moscow, 19-38.

Karpoukhina, A. (1990), *Psychological method of increasing efficiency of work activity*. Knowledge Publishers, Kiev (in Russian).

Karpoukhina, A., One-Jang, J. (2003), "Systems Approach to Psycho-physiological Evaluation and Regulation of the Human State during Performance." Proceedings of the XVth Triennial Congress of the International Ergonomics Association and The 7th Joint Conference of Ergonomics Society of Korea. Japan Ergonomics Society, 6, Korea, 451-454.

Karpoukhina, A., Kokun, O., Zeltser, M. (2008), "Monitoring of Human Psychophysiological Condition as a Method of Increasing of Activity's Efficiency." Conference Proceedings AHFE International Conference. USA Publishing, Las Vegas, USA.

Kokun, O. (2004), Optimization of person's mechanisms of adaptation: psycho-physiological aspect of upholding activity. Millenium, Kiev (in Ukrainian).

Kokun, O. (2006), Psychophysiology. Centre of educational literature, Kiyv (in Ukrainian).

Chapter 65

Evaluation of the Surgeon's Workload from Three Hospitals in Brazil

R.L. Diniz[a] , L.B.M. Guimarães[b]

[a] Department of Design and Technology, Post-Graduation in Environment and health, Federal University of Maranhão, São Luís, MA, Brazil

[b] Department of Industrial Engineering, Federal University of Rio Grande do Sul, Porto Alegre, RS, Brazil

ABSTRACT

This research has as its main objective to evaluate the workload of volunteer surgeons of three hospitals in Porto Alegre, State of Rio Grande do Sul, Brazil, during elective surgeries of low, moderate and high levels of complexity. The objectives were i) to investigate the physical demand level by means of field observation, questionnaires, posture assessment technique (HIGNETT & McATAMNEY, 2000), Heart Rate (HR), Blood Pressure (BP) (systolic and diastolic) and catecholamine (adrenaline and noradrenalin); ii) to suggest a preliminary proposal to physical workload. The results indicated moderate physical workload amongst surgeons. The results suggested that the youngest surgeons (novice) have more mental effort and less physical effort than the oldest surgeons (seniors) who have less mental effort in all levels of surgery complexity but more physical effort during surgeries of low level of complexity. A seat-standing position was indicated as a preliminary proposal to reduce the physical workload.

Keywords: physical workload assessment, elective surgery, surgeon's work, seat-standing position

INTRODUCTION

Hospital ergonomics deals with studies in healthcare settings. It may help prevent injury and promote health, safety, and efficiency, and comfort in health care units to optimize human performance and improve patient health. Health care personnel are exposed to a remarkable array of biological, chemical, physical, and psychological hazards in their working environment (Mirbord et al. 1995). According to KANT et al. (1992) little is known about ergonomic stress in more specific groups of hospital workers. A surgeon's work has a range of complex operations and imposes significant constraints on posture and movements, force transmission and visual perception, a lake of workload (mental and physical). Under these conditions, operating room procedures may contribute to increased personnel injury, technical errors and costs (Berguer, 1996). There are no studies of ergonomics related to the relationships between physical and mental demands and age and experience in the operating room (elective surgeries). All studies are about only mental or only physical workload (movements of the surgical team, visual and manual constraints, the operating room environment, decision making and, surgical instruments) (KANT et al., 1992; MIRBOD et al., 1995; LUTTMANN et al., 1996; BERGUER, 1997; Herron et al., 2001). What can be affirmed in fact is that ergonomics has more to research in the operating room and more challenges with telemedicine and surgical robots, mainly in emergent countries like Brazil.

The aim of this study was to investigate the correlation of physical workload according expertise of the surgeon and surgical complexity and propose ergonomic recommendations. Qualitative (questionnaires and field observation) and quantitative techniques were used (posture assessment, Heart Rate, Blood Pressure and hormonal levels assessments).

Materials and methods

The study population consisted of a group of elective surgeons in Porto Alegre (RS), Brazil. The hospitals involved were Mãe de Deus, Pereira Filho and Santa Rita (Santa Casa Hospital). At the first stage questionnaires were given to 32 surgeons (20 at Santa Rita and 12 at Pereira Filho surgery rooms) that evaluated body discomfort intensity related to physical effort. A verbal rating scale which uses "no discomfort, neutral and, maximal discomfort" to indicate increasing intensities of discomfort was used and, field observations were made (14 thoracic surgeries). At the second stage a quantitative assessment was performed related to physical workload in the operating theatre on 6 surgeons (2 residents, 2 staff and 2 seniors) during 3 types of surgery (minimal, medium and high levels of complexity according to surgical techniques). Heart Rate (HR), Blood Pressure (BP) (systolic and diastolic) and cathecolamines (Noradrenaline, Adrenaline, Cortisol and Adenocorticotrophine), posture assessment technique (REBA – Rapid Entire Body Assessment) (HIGNETT & McATAMNEY, 2000), were applied before, during and after surgeries.

For recording Heart Rate a **POLAR®** Monitor was used and the data was recorded by means of **Polar Precision Performance™** software in 5 seconds; for recording Blood pressure a **CITIZEN™** digital monitor was used with automatic insufflate before, during and after surgeries. Rest values before the surgeries were not considered with a view to eliminating the onset of fatigue in the surgeon before he begins his activities.

Systolic arterial pressure and Diastolic arterial pressure were measured using a digital arterial pressure monitor of the **CITIZEN™** brand CH-491E (system oscillometer), with automatic inflation of the strap until an appropriate level for the internal pump was reached and emptied by an electromagnetic valve. The precision of the apparatus was ± 4 mmHg for pressure and ±5% for the pulse, with extension of the medication of 0 ~ 280mm for pressure of 40 ~ 180 beats/min. The rest levels of systolic and diastolic arterial pressure were registered while at rest before, during and after surgery. The recording of levels at rest) was done on the arm (at the height of the braquial artery) and as during surgery it would not be viable to take further measure of arterial pressure since the surgeon cannot stop what he is doing, the measurement were taken from the leg (at the height of the popliteal artery). The measurements taken during surgery were taken at intervals of fifteen to twenty minutes during procedures of medium and large gravity and at intervals of eight and ten minutes during lesser procedures. In total 5 samples were taken during the bigger operations and three during the lesser ones. The treatment of the data required calculation of the arterial pressure average (approximated from the integral Arterial Pressure) using the following formula (McArdle et al, 1996; Fox & Mathews, 1986): Average Arterial Pressure = Dialtolic Pressure + 1/3 (Systlic P – Dialtolic P), Where: AAP = Average Arterial Pressure; ADP = Average Diastolic; ASP = Average Systolic Pressure.

For the measurement of hormonal levels blood samples were collected minutes before the procedures and also after. Each sample was of 5ml of venous blood taken and conditioned in the following way: in BD Vacutainer™ glass tubes withut reagents for analyses of cortisol and in BD Vacutainer™ glass tubes with K_3 EDTA (anti-coagulant) for analysis of catacholamines (adrenalin, noradrenalin and dopamine). The test tubes, together with the samples were immediately refrigerated in Styrofoam cases with ice and then sent to the lab, **IMUNO Pesquisas Clinicas LTDA,** where they were analyzed. Following the information sent back from the lab, the samples were out in a common centrifuge at 3000RPM for 5 minutes and stored at –80°C. The trials for analysis of catecholamine were done using chromatographic fluid of high resolution with a detection limit of 2pg/ml for adrenalin (CV less than 8,4 %) and 1pg/ml for Noradrenalin (CV less than 9,3%) and with the following values of reference: adrenalin less than 150pg/ml; noradrenalin less than 370pg/ml and dopamine less than 200pg/ml. The blood samples (plasma) were collected by both nurses and doctors (anesthetists or surgeons).

In this research the ratio of ratios was utilized between the samples collected after the surgeries (collection 2) and the samples collected before the surgeries (collection 1) for all of the physiological parameters (Na and A). The data was then tabulated on an Excel plan and plotted on tendency graphs. In the case of the analysis of the relationship between physical load and mental load for the surgeon the ratio Na/A was used as in similar studies (Fibiger et al., 1984; Basset et al., 1987), which suggested that the resulting values of this ratio were between 2 and 3 and the load is essentially mental and if the values were greater than 5 the load was essentially physical. In this case, the evaluation was done considering the ratio between Na/A of the collection 2 (post surgery) and Na/A of collection 1 (before surgery. So, differing from the literature, this test used the "ratio of ratios" (ratio between the before and after Na/A) getting values less than those suggested by the literature meaning that when the ratio was between 0 and 1 the load on the surgeon was essentially mental and when the results were greater than 1 the workload was essentially physical.

The systematic register of the occupational posture of the surgeon was done with a digital video camera mounted on a tripod positioned on the sagittal plane in relation to the surgeon's position. The REBA was used for evaluating the posture assumed by way of EBA software 1.3 (*Neese Consulting Company, 2001*). The REBA technique was applied to the longer duration surgeries. 120 observations were collected from the longer procedures in three different moments: 40 observations in the first 20 minutes of surgery, 40 observations at the approximate middle of the procedures and a further 4 during the final twenty minutes. For the shorter operations REBA was applied during the entire procedure. The interval between observations was 30 of seconds.

Results and discussion

In relation to physical effort it was noted that the surgeons submitted their upper limbs to complex movements, they were sequential and repetitive owing to pre-established surgical techniques (dieresis, prehension, exposition, synthesis, etc...). In the work of a surgeon there are postural demands which are directly related to the amount of time that the position has to be maintained (figure 1).

FIGURE 1. General postures assumed by surgeons

The results from the questionnaires showed that, in a general way, the item of fatigue was perceived as the most problematic between surgeons, followed by discomfort/pain in the arms back, legs and feet.

The graph of the mean heart rate obtained from the heart rate after and during the surgeries show a variability among the surgeons. However, variability was expected not only because of individual differences but also due to the imprecise character of HR (Figure 2).

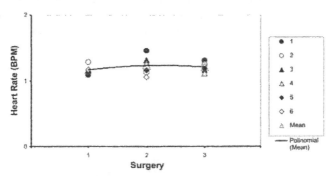

FIGURE 2. Mean HR of the six

However, considering only the heart rate at 5 seconds interval it can be noted that the CF of the expert surgeons are more constant than the CF of the novices, who showed ups and downs more frequently during both low level and high level of complexity surgeries (figure 3). One possible explanation is that the mental effort and emotional state of the novices (who need to make decisions, assume responsibility in situations they have no experience yet) impact the heart rate in any kind of surgery while the same does not happen with the experts, It is important to note that the collected data do not allowed the evaluation of the heart rate variability (VFC) or sinoidal arrhythmia as suggested in the literature (Kalsbeek and Ettema, 1963; Meers and Verhagen, 1972; Meshkati, 1988; Meshkati et al., 1995; Veltman and Gaillard, 1998). However the data can be used as an indicator of mental stress and emotional load, mainly among the surgeons with less experience.

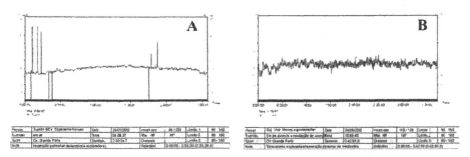

FIGURE 3. Examples of HR of the seniors surgeons (graph A) and novice (graph B)

The rate between the MAP (Mean Arterial Pressure) measurements during rest (after the surgery) and the measurements during the surgeries, taken from the arm (braquial artery) and from the leg (popliteal artery), also showed variability following the results of the heart rate.

The values of the rate Na/A of the novice surgeons, in general, were between 0 and 1 suggesting more mental load than physical load, mainly in the case of high level of complexity surgery (Figure 4). For the most experiment surgeons, the values, in general, were higher than 1 suggesting that they have more physical than mental load, mainly in the lower complexity surgeries, and less mental load, in the higher complexity surgeries in comparison to the novices (Figure 5).

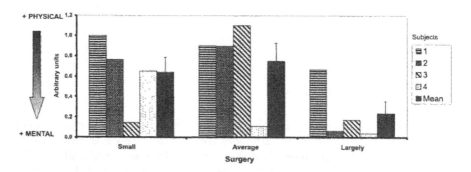

FIGURE 4. Results of the catecholamine measurements of novice surgeons

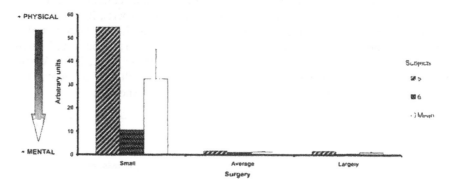

FIGURE 5. Results from the catecholamine measurements of seniors surgeons

It can be said that the novice surgeons present more mental effort than physical, mainly in the high level of complexity surgeries while the expert surgeons present more physical than mental effort, mainly in the low level of complexity surgeries. This suggests that in the low level of complexity surgeries, in the case of the expert surgeons the load component is physical, what can lead to the conclusion that the "easier" the surgery is physical effort is the major factor since the more experient the surgeon is, less mental resources are used in low level of complexity surgeries.

The REBA technique showed that the observed tasks have a medium level of postural risk (scoring between 4 and 7) suggesting a level 2 of action (i.e. it is necessary to better evaluate the postural actions). This result is in accordance to the unsystematic observations and with the results from the reviewed literature (Magalhães et al., 2000; Berguer, 1999; Luttman et al., 1996; Mirbord et al., 1995; Kant et al., 1992), confirming the physical stress inherent in the surgery task. The REBA technique also showed that the expert surgeons (subjects 5 and 6) and the ones with medium experience (subjects 3 and 4 who were staff surgeons) present REBA scores lower than the surgeons with less experience (resident surgeons, subjects 1 and 2). This can be explained by the fact that the most experients have more practice with the surgical techniques and maneuvers and experienced more chirurgical cases, therefore being more adapted to the cases and having more capacitation to overcome problems including the postural ones. It is important to note that subject 5, who is experient and has trainment on postural behaviour during surgeries presented, in all cases, the lower scores. As expected, the low level of complexity surgeries always presented lower scores in comparison to the high level of complexity surgeries (Figure 6).

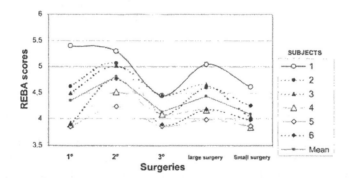

FIGURE 6. Results from the postural evaluation of the surgeons using the REBA technique

From the results, it is clear that the physical workload is part of the surgeon's task. One of the possibilities to reduce this workload is postural alternation, what can be done by providing a seat. Due to the fact that surgeons can not operate in a regular seat position (due to the table and the fact that they need mobility to access the surgical area), the semi-seated position, although not perfect, is a possibility to minimize the effort (Figure 7).

FIGURE 7. Semi-seat chair as an alternative for minimizing the physical effort during surgeries

Final Considerations

This research evaluated the workload of volunteer surgeons of three hospitals in Porto Alegre, State of Rio Grande do Sul, Brazil, during elective surgeries of different levels of complexity. It adds to the other studies reviewed in the literature because, besides the postural analysis, it investigated the relationship among workload (mental and physical), surgery level of complexity (low, medium and high) and the surgeon expertise (experient or novice). The results of catecholamine concentrations, made evident by the rate of Na/A (rate between noradrenalin and adrenaline) showed that the younger and less experient surgeons have more mental than physical load mainly in high level of complexity surgery while the older and more experient surgeons have more physical than mental load mainly in low level of complexity surgery. In this kind of surgery, mental load is minimal for the experient surgeons who, probably due to the age factor, showed physical load. In short, it is possible to assume that the easier the surgery more evident is the physical load and that physical load affect more the experient surgeon while mental load affect more the novices. It is important to note that the small sample of the study probably interfere in the variability of data, mainly in the results of heart rate (HR), arterial pressure (AP), cortisol and ACTH. The variability is also due to the low sensitivity of the techniques used for measuring HR and AP to distinguish mental from physical load. The best indicator was the rate Na/A proposed by Fibiger et al., 1984 e Basset et al., 1987.

From the obtained results, it is clear that physical load impact the surgeons independently of the level of complexity of the surgery what is in accordance to the reviewed literature that points to postural problems during surgery (Magalhães et al., 2000; Berguer, 1999), fatigue and stress (Kant et al., 1992; Mirbord et al., 1995; Luttman et al., 1996; Mcroberts et al., 1999; Nguyen et al., 2001) and physical and mental load (Berguer, 1997; Berguer, 1999; Czyewsaka et al., 1983; Böhm et al., 2001; Bertram, 1991). It is also important to note the technological, organizational, and social factors inherent to developing countries, as is the case of Brazil, which is responsible for the deficiencies in the hospitals and the low wages of the professionals what add to physical and psychological complaints and impact on the overall workload imposed to the surgeons. These professionals have to attend a great number of patients, under time pressure, high number of hour of work without rest and bad technological conditions, leading to adverse consequences such as medical error, low productivity, high level of absenteeism and staff leaving, what makes the professional do not support the workload anymore until he/she changes to another medical specialization. More ergonomic studies in the medical area are needed to improve this socio-technical system in order to force the government to pay more attention to the hospital conditions and, therefore, enhance the Brazilian medical system.

REFERENCES

BASSETT, J. R., MARSHALL, P. M., SPILLANE, R. The physiological measurement of acute stress (public speaking) in bank employees. International Journal of Psychophysiology. Elsevier Science Publishers B.V. (Biomedical Division). N° 5, 1987. pp. 265 – 273.

BERGUER, R. Ergonomics in the operating room. The american journal of surgery. [editorial], vol. 171, 1996, pp. 385 – 386.

BERGUER, Ramon. The application of ergonomics to general surgeons' working environment. Rev. Environmental Health. N°12, 1997. pp. 99 – 106.

BERGUER, Ramon. Surgery and ergonomics. The Archives of surgery. American Medical Association. Vol. 134, 1999. pp. 1011 – 1016.

BERTRAM, D. A. Measures of physician mental workload. In: Anais do Human Factors Society 35th Annual Meeting. Bay Area Chapter: São Francisco, CA. Vol. 2, 1991. pp. 1293 – 1297.

BÖHM, B., RÖTTING, N., SCHWENK, W., GREBE, S., ULRICH, M. A prospective randomized trial on heart rate variability of the surgical team during laparoscopic and convencional sigmoid resection. Arch Surg. American Medical Association. Vol. 136, 2001. pp. 305 – 310.

CZYZEWSKA, E., POKINKO, P., KICZKA, K., CZARNECKI, A. The surgeon's mental load during decision making at various stages of operations. Jornal Europeu de Fisiologia Aplicada. N° 51. 1983. pp. 441 – 446.

FIBIGER, W., SINGER, G., MILLER, A. Relationships between catecholamines in urine and physical and mental effort. International Journal of Psychophysiology. Elsevier Science Publishers B.V. N° 1, 1984. pp. 325 – 333.

FOX, E., MATHEWS, D. Bases fisiológicas da educação física e dos desportos. 3° Ed., Rio de Janeiro: Guanabara, 1986. Pp, 173 – 177.

HERRON, D. M., GAGNER, M., KENYON, T. L., SWANSTRÖM, L. L. The minimally invasive surgical suite enters the 21st centrury. Surgical endoscopy – Ultrasound and Interventional Techniques. Nova Iorque: Springer-Verlag 15: 2001. pp. 415 – 422.

662

HIGNETT, S., McATAMNEY, L. Rapid Entire Body Assessment (REBA). Applied Ergonomics. Elsevier Science Ltd. N° 31, 2000. pp. 201 – 205.

KALSBEEK, J. W. H., ETTEMA, J. H. Scored regularity of the heart rate and the measurement of perceptual load. Ergonomics, n° 6. 1963. pp. 306 – 327.

KANT, I., DE JONG, L., VAN RIJSSEN-MOLL, M., BORM, P. A survey of static and dynamic work postures of operating room staff. Int. Arch. Occupational Environmental Health. N°63, 1992, pp. 432 – 438.

LUTTMANN, A., SÖKELAND, J., LAURIG. Electromyographical study on surgeons in urology. I. Influence of the operating technique on muscular strain. Ergonomics, vol. 39, n°2, 1996. Pp. 285 – 297,

MAGALHÃES, R. A. S. et al. Identificação de riscos ergonômicos no posto de trabalho de médicos-cirurgiões em um hospital universitário. In: I Encontro Pan-Americano De Ergonomia - X Congresso Brasileiro De Ergonomia, Rio de Janeiro. A Ergonomia na Empresa: útil, prática e aplicada. Rio de Janeiro: ABERGO, 2000. p. 22 – 33.

McARDLE, W. D., KATCH, F. I., KATCH, V. L. Sistema Endócrino e Exercício. In: McARDLE, W. D., KATCH, F. I., KATCH, V. L. (Ed.s) Fisiologia do Exercício: Energia, Nutrição e Desempenho Humano. 4ª.ed, Guanabara Koogan, Rio de janeiro. Tradução do original em Inglês: Exercise Physiology: Energy, Nutrition and Human Performance, 4th ed., 1996), 1998. pp.339-367.

McROBERTS, J., GILL, I., RAY, C., WOOD, D. "Stand-up" urologic endoscopy. Urogoly. Adult urology. Digital Urology Journal: Boston, MA.47 (2), 1999. pp. 201 – 203.

MEERS, A., VERHAGEN, P. Sinus arrhytmia, information transmission and emotional tension. Psychological Belgrade, XXII – 1, 1972. pp. 45 – 53.

MESHKATI, N. Heart rate variability and mental workload assessment. In: Hancock, P. A., Meshkati, N. (eds). Human mental workload. Amsterdam: Elsevier Science, 1988. Pp. 101 – 115.

MESHKATI, N. HANCOCK, P., RAHIMI, M., DAWES, S. Techniques in mental workload assessment. In: WILSON, J. R., CORLETT, E. N. (eds). Evaluation of Human Work: a practical ergonomics methodology. Londres: Taylor & Francis Inc., 1995. Pp749 – 782.

MIRBOD, S. M., YOSHIDA, H., MIYAMOTO, K., MIYASHITA, K., INABA, R., IWATA, H. Subjective complaints in orthopedists and general surgeons. Int. Arch. Occup. Environmental Health. New York: Springer-Verlag, n° 67, 1995. Pp. 179 – 186.

NEESE CONSULTING COMPANY. Rapid Entire Body Assessment (REBA) software®. Versão 1.3. 2001.

NGUYEN, T. N., HO, S. H., SMITH, W. D., PHILIPPS, C., LEWIS, C., DE VERA, R. M., BERGUER, R. An ergonomic evaluation of surgeons'axial skeletal and upper extremity movements during laparoscopic and open surgery. The American Journal of Surgery. Excerpta Medica, Inc. n°182, 2001. pp. 720 – 724.

VELTMAN, J. A., GAILLARD, A. W. Physiological workload reactions to increasing levels of task difficulty. Ergonomics. Taylor & Francis Ltd. Vol. 41, n° 5, 1998. pp. 656 – 669.

Chapter 66

The Concept of Task for Non-Production Human-Computer Interaction Environment

Gregory Bedny, Waldemar Karwowski, Inna Bedny

Institute for Advanced Systems Engineering
University of Central Florida
Orlando, FL 32816-2993, USA

ABSTRACT

The task is a critically important concept for the study. Computers are now used beyond production environment for interaction, entertainment and as a source of information and some HCI professionals express an opinion that in such cases concepts of task and goal-directed activity are not applicable. They promote an idea that science of "enjoinment" reject concept of task and goal-directed approach. In this paper we demonstrate that even entertainment related human activity is goal-directed, and task concept is still critically important for its analysis.

Keywords: Task, Goal, Sense of Task, Task Complexity, Task Difficulty, Motivation, Non-Production Environment, Self-Regulation, Functional Mechanism

INTRODUCTION

The nature of work constantly changes. The main changes are recently related to computerization. This evokes new issues in the study of human work associated with informational technology. These issues are known as task analysis of computer-based systems. One weakness common to cognitive approaches in studying computer based tasks is the disregard for emotionally-motivational aspects of human work activity. Usually this problem in ergonomics is reduced to the study of emotional stress and reliability analysis. This approach does not consider positive emotions related to task performance, relationship between emotions and motivation, etc. In industrial/organizational psychology motivation is considered separately from human-information processing approach developed in cognitive psychology. In most cases motivation is described from personality, group dynamics and productivity perspectives.

From activity theory (AT) perspectives emotions play an important role in regulation of activity. One important distinguishing feature of motivation versus emotions is that the motivation directs activity to a specific goal. In self-regulation of activity feedback has both cognitive and motivational effect. Studying activity as a self-regulative system is called functional analysis of activity (Bedny, Meister, 1997; Bedny, Karwowski, 2007).

Emotionally-motivational state while performing a task to a significant degree depends on task significance or emotionally evaluative components of activity. Users actively select required information; interpret it depending on the excepted or formulated task goal and personal significance of the task. SSAT considers activity as a system, which integrates not only cognitive and behavioral but also emotionally-motivational components.

Computing technology with increasing frequency is now used for non-productive purposes: recreation activity, education, games, etc. (Karat, et al., 2004). In order to improve the design of such systems we need a good understanding of what is task, goal, emotionally-motivational aspects of human performance, etc. Recently the concept of affective design has been introduced in ergonomics (Helandar, 2001) with its pleasure-based principles to in ergonomics. Such concepts as emotion, motivation or aesthetics requirements are particularly important for such design. However human information processing should not be separated from emotionally-motivational aspects of activity. A person always emotionally relates to presented information. Hence the emotionally-motivational processes always interact with cognition. Interpretation of information depends on emotionally-motivational aspects of activity. Pleasure-based design is only one aspect of implication of the principle of unity of cognition, behavior and emotionally-motivational components of activity. In traditional design this principle is also critical. Conceptualization of the principle of interdependency of cognition, behavior and motivation can be found in SSAT. Application of emotionally-motivational aspects of affective design requires development of adequate concepts, terminology and that would reflect their relevance to the design process. In this work we are considering the concept of task and its main attributes in the non-productive environment.

EMOTIONALLY-MOTIVATIONAL ASPECTS OF TASK ANALYSIS

In this work we discuss emotionally-motivational aspects of task analysis from functional analysis perspective when activity is considers as a self-regulative system and its main units of analysis are function blocks (for more details see Bedny, Kawrowski, 2007). From the self-regulation analysis perspectives when we study motivation such functional mechanisms or blocks as "goal", "assessment of task difficulty", "assessment of the sense of task "(personal significance of task and its elements or emotionally-evaluative mechanism), "level of motivation" (inducing components of activity), "criteria of evaluation" are particularly important.

Let us briefly consider interaction between such blocks as "assessment of task difficulty", "assessment of the sense of task" and "formation of the level of motivation".

Functional analysis distinguishes between objective complexity of task and subjective evaluation of task difficulty. The subject can evaluate the same task as more or less difficult depending on complexity of task, subject's own past experience, individual features of the subject and even his/her temporal state. The higher is the task complexity the more is the probability that this task will be evaluated as being difficult. Cognitive task demands during the task performance depend on the task complexity. A subject experiences not the complexity of task but its difficulty. A method of quantitative evaluation of task complexity can be found in Bedny, Karwowski (2007). An individual might under or overestimate the objective complexity of the task. For example, a subject can overestimate the task difficulty and task can be rejected in spite of the fact that objectively the subject would be able to perform it. Moreover overestimation of task difficulty produces emotional stress and even if the task is accepted the quality of performance can suffer. On the other hand if a subject underestimates task difficulty he/she can fail to perform the task. Psychological concepts that reflect subject's evaluation of her/his own abilities to perform the required task are called self-efficacy (Bandura, 1987), self-esteem, etc. From a functional analysis perspective functional block "difficulty" depicts cognitive mechanisms of self-regulation that are critical for the motivational process. The block "difficulty" interacts with a number of other blocks in the self-regulation process. In this work we'll consider interaction of blocks "difficulty" and "sense ". Function block "sense" predetermines the significance of the task, its elements, situation and a subjective value of obtaining the desired result of the task elements and of the task as a whole that provides a sense of achievement. Blocks "difficulty" and "sense" have complex relationship and interaction between them influences the process of motivation (figure 1).

666

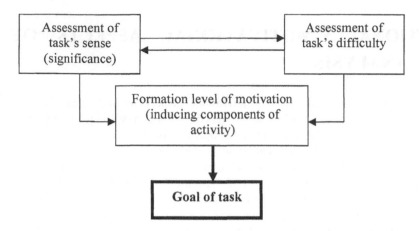

Figure 1. Relationship between functional mechanisms "assessment of task's significance", "assessment of task's difficulty ", "formation level of motivation", and "goal of task"

Interaction of these blocks explains motivation from a totally different point of view in comparison with the existing theories. Let's consider the construct of self-efficacy developed by Bandura (1987). According to Bandura, the stronger the belief in self-efficacy is, the more persistent a person will be in pursuing the desired result. He suggests that people with high personal efficacy set more difficult goals and show greater persistence in their pursuit, while people with low efficacy set lower goals which often negatively influences the motivation and increases the probability of goal abandons in face of adversity. It implies that all motivational manipulations take effect through self-efficacy. From the self-regulation or functional activity analysis point of view, if a person evaluates a goal as very difficult due to her/his low self-efficacy it negatively influences motivation (inducing component of motivation) and increases the probability that the goal will be avoided or abandoned. On the other hand, if the particular goal of task is significant or has high level of positive subjective value, those with low self-efficacy can nevertheless be motivated to strive for such goal.

Motivation among other things depends on a subjective criterion of successful result (see function block "criteria of success"). The subjective standard can deviate from objective requirements, so the satisfaction of goal attainment depends on this criterion as well. If the subject achieves a required goal but her/his level of aspiration exceeds the goal, the subject will not be satisfied by the obtained result. The concept of the subjective standard of successful result is also deemed important in the social learning theory (Bundura, 1977). However, in this theory there is no clear understanding of what is the difference between the goal and subjective standard of success. Subjective standard of success can deviate from goal, particularly at the final stage of the task performance. A subjectively accepted goal can be used as a subjective standard of success. However, goal by itself might not contain enough information to evaluate the result of task performance. This standard has dynamic relationship with the goal and past experience (Meister, Bedny, 1997) and might change during goal acceptance and task performance.

RELATIONSHIP BETWEEN FUNCTIONAL MECHANISMS OF "GOAL" AND "SENSE" IN GAMBLING

"Sense" (significance) is one of the function blocks of self-regulation that interacts with the goal of task. If the goal of task has only positive value for the subject then function block "sense" has homogeneous structure and has only positive significance. On the other hand if the goal of task possesses attributes that have positive and/or negative personal value, then function block "sense" has a heterogeneous structure and includes positive and negative significance. Proportion of these two types of significance determines integrative character of the evaluation of task significance.

For example, in gambling the goal is to win money in a risky situation Achievement of this goal is conveyed also with the possibility of loosing money. The relationship between these two kinds of significance in function block "sense" determines the subject's level of involvement in gambling. Moreover there is also a possibility that a risk addicted person will choose a task with a high probability of failure. For such a person it's important not only to obtain money but also to experience the danger. Elimination of danger immediately eliminates feeling of positive significance in gambling for such a person. Risk addicted people are usually involved in gambling based on some proportion of positive and negative significance associated with mental representation of a probability to loose or win money. Feeling of some level of risk (to loose money) increases feeling of positive significance which in turn triggers inducing components of motivation. After success in a risky situation risk seekers are often get involved in even riskier tasks until they fail. Hence there are individual differences between risk addicted and non-risk addicted people in the selection of activity strategies in risky situations. Some individuals always attempt to be involved in tasks that have some negatively significant factors with various proportion of positive and negative significance. Moreover this proportion can have different value for different individuals. For some individuals such relationship is dynamic and depends on their history of successes and failures.

Considered above aspects of motivation are important for development of computer-based gambling tasks and for non-production tasks in general.
Relationship between positive and negative significance is also important for the analysis of safety because it might produce conflict of motives during achievement

of the task's goal. It is well known that requirements of safety very often contradict with productivity. The factor of productivity is significant for workers because it influences their wages and social status. Therefore workers have high aspiration to increase productivity. However the safety rules are often restrictive. As a result workers face an ambiguous situation. The achievement of high productivity can contradict with the safety requirements. This produces some conflict between positive and negative significance in the task performance. In this case, the worker's strategy would depend on what kind of significance dominates. Often workers ignore high risk with low subjective probability of occurrence in favor of more probable and valuable goal of increasing productivity.

ANALYSIS OF TASK PERFORMANCE IN THE NON-PRODUCTION ENVIRONMENT

Playing and games are important topics in the study of HCI in non-production environment. For our analysis it is useful to consider the role of playing in mental development. Vygotsky's (1978) work gave the most wide-ranging account of psychological characteristics of games and of their role in mental development. The purpose of play is not to achieve a useful result of activity but rather the activity process itself. However, this fact doesn't eliminate goal formation and motivational aspects of such activity. Actions of a child are purposeful and goal directed. For example, an adult demonstrate how to feed a doll. Children have their own experience of being fed. In spite of the fact that a child can't really feed the doll she/he still performs conscious goal directed actions. Moreover imaginative aspects of the play become critical when children operate with a variety of objects that have particular purpose and are associated with these objects' meaning. The child at play operates with the meanings that are often detached from the usual objects (Vygotsky, 1978). For example, a child can take stick and tell that it is a horse and start performing meaningful actions that have some similarity with the rider's actions. Play is associated with pleasure and develops motivational components of activity. Gradually, the play becomes more and more important and a child develops the rules for her/his play. Submitting to these rules makes playing more complex and transfers it into a game. The imposed rules force children to learn how to suppress involuntarily impulses. Hence the childhood games are preceding real adult activities.

With development of computer industry the games became important even for adults. Motivational factor plays a particular role in a game. So, let's consider the stages of motivational process in human activity. According to concept of motivation developed by Bedny and Karwowski (2006) there are five stages of motivation: (1) preconscious; (2) goal-related; (3) task-evaluative; (4) executive or process-related and (5) result-related motivational stage. These stages, according to the principles of self-regulation, are organized as a loop structure and depending on specificity of task, some stages can be more important than others. Depending on task specificity some of these stages and their relationship become more or less important. The first pre-conscious stage of motivation predetermines motivational tendencies. This stage is not associated with a conscious goal but rather with an unconscious set that can be later transferred into a conscious goal and vice versa. The second goal-related motivational stage is important for goal formation and goal acceptance. This stage can be developed in two ways: by bypassing preconscious stage of motivation or through the transformation of an unconscious set into a conscious goal. When the execution of a current task is interrupted and attention is shifted to a new goal, the previous goal doesn't disappear, but is transformed into a preconscious set. It helps a subject to return to an interrupted task, if necessary, through transition of a set into a conscious goal. The third motivational stage is related to evaluation of the task difficulty and significance which has been discussed in the previous sections. The fourth executive or process related motivational stage is associated with executive aspects of task performance. Goal formation, task evaluation (evaluation of task difficulty and its significance and their relationship) and process related stages of motivation are particularly important for understanding of risky tasks, games and development of recreational computer-based tasks. The fifth stage of motivation is related to evaluation of activity result (completion of task). These five stages of motivation can be in agreement or in conflict.

Process-related motivational stage (stage 4) is critical for computer-based games and should be associated with positive emotional-motivational state. Result related stage (stage 5) vary when positive results are combined with negative results producing a combination of positive and negative emotionally-motivational states. At the same time only positive result in computer-based games can reduce interest in the game. A simple game without a risk of loosing can reduce positive aspects of process-related stage of motivation. Hence, complexity of task should be regulated depending on the previously obtained results. Even in gambling when people can loose their money some ratio between success and failure is important. Strength of positive and negative emotions during various stages of motivation is also important. This is particularly relevant for the gamblers since manipulation with process and result related stages of motivation are critical. Of course, other stages of motivation also should be taken into consideration. In the non-productive tasks simplicity or difficulty to obtain a desire result is an important factor. Understanding motivation as a sequence of interdependent motivational stages helps to create a desire motivational state in production and non-production environment.

Sometimes the goal of task-game isn't precisely defined and at the beginning the goal is presented in a very general form. Only at the final stage of the game the goal becomes clear and specific. This is not new. For example, when playing chess the goal can be formulated only in a very general form "to win" or "to tie the game ". A chess player also formulates in advance some hypotheses about his/her possible strategies that are tightly connected with the goal of task. For this purpose a player uses some algorithmic and heuristic rules stored in player's memory. When a chess player selects a possible strategy he/she starts to formulate multiple intermittent goals corresponding to this strategy. The selected strategy can be corrected or totally abandoned depending on the rival's strategies. Clear and specific understanding of an overall goal is possible only at the final stage of the game, just before a checkmate. Even when a goal of task is externally given in a very precise form, a subject can reach this goal by using a variety of strategies and various intermittent goals. Therefore, a goal can not be considered as an end state of the system that a person or a system strives to achieve as has been stated by Preece, et al. (1994, p. 411). A goal of a system and a goal of a person are two totally different concepts.

This understanding of goal is radically different from its description as a clearly defined end state of the system as it is considered by some usability engineers. Karat at al. (2004) stated that only production tasks have a clearly intended purpose or goal. According to these authors HCI field shifts its focus from production environment with its clearly defined tasks and goals to the non-production field, where the main purpose is communication, engaging, education, entertainment, etc. The authors mix the goal of task with the motive and insist that in non-production environment and particularly in games there are no tasks and goals (Karat, Karat and Vergo, 2004, p. 587). However, in the above described example we've shown that the goal is to reach the desired future result of the game. The motive is to obtain pleasurable state while playing and to feel satisfaction after a game. The goal in non-production and production environment can be imprecise at the beginning of the task performance. In designing task in any environment it is very important to find out how initially formed in general terms goal is gradually clarified and specified during farther task performance. As an example we can consider such games as "sudoky", "cheese playing", etc. In non-production situation the same as in production situation not just the final goal of task but also intermittent goals are important. Sub-goals are integrate into a stage of task performance. A goal of task and goals of separate actions are cognitive and conscious entities. There is also a need for a general motivational state that creates an inducing force to induce this sub-goaling process. Comparison of a future hypothetical end-state with an existing state is provided by feedback which is processed in the mental plane. This demonstrates that thinking works as a self-regulative process. A sub-goaling process includes formation of hypotheses that has conscious and unconscious components.

Karat at al. (2004) introduce the so called "science of enjoyment" and substitute such concepts as task, goal, emotions and motivation with the term "value" that just has a common sense meaning. People can enjoy drugs, alcohol, work, sports, etc., depending on various emotionally-motivational factors. Hence, study of enjoyment should be associated with emotions and motivation. Is there a need for a new "science of enjoyment" when psychology already studies emotions and motivation?

CONCLUSION

Some practitioners raise the questions about the future of the task analysis. There are even suggestions to eliminate the concept of task because it ignores motivational forces, or to eliminate this concept just when studying entertainment. Other professionals insist on eliminating the concept of goal in task analysis. From activity theory perspectives we can not agree with such opinions.

Any technology is just means and/or tools of human work or entertainment activity. Hence we need to study the specifics of utilizing such tools or means of work in various kinds of activity. Progression of the technology into the home environment and everyday life can not eliminate task-oriented and goal-directed characteristics of activity. In non-production environment people strive to break down activity flow into smaller segments or tasks which are often self-initiated. Social interaction, learning and playing are always motivated and goal directed as any other type of human activity. Therefore subjects' everyday activity can't be understood without referring to motivational and goal-formation processes.

In SSAT every task includes means of activity, materials and mental tools, work processes and technological or control processes. Hence, the task can be studied from behavioral or technological perspectives. Of course these two aspects of study are interdependent. In this paper we've considered some basic characteristics of task performed in non-production environment. It was demonstrated that goal and task are critically important for the study of any kind of human activity, including playing games, social interaction, entertainment, etc.

REFERENCES

Bandura, A. (1987). Self-regulation of motivation and action through goal systems. In V. Hamilton, and N. H. Fryda (Eds.), Cognition, motivation, and affect: A cognitive science view. Dordrecht: Martinus Nijholl

Bandura, A. (1977). Social learning theory. Englewood Cliffs, NJ: Prentice-Hall

Bedny, G., Mejster, D. (1997). The Russian theory of activity. Current application To design and learning. Lawrence Erlbaum Associates Publishers. Mahwah, New Jersey

Bedny, G., Karwowski, W. 2007, A systemic-structural theory of activity. Application to human performance and work design. Taylor and Francis, Boca Raton, London, New York.

Bedny, G. Z., Karwowski, W. 2006, The self-regulation concept of motivation at work, Theoretical Issues in Ergonomics Science, V. # 4, pp. 13 – 436.

Bedny, G.Z., Harris, S. R. 2005, The systemic-structural theory of activity: Application to the study of human work, Mind, Culture, and Activity: An International Journal, V. 12, # 2, pp. 128-147.

Karat, J., Karat, C-M, and Vergo, J. (2005). .Experiences people value: The handbook of task analysis. In D. Diaper, N. Stanton (Eds.). The handbook of task analysis for human-computer interaction. Lawrence Erlbaum Associates Publishers. Mahwah, New Jersey

Helander, M. G. 2001, Theories and methods in affective human factors design, in M. J. Smith, G. Salvendy, D., D. Harris and R. J. Koubeck (Eds.), Usability Evaluation and Interface Design. V. I of the proceedings of HCI 2001, Mahwah, NJ: Lawrence Erlbaum Associates, pp. 357 – 361.

672

Preece, J. Rogers, Y. Sharp, H., Benyon, D., Holland, S., Carey, T. 1994, Human-computer interaction. Addison-Wesley.

Vygotsky, L.S. 1978, Mind in society. The development of higher psychological processes. Cambridge, MA: Harvard University Press.

Chapter 67

The Concept of Task in the Systemic-Structural Activity Theory

Waldemar Karwowski[1], Gregory Bedny[1], O. Chebykin[2]

[1]Institute for Advanced Systems Engineering
University of Central Florida
Orlando, FL 32816-2993, USA

[2]Academy of Pedagogical Science, South
Ukrainian Pedagogical University, Odessa
Ukraine

ABSTRACT

In this paper we describe the role of task concept in the study of human work and HCI interaction in particular. Every task includes human activity, technological components or means of work, and the object under transformation. The nature of the changes undergone by the object depends on both human activity and the specifics of technological components of work. Hence task analysis may be studied from either the technological viewpoint or from human activity perspectives. In this paper we consider task concept from activity theory viewpoint.

INTRODUCTION

Changing equipment or software characteristics influences the methods of task performance. Hence the development of human-machine systems begins with task analysis. By analyzing the task that is being performed we can estimate the efficiency of its performance. Selection process for complicated jobs and development of job training also requires preliminary task analysis. All of this makes task concept critically important in work psychology and ergonomics. At the same time there is no general agreement on what a task is. Contemporary cognitive approach concentrates its efforts on studying internal mental processes required to support mental work or external behavioral acts. Such approach ignores the concept of goal and motivational components of activity. Moreover in cognitive psychology and in industrial-organizational psychology there is no clear understanding of the concept of goal. For example, Vancouver (1996) gives the following definition of goal: "Goal is an internal representation of desired state of the system". However the goal of the artificial system and goal of the human being are not the same. A person understands what she/he wants to achieve. Hence goal is associated with our consciousness. Goal is not just a representation of a desired state, but also a representation of a future desired state which can be achieved as a result of subject's own activity. For instance, a desired state of the system can be achieved as a result of the system functioning without direct involvement of the subject in its performance. This future state can not be considered as a goal. Subject can understand what he/she wants as "a desired state", but does nothing to achieve this state. Hence such desired state can't be the goal of the subject. General agreement in ergonomics, according to which task is a set of human actions that can contribute to a specific functional objective and ultimately to the output goal of the system, is also ambiguous. This definition does not distinguish between human and non-human goals. There is no clear understanding of actions, motivational forces, etc.

Similarly, there are a lot of problems in understanding the concept of task in the field of human-computer interaction. For example, Diaper and Stanton (2004) wrote that such main concept as goal should be abandoned as unnecessary in task analysis. According to them the term "goal" is one of the major sources of confusion. The entire future of task analysis, according to them, is not clear.

These drawbacks with understanding of task and its basic components bring some scientists to the conclusion that concept of task is redundant. Kaptelinin and Nardi (2006) recommend the concept of engagement as a replacement of task. Without clear understanding of the concept of task accurate task analysis is impossible. In this work we consider the concept of task from the activity theory perspective.

TASK AS A MAIN ELEMENT OF WORK PERFORMANCE

Task is a main element of work that can be seen as a collection of tasks falling within the scope of a particular job title. In work settings tasks usually have some logical organization. This logically organized sequence of tasks is called production process (Bedny, Harris, 2005). From technological perspectives production process in manufacturing can be defined as a sequence of transformation of raw material into finished product. Any production process has two main components: work process and technological process.

In automated systems instead of production process the term operational-monitoring process is used. An operational-monitoring process is a combination of tasks essential to accomplish some automated or semi-automated system function. One important aspect of operational-monitoring process is that the operator is not simply involved in changes of physical material, but also in transformation of information. In operational-monitoring process the sequence of tasks often doesn't have rigorous logical organization. This is especially relevant for HCI task, where a significant percentage of tasks are formulated by users. They decide what has to be done. The task in operational-monitoring process is often a problem-solving one. Operational-monitoring process includes both work process and control process.

The structure of production process and operational-monitoring process are similar. The only difference is that in operational-monitoring process an operator or user interacts with input and output information instead of raw materials and finished products.

HCI process is similar to operational-monitoring process. The main difference being that in the first one is that the computer is the dominant means of work which produces on the screen a variety of tools required for a particular task and mediates human-computer interaction. We don't agree with Kaptelinin (1997) who considers computer as a tool. From our point of view this is not an accurate interpretation of the role of computer in human work.

One can study any task from two main perspectives: technological perspectives and work analysis perspectives. These two approaches are interdependent but have their own specifics. In manufacturing the task is synonymous to production operation. A process engineer is responsible for developing manufacturing process, which includes such steps as citing, drilling, reaming, etc. All these steps are performed in a particular sequence and speed by using special equipment, tools, computers, and so on. Ergonomists/industrial-organizational psychologists on the other hand pay attention to behavioral aspects of human work. It is obvious that equipment configuration, manufacturing process, computer interface influence strategies of human work activity and vice versa. Hence, changes in equipment configuration, computer interface, and technological process will result in a probabilistic way in the new strategies of human work activity.

We define the task as logically organized system of cognitive and behavioral actions utilized by a worker in order to achieve the goal of task (Bedny, Karwowski, 2007).

Leont'ev (1978) and Rubinshtein (1959) divided activity into tasks. Task in turn can be divided into hierarchically organized units such as actions and operations. The last one in addition can be divided into smaller units called functional micro-blocks. Actions can be behavioral or motor and cognitive. Each action is directed to achieve a conscious goal. This goal should be distinguished from an overall goal of task. The goals of actions involved in task performance can be quickly forgotten. In contrast the goal of a task is kept in memory until the task is completed. The same goal of task can be achieved by different courses of actions.

Sometimes during an automatic processing of information the action's goal ceases to be conscious and as a result an action is transformed into unconscious operation. Such operation is a component of a more complicate action. This means that strategy of task performance is changed. For example, when a paraplegic learns to walk again, he is at first conscious of the individual goals of standing, raising the leg, moving the leg forward, and so forth. As she/he learns to walk again, she/he is no longer aware of these individual movements' goals and is aware only of the goal of walking. On the other hand, the complicated motor actions can be disintegrated. If for any reason a walker starts considering consciously individual steps while walking such steps are consciously controlled according to the goals of steps. Such steps become actions (stepping on the Moon). Actions and activity in general are complex self-regulated systems. Their performance depends on strategies of task performance. During task analysis we need to select the most preferable strategies of task performance. For task analysis we also have to select a standardized language of task description. Selection of standardize units of analysis and language of description is prerequisite for any design process.

We utilize the concept of standardize motions and the MTM-1 system for the description of motor actions (Karger, Bayha, 1977). According to our definition a motor action includes no less than two standardized motions integrated by one conscious goal. For instance, a motor action "grasp the part" can include two motions: move an arm and grasp. They can be performed in various modes. Cognitive actions are main units of analysis of mental activity. Usually, in cognitive psychology mental actions are not extracted. Scientists extract only mental processes such as: short-term memory, long-term memory functions, decision-making, etc. Such descriptions are useful but not sufficient. In order to describe the structure of cognitive components of activity utilization of the concept of cognitive actions is required. Psychic processes that are involved in task performance can be the same but associated with them mental actions can be different. This is why the concept of mental action is useful for the description of the task structure. Mental actions can be classified based on various criteria. The main method of cognitive actions' classification is based on dominant psychic process during actions performance. Based on this principle it is possible to distinguish between sensory, perceptual, thinking, decision making and mnemonic actions. Basic criterion for separation of one action from another is the goal of action. The simplest cognitive actions have very short duration. Actions have logical organization. Some actions can be performed simultaneously. Detailed description of actions can be found in works of Bedny and Karwowski (2007).

The type of actions involved in task performance depends not only on goal but also on the task conditions. Leont'ev (1978) stated that a task can be defined as a situation that requires achievement of a goal of task in specific conditions. The elements of task include initial information, the actions required, and the goal of task that organizes all task elements as a whole. There are also some motivational forces associated with the goal that give activity during task performance goal directed character. We discuss this aspect later on in this work. The structure of task is presented on figure 1. From this figure one can see that conditions that are related to motives form a goal. The relationship between the goal and the conditions determines the task performance.

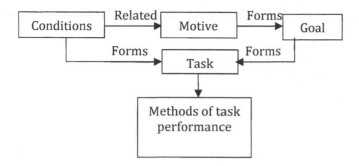

Figure 1. Schema of task formation in structure of activity

There are two major components that determine a particular task. They are called task requirements and task conditions. Anything that is presented to the performer or known by her/him is a condition of the task. Finding the solution (or what we need to achieve) becomes a goal of task after acceptance of the requirements by a subject or can be formulated by a subject independently. We'll show that the subject's goal doesn't exist in a ready form. Only objectively given requirements do. At the later stage these requirements can be transformed into a subjectively excepted goal. Hence, there are requirements given in particular conditions and after transformation of these requirements into a goal the task emerges. Acceptance or formulation of task's goal is closely associated with subjective representation of the task.

THE CONCEPT OF GOAL IN PSYCHOLOGY AND ERGONOMICS

The goal is a critically important element of task. In the US the concept of goal is associated with such scientist as McDougal (1930) and Tolman (1932). McDougal rejected interpretation of human behavior as a stimulus triggered system. He explained human behavior as an activity striving toward an anticipated goal.

Tolman developed purposeful behaviorism. According to him goal is the end state toward which the motivated behavior is directed. Knowledge about responses and their consequences are important in both animal and human behavior. Hence animal and human behavior is flexible in reaching its purpose. However goals of animals do not include consciousness. This is why a low level organization animals' goal is understood by Tolman as a purpose.

Goal is also an important concept in the field of motivation and social psychology. One of the peculiarities of considered theories is that motivation and goal are not distinguished. According to AT motivation and goal are two different entities. In more resent studies the concept of goal and motivation are also not differentiated. Locke and Latham (1990) postulate that goal has both cognitive and affective features. They state that goal can be weak or intensive and consider goal as a motivational component of human behavior. Presumably, the more intense the goal is, the more one strives to reach it. Hence, the goal "pulls" the behavior or activity. Kleinback and Schmidt (1990) describe volition process and claim that goal is inducing behavior (see figure 2).

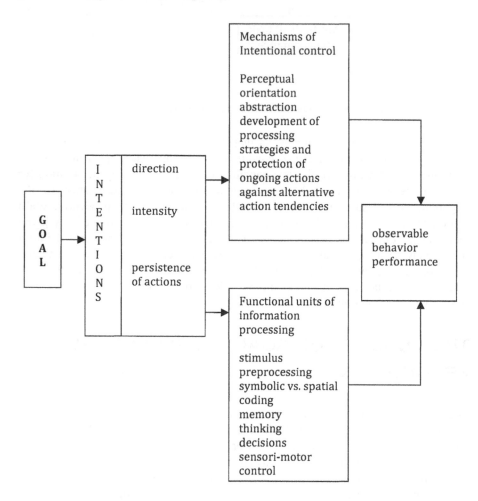

Figure 2. Schema of the volition process according to Kleinback and Schmidt

This figure depicts volition process moving from goal to behavior. So, Kleinback and Schmidt also didn't distinguish between goal and motive. In the field of HCI the goal is defined as a state of a system that human, machine or a system strives to achieve (Preece, et al., 1994). Hence the human and non-human goals are considered as similar phenomenon. This is the cybernetic understanding of goal.

Such theories don't discuss objectively given requirements of tasks and process of goal acceptance, intermittent goals of actions and final goal of task, the difference between goals of artificial system and human goal, etc. Hence some scientists (Diaper, Stanton, 2004) suggest eliminating the concept of goal and some others (Nardi, 1997) suggest to eliminate the concept of task.

The goal should not be confused with motives. Our behavior is poly-motivated and usually includes a number of motives. If the goal includes various motives, according to cognitive psychology, a person pursues not one but multiple "goals-motives" at any given time during the task performance. This is imposable, because one task has only one goal.

Austin and Vancouver (1996) introduce the terms "maintenance goals and attainment goals". Vancouver (2005) reduces human behavior to a homeostatic self-regulated process. Deviation from "maintenance goal" is considered to be an error that should be corrected through negative feedback. However such self-regulated process is more suitable for artificial systems and physiological processes. In human activity self- regulation is goal directed. These authors mix goal of a system with human goal. Several tasks with their specific goals are reduced to one task with one maintenance goal. Any deviation from a so called "maintenance goal" is considered to be an error. However in Vancouver's hypothetical example those are not errors because deviations from some parameters can be considered errors only if they exceed particular range of tolerance which exists for this system. As it can be seen a number of critically important aspects of human work in general and task analysis in particular are concentrated around the concepts of goal and task.

GOAL AS ONE OF THE MAJOR ANTICIPATORY MECHASNISM OF ACTIVITY

In activity theory goal is one of the most important anticipatory mechanisms at the psychological level of the activity self-regulation. It facilitates forecasting of our activity outcome. Goal is a mental model of a desired future result. During ongoing activity this goal becomes more specific and is corrected if needed.

Goal in AT is understood as a conscious image or logical representation of desired future result. An image or mental representation of future result, when a subject isn't directly involved in achieving this result, is not a goal of activity. An imaginative, future result emerges as a goal only when it is a consequence of the subject's own activity. In AT the goal is connected with motives and creates the vector "*motive* → *goal*" that lends activity its goal directedness. We also need to distinguish between the "overall or terminal goal of task" and "partial or intermittent goals of actions and sub-goals of a task".

In AT there are energetic and informational (cognitive) components of activity that are interdependent but still different. Goal is a cognitive component of activity. Motives or motivation is an energetic component of activity. Goal can't be presented to the subject in a ready form but rather as an objective requirement of the task. However these requirements are then comprehended and interpreted by the subject. At the next stage these requirements are compared with the past experience and the motivational state which leads to the goal acceptance process. Subjectively accepted goal doesn't always match the objectively presented goal (requirements). Moreover, very often subjects can formulate the goal independently. As it can be seen goal always assumes some stage of activity, which requires interpretation and acceptance of goal. So, we can conclude that the goal doesn't exist in a ready form for the subject and can't be considered simply as an end state to which the human behavior is directed.

Not just the overall goal of task but also intermittent goals are critical elements of task performance. Sub-goaling process regulates all stages of task performance. Such sub-goals are hierarchically organized. Some sub-goals and goals of separate actions that flash across subject's mind can be quickly forgotten. Subject formulates a number of potential goals that can be also quickly forgotten. Such goals are associated with gnostic hypothesis which often are not verbalized. However some of such hypotheses can be verbalized and become conscious at the farther stages of activity.

CONCLUSIONS

Activity is a coherent system of internal mental processes and external behavior and motivation that are combined and directed to achieve conscious goals. Existence of a conscious goal demonstrates that human activity is behavior specific. Task is the main component of activity. Our lives can be conceptualized as a continuous performance of various tasks. Task is also a main component of work activity. Usually tasks are prescribed in advance. However contemporary production process is very flexible so the task can be formulated by a performer independently. In spite of critical importance of this concept for work psychology, ergonomics and HCI fields there is no clear understanding of the concept of task and its main attributes and some scientists even suggest to eliminate concepts of task and goal.

Historical analysis of the concepts of task and goal demonstrate that they are not sufficiently developed in contemporary cognitive and social psychology. The concept of task usually doesn't include motivational aspects. Some practitioners mix motives with goals of task because human behavior is usually poly-motivated and conclude that the same task can have multiple overall goals. This is the main reason for the rejection of the concept of goal in task analysis. Such interpretation contradicts with AT where one task has only one final goal. If the final goal of task totally changes during performance this means that a subject formulates a new task. An overall goal of task should be separated from goals of individual actions.

The other problem arises from integration of goals with motives. However in activity theory goal is just an anticipatory, cognitive mechanism. Goal can be precise or imprecise. It includes verbally-logical and imaginative components, it can be preliminary defined and later specified more precise. If the goal isn't sufficiently defined at the beginning of task performance it can be specified or modified during task performance. Goal is only a cognitive, anticipatory mechanism of subject's own activity which should be separated from emotionally-motivational or energetic aspects of activity. Goal and motive are interdependent just like information and energy but they are different phenomenon. In production processes task and its goal are usually presented to a worker in a ready form. However, very often, in operationally-monitoring processes and in computer-based tasks in particular the overall goal of task can be formulated by the user independently.

There is yet another aspect of goal in task analysis. Some specialists consider goal as an end state of the object or system toward which the human activity is directed. This is a cybernetic understanding of the goal that is popular among HCI specialists. The shortcoming of such interpretation is that the goal of a technical system isn't distinguished from a human goal. Goal can't be considered as an end state of the system and can't be presented to a subject in a ready form. There are only objectively given requirements of the task which has to be transferred into a subjectively excepted goal.

Analysis of users' performance demonstrates that even in situations when a goal of task is clearly defined a person has to interpret, clarify, reformulate and accept it. After that a person formulates intermittent goals, achievement of which brings him/her closer to the overall goal of task. So, there is one final goal of a task and multiple intermittent goals. Goal of task is associated with motives and creates vector "motives → goal". The task includes initial situation; the accepted or formulated goal that is associated with motives or motivational state as an inducing force; cognitive and behavioral actions required to achievement the overall task's goal and elements of external environment. Every task has requirements and conditions. When requirements are interpreted and accepted by a person they become the goal of task. This transformation process is one of the main means by which the goal and therefore the task is formed

REFERENCES

Austin, J. T., Vancouver, J. B. (1996). Goal constract in psychology. Structure, process and content. *Psychological Bulletin*, 120 (3), pp. 338-375. Bedny, G., Z.,Harris, S., 2005, The systemic-structural theory of activity: Application to the study of human work. Mind, Culture, and Activity, V 12, N 2, pp.128-147.

Bedny, G., Karwowski, W., 2007, A systemic-structural theory of activity. Application to human performance and work design. Taylor and Francis, Boca Raton, London, New York.

Diaper, D., Stanton, N., (200, Wishing on a sTAr: The future of task analysis. In D. Diaper and N. (Eds.) Stanton The handbook of task analysis for human-interaction. Lawrence Erlbaum Associates, Publishers. Mahwah, New Jersey, pp. 603-619

Kaptelinin, V., Nardi, B., B., A.., 2006, Acting with technology. Activity theory and interaction design, The MIT Press, Cambridge, Massachusetts.

682

Kaptelinin, V., 1997, Computer-mediated activity: Functional organs in social and developmental contexts. In B. Naedi (Ed.) Context and consciousness. AT and human-computer interaction. The MIT Press, Cambridge, Massachusetts, London, England.

Karger, D. W., Bayha, F. H., 1977, Engineering work measurement (3rd ed.). New York: Industrial Press.

Kleinback, U., & Schmidt, K. H., 1990, The translation of work motivation into performance. In V. Kleinback, Quast H.-H., Thierry, H., H. Hacker (Eds.), Work Motivation (pp. 27-40). Hillsdale, NJ: Lawrence Erlbaum Associates, Inc.

Leont'ev, A. N., 1978, Activity, consciousness and personality. Englewood Glifts.Prentice Hall

Locke, E. A. & Latham, G. P., 1990, Work motivation: The high performance cycle. In V. Kleinbeck, etal.(Eds.), Work Motivation (pp. 3-26). Hillsdale, NJ: Lawrence Erlbaum Associates

McDougall, W., 1930, Autobiography. In C. Murchison (Ed.). A history of psychology in autobiography. Worcester, MA: Clark University Press.

Nardi, A., 1997, Some reflection on the application of AT. In B. A. Nardi (Ed.). Context and consciousness: AT and human-computer interaction, The MIT Press. Cambrige, Massachusset London,17-44.

Preece, J. Rogers, Y. Sharp, H., Benyon, D., Holland, S., Carey, T., 1994, Human-computer interaction. Addison-Wesley.

Rubinshtein, S.L., 1959, Principles and directions of developing psychology. Moscow: Academic Science.

Tolman, E.C., 1932, Purposive behavior in animals and men. New York: Century.Vancouver J. B., 2005, Self-regulation in organizational settings. A tale of two paradigms, In M. Boekaerts, P. R. Pintrich, M. Zeidner (Eds.) Handbook of self-regulation, pp. 303-342.

Chapter 68

Analysis of Abandoned Actions in the Email Distributed Tasks Performance

Inna Bedny, Gregory Bedny

Institute for Advanced Systems Engineering
University of Central Florida
Orlando, FL 32816-2993, USA

ABSTRACT

In this work we demonstrate that analysis and reduction of abandoned actions is important in increasing efficiency of computer based tasks performance. Cognitive and motor explorative actions that give undesirable results will be referred to in this work as abandoned actions. Gnostic exploratory activity is virtually ignored in the task analysis. Meanwhile, it is an important component of the work activity because of the special role it plays in orienting activity. Gnostic exploratory activities that are carried out in the mental plain are called Gnostic dynamics. Orienting activity mainly precedes executive activity. The orienting activity concept is close to the concept of Situation Awareness in cognitive psychology but it's much wider than the concept of SA. It includes a conscious goal, and not only its verbally-logical but also imaginative components, etc. Emotionally-motivational mechanisms that are virtually ignored in cognitive task analysis play significant role in orienting activity which is a vital part of the human-computer interface.

Keywords: Explorative Activity, Self-Regulation, Goal, Significance, Motivation, Strategies, Abandoned Actions, Efficiency of Performance

INTRODUCTION

From the functional point of view activity is a complex self-regulative process. Let's consider two activity components: gnostic or explorative component of activity, the goal of which is to gather and comprehend information about the external world; and executive component of activity which is directed to the transformation of external environment or object of activity according to its goal. Explorative activity can be divided into external explorative activity, which is interconnected with internal, mental activity. For example, assembling and disassembling objects, transforming information on computer screen, etc. This kind of activity can also be internal or mental where a person operates with words, statements, symbols, and images. In order to make mental exploration more effective a subject might combine these two types of explorative activity.

The consequences of explorative actions are the source of feed-back information used to evaluate the situation at hand. Gnostic activity can be performed at the conscious or sub-conscious levels of self-regulation. From functional perspective a rat in a Skinner box performs explorative activity at the sub-conscious level. This according to systemic-structural activity theory is an example of graduate transformation of explorative stage of behavior into its executive stage. In the beginning the purpose of this behavior might be not to find food but simply to escape from the box. At the second step a rat might start looking for the food. At the final stage this behavior is transformed into executive one (to obtain food by pressing bar). These possible stages are not discussed in behavioral psychology.

Conscious level of explorative activity is a dominant one for human beings. At the sub-conscious level of exploration subjects use mental and motor operations and at the conscious level they use cognitive and behavioral actions. Gnostic activity is an important component of human performance. The more complex and ambiguous the tasks are, the more unexpected are events, the less precise are procedurals components of activity the more important are gnostic activity and explorative actions. In complex and dangerous situations explorative activity often becomes chaotic and is undesirable.

Gnostic or explorative components of activity are especially important in performance of computer based tasks. Computer based tasks are complex and have tremendous diversity. Their procedural components are not well defined. Thus gnostic activity and explorative cognitive and behavioral actions are important component of computer users' activity. The major purpose of explorative actions is not transformation of situation on the screen but rather an extraction of information from the screen that determines possible executive actions. When computer-based task is not clearly conceptualized the user explores the task by attempting to formulate the goal and extract what appear to be the significant aspects of the task. Then the user evaluates the difficulty and significance of the task. This in turn influences user's motivation. For example, if the task is evaluated as an easy and insignificant the motivation to perform this task goes down. In contrast if the task is evaluated as sufficiently difficult and significant the motivation is increased (Bedny, Karwowski, 2007; 2008). User forms a hypothesis about the task's nature based on explorative actions. Very often excessive number of explorative actions is an undesirable factor. Moreover sometimes explorative actions can be a cause of irreversible errors. As can be seen user activity during task performance is a self-regulative system. Analysis of explorative activity during performance of computer based task utilizing methods developed in Systemic-Structural Activity Theory (SSAT) is a purpose of this study.

Algorithmic description of the web-survey task performance allowed us to discover more efficient task organization based on logical consequence of user actions and reduce or eliminate abandoned actions. We also developed quantitative measures for evaluation of performance efficiency based on assessment of abandoned actions. Measures of efficiency derive from evaluation of time of task performance and duration of various types of abandoned actions.

WEB-SURVEY TASK

In most cases computer based tasks do not possess rigorous standardized method of performance as it's observed in the mass production processes. In HCI tasks a user often does not know in advance the sequence of actions he/she has to take and even experienced users have to discover the details of the task performance through exploratory activity. Moreover often users are not aware of all existing constrains in task performance. Hence continuous self-learning through explorative activity is an important component of users' professional activity.

In our study we have chosen a real email distributed task. Several hundred employees have received the same email. This email distributed task is not a high priority task. However users have an obligation to complete this task. It was discovered that low level of motivation in combination with inefficient screens' design for the considered task result in a lot of redundant explorative actions. Users move from one screen to another, return back, and so on. Users' performance included a lot of erroneous actions. Cognitive and motor explorative actions that give undesirable result will be referred to as abandoned actions. During development of HCI tasks designers should strive to reduce or eliminate such actions. It is necessary to take into account that some explorative actions can lead to significant undesirable consequences, and regarded as failures. This type of explorative actions should be analyzed separately.

According to self-regulation concept of motivation (Bedny, Karwowski, 2006) emotionally-evaluative mechanism is important for activity regulation. It influences selection of required information and development of adequate strategies of task performance. In model of activity self-regulation function block "Assessment of the sense of input information" is a part of emotionally-evaluative stage of activity performance. Distributed emails that have low positive or, in some cases, even negative significance are perceived as waist of time and money. The email composer should take into account effectiveness of such emails because it saves companies a lot of money if they design properly.

In this work we demonstrate that not only cognitive but also emotionally-motivational aspects of user activity are important for task analysis. Cognitive aspects of activity should be studied in unity with emotionally-motivational components. The more complex and lengthy supplemental task is the greater the negative emotional effect it has on the user because she/he loses the connection with the main task of her/his work and it becomes more difficult to return to it. "Where was I?" is the first question for user after completing supplemental task. Email distributed tasks are the once that are sent to multiple employees. Sometimes it's done with the regular frequency, once a quarter, once a year, etc. Every employee has to complete the distributed task that contains a questioner. These emails require careful reading and answering a number of questions.

The questions can be wordy and numerous. Until the employee completes the questioner the same email keeps coming back. If such communication is composed without taking into consideration some psychological factors it can cause confusion, loss of time and negative emotional state that effects productivity. We've chosen the email distributed task that has been associated with poor emotional and motivational state, because these conditions influenced cognitive strategies.

QUALITATIVE ANALYSIS OF THE WEB-SERVEY TASK

As an example we have chosen a real web-survey task. More than 5 thousand employees have received the email with the following content (see Figure 1 and Figure 2). Each employee should read this email and fill up the questioner. As can be seen from figures 1 and 2 this email does not fit on one screen. As a result attached file can not be observed without scrolling down. There is an interesting psychological factor associated with the employees' motivation.

This email distributed task is not a high priority task. It is considered by most employees as annoying, with low personal significance because it takes time from the main duties and as a result is accompanied by low motivation. However it's required to complete this survey. The busier the employee is the less he/she is motivated to take on this task. Often performance of such tasks is accompanied by irritation and negative emotional state. From functional analysis perspective this task has negative significance for the employee (Bedny, Karwowski, 2006; Bedny, Karwowski, 2007).

Functional analysis considers activity as self-regulative system. Self-regulation model of activity outlines the following motivational stages: (1) preconscious motivational stage; (2) goal-related motivational stage; (3) task evaluative motivational stage; (4) executive or process-related motivational stage; (5) result-related motivational stage. For the task under consideration the goal-related motivational stage is in conflict with the executive stage because the task has to be completed but it is boring and out of the scope of the main duties.

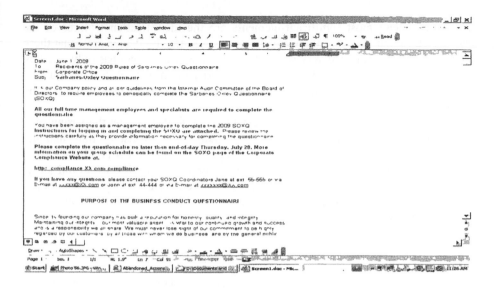

Figure 1. The first page of the email

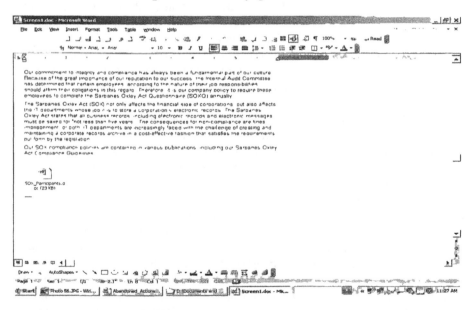

Figure2. The second page of the email

Observation demonstrates that employees try to complete this task as quickly as possible. They select a strategy to move through the content of the email without carefully reading it and to get the questioner just as quickly as possible. For the employees the most significant identification elements of the task are those that provide the direct link to the questioner so they are looking for the links to the questioner. These most significant elements are the identification features or the task attributes. Our analysis of the email and the user activity demonstrate that these attributes are not organized very well. The first page of the email has only one link which immediately attracts attention of the majority of employees. So they quickly click on the first link they see. When the webpage opens up they don't see the expected login screen.

After looking at this screen employees realize that they are on the wrong pass and start asking each other for help. It's usually just a waist of time and the employees decide to go back to the email. Employees don't know in advance what information will be required at the next step. Moreover even if they would know they still have to keep this information in working memory until they open the login screen and key it in. Keeping information in working memory is an undesirable factor for any task. The unnecessary mnemonic actions could be avoided if this information would be presented on the login screen itself.

There are several factors that are leading to the inefficient strategies of task performance. One of them is low subjective significance of the task that results in the negative emotionally-motivational association. The second factor is the inefficient task design where the identification features of the task are not adequate with the strategies of activity. The employees' strategy doesn't include careful reading of the screen content. Developers of the email had two broad goals (objective requirements) in mind: familiarization with the presented information and completing the questionnaire. However, objectively defined goals (requirements) may not coincide with subjectively formulated or excepted goals (Bedny, Karwowski, 2007). Employees rejected the first goal and formulated their own general goal to complete the questionnaire as quickly as possible and return to their primary duty. In accordance with this goal they had developed an adequate strategy of task performance. We can not describe in details the most likely strategy of searching for the necessary information. One only needs to note that users are trying to find it on the screen by trial and error. This contributes to a large number of unnecessary explorative actions.

When performing computer based tasks users often need to read or print a text. Two strategies of reading were discovered: one involves careful reading of the entire text that is similar to reading a book (detail reading); the second strategy is browsing the text when certain parts of text are scanned (user looks at a piece of text, captures a main idea of the fragment and moves to the next piece of text). Browsing allows a user to get quickly acquainted with the main idea of a particular fragment of the text. After browsing the whole text user can return to some parts of the text for more details or move to another screen. Browsing text usually requires simultaneous perceptual actions. Before describing algorithmically process of reading it is necessary to understand relationship between these two strategies.

Sometimes it is necessary to segment reading of text into separate verbal actions. Each verbal action represents an elemental phrase, each of which represents a separate meaningful unit of information. Separate verbal-motor actions determine meaningful typing units such as typing a word or several interdependent words which convey one meaning. This is in line with Vygotsky's (1978) idea about meaning to be one of the basic units of analysis. Similarly cognitive psychology shows a possibility of segmenting verbal activity in verbal protocol analysis (Bainbridge, Sanderson,_1991). According to SSAT (Bedny, Karwowski, 2007) reading and typing normally requires the third level of concentration of attention and has the third level of activity complexity. Sometimes processing the most subjectively significant units of text can be transferred into higher level of complexity. If the text is relatively homogeneous there is no need to extract separate actions. Time spent processing text can be measured according to described above strategy.

GENERAL PRINCIPLES OF ALGORITHMIC DESCRIPTION OF ACTIVITY

In our study we use the concept of human algorithm, which should be distinguished from mathematical or computer algorithm. According to SSAT the algorithmic analysis follows the qualitative stage. It includes the subdivision of activity into qualitatively distinct psychological units and determining their logical organization and sequence. A number of methods can be employed for the qualitative analysis (Bedny, Karwowski, 2007). We utilized objectively logical analysis and some elements of functional analysis where main units are function blocks or functional mechanisms such as goal, sense (significance), motivation, etc. For the algorithmic analysis of activity main units of analysis are cognitive and behavioral actions and members of the algorithm. These units of analysis distinguish a human algorithm from other types of algorithms. Algorithmic description allows to describe the more preferable strategies of activity during task performance (Bedny, Karwowski, 2003).

Every activity varies which is especially true for the performance of computer based tasks performance. In order to analyze activity there is no need to describe all possible strategies of task performance but rather consider the most typical ones and ignore minor variations in the tasks performance. The most representative and important strategy of task performance should be selected. The combined probability of considered strategies should be equal to one. This means that the selected strategies absorb all others strategies. Such approach is justified because the designed activity only approaches the real task performance and describes it with a certain level of approximation as does any model in the design process. Algorithmic description of task in SSAT is a stage of the morphological activity analysis.

In the algorithmic analysis a member of an algorithm is an element of activity which integrates one or several actions united by a higher order goal. It consists of interdependent homogeneous actions (only motor, only perceptual, or only thinking or decision-making actions, etc.) which are integrated by a particular goal into a holistic system. Subjectively, a member of an algorithm is perceived as a component of activity, which has logical completeness. A sense of logical completeness of actions for one member of the algorithm is usually associated with a sub-goal of task that integrates several actions. Sometimes sensing the logical completeness of homogeneous actions can be determined by the capacity of short-term memory. When actions are performed simultaneously or require keeping their order in working memory, due to limits on the capacity of working memory, each member of an algorithm is limited to between one to four homogeneous actions. Classification of cognitive and behavioral actions and methods of their extraction can be found in (Bedny, Karwowski, 2003; Bedny, Karwowski, Sengupta, 2008; and I. Bedny, Karwowski, Chebykin, 2006). Members of the algorithm are classified according to psychological principles and are designated by special symbols.

ANALYSIS OF ABANDONED ACTIONS

In this section we consider abandoned actions for the first main strategy. The most common abandoned actions for a particular task should be presented in algorithmic analysis. Depending on the purpose of the study performance time of abandoned actions can be considered or ignored. The more switching from one screen to the other are performed the more time delays are associated with such switching.

We have developed quantitative measures for assessment of abandoned actions. Measures derive from evaluation of time of task performance and duration of various types of abandoned actions.

The following symbols will be utilized: A - general time for all abandoned actions; A^{α} - time required for afferent abandoned actions; A^{ε} - time required for efferent abandoned actions; A^{l} – time required for abandoned logical conditions; A^{μ} - time required for abandoned actions associated with keeping information in working memory.

The first step in assessment of task performance efficiency is evaluation of task performance time (formula 1).

The first basic strategy has probability $p = 0.9$ and the second one $p = 0.1$. So, the task performance should be determined using the following formula:

$$T = \sum P_i \, t_i ; \qquad\qquad (1)$$

where P_i – probability of the i-th member of algorithm; t_i – performance time of i-th member of algorithm.

Then the task performance time equals:

$$T = P_1 \times Tst_1 + P_2 \times Tst_2; \qquad\qquad (2)$$

where $Tst1$ and $Tst2$ – performance time of the first and the second basic strategies; P_1 and P_2 are probabilities of performing these strategies.

So, in our case the task performance time equals:

$$T = 0.9 \times Tst_1 \ldots + 0.1 \times Tst_2 \quad (3)$$

The time taken for all described abandoned actions can be determined as follows:

$$A = A^\alpha + A^\varepsilon + A^l + A^\mu + A^{th}; \quad A^\alpha = \sum P^\alpha{}_l \times t^\alpha{}_i; \quad A^\varepsilon = \sum P^\varepsilon{}_b \times t^\varepsilon{}_b; \quad A^l = \sum P^l{}_r \times t^l{}_r;$$

$$A^\mu = \sum P^\mu{}_j \times t^\mu{}_j; \quad \hat{A}^{th} = \sum P^{th}{}_k \times t^{th}{}_k;$$

where $P^\alpha{}_i; \; P^\varepsilon; \; P^l; \; P^\mu{}_j - P^{th}$- probability of i-th abandoned action of corresponding type;

$t^\alpha{}_i; \; t^\varepsilon; \; t^l; \; t^\mu; t^{th}$ - time performance of i-th abandoned action of corresponding type.

This time can be obtained based on existing studies or experimentally based on chronometrical analysis.

The next step in the evaluation of the task performance efficiency utilizes the following measures of efficiency:

$$\hat{A} = A/T \; (4); \quad \hat{A}^\alpha = A^\alpha/T \; (5); \quad \hat{A}^\varepsilon = A^\varepsilon/T \; (6); \quad \hat{A}^l = A^l/T \; (7);$$
$$\hat{A}^\mu = A^\mu/T \; (8); \quad \hat{A}^{th} = A^{th}/T \; (9);$$

These are measures of various types of abandoned actions. The less is the value of these measures the more is the performance efficiency. In any given study all these measures or just the most suitable ones should be utilized. When mnemonic, thinking or decision-making activities are performed simultaneously with motor components of activity each type of activity is accounted for separately in the calculation.

The fraction of abandoned actions in the task performance time is:

$$\hat{A} = A/T = 19.79/49.84 = 0.4$$

So, abandoned actions take about 40% of this task performance time.

It's also possible to determine the fraction of perceptual actions, thinking, decision-making and mnemonic abandoned actions in the task performance time by calculating times for various types of abandoned actions: $A^\alpha; A^\varepsilon; A^l; A^\mu; A^{th}$.
We calculate described above measures only for one basic strategy that has probability 0.9. However if it is required we can calculate these measures taking into consideration two basic strategies, one with probability 0.9 and the other with probability 0.1. If two main strategies are considered the formula for the task performance time would be:

$$T = 0.9 \times Tst_1 + 0.1 \times Tst_2$$

We've considered in abbreviated manner the principle of quantitative assessment of explorative actions utilizing developed quantitative measures. A better method of providing information on the screen has been offered based on obtained measures.

CONCLUSIONS

The more uncertain the task is the more complicated is the explorative activity. In conditions when the uncertainty about possible strategies is significantly increased explorative activity can approach the chaotic mode. This situation has been encountered in the study of the web-survey questioner task. The concept of abandoned actions and a method of their analysis are important for studying explorative activity during computer-based tasks performance.

The total elimination of explorative actions is usually not achievable but they can be reduced. It is necessary to reduce actions that include the processes of thinking, decision-making and memory workload. It is desirable for the actions to be performed based on the perceptual information which is recommendation for production tasks and especially tasks performed under stress. In contrast for entertainment tasks we often need to introduce explorative actions including decision-making, thinking, etc. so that the entertainment tasks are more interesting for their users.

The presented above analysis leads to the conclusion that there is a need for quantitative analysis of explorative activity. Study demonstrates that emotionally-motivational components of activity are tightly connected with cognitive components of activity. Humans are not simple logical devises and always have predisposition to events, situation or information. In activity approach design process always includes analysis of emotionally-motivational component of activity in both production and entertainment environments.

REFERENCES

Bainbridge, L., Sanderson, P.(1991). Verbal protocol. In J. R. Wilson, & E, N. Corlett (Eds). *Evaluation of human performance. A practical ergonomics methodology*. Second Edition, Taylor and Francis, pp. 169 – 201.

Bedny, G. Z., Karwowski, W. (2003). A systemic-structural activity approach to the design of human-computer interaction tasks. *International journal of human-computer interaction.* 2, 235-260.

Bedny, G. Karwowski, W. Sengupta, T. (2008). Application of systemic-structural theory of activity in the development of predictive models of user performance. International Journal of Human-Computer Interaction, V. 24, # 3, 239-274.

Bedny, G. Z., Karwowski, W. (2007). A systemic-structural theory of activity. Application to human performance and work design. Taylor and Francis, London, New York

Bedny, G. Z., Karwowski, W. (2006). The self-regulation concept of motivation at work. Theoretical issues in ergonomic science, 7 (4), 413-436.

Bedny, I. S., Karwowski, W., Chebykin, A. Ya. (2006). Systemic-structural analysis of HCI tasks and reliability assessment, Triennial IEA 2006 16th World Congress on Ergonomics, Maastricht, Netherland, e-book.

Chapter 69

Specifics of Becoming a Professional Bank Manager

Oleg Kokun, Aza Karpoukhina

Laboratory of Psychophysiology
G.S. Kostiuk Institute of Psychology APS of Ukraine
2 Pankivska Street, Kyiv, 01103, Ukraine

Ukrainian Academy of Pedagogical Sciences,
205 East Ruth Avenue, Suite 330, Phoenix, AZ 85020, USA

ABSTRACT

The need of psychological support to managers in their work is explained by ever increasing competition they have to face and elevated levels and of tension and stress associated with it. This article relies on Systemic-Structural Theory of Activity to analyze the specifics of managers' professional activities. The authors described the results of research into bank managers' professional growth that could be used as a tool of their psychological support. The research results reflect the influence produced by various external and internal conditions in the bank manager daily activities affecting their professional success as well as crucial factors of their professional development interrelated with this success. The article also comprises the formula to estimate the integral diagnostic index that can be used as the scale of bank managers' overall professional efficiency.

Keywords: Professional Development, Elements of Activity, Managers, Psychological Support

INTRODUCTION

The modern trends of the social development call for an increased level of psychological support in different areas of professional activities. It includes primarily the systematic psychological support throughout all stages of professional development, with continuous upgrade of professional qualification and efficiency accompanied by constant personal perfection. This psychological support must rely on scientifically authentic research into psychological specifics of professional activities and development that individuals undergo in a concrete professional environment. Such support will be most beneficial for occupations that are characterized by high professional pressure, permanent stresses and increased risk of professional burnout. These features apply among others to the work of bank manager whose professional development became the focus of research described in this article.

SPECIFICS OF MANAGERS' PROFESSION

In the most general sense manager can be described as a person who holds a managerial position and masters a set of knowledge and skills that are necessary to manage people and administer organizations in conditions of uncertain and permanently changing environment, in which this organization is functioning (Atwood, 2008; Hosmer, 2008).

One of the modern theories that can be used to perform in-depth analysis of professional activities performed by humans in general and managers in particular is the Systemic-Structural Theory of Activity – SSTA (Bedny, Meister, 1997; Bedny et al, 2000; Bedny, Karwowski, 2007; Bedny, Harris, 2008).

According to this theory the major elements of activity include: 1) subject of activity; 2) task; 3) conditions; 4) tools; 5) method or procedures; 6) object; 7) product or result. It should be kept in mind that activities are organized according to principles of self-regulation and feedback.

In their work managers (*subject of activity*) can pursue the following common *tasks*: tie up with other individuals inside organizations to achieve organizational objectives and economic success; study evolution of demand and supply in the market; plan organization's activities; communicate (establish and develop interrelation among personnel, other institutions and organizations); search for and use necessary means and resources to better achieve organizations' core objectives; develop business plan and control its implementation; manage, supervise and evaluate subordinates' work; control the performance of tasks assigned to subordinates; participate in recruiting and selection process, assist to other employees in their professional growth; set up negotiations, perform client outreach efforts; hold presentations etc.

Conditions that stipulate success of achieving the above tasks can be broken down in external and internal ones. The external conditions include such as: manager's place in the organization's hierarchy, range of his/her authority, other employees' qualification and experience, social and psychological climate inside organization, objective market situation within organization's scope of activity, its financial status, image etc. Internal conditions include managers' professional experience and qualification underpinned by their core professional qualities such as: highly developed managerial and communicational skills; ability to affect the behavior of other persons, build and develop efficient task forces, solve problematic issues in reduced timeframe; the ability to forecast and predict the way things are going to develop; well-developed analytical skills; high self-control, self-assertion, confidence in own decisions; vitality, consistency, responsibility, resolution, spirit of enterprise; ability to follow organizational requirements and regulations; drive for the permanent personal improvement (Atwood, 2008; Mescon et al, 1988).

We also consider that among crucial internal conditions that manager needs to have in order to become and remain successful most important are his work efficiency and psychophysiological state (Karpoukhina, 1990; Karpoukhina, One-Jang, 2003; Karpoukhina et al, 2008; Kokun, 2004, 2006).

It should taken in account that managerial work is characterized by its high intensity, tension, frequent interference of external factors, multiple social relations of different levels, prevalence of direct personal contacts with other people. Besides, it is associated with a multitude of economic, organizational and social-psychological stresses. Subject to these stresses managers become less productive, experience sensation of permanent fatigue, headaches, insomnia, general health deterioration. In many cases they cease to feel satisfaction with their work, experience hostility; sometimes feel themselves incompetent, helpless and exposed to other symptoms of professional burnout.

In the course of their professional activities managers most frequently apply material and internal/ external functional tools. Material tools are primarily those related to various means of accepting, receiving, processing and transferring information. Internal functional tools may include perceptions of activity results, situation analysis, search for new approaches, decision taking process, self-adjustment, self-motivation etc. External tools include behavioral and linguistic expression, clothing and other attributes of exterior image.

While working with their *objects* such as subordinates, clients, customers, partners, co-workers, representatives of controlling organizations, management etc. managers can recur to such *method or procedures* as: negotiations, meetings, consultations, instructions, control, fines, information etc.

The examples of *product or result* that managers try to achieve by their work can include: received financial revenues; number of closed deals; increased reputation and better image of their organization; higher stimuli, discipline, collaboration, professional qualification exhibited by their co-workers; creation of new promising projects etc.

What makes managers' work unique is that in many cases it is difficult to evaluate how well managers accomplish any given task without taking into account the context of their overall work efficiency. Most frequently the manager's productivity is assessed by the end result of his/her professional efforts (*product or result*) that can be achieved in different manners (depending on environment, resources, available possibilities, experience, manager's individual traits etc.).

Professional development means building vocational orientation, competence, socially significant professional qualities and their integration, readiness to constant professional growth, search for better quality and creative performance of tasks in line with personal and psychological singularities (Zeer, 2007). The process of manager's professional growth is shaped by all above elements of his/her activity. Below, we will study in detail specifics of such conditioning based on bank managers' cases.

METHODS AND TARGET GROUP

Methods: questioning; expert evaluation; methods of revealing "communicative and organization abilities" (B. Fedorishin); methodology of "Studying Satisfaction Levels Associated with Profession and Work" (N.Zhurin and E.Ilyin); Maslach Burnout Inventory (MBI); Modification of The General Self-Efficacy Scale (M. Jerusalem and R. Schwarze).

The *target group* for this research was 51 managers of a Kiev bank.

SPECIFICS AND FACTORS OF BANK MANAGERS PROFESSIONAL DEVELOPMENT

Let's first review the summarized questioning and psychodiagnostics results that shed the light on external and internal *conditions* of bank managers' operation. Then we will attempt to determine how they develop such crucial components of their professional development as occupational focus and competence (internal *conditions*) as related to their professional efficiency.

QUESTIONING RESULTS

The results of questioning included in Table 1 show that the managers in the study group revealed deepest satisfaction with relation to sheer professional aspect of their activities. 94 % of all interviewees said they were happy with their occupation. As to the aspects that are of more pronounced social and professional nature such as carrier are social status the satisfaction level dropped to 68 % and 56 %, respectively. The fact that only the third of all managers said that they were happy with their salary rate can be considered as absolutely natural as most of all hired employees tend to be dissatisfied with the salary they receive.

Also the above results corroborate the conclusion that 90 % of all studied managers have high to very high interest in their professional activity, 94 % of them said they like what they do and 97 % take regular efforts to upgrade their professional level. The total picture of the studied group shows that the managers are characterized by a rather high level of their professional focus, especially motivation, presence of appropriate professional values and professional perspective.

Table 1. Level of professional and social/professional satisfaction among managers

Level of Satisfaction	Not happy	Rather unhappy	Rather happy	Totally happy
With his/her occupation	2 %	4 %	49 %	45 %
With salary rate	34 %	31 %	31 %	4 %
With his/her carrier	14 %	18 %	58 %	10 %
With his/her social status	16 %	24 %	36 %	20 %

Let's now consider summarized results of indices that represent factors capable of considerably affecting the managers' professional development and daily operations.

In our case they turned out to be predominantly favorable. The majority of all interviewees said they were on good terms with their colleagues and superiors (see Table 2). Most of them did not feel overstressed during their work and confessed to be in a rather good physical shape.

Table 2. Managers' self-evaluation of their relationships with colleagues and superiors

Relationships	Very bad	Bad	Mediocre	Good	Very good
With colleagues	–	–	6 %	63 %	31 %
With superiors	–	–	16 %	60 %	24 %

RESULTS OF PSYCHODIAGNOSTICS METHODS

Results of study performed by method of revealing "communicative and organization abilities" (Table 3) show that 50 to 60 % of all interviewees have appropriate (high to very high) level of personal qualities that are required for successful professional activity such as communicative and managerial abilities. These specialists tend to quickly adapt to complex situations and new environment, exhibit initiative in communication, act of their own accord in difficult circumstances, are persistent in their doings, aspire to broaden their social circle etc.

Table 3. Level of managers' communicative and managerial abilities

Level	Grade	Abilities	
		Communicative	Managerial
Low	0 – 4	4 %	–
Under low	5 – 8	11 %	9 %
Medium	9 – 12	24 %	41 %
High	13 – 16	31 %	37 %
Very high	17 – 20	30 %	13 %

Another third or so of all managers (24 – 41 %) have intermediate level of such abilities, which can somewhat interfere with their professional growth and success. Although they want to contact people and, in principle, can withstand their standpoint their overall potential is not stable and needs special efforts to be strengthened.

Another relatively insignificant portion of interviewees (9 – 15 %) showed too low for managers level of communicative and managerial abilities, which jeopardizes their professional aptitude. These are people who prefer to stand apart and feel most comfortable in their own company. They tend to feel uneasy among new people or employees, have trouble getting in touch with the others, do not defend their position, feel hurt if offended, rarely do something on their own and avoid personal decisions.

One of professional competence indicators that we used in our research was the professional self-efficacy index according to the General Self-Efficacy Scale (M. Jerusalem and R. Schwarze) (see Table 4) that is used to assess the person's confidence in his/her potential ability of setting up and managing his/her activities undertaken to achieve a certain objective.

Table 4. Level of managers' self-efficacy

Professional self-efficacy	Grade	Number
Low	≥ 19	–
Under low	20 - 24	–
Average	25 - 29	17 %
Above average	30 - 35	64 %
High	36 - 40	19 %

The above results demonstrate that the interviewees normally exhibited the appropriate level of professional self-efficacy: 19 % of them had high self-efficacy level, 64 % above average and only 17 % showed the average results. Remarkably, none of them went to low or below low levels.

Data that we have obtained through the "Studying Satisfaction Levels Associated with Profession and Work" methodology (Table 5) are in complete harmony with the results of questioning we set up to measure for different aspects of managers' professional satisfaction. Only 7 % of the interviewees said they were slightly dissatisfied with their occupation and work and other 21 % exhibited the low satisfaction level. Among the rest 72 % this index ranged between average and high.

Table 5. Level of managers' satisfaction with their profession and work

Satisfaction / Dissatisfaction Level		Grade	Number
Dissatisfaction level	high	$\geq$ -11	–
	average	-6 – -10	–
	low	-1 – -5	7 %
Satisfaction level	high	+1 – +5	21 %
	average	+6 – +10	36 %
	low	$\leq$ +11	36 %

Most managers in the study group also exhibited insignificant extent of professional burnout (equivalent of "emotional depletion") and deformation (equivalent of: "depersonalization", "reduction of personal achievements"). For different components, this level was measured as average or above average only among 15 to 24 % of all managers (see Table 6).

Table 6. Level of professional "burnout" and deformation among managers

Components of professional burnout and deformation	Extent				
	low	under low	average	above average	high
Emotional depletion	36 %	49 %	9 %	6 %	–
Depersonalization	55 %	26 %	15 %	4 %	–
Reduction of personal achievements	34 %	43 %	22 %	2 %	–
Summarized value	28 %	59 %	13 %	–	–

One final comment should be made though with regard to results obtained through questioning and psychodiagnosis. In general they can be viewed as positive (which is corroborated by their rather high level of interviewees' professional qualities and level). However all methods include some single deviations from the bulk of results. Admittedly such deviations should be considered as normal and be disregarded. At the same time it should be noted that the results we have achieved could have been affected (toward their somewhat artificial improvement) by some interviewees' desire to seem in their responses better than they in fact are (i.e., lack of candor). But this phenomenon is absolutely natural and in practical terms is unavoidable when performing similar research inside organizations.

SPECIFICS OF DEVELOPING THE PROFESSIONAL FOCUS AND COMPETENCE AMONG BANKING MANAGERS AT DIFFERENT STAGES OF THEIR PROFESSIONAL DEVELOPMENT

The spread of managers in the study group across different stages of professional development was rather uneven (detailed spread was determined by questionnaire's indicator questions and expert evaluation results). Only 3 of them were at the stage of professional adaptation. Other 13 managers were at the subsequent stage of primary professionalization, 26 at the second professionalization stage and 9 managers at the stage of professional excellence. It seems that such stratification is completely natural since 83 % of all interviewees had professional record of above 5 years and 73 % of more than 10 years (maximum length of service was 37 years).

The specifics of bank managers' professional target and competence at different stage of their professional development can be most explicitly illustrated by results of (Spearman) correlation analysis between interviewees' proven professional record (which to considerable extent reflects their life and professional experience) and results of questioning, psychodiagnosis and expert evaluation (see Table 7).

Table 7. Correlation between managers' professional record and results of questioning, psychodiagnosis and expert evaluation

№	Parameters	Age	Record
1	Satisfaction with salary rate	,33*	,32*
2	Satisfaction with carrier	,24	,28*
3	Relations with colleagues	-,47**	-,52**
4	Health	-,23	-,29*
5	Professional self-efficacy		,27*
6	Depersonalization	,23	,27*
7	Reduction of personal achievements	,29*	,28*
8	Professional independence	,38**	,31*
9	Success in improving professional skills	-,24	-,27*
10	Social and legal competence		-,25
11	Extreme professional competence		-,28*

Note: 1) ** means that correlation is valid at $p \leq 0,01$; * means that correlation is valid at $p \leq 0,05$; if no asterisk correlation is valid at $p \leq 0,1$; empty box means correlation is valid at $p > 0,1$;

2) values 1 - 4 were obtained by questionnaire; 5th by Modification of The General Self-Efficacy Scale; 6 - 7 by Maslach Burnout Inventory; 8 - 11 by expert evaluation.

The above table shows that the age and professional record positively correlate with managers' overall satisfaction with their salary and carrier. In our perspective this is totally logical since professionals tend to get paid better and climb higher the carrier ladder as their professional experience gets richer. Also natural are positive correlations between managers' age and professional record, on one hand, and expert evaluation of their professional independence and professional self-efficacy index (only record), on the other hand, since these qualities are also acquired along with work experience.

In a similar vein, results of correlation analysis truthfully reflect certain deterioration of health with age and a slight tendency of acquiring some indications of professional deformation such as depersonalization and reduction of personal achievements.

The reasons behind other correlations shown in the Table above (all of which are negative), are not in our opinion as evident and self-explanatory and therefore require a more detailed interpretation. For example, negative correlation between age/ professional record and expert evaluation of success in improving professional skills can be explained by an observation according to which the longer an employee works at the same workplace the lesser is his/her desire to improve his/her professional qualities, and the lesser objective possibilities he/she has for this (the higher is professional level the harder are the efforts to improve it). Negative correlation between professional record and expert evaluation of social/ legal and extreme professional competence can be explained by younger employees' ability to considerably easier find their bearings in fickle social and legal environment and keep abreast of headlong events. The most pronounced negative correlation between age/ professional record and managers' self-evaluation of their relations with colleagues can be explained by fastidiousness with regard to co-workers that gets stronger with time the manager spends at his/her work place, by certain professional deformations that accompany managers' carrier growth (if a manager climbs to a position where he/she has to control his/her subordinates and require from them the necessary level of workload it certainly is not a way of making new friends among them).

FACTORS OF BANK MANAGERS' PROFESSIONAL FOCUS, COMPETENCE AND EFFICIENCY

Let's first analyze the valid correlation between bank managers' professional competence that we used in our research and indices that can be considered as favorable or unfavorable to its evolution (see Table 8).

Table 8. Correlation between bank managers' professional competence indices

No	Factors	Professional competence indices					
		Self-efficacy	Expert evaluation of professional competence				
			Social and legal	Special	Personal	Self-compet.	Extreme
1	Relations with colleagues		,57**		,43**	,39**	,56**
2	Relations with superiors		,50**		,48**	,34*	,45**
3	Work fatigue	-,30*					
4	Health condition	,27*					
5	Communication abilities	,25	,25	,29*	,24		
6	Organizational abilities	,44**		,25			
7	Emotional depletion		-,24		-,33**		
8	Depersonalization	-,39**	-,37**	-,25		-,26	-,40**
9	Reduction of personal achievements	-,58**	-,37**				

Note: 1) ** means that correlation is valid at $p \leq 0,01$; * means that correlation is valid at $p \leq 0,05$; without asterisk means that correlation is valid at $p \leq 0,1$; empty box means that correlation is valid at $p > 0,1$;

2) indices of "factors" 1 - 4 were acquired by questionnaire; 5 - 6 by methods of revealing "Communicative and organization abilities"; 7 - 9 by Maslach Burnout Inventory.

From the Table above we can clearly see that four of five managers' professional competence indices (according to expert evaluation) are quite tightly correlated in a positive sense with self-evaluation of their relations with colleagues and superiors. Therefore such relations, on one hand, can be undoubtedly viewed as favorable *factors* to develop the bank managers' professional competence, and on the other hand, considered as a *consequence* and integral part of well-developed professional competence.

It allows to draw the conclusion that there exists an interrelation between the two variables. Another explanation of the link between them is the influence produced by a "third factor", in this case its role is played by such index as communication skills that are undeniably connected with three of five professional competence varieties. It is a well established fact that full-fledged communicative skills enable an individual to build positive relations with other people and also constitute an indispensable component of manager's professional competence and prerequisite of their development.

A considerable number of negative correlations between different types of managers' professional competence with indices of professional burnout and deformation such as emotional depletion, depersonalization and reduction of personal achievements also seems to be a natural phenomenon. These qualities indisputably represent factors that interfere with successful professional development and, in particular, professional competence.

Another positive correlation that seems to be quite intuitive is that between the professional self-efficacy (which is an integral index of managers' professional competence) and communication/ organizational skills that similarly to the case above can be interpreted as both a factor and component of the managers' professional self-efficacy. On the other hand, this index is negatively correlated with depersonalization and reduction of personal achievements that naturally hamper the efficiency of managers' work.

Also we observed that managers with high self-efficacy level subjectively felt themselves less tired and correspondingly graded their health better than the others. This, in our opinion, can be explained by the fact that such managers tend to experience less fatigue at work due to better organization and efficiency of their efforts and better health and physical condition that normally result in higher work efficiency.

Let's now consider valid relations between indices of interviewees' professional focus and indices that can be construed as conditioning factors (see Table 9).

Table 9. Correlations between bank managers' professional focus indices

No	Factors	Professional focus indices				
		I	II	III	IV	IV
1	Relations with colleagues	,38**	,34*		,32*	
2	Relations with superiors	,31*			,28*	
3	Work fatigue					
4	Overall health condition			,24		
5	Communication skills		,25		,24	,35*
6	Organizational skills	,27*				
7	Emotional depletion		-,25	-,24		
8	Depersonalization		-,42**	-,33*	-,43**	-,40**
9	Reduction of personal achievements	-,27*	-,35*		-,35	-,30*
10	Satisfaction with profession and work	,47**	,31*	,38**	,46**	,49*

Note: 1) ** means that correlation is valid at $p \leq 0,01$; * means that correlation is valid at $p \leq 0,05$; without asterisk means that correlation is valid at $p \leq 0,1$; empty box means that correlation is valid at $p > 0,1$;

2) indices of "factors" 1 - 4 were acquired by questionnaire; 5 - 6 by methods of revealing "Communicative and organization abilities"; 7 - 9 by Maslach Burnout Inventory; 10 by methodology of "Studying Satisfaction Levels Associated with Profession and Work";

3) professional focus index: I means interest in own occupation; II means satisfaction with work content; III means satisfaction with carrier growth; IV means feeling

affection to one's profession; V means efforts to raise his/her professional qualification (acquired by questionnaire).

The above table shows that all professional focus indices at sufficiently high level logically correlate with the index measured by using the "Studying Satisfaction Levels Associated with Profession and Work" methodology. In a similar fashion, we can consider natural the considerable number of positive correlations between focus index and factors that favor its development in managers, including relations with colleagues and superiors and level of communication skills.

As in a previous case (of professional competence) we have established a fair number of sufficiently valid correlations between various professional focus types and indices of professional burnout and deformation such as emotional depletion, depersonalization and reduction of personal achievements, which can also be viewed as factors obstructing the managers' professional focus.

The obtained research results allowed us to calculate the integral diagnostic index of bank managers' professional efficiency. To perform these calculations we applied the procedure of the stepwise linear regression analysis. In its course we defined five of the most informative indices that we have used during research (along with corresponding coefficients), which taken together provide a rather precise characterization of managers' efficiency. It turned out that the coefficient of multiple correlation (R) for these five indices that describe internal and external *conditions* of the managers' professional activities with the expert evaluation of these activities' efficiency is rather high ($R = 0,88$). It enabled us to come up with a sufficiently precise regression formula to forecast efficiency of bank managers by using the procedure that we have developed and tested before (Kokun, 2004).

$$DLEA = (2,24*A + 1,22*B + 2,63*C + 0,13*D - 0,17*I + 6,3)*2,73,$$

where: DLEA is diagnostic level of bank managers' efficiency;
A is an index of efforts taken by interviewee to raise his/her professional skills;
B is index of interviewee's relations with colleagues;
C is index of interviewee's relations with superiors;
D is index of interviewee's satisfaction with profession and work;
I is index of interviewee's personal achievements reduction;
6,3 is a constant value.

The indices "A" to "C" were obtained by way of questionnaire (index "A" can vary between 1 and 4; indices "B" and "C" can range between 1 and 5). Index "D" was obtained by applying the "Studying Satisfaction Levels Associated with Profession and Work" methodology (can vary between -17 and +17). Index "I" was obtained by using the Maslach Burnout Inventory (can vary between 1 and 48).

The integral diagnostic index DLEA that was calculated by using the above formula is based on 1 to 100 scale and can be interpreted as follows: between 1 and 20 points means that efficiency of bank manager is low; between 21 and 40 under low; between 41 and 60 average; between 61 and 80 above average and between 81 and 100 this level is high.

This formula can be used to either rank bank managers or to take decision about their promotion, or to move them to another position or define success of professional tests etc.

CONCLUSION

The theory, its methodological premises and practical results of research into specifics of bank managers' professional development can be successfully implemented to build psychological support of their activity. The need of such support becomes evident if not urgent as the business competition gets tougher; the managers' work becomes more stressful and abundant with various symptoms of professional "burnout".

Discovering the extent of impact produced by external and internal *conditions* in managers' activity to their efficiency; studying specifics of this activity; determination of lead factors behind success of the managers' professional growth allow to develop scientifically founded psychological support of their professional activities.

REFERENCES

Atwood, C. (2008), *Manager skills training*. ASTD Press, Alexandria.

Bedney, G., Mester, D. (1997), *The Russian Theory of Activity: Current Application to Design and Learning*. Lawrence Erlbaum Associates, Mahwah.

Bedney, G., Seglin, S., Mester, D. (2000), "Activity Theory: history, research and application." *Theoretical Issues in Ergonomics Science*, 1, 2, 168-206.

Bedny, G., Karwowski, W. (2007), *A Systemic-Structural Theory of Activity: Applications to Human Performance and Work Design*. CRC Press, Boca Raton.

Bedny, G., Harris, S. (2008), "Working sphere/engagement" and the concept of task in activity theory." *Interacting with Computers*, 20, 2, 251-255.

Hosmer, L. (2008), *The ethics of management*. McGraw-Hill Irwin, Boston.

Karpoukhina, A. (1990), *Psychological method of increasing efficiency of work activity*. Knowledge Publishers, Kiev (in Russian).

Karpoukhina, A., One-Jang, J. (2003), "Systems Approach to Psycho-physiological Evaluation and Regulation of the Human State during Performance." *Proceedings of the XVth Triennial Congress of the International Ergonomics Association and The 7th Joint Conference of Ergonomics Society of Korea. Japan Ergonomics Society*, 6, Korea, pp. 451-454.

Karpoukhina, A., Kokun, O., Zeltser, M. (2008), "Monitoring of Human Psychophysiological Condition as a Method of Increasing of Activity's Efficiency." *Conference Proceedings AHFE International Conference. USA Publishing, Las Vegas, USA*.

Kokun, O. (2004), *Optimization of person's mechanisms of adaptation: psycho-physiological aspect of upholding activity*. Millenium, Kiev (in Ukrainian).

706

Kokun, O. (2006), *Psychophysiology*. Centre of educational literature, Kiyv (in Ukrainian).

Mescon, M., Albert, M., Khedouri, F. (1988), *Management*. Harper & Row, New York.

Zeer, E. (2007), *Psychology of trades*. Academic Project, Moscow (in Russian).

Chapter 70

Training Process Design: A Case Study in the Sanitation Sector

Catarina Silva[1], Cláudia Costa[2]

[1]Faculdade de Motricidade Humana
Technical University of Lisbon
Lisbon, Portugal

[2]Oeiras Water Municipal Services
Oeiras, Portugal

ABSTRACT

The present study seeks to examine the importance of contextualized training in the acquisition of safety behaviours of a group of workers in the area of sanitation of a municipalized service. With this training process we intend to develop their working activity skills regarding self-analysis and self-learning so they can also be players in the safety of their own working environment. This approach has its roots in the "ergonomics activity" tradition and follows similar orientations of some work psychology projects: professional training is always designed in constant dialogue with ergonomics work analysis and promotes a participatory transformation process. Through the activity analysis and work accidents reports in the sanitation sector, two major themes were identified that were also used in the developed training: (1) work in confined spaces and (2) adopted working posture and manual cargo handling. The training process was designed based on a model alternating between theoretical expositive sessions and sessions of self-analysis, supported during work, and sessions of collective analysis and discussion (problem-situations, using work activity video analysis for a proper accountability). The training process ended with the formalization of proposals to improve working conditions. All the proposals for improvement were presented, discussed and negotiated collectively with the local manager, the administration and the board of directors. A subsequent analysis confirmed that all the proposals were made. Seven months later we came back to reanalyze work activity and we identified relevant differences in safety behaviors. This was considered as a result of the new activity's risks representation, carried out by the sanitation workers. We concluded that the participation of all the system's stakeholders (workers and people in charge) is very important when designing training procedures aiming at promoting safety.

Keywords: Training, Work analysis, Self-analysis of work, Safety, Sanitation

INTRODUCTION

According to occupational health and safety regulations, the wastewater system's exploitation activities of the public systems' water supply and wastewater disposal (Portaria n°762/2002), have specific risks that are the result of situations such as inadequate atmospheric oxygen, the presence of hazardous gases and vapors, contact with reagents, sewage water, and the sudden increase of flow and flash floods.

The principles expressed in the Occupational Health at Work Directive 89/391/EC, in reality lead to a prevention model that consists of strict compliance to the law by the company and the working and security procedures' by the workers. In this scenario the workers participation in the promotion of safety results simply from their observance of the legal rules or local instructions. Thus, in practice, we start off with principles (the wrong ones) that prevention is something that can be completely organized beforehand, neglecting the potential that the different stakeholders of an organization can collectively have in the construction of safety, and that the strategy of teaching rules suffices for prevention to be successful.

The fact that every year there are work related accidents in the sanitation sector, makes it essential not to look at the prevention problem focusing only on those occurrences, but on the work organization itself and the daily performed activities. It is therefore essential, to promote the workers effective participation (usually the simple cooperation or observance of instructions) of work related transformation projects or of work plans aiming at the promotion of safety.

The approach developed in this project has its roots in the "ergonomics activity" tradition. It has the same scientific orientations as some work psychology training projects. These analyses broach occupational health and safety issues in the activity context and in real exercise conditions. They consider that it is not enough applying rules or laws to obtain better safety at work. The safety organization and promotion is a process, a continuous one, involving all stakeholders and considering the reality of work activities (Frontini & Teiger, 1998, Delgoulet, 2001; Lacomblez & Vasconcelos, 2002, Gonzalez & Teiger, 2003, Vasconcelos et al. 2010).

We therefore consider that the safety's organization and promotion and the occupational risks prevention results, on the one hand in recognizing that all stakeholders play an effective role and, on the other hand, assuming, as a result of the training process, the need to change real working conditions. Therefore, it consists in an approach seeking to articulate ergonomic work analysis, training and action (transformation) in the same project, integrating workers, technical staff and people in charge. The increased number of training participants at different levels of stakeholders is crucial for making the transformation of the work execution conditions feasible (Lacomblez & Vasconcelos, 2009; Vasconcelos, 2010).

We consider the XIVth Congress of the French Language Ergonomics Society, in the early 80th of the past century, a milestone in the emergence of this link between training and action. In the balance of the event, organizers consider training as "an effective way of linking the various stakeholders to technical and organizational choices, and to introduce a participatory approach, allowing those who use the technical devices to evaluate and modify it" (Terssac & Queinnec, 1981).

After that date, the link between training and action has developments in their frames of references (Teiger and Lacomblez, 2006; Lacomblez & Teiger, 2007). But, the reflections, the knowledge and experience achieved, confirm the necessity of assuming that training cannot be isolated from a process of change, sustained by a dialogue among different stakeholders for a wider view of the emerged problems and their possible solutions.

With this logic we will present the design of a project where analysis, action and training was articulated, among a large number of participants and the results are a sign of its success.

ARGUMENTS FOR TRAINING PROJECT IN SANITATION SECTOR

This study (Costa, 2010) was carried out in the Oeiras Water Municipal Services and is focused on the tasks of sanitation workers who are cleaning and clearing domestic waste water and rainwater draining systems, home care branches and sewage boxes, ensuring proper functioning of the municipal sewage network.

The analyzed workers' group is exclusively composed of male gender elements, with an average age of 49 years and average seniority of 20 years. This group has a low educational level (mostly less than 6 years of schooling).

The sanitation sector is very particular about physical, chemical and biological agents that can endanger the workers' safety and health and therefore regards the following specific situations: insufficient atmospheric oxygen and the presence of hazardous gases or vapors, contact with sewage water or the sudden increase of flow and flash floods as risk activities.

Since 2001 there are recorded accidents of a different nature (handling equipment, falling and pitching materials) causing damage or injuries in various body parts.

For these reasons the Division of Human Resources Management of that Municipal service considered developing a training project aimed at promoting safety and health of sanitation workers a priority.

FROM WORK ACTIVITY ANALYIS TO TRAINING

Considering the risks associated with the activity as well as the occurred work-related accidents it was decided to develop a training process, framing it in its context and in work activity. Such a process differs from traditional models of training because they take into consideration the actual conditions of the activity, the experience of the workers and provide a basis of shared knowledge and know-how between different stakeholders.

The difference that marks the success of this training process was using knowledge and non-formalized experience of the workers, as demonstrated in others studies. Thus, "besides training", we tried to create conditions for a "co-training" between stakeholders where the parameters of change were not previously defined, but defined or shaped with the input from all and through everybody's work knowledge. Through the "co-construction" and "conjunction" of this knowledge it will be possible to identify the risks of the work situation, the know-how used to prevent accidents (Cru, 2000) and to acquire new skills and define the strategies for change (Vasconcelos et al., 2010).

In this approach, the ergonomic work activity analysis was the first phase. We tried to understand the overall work situation characterizing it by using observation techniques, interviews and documentary research. We also made videos of the work activity, with the objective of their subsequent use.

This preliminary analysis of work activity was a crucial phase of the project. Through it the trainer (at that stage in the role of work analyst), sought to capture the sense of activities, identify problem situations, understand the group dynamics. In short, all that could contribute to a better orientation of the formative process. This analysis revealed the need to develop two subjects during the training sessions: work in confined spaces and adopted working postures and manual cargo handling.

Based on the acquired knowledge, we designed a training process taking the target population's characteristics (older age and low educational level) into account. The process was designed according to an alternating session's model (table 1): theoretical sessions in class about safety in sanitation work, activity self-analysis sessions in the workplace itself, collective discussion sessions (using the analysis of video recordings of work activity and problem situations) in class and balance and negotiation sessions in class with the people in charge. In total the training process had eleven sessions. All training sessions were recorded on video and later reproduced in a text document for detailed analysis.

Table 1 Training sessions

Subject	Session Type	Context
Occupational Health and Safety	Introduction	Class
Confined Spaces	Self analysis of work	Workplace
	Collective discussion	Class
	Collective discussion	Class
Working Postures and Manual Cargo Handling	Collective discussion	Class
	Self analysis of work	Workplace
	Collective discussion	Class
Development and Negotiation Proposals	Training process analysis	Class
	Presentation and negotiation of the proposals	Class (with the participation of people in charge)
Evaluation	Self analysis of work	Workplace
	Collective discussion	Class

TOWARDS THE EFFECTIVE TRANSFORMATION OF WORK ACTIVITY AND CONDITIONS

According to the workers' opinion, the activity developed by the sanitation sector has characteristics that distinguishes it from other work activities and causes a representation of "dirty work". These characteristics triggered in the workers themselves needs of identity assertion and feelings of difference from other company sectors.

For this reason, the reaction to the training model and the level of participation in the discussions proposed was very positive. The high professional experience and familiarity among group members enhanced spontaneous manifestation of experiences and personal and collective strategies and made the reflections activities about work activity and conditions easier.

The trainer took a leading role in the conduction of the group discussion, guiding and asking the key issues that stimulated workers to reflect about their work activity, allowing for the expression of "not always knowing that one knows" (Lacomblez & Vasconcelos, 2009, p.54) and contributing to building the competences they aimed for.

Four subjects dominated the discussions and reflections in the different sessions:

- Human resources - the constraints that an aging group brings to the daily management of work activities, the shortage of workers to the volume of work involved, the teams organization, etc.

- Material resources - the adequacy and effectiveness of different personal protective equipment and work tools, the shortage of vehicles for responding (on time) to requests, lack of space in the vehicles for transporting all equipments and tools, etc.
- Procedures - a detailed description of the procedures realized during work activities, security procedures (individual and collective) adopted (or not) in relation to the circumstances, etc..
- Training - description of non-formalized training practices among workers, identification of training needs, adequacy of the provided training, etc.

The training process led to the formalization of proposals for improving working conditions. All proposals were presented, discussed and negotiated collectively with the people in charge (including the administration). These implemented the improvement measures for the working conditions as a counterpart for applying the safety working procedures. The transformation proposals were: the recruitment of new workers, the acquisition of personal protective equipment appropriate for the task, the acquisition of individual bags for the storage of equipments and first aids, the purchase of another vehicle and the planning of technical training whenever new equipments were acquired.

Seven months after the end of the training process we reviewed the work situation characteristics. It was our intention to verify if the commitments made by the people in charge were implemented and, in fact, they were. On the other hand the workers showed another awareness regarding their work activity and its risks, assuming another posture with respect to the possibilities of being promoters of their own safety.

Although one cannot establish a direct causal relationship, it's a fact that since the training sessions (September / October 2008), by the end of 2009 there were no work accidents in the group of workers addressed by this project.

REFLECTIONS ABOUT TRAINING PROCESS DESIGN

The results showed the relevance and the potential that a participatory training approach can have, representing a real alternative to "traditional" models of safety promotion.

Apart from the indisputable mark left on the research field by this project, it was possible to influence different categories of workers (execution and decision makers) with another view and ability to act in promoting security. It also demonstrated the value of a "situated" and "timely" training processes.

The mediation role assumed by the trainer proved to be of enormous importance. It leads to individual development, intrinsically linked to explanation and formalization of knowledge, collective development, creating dynamics of experience and knowledge confrontation and mutual enrichment, and social recognition in order to give meaning to that experience within the organization.

With the present study it was also possible to identify the non-formalized know-how, constituting "formalized" work rules according to the group of workers. Know-how focused on the protection of workers who go into the sewage boxes was identified. The knowledge expressed by these workers, the recognition of its relevance and the conversion into security procedures by people in charge and its subsequent dissemination within the company is an undeniable contribution towards health and safety promotion.

This project is currently preparing a reassessment edition for the group of workers involved. We are preparing to extend the project to other professional groups whose accidents increased in recent years.

REFERENCES

Costa, C. (2010), *Formação em Contexto Profissional. Análise da Importância da Formação Contextualizada no Desenvolvimento de Actos Seguros num Grupo de Operadores do Sector do Saneamento*. Tese de Mestrado. Lisboa: FMHUTL.

Delgoulet, C. (2001), "La construction des liens entre situations de travail et situation d'apprentissage dans la formation professionnelle". *Pistes*, 3(2). http://www.pistes.uqam.ca/v3n2/articles/v3n2a2.htm

Frontini, J.-M. & Teiger, C. (1998), "L'apprentissage de l'analyse ergonomique du travail comme moteur de changement individuel et organisationnel. Le cas de la formation des préventeurs en entreprise." *Performances Humaines & Techniques*, n° hors de série, dec.,101-110.

Gonzalez, G-R.; Teiger, C. (2003), "Collective self-analysis of the activity of supervisors in a child day care center." *Proceedings of the congress of the International Ergonomics Association*, 24-29 Aug, Séoul, vol 7, sur CD-ROM.

Lacomblez, M. e Vasconcelos, R. (2002), "Análise guiada do trabalho e desenvolvimento da segurança e saúde no trabalho: contributos, reflexões e desafio." *2° Colóquio Internacional de Segurança e Higiene do Trabalho*, Porto, 33-38.

Lacomblez, M.; Teiger, C. (2007), "*Ergonomia, formações e transformações*". In P. Falzon (ed), Ergonomie. São Paulo:Editora Blucher, pp. 587-601.

Lamcomblez, M.; Vasconcelos, R. (2009), "Análise ergonómica da actividade, formação e transformação do trabalho: opções para um desenvolvimento durável." *Laboreal*, 5 (1), 53-60. http://laboreal.up.pt/revista/artigo.php?id=37t45nSU5471123592231593411

Portaria n° 762/2002, "Regulamento de Segurança, Higiene e Saúde no trabalho na exploração dos sistemas públicos de distribuição de àgua e de Drenagem de àguas residuais." *Diário da Républica* n° 149, I série – B, de 1 de Julho de 2002, pp. 5123-5130.

Teiger, C. Lacomblez, M. (2006), "L'ergonomie et la trans-formation du travail et/ou des personnes. Permanences et evolutions (2)." *Education Permanente*, 166(1), 8-28.

Terssac, G de; Quéinnec, Y. (1981), "Exposé de synthése du XVIe congrés de la SELF". Toulouse, 1980. *Le Travail Humain*, 44(1), 113-121.

Vasconcelos, R. (2000), *Analisar o Trabalho para Formar e Transformer: a Auto-análise do Trabalho ao Serviço da Higiene e Segurança no Trabalho num Contexto de Desenvolvimento e Transmissão de Competências Profissionais.* Tese de Mestrado. Porto:FPCEUP

Vasconcelos, R. (2008*), O Papel do Psicólogo do Trabalho e a Tripolaridade Dinâmica dos Processos de Transformação: Contributo para a Promoção da Segurança e Saúde no Trabalho.* Tese de Doutoramento, Porto:FPCEUP.

Vasconcelos, R.; Duarte, S., D.; Moreira, V. (2010), "Matriosca Project: Work analysis, hands-on training and participative action for accident prevention." *Proceedings of International Symposium on Occupation Safety and Hygiene* 11-12 Feb. Guimarães:Portuguese Society Of Occupational Safety and Hygiene, 542-546.

Chapter 71

Effects of Training Program on Interface Task Performance for Older Adults

Wang-Chin Tsai, Chang-Franw Lee

Graduate School of Design
National Yunlin University of Science and Technology

ABSTRACT

This study is grounded on cognitive learning theory and the representative SRK model "Three Levels of Skilled Performance" suggested by Rasmussen. This study provides two different training programs (declarative knowledge training and procedural knowledge training) to determine the effects on 3 interface task performance (skill-based behavior task, rule-based behavior task, and knowledge-based behavior task). 32 older adults were divided into 2 groups and each group then participated in 2 training programs separately to examine the resulting performance on interface tasks. After training program, participants are requested to conduct Microsoft Media Player interface task on three levels of skilled performance and complete a questionnaire for the subjective satisfaction on the two training programs. Results show that the type of training material has main effects on different interface types operations. It reveals that older adults under declarative knowledge training program develop a better understanding of the knowledge-based and rule-based behavior tasks than procedural knowledge training program. The results are applicable to the development for the instructional design in the interface research filed.

Keywords: Older adults, training program, interface task

INTRODUCTION

Similar to many other countries in the world, the populations of Taiwan is aging rapidly (Ministry of Interior, 2009). This aging trend is also coincident with the dramatic development of computers, Internet and information appliances. Owing to the progress in microprocessor control, there are many electronic appliances and systems used in our daily life. For instance, in many places public transportation tickets can only be purchased at vending machines, and money from a bank account can only be withdrawn at a cash machine. However, older adults often experience problems when using these modern technologies. It has been shown in several studies that, for instance, the older adults have difficulties in using electronic appliances with smaller displayed characters and less effective user interface design. In order to create tools to enable self-reliance for older adults, Human-computer interaction textbooks state that individual differences, such as age, need to be taken into account in the designing process. Designers have to understand their unique physical characteristics, such as their motion, perception, and cognition, and put more emphasis than ever on satisfying their needs. Likewise, in a time when society is becoming more and more reliant on computers and the Internet, it is imperative that older adults also need to learn to use new electronic appliances and digital product so that they can also take advantage of the rich opportunities and resources, increase the autonomy provided by these technologies (Dickinson & Hill, 2007). Older people are able to learn to use computers, internet browsing and technological product effectively with appropriate training(Gagliardi, Mazzarini, Papa, Giuli, & Marcellini, 2008; Ng, 2008); their confidence and ease of use of these technologies are proportional to actual practice and they usually are enthusiastic about learning digital technologies(Gatto & Tak, 2008; Lagana, 2008). As regard to the training and learning, Researchers in aging and training have developed several training theory and design guidelines to help individuals, especially older adults, learn. To ensure the success of older adults with technology, researchers should incorporate these design guidelines into the design of training development.

Training Design for older adults is the systematic development of instructional specifications using adult learning and instructional theory to ensure the quality of training. It is the entire process of analysis of training needs and goals and the development of a sophisticated program to meet those needs(Githens, 2007). It includes development of training materials and programs; and tryout and evaluation of all instruction and trainee feedback. Thus, designing training and instructional programs to promote meaningful learning has been a long-standing challenge. At the most fundamental level, the issues are what to teach and how to teach it. Before developing specific strategies that focus on the transfer of technological skills to novel technological problems that older adults are faced with in everyday interface activities, it is important to investigate whether this transfer can actually be achieved. Otherwise, research should focus on different approaches or programs of reducing older adults' problems with technology: such as, specific training programs for different tasks requirements. The present study focused on the design of training programs to facilitate both the acquisition of interface skills and the development of an adequate representation of the interface structure of a simple and complex task. To support the assertion that training program affects interface tasks acquisition, the following section will present a discussion of patterns of age-related differences in interface tasks acquisition. Then the training programs will be presented followed by a discussion of the theoretical framework of procedural and conceptual knowledge. The model of three levels of skilled performance grounded by Rasmussen's (1983) will be presented followed by to classify our interface tasks. The section will conclude with a discussion of the rationale for the present study.

AGING AND TRAINING PROGRAM

While learning interface tasks, the aging declines may contribute to older adults' need for longer training and poorer performance, in that, older adults may have trouble understanding the spatial structure of the interface. Older adults need to take advantage of their knowledge system for interface skill acquisition and problem solving. Knowledge in the training context can be broadly classified into two main categories: declarative knowledge and procedural knowledge. In the current research, we developed two training programs are grounded in the theoretical concept of procedural and declarative knowledge. Ontologically declarative knowledge relates to the what, where and when aspects of temporal and strategic knowledge domains. It is a static description which captures an insight of the physical world through the medium of words, images, sounds and emotions. For all practical purposes, declarative knowledge can be identified with explicit knowledge or knowledge that can be coded and clearly articulated in textual, graphical or verbal structures of representations (Nickols, 2000). Since declarative knowledge deals with the exposition of facts, methods, techniques and practices, it can easily be expressed, recorded and disseminated in the form of artifacts, written norms and verbal communications to become explicit knowledge assets. Procedural knowledge is related to the procedure to carry an action out. Knowledge about how to do something is procedural knowledge. Procedural knowledge is instruction-oriented. It focuses on how to obtain a result (Turban & Aronson, 1988). The overview provided of the key concepts of declarative and procedural knowledge forms potentially gives rise to the natural question of how these two categories of knowledge could tie in together organically as the integral components of a training program.

In this research, procedural training program presents interface task information in a sequence of "how to" steps or procedures. These procedures are specific goals and sub-goals necessary to complete the given interface task, therefore instilling procedural knowledge. However, declarative training program presents factual task information at each interface state, which is analogous to declarative knowledge. Training presented in declarative form consists of general facts that do not direct translate into procedures. These programs must be interpreted into training development. Both training program have their advantages where procedural training could be less error prone and faster but declarative training is more flexible. And this is depends on different instructional task content and complexity.

THE SKILL, RULE AND KNOWLEDGE BASED CLASSIFICATION

In this research, we use an influential classification of the different types of information processing involved in industrial tasks was developed by J. Rasmussen (1983)of the Risø Laboratory in Denmark. This scheme provides a useful framework for identifying the interface tasks likely to occur in different operational situations, or within different aspects of the same task where different types of information processing demands on the individual may occur. Rasmussen describes the simplest form of task behavior as skill-based behavior (SBB). It is controlled from the lowest level of the cognitive processing hierarchy, and may be characterized as "smooth, automated, and highly integrated" and takes place (critically) "without conscious attention or control". Effective SBB performance relies on heavy feed forward control flows throughout, "depends upon a very flexible and efficient dynamic internal world model", and will usually involve rapid coordinated movements. Examples of SBBs on some interface tasks are simple double click starting or pausing operation for digital product. As for the nature of the information at this level, SBB is described as relying on signals, which are defined as "representing time-space variables from a dynamical spatial configuration in the interface environment". The next level of task behavior complexity is rule-based behavior (RBB). It is controlled by the middle level of the processing hierarchy, and may be characterized as consisting of "a sequence of subroutines in a familiar work situation", where the subroutines follow previously stored rules, again relying primarily on feed forward control. Examples of RBBs on interface tasks are sequenced task decision and system control tasks such as the discrete maneuvering of aircraft or cars. As for the nature of the information at this level, RBB is described as relying on signs to indicate the state of the environment. These are defined as "related to certain features or rules in the environment and the connected conditions for action". Rasmussen also explicitly explains that "the boundaries between skill-based and rule-based performance is not quite distinct" sometimes, varying with both level of training and attention state. Rule-based control is ultimately based upon "explicit know how" - the rules can be explained in words by the person concerned; so if you cannot explain it then it must be skill-based.

The highest level of complexity is knowledge-based behavior (KBB). It is controlled by the highest level of the processing hierarchy, relies upon a "mental model" of the system in question, and in general terms is to be strongly avoided because what it achieves in terms of sophistication it loses in the time it takes. KBB is therefore what you have to turn to only when SBB or RBB are momentarily not up to the task at hand. This means that the goal of a given piece of KBB has to be "explicitly formulated" at the time it is needed, taking into account the nature of the problem and the overall aims of the subject. Examples of KBBs are problem solving and fault diagnosis. For some interface tasks, those are seldom-used or advanced function. Users are required to use their higher knowledge and logical method to complete a task. As for the nature of the information at this level, KBB is described as relying on symbols. These are defined as "abstract constructs related to and defined by a formal structure of relations and processes", and include prior knowledge itself and information processing ability.

HYPOTHESES- PERFORMANCE AT TEST - TRAINING X INTERFACE TASK INTERACTION

Hypothesis: It has been suggested that participants who receive training program that focus on declarative knowledge develop a better understanding of the system structure than participants who receive instructions that focus on procedural knowledge (Zeitz & Spoehr, 1989). Also, Mead & Fisk (1998) found that older adults in a declarative training condition navigated the system structure better than those in the procedural training condition. Therefore, we assume the training program affects the performance of older adults operating different interface tasks. Older participants in declarative knowledge training should produce higher performance on the identification of three types of information processing tasks and take fewer interface error rates for untrained tasks than older participants in procedural knowledge training.

METHODS

EXPERIMENTAL DESIGN

A 2 (training program) x 3 (interface task type) nested design is adopted in this study. Two independent variables are training program: between-group factor (Variable A)and interface task type: within-group factor (Variable B). Training program variable includes declarative knowledge training(A1) and procedural knowledge training(A2). Tasks type variable consist of a skill-based behavior task (B1), rule-based behavior task (B2) and knowledge-based behavior task (B3) of Microsoft Media Player V11.0 control panel. The dependent variable is defined as the averaged complete time for each task from the starting till to the finished stage.

PARTICIPANTS

In this study, we chose 32 participants in accordance with the training program arrangement. The 32 older adults were randomly chosen from the Adult learning Center. Participants, as a function of condition, were 16 older declarative program trainees aged 58 to72 years (M=65.12, SD=6.23), the other 16 older procedural program trainees aged 63 to78 years (M=68.36, SD=4.17). All participants passed both near and far visual acuity with the criterion set at 20/40 (corrected and uncorrected). Those participants were no related Microsoft Media Player experience, but spent 30-60 minutes on internet browsing per week.

PROCEDURE OF INTERFACE TASKS AND TRAINING PROGRAMS

In this study, we use the Microsoft Media Player as a demonstration platform to train older adults on interface skill acquisition and problem solving for 6 interface tasks (including 3 basic tasks and 3 advanced tasks). The training program taught each older adults group separately what to do or how to do it with hard copy tutorials designed by the researcher. The declarative training program presented a conceptual paragraph for the interface task information, but didn't include detailed and sequenced instruction for using it to complete the task. That is, the declarative training program told participant what to do instead of how to do it. Likewise, the procedural training program told the participants how to operate the tasks but didn't explicitly tell them what they were doing. The detailed training program content is shown as the table 1. After participant complete this assigned 30 minute training program, they were requested to perform 6 re-tested tasks based on 2 skill-based behavior tasks, 2 rule-based behavior tasks and 2 knowledge-based behavior tasks(to differentiate the training performance and effectiveness).

Table 1. Sample as two training program content

Basic function training task-Please adjust the playlist in a certain order	
declarative knowledge training	There are 7 songs in the playlist. You can image the playlist is similar to the Chinese abacus. Each song has its position. If you want to change the unit, you need to think how to move in a certain way by selecting up and down function.
procedural knowledge training	S1please select the song/ S2 drug it up and down in a certain direction/S3 see the different orders
Advanced function training task-To create a playlist by selecting songs from specific folder	
declarative knowledge training	The scene is likely you are going to select bottle drinks on a vending machine. You need to think about the whole procedure. How to choose the wanted drinks in different location and category and the drink should appear in a slot on the bottom of the vending machine.
procedural knowledge training	S1 File/S2open file/S3 Select the dropped arrow /S4Select Disc D/S5.Seclect folder title as" Chinese song ". /S6 Select the song named" When Shall You Return? /S7open/S8hear the music。

STATISTICAL ANALYSIS

For the analysis of the data, this study applied Two-way Analyses of Variance (ANOVA) to examine significant differences of the task performance of 2 training program conditions within each group. In addition, the significant differences were analyzed by utilizing the Scheffe Method as the post hoc test for multiple comparisons. Significance was accepted at the level of p<.05, while the degrees of freedom and corresponding probability, or the F-value, were also shown in the statistical test. In all, the statistical analysis was conducted by utilizing the Windows SPSS Statistics 17 Program.

RESULTS

SIGNIFICANT ANALYSIS OF THE TWO INDEPENDENT VARIABLES

The effects of two independent variables, training program and task types, were explored (Table 2). The analysis of variance indicated that both the training program [F $(1, 90)$ = 12.59, p <.05] and the task type [F $(2, 46)$ = 129.08, p <.05] affected performances of the participants. These findings also indicate that the training program and task types have the significant interaction effect among the older adults in the experiment [F $(2, 90)$ = 11.93, p <.05]. Therefore, we continue perform the post hoc multiple comparisons in order to understand the relation between these two variables. The post hoc multiple comparisons were integrated and are shown in tables 3 and tables 4 in this study.

Table 2.ANOVA table and significant analysis of two independent variables

Source of Variances	Sum of square	Df	MS	F	P value
Training program (A)	4875.91	1	4875.91	12.59	.001
Task type (B)	99915.69	2	49957.85	129.08	.000
A * B	9235.96	2	4617.98	11.93	.000
Error (A * B)	34832.78	90	387.03		

*: p < 0.05

Table 3.Results of operational time on task types (unit: seconds)

Source of Variances	SBB Task(B1)	RBB Task(B2)	KBB Task(B3)	P value	Post hoc tests
Declarative training(A1)	20.3(3.6)	38.7(11.4)	74.9(9.6)	.001	B3>B2>B1
Procedural training(A2)	19.2(4.5)	40.7(10.0)	116.9(44.3)	.000	B3>B2> B1

*: p < 0.05

Table 4.Results of operational time on training programs (unit: seconds)

Source of Variances	Declarative training(A1)	Procedural training(A2)	P value	Post hoc tests
SBB Task(B1)	20.3(3.6)	19.2(4.5)	.072	-
RBB Task(B2)	38.7(11.4)	40.7(10.0)	.021	A2>A1
KBB Task(B3)	74.9(9.6)	116.9(44.3)	.003	A2>A1

*: $p < 0.05$

According to the results of the ANOVA, there was a significant interaction between training program and task type on the performance of total times. For both training program, the performance of total time was larger on the knowledge-based behavior task than on the rule-based behavior task and skill-based behavior task (Table 3). Moreover, for knowledge-based behavior task, the performance of total time was the smallest on the declarative training program, followed by the procedural training program. For rule-based behavior task, the performance of total time on the procedural training program was larger than declarative training program (Table 4). There was no significant difference on the skill-based behavior task.

DISCUSSION

Previous research on the acquisition of complex interface skills refers to the notion from a knowledge structure or a memory organization (Hickman, Rogers, & Fisk, 2007; Rogers, Fisk, Mead, Walker, & Cabrera, 1996). In this research, older adults who received declarative knowledge training program showed superior performance on the knowledge-based behavior task than procedural knowledge training program. Thus, presenting declarative information to older adults during training was more important than presenting procedural information. This is different than the previous research (Mead & Fisk, 1998). We assume that knowledge organization consists of both declarative memories, the part of long-term memory where factual information is stored, and procedural memory, the part of memory where knowledge of skills or procedures is stored. As processing continues through a knowledge-based behavior task, the interface tasks of early processing may no longer be available when later processing is complete. As a result, the procedural training would take longer and are more prone to more sequenced steps for older adults compared to declarative training programs. Furthermore, the questionnaire results partly explain the problems encountered by participants during training program. For example, certain subjects under procedural training could not fully understand and memorize the sequenced operational processes and button functions and operated tasks. As a result, for the rule-based behavior task or skill-based behavior task, it has little burden for them. However, when conducting the knowledge-based behavior task, they made several mistakes during operational processes and took a longer operational time. Many of them were furthermore reluctant to memorize the whole sequenced steps.

On the other hand, the declarative training program could provide the number of cognitive nodes when new information adds to existing information. Therefore, if older adults have less experience (i.e., existing information) about computers or digital interface, they may have a greater number of existing cognitive nodes and units to be made stronger with the storage of new information under the declarative training program. This strengthening of cognitive units may account for the development of an adequate training representation.

CONCLUSION

This research summarized the possible application of providing training on the task types, and the issues that one should consider when developing training programs for older adults. Thus, older adults are capable of learning to use new interface tasks. The optimal training program will likely involve a combination of specific task types depending on the complexity level. The present study makes clear the relative benefits of declarative knowledge training program versus procedural knowledge training program for older adults and especially suggests when conducting the knowledge based behavior task, how instructor should incorporate declarative knowledge training program into these tasks.

REFERENCES

Dickinson, A., & Hill, R. L. (2007). Keeping In Touch: Talking to Older People about Computers and Communication. *Educational Gerontology, 33*(8), 613-630.

Gagliardi, C., Mazzarini, G., Papa, R., Giuli, C., & Marcellini, F. (2008). Designing a Learning Program to Link Old and Disabled People to Computers. *Educational Gerontology, 34*(1), 15-29.

Gatto, S. L., & Tak, S. H. (2008). Computer, Internet, and E-mail Use Among Older Adults: Benefits and Barriers. *Educational Gerontology, 34*(9), 800-811.

Githens, R. P. (2007). Older Adults and E-learning: Opportunities and Barriers. *Quarterly Review of Distance Education, 8*(4), 329-338.

Hickman, J. M., Rogers, W. A., & Fisk, A. D. (2007). Training older adults to use new technology. *The Journals of Gerontology Series B: Psychological Sciences and Social Sciences, 62*, 77-84.

Kumar, M, Organizing Curriculum Based upon Constructivism: What To Teach and What Not To, (2006), *Journal of Thought, (41)*, 81-94.

Lagana, L. (2008). Enhancing the Attitudes and Self-Efficacy of Older Adults Toward Computers and the Internet: Results of a Pilot Study. *Educational Gerontology, 34*(9), 831-843.

Mead, S., & Fisk, A. D. (1998). Measuring Skill Acquisition and Retention with an ATM Simulator: The Need for Age-Specific Training. *Human Factors: The Journal of the Human Factors and Ergonomics Society, 40*(3), 516-523.

Mead, S., & Fisk, A. D., (1998), Measuring skill acquisition and retention with an atm simulator: the need for age-specific training, *Human Factors: The Journal of the Human Factors and Ergonomics Society, 40*(3), 516-523.

Ministry of Interior, (2009), Taiwan population statistic report.

Ng, C. (2008). Motivation Among Older Adults in Learning Computing Technologies: A Grounded Model. *Educational Gerontology, 34*(1), 1-14.

Nickols, F. W. (2000). The knowledge in knowledge management. Boston: Butterworth-Heinemann.

Rasmussen, J. (1983). Skills, Rules, Knowledge: Signals, Signs, and Symbols and other Distinctions in Human Performance Models. IEEE Transactions on Systems, Man, and Cybernetics, SMC-13(3), 257-267, as cited in Lehto, M. R. (1997). Handbook of human factors and ergonomics (2nd Ed.)(Gavriel Slvendy Ed.), 1235-1236. New York: John Wiley & Sons.

Rogers, W. A., Fisk, A. D., Mead, S. E., Walker, N., & Cabrera, E. F. (1996). Training Older Adults to Use Automatic Teller Machines. *Human Factors: The Journal of the Human Factors and Ergonomics Society, 38*(3), 425-433.

Turban, E., & Aronson, J. (1988). Decision support systems and intelligent systems. Upper Saddle River, NJ: Prentice Hall.

Zeitz, C. M. & Spoehr, K. T. (1989), Knowledge organization and the acquisition of procedural expertise, *Applied Cognitive Psychology*, 3, 313-336

CHAPTER 72

Face to Face and On-line Learning: Are Emotions so Blended in a Blended Learning Environment?

Parlangeli Oronzo, Ciani Natalia, Marchigiani Enrica

Communication Science Department,
University of Siena
53100 Siena, Italy

ABSTRACT

Recently, research advancements in neuro-psychology have emphasized the role of emotional states in learning, investigating the relationship between emotions and cognition. Emotions are seen to be profoundly involved in educational experiences, both in traditional education and in on-line learning.

This study describes research aimed at analyzing the emotions experienced during a blended English course for Italian adult students offered to employees of the main hospital in Siena, Italy. Emotions experienced by the students during their first approach with both classroom activities and on-line lessons have been assessed and compared. The results of the study show that both the face to face lessons and the on-line learning environment show an apparent differentiation between the expression of negative and positive emotions. On the whole, however, the pattern of results obtained from the two learning environments show significant differences.

Keywords: Emotion, Learning, Blended Learning, Language Learning

INTRODUCTION

Emotion is a complex process with adaptive, communicative and motivational functions (Darwin, 1872) including different aspects as subjective experiences, appraisal of situations, facial and body expressions, physiological responses and action tendencies (Lazarus, 2001; Parkinson, 2001).

Some researchers (Damasio 1994; Ekman, 1992) agree on the existence of primary or universal emotions, such as happiness, sadness, fear, anger, surprise, and disgust, and secondary emotions, like embarrassment, jealousy, guilt and pride, which are socially and culturally determined. Primary emotions are responsive processes that, in the course of evolution, have fulfilled the need for survival, while the secondary emotions are mainly connected with experience and learning, therefore they may be considered as a mixture of the emotions of the first type. Furthermore, within the affective distinction, Damasio (1999; 2003) introduces another concept regarding background feelings such as well-being or malaise, calm or tension, fatigue or energy. These focus mainly on the internal state of the body, whereas the physical state of the body would be the emotion and the perception of that emotion would be the feeling.

Affective and cognitive domains have traditionally been seen as opposite realms. The influence of emotions on learning is, in fact, a relatively new concept and it is still under-emphasized. However, a growing body of literature (Martin and Briggs, 1986; Dirkx, 2001) has begun to underline the central role of emotions in learning tasks. Advances in neuroscience have enabled us to understand that emotions have an influence on reason, or rather, that emotions are indispensable for rationality (Damasio, 1994; Bar-On, 2003; Pessoa, 2008). Other studies (Isen, 2000; Erez & Isen, 2002) demonstrate that a positive emotional state can induce a different kind of thinking that is characterised by more thorough decision making, greater creativity and flexibility in problem solving, a deeper level of awareness, and increased memory and retrieval skills. Furthermore, a positive emotional experience can facilitate the emergence and duration of intrinsic motivation, which is of primary importance in all types of learning. These findings highlight the significant effects of emotions on learning. However, emotions can not only support, but may even impede learning. In particular, negative emotions, as well as positive emotions, need to be regulated by learners. In other instances they would require all the attention resources, thus becoming a variable that would negatively affect learning.

Studies on emotions and e-learning (O'Reagan, 2003) indicate that particular emotions, such as frustration and anxiety, are related to studying on-line. In this case, negative emotions related to the use of technologies that are characterised by a low level of usability and/or which have various drawbacks in communication. On the other hand, positive emotions such as enthusiasm and satisfaction, in this context are mainly due to the content and to the mere use of new technologies. However the majority of these studies highlights the importance of encouraging positive emotions and preventing the emerging of negative emotions. In this paper we describe research aimed at studying the emotions throughout a blended learning course, investigating and comparing the emotions experienced by the students during their first approach with both classroom lessons and with the on-line learning activities.

THE BLENDED LEARNING ENVIRONMENTS

Blended learning can be described as a combination of on-line learning and traditional education delivered in a classroom, implying individual and collaborative activities (Reay, 2001; Rooney, 2003; Sands, 2002; Ward and LaBranche, 2003; Young, 2002). Its use is progressively increasing in higher education due to the potential for taking advantage of the strengths of each environment. So it aims at including the qualities of learning at a distance on the Internet, which is mainly learner-driven and self-paced, maintaining, however, face-to-face contact, which is especially important for strengthening social relationships. Social practices and active participation in activities are considered, in fact, basic for knowledge and skills construction (Brown & Campione, 1990; Wenger, 1998; Bereiter and Scardamalia, 2003). Studies on blended learning (Rovai and Jordan, 2004) highlight in particular the stronger sense of community found in blended learning, in respect to fully on-line courses because of the varied opportunities of interaction among students, as well as with tutors and professors. The possibility of this ongoing interaction increases exchange, communication, socialization, sharing of experiences and feelings, which are all factors that contribute in the construction of knowledge.

In the past, research on emotions and learning dealt primarily with e-learning, and less with blended learning. The literature relates that students learning on-line suffer mainly from confusion, anxiety and frustration, and possibly isolation, possibly due to ambiguity or delay in instructions received by professors (Hara and Kling, 2000). The same authors still highlight that the reason for dropouts in on-line courses very often does not depend on the technology, but rather on the course design and the underlying pedagogical theory. Kerry O'Reagan (2003) indicates specific emotions related to e-learning such as frustration, fear, distress, and even excitement and pride. Frustration is mainly due to technologies that are difficult to use; fear and distress are related to apprehension of the unknown interlocutor; excitement is linked to the use of new technologies, to the possibility to get in touch with new people and to the effectiveness of the learning process; pride is mainly a result of accomplishment.

THE STUDY AND THE LEARNING ENVIRONMENT

The present study aims at identifying the emotions experienced during the first phases of a blended course for language learning. More specifically, it has been designed to compare the different emotions experienced in relation to the face to face lessons and to the on-line learning environment. The study was based on blended English language courses offered by the Siena University Language Centre to the technical, administrative and professional staff of the main Hospital in Siena.

The course we considered comprises face-to-face lessons and on-line lessons in the open source CMS (Course Management System) MOODLE.

Table 1. Heuristic evaluation results

Guidelines	Degree of assessment	Comments: strengths and weaknesses
Visual design	4	Tools and elements are clear and consistent
Navigation	4	Simple and efficient with clear marked exits
Accessibility	4	Free from technical problems
Interactivity	3	The e-learning program provides content-related interaction and tasks that support meaningful learning but material integration is not foreseen.. Glossary and note-taking should be more handy
Help and documentation	5	Users' tasks are clearly explained even with brief explanatory summaries
Learning design	4	The program engages learners in tasks that are closely aligned with the learning objectives
Content	4	The content is organized in manageable units and they are supplied with clear explanation and summaries. Vocabulary and terminology used are appropriate for the learners
Learning and support	4	The course offers tools (note taking, glossary and the steady feedback of the tutor on-line.) that support learning. It includes both individual-based and group-based activities
Assessment	4	The course provides opportunities for self-assessment that advance learner achievement. Automatic assessment does not always provide sufficient feedback to the learner to afford remedial directions, instead the tutor feedback for forum activities is very efficient
Media integration and resources	5	The inclusion of media serves clear pedagogical and motivational purposes. It provides access to resources such as video, audio and links to external web sites that enhance an effective learning process
Motivation to learn	4	The course stimulates further inquiry and provides learner with varied learning activities that increase learning success. Feedback from the tutor is very prompt and useful. The group-based activity is challenging.

It provides a collaborative learning space where data can be shared and it offers different resources: course administration and content management, then, the participant block, a space for personal profiles, forums for asynchronous discussion, news, a glossary, blogs, quizzes, assignments, multimedia integration, and the participant's log.

Within this specific course, there are weekly classroom and on-line activities which are directly related to and integrated with each other. The classroom activities of this course mainly focus on listening, speaking and comprehension skills, also covering grammar and vocabulary both in collaborative and individual tasks. The on-line course is especially designed to improve reading skills and writing on various topics, expanding vocabulary, reinforcing grammar abilities either individually or with group exercises. Within the on-line orientation tutorial, participants are encouraged to express their feelings right from the start so the tutors can help them through the difficult parts of starting the course.

Before the study on learning and emotions we conducted a heuristic evaluation in order to assess the effectiveness and the efficiency of the on-line learning application (see Table 1). Actually, effective, efficient and satisfying interaction, that does not hamper the learning task, is a golden rule for educational applications. An inefficient and unpleasant application could make learners frustrated enough to drop the course. On the contrary learners can take advantage of, and can be motivated to accomplish the learning task through strong interactivity and an effective educational process.

Therefore, two HCI experts conducted an evaluation through the inspection method, driven by a set of specific guidelines aimed at evaluating the usability of e-leaning applications. The usability criteria have been arranged drawing together more general usability guidelines (Nielsen, 1994; Ravden and Johnson, 1989) with more specific ones addressed to learning applications (Ardito et al., 2005; Zaharias, 2009). More specifically the guidelines used in this study refer to the following criteria: visual design, navigation, accessibility, interactivity, help and documentation, learning design, content, learning and support, assessment, media integration and resources, motivation to learn. These criteria have been assessed on a five points scale from strongly disagree to strongly agree. The two evaluators conducted the inspection separately, exploring the interface freely and comparing their findings at the end of the reviewing session. In general, the heuristic evaluation results show that the application is easy to use and the violations found do not hamper the learning process.

METHOD

Participants included 76 employees of the main Hospital located in Siena. They were physicians, nurses, technicians and office workers, who attended blended courses to learn the English language, from October 2009 to January 2010. Students enrolled within the course were asked o fill in a questionnaire during the learning path. The study was conducted by means of interviews and questionnaires.

A preliminary interview was directed to two professors involved in the learning program, in order to gather general information on the course and in particular to try to identify the emotions associated with, and impacting on their experience of blended learning.

The questionnaire was designed to collect general information on the students, their expectations, motivation, familiarity with technology, language expertise and degree of satisfaction. The emotions investigated in the questionnaire were derived from the results that emerged during interviews and from the literature. The resulting set comprised primary emotions (Damasio, 1994) - i.e. joy, fear, anger and surprise - and additional emotions that have been generally considered relevant in learning - i.e. interest, embarrassment, curiosity, gratification, frustration, and boredom. Students' evaluations in relation to these emotions were collected on a seven point scale related to both the face-to-face and on-line learning environments.

RESULTS AND DISCUSSION

Data analyses show that both the face to face lessons and the on-line learning environment bring an apparent differentiation between the expression of negative and positive emotions. While for negative emotions we have obtained mean values that are consistently valued as "irrelevant" (point "4" on the 7 point scale), the positive emotions are instead all on the high end of the scale indicating an expression of relevance (see figure 1).

On the whole, both learning environments thus prove to be able to induce a generally positive expression of emotions. This global positive trend, however, is not always due to a coinciding evaluation of the two environments since four emotions out of ten show significant differences. "Anger" is generally evaluated as an emotion scantly felt, but this is more evidently true for face to face lessons (mean difference=.415; t=2.605; p<.011). On the contrary, "embarrassment" is more intensely felt during face to face lessons, this being most likely due to the intrinsically social characteristic of this environment (mean difference=-.746; t=-2.994; p<.004), which implies a more evident exposition of the Self. Two positive emotions, "joy" and "gratification" are instead felt more intensely during face to face lessons. Again, in this case differences are both significant, respectively being -0.55 (t= -2.260; p<.009) and 0.39 (t=-2.137; p<.04).

It is worthwhile to highlight that "surprise" appears to be an emotion in the middle, between the more positive and negative emotions. This may be ascribed to the experience of this emotion as something that can be related both to negative and positive events.

It should also be noted that "frustration" is usually related to on-line courses (O'Reagan, 2003), however, it was found in this study to have a neutral evaluation and, more interestingly, no significant differences were found between the two environments.

Taking into account correlations between emotions within each teaching environment, different patterns of results emerge when the on-line environment is compared with face to face lessons, thus suggesting non coincidental emotional experiences.

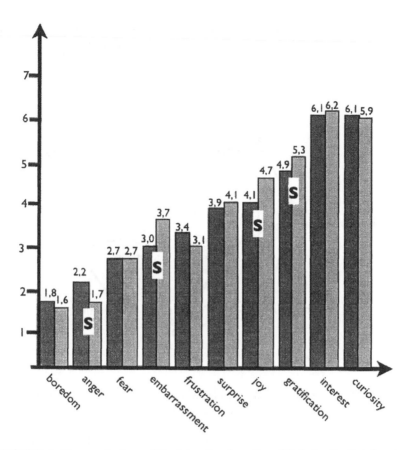

FIGURE. 1. Mean evaluations of the intensity of emotions felt during the first face to face lessons (light gray columns) and the first interactions in the on-line learning environment (dark gray columns).
Differences between the two teaching environments are significant for four emotions: "anger", "embarrassment", "joy" and "gratification".

With regards to the on-line learning environment, nearly all negative emotions are positively correlated (figure 2). The same result is obtained with positive emotions as well, and different negative correlations emerge between emotions with the opposite sign (for instance: "anger" vs. "gratification"). On the whole, these results suggest a complex and articulated emotional experience, consistently felt through the separation of those events bringing a positive reaction from those, on the contrary, eliciting negative sensations. By and large, the notable number of counter-correlations suggests that in the on-line learning environment there is a sort of difficulty in identifying emotions per se, since they are often characterized by the contrary of some other emotion. For instance, the clearly positive emotion "gratification" is counter-correlated with three different negative emotions, namely "fear" ($r=-271$; $p<.036$), "frustration" ($r=-.293$; $p<.023$), and "anger" ($r=-.362$; $p<.004$). This may be seen as an inability to clearly identify each emotion without the conceptualization of its

732

negation. In some cases emotions are thus expressed as if they were on a continuum in which there are two or more extremes of opposite sign.

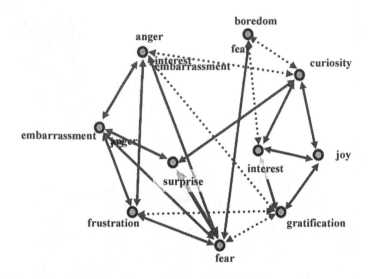

FIGURE 2. Arrows shows correlations between emotions within the on-line teaching environment. Dotted lines are negative correlations.

Results suggest something quite different for the face to face teaching environment (figure 3). In this case, negative emotions are even more evidently separated from positive ones. The first ones are in fact linked by new reciprocal correlations, suggesting that, in this environment, negative emotions are even more integrated and coherently felt and identified.

This is also true for positive emotions since in this environment each positive emotion is correlated with all the other ones of the same sign. This pattern of result may be seen as the expression of a clearer differentiation of what is positive from what is negative. This hypothesis is also supported by the fact that just one counter-correlation is now found, namely between "boredom" and "interest" (r=-.360; p<.004). This result suggests a sharper sensitivity in discriminating positive emotional states from negative ones. Face to face lessons are thus experienced with a more evident understanding of what is able to induce positive or negative feelings. Emotions are better identified in their essence, and positive feelings are not the contrary of something negative, and vice versa. Positive and negative emotions have their own origin, their motivation, their meaning.

It should also be noted that "surprise", although still having an intermediate position between positive and negative emotions, in this case appears to be even more related to negative emotions. This may be due to the fact that, "surprise" is something that in the on-line learning environment is related to events that the student is more eager to explore within the system.

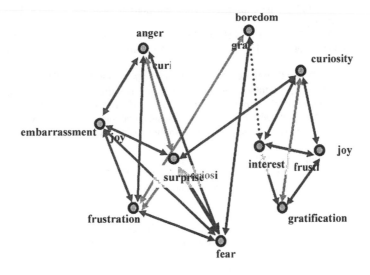

FIGURE. 3 Arrows shows correlations between emotions during face to face lessons. Dotted lines are negative correlations and light gray arrows are correlations that were not significant in the on-line learning environment.

ACKNOWLEDGEMENTS

The Authors would like to deeply thank Linda Mesh who, during the months in which this research lasted, provided us with useful and helpful assistance.

REFERENCES

Ardito, C., Costabile, M. F., De Marsico, M., Lanzilotti, R., Levialdi, S., Rosselli, T and Rossano, V. (2005), An approach to usability evaluation of e-learning applications. Springer-Verlag.

Bar-On, R, Tranel, D., Denburg, N. L. and Bechara, A. (2003). Exploring the neurological substrate of emotional and social intelligence. *Brain*, 126, (8), 1790-1800.

Bereiter, C. and Scardamalia, M. (2003), *Learning to work creatively with knowledge*, in: E. De Corte, L.Verscheffel, N. Entwistle and J.V. Merrienboer (eds.), *Powerful Learning Environments: Unravelling Basic Components and Dimension.* Oxford: Elsevier Science, 73-78.

Brown, A.L., and Campione, J.C. (1990), *Communities of learning or a content by any other name*, in D. Kuhn (ed.), *Contribution to Human Development.* Cambridge: Cambridge University Press, 108-126.

Damasio, A. R. (1994), Descartes' Error: Emotion, reason and the human brain. New York: Grosset/Putnam.

Damasio, A. R. (1999), The Feeling of What Happens: body and Emotion in the making of consciousness, New York: Harvest Edition.

Damasio, A. (2003), *Looking for Spinoza: Joy, Sorrow, and the Feeling Brain.* Orlando,

FL: Harcourt, Inc.

Darwin, C. R. (1872), The expression of the emotions in man and animals. London, John Murray.

Dirkx, J. (2001), The power of feelings: Emotion, imagination and the construction of meaning in adult learning. *New directions for adult and continuing education*, 89, 63-72.

Ekman, P. (1992), Are their basic emotions? *Psychological Review*, 99, 550-553.

Erez, A. and Isen, A.M. (2002), The influence of positive affect on the components of expectancy motivation. *Journal of Applied Psychology, 87*(6), 1055-1067.

Isen, A. M. (2000), "Positive affect and decision making", In: M. Lewis and J. Haviland-Jones (Eds.), Handbook of Emotions (2nd ed., 417-435). New York: Guilford.

Hara, N., & Kling, R. (2000). Students' distress with a web-based distance education course. Information, Communication & Society, 3(4), 557-579.

Lazarus, R. S. (2001), Relational meaning and discrete emotions. In: K. R. Sherer, A. Shorr and T: Johnstone (Eds.), Appraisal processes in emotion: theory, methods, research (pp. 37-67). Oxford, UK

Martin, B. and Briggs, L. (1986), *The Affective and Cognitive Domains: Integration for Instruction and Research*, Educational Technology Publications, Englewood Cliffs.

O'Reagan, K. (2003), Emotion and E-Learning. Journal of Asynchronous Learning Networks (JALN), 7, (3).

Nielsen, J. (1994), Heuristic evaluation. In Nielsen, J., and Mack, R.L. (Eds.), Usability Inspection Methods. John Wiley & Sons, New York, NY.

Parkinson, B. (2001), Putting appraisal in context. In: K.R. Scherer et al. (Eds.), *Appraisal processes in emotion: Theory, methods, research.* Oxford University Press, Oxford, UK.

Pessoa, L., (2008), On the relationship between emotion and cognition. *Nature Review Neuroscience.* 9, 148-158.

Ravden, S. and Johnson, G. (1989) Evaluating the usability of human-computer interfaces. Ellis Horwood, Chichester.

Reay, J. (2001), Blended learning - a fusion for the future. *Knowledge Management Review, 4* (3), 6.

Rooney, J. E. (2003), Blending learning opportunities to enhance educational programming and meetings. *Association Managment, 55* (5), 26-32.

Rovai, A. P., and Jordan, H. M. (2004), Blended learning and sense of community: A comparative analysis with traditional and fully online graduate courses. *The International Review of Research in Open and Distance Learning, 5*(2).

Sands, P. (2002), Inside outside, upside downside: Strategies for connecting online and face-to-face instruction in hybrid courses. *Teaching with Technology Today, 8* (6).

Ward, J., and LaBranche, G. A. (2003), Blended learning: The convergence of e-learning and meetings. *Franchising World, 35* (4), 22-23.

Wenger, E. (1998), *Communities of Practice. Learning, Meaning and Identity.* Cambridge: Cambridge University Press.

Young, J. R. (2002), 'Hybrid' teaching seeks to end the divide between traditional and online instruction. *Chronicle of Higher Education*, A33.

Zaharias, P. (2009), Developing a Usability Evaluation Method for E-learning Applications: From Functional Usability to Motivation to Learn. *International Journal of Human Computer Interaction*, 25 (1), 75-98.

Chapter 73

A Practitioner's Guide to Personalized Instruction

Benjamin Goldberg*, Sae Schatz, Denise Nicholson

Institute for Simulation and Training, ACTIVE Lab
University of Central Florida
Orlando, FL 32816, USA
* Mr. Goldberg now works with RDECOM-STTC, Orlando

ABSTRACT

Adaptive instructional technologies may represent the most effective and efficient computer-based learning tools. However, to garner the benefits of adaptive instruction, such technologies must account for what students know and how their individual characteristics affect the way they learn. This paper (1) reviews the macro-adaptive learning variables that appear to most significantly moderate learning outcomes, (2) discusses practical considerations regarding these variables, and (3) suggests available self-response tools for data collection.

Keywords: Adaptive Instruction, Intelligent Tutoring, Individual Differences

MACRO-ADAPTIVE VARIABLES

Computer-based adaptive instructional systems "attempt to be different for different students and groups of students by taking into account information accumulated in the individual or group student models" (Brusilovsky & Peylo, 2003). Such technologies try to replicate the success of one-on-one (human) tutoring (Bloom, 1984) by tailoring instruction to each student's current knowledge and individual characteristics.

Macro-adaptive tools, specifically, adapt the systems' base states, before learning begins, in response to learners' traits and general attributes. This is in contrast to *micro*-adaptive systems, which adapt the instruction dynamically throughout the episode, in response to learners' performance and behaviors.

Both macro- and micro-adaptation are considered important functions; however, this paper focuses exclusively on variables for macro-adaption. More specifically, this paper presents a brief review of empirical literature aimed at answering the question: *Which macro-adaptive student characteristics can be readily assessed while also providing significant predictive power on key learning outcomes, for adult-oriented instruction, within a computer-based environment?*

DOMAIN-SPECIFIC APTITUDES

Domain-specific aptitudes describe the declarative, procedural, and conditional knowledge one possesses within a field of study (Alexander & Judy, 1988), i.e., prior knowledge, skills, and abilities. Dochy et al.'s (1999) review of 183 classroom-based studies suggests that prior knowledge predicts around 30-60% of posttest variance. These results hold true for adaptive instructional systems as well. For example, Shute (1995) reports that adults' prior knowledge of statistics correlated about 50% with outcome performance on the subject within an ablative intelligent tutor. Along with prior knowledge, prior skill represents an important predictive variable. Domain-specific skills coincide with the *application* of domain knowledge. Individuals' skills can be determined by measuring their real-world achievements. For instance, in a military context, rank may serve as a rough adjunct for a person's skill level (Rabinowitz & McAuley, 1990).

Due to the significant effect these aptitudes have on task performance, practitioners should obtain initial knowledge and skill measures. For measures of knowledge, Dochy et al. (1999) recommend multiple choice tests, open questions, completion tests, recognition tests, and matching tests, because they can all validly measure prior knowledge. For measures of skill, objective practical applications (e.g., scenarios or situational judgment tests) may be used, or evidence of prior skill (e.g., deployment history, awards) may be considered.

GENERAL APTITUDES

General aptitudes include key variables that characterize an individual's mental characteristics, including those *knowledge, skills,* and *abilities* (KSAs) not directly related to a specific instructional domain (Kyllonen & Gitomer, 2002).

General Cognitive Abilities

General cognitive abilities include attributes such as working memory capacity, inductive reasoning skills, information processing speed, and associative learning ability (e.g., Shute, 1995). Often, g is used as a catchall for such cognitive abilities. Meta-analyses within education and training literatures, in general, suggest g explains between 12% and 26% of the variance in performance (Dochy et al., 1999). Many researchers suggest other measures may substitute for g. For instance, Schmidt (1988) argues that ACT or SAT test scores may be used as valid adjuncts of *g*, and that they correlate highly with future test performance (r = .94; 88%) (Gully et al., 1999). Schmidt further suggests that academic history and GPA provide valuable information related to general cognitive aptitudes associated with educational performance (Schmidt, 1988).

However, if prior knowledge is controlled for, then the correlation between intelligence and performance approaches insignificance (Walker, 1987). Studies in adaptive instruction report similar findings. For instance, in Baker et al.'s (2004) middle-school scatter plot tutor, the general level of academic achievement (in lieu of *g*) predicted about 13% of posttest performance; however, academic achievement substantially correlated with prior knowledge (37%), and when both academic achievement and prior knowledge were used as predictor variables, only prior knowledge meaningfully contributed to the regression model (Baker et al., 2004). In light of the seemingly greater predictive power of prior knowledge (as well as the relative ease of measuring it), the authors recommend using prior knowledge instead of collecting measures *g* or even its adjuncts, such as ACT, SAT, or GPA scores.

Metacognition

Metacognition, or "thinking about thinking," relates to an individual's awareness of his/her learning processes (Schraw & Dennison, 1994). High metacognitive ability positively correlates with learning performance. For example, Stimson (1998) reports that, in the traditional classroom setting, measures of metacognitive ability accounted for 31% of the variance in performance. Similarly, with an intelligent tutor, Kim et al. (2006) found that metacognition inversely affected (r = -0.382, 14%) the amount of time fifth graders spent seeking instructional resources (without negatively affecting their performance); that is, students sought only those resources they required. Kim and colleagues report that "Metacognition was found to be the most significant individual characteristic to influence participants' responses..." (p. 1239). In light of metacognition impact on learning outcomes, the authors recommend gathering initial estimates of students' metacognitive abilities. The Metacognition Awareness Inventory (MAI) (Schraw & Dennison, 1994) provides the means for identifying student's planning, reflecting, and reasoning metacognitive skills. The 52-item MAI is freely available for use, and it shows excellent internal consistency, between .88 and .93 (Schraw & Dennison, 1994).

Computer Competency

While computer competency would appear to affect learning via instructional technology, the authors were unable to find empirical results that supported this belief. More often, individuals' *attitude* towards computers influences their performance. This finding is explained in more detail in the attitudes section, below.

CONSTITUTIONAL ATTRIBUTES

Constitutional attributes include overt personal traits, such as gender or age, as well as functional classifications, such as occupational role, class standing, or military rank (Woolf, 2009). They are the simplest data to obtain and can help predict outcome performance due to their correlations with prior knowledge/experiences.

For instance, the age of students commonly correlates with memory capacity and cognitive processing. In a meta-analysis of 75 studies, Verhaeghen and Salthouse (1997) found an average correlation of $r = -0.27$ (7%) between (healthy) adults' age and their working memory capacity. Similarly, Ackerman (2000) administered a battery of tests to college graduates and found that older adults generally performed poorer on cognitive processing tasks ($r = -0.39$; 15%); while they performed slightly better on tests of accumulated knowledge ($r = 0.14$; 2%). While this suggests that age can be used as an indicator for some cognitive functions (see also Dede, 2004), there is a lack of studies on any adaptive instructional systems that empirically test the influence of adult age differences on outcome variables. Thus, the authors cannot readily advise using age as a variable for adult-level instruction.

Gender differences have been linked with instructional variables in the general education literature (for a review, see Beal et al., 2002). However, most adaptive training systems fail to find significant effects of gender on performance. Studies of middle-school students studying mathematics (Baker et al., 2004), college students studying databases (Naser, 2006), or adults studying statistics (Shute, 1995) all failed to find gender–performance effects. However, some children-focused studies have reported effects for gender and tertiary variables, such as boys' and girls' confidence when using an intelligent tutor (Beck et al., 1999) or how boys and girls work with different types of hints (Arroyo et al., 2000). Nonetheless, for adult-oriented instruction gender appears to hold little significance.

Classifications such as occupational role or military rank can predict learning outcomes. Identifying such functional classifiers may be an efficient strategy for gaining insight on students' prior knowledge and skills. However, the relevancy of these classifications varies across domains. For military settings, for instance, Wisher et al. (2001) report that inclusion of military rank added predictive power to the ITS's internal models. However, for other contexts the link may not be so clear-cut. Thus, the authors recommend that adaptive instruction programs collect relevant trainee classifications, as applicable for the specific domain of interest.

AFFECTIVE APTITUDES

Affective attributes relate to both static (i.e., trait) and dynamic (i.e., state) emotions; however, the macro-adaptive approach focuses primarily on the attributes related to individuals' static or trait attributes. Thus

Attitude Towards Computers

Students' attitudes towards computers affect their likelihood to "game" the system and may influence their motivation to use the instructional technology, overall. Roll et al. (2005) describe "gaming the system" as attempting to obtain guidance and correct answers without properly working through the instructional materials. Baker et al. (2004b) found that when grade school students' attitudes towards computers were positive, they were about 6% less likely to game an intelligent tutor and showed an increase in performance of about 10%. It is unclear whether these correlations hold true in adult education; however, given the ease with which these attitudes are collected, it may make sense to determine students' attitudes towards computer before delivering the instruction. To gather these data, the authors recommend using the Game-Based Motivation Questionnaire (Zweig & Webster, 2004), an adapted 7-item survey questionnaire, with an internal consistency of .93.

Motivation

The link between motivation and learning outcomes is well recognized. Intrinsic motivation stems from self-generated rewards, such as the love of learning, and it correlates with positive learning outcomes. Extrinsic motivation, on the other hand, stems from the prospect of outside rewards or punishments (e.g., Deci, 1972). Wolf et al. (1995) conducted an empirical study, in which one group of students took a test that affected their grades while another group took the same test but it did not affect their grades. Wolf and colleagues found correlations between an individual's motivation score and task performance ($r = .351$; 12%), and, not surprisingly, they discovered that many students are motivated by the consequence, i.e., the extrinsic threat of punishment.

In terms of pre-task measures, self-response questionnaires provide a mechanism through which to assess a student's trait motivation. However, such questionnaires have been criticized as static, since most researchers agree that state motivation represents a more critical variable (Vicente & Pain, 1998). Furthermore, the context of adult education varies across many domains—from requisite compliance training to life-long learning courses. Similarly, adults' motivation is likely to vary across setting. For these reasons, the authors shy away from measuring trait motivation, directly, and instead suggest that practitioners consider measuring adults goal orientation instead.

Goal Orientation

Goal orientation describes the way students approach their educational goals. A student's goal orientation can fluctuate between *performance* (e.g. focused on earning a high grade) and *mastery* (e.g. focused on overall learning) (Pintrich, 2000). Mastery goal orientation corresponds with a student's desire to acquire new skills and mastering new situations; in contrast, performance goal orientation is characterized by a student's desire to prove his/her competence and gain favorable judgments. Those exhibiting performance goal orientation aim to avoid the disproving of their abilities and to evade negative judgments (Vandewalle, 1997).

Goal orientation has been included in several adaptive systems (e.g. Baker et al, 2004; Jaques & Vicari, 2005) and is theorized to interact with multiple variables, including self-efficacy, cognitive ability, performance, and knowledge. For instance, Zweig and Webster (2004) found that goal-orientation predicts a considerable level of variance for final grades (11%). Similarly, Bell and Kozlowski (2002) researched goal orientation interactions with TANDEM, a simulation for Naval radar tasks. Their regression analyses revealed that for every unit increase in the mastery orientation scale, there was a 24% increase in self-efficacy, a 14% increase in performance, and a 19% increase in prior knowledge. Given the significant impact of goal orientation, as well as the readily available interventions, the authors recommend collecting this variable. The 21-item Goal Orientation Questionnaire (Zweig & Webster, 2004) can serve this purpose; it is freely available and exhibits high levels of internal reliability (above .80) (Zweig & Webster, 2004).

Self-Efficacy

General self-efficacy is a well-documented correlate of motivation and educational effectiveness ($r \approx .32$; 10%) (e.g. Zimmerman, 2000). Judge et al. (1998) define general self-efficacy as "individuals' perception of their ability to perform across a variety of different situations." General self-efficacy is a strong determining factor of students' effort, persistence, and strategy when performing routine tasks (Jerusalem & Schwarzer, 1993). Research suggests that individuals are more likely to engage in learning activities if they perceive themselves to have confidence for performing the tasks (Vockell, 2008), and results demonstrate that the positive correlation between self-efficacy and instructional performance holds true with adaptive technology. Kim et al. (2006) carried out a post-hoc analysis on grade school students' interactions with an intelligent tutor and found self-efficacy to be highly correlated with how much time students spent using optional resources ($r = 0.34$; 11%) and their outcome performance ($r = 0.345$; 11%). Due to the predictive power of self-efficacy in relation to task performance and other affective attributes, the authors recommended gathering these data. Domain-specific self-efficacy measures are the most appropriate tools (Scherbaum et al., 2006); however, a possible generic instrument is the General Self-Efficacy survey (Jerusalem & Schwarzer, 1993). It contains 10 items designed for the general population, and returns Cronbach's alphas ranging from .76 to .90 (Jersusalem & Schwarzer, 1993).

LEARNING STYLES

Learning styles are theorized differences in how individuals study and retain information. During the 1980's and 90's many classification systems for learning styles were created, and a few research efforts have empirically examined their effect within adaptive systems. For instance, Bajraktarevic et al. (2003) report an 18% increase in task performance between matched and mismatched learning style conditions. Kelly & Tangney (2002; 2003) and Kelly (2008) applied Multiple Intelligences within the intelligent tutor EDUCE, and found a mean posttest gain of 70.7% in the adaptive condition. While this suggests that adaptation to learning styles is profoundly significantly, it must be noted that Kelly (2008) compared the learning styles condition to a learner-controlled condition, and learner control systems notoriously lead to diminished performance. Further, despite a few positive reports, many experts dispute the validity of learning styles, and in the general education literature most empirical evidence suggests that learning styles fail to significantly influence learning (e.g., Coffield et al., 2004). Thus, due to the lack of significant evidence that learning styles have a direct impact on task performance, the authors recommend that practitioners pass over collection of these variables, at least until the literature offers clearer evidence on the impact of learning styles.

LEARNER PREFERENCES

Some adaptive systems allow students to make self-guided decisions about their instructional experiences, such as self-pacing or self-selection of a preferred learning style. Considerable research has demonstrated that this approach most often inhibits instruction because students typically select non-optimal learning approaches (Ghatala, 1986; Pressley & Ghatala, 1990). For example, through a within-subjects design, Kelly (2008) demonstrated that when students receive their *least* preferred pedagogical strategy they showed a greater posttest performance. Mean posttest gains were 77.2% compared to 55.5%, respectively.

As well, learner control permits students to misuse their freedom, and they may attempt to "game" or manipulate the system. Aleven et al. (2004) among others (Baker et al., 2004; Aleven et al., 2003) have demonstrated students' propensity to abuse their control of the training technology. Aleven and his colleagues reported that 72% of their participants used their intelligent tutor in unintended ways, the most prevalent being abuse of the help feature (e.g. flipping through all of the hints). They further showed that frequency of bad behavior was correlated strongly ($r = -0.61$; 37%) with students' overall learning (or lack thereof). While this behavior is most prevalent in children; nonetheless demonstrates little benefit for adults. Consequently, because personal learning preferences have not been found have a significant positive correlation with instructional performance, it is recommended to avoid use of these data for adaptive mechanism selection.

CONCLUSION

Table 1: Summary of macro-adaptive variable recommendations

Attribute	Effect	Predicts	Recommend?
Prior Knowledge/Skill	Large	Performance	Yes – Highly
General Intelligence	Unclear	---	No, *Use Prior Knowledge*
Metacognition	Large	Performance; Help Use	Yes – Highly
Computer Competency	Unclear	---	No, *Use Comp. Attitudes*
Age	Moderate	General Intelligence	No, *Use Prior Knowledge*
Sex	Small	Kids' Tertiary Attributes	No (for Adults)
Functional Classifiers	Moderate	Prior Knowledge Adjunct	Maybe (by Domain)

Computer Attitudes	Moderate	Performance; Gaming	Yes
Motivation	Moderate	Performance	No, *Use Goal Orientation*
Goal Orientation	Large	Performance; Efficacy	Yes – Highly
General Self-Efficacy	Moderate	Performance; Help Use	Yes
Learning Styles	Unclear	---	No
Learning Preferences	Negative	Inhibits Performance	No

Instructional technologies provide a means for applying personalized learning to individuals based on their unique needs and abilities. For effective adaptive instruction, a system must gather knowledge on relevant student traits. These individual differences can be used to model students' aptitude, attitudes, and abilities and then inform delivery of macro-adaptive instructional interventions. This chapter focused on those variables that offer the greatest predictive power related to learning performance, are easily measured, and can inform macro-adaptive interventions. Although this chapter provided only a short synopsis, considerable empirical research underlies the recommendations provided herein. A summary of the authors' suggestions is provided in Table 1.

REFERENCES

Ackerman, P.L. (2000). Domain-Specific Knowledge as the "Dark Matter" of Adult Intelligence, *Journal of Geronotlogy: Psychological Sciences, 55B*(2), 69-84.

Aleven, V., Stahl, E., Schworm, S., Fischer, F. & Wallace, R. (2003). Help Seeking and Help Design in Interactive Learning Environments, *Review of Educational Research, 73*(3), 277-320.

Aleven, V., McLaren, B., Roll, I. & Koedinger, K., 2004, Toward Tutoring Help Seeking: Applying Cognitive Modeling to Meta-Cognitive Skills, *Proceedings of the 7th International Conference on ITS* (pp. 227-239). Alagoas, Brazil.

Alexander, P.A. & Judy, J.E. (1988). The Interaction of Domain-Specific and Strategic Knowledge in Academic Performance. *Review of Educational Research, 58*(4), 375-404.

Arroyo, I., Beck, J.E., Woolf, B.P., Beal, C.R., & Schultz, K.. (2000). Macroadapting Animalwatch to Gender and Cognitive Differences with Respect to Hint Interactivity and Symbolism, *Proceedings of the 5th International Conference on ITS* (pp. 574-583), Montreal Canada.

Bajraktarevic, N., Hall, W. & Fullick, P. (2003), Incorporating Learning Styles in Hypermedia Environment: Empirical Evaluation. *Proceedings of AH2003: Workshop on Adaptive Hypermedia and Adaptive Web-Based Systems.*

Baker, R.S., Corbett, A.T. & Koedinger, K.R. (2004a). Detecting Student Misuse of Intelligent Tutoring Systems, *Proceedings of the 7th International Conference on Intelligent Tutoring Systems* (pp. 531-540), Alagoas, Brazil.

Baker, R.S., Corbett, A.T., Koedinger, K.R. & Wagner, A.Z. (2004b). Off-Task Behavior in the Cognitive Tutor Classroom. *Proceedings of ACM CHI 2004: Computer-Human Interaction* (pp. 383-390), Vienna, Austria.

Beal, C.R., Beck, J., Westbrook, D., Atkin, M., & Cohen, P. (2002). Intelligent modeling of the user in interactive entertainment (pp. 8-12). *Proceedings of AAAI* (SS-02-01).

Beck, J.E., Arroyo, I., Woolf, B.P. & Beal, C.R. (1999). An Ablative Evaluation. *In Proceedings of the Ninth International Conference on Artificial Intelligence in Education* (pp. 611-613), Le Mans, France,.

Bell, B.S. & Kozlowski, S.W.J. (2002). Goal Orientation and Ability. *Journal of Applied Psychology, 87*, 497-505.

Bloom, B.S. (1984). The 2 Sigma Problem. *Educational Researcher*, 13, 6, 4-16.

Brusilovsky, P. & Peylo, C. (2003) Adaptive and intelligent Web-based educational systems. *International J. of Artificial Int. in Ed., 13*(2-4), 159-172.

Coffield, F.J., Moseley, D.V., Hall, E. & Ecclestone, K. (2004). *Learning Styles for Post 16 Learners*. Learning and Skills Research Centre, U. of Newcastle.

de Vicente, A. & Pain, H. (1998). Motivation Diagnosis in Intelligent Tutoring Systems, In B.P. Goettl, H.M. Halff, C.L. Redfield, and V.J. Shute, (Eds.), *Proceedings of the 4th International Conference on Intelligent Tutoring Systems* (pp. 86-95). *Berlin*, Germany, Springer.

Deci, E. (1972). Intrinsic Motivation, Extrinsic Reinforcement, and Inequity, *Journal of Personality and Social Psychology, 22*(1): 113–120.

Dede, C. (2004). If Design-Based Research is the Answer, What is the Question?, *Journal of the Learning Sciences, 13*, 1, 105-114.

Dochy, F., Segers, M., & Buehl, M.M. (1999). The Relationship between Assessment Practices and Outcomes of Studies: The Case of Research on Prior Knowledge, *Review of Educational Research, 69*, 2, 145-186.

Ghatala, E.S. (1986). Strategy, Monitoring Training Enables Young Learners to Select Effective Strategies, *Educational Psychologist, 21*(1&2), 43-54.

Gully, S.M., Payne, S.C., Kiechel, K.L. & Whiteman, J.K. (1999). Affective Reactions and Performance Outcomes of Error-based Training. Paper presented at the meeting of the Society for Industrial and organizational Psychology, Atlanta, GA.

Hothi, J. & Hall, W. (1998). An Evaluation of Adapted Hypermedia Techniques using Static User Modeling, *Proceedings of the Second Workshop on Adaptive Hypertext and Hypermedia* (pp. 45-50), Pittsburgh, PA.

Jaques, P.A. & Vicari, R.M. (2005). A BDI Approach to Infer Student's Emotions in an Intelligent Learning Environment, *Computers & Education, 49*, 360-384.

Jerusalem, M. & Schwarzer, R. (1993). The general self-efficacy scale (GSE). Retrieved from http://userpage.fu-berlin.de/ health/engscal.htm

Judge, T.A., Erez, A. & Bono, J.E. (1998). The Power of Being Positive. *Human Performance, 11*(2/3), 167-187.

Kelly, D. (2008). Adaptive Versus Learner Control in a Multiple Intelligence Learning Environment, *Journal of Educational Multimedia and Hypermedia, 17*(3), 307-336.

Kelly, D. & Tangney, B. (2002). Incorporating Learning Characteristics into an Intelligent Tutor, *Proceedings of the 6th International Conference on Intelligent Tutoring Systems* (pp. 729-738), Honolulu, HI.

Kelly, D. & Tangney, B. (2003). A Framework for Using Multiple Intelligences in an Intelligent Tutoring System, *Proceedings of Educational Multimedia, Hypermedia and Telecommunications* (pp. 2423-2430), Honolulu, HI.

Kim, S., Lee, M., Lee, W., So, Y., Han, C., Lim, K., Hwang, S., Yun, S., Choi, D. & Yoon, M. (2006). Student Modeling for Adaptive Teachable Agent to Enhance Interest and Comprehension (pp. 1234–1241). In E. Corchado et al. (Eds.), *IDEAL 2006, LNCS* 4224.

Kyllonen, P.C. & Gitomer, D.H. (2002). Individual Differences, Encyclopedia of Education, The Gale Group Inc, Retrieved August 20, 2009 from http://www.encyclopedia.com/doc/1G2-3403200312.html.

Lane, H.C. (2006). Intelligent Tutoring System. Whitepaper for the Science of Learning Workshop, Hampton, VA.

Martin, K. & Arroyo, I. (2004). AgentX. *Proceedings of 7th International Conference of Intelligent Tutoring Systems* (pp. 564-572.), Alagoas, Brazil.

Naser, S.S.A. (2006). Intelligent Tutoring System for Teaching Database to Sophomore Students in Gaza and its Effect on their Performance. *Information Technology Journal, 5*(5), 916-.

Pintrich, P.R. (2000). The Role of Goal-Orientation in Self-Regulated Learning, In M. Boekaerts, P.R. Pintrich, & M. Zeidner (Eds.) *Handbook of Self-Regulation.* Academic Press, San Diego, CA.

Pressley, M. & Ghatala, E.S. (1990). Self-Regulated Learning: Monitoring Learning from Text. *Educational Psychologist, 25*(1), 19-33.

Rabinowitz, M. & McAuley, R. (1990). Conceptual Knowledge Processing. In W Schneider & F.E. Weinert (Eds.) *Interactions Among Aptitudes, Strategies, & Knowledge in Cognitive Performance* (pp. 117-133), Springer-Verlag: NY.

Roll, I., Baker, R.S., Aleven, V., McLaren, B.M. & Koedinger, K.R. (2005). Modeling Students' Metacognitive Errors in Two Intelligent Tutoring Systems, *Proceedings of the 10th International Conference on User Modeling* (pp. 367-376), Edinburgh, Scotland.

Scherbaum, C.A., Cohen-Charash, Y. & Kern, M.J. (2006). Measuring general self-efficacy: A comparison of three measures using item response theory. *Educational and Psychological Measurement, 66*(6), 1047-1063.

Schmidt, F.L. (1988). The Problem of Group Differences Ability Test Scores in Employment Selection. *Journal of Vocational Behavior, 33*, 272-292.

Schraw, G. & Dennison, R.S. (1994). Assessing Metacognitive Awareness. *Contemporary Educational Psychology, 19*, 460-475.

Shute, V.J. (1995). SMART: Student Modeling Approach for Responsive Tutoring. *User Modeling and User-Adapted Interaction, 5*, 1-44.

Shute, V.J. & Towle, B. (2003). Adaptive E-Learning, *Educational Psychologist, 38*(2), 105-114.

Stimson, M.J. (1998). *Learning from Hypertext Depends on Metacognition* (Doctoral Dissertation, University of Georgia), Dissertation Abstracts International: Section B: The Sciences and Engineering, 60, 2-B, 851.

Vandewalle, D. (1997). Development and Validation of a Work Domain Goal Orientation Instrument, *Educational & Psych Measurement, 57*(6), 995-1015.

Verhaeghen, P. & Salthouse, T.A. (1997). Meta-Analyses of Age-Cognition Relations in Adulthood. *Psychological Bulletin, 122*, 231-249.

Vockell, E., 2008, *Education Psychology: A Practical Approach.* Retrieved from http://education.calumet.purdue.edu/vockell/EdPsyBook/.

Walker, C.H. (1987). Relative Importance of Domain Knowledge and Overall Aptitude on Acquisition of Domain Related Information, *Cognition and Instruction, 4*, 25-42.

Wisher, R.A., MacPherson, D.H., Abramson, L.J., Thorndon, D.M. & Dees, J.J., (2001). *The Virtual Sand Table* (ARI Research Report 1768), Alexandria, VA: U.S. Army Research Institute for the Behavioral and Social Sciences.

Wolf, L.F., Smith, J.K. & Birnbaum, J.E (1995). The Consequence of Consequence, *Applied Measurement in Education, 8*(3), 227-242.

Woolf, V. (2009). *Building Intelligent Interactive Tutors.* Elsevier, Burlington, MA.

Zimmerman, B.J. (2000). Self-Efficacy: An Essential Motive to Learn, *Contemporary Educational Psychology, 25*, 82-91.

Zweig, D. & Webster, J. (2004). Validation of a Multidimensional Measure of Goal Orientation, *Canadian Journal of Behavioural Science, 36*(3), 232-243.

CHAPTER 74

What is a Scenario? Operationalizing Training Scenarios for Automatic Generation

Glenn A. Martin, Sae Schatz, Charles E. Hughes, Denise Nicholson

Institute for Simulation & Training
University of Central Florida
Orlando, FL 32826

ABSTRACT

Scenario-based training (SBT) is an instructional approach ideally suited to take advantage of the characteristics of simulation. However, it is often difficult for instructors to create scenarios for computer-based simulation; this problem has prompted researchers to begin exploring automated scenario-generation techniques. Yet, for software to automatically generate an efficacious training scenario, it must "understand" the components, and the relationships among components, that comprise effective scenarios. This chapter discusses an approach to operationalizing scenario pieces so that automatic generation software can assemble complete training scenarios from these component parts.

Keywords: Scenario-Based Training, Scenario Generation, Training Objectives, Domain Ontology

INTRODUCTION

Developing scenarios for simulation-based training can be a time-consuming and expensive process. Unfortunately, such inefficiency often results in only a small number of scenarios being developed and constantly reused. Not only does this reduce variety in training, but the loss of variety and limited scope of scenarios also diminishes training effectiveness (Martin et al., 2009). To help address these issues, the authors are investigating processes and tools for improved manual scenario generation, as well as computational approaches for automated scenario generation.

Yet, before any method for scenario generation can be implemented, the scientific community must truly understand what a training scenario is. This understanding includes not only parameters of the scenario, but also a conceptual model of scenarios, an operational definition of scenario complexity, and a framework for linking scenarios to learning objects. Objective, operationalized definitions of these facets will enable software to generate appropriate scenarios for specific trainees, based upon input such as their past performances and their individual characteristics and goals.

This chapter describes the foundational work of developing this conceptual definition of training scenarios. The conceptual approach is based upon the notion of selected "training objectives," coupled with the use of "baseline scenarios" and modifications that offer increased complexity. The training objectives are enumerated based on the trainee audience and provided for user selection. Baselines represent simplified scenarios that depict ideal conditions (no weather, perfect lighting, no surprises, etc.). Additional modifications allow the baselines to be altered and the overall scenario complexity enhanced. Given a trainee profile, the system can assemble combinations of baseline and modifications that support specific training objectives, reach appropriate levels of complexity, and provide adaptive training opportunity for the trainee(s) to further their understanding and performance.

In addition to basic scenario generation based on trainee profiles, the science of training suggests that training scenarios should support varied pedagogical approaches. For example, a given trainee may be presented with a "compare and contrast" scenario where a particular event is specified in two different ways. Or a "disequilibrium scenario" may show a possible worst-case scenario to enhance the understanding of why such training is necessary. The authors will discuss how the scenario generation framework supports such capabilities, and example instructional strategy use cases will be presented.

Finally, the authors will briefly touch on some of the computational approaches that could be used to support automated scenario generation based on the presented conceptual model. A recently re-discovered approach and a newly-developed approach show great promise in their possible use in automated scenario generation. Both of these computational approaches fit particularly well with the proposed conceptual model, and their benefits and limitations will be outlined.

AUTOMATIC SCENARIO GENERATION

For the purposes of this research, *scenario generation* describes the design and development of training episodes for simulation-based instruction. Traditionally, this process is carried out by one or more subject-matter experts, who manually plan out the scenario and then carefully program their plan into a simulation system.

Automatic scenario generation occurs when a computer executes (or assists with the execution of) the scenario design and development process (for a more complete review, see Martin et al., 2009). Yet, what is a scenario? The answer to this question includes not only parameters of a scenario, but also a conceptual model of scenarios, an operational definition of scenario complexity, and a framework for linking scenarios to learning objects. Objective, operationalized definitions of these facets will enable software to generate appropriate scenarios for specific trainees, based upon input such as their past performance and individual characteristics.

TRAINING SCENARIOS

First, a common understanding of the term "scenario" is required. When discussing scenarios, a distinction must first be made between the scenario, itself, and the simulation in which it is embedded. Roughly, simulation supports simulation-based training, which is the use of a virtual environment to support practice. In contrast, scenarios support scenario-based training, which is often used in conjunction with a virtual environment. Thus, scenarios can be thought of as the purposeful instantiation of simulator events to create desired psychological states (Martin et al., 2010).

Next, the difference between a "scenario" and a "situation" must be clarified. Situations refer to instant snapshots, which occur at any given time within an exercise; whereas scenarios can be thought of a series of situations over time (Tomizawa & Gonzalez, 2007). Consequently, a training scenario is a series of simulator events that create specific situations, which facilitate situated learning. Finally, in addition to simply describing the environmental context, training scenarios should include pedagogical accompaniments, such as training objectives and performance measures, in order to facilitate optimal transfer-of-training.

INPUT-PROCESS-OUTPUT

Automated scenario generation can be conceptualized via an Input–Process–Output model (e.g., Hofer & Smith, 1998). Inputs may include specific training objectives and information about the trainees, i.e., information used to "seed" the generation process. Once inputs are received, the software processes them, and then assembles a scenario, constrained by certain predefined heuristics. The software then outputs a composite scenario definition file.

Specifically, the inputs include a preselected training objective, an optional recommended pedagogical approach, and information about the trainees, including the number of trainees, the functional roles they will play in the simulation, and their levels of expertise. Once inputs are provided, the generation system constructs a unique, valid scenario that emphasizes the given training objective and is tailored to the specific trainees' instructional needs. The output composite scenario definition is automatically assembled from pre-existing scenario baselines, "vignettes" that represent an element of a scenario, and pedagogical templates. Specific instances of each are selected based on their goodness-of-fit relative to the training inputs. The scenario is then output and used within the simulation systems for appropriate initialization of the exercise. Conceptually, XML is a good technology to represent the scenario components as it maximizes flexibility and is extensible to different simulation platforms.

SCENARIO BUILDING BLOCKS

In this section, the specific building blocks of a scenario (briefly discussed above) are articulated in more detail. All together these blocks are referred to as *facets* of the scenario; however, each has a specific role to play in the formulation of a scenario.

TRAINING OBJECTIVES

The military formally defines *training objectives* in their Training & Readiness (T&R) manuals. Each objective in the T&R manual is accompanied by a list of "conditions," which describe the context under which the action can be performed. In the automated scenario generation system, these conditions become requirements for elements that must be present in the scenario. For example, to train an artillery gunner to fire upon an enemy convoy, the virtual environment must at least include available supporting arms, munitions, and an enemy convoy to target. Thus, the selection of a particular training objective causes a set of conditions to become "active" (i.e. valid for use in this scenario).

Note that training objectives can be created for any particular domain. While our initial focus is in military exercises, this approach could be used to develop scenarios for other domains such as cognitive rehabilitation or education. However, for any domain, it is important to consider what the training objectives for that domain would be and how it can be specified at a level appropriate for scenario generation.

For example, the training objectives in the military's T&R manuals typically have broad definitions. For example, the *Marine Corps Infantry T&R Manual* describes training objective "0302-FSPT-1302: Employ Supporting Arms" as:

> Given a radio, call signs, frequencies, available supporting arms, equipment, a scheme of maneuver and a commander's intent...achieve desired effect(s) on target that support(s) the ground scheme of maneuver.

Such a description is not sufficiently detailed for automatic scenario generation. Consequently, an approach must be devised to break down training objectives into their component knowledge, skills, and attitudes (KSAs) (see Fowlkes et al., 2010). Training objectives are decomposed into a formal "domain ontology," which resembles a tree-diagram. For instance, the "Employ Supporting Arms" training objective may include KSAs related to spatial and temporal coordination, battlefield sense making, tactical positioning, and communication (just to name a few). KSAs are associated with one another in a hierarchical fashion, with those KSAs that require more coordinated actions listed higher on the diagram.

BASELINE SCENARIOS

Identification of a training objective triggers the construction of a baseline scenario. *Baseline scenarios* are simplified scenarios that include minimal, ideal conditions that support the selected training objective. "Minimal" implies that only those elements required to support the training objective are present. "Ideal" implies that all scenario variables are set to their least complex settings; for instance, weather conditions are set to optimal (e.g., daytime lighting, no precipitation, no wind).

Baseline scenarios can support training, in theory. However, these simple scenarios do not offer particularly beneficial training experiences. At best, they are only suitable for training the most novice trainees in procedural operations. They lack variability and complexity, and they are not well suited for the training of cognitive skills. In order to expand baseline scenarios into more functional training tools, additional scenario elements are needed.

AUGMENTATIONS

The baseline scenario can be enhanced by adding *augmentations*. These elements add complexity to the scenario and can affect aspects of the scenario itself (both entity and weather). Examples include moving the scenario to night, adding an additional target, or making a target hidden from first view. Each of these adds complexity to the generated scenario; however, not all may be applicable to a given set of training objectives. In addition, some may need to be limited. It makes no sense to add the "night" augmentation to a scenario more than once. Therefore, specifications of augmentations must also provide some limitations on quantities allowed.

The adding of augmentations to a baseline begins to collect together an appropriate training scenario. Through adding elements to a scenario, appropriate complexity levels can be supported for the full range of trainees from novices to experts. Much like training objectives, augmentations add requirements to the scenario. Adding an additional target requires that the target be specified (type and position). However, it is important to note that augmentations still focus on the initial *situation* of the training exercise.

SCENARIO VIGNETTES

The baseline and augmentations set a basic initial situation, focused on the environment. In turn, *scenario vignettes* add learning-objective focused content to the baseline. Scenario vignettes are pre-packaged alterations and/or additions to the scenario; they may be both macroadaptive (i.e., predesigned to meet instructional needs of trainees) or microadaptive (i.e., adjusting in real-time to better facilitate the training).

Scenario vignettes are defined as sets of associated *triggers* and *adaptations*. Triggers are any kind of check or comparison that returns a Boolean (true or false) value. They may listen for specific events (e.g. a detonation occurred nearby) or be time-based (allowing time-specific events to occur if desired). When a trigger is determined to be true, its corresponding adaptation is activated. Note that a trigger could have more than one adaptation associated with it. In addition, triggers can be chained to provide "if-then" type logic or even Boolean "and" logic.

Adaptations are alterations made to the current situation within the exercise. They could include entity manipulations (create, kill, move, fire weapon) or environmental manipulations (reduce rain, raise sun). They may be used to adjust the focus of the training (e.g., providing remediation) or they may be used to repair an exercise. For example, if a critical entity is killed, then an adaptation can be used to recreate it and, thereby, facilitate the completion of the scenario (i.e., the training opportunity is not lost). Ultimately, adaptations cause changes to occur within the scenario itself. Together with triggers, they form the basis for adjusting scenarios during run-time (i.e., microadaptation).

While triggers and adaptations form the basis for a vignette, they also provide an additional capability to support dynamic scenario (training) adaptation. This form of scenario adaptation can be paired with machine intelligence or developer-prescribed rules to create new triggers and adaptations on the fly. In other words, in addition to a scenario creator (whether human or automated program) creating vignettes for anticipated events, an adaptation system can dynamically create new triggers/adaptations to satisfy common instructional needs during run-time. For example, if a trainee accidentally gets killed during an exercise, the system may have implicitly prepared a trigger to detect this and then fire an adaptation to re-spawn that trainee (so that the remainder of the training opportunity is not lost). Similarly, if trainees' performance falls outside of a predetermined range, then the system may trigger an adaptation that escalates the training (e.g., introducing the next training objective) or offers remediation.

SATISFYING REQUIREMENTS

Before the scenario can be considered complete, one additional step is required. The training objectives, baselines, augmentations and vignettes may have specified required elements for the scenario. For example, a vignette may require an entity to exist (such as a target). However, these scenario components generally do not specify the type or position of said entity. These details are left to the end.

Requirements can be satisfied either manually, by a user, or via an automated approach. We enumerate the types of requirements that may be necessary, which allows the system to prompt for the specification of that requirement type. For example, knowing that a target is required, the system can prompt for type and position of that entity. Once requirement is satisfied (specified), then the scenario is complete.

INSTRUCTIONAL STRATEGIES

In the previous subsections, scenarios were formalized as a baseline, a set of augmentations, and a set of vignettes (each made up of a set of triggers and a set of adaptations). To transform these elements into *training* scenarios, pedagogical data should be incorporated.

The science of training suggests that training scenarios should support varied instructional approaches. For example, a given trainee may be presented with a "compare and contrast" scenario where a given event is shown in two slightly different ways, which emphasizes the subtle differences between the cases. Or a "disequilibrium scenario" may show a possible worst-case scenario to enhance the understanding of why such training is necessary. Such instructional templates help guide the construction, and potential microadaptation, of a scenario. This is the final (high-level) building block for our conceptualization of dynamic scenario generation and augmentation.

BUILDING TRAINING SCENARIOS

With the formulation of scenarios operationalized, an application can then be built to allow a user to develop a scenario and export it in a way that allows various training applications to initialize (and then execute) the scenario. The authors are participating in a research program to develop such an application with an initial focus on U.S. Marines Fire Support Teams.

The user first selects the trainees' expertise levels (which in turn sets the scenario's complexity range) and training objectives to be trained. Coupled with the trainee profiles and the KSAs related to the training objectives, the baselines, augmentations and vignettes are then filtered to only provide those that support training that desired configuration. The user then selects a baseline, with zero or more augmentations and zero or more vignettes to create a scenario of sufficient complexity (the system ensures that the resulting scenario is within the desired complexity range and requires the user to alter the scenario if not).

Future work will investigate how to provide some automation to the scenario creation process. The user will still provide the initial inputs, but the system itself will then formulate a baseline, set of augmentations and set of vignettes to couple into a scenario fulfilling that request. Currently, procedural models such as shape grammars and L-systems are being considered.

CONCLUSION

In this chapter, an operationalization of scenarios was given. When creating scenarios, it is necessary to formulate a fixed definition for a scenario that can satisfy building a variety of scenarios given specific training objectives. This is more important for creating such scenarios in an automated fashion. In addition, the use of pedagogical templates can further enhance the training effectiveness. They provide for scenarios that can use different lesson types to enhance the learning experience. This further enhances the variety of scenarios that can be provided as well.

ACKNOWLEDGEMENTS

This work is supported in part by the Office of Naval Research Grant N0001408C0186, the Next-generation Expeditionary Warfare Intelligent Training (NEW-IT) program, and in part by National Science Foundation Grant IIP0750551. The views and conclusions contained in this document are those of the authors and should not be interpreted as representing the official policies, either expressed or implied, of the ONR or the US Government. The US Government is authorized to reproduce and distribute reprints for Government purposes notwithstanding any copyright notation hereon.

REFERENCES

Hofer, R. C. & Smith, S. H. (1998, March). Automated scenario generation environment. Paper presented at the Simulation Interoperability Workshop, Orlando, FL. Retrieved from www.sisostds.org.

Fowlkes, J., Schatz, S., & Stagl,K. (2010). Instructional Strategies for Scenario-based Training: Insights from Applied Research. *Proceedings of the Military Modeling & Simulation Symposium (MMS2010)*, Orlando, FL.

Martin, G., Schatz, S., Bowers, C.A., Hughes, C. E., Fowlkes, J. & Nicholson, D. (2009). Automatic scenario generation through procedural modeling for scenario-based training. *Proceedings of the 53rd Annual Conference of the Human Factors and Ergonomics Society*. Santa Monica, CA: Human Factors and Ergonomics Society.

Schatz, S., Bowers, C.A., & Nicholson, D. (2009). Advanced situated tutors: Design, philosophy, and a review of existing systems. *Proceedings of the 53rd Annual Conference of the Human Factors and Ergonomics Society*. Santa Monica, CA: Human Factors and Ergonomics Society.

Stout, R., Bowers, C., & Nicholson, D. Guidelines for using simulations to train higher level cognitive and teamwork skills (pp. 270-296). In D. Schmorrow, J. Cohn, & D. M. Nicholson (Eds.) *Handbook of virtual environments for training and education: developments for the military and beyond.* Westport, CT: Praeger Security International.

Tomizawa, H. & Gonzalez, A. Automated scenario generation system in a simulation. *Proceedings of Interservice/Industry Training, Simulation and Education Conference*, Orlando, FL: National Defense Industrial Association.

CHAPTER 75

Exploring Learning and Training Abilities Using Assistive Technologies

Guy A. Boy, David Lecoutre, Laurent Mouluquet, Anil Raj

Florida Institute for Human and Machine Cognition
40 South Alcaniz Street
Pensacola, Florida 32502, U.S.A.

ABSTRACT

This paper presents the development of a training protocol for learning to read text with the BP-WAVE-II. The various learning tasks tested in the development of this paradigm could be applied to other situations in order to train BP-WAVE-II users to adapt to operations in unknown environments. Other types of assistive technologies increase situation awareness for the blind, such as the white cane and the guide dog. Unlike the BP-WAVE-II, these alternative systems are costly and require important environment modifications: (1) Braille representations; (2) street traffic lights equipped with speakers (audio CFS) that tells bind people when to cross the street; and (3) pedestrian crosswalks and sidewalk transitions equipped with grooves or bumps to inform the blind about changes in the nature of the ground (tactile CFS).

Keywords: Assistive technologies, impaired people, training protocol, reading.

INTRODUCTION

Individuals such as wounded military service members or civilians who recently became visually impaired require special attention with respect to their lost abilities such as vision or audition. Until their injury or illness, they possessed well-developed visual processing capabilities that enabled them to recognize previously learned external objects rapidly. The sensory substitution approach can recover this ability to pattern-match information using these previously stored visual memories by using data from sensors other than their eyes. We also could legitimately think that they create new sets of patterns and tactile "image" memories. The more general question relates to the brain's mechanism of re-adaptation to an alternate sensory input.

Bach-y-Rita and his colleagues (1998) already showed that sensory substitution systems can deliver visual information from a digitized video stream to an array of stimulators in contact with the skin of one or several part of the body, and the tongue in particular. The tongue is very sensitive and enables good electrical contact (Kaczmarek, 2005; Ptito etal., 2005). He notes that "We do not see with the eyes; the optical image does not go beyond the retina where it is turned into space-time nerve patterns (of impulses) along the optic nerve fibers. The brain then recreates the images from analysis of the impulse patterns." (Bach-y-Rita et al., 1998).

While assistive technologies have been developed for sensory deficits (Capelle et al., 1998; Lenay et al., 2003), there is very few available data on optimal learning and training with these technologies. Indeed, when a natural sensory input such as eyes is replaced by an artificial input such as the combination of electro-tactile stimulation and human skin, the brain needs to learn how to interpret such new sensory input. Our main objective was to develop a principled approach to training for assistive technologies.

WHAT IS THE PROBLEM?

The multiple channels that carry sensory information to the brain, from the eyes, ears and skin, for instance, are set up in a similar manner to perform similar activities to enable understanding of the local environment. Sensory information sent to the brain is carried by nerve fibers in the form of patterns of impulses, and the impulses route to the different sensory centers of the brain for interpretation. Substituting one sensory channel for another can result in correct re-encoding of the information by the brain. The brain appears to be flexible in interpreting sensory signals. It can be trained to correctly interpret signals sensed through the tactile channel, instead of the visual channel, and process it appropriately (Kaczmarek et al., 1995). It just takes training and accurate, timely feedback from the sensory channel in response to user inputs (Bach-y-Rita & Kercel, 2002).

Using the BP-WAVE-II[1], there are three sensory data-processing layers between the external environment and the brain. First, the camera provides digitized visual information. The second layer is the physical array itself that provides electrical stimuli to the tongue. The third layer is the tongue itself, equipped with nerve fibers, that transfers the stimuli to the tactile-sensory area of the cerebral cortex, the parietal lobe. The parietal lobe usually receives tactile information, the temporal lobe receives auditory information, the occipital lobe receives visual

---

[1] BrainPort® Wearable Aid for Vision Enhancement (BP-WAVE-II).

information and the cerebellum receives balance information. The frontal lobe is responsible for many higher brain functions. Obviously, the brain needs training to process signals normally processed in the occipital lobe connected to the parietal lobe in order to interpret artificially-provided tactile representations of visual information.

There are many ways to approach augmented perception and cognition via assistive technologies. We chose to represent human and machine cognition using the cognitive function paradigm (Boy, 1998). A cognitive function is represented by three attributes, a role, a context of validity and a set of useful resources. We will consider five types of human cognitive functions related to vision, i.e., frame catching, pathway of data, localization, data processing and recognition. Frame catching involves the pupil, cornea, iris and lens, as well as the cones and rods photoreceptors cells of the retina. These components have their own cognitive functions, which may be impaired leading to a total or partial blindness. There are cognitive functions on the machine side (assistive technology), i.e., the BP-WAVE-II. In the camera layer, there are several types of cognitive functions such as the ability to manipulate contrast, light intensity, zoom and so on. In the array layer, two types of cognitive functions can be modeled such as image resolution related to the number of points and the distance between points, and the tongue affordance to accept a large number of points. Finally, there are other cognitive functions related to supervision that needs to be provided to synchronize and support the use of the BP-WAVE-II, e.g., verbal indications or instructions in real-time, various kinds of training, and so on. This can be represented by the triplet *Human-Supervision-Machine* (HSM), where supervision is the gap between human and machine that an impaired user needs in order to reconstruct some kind of natural perception.

Consequently, the experimental problem is as follows. Would it be possible to better understand the way HSM cognitive functions work together to insure a better human-machine adaptation while minimizing supervision? In order to specify this problem statement in more concrete details, we focused on letter size, complexity, color and contrast in order to determine the optimal stimulation on the tongue. In addition, light intensity, external contrasts, familiarity of targeted objects, static objects vs. motion, and complexity of the scene were considered.

METHOD

The first requirement was to develop scenarios appropriate for representation using the BP-WAVE-II, i.e., we had to define the distance between the camera and the object, size and thickness of items presented to the user, brightness, glint, and contrast. We simplified the environment to a 22-inch computer screen with white letters on a black background using a distance between the head mounted video cameras and the display of 80 cm promote efficiency and rapid success. Interaction revealed that all capitals Arial regular text at a size of 400 points provides a good signal quality on the tongue when using the BP-WAVE-II.

Initial familiarization with the BP-WAVE-II is required to learn to operate the system independent of its use as a vision substitution interface. Users operate the controls while presented with simple shapes before other more complex shapes can be presented. In addition, it is easier to feel a moving object, e.g., a line being drawn versus a static one. Once familiar with moving objects, a user can learn simple static shapes such as circles, triangles and stars, which can be presented before differentiating multiple objects. This step is crucial to increase user's awareness of spaces between items, i.e., the way users feel and understand how to avoid interferences between letters. For example, interference may occur when a user may see half of a letter on the left part of the display, and half of the next letter on the right part of the display. This can result in a cognitively demanding disturbing situation for a novice.

Initially, we envisioned training with 36 alphanumeric characters, but because numbers can be more difficult to perceive than letters, we focused on designing the training paradigm using only capital letters, which were categorized into four groups after several iterative trials: easy (C, I, J, L, O, U, T); medium (D, E, F, H, P, V, Y); hard (A, B, M, N, K, X, Z); very hard (G, Q, R, S, W). These results could be interpreted as follows. First, there are simple lines such as vertical and horizontal lines, and complex lines such as diagonal and intersected lines. Second, complexity comes from the number of lines in a letter. Third, circles are easy to recognize, but when augmented with other shapes they can become extremely difficult to identify. Fourth and more generally, letter-reading complexity comes from the combination of various shapes. In addition, we found out that 4 presents the frequency of each letters in English. We can see that the easy and medium categories are the most commons letters. This is encouraging for BP-WAVE-II use.

A software presentation control program was written that presents a letter on the computer monitor. If the participant identifies it correctly, another letter is presented, else he or she is informed of the error and prompted to try again. After a second letter recognition failure, the system informs the participant, which letter was presented. For each trial, user's reaction time can be measured, and subjective assessments of the perceived difficulty can be recorded.

First, users would first be presented easy letters. After 80% reading success, they will proceed to the next category of letters. In case of failure in a letter category, this whole category is presented again. This process is used for the first two categories of letters, i.e., easy and medium. The series of letters presented during the test include redundancies in order to prevent participants from predicting the possible next letter, i.e., there are more items to guess than letters to be learned. For example, in easy trials where users learn 7 letters, a typical quiz is the following sequence of 10 items "C, I, J, I, C, L, O, T, U, T". So even if the letter "C" is indentified, it could be presented again for recognition.

For the hard and very hard categories of letters, the process consists of a series of quizzes, e.g., the three letters DPF were presented on the same screen and the participants asked to find the letter F. A quiz consists of asking for a wrong letter in a series, e.g., DPF is presented and the user is asked to find the letter E. For the very hard letters, a quiz consisted for example in presenting the series SZXSK and asking to find the letter S.

Finally, participants will be asked to read a word such as ALBUM and reading commonly used English words such as AND, THE, OF, TO, IN, IS, WITH and so on. Potential users of this training paradigm must possess sufficient cognitive and communicative capabilities to understand the instructions, control the BP-WAVE-II and respond to the stimulus. Once this training protocol was designed, we elicited human, machine and supervision cognitive functions. Cognitive functions were categorized with respect to the following dimensions: ease of learning; ease of retention; and level of required supervision.

RESULTS

Our research team used a four-trial evaluation procedure to design the protocol after a familiarization phase (detection of basic movements and shapes before starting reading letters, see Table 1). Each evaluation starts with a training period and ends with a simple test or a quiz. **Easy and medium trials** are essential for a participant to learn how to recognize a single letter through his or her tongue. Only one item was presented on the screen in order to keep workload as low as possible. Novices with the BP-WAVE-II tend to have difficulties with spatial recognition; they acquire these skills with experience. **Hard trials** involve reading several letters in order to prepare for reading words. Two trials are presented for identification of a single letter among three letters. Then, three trials are presented for identification of three letter words. In the **very hard trials**, two trials devoted to the identification of a given letter that may or may not be present in a set of five letters, and two trials for the reading of a word of five letters. The participant is awarded one point for each right answer, and the final score is the sum of all right answers.

After experiencing all these trials, an individual should be capable of successfully reading most common English words, defined as recognizing a 3-letter word in less than 5 sec. In that case, this means that a user performed good pattern matching. Alternate procedures were considered, for example instead of presenting fixed (already written) letters, a letter could be dynamically drawn. This was motivated by the fact that when a letter is manually traced on the skin of someone, it is usually recognized. This is a question of memory of the letter shape associated with the motor movement. Initial pilot testing using the BP-WAVE-II found that while the letter being drawn was felt, the letter was seldom identified. The use of the camera zoom was also tested to better understand if it could improve the identification of letter shapes, and increase the awareness of users of zoom capabilities. During letter recognition tasks, zoom in – zoom out movements could be appreciated but not letter shapes. Zoom proved useful for exploration of specific parts of letters, e.g., in order to distinguish letters O and Q, zoom is useful to identify the small line at the bottom of the circle of the letter Q.

Ease of learning. Learning of letters depends on user familiarity with the system, and on letter complexity (i.e., easy, medium, hard and very hard categories of letters). Initial pilot testing revealed an average learning time of 45 sec ± 15 sec per letter, which drops to near 20 sec with experience (i.e., at least 5 hours of experience in reading with the BP-WAVE-II). With 45 min of effective training (± 15 min), which is approximately 1 hour and 30 min of overall use, the researchers were able to read a 3-letter word in less than 4 minutes. A first important milestone occurs after approximately 200 min training time when a user manages to easily read small words using the BP-WAVE-II before proceeding to reading longer words. At this point the researchers were able to read a word in less than 15 sec. As a general standpoint, the more training, the faster the reading rate with the BP-WAVE-II. After 2 to 3 hours training, interpretation of simple patterns and words presented on the tongue was possible. It must be acknowledged that we still do not know how much training is needed for participants to read fast enough to process an entire sentence at a reasonable pace. A learning plateau occurs when individuals stabilize their performance. We believe that with more experience an expert user will use strategies to read a group of letter instead of one letter at a time. This is a matter of context awareness that should enable readers to recognize words instead of always recomposing words from the recognition of single letters. Imagine when you started learning at school; in the beginning, it took a lot of time for you to read a single word, because you were probably trying to read each letter and then recognize or identify the word. Normal readers do not even read each letter of a word; they are able to understand and reconstruct a sentence even if some words or letters are missing, misspelled or duplicated. We faced the same problem and we expect to find the same solutions using the BP-WAVE-II. We believe that users can learn faster because their brains were already equipped with appropriate cognitive functions that enable them to read efficiently.

Ease of retention. Learning is the ability to retain facts and figures in memory. We found out that not using the BP-WAVE-II for intervals of many days did not affect recognition accuracy or reaction. As already said, the more training, the faster one can read a word.

Levels of required supervision. Initially users manifest three kinds of focus problems: targeting the screen, focusing on the letter and adjusting the appropriate zoom on the letter. This is a general problem of lack of feedback, resolved with time and experience. Supervision required appears variable with individuals as some individuals try to "Feel borders", "Feel specific parts of a shape", "Change level of zoom", and so on. Four types of useful advices can be given to the users: orientation, technical settings, feedback, and alternative strategies. **Orientation**. Assisting users to appropriately set the zoom and recapture the area of interest (i.e., the center of the screen, the first letter of a word or a series). **Technical settings**. Helping users set the intensity of the display and the orientation of the camera. **Feedback**. Since there are different ways of recognizing shapes, users should explore all possible feedback methods using the BP-WAVE-II. This includes following the shape, moving the head from left to right, top to bottom, trying to draw the letter mentally, and finding various curves of each letter. For example, we know that the letter E is made of one vertical line and three horizontal lines. What is important in the pattern matching is the fact that the E is the only case in this class. It is thus necessary for users to create reference points for each letter. These reference points are typically anticipating recognition patterns. Supervisory guidance can be provided to those who cannot locate important reference points on their own. **Alternative strategies**. There are two methods for reading an item: following the borders of a letter as if it was being written; and placing the global shape of the letter on the tongue. Users use the former at the beginning of the training and the latter when they start to know how to recognize a letter pattern.

ANALYSIS

Easy letters are found faster than complex ones. When the complexity increases reading time increases, but progressively decreases with experience acquisition. The very hard group of letters requires less reading time than a hard one and so on. Therefore, user experience improves the learning ability of users to read new letters and words. Two issues forced us to modify the initially planned protocol:

- Other BP-WAVE-II training protocols inform users about the letter presented, and asked them to feel it (using large three dimensional physical models). This procedure was found to be unproductive because it was very difficult to verify the user's acuity when feeling of the letter.
- Some users spend few seconds (about 5 sec) to learn an item, then they spend a lot of time on another item (1 min), skewing total test response time during the quiz. Others spent more time (50 sec) during the learning part; they actually took the time to feel the different patterns. These users were more efficient at recognizing letters afterwards (10 sec). It could be reasonably inferred that some may overestimate BP-WAVE-II capacity for providing appropriate information users need to find, associate and integrate novel patterns in order to be able to perform good pattern matching.

The following learning steps were identified. **Acquisition**, or assimilation in Piaget's terminology (1985), is the process of learning a new behavior or skill. It can be measured by the time that users take to learn the patterns of an item. This time differs from one user to another. We observed that when providing appropriate supervision, such as orientation or coaching alternative strategies as mentioned above, skill acquisition improves. **Execution adaptation and stabilization** (maximum efficiency), or accommodation in Piaget's terminology, is the process of refining and automating a new cognitive skill. This is the capacity to read a word with ease and recognize interferences, e.g. camera artifacts, distractions in the visual field, etc. **Retention** is the power of retaining and recalling past experience. At this learning step, it appears that cognitive performance remains unaffected following a lack of stimulation and/or spacing among periods of BP-WAVE-II use. Retention was very good, especially with a few exercises to go back to previous performance levels. Retention is better when both acquisition and adaptation are well done. **Endurance** is the ability of using an acquired and adapted skill over a long period of time without affecting results. Users often need a break after more than 45 minutes. Physically speaking, the camera gets warm and puts pressure on the user's head. Cognitively, concentration and mental workload associated with the adaptation task contributes to fatigue. **Transfer** is the capacity of transferring skills, e.g., after being able to read upper case letters, users are likely to learn how to read lower case letters faster. Such a transfer is not immediate; users should learn how to recognize a few patterns and reuse these patterns by analogy. For example, changing font style leads to learning a new pattern. This can be a limit of the device, but expert users can easily recognize a letter even if its font was changed. It actually depends on the distance in shape between both fonts. For example, the difference between upper-case "O" and lower-case "o" is insignificant, whereas the difference between "G" and "g" is very significant. **Application** is the ability to use a new cognitive skill in everyday life or outside its learning context. The environment for administration of this protocol should be set up to maximize the performance of the user. Transition to real world conditions with variable lighting, contrast, reflections, and so on, would be helped with system technology improvement.

DISCUSSION

Human vision obviously provides far better capability than the use of the current BP-WAVE-II system. This is due to several reasons that can be expressed in terms of cognitive functions on the human side (CFH), the supervision side (CFS), and the machine side (CFM). At this point in time, the BP-WAVE-II training cannot occur without supervision. The main goal of further studies and developments is to transfer some of the cognitive functions from the supervision to the machine, i.e., the BP-WAVE-II (CFS→CFM). However, supervision shouldn't be ignored because it enables increasing users' trust using the BP-WAVE-II. The more trust, the less supervision would be needed.

We should add richer **feedback** (CFS→CFM) to increase user's situation awareness (CFH), i.e., it would be good if the system could identify the reading process context and adjust appropriately. With the current system, the only available feedback is the central vision context, and not the periphery. Adding peripheral visual cues using sensory substitution technology presented on another part of the body could enhance the users understanding of the BP-WAVE-II stimuli. For example, when someone is reading, the system should be able to follow the reading process and dynamically follow up on the next word and line. The system should be able to interpret the user's point of regard through gaze and accommodation tracking.

In human vision, the eye is able to **focus** on near and far objects (CFS→CFM). BP-WAVE-II supports optical and digital zoom instead but without strong direct zoom ratio/level feedback (the current system relies on tactile feedback through the fingers controlling the zoom level). Since users are not able to read letters lower than 18-point font and without a good contrast, then BP-WAVE-II cannot be used for reading most texts of our everyday life. This problem could be mitigated with increases in the resolution of the tongue array.

The system should provide **auto-focus/orientation** (CFS→CFM) when the scene is not presented orthogonal to the vision axis, the system should be able to read a word, identify each letter, and present it to the user. While sighted people use this cognitive skill automatically (CFH), BP-WAVE-II users experience difficulty probably due to its low resolution and lack of orientation feedback (two CFM to be improved), e.g., if the letter E is presented and it appears that the item is not straight, it is not directly clear if the letter is rotated or a different letter with angled lines.

Frame catching (CFM) needs adjustments, such as feedback of the gaze position. In addition, the concavity of the lens is not corrected. A straight line can be perceived as a curve in many cases. Sighted people are able to promote the saliency of part of the scene and filter out the surroundings in order to concentrate on the main task of processing purposeful data. Instead, the BP-WAVE-II acquires the whole scene without any semantic filtering. The BP-WAVE-II, therefore, requires cognitively more complex image processing. Consequently, a new machine cognitive function should be developed to perform **appropriate image processing** to improve the saliency of the visual point of regard. Various filters (e.g., infra red vision, edge detection, night vision, sonar vision, hollow shape) could be added on the camera that BP-WAVE-II user could switch as much as they want. In addition, peripheral vision sensing that provided a three-dimensional mapping (CFM) of the environment around the user that can be sent to the body, either on the tongue or chest for example, could improve situation awareness.

CONCLUSIONS

In this paper we present the development of a specific protocol and its applications to the exploration of learning and training abilities using the BP-WAVE-II assistive technology system. We designed the protocol based on observation and analysis of the various distinctions among human, supervision and machine (HSM) cognitive functions in order to propose a possible training solution that is likely to improve current BP-WAVE-II learning and use. In contrast with previous work, we did not only focus on how users feel the various ways of presenting simple letters such as T or C. We clustered the letters with respect to their level of difficulty and frequency in the English language. We then developed the HSM method that users practiced to learn how to read using the BP-WAVE-II.

Appropriate supervision facilitates users' learning and reading performance. For example, there is a learning asymptote for getting close to 100% success at about 5 hours of user experience in a reasonable reading time. However, current BP-WAVE-II technology (CFM) is not mature enough to provide good reading quality in the everyday life. Users (CFH) are still too dependent on supervision (CFS). This is why sensory substitution should be augmented to remove supervision as much as possible. The VideoTact, as an example of a technology for context augmentation of BrainPort® stimuli is already under investigation. This work revealed that the tongue array resolution may be suboptimal, but we still do not know how many pixels the tongue is able to manage. We clearly observed that movements are felt through the tongue, but they are not always recognized. Only simple shapes and simple movements are easily recognized. Therefore, BP-WAVE-II functions should be augmented to facilitate interaction and situation awareness.

Human variability is tremendous and vision impaired individuals are not likely to adapt in the same way as our blindfolded researchers. Whether we should imitate human cognitive functions in the system, or augment the BP-WAVE-II by extending the machine's cognitive functions to sense infrared, sonar, radar, etc.. We are currently continuing to investigate in these directions. More generally, this research effort enabled the development of a human-centered methodology for the co-adaptation of both user experience and technology. This methodology applied to the BP-WAVE-II is based on a multidisciplinary approach, as well as cross-fertilized competences in neuroscience, psychology, computer and electrical engineering, and HCI. From an HCI point of view, it enables user experience design, and consequently the construction of training protocols, through the incremental elicitation of HSM cognitive functions.

REFERENCES

BP-WAVE-II (BrainPort® Wearable Aid for Vision Enhancement), Wicab, Inc., Middleton, WI, USA: http://vision.wicab.com.

Bach-y-Rita, P., Kaczmarek, K.A., Tyler, M.E. & Garcia-Lara, J. Form perception with a 49-point electrotactile stimulus array on the tongue: A technical note. *Journal of Rehabilitation Research and Development*. Vol. 35, 4, October, (1998), 427-430. www.rehab.research.va.gov/jour/98/35/4/bachyrita.htm.

Bach-y-Rita, P. and Kercel, S.W. (2002) Sensori-'motor' coupling by observed and imagined movement. *Intellectica* 35, 287–297.

Boy, G.A. Cognitive function analysis for human-centered automation of safety-critical systems. *In Proc. CHI'98*. ACM Press (1998), 265-272.

Capelle, C., Trullemans, C., Arno, P., & Veraart, C. A real-time experimental prototype for enhancement of vision rehabilitation using auditory substitution. *IEEE Transactions Biomedical Engineering*, **45** (1998), 1279-1293.

Kaczmarek, K. A. Sensory augmentation and substitution. In J. D. Bronzino (Ed.), *CRC Handbook of Biomedical Engineering* (1995), 2100-2109). Boca Raton, FL: CRC Press.

Kaczmarek, K. A. Tongue Display Technology. Technical Report. University of Wisconsin, Aug. 18, 2005.

Lenay. C., Gapenne, O., Hanneton, S., Marque, C. & Geouelle, C. Sensory Substitution: limits and perspectives. *Touching for Knowing, Cognitive psychology of haptic manual perception* (2003), 275-292.

Piaget, J. (1985). *Equilibration of cognitive structures*. University of Chicago Press.

Ptito, M., Moesgaard, S.M., Gjedde, A. & Kupers, R. Cross-modal plasticity revealed by electrotactile stimulation of the tongue in the congenitally blind. *Brain* 128(3), (2005), 606-614.

VideoTact. ForeThought Development, LLC, Blue Mounds, WI, USA: http://www.4thtdev.com

Chapter 76

The Disadvantageous but Appealing Use of Visual Guidance in Procedural Skills Training

Nirit Yuviler-Gavish, Eldad Yechiam

Faculty of Industrial Engineering and Management
Technion - Israel Institute of Technology
Haifa, Israel, 32000

ABSTRACT

The use of visual guidance tools in procedural tasks training is considered to improve performance. However, facilitating training with visual guidance tools may also impair skill transfer, because of the possibility of inhibiting active task exploration. The present work examined four different kinds of visual guidance tools which can facilitate training: visual pointing aid; visual pointing aid combined with drawing options; observational learning in the first part of the training; and dyad trainer-trainee performance. These tools were examined by using a 3-D puzzle task and different training settings (computerized training, AR training, etc.). Three experimental studies demonstrate that although these tools were preferable for trainers and trainees, and required less cognitive load during training, they impaired non-supervised performance. These findings suggest that when designing training platforms, using sophisticated visual communication aids should be considered very carefully given their potentially negative effects on active exploration processes and training quality.

Keywords: Procedural Skills, Training, Visual guidance, Augmented Reality, Skills Transfer.

INTRODUCTION

Within the framework of the SKILLS Integrated Project (Skills Consortium, 2008), one of the domains that were selected is designing a training simulator for Industrial Maintenance and Assembly (IMA) tasks. Because of the comprehensive workflows of assembly and maintenance steps which are commonly required from the technician, the main focus of IMA tasks that were selected within the project, and hence – the main focus of the training simulator, was procedural skills transfer. The current work is part of the research effort that has been done to address the issue of procedural skills transfer in the IMA simulator.

The use of effective trainer-trainee communication during training of procedural skills can facilitate learning and improve skill transfer. Trainers with task-related experience can improve the learning process by enhancing trainees' knowledge on the nature and best heuristics for the task, expanding their exploration process and exposure to task variants and giving the necessary feedback and knowledge of results during the performance of the task. Trainer-trainee communication can take several formats and make use of different training aids. For example, the trainee can be fully active in the performance of the trained task or can be more passive and observe the trainer's actions. The trainer can use visual pointing devices, computer graphics objects (2D or 3D), audio signals, haptic feedback, etc. as additional aids accompanying verbal communication. The latter trainer-trainee communication modes are commonly used to facilitate training because they make the process of transferring the necessary information easier for both trainers and trainees.

The focus of the present work is on the use of visual guidance tools in trainer-trainee communication in the process of procedural skills learning. Two main theories claim that using the visual and auditory channels together for presenting information improves training effectiveness compared to the use of a single modality. The first theory is the Cognitive Theory of Multimedia Learning (Mayer, 2001) which postulates that having two modalities available during training enables a trainee to build two different mental representations - a verbal model and a visual model - which enriches the stored memory. This theory is supported by finding in multimedia studies showing that adding pictures and animations to vocal narrations improves success rate in subsequent knowledge tests (Mayer, 1989; Mayer & Anderson, 1991, 1992). A second theory supporting the use of multiple modalities during training is Cognitive Load Theory (Sweller et al., 1998; van Merrienboer & Sweller, 2005), which asserts that using multiple sensory channels reduces the cognitive load on working memory. These arguments were also supported by research findings (Chandler & Sweller, 1991; Penney, 1989).

Visual guidance tools can also be used to enable observational learning. In observational learning, the trainee is not active in performing the task, and his learning is a consequence of observing his trainer's performance. Evidences show that this training method can be effective (Bandura & Jeffery, 1973; Wouters, Tabber & Pass, 2007; Wulf & Shea, 2002). Bandura (1986) stated that while observing an expert performing a task, learners can construct adequate cognitive representation of the performance strategies, which in turn enables them to rehearse the task mentally or physically, and by doing so refine both their initial representation and their behavior. Observational learning is considered to be effective especially in the first stages of training. Performing a complex task typically impose high demands on novices' working memory, which in turn can lead to an inability to process essential information. According to Wouters et al. (2007) observing an expert performs the task at the beginning of the training may release the necessary cognitive resources. A variant of observational learning is a dyad training protocol, in which the trainer and the trainee are performing the task together during training. Shebilske et al. (1992) demonstrated that in a complex computer game task, trainees that practiced in pairs, each controlling half of the task and both viewing each others' activities on a computer monitor, performed as well as individual trainees on 10 test sessions, despite having only half as much hands-on experience during practice sessions.

However, facilitating training with visual guidance tools may also harm skill transfer, because of the potential inhibition of active exploration of the task. Active exploration is considered a key part in the development of situational awareness (i.e., the mental model of the spatial environment) (Neisser, 1976), which enables the learner to perform the task independently, and is also important for using features of the spatial environment for problem solving and dealing with errors (Fu & Gray, 2006; Yechiam et al., 2004). Active exploration naturally takes place if transferring the information about the task during training is accompanied with some difficulties, forcing the trainee to independently explore the task. However, when such difficulties are reduced, active exploration may not take place. Visual guidance tools may impede active exploration because they provide the trainer with an option to guide the trainee in specific movements and thus inhibit the trainee's active exploratory responses. Moreover, visual guidance tools may be highly attractive to trainers (and trainees as well) exactly because of the apparent success of performing these "micro-instructions". The use of visual guidance may therefore be both appealing and disadvantageous.

EXPERIMENTAL STUDIES

GENERAL OVERVIEW

Three studies were conducted in order to investigate the effects of four visual guidance tools on facilitating trainer-trainee communication and procedural skills acquisition: visual pointing aid; visual pointing aid combined with drawing options;

observational learning in the first part of the training; and dyad trainer-trainee performance. Despite theories and evidences which support the contribution of these tools to better skill transfer, some findings from our studies demonstrated the opposite effect. In addition, three different combinations of training platform and tested task performance conditions were tested: computerized training for computerized (virtual) task performance; computerized training for both computerized task and real-world task performance; and Augmented Reality (AR) training (real world task training enhanced with additional computerized aids) for a real world task performance.

Study 1 examined the addition of visual pointing aid to vocal guidance in a computerized training for computerized task performance. Study 2 examined the addition of visual pointing device combined with drawing options to vocal guidance in AR training for a real world task. Study 3 examined the effects of observational learning in the first part of the training and dyad trainer-trainee performance in a computerized training for both computerized and real-world task performance (see Table 1). All participants were undergraduate engineering students from the Technion who received reward based on their success in performing the experimental tasks.

Table 1: Summary of visual guidance tools and training conditions in the different studies

		Study 1	Study 2	Study 3
Visual guidance tools	Visual pointing aid	X		
	Visual pointing aid + drawing option		X	
	Observational learning in the first part			X
	Dyad trainer-trainee performance			X
Training platform	Computerized training	X		X
	AR training		X	
Tested task performance	Computerized task	X		X
	Real-world task		X	X

The task that was trained in all studies was a three-dimensional puzzle. This task had two versions. One was computerized (Shuzzle: http://leweg.com, Figure 1). In this task, puzzle pieces must be moved in a virtual three-dimensional space so as to complete a shape indicated by a wire frame, using the mouse and the keyboard. The second version was a real-world puzzle, identical to the computerized one. In this version colored puzzle pieces must be put in a certain places so as to complete a shape indicated by a solid shape (see Figure 2). Although the 3-D puzzle task may demand not only procedural skills, but also other skills such as spatial perception and mental rotation, it appears that procedural skills were the most dominant in it. For example, although there were at least ten different alternatives to solve the puzzle (i.e., different configurations of the shapes), about 89% of the trainees in Study 1 who succeeded in the test chose the alternative that was demonstrated to them during the training phase, suggesting that training was based on procedural memory of the allocation of the shapes during puzzle solving.

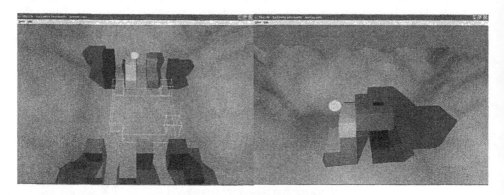

Figure 1. The computerized version of the 3-D puzzle task.

Figure 2. The real-world version of the 3-D puzzle task.

STUDY 1: THE EFFECT OF VISUAL POINTING AID IN COMPUTERIZED TRAINING FOR COMPUTERIZED TASK PERFORMANCE

Study 1 examined the hypothesis that the use of visual pointing aid will impair non-supervised performance, and will inhibit the natural process of active exploration of the task by the trainee. Seventeen participants (16 males and one female) served as trainers in the 3-D computerized puzzle task, after being familiarized with the task and acquiring the necessary level of expertise. Each trainer trained two trainees, each in one experimental condition, in a randomized order. In one condition (Vocal, 9 males and 8 females) trainers could only give vocal instructions. In the second condition (Pointer, 9 males and 8 females) trainers had the additional option of using a visual aid: Using a mouse pointer, trainers in this condition could point out positions and trajectories on the trainees' computer screen. In reality, 94% of the trainers used the visual pointer. After the training session, which was not limited on time, each trainee performed the same 3-D computerized puzzle alone, and a subsequent new 3-D computerized puzzle. The Trainees received a bonus payment for fast performance in these non-supervised tests, and the trainers received similar bonuses according to the success of their trainees. In addition to the performance measures, the participants' Heart Rate (HR) and Peripheral Arterial Tone (PAT) were monitored, enabling to assess their cognitive load in the different experimental conditions (previous examinations of the PAT measure include Iani et al., 2004; Iani et al., 2007; Lavie et al., 2000).

The results indicated that although there were no significant differences in training time between the Pointer and Vocal conditions, performance in the non-supervised test on the same 3-D puzzle was poorer in the Pointer condition. Only 64.7% of the participants in Pointer condition were able to solve the puzzle on their own, compared to 94.1% in the Vocal condition. In addition, participants in the Pointer condition took much longer to complete the test. Furthermore, it seems that training effort in the Pointer condition was less effective than in the Vocal condition, since no correlation between training time and performance time was found for the Pointer condition. In contrast, a negative correlation was found for the Vocal condition, indicating that more time dedicated for training was correlated with greater success in test performance for that condition. Performance in the new 3-D puzzle task did not show differences between the Vocal and Pointer conditions, either in terms of success rate or performance time (this is another indication of the important role of procedural skills in this task, which are not easily transferrable to a new task). Analysis of arousal using the PAT measure revealed that trainees' average level of arousal in the Pointer condition was lower than in the Vocal condition, suggesting that the use of the pointer reduced the cognitive efforts invested in learning the task. Additionally, analysis of the participants' verbal protocols showed that the frequency of specific command words (such as "do this" and "pick that") was associated with longer performance times on the non-supervised tests, indicating an association between passive visual guidance and performance decrements.

STUDY 2: THE EFFECT OF VISUAL POINTING AID AND DRAWIND AID IN AR TRAINING FOR REAL-WORLD TASK PERFORMANCE

In Study 2 AR training replaced the computerized training, and both training and test were performed with the real-world version of the 3-D puzzle task. Trainees' actions were recorded on-line using a web camera, and trainers could see them on their computer screen. Another computer screen with the same display was put in front of the trainees, enabling them to receive visual instructions from the trainers. This study, similarly to Study 1, involved two main conditions. In one condition (Vocal, 8 males and 8 females) trainers could only give vocal instructions. In the second condition (Visual, 6 males and 10 females) trainers had the additional option of providing visual instructions using a mouse pointer and a drawing program with two optional colors.

The results showed that similarly to Study 1, despite no significant differences in training time, participants in the Visual condition had much lower success rates in the non-supervised test: Only 62.5% of the participants in the Pointer condition were able to solve the puzzle, compared to 93.8% in the Vocal condition. In addition, participants in the Visual condition took much longer, on average, to complete the test. These results again demonstrate a negative effect of visual aids on later performance in an AR training environment.

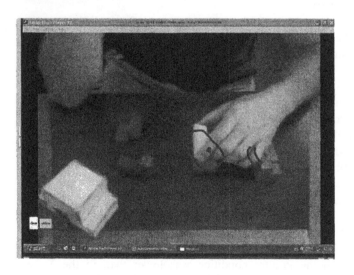

Figure 3. AR training with a pointer and a drawing option.

STUDY 3: THE EFFECT OF OBSERVATIONAL LEARNING IN COMPUTERIZED TRAINING FOR COMPUTERIZED AND REAL-WORLD TASK PERFORMANCE

Two variants of observational learning enabled by visual guidance tools were examined in this study. The first is observational learning at the beginning of the training, in which the trainer was instructed to begin his training with an explanation or demonstration, while the trainee stayed passive and could not use the mouse. The second is dyad performance of the trainer and the trainee during training, in which both of them could use the mouse and move the shapes. Nineteen participants (12 males and 7 female) who were highly experienced in the 3-D computerized puzzle task served as trainers. Each trainer trained four trainees in a 2×2 experimental design, with the order being randomized (altogether 76 trainees, 39 males and 37 females, were evenly distributed among the four experimental conditions). The training variables were the addition (or absence) of the preliminary observational learning part, and the addition (or absence) of dyad performance during training. After the training session, both trainers and trainees filled in subjective evaluation questionnaires about the training. Then, each trainee performed the same 3-D computerized puzzle alone followed by the real-world 3-D puzzle. Trainers and trainees' HR and PAT were monitored.

Training time was longer for the preliminary observational learning condition, but the additional training time had no effect on performance in the computerized or in the real-world test, hence training was less effective in this condition. The dyad performance condition lowered mental effort during training for both trainers and trainees. This was demonstrated in the PAT measure and in the subjective evaluations: Both trainers and trainees thought that the task was easier for the trainees in the dyad performance condition, and trainers felt that it was easier to train in the dyad performance condition. No significant differences were found in the computerized test. However, performance in the real-world task was inferior for the dyad performance condition, which led to lower success rates. In addition, trainers' effort (indicated by the PAT measure) was less effective in the dyad performance condition than in the no-dyad performance condition, since no correlation between the PAT measure time and performance time was found for the dyad performance condition, but a negative correlation was found for the no-dyad condition.

CONCLUSIONS

The current work examined the value of four types of visual guidance tools for facilitating training. These tools were examined by using a 3-D puzzle task and different training settings (computerized training, AR training, etc.). Three experimental studies clearly showed that these tools were preferable for trainees and trainers (for example, in Study 1, although the employment of the pointer was optional, only one of the 17 trainers did not use it) and demanded less cognitive effort during training. However, they apparently inhibited the active exploration of the task. As a result, using these tools led to less effective training and inferior performance measures. Contemporary technology offers a variety of visual guidance tools to facilitate trainer-trainee communication in Virtual Reality (VR) and AR training platforms. However, as pointed by Stedmon and Stone (2001), training systems development should be led not only by technological considerations, but also by human factors and training models. The findings from the current work suggest that when designing training platforms, using sophisticated visual communication aids should be considered very carefully given their potential negative effect on active exploration processes and transfer capabilities.

Applications within the SKILLS project (Skills Consortium, 2008) can be implemented in the IMA training simulator focusing on procedural skills acquisition, which reflects the operator's ability to obtain a good representation of how to perform each of the steps of a task and the correct order of performing them, according to their hierarchic organization. This simulator is developed by means of VR and AR training systems. In the VR training system, the trainee is immersed within a virtual environment, and the assembly and maintenance procedures are simulated via VR-interaction paradigms using 3-D and haptic interface technologies. In the AR training system, the trainee is performing assembly operations on the real machines using the real instruments for interaction, and is guided via a mobile training system including AR visualization as well as tactile and audio feedback. Both platforms use trainer-trainee communication to enhance procedural skills transfer. The current work suggests that visual guidance tools within these frameworks should not enable passive guidance of hand movements and visual trajectories.

REFERENCES

Anderson, J.R. (1982), "Acquisition of cognitive skill." *Psychological Review,* 89(4), 369-406.

Bandura, A. (1986), *Social Foundation of Thought and Action: A Social Cognitive Theory.* Englewood Cliffs, NJ: Prentice Hall.

Bandura, A., and Jeffery, R.W. (1973), "Role of symbolic coding and rehearsal processes in observational learning." *Journal of Personality and Social Psychology*, 26, 122-130.

Chandler, P., and Sweller, J. (1991), "Cognitive load theory and the format of instruction." *Cognition and Instruction,* 8, 293-332.

Fu, W.-T., and Gray, W. D. (2006), "Suboptimal tradeoffs in information seeking." *Cognitive Psychology,* 52, 195-242.

Gupta, P. and Cohen, N. J. (2002), "Theoretical and computational analysis of skill learning, repetition priming, and procedural memory." *Psychological Review,* 109(2), 401-448.

Iani, C., Gopher, D., Grunwald, A. J., and Lavie, P. (2007), "Peripheral arterial tone as an on-line measure of load in a simulated flight task." *Ergonomics,* 50, 1026-1035.

Iani C., Gopher D., and Lavie, P. (2004), "Effects of task difficulty and invested mental effort on peripheral vasoconstriction." *Psychophysiology,* 41, 789-798.

Lavie, P., Schnall, R. P., Sheffy, J., and Shlitner, A. (2000), "Peripheral vasoconstriction: A new physiological marker of REM sleep." *Nature Medicine,* 6, 606.

Mayer, R. E. (1989), "Systematic thinking fostered by illustrations in scientific text." *Journal of Educational Psychology,* 81, 240-246.

Mayer, R.E., and Anderson, R.B. (1991), "Animations need narrations: An experimental test of dual-coding hypothesis." *Journal of Educational Psychology,* 83, 484-490.

Mayer, R.E., and Anderson, R.B. (1992), "The instructive animation: Helping students build connections between words and pictures in multimedia learning." *Journal of Educational Psychology,* 84, 444-452.

Neisser, U. (1976), *Cognition and Reality.* San Francisco: Freeman.

Penney, C. (1989), "Modality effects and the structure of short-term verbal memory." *Memory and Cognition,* 17, 398-422.

Shebilske, W.L., Reigan, J.W., Arthur, W., and Jordan, J.A. (1992), "A dyadic protocol for training complex skills." *Human Factors,* 34, 369-374.

SKILLS consortium (2008), *Beyond Movement: The History and Future of Gesture Analysis.* Alinea publishing house.

Stedmon, A.W., and Stone, R.J. (2001), "Re-viewing reality: human factors of synthetic training environment." *International Journal of Human-Computer Studies,* 55, 675-698.

Sweller, J., van Merrienboer, J.J.G., and Paas, F.G.W.C. (1998), "Cognitive architecture and instructional design." *Educational Psychology Review,* 10, 251-296.

van Merrienboer, J.J.G., and Sweller, J. (2005), "Cognitive load theory and complex learning: Recent development and future directions." *Educational Psychology Review,* 17, 147-177.

Wouters, P., Tabbers, H.K., and Pass, F. (2007), "Interactivity in video-based models." *Educational Psychology Review,* 19, 327-342.

Wulf, G., and Shea, C. H. (2002), "Principles derived from the study of simple skills do not generalize to complex skill learning." *Psychonomic Bulletin & Review,* 9, 185-211.

Yechiam, E., Erev, I., and Parush, A. (2004), "Easy first steps and their implication to the use of a mouse-based and a script-based strategy." *Journal of Experimental Psychology: Applied,* 10, 89-96.

Chapter 77

Shortening the Expertise Curve: Identifying and Developing Cognitive Skills in Board Operators

Danyele Harris-Thompson, David A. Malek, Sterling L. Wiggins

Applied Research Associates, Klein Associates Division

ABSTRACT

Many process control plants are struggling with the dual challenges of increased regulatory requirements for safe and efficient operation and the expected retirement of a large population of its experienced workforce. Bringing new operators up to speed quickly to maintain safe and efficient operations is complicated by two factors. First, it is not clearly understood what constitutes an "expert" operator. Second, improvements in plant reliability and automation have greatly increased safety and efficiency. However, they also have reduced the number of abnormal events; thus reducing the number of opportunities operators have to build experiential knowledge. This paper details an effort that introduced Decision Making Exercises (DMXs), a cost-effective cognitive training solution, to the process control industry as a way to meet these challenges.

We applied cognitive-based methodologies to create "expert profiles" of Pipeline Analysts, Fluid Catalytic Crackers, and Crude Unit Operators at a large Gulf Coast Refinery and at one of the largest pipeline operator facilities in the U.S. Using expert stories and lessons learned, we generated DMXs to support the development of skills that experts noted as critical to managing specific types of events.

Keywords: expertise, decision making, Decision Making Exercises, process control, operator performance, scenario-based training, blended learning

INTRODUCTION

It is fall and the refinery has been busy. A feed tank that had been taken down for service has recently come back on line. The gas oil hydrotreater that was offline has been returned to service. A turnaround was completed last week and it went very smoothly. The plant has been consistently meeting its barrel target of 40k barrels/day. The temperature has been unusually hot and humid for weeks. It is late afternoon and you have had an uneventful shift when you notice an increase in the reactor and regenerator pressures. The reactor temperature has increased, too. You look at the history on these units and see the temperature and pressure increases happened recently. You monitor the temperature and pressures over the next minute as they continue to rise. You check the heater passes, but only see very small fluctuations. The charge rate and line pressure are normal and have been all shift, yet the pressure continues to increase in the reactor and the main column. The reactor temperature continues to increase as well. You have 1 minute. What is your assessment of the situation?

You have just read a portion of an exercise designed to improve decision-making skills of process control operators. The situation is a critical one; and before the operator can determine how to fix the problem he/she must first identify what the problem is.

If you are unfamiliar with process control, you may have a difficult time figuring out what is going on. If you have a process control background you probably have an idea based on your own experience. Someone with a similar background, but different experiences, will likely have a different interpretation and solution to the challenge. The opportunity to learn from the experiences and expertise that motivate responses to critical events is one of the key benefits offered by Decision Making Exercises (DMXs). DMXs (cognitive-based scenarios most often delivered through a paper-and-pencil, facilitated format) capture operator expertise and provide a forum for building expertise in a flexible, cost-effective format. We adapted this scenario-based training tool originally used to develop expertise in the Department of Defense, healthcare, and other domains to provide a cost-effective solution to the loss of expertise in process control.

Many process plants are struggling with the dual challenges of increased regulatory requirements for safe and efficient operation and the expected retirement of a large population of its experienced workforce. This translates into the industry's loss of key operational expertise at a time when it is most critical (Strobhar & Harris-Thompson, 2009). The need to bring new operators up to speed quickly while maintaining safety and efficiency is complicated by two things. First, developing skilled performance beyond basic procedural knowledge requires operators to make decisive judgments and assessments. However, it is not clearly understood what constitutes an "expert" operator. Many organizations do not know how to tap into and make explicit the tacit knowledge of their most experienced operators much less develop training that effectively targets the rapid development of this expertise (such as decision-making skills, attention management, and problem detection). Second, building expertise on a unit has typically come from experiencing numerous events and building a mental model of what those events look like (in terms of changes in the unit, pressures, or outputs) and some appropriate responses. Improvements in plant reliability and automation have greatly increased safety and efficiency. They also have reduced the number of abnormal events; thus reducing the number of opportunities operators have to build experiential knowledge. We addressed these challenges by identifying characteristics of expertise for process control console operators and then developing a cost-effective scenario-based training application to begin building those characteristics more rapidly in novices.

EXPERTISE IN PROCESS CONTROL

The underlying theory that drives DMXs concerns the nature of expertise in performance and how expertise develops. Hoffman (1998) defines an expert as one whose judgments are uncommonly accurate and reliable, whose performance shows consummate skill and economy of effort, and who can deal effectively with certain types of tough cases. Klein (1998; 1993) describes the keen intuitive ability that we credit to experts as the process of rapidly integrating information from a large array of accumulated experiences to size up a situation; selecting a course of action through recognition; and then assessing that course of action through mental simulation. Klein (1993) states that expertise requires a rich store of experiences to operate successfully and consists of facts and causal relationships being linked in terms of:

- Cues: If I see this, it means this larger pattern probably exists in the situation.
- Expectancies: In that pattern, I've usually seen things unfold in this way.
- Goals: It is important in this type of situation to do this.
- Typical actions: I have seen this goal achieved by doing the following.

Often, the assumption is that expertise is inherent in senior operators or developed through the execution of tactics and procedures. However, factual knowledge that is not acquired through experience – real or surrogate – is often not accessible during later performance and, therefore, may not contribute to advanced learning or performance. Simply knowing a lot of information or rules about situations will not meet the two conditions necessary for a novice to gain expertise. First, novices must acquire information in a manner that makes it mentally accessible in the appropriate situations, (i.e., the information must be learned as it occurs in real situations). When information is appropriately indexed, a situational cue is recognized immediately and it calls forth the associated information appropriate for conditions of similar situations. The same piece of information may or may not be a cue in situations with different conditions (in the scenario that began this paper the "unusually hot and humid" temperature is a cue that, when combined with recent tank maintenance, should have some relevance to the trainee. In a different context, the temperature may be irrelevant.).

Second, the novice must practice recognizing cues, expectancies, goals, and actions in context. *Knowing* what is typical in a domain and *using* that information to act in a situation are two different types of knowledge. Expertise is based on largely unconscious recognition. Therefore, one cannot "learn about" decision making or planning to become an expert decision maker or planner in a domain. One must practice making decisions in a context that provides the elements experts attend to. Decision Making Exercises provide repetitious opportunities for novices to develop expert skills in a safe context.

PROFILES OF OPERATOR DECISION MAKING

Supporting the development of operator expertise began with defining what expertise "looks like" in the process control domain. With the guidance of the customer,[1] we identified three positions to focus on: Pipeline Analysts, Crude Unit Operators, and Fluid Catalytic Crackers.

Using Cognitive Task Analysis (CTA) methodologies[2] we conducted interviews (n=17) and observations at a large Gulf Coast refinery and at one of the largest pipeline operator facilities in the United States. Interviewees defined the primary tasks for their positions and prioritized which tasks required the most judgment and decision making. We asked each interviewee to recall a critical, non-routine situation in which he/she was the primary decision maker. This specific incident provided us with a salient example to discuss and revealed the expertise required to perform in this environment.

These expert stories serve several purposes. First, they provide a glimpse of expertise in context. Second, they provide case studies that can support learning across the organization. Finally, these cases became the foundation of the DMXs. We created scenarios that were realistic and relevant to the operators' performance environment because they were based on events that actually happened on their units versus generic situations.

This phase of work offered interesting insights regarding operator expertise (Harris-Thompson, Long, O'Dea, Gabbard, & Sestokas, 2007).[3]

PIPELINE ANALYST

Pipeline Analysts form expectancies of how their shift will play out by considering

---

[1] This work was conducted for the Center for Operator Performance, a consortium of academic and process companies that research generic issues in human factors and process operator performance.

[2] CTA is a family of methods that document the cognitive processes behind the behaviors and judgments related to a task (Crandall, Klein, & Hoffman, 2006).

[3] Findings reflect positions at two facilities. They are not meant to stand as an industry-wide representation; although some of the findings may be generalized across other facilities.

at sufficient depth how events will unfold over a 6 to 12 to 24 hour time period. These expectancies allow the Pipeline Analyst to develop a baseline against which to compare events as they unfold, more quickly detect anomalies, and make faster decisions. Decision making was cited as an extremely important skill: recognizing what will happen and when, where to go for critical information, what screens to reference in the control systems, and what help is available from whom. Experts stated that novices have not been exposed to enough realistic events to develop confidence to act in a thoughtful and deliberately paced way. Because the experiences do not occur with enough frequency (and training does not fill the gap), novice Analysts will not have developed these strategies as completely.

Expert Pipeline Analysts can quickly recognize when instruments are not in agreement or that human sources of information disagree with their expectations. This can be an important strategy for anomaly detection or event prevention. For example, when a discrepancy exists between a schedule and a meter reading an expert will recognize the issue, cross-check, and de-conflict information. A novice may do neither or be reluctant to challenge incorrect resources.

Crude Unit Console Operator

Optimizing production is a key goal for Crude Unit Console Operators. This involves knowing which system constraints may limit production, pushing the limits on the constraints, and working around the constraints to reach daily targets. This calls for a process of continuous adjustment or 'tweaking' the system to get the maximum out of it. Spotting ways to remove constraints or work around them is an important skill.

Crude Unit Console Operators use the schematic system to monitor the plant. They use the alarms extensively to alert them when rates and levels begin moving in a particular direction or nearing the edge of the set parameter. They tighten the parameters so they are alerted when the system goes outside these tight limits. The different approaches seem to be connected to how each group is trained.

Expert Operators recognize that the unit is made up of several interdependent units, but they see it as one system. For example, the more expert Crude Unit Console Operators said they often do not have to think sequentially about how an action is going to affect the rest of the system. Rather they "know" how an action or adjustment will affect the whole system. This understanding is developed through observing and developing an understanding of how the system has reacted in the past during different events. They also actively seek information from inside and outside parties to determine how their actions may affect the plant state. This proactive stance helps them catch issues early and make adjustments to head off an upset.

Fluid Catalytic Cracker (FCC) Console Operator

Fluid Catalytic Cracker Console Operators form two main categories of expectancies: about the unit and about how external circumstances will affect the unit. What appears to be different for the FCC Console Operators is the order in which they gain expertise around expectancies. Operators have formed expertise around expectancies of external circumstances first, then expectancies about the unit during a crisis, and finally expectancies about the unit during normal operations. One reason for this sequence appears to be that current experts can utilize their mental models of how to do their work during crisis situations (because they are operating the system manually), but are still trying to form their mental models about how the process controls optimize or run the unit during normal operations.

One of the biggest aspects of expertise for FCC Console Operators is the ability to adapt their performance to handle crisis situations. Part of this expertise comes from personally experiencing those situations and working through identifying, diagnosing, and resolving the issues. The combination of personally experiencing the situation and seeing how the system reacts allows them to anticipate what will happen next.

The Training-Performance Mismatch

The profile summaries highlight an important performance requirement – proactive behavior. Maintaining safety and efficiency hinges largely on the operators' abilities to spot and prevent issues from arising or worsening. This includes planning, scheduling, monitoring for prevention, and considering the impact of one's own actions on downstream operations. Considering this, it was interesting to hear from many operators that their simulation-based training did more to develop skills for reactive response than proactive prevention. The simulators provided a key training benefit by immersing trainees in a similar physical environment. However, there were deficiencies in the simulators that were not being addressed through other training (findings related to simulators are based on data collected in this effort; not documented as a standard for simulators across the industry).

The simulator trained Crude Unit and FCC Console Operators on baseline skills. Many operators do not gain experience managing critical events until they occur on the job. Because of this, some Crude Unit and FCC Console Operators did not feel fully prepared to manage an upset if it occurred during their shift.

For Pipeline Analysts, the simulators were primarily used to train critical events. As one Analyst stated, when they step into the simulator, they know it will be a critical event. The problem is stated very clearly: "there is a leak, how do you fix it?" What the simulator does not develop is the skill to recognize small changes in the system, understand the causes of problems, and troubleshoot effectively (or work proactively to prevent the problem).

The second phase of work addressed this training - performance mismatch by providing low-cost DMXs to supplement existing training and close the gap between the mastery of procedural knowledge and the development of expertise.

USING DECISION MAKING EXERCISES

By this point, your one-minute challenge from the introduction is up. What do you think is happening in the situation? Consider the next part of that same situation.

You have the outside operator to check the gas oil. The operator reports that the feed looks normal. The crude temperature has been 350 degrees for the entire shift. You confirm that the gas oil hydrotreater is operating correctly. You reduce the feed to the reactor, but it does not stop the pressure increase in the reactor and the main column or the reactor temperature increase. You have three minutes. What do you do now?

Decision Making Exercises have emerged as an effective tool for training high-level thinking skills because they pose realistic challenges to which trainees must respond. This enables intense focus on specific learning points and sharing of expertise across trainees. The experience allows novices to develop recognition of salient situational cues, expectancies, typical goals, and typically successful actions. This recognition helps operators diagnose a situation and quickly generate a course of action.

Three characteristics distinguish DMXs from other forms of training. First, procedural training often ends when the "right" solution is reached. Often, it is not easy to verify how the trainee reached the right solution (through knowledge or luck?) until the trainee becomes an operator and faces the situation in real time. In a DMX, the learning event continues beyond the first solution, as the trainees get more data that may verify their situational assessment or make them rethink it completely. Second, traditional training often gives trainees unrealistic time frames to figure out a problem. DMXs mimic, as closely as possible, the time constraints that an operator would face in the real world. If an operator only has three minutes on the job to interpret data and carry out actions to prevent a problem from escalating, then training should give the trainee three minutes; not 15 minutes or more. Finally, DMXs are designed to be facilitated using the Socratic Method, uncovering the "why" behind decision-making. Skilled facilitators explore how trainees distinguish between relevant and irrelevant data; aggregate data; assess situations; understand how to use systems; and recognize when data can mislead them.

DMXs in Refinery Training

DMXs are typically delivered to a mixed group of expert and novice participants. The value is in having more experienced workers share their expertise in a specific situational context. This is very important to note. There is a general assumption that putting experienced and less-experienced workers together at a console will naturally facilitate learning. Unfortunately, this is usually not realized because experts typically find it difficult to articulate their expertise. In the DMX session, experts and novices think through the problem together and discuss not just their actions, but what they are noticing, why it is important, and how it is contributing to their decision making. Participants bring different perspectives and responses to the session. This facilitates a greater value of learning than training that is structured around passive interactions (the trainer talks while the trainee listens).

This project allowed us to demonstrate the flexibility of DMXs. The refinery we worked with incorporated the DMXs into its existing one-to-one training format. The trainee, either novice or expert, goes through the exercise as he/she would in a group format. The important component (as with any training) is the facilitator. The facilitator of the DMXs is experienced in managing the type of challenge the DMX poses to the trainee. In events when the trainer does not have this experience, an expert operator supports facilitation of the exercise. This allows the discussion around situation assessment and decision making to have the same interaction and context-specific learning as a group session.

The effectiveness of DMXs in supporting the development of expert skills such as situation awareness has been evaluated with small-unit military leaders (Baxter, Harris, & Phillips, 2004). In this first application in the refinery, an ROI evaluation has not yet been conducted. Anecdotal evidence, however, points to a number of training and organizational benefits.

Uncovering Mental Models

Simulators do a good job of recreating operators' system interfaces. Novices can learn the range of data the system provides, the technical classification of that data, and how to use the controls to manage the larger unit. But, simply watching a trainee navigate the system interface and adjust controls in the simulator will not allow a trainer to understand the trainee's mental model of the system. The trainer will know which actions a trainee took, but not necessarily why. In the past, we have heard trainers question why a trainee got something wrong in the simulator, but during the debrief the trainee knew exactly what should have been done. The reason is because what the trainer observed as an "incorrect response" from the trainee in the simulator was often the result of an incorrect assessment of what was happening in the situation. The action may have been appropriate for the situation as the trainee understood it. Once the trainer provided an explanation of the situation during the debrief, the trainee provided a situational-appropriate ("correct") response.

The Socratic questioning used when facilitating DMXs uncovers why the trainee took a particular action. For each action the trainee takes, the trainer can ascertain the cues and information the trainee was paying attention to, the trainee's goals, and what the trainee expected to happen as a result of his/her actions. This approach makes the trainee's mental model and span of control of the system transparent.

Identifying Skill Development Needs

The greatest benefit the refinery has realized from the DMX training is the knowledge they have gained about their younger operators. The sessions have allowed trainers to target content to address specific performance weaknesses that trainers and managers did not know existed before the operator participated in the DMX training. We heard examples where operators who were considered novices have proven to be "sharper than expected" in the DMX session; demonstrating a comprehensive understanding of the systems they managed, the larger unit, and the downstream effects of their actions. Interestingly, we also heard examples of operators who were considered to be more expert struggling with the challenges presented in the DMX. Perhaps a long tenure on the system led to latency in skills (which was not being supported through refresher training); or, the operator had not experienced the types of situations that would expose skill deficiencies on the console.

Using DMXs to document skill gaps can provide a resource and efficiency benefit; allowing the refinery to focus training resources to close specific skill gaps rather than "broad brushing" everyone with the same content.

Maximizing Simulators

Though DMXs are not intended to be a replacement for simulators, they can enhance their value. The physical fidelity of simulators is important for some learning objectives, such as the assessment of visual cues in the environment. However, physical fidelity is not a substantial requirement for developing cognitive skills such as problem detection.

The key consideration when training for decision-making contexts is its *cognitive authenticity* (Ross, Halterman, Pierce, & Ross, 1998; Ross & Pierce, 2000), which is the emulation of the features that an expert would perceive in the performance environment that support perception and decision-making. DMXs provide richer content options for simulators. The trainee can be presented with a challenge (such as the example scenario presented in this paper), make adjustments to the system based on his/her assessment of the situation, and see the system response to his/her action through the simulation. This will allow trainers to set up the simulator to target challenging areas, such as pumps or compressors; making each operator's time in the simulator more efficient.

Long-Term Cost Savings

Prior to adopting DMXs in their training, the Gulf Coast refinery brought operators in on their off days to spend a half day in the "generic" simulator. This required organizational resources, as well as a personal commitment by the operators. Currently, the refinery is using DMXs to restructure this training design. Rather than spending their off-time in the simulator, operators begin their regular shift one hour earlier and participate in a DMX session before starting the shift. As DMXs are rolled out to more positions, training on off days will occur less frequently.

SUMMARY

In summary, DMXs are cost effective scenarios that can support training in individual and team structures; provide content for classroom-based training; and enhance simulators. The format of these exercises lends itself to easy modification creating opportunities for more diversity in training content (e.g., focus on different areas of skill development; develop unit- or system-specific expertise) and create different levels of complexity (vary challenges for novice to expert participants). This can be done faster, easier, and with less cost than it would take to rewrite classroom curriculum or modify simulators.

REFERENCES

Baxter, H. C., Harris, D., & Phillips, J. K. (2004, Dec. 6-9). *Sensemaking: A cognitive approach to training situation awareness.* Paper presented at the Interservice/Industry Training, Simulation, and Education Conference, Orlando, FL.

Crandall, B., Klein, G., & Hoffman, R. R. (2006). *Working minds: A practitioner's guide to Cognitive Task Analysis.* Cambridge, MA: The MIT Press.

Harris-Thompson, D., Long, W. G., O'Dea, A., Gabbard, S. R., & Sestokas, J. M. (2007). *Center for operator performance pilot study: The nature of expertise in the control of continuous processes* (Final Report prepared for the Center for Operator Performance). Fairborn, OH: Klein Associates Division, Applied Research Associates.

Hoffman, R. R. (1998). How can expertise be defined? Implications of research from cognitive psychology. In W. F. R. Williams, J. Fleck (Ed.), *Exploring expertise* (pp. 81-100). New York: MacMillan.

Klein, G. (1998). *Sources of power: How people make decisions.* Cambridge, MA: MIT Press.

Klein, G. A. (1993). A recognition-primed decision (RPD) model of rapid decision making. In G. A. Klein, J. Orasanu, R. Calderwood & C. E. Zsambok (Eds.), *Decision making in action: Models and methods* (pp. 138-147). Norwood, NJ: Ablex.

Ross, K. G., Halterman, J. A., Pierce, L. G., & Ross, W. A. (1998). *Preparing for the instructional technology gap: A constructivist approach.* Paper presented at the 1998 Interservice/Industry Training, Simulation, and Education Conference, Orlando, FL.

Ross, K. G., & Pierce, L. G. (2000). Cognitive engineering of training for adaptive battlefield thinking. In *IEA 14th Triennial Congress and HFES 44th Annual Meeting* (Vol. 2, pp. 410-413). Santa Monica, CA: Human Factors.

Strobhar, D. A., & Harris-Thompson, D. (2009). Build operator expertise faster, *ChemicalProcessing.com.*

Chapter 78

Using Virtual Reality for Interior Colors Selection and Evaluation by the Elderly

Cristina Pacheco[1], Emília Duarte[1], Francisco Rebelo[2], Júlia Teles[3]

[1] UNIDCOM/IADE – Superior School of Design,
Av. D. Carlos I, 4, 1200-649 Lisbon, PORTUGAL

[2] Ergonomics Laboratory, FMH/Technical University of Lisbon
Estrada da Costa, 1499-002 Cruz Quebrada, Dafundo, PORTUGAL

[3] Mathematics Unit, FMH /Technical University of Lisbon
Estrada da Costa, 1499-002 Cruz Quebrada, Dafundo, PORTUGAL

ABSTRACT

The aim of this study was to assess the potentiality of Virtual Reality as a technique to be adopted for color selection and evaluation purposes, by the elderly, in interior design studies. Two processes were compared: Paper-based illustrations and Virtual Environments. Twenty users, males and females, with age over 60 years, were requested to evaluate four bedrooms, colored with four distinct colors, after observing it in both display mediums and choose the preferred solution. The results revealed no statistical significant differences for the selected color and subjective evaluation made to the environments, in function of the display medium. However, significant differences were verified regarding variables as realism, interaction level, sense of presence, chromatic fidelity, general interest and amusement.

Keywords: Virtual reality, interior design, color, elderly

INTRODUCTION

The purpose of the study was to assess the potentiality of Virtual Reality as technique to be adopted for color selection and evaluation purposes, by the elderly, in interior design studies.

Previous research has demonstrated that environmental stimuli such as light and color can affect both mood and/or behavior (e.g. Knez, 2001, Knez & Kers, 2000, Kuller, Ballal, Laike, Mikellides & Tonello, 2006, Baron, Rea & Daniels, 1992, Dijkstra, Pieterse & Pruyn, 2008, Yildirim, Akalin-Baskaya & Hidayetoglu, 2007). It has generally been concluded that blue environments evoke better feelings than orange environments do (e.g.Valdez & Mehrabian, 1994). Stone (2003) suggests that blue is a calming color and red is a stimulating color, which may interact with other environmental factors. Jacobs and Suess (1975) found that subjects exposed to warm red and yellow colors reported higher levels of anxiety than did the subjects exposed to cool blue and green colors. Kwallek and Lewis (1990) found that individuals who worked in a red environment, as opposed to white or green environments, revealed lower amount of confusion. Babin, Hardesty & Suter (2003), in a study about store design and shopping intentions, suggest that it is generally expected that violet/blue interiors will produce higher levels of positive affective tone and increased purchase intentions than red/orange interiors. Furthermore, color affects the way people evaluate some environmental features such as dimensions and obstacles (e.g. Acking & Kuller, 1972) and detect a given stimuli present in cluttered environments (e.g. Jansson, Marlow & Bristow, 2004).

The effects of the physical aspects of the environment are of particular importance for the elderly, whose sensorial and motor capacities are diminished in consequence of the aging process. Also it was found an effect of age on evaluation of illuminance and the color temperature of the lighting (e.g. Knez & Kers, 2000).

As consequence of an increasing life expectancy, elderly is a growing group in our societies that consequently are getting older. It is estimated that by 2030 the percentage of the population older than age 65 ranges from 12 to 24 percent (Fisk et al., 2004). Such change in demographics is placing new challenges to all players involved in the design of our environments, since inclusive design solutions are becoming even more critical for our societies. Embracing older users' needs in the projects not only reinforce their inclusion, but also will increase the quality of life for all users in general.

A user-centered design approach can help designers to produce more usable, accessible, adequate and safe solutions. Such approach involves having potential users participating in the design process, in diverse iteration moments, from the early stages of the process until an adequate solution is found. For obvious reasons, testing environment variables such as color, in real settings, is not easy to do. Alternatively, simulation has already shown that is a good way to introduce end-users to the design process (Hunt, 1993). But, real scale models are expensive, time and space consuming, which turns such option incompatible with most research resources. Therefore, paper-based prototypes (e.g. drawings, photographs), physical mock-ups or digital prototypes (e.g. CAD) have been the most common options for presenting the solutions to the potential users. But, nowadays, thanks to recent technology developments, Virtual Reality (VR) is becoming an alternative with good potential for promoting user testing in diverse research areas (Burdea & Coiffet, 2003).

Virtual Reality advantages, as a simulation process, can be grouped in: availability, safety, and provision of data. Regarding availability, VR can allow access to almost all locations (e.g. private homes, restricted access buildings), under specified environmental conditions (e.g. day, night, smoke), in a repeatable and systematic way, without the time and the costs required if it was a real setup. Concerning safety, VR provides a mean for experience the environment in a safe mode, even if exposed to controlled critical conditions (e.g. egress during a fire). If used as a resource for training, it may contribute to reduce accidents and injuries. Finally, about provision of data, VR simulation provides the opportunity to collect data, with good accuracy that is not easily available in the real world settings. For example, VR allows the manipulation of design variables (e.g. color, light, dimensions and materials), in a relatively simple but accurate way, which would not be possible with other research processes. Also, while interacting with a Virtual Environment (VE), participants can be involved in a scenario, with a list of tasks to perform, which is essential for the ecological validity of the research. Virtual Reality can be used from the earlier stages of the design cycle, in order to find as must unconformities as possible, before spending too much time and money with solutions that might not be worthwhile.

Considering the just exposed and the fact that few studies have investigated the interaction of the elderly with VR, this study had the main goal of determine the potentiality of VR for interior color selection and evaluation by the elderly. Two design display mediums were compared: Paper-based illustration and Virtual Environment. Twenty users, males and females, with age over 60 years, were requested to choose and evaluate four bedrooms, colored with four distinct colors, after observing them in both display mediums.

METHOD

PARTICIPANTS

Twenty subjects, 7 (35.00%) males and 13 (65.00%) females, with ages ranging from 60 to 85 years old (mean = 69.45, SD = 8.672) participated in this study. Participants were randomly distributed by two groups of 10 subjects each. Each group of individuals was assigned to both experimental conditions in a reverse order. This procedure was adopted to evaluate the procedures sequence potential influence on the results.

APPARATUS

The experimental setup consisted of two main areas (Room 1 – Ergonomics Laboratory and Room 2 – VR room). Both rooms were equipped with desks and chairs, since the participants carried out the procedures seated. In Room 1 participants received the initial explanation, underwent the procedure involving paper-based prototypes (illustration) and answered to the questionnaires. In Room 2, participants performed the procedure involving the VR simulation (see Figure 1). The VR room was darkened and protected from outside noise.

The devices used included: 1 magnetic motion tracker from Ascension-Tech, model Flock of Birds for monitoring the head movement; a joystick from Thrustmaster as a locomotion device; a Head-Mounted-Display (HMD) from Sony, model PLM-S700E; wireless headphones and a graphics workstation.

The VE was presented at a resolution of 800 x 600 pixels, at 32 bits, with a FOV 30°H, 18°V and 35° diagonal. The participants' viewpoint was egocentric. The VE displayed in the HMD was also simultaneously displayed in a second monitor. Thus, the researcher could see the same image as the participants did. All participant sessions were videotaped.

The *ErgoVR* system, developed in the Ergonomics Laboratory, of the FMH - Technical University of Lisbon, allowed not only the display of the VE but also the automatic collection of data such as the duration of the simulation, distance and path taken by the participants, among others.

THE VIRTUAL ENVIRONMENT

For this study two VEs were developed, one for training and other for the experimental session. The training VE consisted of two rooms with 65 m^2 (5x13m), without windows, connected by a door and containing a number of obstacles (e.g. narrow corridors, pillars, etc.), requiring some skill to be circumvented. The purpose of this practice trial was to get the participants acquainted with the setup and to make a preliminary check for any initial indications of simulator sickness.

The VE designed for testing, that is, the experimental environment, consisted of a distribution hall, with direct access to four bedrooms, corresponding to four distinct color environments (see Figure 2). Each room has an outside window with curtains and the following equipment: 1 bed, 2 bedside tables, 2 table lamps, 1 ceiling lamp, 1 sofa, 1 wardrobe, 1 dresser and 1 television set.

The base structure of the VE was designed using *AutoCAD® 2009*, and after imported into *3D Studio Max® 2009* (both from Autodesk, Inc.). The scenario was then exported using a free plug-in called *OgreMax*, to be used by the *ErgoVR* system.

Figure 1. Participant during the simulation

Figure 2. Experimental VE floor plan

THE PAPER-BASED MEDIUM

Four images, gathered from the simulation, and showing distinct points-of-view of the bedrooms, were selected for the paper-based condition. Such images were printed with photographic quality and applied to paperboards with 42 x 59.40 cm. Since it was not possible to use color analysis equipment (e.g. spectrophotometer) in the VE displayed in the HMD, several subjective analysis, made by a group of 6 specialists (researchers at the Ergonomics Laboratory), were conducted to ensure the maximum color equivalence between the stimuli displayed in both mediums.

DESIGN OF THE STUDY AND PROCEDURE

The study used a within-participants design with 2 experimental conditions: Paper-based illustrations, and Virtual Reality. The process used to display the VE for user testing characterizes these conditions. The scenario used for testing (bedroom) was the same for both conditions, varying in the color schema (white, beige, green and blue). Such colors resulted from previous questionnaires applied to end-users, heuristics and from literature, not discussed in this paper.

The procedure was divided in three major steps: pre-experimental training, experimental session, and post-hoc questionnaire.

In the pre-experimental training, participants were given a brief explanation about the study and were introduced to the equipment. The Ishihara Test (Ishihara, 1988) was used to detect color vision deficiencies. Next, participants were asked to sign a consent form and advised they could stop the simulation at any time. The pre-experimental training ended as soon as the participants presented an adequate level of navigation skill (i.e. were able to circumvent obstacles and move without colliding with the walls) and expressed that they were comfortable with the interaction devices to start the simulation.

In the experimental session it was asked each participant to observe the 4 chromatic environments, in paper-based and the VR condition, according to the sequence they were assigned and, in the end, choose the one they liked most. After the choice made, they had to answer a questionnaire, after which they executed the other condition and repeat all the procedure (observation, choice and evaluation by questionnaire). In this case, participants should indicate if the choice they made was the same of the previous condition or if they desired to change the chosen chromatic environment. There were no time restrictions and it was possible to review each chromatic environment as many times as wanted.

RESULTS

All the participants ended the procedure and there were no cases of simulator sickness.

The Wilcoxon-Mann-Whitney Test tested the existence of an eventual influence of the sequence of the experimental condition, regarding the chromatic choices and evaluations made by the participants, for a significance level of 5%. The results show that there is no statistically significant difference, both in chromatic choices or the responses to the questionnaires, in function of the sequence of the application of the experimental condition. In this sense, the sequence of the application was ignored in the remaining statistical tests.

It was carried out the Chi-square test of independence between the factors *display medium* (Paper-based and VR) and *color* (white, beige, green and blue). No statistically significant association between the two factors was found, either for the first or the second experimental condition to which the participants belonged. The fact that no statistically significant association was found could be due to the small sample size. Yet, empirically, it was noted some differences in the choices of the chromatic environments depending on the experimental condition (see Table 1).

Table 1 Crosstab for color environment and experimental condition

	Color environment			
	White	**Beige**	**Green**	**Blue**
Virtual Reality condition	5 (25.00%)	7 (35.00%)	4 (20.00%)	4 (20.00%)
Paper condition	7 (35.00%)	6 (30.00%)	3 (15.00%)	4 (20.00%)
Both	12 (30.00%)	13 (32.50%)	7 (17.50%)	8 (20.00%)

There were 4 cases in which the participants wished to change the first choice, after viewing the chromatic environment in the second experimental condition (all had been exposed in the first place to paper-based condition). The green and beige were the cases where there were more changes related to the experimental condition. The results also show that in the paper-based condition, white was the most chosen color while in the VR condition, beige and green were favorites. The choice of blue was similar in both display mediums and chosen by the same participants.

In addition, participants were asked to make a subjective rating of the chromatic environments, in both conditions, using a semantic differential scale (7-point scale from -3 to 3). It was used 8 antagonistic pairs of adjectives (cold-hot; dark-light; sad-happy; little-big; heavy-light; repulsive-attractive; aggressive-peaceful and conservative-extravagant). The hypothesis that this rating could be different between the two experimental conditions was tested with the Wilcoxon-Mann-Whitney test, to a value of 5% significance. When comparing the two allocation sequences of the experimental conditions, the results show that there are no significant differences in the ratings of the 8 semantic differentials that characterize the environments.

In the post-hoc final questionnaire participants were asked to compare the two experimental conditions regarding variables as realism, interaction, sense of presence, color fidelity, interest and enjoyment. The Binomial Test was used to assess the existence of an eventual influence of the display medium on the variables just mentioned. The inferential statistical analysis indicates significant differences, between both display mediums, for all variables (realism: $p = 0.001$, interaction: $p < 0.001$; presence: $p < 0.001$; color fidelity: $p = 0.029$; interest: $p < 0.001$ and amusement: $p < 0.001$). The majority of the respondents considered the VR medium as the one that produced higher levels of realism, interaction, presence, color fidelity, interest and enjoyment.

CONCLUSIONS

This study aimed to assess the potentiality of Virtual Reality as a technique to be adopted for color selection and evaluation purposes, by the elderly, in interior design studies. It also had the purpose to determine the extent to which subjects over 60 years old would be able to interact with the interfaces used in immersive Virtual Reality and how they would rate such experience. For such purpose, a task of chromatic environments selection and evaluation was designed and two display mediums were compared: Paper-based illustration and Virtual Reality.

Twenty users, males and females, with age over 60 years, were requested to evaluate four bedrooms, corresponding to four chromatic environments, after observing them in both display mediums and, after, choose the preferred color.

The results show no statistically significant differences in the colors selected and subjective evaluation made to the color environments, in function of the display medium (Paper-based and Virtual Reality). This fact can be due to the small sample size. Yet, some differences were empirically observed in the choices, in function of the display medium. Such result, indicating that the chromatic environment was evaluated in an equivalent manner in both display mediums can be understood as a positive outcome, since it could mean equivalence between processes. However, results reveal that both display mediums were evaluated, with a significant difference, regarding variables such as realism, interaction, presence, color fidelity, interest and amusement.

This study shows that VR can be applied in studies involving the elderly. All participants completed the procedure and have not been verified cases of simulator sickness. Also, time spent on the training phase and the navigation skills demonstrated by this sample of elderly users, was similar to the verified in previous studies with younger users (university students).

Studies with larger samples that compare different age groups could contribute to a greater knowledge about age differences regarding color issues. Also, further research regarding issues as pictorial realism, quality of the images displayed by VR displays, interaction and locomotion devices is necessary.

REFERENCES

Acking, C. A. & Kuller, H. (1972). The Perception of an Interior as a Function of its Colour, *Ergonomics, 15*(6), 645-654.

Babin, B.J., Hardesty, D.M. & Suter, T.A. (2003). Color and shopping intentions: the intervening effect of price fairness and perceived affect. *Journal of Business Research*, 56, 541–51.

Baron, R.A., Rea, M.S. & Daniels, S.G. (1992). Effects of Indoor Lighting (Illuminance and Spectral Distribution) on the Performance of Cognitive Tasks and Interpersonal Behaviors: The Potential Mediating Role of Positive Affect. *Motivation and Emotion, 16*(1), 1-33.

Burdea, G., & Coiffet, P. (2003). *Virtual reality technology* (2nd ed.). John Wiley&Sons, Inc.

Dijkstra, K., Pieterse, M.E. & Pruyn, A. Th. H. (2008). Individual differences in reactions towards color in simulated healthcare environments: The role of stimulus screening ability. *Journal of Environmental Psychology, 28*, 268–277

Fisk, A. D., Rogers, W. A., Charness, N., Czaja, S. J. & Sharit, J. (2009). *Designing for Older Adults: Principles and Creative Human Factors Approaches* (2nd ed.): CRC Press.

Hunt, M. E. (1993). Research for aging society, in *Environmental Simulation*. Robert W. Marans and Daniel Stoklos (Eds). New York: Plenum Press, p.98.

Ishihara, S. (1988). *Test for Colour-Blindness* (38th ed.). Tokyo: Kanehara & Co., Ltd.

Jacobs, K. W. & Suess, J. F. (1975), Effects of four psychological primary colors on anxiety state, *Percept. Mot. Skills 41*, 207-210.

Jansson, C., Marlow, N. & Bristow, M. (2004). The influence of colour on visual search times in cluttered environments. *Journal of Marketing Communications, 10*(3), 183-193.

Knez, I. & Kers, C. (2000), Effects of indoor lighting, gender, and age on mood and cognitive performance. *Environment and Behavior, 32*, 817–831.

Knez, Igor (2001). Effects of colour of light on nonvisual psychological processes. *Journal of Environmental Psychology, 21*, 201-208.

Kuller, R. Ballal, S., Laike, T. Mikellides, B. & Tonello, G. (2006). The impact of light and colour on psychological mood: across-cultural study of indoor work environments. *Ergonomics, 49*(14), 1496-1507.

Kwallek, N. & Lewis, C.M. (1990). Effects of environmental color on males and females: a red or white or green office. *Applied Economics, 21*, 275–8.

Stone, N.J. (2003). Environmental view and color for a simulated telemarketing task. *Journal of Environmental Psychology, 23*, 63–78.

Valdez, P. & Mehrabian, A. (1994). Effects of color on emotion. *Journal of Environmental Psychology, 123*(4), 394–409.

Yildirim, K., Akalin-Baskaya, A. & Hidayetoglu, M. L. (2007). Effects of indoor color on mood and cognitive performance. *Building and Environment, 42*, 3233–3240.

Chapter 79

Human Interaction Data Acquisition Software for Virtual Reality: A User-Centered Design Approach

Luís Teixeira[a], Francisco Rebelo[a], Ernesto Filgueiras[a,b]

Ergonomics Laboratory
[a] FMH – Technical University of Lisbon – Portugal
[b] Beira Interior University – Covilhã – Portugal

ABSTRACT

This paper presents a Virtual Reality (VR) system – *ErgoVR* – that is being developed to allow the visualization and measurement of Human behavior interaction's variables in virtual environments, to be mainly used in studies in Ergonomics in Design. The first section of the paper justifies this system's creation compared to the commercial solutions available. The second section describes the development of *ErgoVR*, in particular the definition of its concept and implementation in a User-Centered Design perspective. As the most important feature of *ErgoVR*, stands the automatic data collection of variables of behavioral interaction such as: dislocation paths, trajectories, collisions with objects, orientation of the field of view and occurrence of events triggered by user's actions. *ErgoVR* is being developed and validated in studies at the Ergonomics Laboratory of FMH - Technical University of Lisbon and serves as a tool for two research projects, supported by a grant from the Portuguese Science Foundation.

Keywords: Human interaction, Virtual Reality, Automatic data collection, Ergonomics, Design.

INTRODUCTION

Virtual Reality (VR) is a technology that can be used in behavior evaluation studies that can replace conventional methods of research such observation of behavior in natural settings (e.g. Drury, 1995; Westbrook, Ampt, Kearney & Rob, 2008). The main disadvantages of these methods are that is not ethical to place an individual in hazardous situations for research purposes and the development of believable experimental scenarios is difficult and has usually high financial costs associated. Virtual Reality has the possibility to overcome, or minimize, those limitations.

Although some studies use VR to evaluate human interaction, for example in safety signs in emergency evacuation studies (Ren, Chen, & Luo, 2008; Smith & Trenholme, 2009; Tang, Wu & Lin, 2009), it is not evident the use of a system that collects, in an automatic manner, data related with human interaction that can be used for Ergonomics in Design. Commercial VR solutions (e.g. *Virtools*, *WorldViz Vizard*) that can be integrated with different hardware equipment do not allow data collection of human interaction without the development of additional software. Moreover, they are quite expensive, taking into account most of the research budgets.

Considering that the Virtual Reality unit at the Ergonomics Laboratory of FMH – Technical University of Lisbon has been developing a pilot facility to support the development of Ergonomics in Design studies and the above mentioned limitations, was defined the main concept of *ErgoVR* as a system that is able to automatically collect data regarding human behavior interaction. This system is to be used mainly in Ergonomics in Design studies that centers in the development of new products and systems intended to be safer and more comfortable for the users and promoting, at the same time, the efficiency of those systems.

This system was developed in a User Centered Design (UCD) perspective accordingly to ISO 13407 standard (ISO, 1999), involving the potential users throughout the development process. In this sense, a UCD approach involves the participation of the potential users in design decisions, according to the difficulties that they express during the interaction with several prototypes of the product. In the earlier stages, users may be involved in the evaluation of use scenarios, paper reproductions or partial prototypes. As the design solutions become more elaborate, the evaluations can be based on more complete and concrete versions of the product. According to McClelland and Suri (2005), the elements normally tested are: (1) descriptions of concepts of the product on paper, (2) partial prototypes or simulations, (3) complete prototypes and (4) complete products.

On the remaining of this article it is described the development of *ErgoVR*, in particular the definition of its concept and implementation in a UCD perspective.

THE *ERGOVR* SYSTEM

As mentioned, the *ErgoVR* system was developed in a User-Centered Design perspective, according to ISO 13407.

User-Centered Design is an iterative process that has been used successfully in product design (Kanis, 1998; Chen, Sato & Lee, 2009). The process has four major stages. The process begins with the identification of the need of this process. After the need is identified, the first stage starts which is the understanding and specification of the context of use. The second stage is where the user and organizational requirements are specified. In the third stage, design solutions are produced. In the fourth stage, designs proposed on the previous stage are evaluated against the requirements. If there are changes to be made to the system the iterative cycle is completed, passing to the first stage with the new changes to be made. The cycle ends on the fourth stage when the requirements are satisfied.

The different stages of the UCD process related to the development of the *ErgoVR* system are presented below.

UNDERSTAND AND SPECIFY THE CONTEXT OF USE

In two brainstorming meetings, six specialists with expertise in the fields of ergonomics, engineering, psychology, design and architecture, outlined the context of use of the *ErgoVR* system. Since the context of use is given by the characteristics of the users, tasks and the physical and organizational environments, those were also defined and are as follow:

a) The characteristics of the possible users

The system can be used by university students, Masters and PhD students of courses of Ergonomics, Engineering, Psychology, Design and Architecture.

b) The tasks that the users perform

The system can be used for research works mainly in the fields of Design and Ergonomics to support the research community and in development and evaluation of professional or consumer products. For the research community, *ErgoVR* can help to develop studies related for example with: behavior differences between genders, age, education, cultural background, previous knowledge, among others in different contexts of use. Related to the development or evaluation of products, it is possible to assess different aspects of interaction with the product, including emotion, perception, and cognition, among others.

c) The environment in which the users use the system

The system is to be used in a university environment where activities such as teaching, research and scientific and professional community support take place.

SPECIFY THE USER AND ORGANIZATIONAL REQUIREMENTS

The second stage of the UCD process is where the requirements for the system are defined in a general manner.

The requirements were discussed by the same group of specialists, in three focus group meetings. The requirements collected had taken into consideration the overall performance and feasibility of the task in a timely manner.

The users can use the system in a pilot facility, using specific hardware and the *ErgoVR* software, whose general characteristics are:

- Characteristics of the pilot facility – adjustable lighting so that any external stimuli are minimized, temperature around 22 degrees Celsius (around 71.6 degrees Fahrenheit) and sound intensity level below 40 dB. The pilot facility must have image recording equipment to record the external behavior of the participants;
- Characteristics of the hardware – immersive VR equipment, preferably a stereoscopic solution; a computer with a graphics cards capable of present complex virtual scenes at a constant high velocity, so that the effects of simulation sickness that the participants might experience can be reduced; motion sensors for body movement capture; equipment to use as dislocation interfaces in the virtual world and a device to interact with objects in the virtual world;
- Characteristics of the software – low cost system, preferably free for research use that works at least on Microsoft® Windows and can interact with the VR equipment in the pilot facility. It also must gather automatically data about the human interaction, present and extrapolate information from that data in different formats.

PRODUCE DESIGN SOLUTIONS

The potential design solutions are produced using the state of art, the experience and knowledge of the participants and the results of the analysis of context of use.

a) Use of existing knowledge to develop design proposals with multi-disciplinary input

Taking as its starting points the results of earlier phases, the same group of specialists proposed solutions to the development of *ErgoVR*. It was also taken into account the characteristics of the main commercial programs used for VR simulations such as *3DVIA Virtools* and *WorldViz Vizard*, although they are expensive solutions.

b) Make the design solutions more concrete using simulations, models, mock-ups, etc.

At this point more specific characteristics were defined the *ErgoVR* system.

The pilot facility consists of a room without windows to the outside, can be darkened and is protected from outside noise. The temperature, humidity and air quality of the room can be adjusted taking into account the season and type of use intended. The pilot facility has four Bosch WZ18 video cameras to record the participant's activity that are recorded synchronized with the images from the virtual environment.

The hardware components are: (1) a computer, (2) two Head-Mounted Displays (HMDs), (3) motion sensors, (4) a VR data glove and (5) some interaction devices to be used for navigation.

1) The computer has a graphics card (NVIDIA® Quadro® FX4600) that is capable of displaying complex scenes at a constant refresh rate. It can also display in full screen a virtual world from the two graphics outputs. The computer has a Intel® Quad-Core processor and 8 Gigabytes of RAM;

2) For presenting the Virtual Environments (VE) to the participant, there are two HMDs. One is a Sony® PLM-700S and the other is a Sensics® piSight, model 145-41b with stereoscopic capabilities and a wide horizontal field of view (144° instead of the 28° presented by the Sony model);

3) For capturing the movement of the participant is possible to make use of three Ascension-Tech® Flock of Birds magnetic sensors. Although this type of sensors is sensitive to metal has high precision and high frequency (around 144Hz);

4) To interact with the VE, it is possible to use a 5DT Data Glove 5 Ultra that detects the flexing done by the fingers and hand of the participant;

5) The devices that can be used for navigation currently available are the keyboard and mouse combination, joysticks, a 3Dconnexion SpaceNavigator 3D mouse and the Nintendo® Wii Balance Board.

ErgoVR uses several free and open-source code libraries to accomplish the proposed requirements. *ErgoVR* is composed by several components: (1) graphics engine, (2) physics engine, (3) sound engine, (4) device control, (5) event system, (6) log system, (7) log viewer, among others.

1) The graphics engine is what allows to present visual images to the participant. This engine saves information of which objects exist in the current environment and their relative position. With that information, is presented to the participant an image in the output device that is being used at the moment;

2) The physics engine allows the detection of collision with objects and introduces physical behavior to the object (e.g. if the user collides with a ball inside the virtual environment, the ball will react accordingly);

3) The sound engine allows presenting to the participant sounds with effects such as 3D origin of the sound, occlusion and obstruction by objects. The

sound can be presented to the participant in a headphone set or in speakers;

4) The device control (motion sensors and navigation interfaces) creates a bridge between the hardware's input (e.g. position/orientation of the motion sensor) and the software which uses the data received by the hardware and creates a response in the system where there is a correspondence between the external movement of the participant and its virtual representation;

5) The event system allows the placement of invisible triggers in the virtual environment so that specific events occur in response to the participant's actions. For example, it is possible to activate an animation of a door opening when the participant passes over a specific area in the virtual environment or when the participant interacts with a specific object (e.g. a button) a sound can be played or a particle system started;

6) The most important component of the *ErgoVR* system is the log system that records data automatically. Data is recorded around 60 times per second (although configurable) about the participant's position, orientation of its field of view and occurred events during the simulation (e.g. collision with some object, activated trigger, detection of direction taken in an intersection, detection if the participant looked at a specific object, detection of sudden movements);

7) There are two log viewers. The first one allows reviewing the simulation made by the participant from three different points of view (the same view that the participant had, a third person view where a representation of the participant is on-screen and a top view where is shown a line representing the participants' path). The second viewer allows the visualization of the simulation's data, the path of the participant and extrapolates data from the recorded one. From the recorded data it is possible to extrapolate the following variables: number of collisions with objects and together with other information it is possible to obtain an indicator on how comfortable the participant was with the navigation devices; the activation of specific triggers can show the actions made by the participant and delimit different areas in the virtual environment for later analysis; the detection of the choice of direction can be used, for example, to verify the compliance with direction signs present in the virtual environment; the detection if the participant looked at a specific object is an indicator of its cognitive process since the fixation time may be taken as an indicator of the level of attention of the participant for that particular object; the detection of sudden movements can be used to know if there were a hesitation or if the participant got scared with some event and had a spontaneous reaction, what could be an indicator of immersion according to Held and Durlach (1987, 1992).

c) **Present the design solutions to users and allow them to perform tasks (or simulated tasks)**

For the first pilot tests to evaluate the pilot facility, hardware and the *ErgoVR* system, it was created a simple scene for simulation. This scene consisted in a 25x15 meter space, with four rooms and a cross corridor that separated them. Each room was around 9.5 x 6 meters and a different theme associated (reception desk, meeting room, waiting room and an office).

There were five participants in the pilot tests and they were told to explore the environment freely between 5 to 7 minutes. During each simulation, the *ErgoVR* system was collecting data automatically for later analysis. After the simulation, each participant was asked for commentaries, suggestions and for the difficulties encountered during the simulation.

With the outputs from the *ErgoVR* system and the results from the interviews to the participants, the same group of specialists had two meetings to analyze the data and propose modifications to the system if necessary.

d) Alter the design in response to the user feedback and iterate this process until the human-centered design goals are met

With the information generated after the meetings mentioned in the last topic, changes were made to the system to correct some of the problems noticed by the participants and by the group of specialists. Some changes were done regarding the sensitivity of the joystick as a navigation device. The possibility to interact with objects present in the environment was one change that had to be made, although not considered as a priority in the first iteration. Another change made was regarding the output data of *ErgoVR*. It was considered important to present the path made by the participant in a top view of the environment to ease the analysis of the data.

After the changes made, a new set of tests were performed, in the same conditions of the pilot tests mentioned earlier (five participants and post-interviews). The difference regarding the previous tests was that it was now possible to interact with some buttons in the environment in order to perform some actions.

EVALUATE DESIGNS AGAINST REQUIREMENTS

At the end of each cycle of the iterative process, the implemented solutions are considered against the requirements and the results obtained from the pilot tests. With the considerations resulting from this step new changes and requirements to the system are generated.

At the moment of this writing, improvements are still necessary to the navigation systems in the virtual environments through solutions like the use of the *Nintendo® Wii Balance Board* and the use of motion sensors placed in the inferior members of the participant, to be used as an alternative navigation method. Although the system allows some interaction with object in the environment, there is a need for a greater interaction level (e.g. grabbing/using objects, open/close doors). Such improvements are to be addressed in the next iteration.

CONCLUSION

Virtual Reality is a technology that enables a considerable potential for researchers and professionals to develop or evaluate design solutions. One of the main advantages of VR in this process of developing or evaluating is the possibility to involve the potential users of a product or system, in a User-Centered Design approach, since VR allows developing interactive virtual environments. However, monitoring the interaction behavior in virtual environments is still not a simple process because there are no applications that have been designed exclusively for this purpose. Current VR solutions are mainly focused on ways to enhance visualization of information of new products and systems. In this sense, the User-Centered Design methodology that has being used for the development of *ErgoVR*, allows a considerable success level in a way that integrates systematically the contributions of the future users of this system. As said by Kontogiannis & Embrey (1997), the system has not been designed for the users but with the users.

As the most important feature of *ErgoVR* for studies in the areas of Ergonomics in Design, stands out the possibility to place participants in a hazardous situation inside the virtual environment without compromising their safety. The measurement and analysis of the participant's activity in an automatic manner by *ErgoVR* will allow quantifying, for example, trajectories, direction of the field of view and the occurrence of events triggered by user actions, which could be associated with dangerous behaviors.

Some studies already conducted in the Ergonomics Laboratory of FMH-UTL, in particular to study behavior consonance with warnings and problems, show that a virtual environment can be an effective and relatively inexpensive alternative to studies in the real world. This approach promises to be particularly useful for studies involving hazardous situations (e.g. fires), extreme context of use (e.g. on top of a mountain, underwater, war contexts), expensive products or locations (e.g. skyscrapers, underground stations, airplanes, houses) or products for private use, among others.

ACKNOWLEDGMENTS

This research project was supported by a R&D grant from the Portuguese Science Foundation (FCT) under the project PTDC/PSI/69462/2006.

REFERENCES

3Dconnexion SpaceNavigator. Retrieved February 20th, 2010, from
  http://www.3dconnexion.com/products/spacenavigator.html
3DVIA Virtools. Retrieved February 20th, 2010, from
  http://www.3ds.com/products/3dvia/3dvia-virtools/
Ascension-Tech, model Flock of Birds. Retrieved February 20th, 2010, from
  http://www.ascension-tech.com/realtime/RTflockofBIRDS.php

Autodesk 3ds Max. Retrieved February 20th, 2010, from
http://usa.autodesk.com/adsk/servlet/pc/index?id=13567410&siteID=123112

Chen, C.-H., Sato, K., & Lee, K.-P. (2009). Editorial: Human-centered product design and development. *Advanced Engineering Informatics, 23*(2), 140-141.

Drury, C. G. (1995). Methods for direct observation of performance. *Evaluation of Human Work: A Practical Ergonomics Methodology*.

Held, R. M., & Durlach, N. I. (1987). Telepresence, time delay and adaptation. In S. Ellis, M. Kaiser & A. Erunwald (Eds.), *Pictorial Communication in Virtual and Real Environments*. London: Taylor & Francis.

Held, R. M., & Durlach, N. I. (1992). *Telepresence. Presence: Teleoperators and Virtual Environments, 1*(1), 109-112.

ISO (1999). *Human-centred design processes for interactive systems*. ISO 13407. International Organization for Standardization.

Westbrook, J., Ampt, A., Kearney, L. & Rob, M. (2008), *All in a day's work: an observational study to quantify how and with whom doctors on hospital wards spend their time*, Med. J. Aus. 188 (9) 506–509.

Kanis, H. (1998). Usage centred research for everyday product design. *Applied Ergonomics, 29*(1), 75-82.

Kontogiannis, T., & Embrey, D. (1997). A user-centred design approach for introducing computer-based process information systems. *Applied Ergonomics, 28*(2), 109-119.

Lee, N. S., Park, J. H., & Park, K. S. (1996). Reality and human performance in a virtual world. *International Journal of Industrial Ergonomics, 18*(2-3), 187-191.

Nintendo Wii Balance Board. Retrieved February 20th, 2010, from
http://www.nintendo.pt/NOE/pt_PT/systems/acessrios_1243.html

McClelland, I., & Suri, J. F. (2005). Involving people in design. In J. R. Wilson & E. N. Corlett (Eds.), *Evaluation of human work. A practical ergonomics methodology* (3rd ed., pp. 241-280). Boca Raton, FL: CRC Press.

OgreMax. Retrieved February 20th, 2010, from *http://www.ogremax.com*

Ren, A., Chen, C., & Luo, Y. (2008). Simulation of Emergency Evacuation in Virtual Reality. *Tsinghua Science & Technology*, 13(5), 674-680.

Sensics piSight. Retrieved February 20th, 2010, from
http://www.sensics.com/products/pisightSection/

Smith, S. P., & Trenholme, D. (2009). Rapid prototyping a virtual fire drill environment using computer game technology. *Fire Safety Journal*, 44(4), 559-569.

Tang, C.-H., Wu, W.-T., & Lin, C.-Y. (2009). Using virtual reality to determine how emergency signs facilitate way-finding. *Applied Ergonomics, 40*(4), 722-730.

Watanuki, K. (2009). Development of virtual reality-based universal design review system. [Proceedings Paper]. *Journal of Mechanical Science and Technology, 24*(1), 257-262.

WorldViz Vizard. Retrieved February 20th, 2010, from
http://www.worldviz.com/products/vizard/index.html

Chapter 80

Virtual Reality in Wayfinding Studies: A Pilot Study

Elisângela Vilar, Francisco Rebelo

Ergonomics Laboratory
FMH - Technical University of Lisbon – Portugal

ABSTRACT

Disorientation has many costs, it can lead to physical fatigue, stress, anxiety, frustration and all of them threaten the well-being, limit mobility and can also place the people's safety at risk. In this context, this pilot study was focused in the environment's cues, mainly in guidance systems. Its main objectives are: i) verify the use of Virtual Reality (VR) and immersive environment interaction for data collection related to the Human performance in wayfinding tasks (finding a destination within a building), ii) validate the use of the ErgoVR System for wayfinding studies purpose and iii) evaluate the sense of presence reported by users after the interaction with the virtual environment. ErgoVR System proved to be effective in the register of relevant variables for assessing the performance of Human Wayfinding. Preliminary data suggest also that the performance in accomplishing orientation and navigation task within a building is affected by the existence of a guidance system and that, when using the horizontal guidance system, the participants walked shorter distances to reach their destination.

Keywords: virtual reality, wayfinding, presence.

INTRODUCTION

The disorientation threatens the welfare and limits people mobility. Not knowing where we are or how to get where we want to go is a very stressful and frustrating experience and it has physical and psychological negative effects. Wayfinding difficulties may lead people to avoid places such as shopping malls, museums and convention centres. It can make people late for important occurrences such as business meetings or flights, which may cause loss of opportunity and money (Carpman et al., 2002). Additionally, during emergency situations, badly designed buildings and guidance systems for wayfinding are also a potential danger for people due the fact that they can arise as an increased problem to an extreme stress situation (Raubal, 2001). Otherwise, a place with an ease wayfinding can provide good sensations and the wish to visit this place again (Cubukcu, 2003).

During the last years, the indoor wayfinding problems have been considered also by organizational administrators, interior designers, architects and planners as a key to improve well-being. Thus, the study of individual movement to improve indoor wayfinding process may contribute to increase the visitor's satisfaction, to intensify visitation, to reduce the visitor's anxiety and producing physical, economical and social benefits.

According to Norman (1989), when people try to find their way to a destination in an unfamiliar environment, they look for external information that will complement their orientation and navigation processes in this unfamiliar environment. Many of the information that people need to reach a destination are in the world (knowledge in the world) and the human's mind is perfectly able to understand this world. It is what made the people able to orientate and navigate themselves in environments that they never interact before.

The analysis of the human interaction with the built environment is beyond of to obtain the users' subjective opinion. It is necessary to understand which the consequences of the aspects related to the orientation and navigation during the human interaction with the building are, in way to promote safety, comfort and effectiveness of the system Human-Environment. Thus, the Human must be considered as integrant and central part of the system, taking in consideration its characteristics, needs and limitations.

The literature suggests that environment's configuration aspects have significant cognitive consequences in human wayfinding (Kim, 2003). According to some authors, the spatial layout of the built environment can also influence the accuracy of cognitive representations of real-world spatial information (Appleyard, 1996). Thus, higher levels of configuration understanding are generally associated with more efficient wayfinding performance. However, buildings where the overall configuration is confusing or hard to imagine can be considered environment where the users are more disoriented (O'Neill, 1991 and Weisman, 1981). The signage is a component with large importance in wayfinding process as many place and situations are presented, at the beginning, as a maze or a complex space (and especially unknown) where the architecture, the lighting and the interior design are not able by themselves be intelligible and usable.

According to Smitshuijzen (2007), the traditional methods of guiding or directing are basically done by grouping and repeating the destinations and adding an arrow to show the right way for all destinations. This method is still far the most used although there is another (slightly archaic) method based on making a more or less continuous line on the floor, the wall, or the ceiling that leads from the starting point to the final destination. Different colors (or types of lines) can be used for different destinations. This method has been applied in a few signage projects for buildings (notably a hospital), but it has far too many limitations to be seriously considered in most cases. These limitations are based mainly in the use flexibility, maintenance and visibility when putted in places with a high flow of people.

Many researches were developed in the wayfinding area, but the studies which are based on the influence of the environment physical conditions on human wayfinding are still few and insufficient (Cubukcu, 2003). A pertinent methodological issue is the interaction environment. Some studies used the real world as interaction's environment to analyse the user's behavior (i.e. Kim, 2005; Blajenkova, 2005; Sohlberg, 2007). In these studies people who never interacted with a space (urban or building) have to move around the environment in order to reach some established points or to point the direction of some landmarks in the environment. The main advantage of this approach is that the studies' participants have the real perception of the environment mainly regarding to materials, barriers, people's movement in environment, the anxiety in using a space for the first time, sounds, light and many other characteristics related to the reality.

However, interacting with real world has also many constraints. The main problem for a research in human wayfinding is to control the variables that can influence people orientation and navigation processes as: light, people moving around the environment, noise, eye direction, among others. Another problem is related to safety. When interacting with real world people is submitted to constraints that can put human physical a mental integrity in risk. The behavioral data (those related to users feelings and frustrations) is also very difficult to acquire during the interaction with real environments due the fact that most of times users' verbalizations are missed during this interaction process.

Another interaction environment used in human wayfinding research is the bi-dimensional image of interior spaces, which, in many times, are presented as slides projected in a screen, photos and/or video (Raubal, 2001 and Omer & Goldblatt, 2007). These images are taken from an environment which participants never interacted before and represent decision points in pre-defined routes from where they have many options for choosing a way. Generally a task- where participants must reach a specified destination – is given to the participants.

This methodological approach is useful in way that it represents low financial costs, it does not put in risk the participants' physical integrity and can be applied in any place which few resources. Thus, it can involve a large number of participants with different characteristics.

However, the realism is very poor and depends of images quality and evaluators choices which may direct the user's field of view. Thus, the route chosen by users may be conditioned by images elements and view angles, besides the users are being always directed to go ahead due the fact that they do not know what is behind themselves. From psychology point of view, space around is not seen as equal; front is most important, back next most important and sides least important (Hunt, & Waller, 1999).

Another point to consider is that when people interact with an area for the first time they can show some fear or anxiety that may reflect in navigation errors. Interacting with unfamiliar places through photo can exclude this anxiety factor.

Recently, some researches used virtual reality-based environments to study human wayfinding (Omer, & Goldblatt, 2007; Blackman et al., 2003; Cubukcu, & Nasar, 2005; Moffat et al,. 2001; Morganti et al., 2007; Umemura et al. 2005).

The Virtual Reality (VR) is emerging as an important tool to overcome ethical and methodological constraints. One of the main advantages of VR is flexibility. The use of VR allows the researchers manipulate systematically the environment's layout, and different kinds of interactions can be designed in order to create suitable experimental conditions. It allows changing scenes according to the research's needs with low time and financial costs. VR also allows monitoring and recording for further evaluation the behaviors through which an explorer gains spatial knowledge (Morganti et al., 2007). The use of this technology allows registering, in a precise way, the user's path into a building, also making possible to observer short actions as small change directions, hesitations, glaze direction among others that in the real-world will be difficult to acquire.

METHODOLOGY

The main goals of this study are to verify the use of Virtual Reality (VR) and immersive environment interaction for data collection related to the Human performance in wayfinding tasks (finding a destination within a building), to validate the use of the ErgoVR System for wayfinding studies purpose and to evaluate the sense of presence reported by users after the interaction with the virtual environment. Thus, in order to reach these objectives we developed a protocol based in the evaluation of user's wayfinding performance in two guidance systems during the accomplishment of a wayfinding task (find an specific destination into a building).

Two guidance systems were tested: i) vertical (composed by signage place at the walls with arrows that indicate the way to all destinations) and horizontal (with continuous lines on the floor that leads from the starting point to the final destination). An Immersive VR was used as interaction environment in way to allow simulating the proposed experimental conditions. Thus, the user's evaluation of the presence level into the simulation was also considered in the study's objectives.

VIRTUAL ENVIRONMENT (VE)

A VE was developed in way to reach the study's objectives. It was made from a requirement program generated in brainstorming meetings with experts in Ergonomics, Architecture and Design.

Firstly, a 2D project, which was the base structure if the VE, was designed using software AutoCad 2008® and exported 3D Studio Max®, both from Autodesk, Inc. in order to model the 3D environment. Several objects, such as furniture, and general properties, such as color, texture, light, among others, were created in order to generate a realistic scenario. The scenario was then exported using a free plug-in called OgreMax 1.6.23, to be used by the software ErgoVR.

The VE consists in a symmetric plan formed by a rectangle divided into 8 rooms (12 x 12 meters each) interconnected by symmetrical axes of corridors, 2 meters wide, and circumvented by another corridor (Figure 1).

A small environment with a room and a small corridor was developed the same way as VE for the experimental tests. This environment was used as a training area, where the participants can move themselves freely in order to become familiar with the equipment and with their movements inside the simulation.

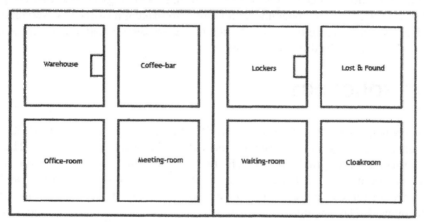

Figure 1. A top down view of the VE plan.

APPARATUS

The experimental tests were made using a Head-Mounted-Display (HMD) from Sony, model PLM-S700E with headphones. The body and head movements were controlled separately in way to give the users a higher level of autonomy and to bring near the natural actions. Thus, as a location device was used an USB joystick from Thrustmaster and the head motion was controlled using magnetic motion tracker from Ascension-Tech, model Flock of Birds.

The experimenter, seated in the surroundings, visually monitored participants to take notes. All experiments were videotaped. The VE displayed in the HMD was being displayed, simultaneously, in a screen placed on the desk. Thus, the researcher could see the same as the participants.

EXPERIMENTAL CONDITIONS

Three experimental conditions were defined in order to test the human wayfinding performance in two-guidance system (Figure 2).

•Neutral (N): In this condition all information and objects that can help the participant when accomplishing the wayfinding task were extracted from the VE as: pictures, fire-extinguishers, furniture, emergency signs. Only the signs with the information about the rooms' use were permitted.

•Horizontal (H): In this condition, a horizontal guidance system (with continuous lines that lead to the destinations) was inserted. There were inserted into the environment elements that increase the sense of presence as: pictures, fire-extinguishers, furniture, emergency signs.

•Vertical (V): In this condition, a vertical guidance system (signs on the wall with the name of all destinations and with arrows that points the way to all destinations) was inserted. There were inserted into the environment elements that increase the sense of presence as: pictures, fire-extinguishers, furniture, emergency signs.

Figure 2. A user's view of the horizontal guidance system (left image), and of vertical guidance system (right image).

PARTICIPANTS

Eighteen subjects, 11 males and 7 females, aged between 15 and 53 years (mean age = 26.44 years) participated in the study. Five (27.77%), all men, were videogame players (VGP) and thirteen (72.22%) were non-videogame player (NVGP). The selection criteria were that they had played action videogames at least twice a week for a minimum of 1 hour per day, and had done it so in the previous 2 months. The participants were randomly assigned to each of three experimental conditions as a function of the guidance system (without signage, horizontal system and vertical system).

PROCEDURE

The experiment used a between-subjects design. The experiment consisted of one preliminary practice trial and three experimental conditions (neutral, horizontal and vertical). The participants, randomly assigned to one experimental condition, were unaware of the real objective of the experiment and were asked to evaluate new software for VR simulation. They were told that they ought to fulfill some tasks as accurate and fast as possible.

The participants received a brief explanation about the experiment and they were asked to sign a form of consent and advised they could stop the simulation at any time they wanted to. After, they were introduced to the equipment and started in a practice trial.

The main objectives of the practice trial were: i) Present a simulation to the participants, in order to get them acquainted with the setup and with their movement inside the simulation; and, ii) Check for eventual simulator sickness case.

For this practice trial, participants were encouraged to freely explore and navigate the virtual room as quickly and efficiently as they could. When they felt they were able to control the navigation devices and felt relaxed or comfortable with the equipment they should declare it, in order to begin the experimental session. The trial VE was composed by a room with some obstacles (an object in the middle and narrow corridors) and was developed in the same way of the experimental VE.

After the practice trial, participants were assigned to one of the three experimental conditions. The same wayfinding task was given for all experimental conditions. Thus, participants were asked to reach the lockers-room as quick as possible and when they find their destinations they must say it aloud. A post task questionnaire was given, requesting subjects to evaluate their sense of presence and immersion level while in simulation.

RESULTS

The mean distance traveled by the entire sample inside the VE was 290,71m. The participants engaged in Neutral condition traveled a greater distance (564,53m) than the two other conditions. In the Horizontal Condition participants traveled a smaller distance (143.41m) than in the Vertical condition (164,19m).

In order to evaluate if the distance traveled into the simulation had a significant influence in the 3 experimental conditions, a non-parametric Kruskal-Wallis test were carried out. We used an error probability type I (α) of 0,05. The Kruskal-Wallis was made using the software SPSS (v.17, SPSS Inc. Chicago, IL). The results revealed that there is a statistical significant difference among 3 experimental conditions. The results revealed no statistical significant differences between the distances for the horizontal and vertical conditions (p-value = 0,558).

Related to the simulation mean time (estimated time and real time), from each experimental condition, the participants were able to estimate, in an accurate way, the time they spent inside the VE (real time mean = 03:52; estimated time mean =

04:01). The mean time for simulation length was longer in Neutral condition (07:32) than in the Horizontal and Vertical conditions. The participants had fastest times (02:00) for the Horizontal condition.). However, according to the Kruskal-Wallis, the results revealed no statistical significant differences between the times for horizontal and vertical conditions (p-value = 0,914). The participants were asked to classify the duration the duration of the simulation, in a 7 points scale (1 - very short to 7 - very long), participants (N 18) classified it as being medium (Mode = 5). The participants who interacts with the vertical condition had considered the duration as being shorter (Mode = 4).

Related to the mean number of pauses from each experimental condition, the mean number of pauses by the entire sample inside the VE was 7,17. The participants engaged in Neutral condition made the highest number of pauses (13,33).

The average velocity from each experimental condition is displayed are measure in meter per second (m/s). The average speed into the simulation for the entire group was 1,45 m/s. The participants were faster in the Vertical condition (1,67 m/s) than in the Horizontal (1,38 m/s) and Neutral (1,30 m/s) conditions. However, according to the Kruskal-Wallis, the results revealed no statistical significant differences between the times for horizontal and vertical conditions (p-value = 0,567).

The post-task questionnaire used a seven-point scale format. Therefore, the scale included a midpoint anchor and the anchors were based on the content of the question (e.g. 1- not compelling, 4 - moderately compelling, 5 - very compelling). The instructions asked respondents to place an "X" in the appropriate box of scale. This questionnaire was formulated in order to investigate questions related to sense of presence evaluation and enjoyment.

Regarding to the levels of presence the entire group presented high presence level. The Vertical condition displayed higher levels of presence, than Neutral and Horizontal conditions. In question (that was: what degree did you feel confused or disoriented at the beginning of breaks or at the end of the experimental session?), participants reported more disorientation in Neutral condition (Mode 4). When asked if they felt themselves inside the VE during the simulation, participants reported higher level of presence in Vertical (Mode 6) and Neutral (Mode 6) conditions than in Horizontal condition (Mode 3). When asked about the visual aspects of the environment participants reported higher levels in Horizontal condition (mode 6) than in vertical (Mode 5) and Neutral (Mode 4) conditions.

Kruskal-Wallis Tests were carried out on each question, with experimental condition (Neutral, Horizontal and Vertical) factor. The results revealed only the Question 16 (Were you involved in the experimental task to the extent that you lost track of time?) presented a statistical significant difference, in function of experimental condition (p-value = 0.012).

CONCLUSIONS

The main objective of this study was to analyze the difference between two types of indoor guidance systems. The use of the VR as interaction environment and the level of presence into the simulation were also considered.

VR technology allows researchers to control and manipulate characteristics of the physical environment. Designers and planners can use the technology to test and refine designs and thus to understand the physical, environmental requirements to ease wayfinding difficulties for different populations. The findings regarding to the participants' sense of presence into simulation suggest that, while interacting with the simulation, participants are able to transport themselves into the virtual world, acting as they were in the real-world. However, some limitations related to the used equipment may have limited the users' sense of presence, mainly the HMD which have a reduced field of view and no stereoscopic view. The use of a joystick as motion controller can also have reduced the immersion and the performance.

The study confirms the ErgoVR's system efficacy in the collection of objective measures to investigate the human wayfinding performance before or after the guidance system implementation.

The findings suggest that the existence of a guidance system increase the wayfinding performance in VE. The task execution time was similar for both groups (horizontal and vertical conditions). However the participants in vertical condition presented a higher distance traveled into the simulation than the participants in horizontal condition, with a high average speed. It can suggest a higher confidence grade while using a guidance system that is the most usual for buildings.

A potential drawback of this study may be the reduced size of sample and the use of rating scales to evaluate sense of presence and performance since it requires non-parametric data analysis.

Improved wayfinding has particular importance for airports, colleges, hospitals, office buildings, museums, libraries, shopping malls, entertainment parks, transport stations, resorts and convention centres. Thus, it is important to gain a better understanding of the effects of guidance systems for indoor wayfinding on users' behaviour.

Acknowledgements: A Portuguese Science Foundation (FCT) grant (PTDC-PSI-69462-2006) supported this research.

REFERENCES

Carpman, J. R., & Grant, M. A. (2002). Wayfinding: a broad view. In R. B. Bechtel & A. Churchman (Eds.), Handbook of Environmental Psychology (1 ed., pp. 427 - 442): Wiley, John & Sons, Inc.

Raubal, M. (2001). Agent- Based Simulation of Human Wayfinding: a perceptual model for unfamiliar buildings. Unpublished PhD., Viena University of

Technology, Viena.

Cubukcu, E. (2003). Investigating Wayfinding Using Virtual Environments. Unpublished Dissertation, Ohio State University, Ohio.

Norman, D. A. (1989). The Design of Everyday Things. New York: DoubleDay.

Kim, Y. O. (2001). The Role of Spatial Configuration in Spatial Cognition. Paper presented at the 3rd International Space Syntax Symposium, Atlanta.

Appleyard, D. (1969). Why Buildings are known: A Predictive Tool for Architects and Planners. Environment and Behavior, 1, 131-156.

O'Neill, M. J. (1991). Evaluation of a Conceptual Model of Architectural Legibility. Environment and Behavior, 23(3), 259-284.

Weisman, J. (1981). Evaluating Architectural Legibility: Way-Finding in the Built Environment. Environment and Behavior, 13(2), 189-204.

Smitshuijzen, E. (2007). Signage Design Manual (1st ed.). Baden, Switzerland: Lars Muller.

Blajenkova, O., Motes, M. A., & Kozhevnikov, M. (2005). Individual differences in the representations of novel environments. Journal of Environmental Psychology, 25(1), 97-109.

Sohlberg, M. M., Fickas, S., Hung, P. F., & Fortier, A. (2007). A comparison of four prompt modes for route finding for community travellers with severe cognitive impairments. Brain Injury, 21(5), 531-538.

Omer, I., & Goldblatt, R. (2007). The implications of inter-visibility between landmarks on wayfinding performance: An investigation using a virtual urban environment. Computers, Environment and Urban Systems, 31(5), 520-534.

Hunt, E., & Waller, D. (1999). Orientation and Wayfinding: a review. University of Washington.

Blackman, T., Mitchell, L., Burton, E., Jenks, M., Parsons, M., Raman, S., et al. (2003). The accessibility of public spaces for people with dementia: a new priority for the 'open city'. Disability & Society, 18(3), 357-371.

Cubukcu, E., & Nasar, J. L. (2005b). Relation of Physical Form to Spatial Knowledge in Largescale Virtual Environments. Environment and Behavior, 37(3), 397-417.

Moffat, S. D., Zonderman, A. B., & Resnick, S. M. (2001). Age differences in spatial memory in a virtual environment navigation task. Neurobiology of Aging, 22, 787-796.

Morganti, F., Carassa, A., & Geminiani, G. (2007). Planning optimal paths: A simple assessment of survey spatial knowledge in virtual environments. Computers in Human Behavior, 23(4), 1982-1996.

Umemura, H., Watanabe, H., & Matsuoka, K. (2005). Investigation of the model for path selection behavior in open space. Electronics and Communications in Japan (Part III: Fundamental Electronic Science), 88(1), 18-26.

Cubukcu, E., & Nasar, J. L. (2005a). Influence of physical characteristics of routes on distance cognition in virtual environments. Environment and Planning B: Planning and Design, 32(5), 777-785.

Chapter 81

Behavioral Compliance in Virtual Reality: Effects of Warning Type

Emília Duarte[1], F. Rebelo[2], Júlia Teles[3], Michael S. Wogalter[4]

[1] UNIDCOM/IADE – Superior School of Design,
Av. D. Carlos I, 4, 1200-649 Lisbon, PORTUGAL

[2] Ergonomics Laboratory. FMH/Technical University of Lisbon
Estrada da Costa, 1499-002 Cruz Quebrada, Dafundo, PORTUGAL

[3] Mathematics Unit. FMH /Technical University of Lisbon
Estrada da Costa, 1499-002 Cruz Quebrada, Dafundo, PORTUGAL

[4] Psychology Department. North Carolina State University
640 Poe Hall, Raleigh, NC, 27695-7650, USA

ABSTRACT

Virtual Reality (VR) is used to examine the effect of different warnings on behavioral compliance. Sixty university students performed a virtual end-of-day routine security check and interacted with four workplace ISO type warnings and three posted signs. The scenario was designed so that warning presentation was not pre-cued or expected. Other signs, however, were pre-cued; these were expected because they were part of the instructed tasks that were carried out. Participants were randomly to static vs. dynamic conditions. Behavioral compliance was measured according to whether participants followed the directive to press particular panel buttons. Data demonstrate that dynamic warnings produce higher behavioral compliance than static ones, but there were no dynamic vs. static differences for the pre-cued posted signs. Implications arising from the use of this technique and resultant findings are discussed.

Keywords: Safety warnings and signs; behavioral compliance; virtual reality

INTRODUCTION

Safety-sign and warning effectiveness depends on a series of events. Several cognitive models have been proposed to explain how the processes occur (e.g., Lehto and Miller, 1986; Rogers, Lamson, & Rousseau, 2000; Wogalter, DeJoy, & Laughery, 1999; Wogalter, 2006). Three main stages of warning processing that most models include are noticing, comprehending and complying. Compliance is considered the ultimate outcome measure of warning success since for compliance all or most of the stages will have been successfully processed. But from the point of view in conducting research, compliance is difficult to investigate because it is limited by methodological difficulties and ethical constraints. One main difficulty is that research participants cannot be exposed to real hazards. Another difficulty is that producing a realistic scenario that has no actual risk is expensive in terms of money, time and effort. Consequently, even though there has been a substantial body of research on the topic of warnings, relatively few studies have measure actual behavioral compliance.

Fortunately, new technology and techniques have become available that could change the situation with respect to warning compliance research. Virtual Reality (VR) may help to overcome some of the main constraints since it can simulate adequate, but safe, contexts of Virtual Environments (VE) for use in warnings research. High quality VEs can promote ecological validity, while allowing good control over experimental conditions. Thus VR combines the best aspects of laboratory and field research, and allow the simulation of hazard-associated emergency situations while keeping the participants safe from actual harm.

To date, few studies have used VR in warnings research and the majority of them have mainly focused on exit signs (e.g., Glover & Wogalter 1997, Shih, Lin, & Yang, 2000; Tang, Wu, & Lin, 2009). Research on exit signs has provided valuable information, and in particular, has demonstrated the ability to measure sign manipulations on compliance. However, to fully explore the utility of VR in warning research, other types of signs should be tested. There are two main reasons for this need. One is to determine if VR as a method would provide a means to measure behavioral compliance to warnings. The second reason is to determine if the method would be adequately sensitive to pick up differences between manipulated warnings and resemble results similar to those recorded in real behavioral-compliance situations.

In the present research, the VE was a company headquarters and the scenario was an end-of-day routine security check. In order to carry out the required tasks (including the behavioral compliance aspects), participants had to press buttons associated with the warnings and posted signs shown in the VE.

In warning research literature, there are several behavioral compliance studies showing effects of sign type (e.g., Wogalter & Young, 1991, Wogalter, Kalsher, & Racicot, 1993). One fairly strong and consistent finding is that dynamic presentation produces greater compliance than static presentations (e.g., Wogalter et al., 1993).

Static signs are traditionally made of paper, metal or plastic, and generally, the method of communication is passive. In contrast, dynamic signs usually use more advanced technology that can be multimodal and customized. Recent articles suggest that technology-based warnings can be more effective than the traditional solutions (e.g., Wogalter & Conzola, 2002; Mayhorn & Wogalter, 2003; Smith-Jackson & Wogalter, 2004; Wogalter & Mayhorn, 2005) since they have features that can enhance the warnings in a number of ways, such as making them more noticeable and more resistant to habituation. As a result, dynamic warnings raise levels of compliance over warnings without dynamic properties, i.e., static.

The safety warnings and signs used in the present research are symbol-based type signs consistent with the International Organization for Standardization's (ISO) 3864-1 (ISO, 2002) standard. The signs had both a symbol and text. Accordingly to ISO 9186 (2001) standard, a safety sign is "a general safety message obtained by a combination of color and geometric shape and which, by the addition of a graphical symbol, gives a particular message" (ISO, 2001). Traditionally, a symbol alone format of ISO 3864-1 is used, without text, due to multilingual considerations. Recent efforts have been taken to harmonize with American National Standards Institute (ANSI) Z535 (2002) standards, and now ISO signs can be used together with text panels.

In the present study, there were two kinds of visual displays in the VE: they were designated as "Warnings" and "Posted Signs." Warnings contained an explicit safety message communicating information about a hazard, the hazard's consequences and also providing guidance on how to avoid the hazard. The warnings were the main target of behavioral compliance evaluation. The warnings and scenario that they were embedded were designed so they would not be expected by participants, as they were not pre-cued ahead of time by the task instructions that participants carried out. Nevertheless they were placed coherently within the VR scenario. The second type of visual display were posted signs. These posted signs had a different role than warnings. First they looked different (as will be described later) and they identified a safety device (e.g., a gas valve). Also, the content of the posted signs was mentioned within the instructions given to participants on the tasks they were to perform, and thus, the contents were pre-cued or expected prior to them being viewed in the VE.

METHOD

PARTICIPANTS

Sixty university students, 30 male and 30 female, aged 18 to 35 years old (mean age= 21.15, SD = 3.107) participated. They had no previous experience with navigation in VEs. Participants had normal sight or had corrective lenses and no color vision deficiencies. They reported no physical or mental conditions that would prevent them from participating in a VR simulation. All participants completed an informed consent form. Participants were randomly assigned to one of two experimental conditions (static vs. dynamic) each comprised of 30 individuals with an equal number of males and females in each condition.

APPARATUS

The used apparatus comprised 2 magnetic motion trackers from Ascension-Tech, model Flock of Birds for monitoring head and left hand movements; a joystick from Thrustmaster as a locomotion device; a Head-Mounted-Display (HMD) from Sony, model PLM-S700E; a Monocular Laptop HMD Mountable eye-tracking system from Arrington Research (Part Number MAE06); wireless headphones and a graphics workstation.

The VE was presented at a resolution of 800 x 600 pixels, at 32 bits, with a FOV 30°H, 18°V and 35° diagonal. The speed of movement gradually increases from stopped, to an average walk pace (1.2 meters per second) to a maximum speed around 2.5 m/s. The participants' viewpoint was egocentric. Participants were seated at a desk for the duration of the session. An example participant during an experimental session is shown in Figure 1. All participant sessions were videotaped. The VE displayed in the HMD was also simultaneously displayed in a second monitor. Thus, the researcher could see the same image as the participants did but also saw it superimposed with information on gaze and fixation time.

THE VIRTUAL ENVIRONMENT

The VE consisted of a company headquarters, with 4 rooms (meeting room, laboratory, coffee-shop and warehouse), each measuring 12 x 12 meters in size. The rooms were interconnected by 2 symmetrical axes of corridors, 2 meters wide, and circumvented by another corridor with an exit. The layout can be seen in Figure 2. In terms of visual and auditory complexity or pollution, the VE can be roughly classified as being uncluttered.

The base structure of the VE was designed using AutoCAD® 2009, and then it was imported into 3D Studio Max® (both from Autodesk, Inc.). The scenario was then exported using a free plug-in called OgreMax, to be used by the ErgoVR system (developed by the Ergonomics Laboratory of the FMH / Technical University of Lisbon).

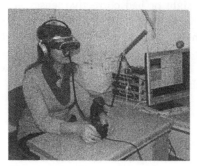

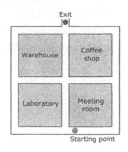

Figure 1. Participant during the simulation

Figure 2. The VE floor plan

DESIGN OF THE STUDY

The experiment used a between-subjects design with two experimental conditions: (1) Static, and (2) Dynamic.

(1) Static: VE with color ISO type warnings and posted signs with a size of 30 x 40 cm, without abrasion marks.

(2) Dynamic: VE with color ISO type warnings and posted signs displayed in illuminated panels, with a size of 30 x 40 cm, augmented with 5 flashing lights and an alarm sound activated or deactivated by proximity sensors. The flashing lights, 3 on the top and 2 on the bottom, were 4 cm diameter, orange colored and with a flash rate of 4 flashes per second, with equal intervals of on and off time (Sanders & McCormick, 1993). The flash was twice as bright as the background. The sound was an alarm beep. An example of a sign in both dynamic and static versions can be seen in Figure 3.

Figure 3. VE images showing the "Inhalation hazard" warning in the (1) static condition on the left and the (2) dynamic condition on the right.

PROCEDURE

The procedure was divided in three major steps: pre-experimental training, experimental session, and post-hoc questionnaire.

In the pre-experimental training, participants were given a brief explanation about the study and were introduced to the equipment. They were unaware of the real objective of the research by being invited to evaluate new software for VR simulation. At the outset, they were asked to sign a consent form and were advised they could stop the simulation at any time. The Ishihara Test (Ishihara, 1988) was used to detect color vision deficiencies. Participants were shown a practice VE using the same equipment used in the experimental session. The practice VE consisted of 2 rooms containing some obstacles (e.g., doors, narrow corridors, pillars, etc.), requiring some skill to be circumvented. The purpose of this practice trial was to get the participants acquainted with the setup and to make a preliminary check for any initial indications of simulator sickness. In this practice trial, participants were told to freely explore and navigate the virtual room as quickly and efficiently as they could. They were told that when they felt that they were able to control the navigation devices and felt comfortable with the equipment that they should say so aloud. When they did, the experimental session started shortly thereafter.

Participants took part in one of two experimental conditions. The given scenario was a series of end-of-day routine security checks that simulated the closing up of a company's facility at the end of a workday. In the VR simulation, participants were to fulfill several tasks inside the VE, involving entering into each one of the main four rooms in the following order: Meeting room, Laboratory, Coffee Shop and Warehouse. In the VE, several warnings and posted signs were placed on walls of the rooms. The entire experiment and the content of the warnings and posted signs as well as all experimental instructions (print and oral) were communicated in Portuguese language. The English translations are given for the purposes of communicating this report. The warnings and posted signs are displayed in Figures 4 and 5.

(1) Mandatory to disconnect before leaving the room (blue)	(2) Caution, laser in operation, do not enter before turning it off (yellow)	(3) Sound warning, mandatory to warn before entering (blue)	(4) Inhalation hazard, start air extractor before entering the room (yellow)

Figure 4. Static version of ISO type warnings. Note that the warnings were not pre-cued by task instructions prior to being exposed in the VE.

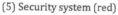

(5) Security system (red)

(6) Gas valve (red)

(7) Cut-off energy to machine room (green)

Figure 5. Static version of ISO type posted signs. Note that the contents of the posted signs were pre-cued by the task instructions.

The written instructions were posted in a projection screen and whiteboards placed inside the rooms in the VE. Once the simulation started, no dialogue between participants and the researcher occurred.

Participants were told they should start the procedure by getting to the Meeting Room. Inside the Meeting Room, the following instructions were displayed: "Check for water cups on the top of the tables. If you find any, please leave a message in the Coffee Shop. Then go to the Laboratory room and turn on the security system." Near the exit door, above an ambient music button switch on a wall was Warning 1 – "Mandatory to disconnect before leaving the room."

Outside the Laboratory, close to the entrance door and above a button switch was Warning 2, which stated "Caution, laser in operation, do not enter before turning it off." However, an "out of order" label was stuck on the button. Inside the Laboratory, Sign 5 was present – "Security system." The instruction displayed inside this room was: "Now, go to the Coffee Shop and turn the gas off."

Outside the Coffee Shop, close to the entrance door was Warning 3, which stated "Sound warning, mandatory to warn before entering" was displayed. Inside the Coffee Shop, Sign 6 – "Shut-off gas" was present together with a button switch. The instruction given was "After the gas is shut-off, go to the Warehouse and cut the energy to the machine room."

Outside the Warehouse, whose door was closed, was Warning 4, which stated "Danger, inhalation hazard, start air extractor before entering the room." There was only one way to open the door, which was by pressing the button adjacent to the warning. Inside the Warehouse was a button switch adjacent to Sign 7, which stated "Cut energy to machine room." Since this was the last room, no further instructions were given here. After entering the warehouse, or following 5 minutes after entering the corridors leading to the warehouse, an explosion occurred, followed by a fire in the Warehouse and in the adjacent corridors. A fire alarm could be heard and flames and smoke blocked all the corridors leaving only the exit route clear. The simulation ended when the participants reached the exit.

After completing the simulation, participants answered a questionnaire concerning their experience in the simulation. These post-hoc data are not described in this article.

RESULTS

The main dependent variable was behavioral compliance, which was defined as the extent to which the participant adhered to the warnings and signs and was measured by the times participants pressed the buttons as directed by the warnings.

Table 1. Frequency of pressing buttons (behavioral compliance) as a function of warnings and posted signs by (1) static vs. (2) dynamic conditions.

Warnings (1-4) and posted signs (5-7)	(1) Static		(2) Dynamic	
	freq	%	freq	%
(1) Mandatory to turn off before leaving the room	12	40.00	29	96.70
(2) Caution, laser in operation, do not enter before turning it off	1	3.33	14	46.70
(3) Sound warning, mandatory to warn before entering	7	23.33	22	73.33
(4) Inhalation hazard, start air extractor before entering the room	23	76.67	30	100.00
(5) Security system	21	70.00	23	76.70
(6) Gas valve	30	100.00	29	96.67
(7) Cut-off energy to machine room	21	70.00	23	76.70

The influence of sign type (static vs. dynamic) on behavioral compliance (number of button pressed associated with warnings and posted signs) was evaluated at a significance level of .05 using t-tests. The analysis showed a significant effect of warning type (static vs. dynamic) on behavioral compliance to warnings ($p < 0.001$) but no effect was found for the posted signs ($p = 0.17$).

The Binomial Test was used to compare the two experimental conditions for each warning. The test assessed whether the proportion of participants who pressed the buttons in the two experimental conditions was equal. The analysis revealed that were significant differences between the static and dynamic conditions for the warnings 1, 2, 3, and 4 ($p < 0.01$) but not significant for the posted signs 5, 6, and 7 ($p > 0.10$). Compliance was higher dynamic warnings than static warnings.

CONCLUSIONS

Virtual Reality was used to simulate an interaction context with warnings signs with the purpose of evaluating behavioral compliance. University students performed an immersive virtual end-of-day routine security check and interacted with ISO type warnings and signs. Previous research indicated that dynamic warnings produce greater compliance behavior than static warnings. In the present study, the dynamic warnings and posted signs had simulated flashing lights whereas the static warnings and posted signs did not, and behavioral compliance was measured (pressing buttons associated with the directives of the signs). The results showed that dynamic warnings produced greater compliance than static ones. This result corresponds with actual behavioral compliance research with warnings that has found that dynamic warnings produced higher compliance than similar static ones. The dynamic features presumably make them more salient and increasing the likelihood that they be noticed and heed to them given, which resulted in the increased compliance compared to static (less salient) ones. But such effects were significant only in the cases of warnings. Warnings in this study were not pre-cued by the task instructions. They just appeared in the course of performing the tasks within the simulation. This was not the case for the posted signs. The posted signs were directly tied to the instructions and were "expected" postings at places participants were specifically instructed to go to. When the posted signs were targets of specific tasks then there was no difference between the two experimental conditions. The pre-cued posted signs did not need to be salient to be noticed. Also note that performance was relatively high in the posted sign conditions probably because their presence and information content were pre-cued. The high level of performance may have produced a ceiling effect that also prevented finding significant improvement for the dynamic posted signs over the static ones without more participants in conditions. Future data analysis such as time spent, paths taken and diverse subjective ratings could reveal other differences between conditions.

Further research is needed on the impact of other aspects of VR such as navigation devices and interaction. In this study, participants "flew" through the VE, since they were actually seated at a desk, moving by the means of a joystick. Also, participants had a rather limited ability to interact with the VE, since they could not manipulate objects as one does in the real world (e.g., to open doors and operate different kinds of machine controls).

The present research has implications for the use of VR to study the effectiveness of warnings in different settings and tasks. This kind of use is in its infant stages. This study affirms that there is good potential for it to serve as a technique to investigate warning effectiveness, particularly behavioral compliance. VR offers a way to overcome several key constraints that have thus far limited warning compliance research.

ACNOWLEDGEMENTS

A PhD scholarship from Portuguese Science Foundation (FCT) supported this study (SFRH/BD/21662/2005), which was developed in the scope of a Research Project financed by FCT (PTDC-PSI-69462-2006).

REFERENCES

American National Standards Institute (2002). Z535. *Accredited Standards Committee on Safety Signs and Colors. Z535.1-5*, National Electrical Manufacturers Association, Rossyln, VA.

Glover, B. L., & Wogalter, M. S. (1997). Using a computer simulated world to study behavioral compliance with warnings: Effects of salience and gender, *Proceedings of the Human Factors and Ergonomics Society 41st Annual Meeting* (pp. 1283-1287). Santa Monica: Human Factors and Ergonomics Society.

International Organization for Standardization (2001). Graphical Symbols - Part 1: Test Methods for Judged Comprehensibility and for Comprehension, *ISO 9186:2001*. Geneva, Switzerland: International Organization for Standardization.

International Organization for Standardization (2002). Graphical Symbols - Safety Colors and Safety Signs. Part 1: Design Principles for safety Signs in Workplaces and Public Areas, *ISO 3864-1*. Geneve, Switzerland: International Organization for Standardization.

Ishihara, S. (1988). *Test for Colour-Blindness* (38 ed.). Tokyo: Kanehara & Co., Ltd.

Lehto, M. R., & Miller, J. M. (1986). *Warnings: Volume 1. Fundamentals, design and evaluation methodologies*. Ann Arbor, MI: Fuller Technical.

Mayhorn, C. B., & Wogalter, M. S. (2003). Technology-based warnings: Improvising safety through increased cognitive support to users, *15th International Ergonomics Association Congress* (Vol. 4, pp. 504-507).

Rogers, W. A., Lamson, N., & Rousseau, G. K. (2000). Warning research: An integrative perspective. *Human Factors, 42*(1), 102-139.

Sanders, M. S., & McCormick, E. J. (1993). *Human factors in engineering and design* (7th ed.). New York: MacGraw-Hill.

Shih, N.-J., Lin, C.-Y., & Yang, C.-H. (2000). A virtual-reality-based feasibility study of evacuation time compared to the traditional calculation method. *Fire Safety Journal, 34*(4), 377-391.

Smith-Jackson, T. L., & Wogalter, M. S. (2004). Potential uses of technology to communicate risk in manufacturing. *Human Factors and Ergonomics in Manufacturing, 14*(1), 1-14.

Tang, C.-H., Wu, W.-T., & Lin, C.-Y. (2009). Using virtual reality to determine how emergency signs facilitate way-finding. *Applied Ergonomics, 40*(4), 722-730.

Wogalter, M. S. (2006). Communication- Human Information Processing (C-HIP) Model. In M. S. Wogalter (Ed.), *Handbook of warnings* (pp. 51-62): Lawrence Erlbaum Associates, Inc.

Wogalter, M. S., & Conzola, V. C. (2002). Using technology to facilitate the design and delivery of warnings. *International Journal of Systems Science, 33*(6), 461-466.

Wogalter, M. S., DeJoy, D. M., & Laughery, K. R. (1999). Organizing framework: A consolidated communication-human information processing (C-HIP) model. In M. S. Wogalter, D. M. DeJoy & K. R. Laughery (Eds.), *Warnings and Risk Communication* (pp. 15-24). London: Taylor & Francis.

Wogalter, M. S., Kalsher, M. J., & Racicot, B. M. (1993). Behavioral compliance with warnings: Effects of voice, context, and location. *Safety Science, 16*(5-6), 637-654.

Wogalter, M. S., & Mayhorn, C. B. (2005). Providing cognitive support with technology-based warning systems. *Ergonomics, 48*(5), 522-533.

Wogalter, M. S., & Young, S. L. (1991). Behavioural compliance to voice and print warnings. *Ergonomics, 34*(1), 79 - 89.

Chapter 82

A Methodological Proposal to Evaluate the Postural Response in Virtual Reality

Ernesto Filgueiras[a,b], Francisco Rebelo[b] and Emília Duarte[b,c]

Ergonomics Laboratory
[a] FMH - Technical University of Lisbon – Portugal

[b] Beira Interior University - Covilhã – Portugal

[c] UNIDCOM / IADE - Superior School of Design

ABSTRACT

This paper describes a methodology to evaluate the postural and interaction behaviors, which was applied in a study involving Virtual Reality, with the intent to relate the postural responses (real world) with actions inside the virtual environment (VE). Twenty-four subjects, seated and using an immersive VR system, were asked to interact with a virtual indoor environment in order to accomplish predefined tasks. The simulation was composed by two scenarios (Pre-emergency and Emergency egress). In the "Pre-emergency" scenario there were instructions inside the VE that subjects had to follow. In the "Emergency egress", subjects were exposed to an explosion, followed by a fire and should leave the virtual building as soon as possible. The methodology allowed collecting necessary data to obtain results regarding to the analysis of the individuals postural responses during an interaction with immersive VR system. These objective measures can be an useful contribute when associated to subjective measures to study sense of presence. The findings indicate that women have more postural responses changes, in while interacting with the VR, than man. The postural responses for the head did not reflect the participant's behaviors into the virtual environment.

Keywords: Postural responses, Virtual Reality, Observation Methodology.

INTRODUCTION

When in Virtual Environments (VE), people experience many degrees of presence and typically divide attention between the real and virtual worlds (Singer et al., 1997). According to Slater, Usoh & Steed, (1994), presence refers to the psychological sense of 'being there' in the virtual environment. However, presence could vary depending on many factors such as: hardware and software and the level of immersion provided by technology; and the individual propensity for using imagination.

Different techniques have been used to measure presence in VEs but three categories of measures can be distinguished: subjective, behavioral and physiological. The subjective measures rely on self-assessment by the user and is investigated using, generally, post-immersion questionnaires. Behavioral measures examine actions or manners exhibited by the user that are responses to objects or events in the virtual environment (e.g. startle reaction); and, physiological methods attempt to measure presence by gauging changes in the subject's heart rate, skin temperature, skin conductance, breathing rate, among others.

The major disadvantage of subjective measures when using questionnaires is that they are post immersion, during prolonged exposure to a VE. Participants may become fatigued or bored, what would influence their responses. Also self-reports may be compromised due to memory erosion. Behavioral measures are more shielded from subject bias than subjective measures and are their lack of intrusion into the virtual experience. However, behavioral measures also have disadvantages. One of then is the inability to know if a certain behavior was caused by the experimental condition. For instance, a participant losing balance during the simulation could be due to the content of the VE or caused by the VR system itself. Physiological measures are more objective than subjective measures and most of the behavioral measures. Although, the physiological levels vary widely from person to person and some of them can only be acquired in an intrusive way (Insko, 2003).

An aggregate measure, comprising subjective and objective components, could be used in order to overcome some of the limitations that can arise from the use of objective or subjective measures separately. A good measure of presence will aid human factors specialists in investigating the relationship between presence and task performance, and may aid the general understanding of the experience of presence (IJsselsteijn, Ridder and Freeman, 2000).

Many of presence research has focused on definition and ideas for measurement (e.g. subjective reports, ratings scales, comparison-based-prediction, behavior and physical measures) and there are several empirical studies of contributing factors for subjective procedures (Sadowski and Stanney, 2002). In many others, the major interest is on body postural responses as an important paradigm in VEs (Slater, Steed and McCarthy, 1998; Freeman et al., 2000). However, ensuring that the physiological, cognitive or postural responses are "directly related" to the level of presence being experienced in VE, is a very difficult issue (Baños et al., 2000).

In this way, the main focus of this paper is to assess the external behavior, during an immersive VR simulation, through a methodology based on video observation that relates, simultaneously, the VR events with the body postural responses. This methodology was developed in the scope of a PhD project focused on Ergonomics Analysis through video observation techniques in real situation that involve sitting work, developed in the Ergonomics Laboratory of FMH - Technical University of Lisbon.

A case study using this methodology in the VR context will be presented. In this way, this study also had as secondary objective to verify the use of the developed methodology in other interaction contexts, as VR simulations.

METHODOLOGY TO CLASSIFY AND MEASURE USER'S ACTIVITY INSIDE THE VE AND POSTURAL RESPONSE

As previously mentioned, the methodology used for this study was originally developed for the Ergonomics Analysis of seated work, and it is based on hierarchical observation of events (Wang, Hu and Tan, 2003). Therefore, this methodology allows the classification of a set of behaviors in hierarchical categories, through video analysis. Hierarchical categories are understood as the pre-defined events, which can be observed and ranked, for instance: task, activity and postures.

The analysis that is carried out with the support of this methodology should be considered a macro analysis.

Five steps compose the methodology, described next:
1. Analysis of reference situation;
2. Definition of categories;
3. Video recorder;
4. Video analysis;
5. Data analysis.

1. Analysis of reference situations: This step consists of studying similar work situations, which can be based on reference literature, in observation of other work situations and ergonomics analysis of the work situation under study.

2. Definition of Categories: From the analysis of the reference situations a list of categories is created, They are re-checked so that they can be grouped in an hierarchical order, until it is obtained a final categorization that will be used the video recorder's definition and for the video analysis. The previous establishment of a hierarchy to combine all interaction categories according to the objective of study is important. This procedure will avoid eventual duplication of categories.

3. Video recorder: It is defined in this step the type and number of cameras to be used, their location (sagittal, coronal and transverse plans) and the recording

duration depending of the objectives of the study and the results from previous steps.

4. Observation Video Analysis: The video analysis is based on time samples configured according to the results of the Definition of Categories step. The video remains in loop for the duration of the period of time configured until the observer decides to move to the next sampling. For example, for a 8 hour work situation it can be defined samples of 5 seconds each in intervals of 30 seconds. This method reduces common analysis errors to the methods of continuous direct observation (Xiao & Mackenzie, 2004). A software specifically developed for this purpose was used to classify the behavior hierarchically and to control the time intervals for the videos.

5. Data analysis: Each moment of observation, or time sampling (period of time that can be adjusted according to the needs of each study), was analyzed in order to gather data related to the categories previously defined (task, activity and posture), for each participant. Such moment, with all its data, is defined as "Event". In this step, the collected data (i.e. frequency of events occurrence) can be analyzed isolated or in conjunction, to make comparisons.

CASE STUDY

The case study presented in this paper used data from a VR simulation, which occurred in the scope of a research project being conducted at the Ergonomics Laboratory of the FMH – Technical University of Lisbon. The data selected for this study includes the activities made by the study participants inside the VE and their external postural behavior. Considering the specificities of the available data, the methodology described in the previous topic, had to be adapted through the use of new interaction categories. Therefore, this case study allowed to verify the methodology adequacy for VR studies that may benefit from postural response evaluation.

For this study it were considered 3 major categories types: interaction scenarios, activities and postural behaviors. The interaction scenarios and activities categories describe the events inside the VE. The postural behavior categories correspond to external events.

- Interaction scenarios categories: two scenarios were defined - "Pre-emergency" in a workplace VE and "Emergency Egress" after an explosion/fire.
- Activity categories: six categories of activities, representing actions made by the participant inside the VE (*Move Forward*, *Move Backwards*, *Reading*, *Manipulation*, *Turning* and *Pause*). Pause was considered when the participants were standing without any visible activity.
- Postural behavior categories: six categories, which can occur in six body segments, were considered for each activity category.

The activity and postural behavior categories can be seen in Table 1).

Table 1 Activity categories inside the VE and postural behavior categories, for the external events.

Codes	Activity categories	Postural Behavior categories					
		HEAD	TRUNK	Left ARM	Right ARM	Left LEG	Right LEG
1	Move Forward	Neutral	Neutral	Arm supported		Neutral	
2	Move Backward	Flexion	Extension	Arm/hand supported		Flexion	
3	Reading	Extension	Hyper-extension	Elbow supported		Extension	
4	Manipulation	Torsion	Inclination	Hand supported		Crossed	
5	Turning	Inclination	Flexion	Without support		Suspended	
6	Pause	With support	Torsion	Forearm supported		With Support	

SAMPLE

Twenty-four participants (11 males and 13 females), aged between 18 and 35 years old (mean= 20.7; SD=2.2) participated in the study. All participants interacted individually with the VE. All participants were asked to sign a form of consent and were advised they could stop the simulation at any time they wanted to.

APPARATUS

The VR system used in this study consisted of the following equipments: i) Two magnetic motion trackers from Ascension-Tech, model Flock of Birds. One positioned on the top of the user's head, to measure the head motion to control the camera (user point of view) inside the VE, and another, attached to the user's wrist, used to measure the hand movements, related to manipulations activities in VE; ii) A joystick from Thrustmaster, used as the body locomotion device inside the VE; iii) A Head-Mounted-Display (HMD) from Sony, model PLMS700E; iv) Wireless headphones Sony, model MDR-RF800R; v) Three infrared Bosch cameras, model WZ18, used to record the postures; vi) A Bosch Multiplexer, model DVR-8K (8 channels), to joint the images from VE and postures, in a single image.

PROCEDURES

The participants were introduced to the equipment and started in a practice trial, with the objective to get them acquainted with the equipment and their movement inside the simulation and to check for eventual simulator sickness cases. Participants remained seated during the entire procedure. In the practice trial, participants were encouraged to freely explore and navigate through a virtual room as quickly and efficiently as they could. When they declared that they were able to control the navigation devices they were assigned to the session.

For the "Pre-emergency" scenario category, the participants were request to fulfill several tasks related to an end-of-the-day routine check (e.g. turn on the security system and turn the gas off). After fulfilling the predefined tasks an explosion following by a fire occurred. The "Emergency Egress" scenario begins at this point. Participants are expected to leave the building as fast as possible.

PROCEDURES TO CAPTURE AND PROCESS THE DATA FROM VR AND EXTERNAL CAMERAS

Three cameras, strategically located in order to capture images from the sagittal, coronal and transverse plans captured videos for each participant. A fourth video collected the participants' egocentric point of view (first person) inside the VE. All four videos were sized to 1/4 of the screen, to fit together in a normally sized frame, using a multiplex.

The observation video analysis was done using software developed for this purpose in the scope of a PhD project in the Ergonomics Laboratory of FMH - Technical University of Lisbon. The videos from each participant were analyzed separately. Samples with 3 seconds of video (in loop) with an interval of 5 seconds between each sample were used. In this way, for each 3 seconds of video only one analysis was accomplish. As the study was a macro-analysis, an important observation criterion for the postural analysis was that only the visible postural changes were considered, while small movements were not considered.

RESULTS

This section has the objective to present and discuss the major results of this case study, in order to verify the methodology potentialities for VR studies, particularly:

- The relationship between the scenarios and activity categories in VE and the postural behavior categories;
- The eventual existence of gender differences in postural behavior categories while interacting with a VR simulation.

The data for all participants (n = 24, mean = 8 minutes/participant) correspond to 2157 samplings (110 samples for each participant). The entire sample spent, in mean, 113 minutes (62.3 %) in the first scenario category - "Pre-emergency", and 67 minutes (37.7 %) in the second scenario - "Emergency Egress". The data regarding activity categories, for both scenarios categories, is showed in Table 2.

When considering both "Pre-emergency" and "Emergency Egress" activity categories, the most frequent occurrences were *Move Forward* (mean=52.3; SD=17.6), representing 55.8 % of all activities, followed by *Manipulation* (mean=13.3; SD=6.2) with 14.2 % of all activities and *Turning* (mean=11.6;

SD=5.4), representing 12.4% of all activities. The less frequent activity category was *Pause* (mean=1.9; SD=1.0), representing 2 % of all activities. These results were expected since the most frequent activities are those that are indispensable for the accomplishment of the required tasks in the VE.

Table 2 shows the means, standard deviations and occurrences percentage for VE activity categories occurrences in function of scenarios categories.

Table 2 Activities categories occurrences in function of scenarios categories

Codes	Activities Categories	Scenarios Categories								
		Pre-emergency			Emergency egress			TOTAL		
		M	SD	%	M	SD	%	M	SD	%
1	Move Forward	29.3	9.1	49.8	23	11.4	65.9	52.3	17.6	55.8
2	Move Backwards	2.5	1.7	4.3	1.1	1.0	3.1	3.6	2.2	3.8
3	Reading	9.3	3.2	15.7	1.7	2.9	4.9	11.0	4.1	11.7
4	Manipulation	7.0	2.0	11.9	6.3	5.7	18.2	13.3	6.2	14.2
5	Turning	9.0	4.0	15.2	2.7	2.3	7.7	11.7	5.4	12.4
6	Pause	1.8	0.9	3.0	0.1	0.3	0.2	1.9	1.0	2.0

When the occurrences in the "Pre-emergency" and "Emergency Egress" scenarios were analyzed separately, *Move Forward* (mean=52.3; SD=17.6), with 55.8% of all occurrences, was again the most frequent activity observed. However, differences regarding the second most frequent activity, for both scenarios, were verified. In "Pre-emergency" scenario the second most frequent activity was *Reading* (mean=9.3; SD=3.2), while in the "Emergency Egress" it was *Manipulation* (mean=6.3; SD=5.7). The results attained in the "Emergency Egress" scenario were not expected since, unlike the "Pre-emergency", no requirements involving the *Manipulation* categories were placed in this scenario. Thus, this result can be explained by the fact that most of the participants tried to extinguish the fire, instead of getting out of the building.

Table 3 shows the means, standard deviations and occurrences percentage for postural categories frequency of occurrence, for each body segment, in function of the defined scenarios categories. The most import data reveal that neutral posture was the most frequent for the head and trunk body segments. The upper limbs remained most of the time supported in different ways, dependent of the task that they need to do as, for example, interactions with the joystick with the right arm (100% of sample) or resting the left arm on the top of the table. For the legs, an alternation between neutral and flexion postures for both legs was verified.

If considering the two scenarios categories separately some differences can be found. The frequency of neutral posture of the trunk diminished, with an increase of the flexion posture, in the "Emergency Egress" when compared to "Pre-emergency". Also, for the left leg, postures modification rates were higher in the "Emergency Egress". For the other body segments no evident differences regarding postures changes were found.

Table 3 Postural categories occurrences for each body segment in function of the scenarios categories (n=24)

YR scenarios	Codes	POSTURAL BEHAVIOR CATEGORY											
		HEAD		TRUNK		LEFT ARM		RIGHT ARM		LEFT LEG		RIGHT LEG	
		M (SD)	%	M (SD)	%	M (SD)	%	M (SD)	%	M (SD)	%	M (SD)	%
Pre-emergency	1	28 (15)	47	43 (26)	75	0.2 (1)	0.4	0 (0)	0	28 (30)	47	25 (30)	43
	2	11 (11)	17	7 (18)	13	50 (12)	85	0 (0)	0	28 (28)	48	31 (28)	53
	3	1 (1.5)	2	0 (0)	0	0.2 (0.5)	0.3	48 (24)	81	2 (11)	4	2 (11)	4
	4	19 (12)	32	0.3 (1)	1	0.5 (2)	0.9	1 (3)	1	0 (0)	0	0 (0)	0
	5	0 (0)	0	7 (16)	12	8 (4)	14	10 (23)	17	0 (0)	0	0 (0)	0
	6	0 (0)	0	0 (0)	0	0 (0)	0	0 (0)	0	0 (0)	0	0 (0)	0
Emergency egress	1	16 (10)	45	23 (19)	66	1 (5)	3	0 (0)	0	19 (21)	54	16 (20)	45
	2	9 (8)	25	4 (10)	10	24 (13)	70	0 (0)	1	9 (13)	25	18 (21)	52
	3	1 (2)	3	0.3 (2)	1	0.6 (2)	2	27 (16)	77	4 (10)	11	1 (6)	4
	4	9 (7)	27	1 (3)	3	3 (8)	7	1 (5)	3	0 (0)	0	0 (0)	0
	5	0 (0)	0	7 (15)	19	6 (6)	18	6 (18)	18	3 (15)	9	0 (0)	0
	6	0 (0)	0	0 (0)	0	0 (0)	0	0 (0)	0	0 (0)	0	0 (0)	0
total Interaction	1	44 (21)	46	68 (40)	72	3 (11)	3	0 (0)	0	47 (48)	50	41 (49)	44
	2	20 (18)	21	11 (28)	11	69 (23)	73	0.4 (1)	0.5	42 (44)	45	49 (46)	52
	3	2 (3)	2	0.3 (2)	0.4	0.7 (2)	0.7	75 (35)	80	5 (18)	5	4 (17)	4
	4	28 (17)	30	1 (4)	1.5	4 (9)	4	2 (8)	2	0 (0)	0	0 (0)	0
	5	0.04 (0.2)	0.05	13 (30)	14	18 (17)	19	17 (39)	18	0 (0)	0	0 (0)	0
	6	0 (0)	0	0 (0)	0	0 (0)	0	0 (0)	0	0 (0)	0	0 (0)	0

Regarding to the relationship between the postural responses and the participants' gender, the findings show that the females had more postural changes (64,4 %) than the males (35,6 %). Table 4 shows the occurrences frequency and percentage for Postural responses occurrences by gender, in function of scenarios categories.

Table 4 Postural responses occurrences and sample gender, in function of scenarios (n=24)

Virtual Reality scenarios	Gender					
	MALE		FEMALE		TOTAL	
	Occu	%	Occu	%	Occu	%
Pre-emergency	512	39	810	61	1322	
Emergency Egress	256	31	579	69	835	100
Both scenarios	768	35.61	1389	64.39	2157	

When the distribution of the postural behavior occurrences is related to the gender of the participants it is seen that males presents more postural changes than females for the "Pre-Emergency" scenario. The opposite happened when the "Emergency Egress" scenario was analyzed, where females had more occurrences in postural behavior than males. The differences found between males and females, regarding the postural changes values, might not be explained just by the ratio between both groups (1.18 females for each male). More research about this subject can help to clarify eventual differences between both genders regarding postural behavior in VR simulations.

CONCLUSION

The data for postural responses obtained with this methodology provide a potential measure to study the relationship between the participants' actions inside the VE and their postural behavior. In this way, it may also represent a good complement measure to analyze the presence once this is an easy to use methodology, which requires few resources and is not intrusive. However, it is also possible that there is a mismatch between postural measures and post-test measures of presence (Frederickson & Kahneman, 1993).

For this case study, the relationship between VR actions and postural responses were not very strong. There are several possible reasons for this. One of them was the fact that the participants remained seated and using the joystick during all the VR interaction; another fact was that the VE did not have stimulus specifically designed with the intention of provoking postural responses (e.g. throw objects in participants direction). The findings from this study suggest new challenges for future researches. For example, this methodology can be used to compare game expertise vs. postural responses, VR interaction duration vs. postural response or different experimental setups (e.g. sitting and standup, using a different type of locomotors based device) vs. postural response.

ACKNOWLEDGMENTS

This research was supported by the Portuguese Science and Technology Foundation (FCT) grants (PTDC-PSI-69462-2006 and (SFRH/BD/47481/2008)

REFERENCES

Baños, R. M., Botella c:, Garcia-Palacios, A., Vilia, H. Perpiña. C., Alcaniz, M. (2000). Presence and reality judgment in virtual environments: A unitary construct? *CyberPsychology and Behavior*, 3(3), 327-335.

Frederickson, B. L., & Kahneman, D. (1993). Duration ne- glect in retrospective evaluations of affective episodes. Jour- nal of Personality and Social Psychology, 65, 45–55.

Freeman, J., Avons, S., Meddis, R., Pearson, D., & IJsselsteijn, W. (2000). Using behavioral realism to estimate presence: A study of the utility of postural responses to motion stimuli. Presence: Teleoperators & Virtual Environments, 9(2), 149-164.

IJsselsteijn, W., de Ridder, H., Freeman, J., & Avons, S. (2000). Presence: Concept, determinants and measurement. *Proceedings-Spie* The International Society For Optical Engineering, 520-529.

Insko, B. E. (2003). Measuring Presence: Subjective, Behavioral and Physiological Methods. In Being There: Concepts, Effects and Measurement of User Presence in Synthetic Environments, Riva, G., Davide, F., and Ijesslsteon, W. A. (eds.) Ios Ptress, Amsterdam, The Netherlands.

Sadowski, W., & Stanney, K. (2002). Presence in virtual environments. In K. M. Stanney (Ed.), Handbook of virtual environments: Design, implementation, and applications. Mahwah, NJ: Lawrence Erlbaum Associates, Inc. 791-806

Singer, M., & Witmer, B. (1997). Presence: Where are you now? In: M. Smith, G. Salvendy, & R. J. Koubek (Eds), *Deign o computing systems: Social and ergonomic considerations.* Amsterdam: Elsevier. 885-888

Slater, M., M. Usoh, A. Steed (1994) Depth of Presence in Immersive Virtual Environments, Presence: Teleoperators and Virtual Environments, MIT Press 3(2), 130- 144.

Slater, M., Steed, A., McCarthy, J., & Maringelli, F. (1998). The influence of body movement on subjective presence in virtual environments. Human Factors.

Wang, L., Hu, W., & Tan, T. (2003). Recent developments in human motion analysis. Pattern Recognition.

Xiao, Y., & Mackenzie, C. (2004). Introduction to the special issue on Video-based research in high risk settings: methodology and experience. Cogn Tech Work, 6(3), 127-130.

CHAPTER 83

Skill Modeling and Feedback Design for Training Rowing with Virtual Environments

Emanuele Ruffaldi, Alessandro Filippeschi,
Carlo Alberto Avizzano, Massimo Bergamasco

PERCeptual Robotics Lab,
Scuola Superiore Sant'Anna / CEIICP
Pisa, 56127, ITALY

ABSTRACT

This chapter presents the models of rowing skill and the feedback design principles behind the rowing training system SPRINT developed in the context of the SKILLS project. This system has been designed for training multiple sensorimotor and cognitive abilities of rowing, addressing efficiency of the technique, coordination and energy management. SPRINT is characterized by a portable mechanical platform that allows realistic motion and force feedback that is integrated with real-time simulation and analysis of athlete performance. The system is completed by training accelerators based on visuo-haptic feedbacks.

Keywords: Training Platform, Skills, Multimodal, Virtual Reality, Rowing

INTRODUCTION

The creation of a multimodal system for training a specific skill is a task that starts from the analysis of the skill itself, the identification of the relevant tasks, the associated skill components and the relevant variables for sensori-motor or cognitive training. This analysis phase allows taking decisions in the context of the training platform, in particular relatively to the simulation part of the platform and the training feedback part. The former part of the training system provided by means of virtual environment technologies, deals with the stimuli that are part of the real task and is constrained by realism requirements and technological limits. The latter part deals with the identification of the types of multimodal stimuli that can be adopted for training in the virtual environment. For a given training system, which are the correct stimuli that can be provided? How we can exploit at most features of Virtual Environments?

There are indeed several aspects of Virtual Environments and multimodal technologies that can potentially improve the training of motor tasks. This work discusses the aspects connected to sport training by introducing a training system for rowing that has been developed in the context of the EU SKILL Integrated Project (http://www.skills-ip.eu/). This system is called SPRINT (Ruffaldi 2009a). The SPRINT system is a platform for in-door rowing training that is aimed at improving specific sub-skill of the general rowing skill. The system has a basis of simulation component that recreates the motion of the user and provides a haptic feedback of the interaction with the water. Over the simulation basis is built a set of exercises based on multimodal feedback for training specific sub-skills, in particular: technique, strategy and coordination. The capturing is performed not only in real-time but in particular during the system design phase for characterizing the model of experts respect given sub-skills. The model is then used, in a user personalized form, for training the user during the execution of the specific training exercises.

This work discusses the modeling of the skill, and the how the feedback is selected and integrated in the multimodal experience of the training system. In particular we discuss the capturing system and the design of feedback based on visual and vibrotactile technologies.

ROWING SKILL

Rowing is a sport that can be can be competitive or recreational, where 1 or 2, 4 or 8 athletes row in a boat on flat water, facing backwards and using oars to propel the boat forward. Rowing is a demanding sport requiring both physical strength and cardiovascular endurance, and involves both lower and upper limbs as well as the trunk, determining the use of almost all human muscles. Rowing task is cyclical, since mainly consists in repeating efficient oar strokes for 6-8 minutes. The most effective measure of efficiency is the distribution along time of the ratio between power output and resulting velocity. For obtaining such efficiency several abilities are involved like cyclical repetition of the most efficient gesture, management of the posture, perceptual abilities for controlling external stimuli, coordination between limbs and with partners, and management of the energy expenditure during the race. In the context of a training system for rowing it is fundamental to correctly identify the above aspects and formalize them with the help of mathematical formulation. In particular after analyzing, with the help of expert coaches and rowing manual, the typical errors and tasks in rowing we have selected the elements that are more relevant for training intermediate rowers. In particular the factors affecting the efficiency of the technique, like errors in the entrance and exit from the water, coordination between limbs, and finally the profile of effort/velocity adopted during the race.

SPRINT SYSTEM

The SPRINT system has been designed around a mechanical platform that allows reproducing out-door features of rowing movements. Both sculling and sweeping tasks are possible and the platform can be adapted to different users. The complete experience of the system is show in Figure 1, in which is presented the mechanical platform in blue, with the force feedback fan highlighted, and the frontal LCD display for the visual feedback. The platform is completed by the sensing components and vibrotactile actuators placed around the wrists of the user (Ruffaldi 2009b). The major difference respect existing commercial solution like Concept2 is first in the possibility of performing a realistic motion, then the use data analysis techniques for characterizing the performance of the user and finally in the integration of multimodal feedback for training the athlete in different aspect of the rowing skill. Respect existing immersive solution like (von Zitzewitz 2008) this system has been designed to be adopted in real rowing contexts and providing the required flexibility for training athletes.

The motion of the user is acquired by means of encoders recording position of the oars in the two main axes, and one infrared sensor that tracks the seat displacement, at the rate of 120Hz. The integration of this information with a simplified kinematic model of the user allows identifying the most relevant motions of the user's body (Filippeschi 2009, Garland 2005). The kinematic model has been validated and corrected by an acquisition session inside a motion capture system (VICON by OMG plc) in which the information obtained from markers has been related with the sensing data of the SPRINT system.

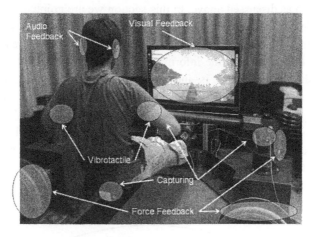

Figure 1 The user's experience in the SPRINT system

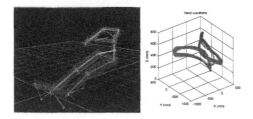

Figure 2 Acquisition session inside the VICON system
for validation of user's motion

Depending on the task being performed over SPRINT, additional sensors can be integrated, like biometric sensors for muscular activity (namely surface ElectroMyo Graphic sensors), Heart Rate sensing, Oxygen consumption devices and detailed motion capturing based on cameras. All the above information is being collected by a single computer that records synchronized data for later analysis and computes the proper feedbacks depending on the training scenario. The software core of the SPRINT system is the simulation of the rowing experience, dealing with the kinematic model of the user, the interaction of the oar with the water and the resulting motion of the boat. A correct simulation of the motion in the environment is motivated by providing the athletes a correspondence between their self-motion and the motion in the environment. In addition the simulation allows to realistically compute energetic parameters of the motion and to experiment the consequences of motions and errors on the rowing experience.

The knowledge about rowing skill and design constraints has allowed designing the SPRINT system as discussed above. In this section we are going to discuss how the different feedback components have been designed and how they participate at different training scenarios. The overall feedback design and the relationship with training have been based on the information flow model depicted in Figure 3. In this model the Environment is intended as the combination of the Virtual Environment, in which the simulated action is taking place, and the Real Environment, that is perceived by the user. Effectively every Virtual Entity is associated to one or more Physical instantiations in the form visuo-haptic feedback. On the left part of this model we have the Digital Trainer that has the role of coordinating the performance evaluation and providing the correct feedback depending on the scenario.

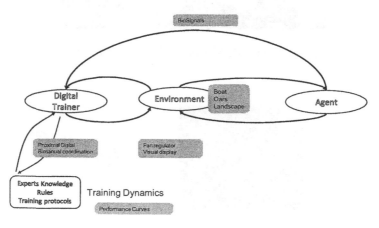

Figure 3 Information flow model of the interaction between the User and the Environment.

FEEDBACK FOR SIMULATION

The first type of feedback we are taking into account is oriented to the realism of the rowing experience, with the two fold objectives of not introducing disruptive effects in the training process and keep the user immersed in the training task. We have addressed two aspects in particular, the first is the force feedback while the second is the visual feedback.

Haptic Feedback

One of the main features of the SPRINT system is the capacity of providing a force feedback that has characteristics similar to the ones of the real rowing task. In particular we have adopted two energy dissipaters of the Concept2 ergometers that have been connected to the oars by means of a mechanical transmission. Such setup has been characterized using force sensor and compared against known profiles of forces measured on boat. The device is effectively a haptic interface because it produces but a force resistance depending on the user task, although in this version of the system it is not possible to tune such resistance during the task execution. The system measures the resulting resistance force from the speed of the fan of the dissipater that has calibrated by means of a force sensor applied on the transmission.

Visual Feedback

The visual feedback is provided by a Virtual Environment in which the user is placed with the boat in a typical rowing basin surrounded by various types of objects. Depending on the scenario the basin is populated with floats, that give the user distance references or the boats with opponents, whose avatars have a motion that is procedurally generated based on behavior schemes. A fundamental aspect in this part of the system is the type and position of the virtual camera that determines the point and field of view. In terms of camera position there are several possibilities (see Bailenson 2008 for an evaluation): the most immersive is first person with or without representation of the athlete limbs, then we have third person just behind the avatar of the athlete, and finally we have a third person facing the avatar and aerial point of views. In the first two cases the visual flow is the same as in the real scenario, although the presence of the avatar of the second case covers most of the visual field. Finally the last two points of view are more adequate for gaming, being easier for reaching the opponents. In the SPRINT system we have adopted the first solution, without the display of limbs, because they are only introducing occlusion in the view. The decision about the point of view is relevant for providing a match between simulated and perceived distances and velocity. For a discussion about distance perception see also (Vanni 2009). Respect other types of simulators, like racing and flight simulators, the perception of distance and velocity is provided both by the physical motion over the device, associated to the resulting fatigue, and by the visual component. For removing the introduction of scaling factors we have decided to follow a projection metaphor that is more common to augmented reality system rather than typical virtual environments: the display screen becomes a window to the virtual world, resulting in a good overlapping between real and virtual oars, and having the tip of the boat in front of the athlete. For obtaining this type of view (Figure 4) we have adopted the same type of perspective corrections that is used in immersive stereographic display, that take into account screen size and position of the user with respect to the projection screen. This projection could take advantage of the tracked position of the user head, but it can be roughly derived from the seat position, and in the context of single channel display the error is not noticeable. We have not considered, for this setup, the adoption of stereo projection because the device has been designed for prolonged training and we wanted to remove elements that could visually fatigue the athlete. Finally it is worth mentioning the fact that the visual refresh rate is 60Hz with a mechanism for reducing the latency between the captured data at 120Hz and the video synchronization signal.

Figure 4 Particular of the point of view from the position of the athlete.

TRAINING FEEDBACK

The design of an immersive system, involving both cognitive and sensorimotor activities, poses challenges in the definition of the way training stimuli are directed to the user. In particular it is important to not overflow a given sensori channel and at the same time to not introduce dependency for a given feedback stimulus. Among the several channels available we have selected the visual and the haptic ones, and in particular for the haptic we have reduced the training feedback to the aspects of vibrotactile stimuli, because the haptic channel is already fully involved in the main task of rowing. The description of the design space for feedback can be performed by sensory channel or by task; in this case we will describe it by channel in parallel with the above description about simulation feedback.

Visual Feedback

The role of the visual feedbacks is multiple, and while their goal is in general to improve the performance of the user toward some specific goal, some act directly others indirectly to the action of the user. In addition in the specific case of rowing the behavior of the user can be modified only periodically at every stroke, or for energy related behaviors with slower rate, and for this reason it is important to select the correct type of feedback. The quantitative indication of power output is an example of direct information that cannot be easily estimated by the user, while velocity can be indirectly perceived from the motion in the virtual environment.

The other important aspect for indicators is the way they are placed on the virtual environment. For example if it is necessary to keep the distance to the opponent that is behind the user a non-immersive solution is to present an horizontal gauge layered over the screen while an immersive solution, selected in one of the strategy training, is to display an arrow over the basin connecting athlete's boat and the one of the opponent behind. Virtual Environments allow anyway to adopt feedback mechanisms that act at the level of the environment and that are more subtle. In particular in the case of the study of inter personal coordination between athletes a direct feedback can be shown layered over the screen, while an indirect one acts on the speed of the boat, without being directly correlated with a precise simulation of the synchronization between the athletes. In general anyway visual feedback selection should follow the rules of visual encoding of information. For example expressing quantitative values with shades of gray makes comparison difficult, but when such shades are animated depending on error it can provide a good estimate of the amount of error. The result of this consideration can be applied to a feedback scheme that gives color to the avatar of the peer during interpersonal rowing.

Haptic Feedback

The haptic feedback that has been selected for training is provided by a set of vibrotactile bracelets that, with the use of two (or four) motor for each wrist provide directional or timing indications to the effort of the user. This type of feedback is currently being investigated in comparison with the visual one for assessing the capacity of training for specific aspects of the technique like the motion pattern. In this context the tactile gives indications of error in the trajectory of the arms (Ruffaldi 2009b).

TRAINING AND EVALUATION STUDIES

The system described above is currently in the phase of evaluation, with specific focus on multimodal contribution to technique optimization and energy management training. In these experiments intermediate and naïve rowers are being first assessed over the system in terms of their skill level and then there are being trained in one of the sub-skills discussed above. Specifically we are focusing on gestural procedure, motion coordination (between arms and legs) and timing for novel-intermediate rowers, addressing the typical errors found at this level. A series of visual and vibrotactile accelerators are used for indicating timing errors and corrections in the trajectory performed. A second study is related to the energy management in 2000 race, characterized by the training for adopting a given velocity profile that manages at best the energy of the athlete. In this case the feedback is based on the use of avatar's opponent for stimulate user's strategy. Finally a third study is focused on inter personal coordination in team rowing based on visual feedback and behavior of avatar's peer for stimulating synchronization.

During the development of the system, for obtaining a first understanding of the quality of the system with real rowers we performed a qualitative assessment of the system asking a group of intermediate rowers to practice with it and express judgments on its capacity of being adopted as a training device for technique and strategy. These comments allowed understanding the qualitative correspondence with boat rowing and the capacity of the system to keep the presence during the training experience. Each rower in the group has been first asked for a self-assessment of its level of expertise in terms of years of practice and types of competition in which participated. Then we asked 11 questions with answers expressed in a 7-point Likert scale. We asked information about the system in terms of stability, perceived capacity of rowing, water resistance in scull and sweep rowing, perceived realism of the virtual environment and effectiveness of displayed real-time information about the rowing performance. The group was of 8 intermediate-expert rowers, all male; with a mean age of 18 years, practicing rowing between 4 and 13 years, two of which competing at international level, 5 at national and one regional. The presence of the virtual environment, compared to commercial training systems in which is it absent, was making the experience more pleasant although at that time the quality of the landscape was not considered sufficient. From the questionnaire it emerged that perceived resistance was fine in scull rowing, while it was too strong for scull rowing, confirming, in this way, the mathematical modeling. This information provided a first insight for updating the strength of the system for the scull configuration. Another force related aspect discussed above that has been considered is the realism of the entrance in the water that at that time was not considered convincing.

CONCLUSIONS

The elements discussed above, although having some aspects specific to rowing, could be easily extended to other training activities in which there is a sensorimotor component, and for which the training feedback has to be accurately selected. There are several aspects that have not yet be addressed: investigate how to balance between different type of stimuli in a given protocol, investigate the role of audio, how the motivational component impacts on the training efficiency and finally the training of teams with multiple SPRINT systems. Additional information and material on the project are available on the website http://www.skills-ip.eu/row .

ACKNOWLEDGMENTS

The activities described in this paper have been carried on with the financial assistance of the EU which co-funded the project SKILLS: "Multimodal Interfaces for capturing and Transfer of Skill"- IST-2006-035005 (www.skills-ip.eu). We also wish to thank people involved in these developments: Vittorio Spina , Oscar Sandoval and Antonio Frisoli from Scuola Superiore S.Anna, Pablo Hoffmann from Aalborg University, Benoit Bardy and Sebastien Villard from University of Montpellier I, Daniel Gopher and Stas Krupenia from Technion.

REFERENCES

Bailenson, J.; Patel, K.; Nielsen, A.; Bajscy, R.; JUNG, S. & Kurillo, G. (2008) The effect of interactivity on learning physical actions in virtual reality Media Psychology, Routledge, 2008, 11, 354-376

Cabrera D. , Ruina A. and Kleshnev V. (2006) "A simple 1+ dimensional model of rowing mimics observed forces and motion", Human Movement Science, vol. 25, pp. 192-220

Filippeschi, A.; Ruffaldi, E.; Frisoli, A.; Avizzano, C. A.; Varlet, M.; Marin, L.; Lagarde, J.; Bardy, B. & Bergamasco, M. (2009), "Dynamic models of team rowing for a virtual environment rowing training system", in The International Journal of Virtual Reality, vol. 4, pp. 19-26

Garland S.V. (2005) "An analysis of pacing strategy adopted by elite competitors in 2000m rowing", Journal of Sports Medicine, vol. 39, pp. 39-42

Ruffaldi, E.; Filippeschi, A.; Frisoli, A.; Avizzano, C. A.; Bardy, B.; Gopher, D. & Bergamasco, M. (2009), "SPRINT: a training system for skills transfer in rowing", ed. Gutiérrez, T. & Sánchez, E. (ed.), in Proceedings of the SKILLS09 International Conference on Multimodal Interfaces for Skills Transfer, Bilbao Spain

Ruffaldi, E., Filippeschi A., Frisoli A., Sandoval O., Avizzano C.A., Bergamasco M. (2009) "Vibrotactile Perception Assessment for a Rowing Training System", in Proceedings of the third IEEE Joint conference on Haptics, World Haptics, Salt Lake City

Vanni, F. e Ruffaldi, E.; Avizzano, C. A. & Bergamasco, M. (2008) "Large-scale spatial encoding during direct and mediated virtual rowing", 1st International Symposium on Neurorehabilitation, Valencia, Spain

Von Zitzewitz, J. and Wolf, P. and Novakovi, V. and Wellner, M. and Rauter, G. and Brunschweiler, A. and Riener, R. (2008), "Real-time rowing simulator with multimodal feedback", Sports Technology, vol. 1, 6, 2008

Chapter 84

An Enactive Approach to Perception-Action and Skill Acquisition in Virtual Reality Environments

B. Bardy, D. Delignières, J. Lagarde, D. Mottet, G. Zelic

Movement To Health Laboratory
Montpellier-1 University, France

ABSTRACT

We describe in this chapter basic principles underlying human skill acquisition, and envisage how virtual environment technology can be used to shorten the route toward expertise. The emphasis is put on *coordination* — between segments, muscles, sensory modalities, and between the agent and the environment — and on enaction, a form a knowledge that is gained by doing, i.e., through the active interaction between action and perception, in real or virtual environments. Examples are taken from the SKILLS FP6 european integrated project.

Keywords: Enaction, skill acquisition, perception-action, virtual environment

SKILLED COORDINATED BEHAVIOR

Generally speaking, skill can be defined as the capacity acquired by learning to reach a specified goal in a specific task with the maximum of success and a minimum of time, energy or both. This simple definition available in any textbook on motor control and learning suggests that skill cannot be considered as a general and abstract ability, but rather as a specific and learned capacity operating in an limited ensemble of situations. A number of criteria can be analyzed for evaluating the level of skill, and have to be integrated in a formal definition of skill.

1. The accuracy of the outcome, with respect to the assigned goal: this criterion is the most commonly used, and measured in terms of errors to the target goal (spatial and /or temporal).

2. The consistency of responses over successive trials: Skilled movement outputs are stable and reproducible. Stability can be traditionally measured by assessing for instance the standard deviation of a set of successive outcomes. More contemporary approaches however focus not only on the amplitude of variability but also on its temporal structure, suggesting that consistency should not be understood in terms of 'behavioral stereotype' but, as proposed by Bernstein (1967), as "repetition without repetition", i.e., as the propensity of experts to regulate small details of the ongoing actions in order to reach a stable performance.

3. Efficiency, i.e. the ability to reach the desired goal at minimal cost: Efficiency can be understood at various levels, including the cognitive level (the use of automatic processes allowing a decrease in mental load), the metabolic level (a reduced metabolic cost in expert to reach the same performance), or the neuro-muscular level (reduction of co-contraction and more phasic muscular activation in experts). These various levels are not independent from each other and generally reveal the same trend toward economy as skill acquisition progresses.

4. Flexibility and adaptability: Skilled behavior copes with endogenous and exogenous uncertainties. This flexibility suggests that skill is not specific to a particular task, but rather to an ensemble of similar tasks, which raises the fundamental problem of skill generalization and transfer. Flexibility and adaptability are crucial and complementary aspects to the stability properties described above. Variability is often functional, not necessarily reflecting noise in the neuro-muscular system for instance, but as a consequence of continuous and functional adaptations operating at the level of perception-action loops.

5. The elaboration of a coordinative structure: Skills are characterized by a delicate spatio-temporal organization of sub-movements, including task specific synergies among body segments, joint angles, and end-effectors, and also between the agent and the environment. A skilled movement is never performed in the exact same way, and always possesses a certain degree of variability; however a pattern of organization — order — is common to each sample. The spatio-temporal patterns of coordinated motor activity are low dimensional, due to the dynamic interactions of the many degrees of freedom of the system, and accordingly can often be described by one or few collective variables. To be proficient, or 'dexterous', the coordinative structure (e.g., Turvey, 1990) has to resist unavoidable internal and external perturbations, and thus must possess stability.

PERCEPTION-ACTION AND SKILL LEARNING

The components of a skilled action briefly outlined above include the set of end effectors and segments-joints effectors, but also include the perceptual systems that are used to guide task performance. Human and other biological organisms are informationally coupled to their environments, in various task-specific perception-action cycles (e.g., Kugler & Turvey, 1987). On the perception-action side, complementary markers of skill performance and skill acquisition are of interest to build efficient virtual reality platforms.

1. Movements are sources of information. Physical properties of the agent - environment system relative to its layout have specific consequences on ambient energies. Although not depending on movement for their existence, these invariant patterns within energies emerge unequivocally only with a flux, "the essentials become evident in the context of changing nonessentials" (Gibson, 1979). Movements themselves are structured and have unequivocal consequences in the informational flow. Accordingly, information resides in the stimulation, that is, in the structure of ambient energies (e.g., optics, acoustics, etc.) that stimulate our perceptual systems.

2. Information, when perceived, is used to regulate movements. In natural or virtual interaction, (multisensory) stimulation is specific to reality, indicating that a lawful relationship (i.e., 1:1) exist between both parts. The specification relation can be formalized in the form Force = f(flow), meaning that the net force acting on the observer (external + internal forces) is specified in the structure of energy flows. First assumed to be modal (e.g., within the structure of one single energy), this equation is now considered to be intermodal (Stoffregen & Bardy, 2001). According to the circular causality principle, motor performance is achieved by the means of control laws. These laws are relations between the informational (perceptual) variables and the free parameters of the action system (motor variables) that are relevant for the ongoing action (Warren, 1988).

3. Learning consists in discovering and optimizing the information-movement coupling. Because information is both created by the movement and regulating the movement, finding the coupling function between the two components is one key problem that learners have to solve. For instance driving a car on a turning road implies maintaining the current direction of the car aligned continuously with the direction of the road. The current direction of the car being given by the optical focus of expansion at the driver's eye, a simple and efficient way of accomplishing this action (going from A to B on a turning road) is to move the steering wheel so as to maintain the focus of expansion in the middle of the road. Learning how to drive a car on a turning road imply discovering and stabilizing the coupling function between movements of the hands on the steering wheel and the optic flow created by these movements.

4. Learning implies stabilizing control laws. Control laws are expressed in the form: $\Delta fint = g\ (\Delta flow)$, where $\Delta fint$ refers to the changes in internal forces applied by the observer, and $\Delta flow$ to the corresponding changes in flow energy (e.g., optical or inertial flow) specifying the (changes in) relationship between the observer and the environment. In the car driving example, the control law for driving a car on a turning road can be written in the form: Intensity of force on the steering wheel = f (focus of expansion / road). This example is a simple one. There

are many control laws that need to be simultaneously stabilized during the acquisition of skilled behaviors — both within and between perceptual modalities.

SKILL DECOMPOSITION

As evidenced in the previous two paragraphs, the various transformations in coordination and perception-action that accompany the acquisition of expertise involve many skill elements, both at sensori-motor and cognitive levels. A specific skill — e.g., juggling, rowing, drilling, reaching, tracking, assembling etc. — can thus be decomposed in a series of skill elements. Within the SKILLS European project (www.skills-ip.eu), several sensori-motor and cognitive skills have been identified throughout six VR training platforms. Table 1 below summarizes these main skill elements call "sub-skills".

Table 1 Skill elements at both sensori-motor and cognitive levels

Sensori-motor sub-skills	
Balance and postural control	The regulation of posture (segments, muscles, joints, etc...) and balance (static, dynamic, etc...) that allows the distal/manual performance to be successfully achieved. It is captured by inter-segmental and inter-muscular coordination, as well as center-of-pressure variables.
Bimanual coordination	The functional synchronization in space and time of the arms/hands/fingers. Bimanual coordination is captured by the relative phase between the coordinated elements and its stability.
Hand – eye coordination	The synchronization of eye / gaze / effector with reference to the main information perceptually detected. It is assesses by gain, relative phase, and in general coupling variables between eye, gaze, and hand.
Interpersonal coordination	The coupling between two or more persons. It emerges from a nexus of components including sociality, motor principles, and neuroscience constraints. It is assessed by the relative phase between persons.
Perception-by-touch	The coetaneous component of the haptic modality. Various receptors embedded in the skin provide information about mechanical properties (*vibration*, *compliance* and *roughness*), *temperature* and *pain*. It is evaluated by psychophysical methods.
Prospective control	The anticipation of future place-of contact and time-to-contact based on spatio-temporal information contained in optic, acoustic, or haptic energy arrays. It requires the coupling between movement parameters and information contained in various energy arrays, and is measured by time-to-contact and related variables.

Proximo-distal coupling	The spatio-temporal coordination of proximal, gross components with distal manipulatory components. It refers to the organization of the body underlying arm movements, or to the synergy between arm postures and hand movements. It is assessed by cross-relational variables.
Respiratory-movement coupling	The synchronization of breathing and movement (segments, muscles, joints, etc...) that allows efficient performance. It is measured by amplitude, phase and frequency synchronization patterns.
Fine force control	The online regulation of the internal forces applied on the surface to successfully reach the goal (drilling, pasting, navigating, etc...). It depends on the properties of the surface in relation to the forces developed by the effectors, and is evaluated by the ratio between the two.
Cognitive sub-skills	
Control flexibility and attention management skills	The ability to change response modes and performance strategies, to apply and manage new attention policies in order to cope with task demands and/or pursue new intentions and goals. It is measured by adjustments to changes in task demand and attention allocation.
Coping strategies and response schemas	A vector of importance or attention weights computed over the many sub-elements of a task, which are associated with the achievement of a specific goal. They are measured by the number and type of strategies to cope with variations in task demands and changes of intention.
Memory organization, structure and development of knowledge schemas	Level of formulated and organized multi hierarchy, task specific memory and knowledge bases that facilitate encoding, retrieval and the conduct of performance. Measured by speed and accuracy of encoding, response and decision-making performance, and number, diversity and speed of generating alternative solutions.
Perceptual Observational	The ability to detect, sample and extract task relevant information from the environment and perceive static patterns and dynamic regularities. Measured by speed, efficiency, amount of conscious supervision, and use of higher-level structures and redundancies.
Procedural skills	Sequences of ordered activities that need to be carried out in the performance of tasks. Performance of every task can be subdivided into a large number of procedures, the competence in the performance of which is developed with training. Evaluated by speed and efficiency of performance, type of supervision (un/conscious).

The first nine sub-skills (upper section of Table 1) are from the sensori-motor repertoire while the last five sub-skills (lower section of Table 1) are from the cognitive repertoire. In general, sensorimotor sub-skills are skills that relate to the relationship between perceptual components and motor components, and cognitive skills are related to higher-level cognitive activities that orient, formulate, monitor and regulate the sensori-motor performance. The distinction is partly arbitrary as the two categories are largely interdependent. This is because of the natural embodiment of cognitive phenomena into sensori-motor dynamics, an embodiment at the heart of the SKILLS project. Although there are several views on embodied cognition (see Wilson, 2002 for a review), the term generally refers to the basic fact that a) cognition is largely for action, and b) off-line cognition (cognition decoupled from the environment) is largely body-based. Consistent with these claims is the fact that perceptual inputs (e.g., vision) can elicit covert motor representations in the absence of any task demand. There is also increasing evidence coming from brain imaging studies that the perception of objects automatically affords actions that can be made towards them (Grezes & Decety, 2002). Similarly, the fact that when individuals observe an action, an internal replica of that action is automatically generated in their premotor cortex (Buccino et al., 2001), suggests that embodied cognition plays a role in representing and understanding the behavior of conspecifics, such as in learning from imitation.

Hence perception is not a perceptual process, preceding symbolic representations of actions to be performed. What one perceives in the world is influenced not only by, for instance, optical and ocular-motor information, but also by one's purposes, physiological state, and emotions. Perception and cognition are embodied; they relate body and goals to the opportunities and costs of acting in the environment (Varela et al., 1991; Proffit, 2006). The taxonomy of skill elements reported in Table 1 is only a convenient way to help researchers and engineers to elaborate technological tools able to accelerate the learning of these skill elements.

ENACTING SKILL ACQUISITION IN VR

Contemporary multimodal virtual reality simulators can be used to accelerate the acquisition of complex skills. It is the aim of the SKILLS EU integrated project to develop training accelerators and training protocols able to speed up the mastering of the various skill elements described in the previous section. What these demonstrators have in common is their potential to *enact* useful — sensori-motor or cognitive — on line or off line information at given moments in time and in a training context, in order to take advantage of the well-known general principles underlying skill acquisition in real life and transfer them to virtual or mixed contexts. In this final section we present three basic examples of how this enactment can be realized (see the SKILLS related chapters for more examples).

DESTABILIZING SPONTANEOUS COORDINATION MODES

As stated previously, learning occurs on the basis of pre-existing coordination modes (between segments, between muscles, or between the agent and the environment. These initial solutions attract the spontaneous behavior, and seem highly resistant to change. The initial attractors of the system exhibit a strong stability, preventing the exploration of the workspace and the adoption of other behaviors more efficient in the long run (Nourrit et al., 2003). A solution for shortening this initial stage is to find tricks helping the destabilization of initial attractor states. This hypothesis was initially examined by Walter and Swinnen (1994) in a bimanual coordination task. In participants who had to learn a 1:3 frequency ratio, they found that slowing down the movement facilitated learning. In oscillation skills, movement frequency is indeed often considered as a control parameter of the coordination (e.g., Kelso, 1984). This adaptive tuning was supposed to reduce the exclusiveness of the spontaneous 1:1 coordination mode, and to allow an effective exploration of other less natural ratios. A related method is currently used in the SKILLS project with virtual juggling of a three-ball cascade (see Figure 1). In this ongoing experiment, jugglers are guided by on line vibro-tactile and auditory information contributing to breaking the spontaneous 1:1 coordination between hands and transforming it into more complex, multi-rhythmic and multimodal, 2:3 ball-hands coordination. Together, these scenarios indicate that control parameters and on line information can be used to accelerate learning through an information-induced destabilization of the initial coordination state.

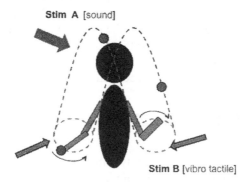

FIGURE 1. Online multimodal information (auditory, vibrotactile) used during virtual juggling in order to destabilize initial 1:1 bimanual coordination and learn a more complex 2:3 multi-rhythmic hand-ball coordination.

INCREASING PATTERN AVAILABILITY

A complementary principle is obviously to increase the availability of the expert coordination mode. Here also, one solution is to use adequate control parameters. This procedure was explored in the domain of rehabilitation. For example Wagenaar and Van Emmerick (1994) showed that during walking at preferred speed, healthy participants had their pelvis and trunk moving out-of-phase, while Parkinson's patients had their pelvis and trunk moving mostly in-phase. When velocity was increased, differences between healthy controls and patients decreased. This result suggests that an appropriate tuning of a relevant control parameter can induce the emergence of the desired coordination pattern during rehabilitation. This example is of interest for the acquisition of complex skills because it reveals that accelerators can be very simple when the dynamics of the perceptuo-motor workspace is known (location of attractors and repellors, disappearance of attractors in patients etc…). It also reveals that slowing down a movement during the acquisition phase is not always a good solution for accelerating learning.

When the desired pattern is not potentially available, and does not emerge as a result of control parameter manipulation, the pattern can nevertheless be required by means of instructions or by presenting a model. In the first case the pattern is described by verbal explanations, in the second case a model is directly provided (by a real or a virtual trainer). For example, Faugloire et al. (2009) demonstrated that on-line postural feedback, by means of a Lissajous plot (position vs. position), allows a quicker learning of a complex postural coordination pattern. In this experiment participant had to learn a coordination pattern with a 90° phase offset between the ankles and the hips. In this case, the expected Lissajous plot is an ellipse (see Figure 2).

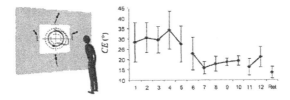

FIGURE 2 Accelerator for learning a new postural pattern (left); Constant error (required - produced pattern) during the acquisition period (right). Data are averaged over participants (Faugloire et al., 2009).

In the initial trials, participants were attracted towards spontaneous coordination modes, especially in phase and anti-phase, yielding straight lines in the Lissajous plots. With time and practice, the competition between the spontaneous patterns and the new pattern progressively turned in favor of the new pattern, with reminiscent presence however of the intrinsic dynamics of the postural system. Interesting for our purpose is the complementary finding that learning persisted over time, suggesting that the accelerator had long-term effects. Similar methods are currently employed within the SKILLS project in various demonstrators (e.g., rowing and juggling).

GUIDING THE EXPLORATION OF THE WORKSPACE

According to Newell et al. (1989), learning is the result of an active exploration of the workspace defined by organism, task and environment constraints interacting with each other. This workspace can be conceived as a landscape characterized by

850

zones of stability (especially that corresponding to the expert pattern). Novices have to explore this landscape in order to discover the optimal zones (i.e. the "valleys", see Figure 3). This exploration can be guided by on line information, especially concerning task requirements, and efficiency gradients. Providing augmented feedback can help learners in their search for optimal solutions. This augmented feedback should be particularly efficient when targeted onto the order parameter representing the macroscopic controlled variable in the task. Enriched multimodal information, using various channels (auditive, visual, haptic,.) can also be delivered in order to guide a more efficient exploration of the workspace.

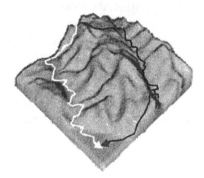

FIGURE 3 Exploring the workspace during learning a new perception-action skill in search for optimal solutions (attractors in the workspace). Multimodal information can be given online to the trainee to find the optimal solution.

CONCLUSION

Demonstrators can be conceived as technological implementation of learning accelerators. The above analysis suggests that according to the nature of the task at hand, but also to the level of advancement of the learning process, different accelerators can be selected. As such, a precise evaluation of the goal of the demonstrator (helping novices, training experts, etc.), and a deep analysis of the task (nature of the workspace, spontaneous coordination mode, expert behavior to be learned) should be undertaken before conceiving the appropriate implementation.

ACKNOWLEDGMENTS

This research was supported by SKILLS, an Integrated Project (IST contract #035005) of the Commission of the European Community. We thank Carlo Avizzano, Massimo Bergamasco, Daniel Gopher, Emanuele Ruffaldi, Eldad Yechiam, and the various SKILLS demonstrator teams for their contribution to the SKILLS unified framework (www.skills-ip.eu).

REFERENCES

Bernstein, N. (1967), *The co-ordination and regulation of movement*. Elmsford, NY: Pergamon Press.

Buccino, G., Binkofski, F., Fink, G.R., Fadiga, L., Fogassi, L., Gallese, V., Seitz, R. J., Zilles, K., Rizzolatti, G., and Freund, H.J. (2001), "Action observation activates premotor and parietal areas in a somatotopic manner: an fMRI study". *European Journal of Neuroscience*, 13, 400-404.

Faugloire, E., Bardy, B.G., & Stoffregen, T.A. (2009). "(De)Stabilization of required and spontaneous postural dynamics with learning". *Journal of Experimental Psychology: Human Perception and Performance*, 35, 170-187.

Gibson, J.J. (1979), *The ecological approach to visual perception*. Boston (MA): Houghton Mifflin.

Grezes, J., and Decety, J. (2002), "Does visual perception of object afford action? Evidence from a neuroimaging study". *Neuropsychologia*, 40, 212-222.

Kelso, J.A.S. (1984). "Phase transitions and critical behavior in human bimanual coordination". *American Journal of Physiology: Regulatory, Integrative, and Comparative*, 15, R1000-R1004.

Kugler, P.N., and Turvey, M.T. (1987), *Information, natural law, and the self-assembly of rhythmic movement*. Hillsdale, NJ: Erlbaum.

Newell, K.M., Kugler, P.N., van Emmerik, R.E.A., McDonald., P.V. (1989). "Search strategies and the acquisition of coordination". In S.A. Wallace (Ed.), *Perspectives on the coordination of movement* (pp. 85-122). North-Holland: Elsevier.

Nourrit, D., Delignieres, D., Caillou, N., Deschamps, T., and Lauriot, B. (2003), "On discontinuities in motor learning: a longitudinal study of complex skill acquisition on a ski-simulator". *Journal of Motor Behavior*, 35, 151-170.

Proffit, D.R. (2006), "Distance perception". *Current Directions in Psychological Science*, 15, 131-135.

Stoffregen, T.A., and Bardy, B.G. (2001), On specification and the senses. *Behavioral and Brain Sciences*, 24, 195-261.

Turvey, M.T. (1990), "Coordination". *American Psychologist*, 45, 938-953.

Varela, F.J., Thompson, E., & Rosch, E. (1991), *The Embodied Mind: Cognitive Science and Human Experience*. Cambridge, MA: The MIT Press.

Wagenaar, R.C., and van Emmerik, R.E.A. (1994). "The dynamics of pathological gait: Stability and adaptability of movement coordination". *Human Movement Science*, 13, 441-471.

Walter, C.B., and Swinnen, S.P. (1994), "The formation of "bad habits" during the acquisition of coordination skills". In S.P. Swinnen, J. Massion and P.Casaer (Eds.), *Interlimb coordination: Neural, dynamical and cognitive constraints* (pp. 491-513). San Diego, CA: Academic Press.

Warren, W.H, (1988), Action modes and laws of control for the visual guidance of action. In O.G. Meijer and K. Roth (Eds.), *Complex movement behaviour: 'The' motor-action controversy* (pp. 339-380). Amsterdam: North Holland.

Wilson, M. (2002), "Six views of embodied cognition". *Psychonomic Bulletin and Review*, 9(4), 625-36.

CHAPTER 85

A VR Training Platform for Maxillo Facial Surgery

Florian Gosselin [1], Christine Mégard [2], Sylvain Bouchigny [2],
Fabien Ferlay [2], Farid Taha [3], Pascal Delcampe [4], Cédric d'Hauthuille [5]

[1]CEA, LIST, Interactive Robotics Laboratory
18, route du Panorama, BP6, Fontenay aux Roses, F 92265, France

[2]CEA, LIST, Sensory and Ambiant Interfaces Laboratory
18, route du Panorama, BP6, Fontenay aux Roses, F 92265, France

[3]Amiens Univ. Hospital, Department of Oral and Maxillofacial Surgery
Place Victor Pauchet, Amiens, F 80054, France

[4]Rouen Univ. Hospital, Department of Oral and Maxillofacial Surgery
1, rue de Germont, Rouen, F 76031, France

[5]Nantes Univ. Hospital, Department of Oral and Maxillofacial Surgery
1, place Alexis Ricordeau, Nantes, F 44093, France

ABSTRACT

Maxillo Facial Surgery procedures are often complex and difficult to train for novice surgeons. Traditional teaching methods include supervised practical training during real surgeries. The trainee's progression is monitored by a senior surgeon who leads the procedure and progressively involves the trainee in the surgery. With such a method, the acquisition of surgery skills is long and depends on the cases the trainee is confronted to. In order to improve training efficiency, we propose to design and use a new skills training VR system composed of a multimodal platform and dedicated training exercises. The aim of this paper is to introduce this system and its main components.

Keywords: Training platform, Skills, Haptics, Multimodal, Maxillo Facial Surgery

INTRODUCTION

The general objective of Maxillo Facial Surgery is to reshape the human face to correct either malformation, results of trauma or tumours. As face features very complex, delicate and intricate structures, such procedures are difficult to perform. Generally, accessing the operating field is an issue in itself and surgeons have only a limited vision of the site. They must thus rely on multisensory information to perform the surgery. It is therefore very difficult to transfer, mostly because critical sub-tasks are mainly based on haptic feedback which is particularly difficult to teach with traditional observational methods.

Training could thus be improved with the use of a dedicated training platform. The objective is not to substitute the traditional teaching composed of theoretical aspects and practical training during real surgeries, but to complement it. With a training system, we can expect more repetitions in the same time period, independently of the availability of patients, hence a shorter training time (Kneebone, 2003). Moreover, training can be focused on the critical steps of the task, and adapted to the level of the trainee.

While some simulators were developed in the past for temporal bone surgery (Morris et al., 2006) (Sewell et al., 2004), we decided to focus on a very delicate operation called the Epker osteotomy (see Figure 1). In this surgery, the front part of the lower mandible is first separated (corticotomy) then reassembled in a better position (osteosynthesis). During this procedure, the alveolar nerve hidden in the mandible must not be damaged. This nerve is responsible for the sensitivity of the face and any injury would be very handicapping for the patient as it is irreversible. Therefore the operation is very stressful, even for expert surgeons with years of practice.

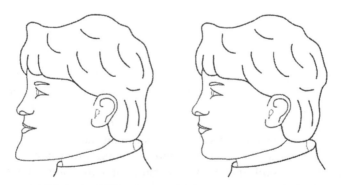

Fig 1: Patient before and after Epker surgery.

Epker Osteotomy is selected because it is considered by surgeons as representative of delicate interventions requiring highly sensitive skills, mainly in haptics and audition (Müller-Tomfelde, 2004), which are very difficult to teach with traditional methods and could greatly benefit from the use of multimodal interfaces, robotics and virtual environments technologies (Morris et al., 2007) (Ström et al.).

While surgical expertise is multiple (theoretical knowledge, technical skills, decision making, communication, and leadership skills, (Moorthy et al., 2003)), we concentrate on sensori-motor skills and put the stress on the most critical steps of the surgery.

SYSTEM SPECIFICATION

METHODOLOGY

Our aim is to provide a skills training system allowing to ease the transfer of skills from senior operators to novices for some of the basic skills that underlie the Epker surgery. To reach this goal and propose an efficient training, we followed a user-centered approach to get a deeper understanding of the surgery and identify its critical steps and its major current teaching difficulties. Three French expert surgeons working in different institutions were interviewed and results were analyzed using the protocol described in (Ericsson and Simon, 1993). A task analysis was also performed during an operation performed by one of the interviewed expert surgeons. The surgery was filmed and analyzed and the verbal exchanges were transcribed verbatim. Results include a task model, the identification of the critical steps and main strategies, and major differences between novices and experts. These results were discussed and validated with the surgeons and served as a basis for the focusing of the training system. The same kind of observations was performed by the Technion in Israel with maxillo-facial surgeons from the Rambam Hospital in Haifa and gave similar results.

IDENTIFICATION OF THE CRITICAL STEPS OF THE TASK

The principle of the Epker osteotomy is to drill the most resistive cortical parts of the maxilla and use natural weaknesses of the bone during fraction to proceed to the distraction of the mandible while protecting the integrity of the alveolar nerve. It is performed in teams. The main surgeon performs the surgical act alone, but needs typically one or two assistants to insure visibility of the very narrow operating field (retraction of the mouth tissues, vacuum of the blood and water off the mouth) plus a nurse who prepares the instruments. The critical steps obtained from the interviews with surgeons are the following (for this project we did not consider osteosynthesis - fixation of the bone in the correct position – which is less difficult to train):

- localization of the Spine of Spix: the Spine of Spix is a small bone protrusion were the alveolar nerve enters in the lower maxilla. It is crucial to locate it precisely to be able to protect the nerve (directly exposed in this region) before drilling. Although vision can help as in some cases the Spine of Spix can be seen in the depth of the operating field, its localization is mainly obtained through controlled haptic exploration of the bone surface during periost retraction. According to the interviews, the surgeons feel "a valley" or "a hole". This procedure is hard to learn and practice for novices as the tool must be constantly maintained in contact with the bone and with a correct angle. Moreover, it is hard for experts to explain what should be felt in case of success, even more as the haptic signature of the Spine of Spix varies according to the specific anatomy of each patient;

- drilling some of the fraction lines: the Epker surgery requires the drilling of the cortical layer of the bone without or with as less as possible overshoots in the underlying spongious bone hiding the nerve. The tool position must be constantly finely controlled while the resistance of the environment varies according to the progression of the task and despite of parasitic movements produced by the resistance to the rotation of the burr. Moreover, the drilling lines must be performed on different complex shapes, with different orientations and tool access possibilities. This skill is therefore very difficult to train. The most critical drilling lines are: 1- the first corticotomy line near the Spine of Spix which is judged difficult as it is necessary to keep removing the muscular insertions at the same time and because the operating field is very narrow, 2- the second lines on the crest of the mandible as the surgeons must follow complex shapes when moving from the internal part of the mandible near the Spine of Spix to the outer one along the teeth, 3- the third line at the basilar edge as the position of the nerve is close to the cortical and the access is very restricted (the access is possible only inside the mouth to keep the integrity of the cheek);

- distraction (separation of the inner and outer parts of the mandible): successful distraction requires that the drilling lines are completely finished (even a small remaining attachment can produce a bad fracture) and that the gesture is sufficiently frank to orient the fracture on the right line. The distraction is more difficult in young patients as in this case the bone does not fracture abruptly ("green bone") and the surgeon must be very cautious when separating the two valves. With older patients, bone fractures as "glass", but as a matter of fact this surgery is mainly performed on teenagers or young adults.

The interviews also revealed major differences between novices and experts during these critical steps:

- novices tend to localize the Spine of Spix using visual indices while expert surgeons rely more on haptic information as vision is limited;
- novices tend to use an insufficiently firm grasp of the instruments, producing more unstable movements of the drill compared to experts. Moreover, novices

tend to proceed to a superficial and incomplete drilling of the cortical in order to avoid entering the spongious bone and impairing the nerve. The risk of such an incomplete drilling is an undesirable multiple fracture of the maxilla;

- during the distraction, a safe gesture is performed when the tools exert the same amount of forces on both sides of the maxilla. The gesture must not be "shy". This seems to be a real challenge for young surgeons.

PURPOSE OF THE TRAINING PLATFORM

Considering these results we decided to focus on a skills training system for a single surgeon allowing to ease the transfer of some of the basic skills involved in the Epker surgery from senior operators to novices. The proposed training protocol will focus on localizing the Spine of Spix and drilling some of the fraction lines.

Skills training is the main focus of the SKILLS European project (www.skills-ip.eu). Within this project, a large amount of work is dedicated to the identification of key issues in VE based training, among which the identification and classification of trained sub-skills. Our study has shown that activities as diverse as juggling, rowing, performing surgery, maintaining assembly lines or manipulating robots require mastering of some common skills. The main trained skills involved in the surgery are:

- perception-by-touch: the detection of Spine of Spix requires a haptic exploration of the bone surface until a change in bone topography is felt. Similarly, correct drilling requires a haptic perception of the change in bone compliance from hard to spongy;
- fine position and force control: this is required to hold and move the tools correctly during tissues removal and detection of the Spine of Spyx. This is also required to drill the corticotomy lines on different complex shapes without any overshoot or uncontrolled movement;
- multimodal feedback management: the information on task progression is redundant (haptic, visual and auditive) and all modalities are used by experts. E.g. during drilling the change from cortical to spongy bone is related to both a change in bone compliance as well as to a visual change in bone color and to a change in sound produced by the drill;
- bi-manual coordination, postural control, and proximo-distal organization: the capacity to perform coordinated and controlled hand movements is related to the capability to adapt posture (especially non-dominant arm) for increased stability of gestures.

These skills are acquired by repeated practice, allowing progressive development of adapted, accurate, stable and efficient perception action loops necessary to perform the surgery. This is also the goal of the SKILLS project to study how this kind of enacted training can be performed in VR environment. A model of the task and environment is needed therefore. As no task is purely unimodal, such environment must be multimodal in order to capture and render all aspects of the activity. VR data can then be used to analyze the user's performances. This requires a high performance Digital Repository able to store tasks related data in real time and to further analyze them to allow displaying feedbacks to the trainee and monitoring the progression towards expertise for the trainer. Taking this information into account, it is also possible to adapt the training content and organization to the capabilities of the user.

Considering the Epker surgery, the reproduction of the actions performed during the surgery requires a multimodal training platform featuring haptic, visual and audio feedback.

SPECIFICATION OF THE TRAINING PLATFORM

In order to get quantitative data about the surgery, we performed a data acquisition campaign during real surgery procedures. The aim is both short term to derive the specifications of the training platform (e.g. range of motion necessary to perform the surgery, amount of forces and torques applied on the tool, ...) and long term to try to extract skilled performance signatures from the raw data in order to be able to focus the training on those aspects. The data acquisition was performed on dead bodies in the Anatomy Laboratory of CHU Rouen. Three Epker osteotomies were performed on both sides of human heads. Each of the three surgeons involved in the study performed the complete procedure (exposition, drilling, fracture) on one side of 2 different patients.

The acquisition platform includes an ART Fast Track 2 motion capture system, an ATI Mini 40 SI80-4 force sensor, two Analog Device ADXL330 accelerometers, EMG electrodes on the surgeon's right arm, 2 Superlux microphones and 2 mini DV plus a Nikon D60 cameras.

The acquired data show that the drill moves within a 200mm cube which defines the haptic rendering workspace. The maximum force (torque) continuously applied by the surgeons is 12N (0.6Nm) while the peak force (torque) reaches 25N (1.5Nm). The maximum stiffness obtained from position and force is around 13kN/m. Finally, the surgeon's hand moves up to 2mm/s during the drilling of the bone while this speed can go up to 0.9m/s when the surgeon is repositioning the tool above the mandible. These values will be used as design drivers for the platform's haptic interfaces. On the other side, the movements of the surgeon's head are small and remain above the patient's head. It is thus possible to implement the visual feedback with a fixed screen placed between the user's hand and head.

MULTIMODAL PLATFORM FOR THE TRANSFER OF MFS SKILLS

The previous results were used to develop a bi-manual multimodal training platform (see Figure 4). The core of the system is composed of two haptic devices used to capture the surgeons' hands movements and to render the tools-bone interaction forces. In the field of surgery, 6DOF capability, dexterity and transparency are key issues to allow fine and dexterous interaction in small volumes. To cope with these requirements, we used hybrid haptic interfaces composed of a 5DOF parallel structure with an additional axis in series which combine large workspace of serial and transparency of parallel robots (Gosselin, 2000). For the manipulation of the drill (right hand on the platform), a new device was designed and optimized to cope with the MFS design drivers in terms of workspace, force and stiffness. This interface incorporates a new piezo actuator based vibrating handle with a bandwidth of 1000Hz which was designed and manufactured to allow the reproduction of the high frequency vibrations produced by the surgery drill. For the manipulation of the retractor or vacuum cleaner (left hand on the platform), we used a pre-existing reconditioned device previously developed for Minimally Invasive Surgery (Gosselin et al., 2005) which fits with mouth tissues interaction requirements. Finally, a tangible element representing the chin is provided to support the surgeon's hand during the operation. The haptic devices are positioned relative to this element in order to allow optimal haptic rendering in the operating zone. Visual feedback is provided by a 3D screen inserted between the surgeon's head and hands. Its placement is also optimized to allow visuo haptic rendering coherency. Finally, surgery sound is synthesized and reproduced. Details on sound analysis and synthesis can be found in (Hoffman et al., 2009).

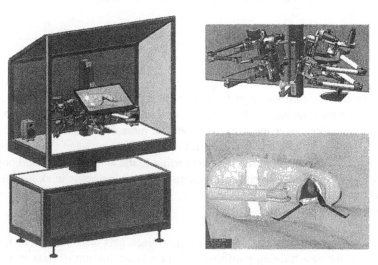

Fig 4: Multimodal training platform for the transfer of MFS skills.

All those elements are driven by a mixed Flash/Adobe AIR architecture for the management of the training progression and the XDE real time simulation framework developed at CEA-LIST for tools-tissues interactions and bone drilling. It makes use of a continuous surface representation based on distance fields (implicit representations of the surface) which is displaced upon drill-bone interaction as a function of parameters including the force applied on the tool at the preceding time step, the drill rotation speed, the angle between the tool and the surface normal and the local bone density. After the surface displacement, we solve for the geometric contact with a constraint based method yielding at the same time the forces to apply to the tool during the next step. This method is still an explicit approach, but has the advantage of being much more robust for large penetration than traditional point-voxel or voxel-voxel based methods (Morris et al., 2006) which introduce numerical instabilities in this case due to poorly continuous geometrical contact models.

The third aspect of the platform is the training program. While in real surgery the trainee has to manage the whole task and its complexity, we aim at focusing on elementary steps. For example, to learn drilling the Epker lines, we will first propose to train punctual drilling on a part with a hard envelope covering a softer core, then to drill straight and curved lines on a plain part with the same characteristics, then on more complex shapes, with or without perturbances (for example a hard element figuring the teeth roots). These simple tasks should allow focusing on specific skills: first exercising perception by touch (detection of the transition from hard to softer material) then fine position control under varying force conditions. A linear progression is proposed based on the logic followed during apprenticeship in which the learner is progressively involved in the surgery with increasing difficulty. Each exercise is unlocked once the trainee has successfully performed the former one three times in a row.

DISCUSSION

Traditional teaching methods in surgery rely on real practice on patients whose availability and variability cannot be controlled. On the contrary, the proposed VR based training methodology will allow to implement a controlled training protocol with more repetitions over the same time period, more focused on the critical steps of the operation, and with a greater amount of variability between the use-cases that can be delivered in a proper order and according to the sensori-motor and cognitive capabilities of the trainee. Such repeated enactive training helps in skill acquisition as enactive knowledge is direct, intuitive, fast, embodied into common action-perception behaviours.

Moreover, contrary to real surgery where the trainee gets only subjective feedback from his mentor, our training platform gathers a large amount of data in real time (positions, forces, ...) which can be used to monitor users progression and provide quantitative feedback to the user, both formative and summative (Tsuda et al., 2009), like for example:

- display of forces and torques applied on the tool: too much effort can be very dangerous as they can produce uncontrolled overshoots in the spongy bone with associated nerve injuries while too little efforts can result in incomplete drilling lines. Display of the applied forces can help the trainee to learn how to control them;
- display of visual feedback from an "impossible" anatomical point of view: Epker surgery is performed in a very small operative field with limited visual access. Viewing the result of the task from another point of view (or in transparency) can help understanding the task and construct an enactive model of it (e.g. shape associated to haptic sensations during bone exploration in the search for the Spine of Spix, overshoot during drilling);
- display of performances related data: proficiency-based training has been established as an optimal method for teaching technical skills (Scott et al., 2008). This method uses expert-derived performance related to task requirements (precision related to the ideal trajectory and time to perform the task) as training end points.

Finally, we will try to further refine our training methodology using the results of recent studies on compliance management and vibration perception (Cohen et al., 2009) (Lev et al., 2009). The experiments aim to give a psychometric scaling of compliance and vibration discrimination and to evaluate the effects of multimodal cues on perception. These experiments are providing an input on the discrimination thresholds, sensitivity and differences between the uni-modal and multimodal task perception and performance.

CONCLUSIONS

In this paper, we introduced a new multimodal platform for the training of Maxillo Facial Surgery skills. The design rationales and methodology were first explained, followed by a presentation of the platform itself, and its potential benefits in training. In the near future, we will proceed to its tuning and implementation of training exercises, which will allow testing its efficiency.

ACKNOWLEDGMENTS

The authors gratefully acknowledge the support of the SKILLS Integrated Project (IST-FP6 #035005, http://www.skills-ip.eu) funded by the European Commission. They also wish to thank people from the Sensory and Ambient Interfaces Laboratory and Interactive Simulation Laboratory at CEA-LIST who contributed to the development of the platform. They thank Haption (www.haption.com),

Technion, the Israel Institute of Technology, EDM Lab from Univ. Montpellier 1, Acoustics from Aalborg University, OMG (www.omg3d.com) for their work (lending of material and contribution to the specification of the platform).

REFERENCES

Cohen A., Weiss K., Reiner M., Gopher D. and Korman M. (2009). "Effects of practice and sensory modality on compliance perception", Proc. Int. Conf. on Multimodal Interfaces for Skills Transfer, Bilbao, Spain.

Ericsson K.A. and Simon H.A. (1993). "Protocol analysis - Revised edition: verbal reports as data", MIT Press.

Gosselin F. (2000). "Développement d'outils d'aide à la conception d'organes de commande pour la téléopération à retour d'effort", PhD dissertation., University of Poitiers, 358p.

Gosselin F., Bidard C. and Brisset J. (2005). "Design of a high fidelity haptic device for telesurgery", Proc. IEEE Int. Conf. on Robotics and Automation, Barcelona, Spain, pp 206-211.

Hoffmann P., Gosselin F., Taha F., Bouchigny S. and Hammershøi D. (2009). "Analysis of the drilling sound in maxillo facial surgery", Proc. Int. Conf. on Multimodal Interfaces for Skills Transfer, Bilbao, Spain.

Kneebone R. (2003). "Simulation in surgical training: educational issues and practical implications", Med. Educ., 37, pp. 267-277.

Lev D., Korman M., Cohen A., Reiner M. and Gopher D. (2009). "Psychometric effects of experience with vibro-tactile detection task in a standard haptic virtual environment", Proc. Int. Conf. on Multimodal Interfaces for Skills Transfer, Bilbao, Spain.

Moorthy K., Munz Y., Sarker S.K. and Darzi A. (2003). "Objective assessment of technical skills in surgery", BMJ, 327, pp. 1032-1037.

Morris D., Tan H., Barbagli F., Chang T. and Salisbury K. (2007). "Haptic feedback enhances force skill learning", Proc. IEEE World Haptics, Tsukuba, Japan, pp. 21-26.

Morris D., Sewell C., Barbagli F., Blevins N., Girod S. and Salisbury K. (2006). "Visuohaptic simulation of bone surgery for training and evaluation", IEEE Transactions on Computer Graphics and Applications, Volume 26, No. 6, pp. 48-57.

Müller-Tomfelde C. (2004). "Interaction sound feedback in a haptic virtual environment to improve motor skill acquisition", Proc. Int. Conf. on Auditory Display, Sydney, Australia.

Scott D.J., Ritter E.M., Tesfay S.T., Pimentel E.A., Nagji A. and Fried G.M. (2008). "Certification pass rate of 100% for fundamentals of laparoscopic surgery skills after proficiency based training", Surgical Endoscopy, Volume 22, No. 8, pp. 1887-1893.

Sewell C., Morris D., Blevins N., Barbagli F. and Salisbury K. (2004). "An event-driven framework for simulation of complex surgical procedures", Proceedings MICCAI VII, Rennes, France. Springer-Verlag Lecture Notes in Computer Science, Volume 3217, pp. 346-354.

Ström P., Hedman L., Särna L., Kjellin A., Wredmark T. and Felländer-Tsai L.. "Early exposure to haptic feedback enhances performance in surgical simulator training: a prospective randomized crossover study in surgical students", Surgical Endoscopy, Volume 20, No. 9, pp. 1383-1388.

Tsuda S., Scott D., Doyle J. and Jones D.B. (2009). "Surgical skills training and simulation, Current Problems in Surgery", Volume 46, No. 4, pp. 271-370.

Chapter 86

Digital Management and Representation of Perceptual Motor Skills

Carlo Alberto Avizzano, Emanuele Ruffaldi, Massimo Bergamasco

PERCeptual RObotics Laboratory
Scuola Superiore Sant'Anna / CEIICP
Pisa, 56127, ITALY

ABSTRACT

In this chapter we address the issue of combining the design of interactive virtual environments with complex human data analysis schemas. A guideline of such a design in order to improve usability, efficacy and intuitiveness of the data content is also being addressed. The work presents a new system to manage and analyze the experimental data which are collected over time across experimental trials which involve different subjects and may span across years. A complete data workflow has been setup within the system. The system supports data acquisition and organization which is based on a self organizing, graphically edited digital format. The system architecture is interoperable by multimodal virtual environment and provides facilities to interact in real-time with complex data formats such as haptic, audio and video streams. e architecture is portable across most operating system and makes use of a standard database format allowing simple data sharing.

Keywords: Multi Modal Interaction, Digital Representation of SKILLS, Digital Repository

INTRODUCTION

The growing complexity of systems is increasing the complexity of experimental data management. In some research disciplines new tools for automatically handling of data are appearing such as biological data (Pacific, 2008), statistical data (Reading, 2003), chemical data (IDBS, 2010). Usually, dedicated programs, commonly known as Laboratory Information Management Systems (LIMS), handle data management from the early design up to publishing results. Among these the BioArray Software Environment – BASE – (Saal et Al., 2002) demonstrated the management of microarray data in genetic research.

Experiments within Virtual Environment (VE) commonly provide user with a huge amount of data. The adoption of database systems in handling data for Haptic Interfaces was pioneered by Yokokoji and Henmi (Yokokoji, 1996; Henmi 1998). Their system stored human motion as a reference for later haptic guidance. In such a system the sensori-motor actions were collected from experts to a database. Yokokoji's database simply recorded force and position data to be forwarded to a specialized impedance controller. By following the Yokokoji approach several authors have later tried to integrate the basic database representation with more advanced systems which included elements of artificial intelligence: Sano (Sano, 1999) proposed to integrate these system with a set of Neural Networks which can model and generalize over task variability; Calinon (Calinon, 2006) employed a structured approach to cope also with non stationary properties of human actions. In this case, PCA was used to extract the relevant component of motions while mixture of probability density functions described the correlation among inputs/outputs.

However, the application of LIMS architecture to virtual environment is not suitable. LIMS systems consider the storage and analysis process as a component independent from the experiment itself, while VEs involve a deeply interaction between collected data and human behavior. This is even truer when considering recent developments on VE systems that include high level artificial intelligence to modify the response of the VE itself.

The system introduced by this chapter deals with a digital representation system that enables the capture, storage analysis and real-time interaction of human motion data. The system supports complex data structures, long term management and manipulation of the real-time feedback required by the interaction in the environment.

SYSTEM ARCHITECTURE

The proposed Data Management System is designed along two orthogonal issues:

- The data content: data content relates the formats that will be employed and ensure that whoever is storing/retrieving information from the system feed it (or is feed) with a minimum requirement of information. To this purpose a specific (extensible) standard has been developed in order to make coherent the data organization: Multi Modal Data Format (Ruffaldi E., 2009);

- The data management: a set of portable tools that supports the storage and retrieval of data: the Digital Repository (Avizzano, 2009).

The storage engine, SQLite, was preferred with respect to other standard formats like HDF-5 (Chilan, 2006) for its implicit capability and flexibility of structuring data by tagging, linking, and attributes, and the additional features offered by relational features.

The Multi Modal Data Formats (MMDF) considers that information can be grouped in classes accordingly to an efficient R-Tree hierarchy (Guttman 1984). Six levels of hierarchy have been predefined in order to facilitate the organization of the data gathered from any interactive session.

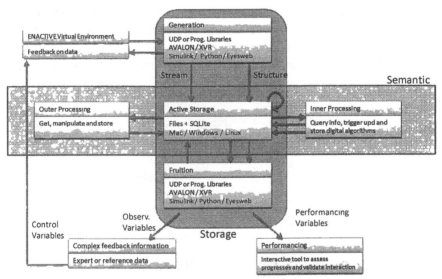

Figure 1 Digital Repository functionalities

All the available tools are built on the Datakit, and engine which embeds the data storage and offers a unique level of C-functionalities available to different development languages. The choice of using a Datakit ensures that portability and maintenance of repository functionalities in much faster. The Datakit is available as a library on Linux, Mac and Windows operating systems, and its calls have been exported to C, Python, Matlab Simulink and 3D interaction toolkits such as XVR (Carrozzino, 2005) and AVALON (Behr, 1998).

Digital Repository tools is complemented with three development interfaces: a Python Library, which is intended to implement by Python code higher level processing on data and complex GUI interactions; A Simulink ® library that combines the storage process with a unique graphical interface that simplifies the process of creating data structure in the database in the process of interconnecting three of graphical blocks in its interface; A Matlab extension class library that links the Digital Repository tools with those numerical functionalities available from the Matlab community.

USING THE DIGITAL REPOSITORY

Most of the Digital Representation functionalities have been implemented as a Simulink extension (a 'toolbox'). The toolbox makes use of Python calls to map digital representation concepts into intuitive graphical interactions. The root of the integration is a basic block that masks the access to specific locations of MMDF structure. Additional nodes (the Data Entities) can be cascaded to the database root by creating a graphical tree. The toolbox provides to map such a tree into a respective relational hierarchy in the database. This is possible because each Data Entity can simultaneously act as a child (using its input ports) and as an ancestor (using the specific output port). The toolkit deeply interacts with the Simulink initialization and simulation procedures in order that this graphical interconnection can be mapped into a proper structural tree before any simulation begins, and with a consequent zero-overhead in runtime (Avizzano, 2009b).

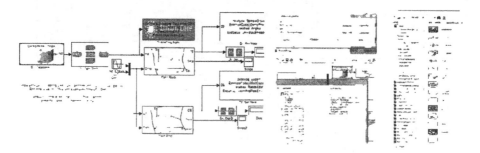

Figure 2. Tree structure (left) and GUI (right) examples

In order to implement the data tagging, several additional types of storing blocks are available: Meta data blocks add informational properties to the stored data; Parameters Data Blocks store information related to parameters required for the experiment and/or the virtual environment; Triggered data Blocks store information related to event related interactions. Additional tools complements the basic Simulink toolkit, among the most relevant ones: A code editor, An experiment name generator.

In the development of the Digital Representation toolbox, we focused on specific research contexts as defined in the SKILLS demonstrators (Bergamasco, 2006; Bardy et al. 2008): a set of multimodal virtual environments (Ruffaldi, 2009; Tripicchio, 2009) dedicated to the acquisition of expertise and to the transfer of to novel and intermediate users.

Such project has several research phases which can be considered to be in common with most experiments that require data processing phases of knowledge management: Real time data acquisition and storage; Pre-processing, filtering and segmentation of data streams; Data clustering, labeling and performance indicators; Model creation and data regression/calibration to fit on models; Realtime delivery of protocols and historical view of performance.

THE JUGGLING ANALYSIS

In order to facilitate the understanding of concepts introduced by the Digital Repository, we will consider a case study based on the capture and analysis of data related to "three ball cascade" in juggling. A set of data has been captured from a group of experts in one development center using a high precision optical tracker. Data captured from experts need to be imported in the digital repository and analyzed to indentify information to be used in later training analyses.

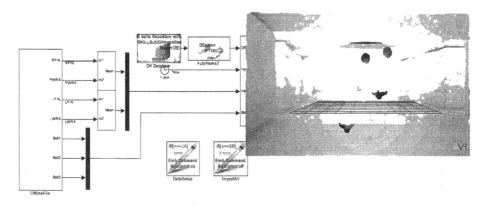

Figure 3. Schema for data acquisition and VR rendering

Acquisition of data was achieved through the scheme represented in figure 3 (left), where two major scripts controlled the data acquisition process. The script named Data Setup provides to setup the simulation process and to decode in real-time the information stored in the storage files of the Optical Tracker. The script named Import All provides to run between all files coming from the experts and run the simulation interactively in order to store all trajectories in the database. The rest of the Schema provides to reorganize data information into information relevant for the Juggling, and encode it in variables which need to be observed for simulation and performance estimation. The tree-architecture is implemented in the StorageRoot block which is organized accordingly to the data format set for the Juggling experiments. On the right part of the same structure, the stored data have been read back from the digital repository in order to play-back and analyze in a virtual environment the sequence of motion. We consider the possibility to reply, analyze and change the point of view of any previous experiment a relevant tool to understand the causalities of data and actions.

868

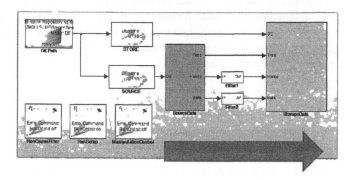

Figure 4. Schema for data filtering and segmentation

Once all data have been stored a specific filtering and segmentation policy is be applied to data. Filtering in this case has been performed using an ad-hoc Piece-Wise-Polynomial (PWP) approximation which allows a reduced set discontinuities on acceleration data and removes outliers by considering the effect of data removal on the statistical properties of the whole regression. The use of PWP has been demonstrated to be effective when using data coming from camera capture, for instance in basketball segmentation. The choice of this policy for Juggling was highly motivated by the following factors: a) PWP do not introduce delay on filtering; b) the introduction of discontinuity points allows to remove the high frequency cut introduced by linear filters; c) an implicit segmentation rule automatically outcome from best fitting optimization and can be verified onto the dynamic of the process; d) Polynomial expressions optimally describe the balls flying phases and allows to reduce the complexity of data description.

In figure 4 the schema to recover data imported in the previous step (DK Path), to iterate all over data (run Simulation), to apply PWP best fitting (non causal filter), and to control the writing back to the database (Manipulation, Control and Storage Data) is represented. Figure 5 shows a comparison between PWP approach and an optimal linear filtering when the Z acceleration is computed from experimental data. The PWP decomposition, not only properly fits the given data, but also matches with a-priori information such as the correct determination of gravity during the flying phases.

The filtering and segmentation process also offer a support for the reduction of problem dimensionality. In the specific example it maps arcs of trajectories into 3rd order polynomials which can be described by only four parameters each (three coefficients and the starting time).

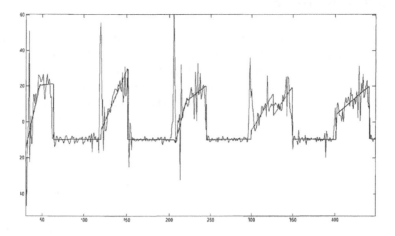

Figure 5. The 2nd derivative of balls acceleration computed on original data and with the implemented PWP regression.

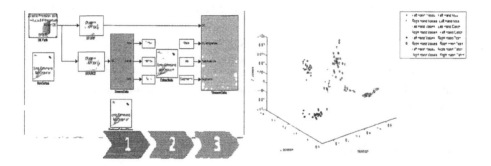

Figure 6. Data analysis.

Once the relevant information has been extracted from trajectories and encoded into a small subset of parameters, data can be compared across experiments. This will allow determining if the there exists a control rule that underlies the motion control.

This is accomplished according to the computing schema shown in figure 6. The schema takes in inputs the parameters extracted from the previous step (filtering and segmentation), applies, if possible, any additional data tagging, such as the procedural decomposition deriving from task analysis, then creates nMaps one for each tasks, which summarize the parameters changes before and after each segment. The plotting includes an iteration among all the available sets of acquired examples, and whenever reasonable classification tools to identify task related properties (in figure 6 a simple classification by task analysis is shown).

A JUGGLING MODEL?

The result of classification/regression on dataset can be used for both devising dynamic relationship underlying the task control, as well as for calibrating on the same properties any external model that replicates similar behavior.

This is generally achieved into with two reference models, the first one that extracts metrics from real-time and batch experiments, the second one that employs the optimal parameters to drive a closed loop model as a digital trainer in the virtual environment.

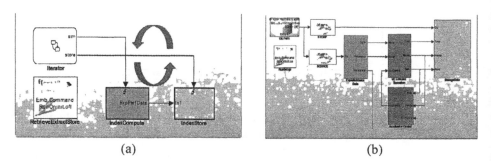

(a) (b)

Figure 7. Model to compute performance (a) and deploy a training experiment (b)

In figure 7(a), an example of complex data analysis for the evaluation of learning curves is shown. The schema makes use of triggering events to store data extracted by complex samples. Finally figure 7(d) shows how this information can be integrated into a real-time feedback trainer. The evaluation of these two models will be shown at the conference.

CONCLUSIONS

A new tool to handle simulation data for complex experiment analysis has been presented. The tool can manage complex experiments carried out by several researchers, at different sites. It encodes the information in a self documented and portable structure. Data management allows simultaneous organization and analysis of data trough a structured data format called multi modal data format.

The system allows real-time as well as batch data analyses, including data collection, filtering, processing and the construction of data driven models. The overhead of the system is minimal and the tools allow portability across several operating systems.

ACKNOWLEDGEMENTS

The activities described in this paper have been carried on with the financial assistance of the EU which co-funded the project SKILLS: "Multimodal Interfaces for capturing and Transfer of Skill"- IST-2006-035005 (www.skills-ip.eu). Authors are grateful to the EU for the support given in carrying on this research.

REFERENCES

Avizzano CA, Ruffaldi E, Bergamasco M (2009) Digital Representation of SKILLS for Human Robot Interaction, Robot and Human Interactive Communication Proc.: 32–37.

Avizzano CA, Ruffaldi E, Bergamasco M (2009b) Digital Representation of SKILLS, SKILLS Int.Work.

Bardy BG, Avizzano CA, et Al. (2008) Introduction to the Skills Project and its Theoretical Framework. IDMME-Virtual Concept, Springer-Verlag.

Behr J, Froehlich A (1998) AVALON, an Open VRML VR/AR System for Dynamic Application. Computer Graphics Topics 10: 28-30.

Bergamasco M. (2006) The SKILLS EU Integrated Project. EU-IST-Call6 FP6-2006-35005.

Buneman, P.; Khanna, S. & Tan, W. (2000), Data provenance: Some basic issues in Lecture notes in computer science, Springer, 2000, 87-93

Carrozzino, M.; Tecchia, F.; Bacinelli, S.; Cappelletti, C. & Bergamasco, M. Lowering the development time of multimodal interactive application: the real-life experience of the XVR project. ACM SIGCHI International Conference on Advances in computer entertainment technology, 2005, 273

Calinon S, Guenter F, Billard A (2006) On Learning, Representing and Generalizing a Task in a Humanoid Robot. IEEE Transactions on Systems, Man and Cybernetics 37: 286-298.

Chilan, C.; Yang, M.; Cheng, A. & Arber, L. (2006), Parallel I/O performance study with HDF5, a scientific data package, The HDF Group

Encyclopaedia Britannica (2009) Computer Assisted Instructions. Online. http://www.britannica.com/EBchecked/topic/130589/computer-assisted-instruction

Gibson E (1969), Principles of perceptual learning and development, Prentice Hall, New York,USA.

Guttman A (1984) R-trees: A dynamic index structure for spatial searching. ACM SIGMOD Proc.:47-57.

Henmi K, Yoshikawa T (1998) Virtual lesson and its application to virtual calligraphy system. Robotics and Automation Proc.

IDBS (2010) ChemBook E-WorkBook for Chemistry, http://www.idbs.com/eln/ChemBook/

Owens M (2006), The Definitive Guide to SQLite. ISBN: 978-1-59059-673-9. Apress.

Pacific Northwest National Laboratory (2008), Experimental Data Management System https://biodemo.pnl.gov/biodemo/

Reading University (2003), Data Management Guidelines for Experimental Projects, http://www.ssc.rdg.ac.uk/publications/guides/topdmg.html

Ruffaldi E., Lippi V.(2009) The Multi Modal Data Format Specification. SKILLS IP: Digital Representation. IST-FP6-2006-35005.

Ruffaldi E (2006) Haptic Web: Haptic Rendering and Interaction with XVR. XVR Workshop.

Ruffaldi E, Filippeschi A, Frisoli A, Sandoval O, Avizzano CA, Bergamasco M (2009) Vibrotactile perception assessment for a rowing training system. World Haptics Proc. IEEE CS: 350-355.

Saal LH, Troein C, Vallon-Christersson J, Gruvberger S, Borg A, Peterson C (2002) BioArray Software Environment (BASE): a platform for comprehensive management and analysis of microarray data. Journal Genome Biology. 3.

Sano A, Fujimoto H, Matsushita K (1999) Machine mediated training based on coaching. IEEE System, Man, and Cybernetics Proc: 1070-1075.

Tripicchio P, Ruffaldi E, Avizzano CA, Bergamasco M (2009) Control strategies and perception effects in co-located and large workspace dynamical encountered haptics. World Haptics Proc. IEEE Computer Society: 63-68.

Warren W (2006) The dynamics of perception and action. Psychological Review 113: 358-389.

Yokokohji Y, Hollis R, Kanade T, Henmi K, Yoshikawa T (1996) Toward machine mediated training of motor skills. Skill transfer from human to human via virtual environment. Robot and Human Communication Proc.: 32-37.

CHAPTER 87

Haptic Technologies in the Framework of Training Systems: From Simulators to Innovative Multimodal Virtual Environment Systems

Massimo Bergamasco, Carlo Alberto Avizzano, Emanuele Ruffaldi

PERCRO, Scuola Superiore Sant'Anna / CEIICP
Pisa, 56127, ITALY

ABSTRACT

Haptic technologies are complex robotic systems devoted to replicate force and tactile stimuli on human body parts such as hands, arms or feet during interaction procedures in Virtual Environments or in Tele-operation conditions. The application of different types of Haptic Interfaces for training purposes are presented in the specific frameworks of simulators, simulator-based training systems, multimodal training and shared multimodal training systems. The importance of integrating Haptics technologies in future innovative Virtual Environment based training systems is evaluated in light of their effective contribution to specific tasks in which paradigms involving contact functionalities are required.

Keywords: Haptics, Skills, Multimodality, Virtual Reality

INTRODUCTION

Today haptic technologies represent a relevant area of the robotics research and find their natural and largest applications in the fields of Virtual Environments (VE) and Tele-operation. Haptic systems are designed to convey contact information to parts of the human operator's body (hands, arms, feet, etc.) in response to specific manipulative or exploratory actions performed in VE or in remote real scenarios. From this general functionality, the term "Haptic Interface" (HI) has been utilized in the course of the recent years (Hayward, 2004).

In general terms, the common functionality of a haptic interface or of a force feedback device is that of being able to generate forces at the level of the human limb where these forces are required or simulated. Usually these places are the palm's surface of the human hand when manipulative operations must be controlled, or other parts of the human body, such as the anterior or posterior aspects of the trunk in the case of whole body motion haptic interfaces.

The design of haptic interfaces represents an innovative area of research per se, since, although haptic interfaces are very similar to serial or parallel robots, their design must address specific issues that are very different from the functional specifications of conventional robots. In particular, the main features characterizing a haptic interface are the following (Barbieri, 1991): *transparency, fidelity and natural movement* of the user's hand or body.

To exert forces on the human limbs, when the user intends to manipulate objects in the remote or virtual operational space, as far as the interaction with the haptic interface is concerned, two possible operative conditions are possible in the control space: in the first the HI is always maintained in contact with the user's limb during the control of the operation, in the second, instead, the HI enters in contact with the user's limb only at the instant of time in which the generation of forces is required.

The first category comprehends the large majority of present haptic interfaces, i.e. robotic devices that are external with respect to the user's body and that are usually grasped at their end-effector during the whole duration of the interaction task. A particular type of "Always in Contact" devices, named external always in contact, refers to those systems whose end-effector is shaped as the effective tool they intend to emulate, i.e. laparoscopic masters, stylus-like appendices, etc. Implicitly, exoskeletons systems, although possessing different characteristics from the general Always in Contact devices, belong to this first category.

The second category represents a completely different concept of haptic interface: here the robotic system is controlled to follow the user's hand or limb at a certain distance except when a contact force is required given the specific hand position and orientation at the specific instant of time. At that exact instant of time, the "Encountered Type" haptic interface is controlled in order to enter in contact with the human limb and generating the required force vector.

A particular type of haptic interface, so far not largely developed in the scientific literature but with a promising development in the near future for training purposes, is the so-called "Whole-Body Motion Haptic Interface", which functionality is that of being able to generate forces of large magnitude on specific locations of the user's body, in particular the trunk or the back (Checcacci, 2003). Among the applications allowed by this type of interface we can mention the tele-operation of humanoid robots in dangerous environments.

HAPTIC SYSTEMS FOR VE APPLICATIONS

A complete haptic system is far more complex than the pure mechanisms which topology has been described above. By analyzing the complete hardware and software technologies subtending the system "haptic interface" utilized for controlling the physical interaction with VE, we will discover that the electromechanical system represents only a small part of the complete system. As depicted in Fig. 1 (right), a complete haptic system requires the implementation of several software components capable of insuring all the functionalities that allow the user to realistically interact with the virtual entities represented in the VE (e.g. virtual objects, etc.).

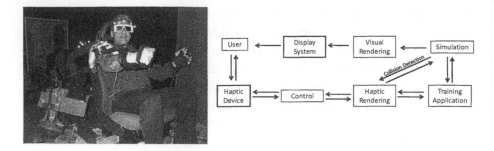

Figure 1 Example of Exoskeleton System for driving simulation (left) and diagram of complete Haptic System (right). In the diagram the user interacts with the system by means of two hardware elements for haptics and visual feedback. The Control module drives the haptic interface actuators under the computation performed by the Haptic Rendering module. The simulation manages the overall behavior of the virtual objects depending and provides contact information to the Haptic Rendering, for feedback, and status to the Training Application.

At present, only few of the above types of haptic interfaces are utilized in the framework of simulators or training systems design. In order to describe how haptic devices or, more general, haptic systems are exploited for training, it is worth to analyze the functionalities required to a haptic interface in the large spectrum of simulators and training systems.

TRAINING SYSTEMS

The use of digital technologies in the framework of training methodologies is largely increasing due to the recent exploitation of specific components of VE. We can distinguish between:

1. Training on Simulators
2. Simulator-Assisted Training
3. Multimodal Training Systems
4. Simulator-Based Training

Training on Simulators

Vehicle or flight simulators are aimed at recreating the same operating conditions as those encountered by the user in a real operation. In fact, the development of such simulators systems was based on Thorndike theory of "similarity" stating that "Learning is the result of associations forming between stimuli and responses" (Thorndike, 1921).

In these cases the user is located in the replica of a real cockpit and is asked to execute the same control operation as during a real operation. The user then interacts with physical commands, sometimes exact replicas of the same levers or buttons, etc., he/she would find on a real vehicle. Motion-based simulators are able to acquire such commands, as generated through real primary controls by the user, and perform adequate movements of the cockpit by replicating an approximation of velocity and accelerations trajectories as the real vehicle.

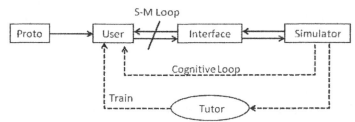

Figure 2 Simulator Block Diagram. On the left the training (Proto)col provided to the User that has a Sensori-Motor (S-M) loop with the Interface. At the same time the Simulator is connected to the User by means of the interface or directly to the User by means of the Cognitive Loop. Finally the human Tutor supervises the whole process.

The use of a real cockpit, integrating real physical primary controls, implicitly allows the user's sensation of contact with the primary controls to be exactly the same as in a real vehicle (Angerilli, 2001). Visual information is provided by computer graphics images generated in real-time by graphical workstation.

.

Figure 3 Specialized haptic device for providing force feedback in driving simulator. These systems provide a flexible and realistic force feedback. Compare them with the general haptic interface used in Fig. 1 left.

Another example of the use of haptic devices in simulators relates to surgery applications, in which training can be performed by utilizing VE systems in which the novice surgeon can exercise in the surgical operation by directly interacting with a mock-up of the human body part. Haptic interfaces, mimicking the surgical tools, at least for the part where grasping is performed, are used to transmit the forces of contact between the tip of the surgical tool and the body's tissue (Raspolli, 2003 and Okamura, 2009). In terms of flexibility, i.e. the capability to reconfigure the system for different vehicles or applications, simulators systems of this kind are very poor. In order to augment the flexibility of simulators systems, VE technologies can intervene by virtualizing the cockpit and all primary commands. In this framework a virtual model of the internal parts of the vehicle can be used and visualized on an immersive visualization system, e.g. a CAVE. As far as sensations of contact with primary commands is concerned, this can be achieved through the use of haptic controls, i.e. haptic interfaces possessing the same appearance and external shape as the real primary commands, but controlled by the computer in order to generate the required level of contact and operative forces to the user for the achievement of a realistic behavior. The same approach is used nowadays for the control commands inside different types of vehicles and it is called drive-by-wire, although in drive-by-wire cases there is no force or tactile feedback as in the haptic controls.

Simulator-Assisted Training

An higher level of interaction and a more sophisticated level of assist in the training protocol can be obtained by inserting a logical module that allows the interpretation of the user's commands and, based on that, be able to adequately modify the training protocol even in real-time.

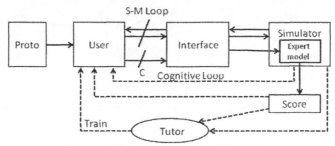

Figure 4 Simulator Assisted Training,
where the Expert Model is used only for the scoring

Different examples of this type of training systems can be found in literature: for example rehabilitation of upper limbs (Frisoli, 2007), training writing based on reactive robots (Avizzano, 2002) and cobots (Faulring, 2004).

Multimodal Training Systems

The training utilizing Multimodal Training Systems (Fig. 5) differs from Simulator-Assisted Training since the information derived from the simulator system is in this case utilized by the User for self-assessment and by the Tutor in order to modify the training protocol. The Project SKILLS (Bergamasco, 2006) exploits such a training methodology for each of the developed demonstrator systems, for instance (La Garde, 2009 and Ruffaldi, 2009).

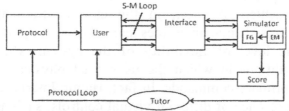

Figure 5 Multimodal Training System,
where the Expert Model is used for the Feedback Generation

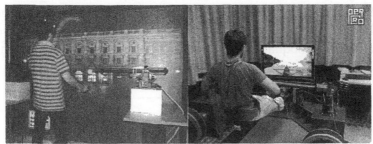

Figure 6 Two examples of systems for training based on the multimodal training schema. On the left the Haptic Juggling with co-located feedback, and on the right the Rowing System providing realistic water feedback.

Simulator-Based Training

In a Simulator-Based Training system, the tutor is substituted by a Learning Model that, based on performance data generated by the simulator, is able to automatically modify the training protocol according to specific performance parameters (Fig. 7).

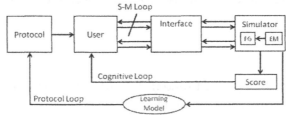

Figure 7 Schematic of the Simulator-Based Training in which both the Cognitive Loop and the Protocol Loop are computed and tuned by the system

CONCLUSIONS

The use of Haptic Interfaces in different cases of Training Systems has been presented. As seen in the case of Simulator-Assisted Training systems, haptic interfaces can play an important role for increasing the level of flexibility of the system when there are operating conditions in which the user must physically interact or operate with primary commands. In fact, Haptic Interfaces can easily simulate the presence of different virtual controls, such as gear-shift, steering wheels or handles, thus allowing the use of the same simulator systems as a training system on several vehicles.

On the other hand the use of Haptic Interface can also enhance the level of functionality of the simulator systems, as seen in the case of Multimodal Training Systems. Here the Haptic system can generate a large spectrum of contact or kinesthetic feedback to the human operator, thus allowing a more sophisticated tuning of the Training Protocol. This has been achieved in the framework of the SKILLS project, in which e.g. Haptic Interfaces have been utilized for the implementation of the resistance force on the oars in the rowing scenario, and for replicating contact conditions on the human hands during juggling operations in VE.

We believe that further sophistications in the developments of Haptic Interfaces, both at the level of electro-mechanical hardware as well as of control algorithms, can surely increase the exploitation of haptic feedback for training purposes.

ACKNOWLEDGEMENTS

The activities described in this paper have been carried on with the financial assistance of the EU which co-funded the project SKILLS: "Multimodal Interfaces for capturing and Transfer of Skill"- IST-2006-035005 (www.skills-ip.eu).

REFERENCES

Angerilli, M., Frisoli, A., Salsedo, F., Marcheschi, S., and Bergamasco, M. (2001). *Haptic simulation of an automotive manual gearshift*. In Proceedings of IEEE Ro-Man02, Berlin, Germany.

Avizzano, C. A., Solis, J., and Bergamasco, M. (2002). *Teaching to write Japanese characters using a haptic interface*. In Proceedings of the 10th Symposium on Haptic Interfaces for Virtual Environments.

Barbieri, L., and Bergamasco, M. (1991). *Nets of tendons and actuators: an anthropomorphic model for the actuation system of dexterous robot hands*. In Fifth International Conference on Advanced Robotics, ICAR, 357-362.

Bergamasco, M. (2006). *SKILL Integrated Project IST-2006-035005.* http://www.skills-ip.eu.

Checcacci, D., Hollerbach, J. M., Hayward, R., and Bergamasco, M. (2003). *Design and analysis of a harness for torso force application in locomotion interfaces.* In Proc. of the EuroHaptics International Conference, Dublin, Ireland, 6-9.

Faulring, E. L., Colgate, J. E., and Peshkin, M. A. (2004). *A high performance 6-DOF haptic cobot.* In IEEE International Conference on Robotics and Automation, Barcelona, Spain, 1980-1985.

Frisoli, A., Borelli, L., Montagner, A., Marcheschi, S., Procopio, C., Salsedo, F., et al. (2007). *Arm rehabilitation with a robotic exoskeleleton in Virtual Reality.* In IEEE 10th International Conference on Rehabilitation Robotics, 2007. ICORR 2007, 631-642.

Hayward, V., Astley, O. R., Cruz-Hernandez, M., Grant, D., and Robles-De-La-Torre, G. (2004). *Haptic Interfaces and Devices.* Sensor Review, *24*, 16-29. Emerald Group Publishing Limited.

Lagarde, J., Zelic, G., Avizzano, C. A., Lippi, V., Ruffaldi, E., Zalmanov, H., et al. (2009). *The Light Weight juggling training system for juggling.* In Proceedings of SKILLS09, Bilbao, Spain.

Okamura, A. M. (2009). *Haptic feedback in robot-assisted minimally invasive surgery.* Current opinion in urology, *19*(1), 102

Raspolli, M., Avizzano, C. A., Facenza, G., and Bergamasco, M. (2005). *HERMES: an angioplasty surgery simulator.* In World Haptics 2005, *Pisa, Italy,* 148-156.

Ruffaldi, E., Filippeschi, A., Frisoli, A., Avizzano, C. A., Bardy, B., Gopher, D., et al. (2009). *SPRINT: a training system for skills transfer in rowing.* In Proceedings of SKILLS09, Bilbao, Spain.

Thorndike, E. (1921). *The Fundamentals of Learning.* New York: Teachers College Press.

CHAPTER 88

Skill Training in Multimodal Virtual Environments

Daniel Gopher, Stas Krupenia, Nirit Gavish

Center for Work Safety and Human Engineering,
Technion, Haifa, 32000, Israel

ABSTRACT

Multimodal, immersive, virtual reality (VR) techniques open new perspectives for perceptual-motor skill trainers. They also introduce new risks and dangers. The chapter describes the benefits and pitfalls of multimodal training and the cognitive building blocks of a multimodal, VR training simulators.

Keywords: Skill training, virtual reality, simulators, perceptual motor skills

INTRODUCTION

With the growing complexity of systems and their operation environments, the required duration of training and increased costs of errors, on the job training became difficult or impossible and alternative simulation platforms have been developed to enable training. Flying, driving, space operation, surgery, power plant and process control are salient examples for tasks for which training simulators were developed. With the advance of computer technology simulators become more and more hybrid and system dynamic, visual field and audition have been increasingly driven and generated by computers. Contemporary developments in sensors and display capabilities and the exponential increase in computation speed and storage capacity led the way to the development of multimodal virtual environments. In these environments, the operator is immersed, experience multimodal sensations and interacts with virtual objects including other humans (Riva 2006). Vision and audition have always been in the study and design of simulators. The new important addition is the inclusion of haptics: the ability to feel and exercise force, touch, texture and kinematic. Haptic technology is developing rapidly and haptic interfaces are now been incorporated in many virtual worlds. It is reasonable to expect that the multimodal, virtual reality platforms a will dominate the next generation of training simulators. From a training and cognitive psychology vantage points this development carries with it some exciting prospects and serious dangers. These are the topics of the present chapter.

At the outset of this discussion it is important to emphasize that topics of training and transfer of skills have not been in the focus of interest for the present research and development of virtual environments. When studying human behavior, task performance and interaction in virtual environments, the interest has been in "virtualization" and the concepts of "presence" and "immersion". Sanchez-Vives and Slater (2005) propose the following definitions for the three concepts: **Virtualization**: "the process by which a human viewer interprets a patterned sensory impression to be an extended object in an environment other than that in which it physically exists" (p 332).

Presence: "… presence is the sense of being in a VE rather than the place in which the participant's body is actually located …… .. A fundamentally different view is that presence is "...tantamount to successfully supported action in the environment. The argument is that reality is formed through actions, rather than through mental filters and that "...the reality of experience is defined relative to functionality, rather than to appearances" (P 333). **Immersion:** "...a person is immersed in an environment that is realized through computer-controlled display systems, and might be able to effect changes in that environment" (P 332).

One aspect in the evaluation of virtual environments is subjective feeling of "presence" and "immersion", which is measured by a variety or rating scales. A second approach is by the use of behavior and performance measures, taken in relatively short duration interactions with the VR environment. The main interest is the correspondence between behavior in the real and virtual environments.

When evaluating the value and possible contribution of VR technologies to the training of skills. Presence and Immersion are related but not the prime focus concern. The value of a training system is judged by its ability to provide relevant experience and guidance for the acquisition of a designated skill, as well as facilitating its transfer from VR training to task performance in the real environment. Relevance and transferability are therefore the crucial criteria for the value of a training system, which are very different from Presence and Immersion. High level of Presence and Immersion may be important motivators and acquisition augmenters if there is a good match between the VR and real life experiences. If there are mismatches or diversions, they may be harmful. High subjective presence formed in spite of experiential diversions may lead to illusionary conjunctions, reduced or even negative transfer. When developing training simulators, it is sometime important to create deliberate diversions and reduce fidelity to avoid involuntary and unconscious illusionary conjunctions. The study of the value and prospects of VR technologies for training should therefore follow its own track and focus on topics related to the acquisition and transfer of skills.

This chapter focuses on the benefits and pitfalls of multimodal experience in the acquisition of perceptual motor skills and its implementation in a VR training platform.

THE SIGNIFICANCE OF MULTIMODAL EXPERIENCE

For humans and most organisms, potential sensory stimulation is characterized by simultaneous changes in multiple forms of ambient energy. (Stein & Meredith, 1993). Ernst and Bülthoff (2004) suggested that no single sensory signal can provide reliable information about the three-dimensional structure of the environment in all circumstances, When the varieties of multimodal cues are registered by the sense organs, the information must be integrated by neural structures to form a coherent and holistic representation of the object. It thus appears that our perceptuo-motor abilities follow from crossmodal processing of our surroundingsFor training purposes it is important to note that presenting information in one sensory modality (such as sound) can immediately draw a person's attention to other sensori stimuli presented in the same location (such as tactile sensations; Spence et al., 2004).

Comparative analysis of the interaction patterns involved in redundant and non-redundant cue processing provides evidence for the robustness of the principle of crossmodal neural synergy that applies regardless of the stimulus content. In addition, a comparative analysis provides evidence for the high flexibility of the neural networks of integration that are sensitive both to the nature of the perceptual task and to the sensory skill of the individual in that particular task (Fort, Delpuech, Pernier, & Giard, 2002). Interestingly, perceptual phenomena associated with single sensory channel processes (such as change blindness and change deafness), can also occur when multiple modalities are presented (Auvray, Gallace, O'Brien, Tan, & Spence, 2008). Several studies have shown that using two modality combinations, visual, auditory or haptic increase performance compared to using each of the modalities separately (Doyle & Snowden, 2001, Murray et al., 2005). A tri-modal combination of auditory, visual and haptic stimuli was detected faster than the shortest of the bi-modal combination (Diederich & Colonius, 2004). Nevertheless, it is important to notice that multimodal experience is not equivalent to the sum of modalities' experience. Vision is often considered as the prime and preferred sensory modality for humans, and as such it can dominate the experience. (Burr and Alais 2006, suggested that visual information does not simply dominate over auditory information, but how multimodal information is combined depends on the reliability of the stimulus inputs, and the more reliable input often dominates. Bresciani, Dammeier, and Ernst (2008) suggested that vision alone plays a minor role in feeling the contact with objects, at least when touch and sound are available, and that audition could be used to enhance the feeling of contact if it is appropriately coupled with touch.

TRAINING IN MULTIMODAL ENVIRONMENTS

Multimodal displays support flexible efficient communication, they are easy to learn, can be used in challenging situations, and people enjoy using them (Oviatt, 2002). Nevertheless, using multimodality displays for training is not straightforward. For example, the dominance of visual feedback in simple motor tasks, discussed above, was reflected in training studies. Training under conditions of combined visual and proprioceptive Knowledge of Results, reduced the efficiency of proprioceptive feedback in a later test during which proprioception was the only available feedback source. (e.g.Adams, Gopher, & Lintern, 1977; Yechiam and Gopher (2003). Virtual spatial sounds can also be used in multimodal contexts to supplement visual information in searching tasks (Bolio, D'Angelo, & McKinley, 1999).

Advantages of training in simulated and virtual environments

Simulated and virtual environments are being increasingly used for teaching and training in a range of domains including surgery (Howell et. al., 2008;), aviation (Blake, 1996), anesthesia (Gabba, 2005), rehabilitation (Holden, 2005), and driving (Godley, Triggs, & Fildes, 2002). These simulated and virtual worlds can be used for acquiring new skills or improving existing ones (Weller, 2004). A review by Holden (2005) on the current methods of VR applications for physical rehabilitation identified that in most cases movements learned in a virtual environment transfer to equivalent real world tasks and in some cases to other untrained tasks. Together, the research reviewed above demonstrates the value and need for successful and efficient training simulators.

Multimodal interfaces succeed in creating a stronger sense of presence by better mimicking reality (Romano & Brna, 2001).The sensorial richness of multimodal environments translates into a more complete and coherent experience of the virtual world and therefore the sense of being present in the virtual realm is stronger (Witmer & Singer, 1998). The experience of being present is especially strong if the virtual world includes haptic (tactile and kinesthetic) sensations (Reiner, 2004). Additionally, navigation in VEs represents a type of spatial task in which performance can be enhanced by employing spatial sound (Gonot, Chateau, & Emerit, 2006). One serious reason for the increase in subjective fidelity when being in a multimodal virtual environment originates from the congruency between perceptual modalities, i.e., a certain pattern of stimulation within the global array, which resembles that occurring in the real situation (e.g., Stoffregen & Bardy, 2004). However, one has to be careful with the concepts of presence and fidelity when interacting with a multimodal interface

Risks of using multimodal virtual reality and augmented reality feedback in training

If fidelity cannot be preserved or is hard to achieve, it is much better to avoid the use of the VR instantiation, or alternatively, a particular care must be taken to develop a training program that identifies the VR to real world mismatch and provides compensatory training mechanisms. For example, the standard view of

U.S. Air Force is that platform motion is not recommended in flight simulators for centerline thrust aircraft (Cardullo, 1991) because for military planes performing large and fast maneuvers, the motion errors in the moving base simulators and those of actual flight are considerable and result in negative transfer because of illusionary conjunctions. Under these conditions, fixed base simulators are much preferred, because the distinction between the simulator and the world is a clear and easy to observe. Additionally, the artificially constrained field-of-regard (the entire visual field available for sampling) found in most high-fidelity flight simulator can cause pilots to adopt novel visual strategies particular to the simulator (Kaiser & Schroeder, 2003). The novel viewing strategies are unlikely to transfer to the real world and therefore real world performance will not match VR performance. Impoverished environments can sometimes be created to emphasize certain components of the virtual world, to overcome the dominance of one modality over another, and to help the trainee to develop certain sensitivities, capabilities and modes of behavior that are otherwise suppressed in the real life operation conditions. For example, Acquisition of touch typing skills, based on proprioceptive information feedback from the hand and fingers, does not develop without deliberate, long, and tedious training (Wichter et al., 1997). A secondary task paradigm, in which the visually guided typing strategy was made less attractive, led to a faster acquisition of touch typing skills and higher performance levels both at the end of training and in the retention tests (Yechiam & Gopher, 2003).

An important advantage of the VR technology is the ease of providing augmented sensory feedback, visual guidance, auditory directors or augmented haptic cues. The greatest potential danger of AR applications is that performers become increasingly dependent on features of the VR devices which may inhibit the ability to perform the task in the AR-feature's absence or in the case of technology failure. For example, during laparoscopic surgeries, surgeons may be required discontinue the procedure and move to traditional, manual surgery. Retention of manual surgery capabilities and the switch between procedures are not trivial. AR features that would not be present in should be used with care and training should include phases that reduce the dependence of these features. Developing dependence, or at least reliance, on VR features that do not exist in the real environment or are very different from their real world counterparts can result in negative transfer to the real world.

DEVELOPING TRAINING IN MULTIMODAL ENVIRONMENT

To illustrate the application of these principles and concerns in the development of training platforms, we briefly describe platforms developed within the European Community 6th framework project "Skills", "Multimodal Interfaces for Capturing and Transfer of Skills". The main building blocks of a training platform and a training program are:

1. A clear specification of the task to be learned, the skills to be acquired, the objectives of training, and the designated criteria of graduation.

2. Design of task scenarios, task versions and difficulty manipulation that best represent typical encounters and key requirements of the task. Richer and diverse training environment affords the development of a more flexible and higher level competence (Bjork and Schmidt, 1992; Gopher, 2007).

3. Identification of key response and performance measures as well as progress criteria to evaluate trainee progress on relevant aspects of task performance and enhanced competence.

4. Definition of desired feedback indices and knowledge of results information to be given to trainees, as well as their frequency and mode of presentation.

When training is conducted as a preparatory stage, in a separate environment, or in a simulator, there is fifth important consideration:

5. Transfer of training. The relevance of the training experience in the learning environment determines the level of transfer from training to actual task performance.

Topics 2-4 have been instantiated through the development of accelerators and training protocols. The term **accelerator** is used to refer to variables that were introduced and implemented to facilitate, assist and improve learning. The term **training protocol** is employed to describe training schedule, duration, selected tasks scenarios, difficulty manipulations and their order of presentation. Six training platforms have been developed and are presently evaluated: Rowing (ROW), Juggling (JUG), Maxillo-Facial Surgery (MFS), Upper Limb Rehabilitation (ULR), Industrial Maintenance and Assembly (IMA), Programming by Demonstration (PBD) and. The IMA and PBD demonstrators include both Virtual Reality (VR) and Augmented Reality (AR) platforms; the ULR demonstrator includes Exoskeleton (Exos) and Bimanual (BM) systems. Table 1 presents a brief summary of the platform focus and their associated accelerators.

Across the six platforms there are a total of 24 implemented accelerators. Ten of which, the dominant category, are feedback indicators Ten of which, the dominant category, are feedback indicators. These capitalize on the elaborated measurement and capturing techniques incorporated in each of the demonstrators, to provide trainees with experiential, on line feedback on their performance which could have not been presented otherwise. In most cases, the information provided to the learner is enacted by the learner him/herself (e.g. energy consumption in rowing), and feedback relate to the discrepancy between current and target values. Feedback indicators vary in their modality of presentation (visual, tactile, auditory), their time mode (continuous versus intermittent) and reference point (Trainee performance level, hitting boundaries or constraints, correspondence to expert or optimal performance models).There are three important tests for the value of feedback based accelerators are: a) The relevance of the information and type of guidance to learning; b) the ability to improve learning without developing dependence on the feedback presence, which will degrade performance immediately once this feedback is removed; c) the presence of feedback should not distract or interfere with the regular modes of performing the trained task. These aspects will be examined in the platforms evaluation studies.

Table 1- Six Multimodal VR training platforms and designed accelerators

	Training Focus	Targeted population	Accelerators FD – Feedback, Cog. Aid –Cognitive aid HP -Haptic
ROW	Acquisition of basic rowing skills, effort and energy manage-ment, interpersonal coordination	Novice and Intermediate rowers	On line Visual spatial trajectory of rowing pattern (**Fd**) On line Vibration directive of rowing pattern (**Fd**) Adjustable auditory pacer of the locomotors/respiratory coupling (**Rhythmic Pacer**) Visual director of energy expenditure (**Fd**) Visual and haptic information of interpersonal coordination (**Rhythmic Pacer**)
JUG	Attention management of multiple moving objects, spatial temporal relationship, bimanual rhythmic coordination	Novice Jugglers	Tactile–auditory rhythm trainer of juggling coordination (**Rhythmic Pacer**) Training at slow and gradually increasing task speed (**Task processing time**) Systematic exploration of the spatial temporal components of the K dwell ratio (**Control strategy**)
MFS	Fine control of force application, use of fine graded touch and visual information	Trained surgeons	Feedback on forces and torques applied to the tool (**Fd**) Visual feedback on performance from an "impossible" anatomical point of view (**Fd**) Performance feedback relative to optimal performance lines (**Fd**) Multimodal feedback to enhance sensitivity to compliance and vibration change (**Fd**)
IMA	Acquisition of proce-dural skills in virtual environment and via a remote augmented reality training	Technicians and machine operators	Including haptic in 3D VR training (**Hp Enact**) Adding abstract representation to enaction (**Cog. Aid**) Introducing direct visual aid (pointer) (**Vis. Director**) Adding images of parts (**Cog. Aid**) Adding rotational haptic hints (**Hp Enaction**) Augmenting enaction by theoretical instructions (**Cog. Aid**)
PBD	Exploring and adapting behavior to the motion and compliance constraints of a robotic arm	PBD robot operators	On line indicators of approaching singularity (Fd) Voluntary exploration of singularity (Control strategy). Haptic exploration of compliance parameters setting (Hp. Enaction)
ULR	Using robotic technology and VR to expand rehabilitation options / interaction	Patients undergoing limb control physiother.	Task selection On line continuous feedback (**Fd**) Motion adaptation (**Fd, Motivation**)

In many cases the study of accelerators and their comparative effect present interesting theoretical conjectures and contrasts. For example, cognitive aids of different types, which have been implemented in the IMA training platform, are natural accelerators for the training of procedural and memory based skills, the focus of this demonstrator. The addition of haptic information as an accelerator provides an interesting test for the influence of motor enaction on the organization of knowledge in memory and its activation. Recent studies have showed the divergence of semantic based and action based representations in memory (e.g. Koriat & Perlman-Avnion 2003). The rowing and juggling platforms include an interesting evaluation of the use of rhythmic pacers as accelerators, the Juggling and PCB platforms both test accelerators based on executive control training approaches (Gopher 2007). These are some of more general and deeper scientific questions and contrasts that emerge and be addressed when assessing the accelerators in the different proposed evaluation plans.

Assessing the value of the developed platforms for skill acquisition and the best ways of applying them in training is a multifaceted task. There are four basic evaluation aspects that need to be examined: 1) A comparative evaluation of the differential experience of performing the same tasks on the VR platform and in the real world; 2) Evaluation of the contribution of accelerators; 3) Assessment of training protocols that will maximize learning and skill acquisition on a platform; 4) Transfer of training studies. The first type of evaluation aims to identify the similarities and differences between performing the same tasks (e.g. rowing, juggling, drilling) in the real world and in the VR training platform. Such an assessment is crucial to better understand the differences between acquiring a skill in the virtual and in the real environment, and the possible implications of these differences on the use of the VR platform in training and transfer. The main question is whether expert performers can comfortably employ the same form of behavior and execute their acquired skills, in the virtual as in the real environment. This question extends much beyond the subjective feeling of immersion or presence.

In conclusion, this chapter clearly shows that from a human performance as skills acquisition perspective, the new multimodal VR technologies offer new and exciting potential for the development of simulators and the training of complex skills. At the same time, they have serious dangers and pitfalls that should be carefully evaluated and avoided.

ACKNOWLEDGMENTS

This work is conducted under SKILLS Integrated Project (IST-FP6 #035005, http://www.skills-ip.eu) funded by the European Commission.

REFERENCES

Adams, J. Gopher, D., & Lintern, G. (1977). Effects of visual and proprioceptive feedback on motor learning. *Journal of Motor Behavior, 9*, 11-22.

Auvray A., Hartcher- O'Brien, J., Tan, H.Z., & Spence, C. (2008). Tactile and visual distracters induce change blindness for tactile stimuli presented on the fingertips. *Brain Research, 1213*, 111-119.

Blake, M. (1996). The NASA advanced concepts flight simulator - A unique transport aircraft research environment. *Paper presented at the AIAA Flight Simulation Technologies Conference.*

Bolia, R. S., DAngelo, W. R., & McKinley, R. L. (1999). Aurally aided visual search in three-dimensional space. *Human Factors*, 41(4), 664–669.

Brescani, J-P., Dammier, F., & Ernst, M., O. (2008). Tri-modal integration of visual, tactile and auditory signals for the perception of sequences of events. *Brain Research Bulletin, 75*, 753-760.

Burr, D., & Alais, D. (2006). Combining visual and auditory information. *Progress in Brain Research, 155*, 243-258.

Diederich, A., & Colonius, H. (2004). Bimodal and trimodal multisensory enhancement: Effect of stimulus onset and intensity on reaction time. *Perception and Psychophysics, 66*(8), 1388-1404.

Doyle, M. C., & Snowden, R. J. (2001). Identification of visual stimuli is improved by accompanying auditory stimuli: The role of eye movements and sound location. *Perception, 30*(7), 795–810.

Ernst, M., O., & Bülthoff, H., H. (2004). Merging the senses into a robust percept. *Trends in Cognitive Science, 8*(4), 162-169.

Fort, A., Delpuech, C., Pernier, J., & Giard, M-H. (2002). Early auditory—visual interactions in human cortex during nonredundant target identification. *Cognitive Brain Research, 14*, 20-30.

Gaba, D. M. (2005). The future vision of simulation in health care. *Journal of Quality and Safety in Health Care, 13*(S1), i2-i10.

Godley, S. T., Triggs, T. J., & Fildes, B. N. (2002). Driving simulator validation for speed research. *Accident Analysis & Prevention, 34*(5), 589-600.

Gonot, A., Chateau, N., & Emerit, M. (2006). Usability of 3D-sound for navigation in a constrained virtual environment. In *Proc. 120th Convention of the Audio Engineering Society* Paris, France. preprint 6800.

Holden, M. K. (2005). Virtual environments for motor rehabilitation: review. *CyberPsychology & Behavior, 8*(3), 187-211.

Howell J.N., Conatser R.R., Williams R.L., Burns J.M., and Eland, D.C. (2008). The virtual haptic back: a simulation for training in palpatory diagnosis. *BMC Medical Education. 8*, 14.

Kaiser, M.K., & Schroeder, J.A. (2003). Flights of fancy: The art and science of flight simulation. In P.M. Tsang & M.A. Vidulich (Eds.), *Principles and Practice of Aviation Psychology*. Lawrence Erlbaum: NJ.

Koriat, A., & Pearlman-Avnion, S. (2003). Memory organization of action events and its relationship to memory performance. *Journal of Experimental Psychology: General, 132*(3), 435-454.

Murray, M.M., Molholm, S., Michel, C.M., Heslenfeld, D.J., Ritter, W., Javitt, D.C., Schroeder, C.E., & Foxe, J.J. (2005). Grabbing your ear: rapid auditory-somatosensory multisensory interactions in low-level sensory cortices *Cerebral Cortex, 15*(7): 963–974.

Oviatt, S. (2002). Multimodal Interfaces. In J. Jacko & A. Sears (Eds.), *Handbook of Human-Computer Interaction* (pp. 286-304). Mahwah: New Jersey: Lawrence Erlbaum.

Reiner, M. (2004). The role of haptics in immersive telecommunication environments. *IEEE Transactions on Circuits and Systems for Video Technology, 14*(3), 392–401.

Riva, G, (2006). Virtual Reality Wiley Encyclopedia of Biomedical Engineering, John Wiley & Sons, Inc.

Romano, D.M., & Brna, P. (2001). Presence and reflection in training: Support for learning to improve quality decision-making skills under time limitations. *CyberPsychology and Behavior, 4*(2), 265–278.

Sanchez-Vives, M.V. Slater, M. (2005). From presence to consciousness through Virtual Reality. *Nature Reviews, 6, 332-339*

Spence, C., Pavani, F., & Driver, J. (2004). Spatial constraints on visual-tactile cross-modal distractor congruency effects. *Journal of Cognitive, Affective, and Behavioural Neuroscience, 4*(2), 148-169.

Stein, B.E. & Meredith, M.A. (1993). *The merging of the senses*. Cambridge, MA: MIT Press.

Stoffregen, T.A., & Bardy, B.G. (2004). Theory testing and the global array. *Behavioral and Brain Sciences*, 27, 892-900.

Weller, J. M. (2004). Simulation in undergraduate medical education: bridging the gap between theory and practice. *Medical Education, 38*, 32-38.

Wichter, S., Haas, M., Canzoneri, S., & Alexander, R. (1997). Keyboarding Skills

Yechiam, E., & Gopher, D. (2003). A strategy based approach to the acquisition of complex perceptual motor skills. *Presented at the 47th Annual Meeting of the* Human Factors society